The Eukaryotic Nucleus

Molecular Biochemistry and Macromolecular Assemblies

Volume 1

The Eukaryotic Nucleus

Molecular Biochemistry and Macromolecular Assemblies

Volume 1

Phyllis R. Strauss

Samuel H. Wilson

THE TELFORD PRESS, INC.
Post Office Box 287, Caldwell, New Jersey 07006

The Eukaryotic Nucleus / [edited by] Phyllis R. Straus, Samuel H. Wilson
p. cm.
Includes bibliographical references.
ISBN 0-936923-32-6 (v. 1) : $65.00. —
ISBN 0-936923-33-4 (v. 2) : $65.00. —
ISBN 0-036923-39-3 (pbk. : v. 1) : $32.50. —
ISBN 0-936923-40-7 (pbk. : v. 2) : $32.50
1. Cell nuclei. 2. Eukaryotic cells. I. Strauss, Phyllis R., 1943- .
II. Wilson, Samuel H., 1939- .
[DNLM: 1. Cell Nucleus. 2. Cells QH 603.E8 E873]
QH595.E89 1990
574.87'32—dc20
DNLM/DLC
for Library of Congress 89-20557
CIP

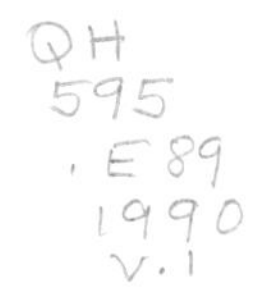

CONTENTS VOLUME 1

CONTRIBUTORS

Hiroyoshi Ariga
The Institute of Medical Science
The University of Tokyo
4-6-1 Shirokanedai
Minato-ku
Tokyo 108
Japan

Robert A. Bambara
Department of Biochemistry and the Cancer Center
University of Rochester School of
Medicine and Dentistry
Rochester, NY 14642

Earl F. Baril
Worcester Foundation for Experimental Biology
Shrewsbury, MA 01545

Kenneth L. Beattie
Department of Biochemistry,
Baylor College of Medicine,
One Baylor Plaza
Houston, TX 77030

James B. Boyd
Department of Genetics
University of California
Davis, CA 95616

Judith L. Campbell
The Braun Laboratories
California Institute of Technology
Pasadena, California 91125

Thomas J. Daly
Department of Pharmacological Sciences
State University of New York
Stony Brook, NY 11794

Errol C. Friedberg
Laboratory of Experimental Oncology
Department of Pathology,
Stanford University School of Medicine
Stanford, CA 94305

Paul V. Harris
Department of Genetics
University of California
Davis, CA 95616

Robert J. Hickey
Worcester Foundation for Experimental Biology
Shrewsbury, MA 01545

Sanae M. M. Iguchi-Ariga
The Institute of Medical Science
The University of Tokyo
4-6-1 Shirokanedai
Minato-ku
Tokyo 108
Japan

Ming-Derg Lai
Institute of Biochemical Sciences,
College of Science,
National Taiwan University,
Taipei, Taiwan,
Republic of China.

Lawrence A. Loeb
The Joseph Gottstein Memorial Cancer Research Laboratory
Department of Pathology SM-30
University of Washington
Seattle, WA 98195

Donal S. Luse
Dept. of Molecular Genetics
Biochemistry and Microbiology
Univ. of Cincinnati College of Medicine
231 Bethesda Ave.
Cincinnati, OH 45267-0524

Rogelio Maldonado-Rodriguez
Departamento de Bioquimica,
Escuela Nacional de Ciencias Biologicas,
I.P.N.,
Prol. De Carpio y Plan de Ayala,
Mexico, D.F. 11340

Linda H. Malkas
Worcester Foundation for Experimental Biology
Shrewsbury, MA 01545

Ralph R. Meyer
Department of Biological Sciences
University of Cincinnati
Cincinnati, OH 45221

Thomas W. Myers
Department of Biochemistry and the Cancer Center
University of Rochester School of
Medicine and Dentistry
Rochester, NY 14642

Fred W. Perrino
The Joseph Gottstein Memorial Cancer Research
Laboratory
Department of Pathology SM-30

University of Washington
Seattle, WA 98195

Charles E. Prussak
Eukaryotic Regulatory Biology Program
Department of Medicine, M-013G
School of Medicine
University of California, San Diego
La Jolla, CA 92093

Louise Prakash
Department of Biophysics
University of Rochester School of Medicine
Rochester, NY 14642

Satya Prakash
Department of Biology, University of Rochester
River Campus Station
Rochester, NY, 1472

Anthony J. Recupero
Department of Biological Sciences
University of Cincinnati
Cincinnati, OH 45221

Michael P. Reed
Department of Biological Sciences
University of Cincinnati
Cincinnati, OH 45221

Diane C. Rein
Department of Biological Sciences
University of Cincinnati
Cincinnati, OH 45221

Ralph D. Sabatino
Department of Biochemistry and the Cancer Center
University of Rochester School of
Medicine and Dentistry
Rochester, NY 14642

Kengo Sakaguchi
Department of Genetics
University of California
Davis, CA 95616

Esteban E. Sierra
Department of Pharmacological Sciences
State University of New York
Stony Brook, NY 11794

Paul T. Singer
Department of Pharmacological Sciences
State University of New York
Stony Brook, NY 11794

K. C. Sitney
The Braun Laboratories
California Institute of Technology
Pasadena, CA 91125

Patrick Sung
Department of Biology, University of Rochester
River Campus Station
Rochester, NY, 1472

Ben Y. Tseng
Eukaryotic Regulatory Biology Program
Department of Medicine, M-013G
School of Medicine
University of California, San Diego
La Jolla, CA 92093

Peter C. van der Vliet
Laboratory for Physiological Chemistry
State University of Utrecht
The Netherlands

Teresa S.-F. Wang
Laboratory of Experimental Oncology
Department of Pathology
Stanford School of Medicine
Stanford Universty,
Stanford, CA 94305

Samuel H. Wilson
Laboratory of Biochemistry
National Cancer Institute
National Institutes of Health
Bethesda, MD 20892

Cheng-Wen Wu
Department of Pharmacological Sciences
State University of New York
Stony Brook, NY 11794

Felicia Y.-H. Wu
Department of Pharmacological Sciences
State University of New York
Stony Brook, NY 11794

Guang-Jer Wu
Department of Microbiology and Immunology
Emory University School of Medicine
Atlanta, GA 30322

1

ADENOVIRUS DNA REPLICATION *IN VITRO*

Peter C. van der Vliet

INTRODUCTION

The replication of adenovirus DNA (Ad DNA) is a very efficient nuclear biosynthetic process. Within a short period, between 8 and 48 hrs after infection, 10^5 - 10^6 new, double stranded progeny molecules are synthesized, each about 36000 basepairs long. This represents a total of at least 3.6×10^9 basepairs, almost equalling the amount of cellular DNA present in the uninfected human host cells. Assuming a linear rate of synthesis and a replication time of 20 min. per molecule (Bodnar and Pearson, 1980), one can estimate that at any moment during replication between 800 and 8000 replicating molecules are present. This requires an extensive pool of replication proteins, in particular those which are required in stoichiometric amounts like the viral DNA binding protein.

For these reasons adenovirus DNA replication has been experimentally accessible to *in vitro* studies (Challberg and Kelly, 1979). It represented the first cell-free system from higher eukaryotic cells in which initiation of DNA replication could be studied and served as the prototype for similar cell-free replication systems employing SV40 and polyomavirus DNA. The Ad *in vitro* system has permitted complementation experiments, leading to the purification of both viral and cellular proteins required for optimal DNA replication, and the system can now be reconstituted to a large extent, replicating at a rate close to that observed *in vivo* (Nagata *et al.*, 1983a).

Several new aspects of eukaryotic DNA replication have resulted from these studies. Proof for the long anticipated protein-priming mechanism has been obtained, a core origin-sequence has been defined as well as an auxiliary region which is capable of binding cellular proteins. Most interestingly, the sequence motifs in this auxiliary region are also used in many cellular genes for optimizing transcription. This has led to the concept of "transcription factor activation", strengthening the evolutionary link between transcription promoters and enhancers on the one hand and replication origins on the other hand.

Several recent reviews have appeared on Ad DNA replication (Sussenbach and Van der Vliet, 1983; Kelly, 1984; Campbell, 1986; Tamanoi, 1986; Van der Vliet *et al.*, 1988). This chapter will briefly recapitulate our knowledge of the DNA replication mechanism and will describe in more detail the various proteins that are required for optimal replication *in vitro*. Unless otherwise stated, results for the best studied serotypes, Ad2 and Ad5, will be described.

OUTLINE OF THE DNA REPLICATION MECHANISM

The human adenovirus genome consists of a linear double-stranded DNA of 35-36 kbp. The 5′ termini are covalently linked to a viral protein, the terminal protein (TP) by a phosphodiester linkage between the β-OH group of a serine residue and the 5′-OH group of the first nucleotide, dCMP. The DNA contains an inverted terminal repetition (ITR). The lengths of the ITR in human adenoviruses vary between 103 and 163 bp, depending upon the serotype. The ITR includes the origin of replication.

Early studies of DNA replication employed the classical approach of pulse labeling to identify replicative intermediates. Analysis of these molecules by sedimentation and density gradient centrifugation and by electron microscopy showed the presence of two types of intermediates, one in which a single strand of variable length extruded from an intact double-stranded molecule and one in which a partially double-stranded, partially single-stranded intermediate was found (Sussenbach *et. at.*, 1972; Ellens *et. al.*, 1974; Lechner and Kelly, 1977). On the basis of this a model was proposed in which replication starts at either end of the molecule, displacing one of the parental strands. The single displaced strand is subsequently converted into a duplex by complementary strand synthesis (Figure 1). Essential in this model is a continuous strand displacement mechanism as well as intrastrand hybridization of the ITR sequences to form a partially double-stranded panhandle structure, reshaping the same origin as used for displacement synthesis. Subsequent analysis of replicative intermediates by restriction enzyme digestion confirmed the presence of origins at or very near both ITR's (Schilling *et. al.*, 1975; Sussenbach and Kuijk, 1978; Weingärtner *et. al.*, 1976; Horwitz, 1976) and were in complete agreement with the proposed displacement mechanism.

For several years, the mechanism of initiation has been obscure. Models proposed for other DNA molecules like selfpriming (AAV) or concatemerization (T7) had to be discarded on experimental grounds. The presence of the TP bound to viral DNA as well as to replicative intermediates led to a protein-priming model in which the TP serves as a primer (Rekosh *et. al.*, 1977). Finding conditions for *in vitro* initiation (Challberg and Kelly, 1979) enabled experimental tests for this model and detailed analysis demonstrated that this initiation mechanism is essentially correct, but that a 80 kDa precursor of the terminal protein (pTP) rather than the terminal protein itself is used for priming (Challberg *et al.*, 1980, Lichy *et al.*, 1981; Tamanoi and Stillman, 1982). Further dissection and reconstitution experiments, as well as mutant complementation studies, led to the discovery of other proteins required for optimal DNA synthesis, like the viral DNA polymerase and three cellular proteins, nuclear factors I, II and III.

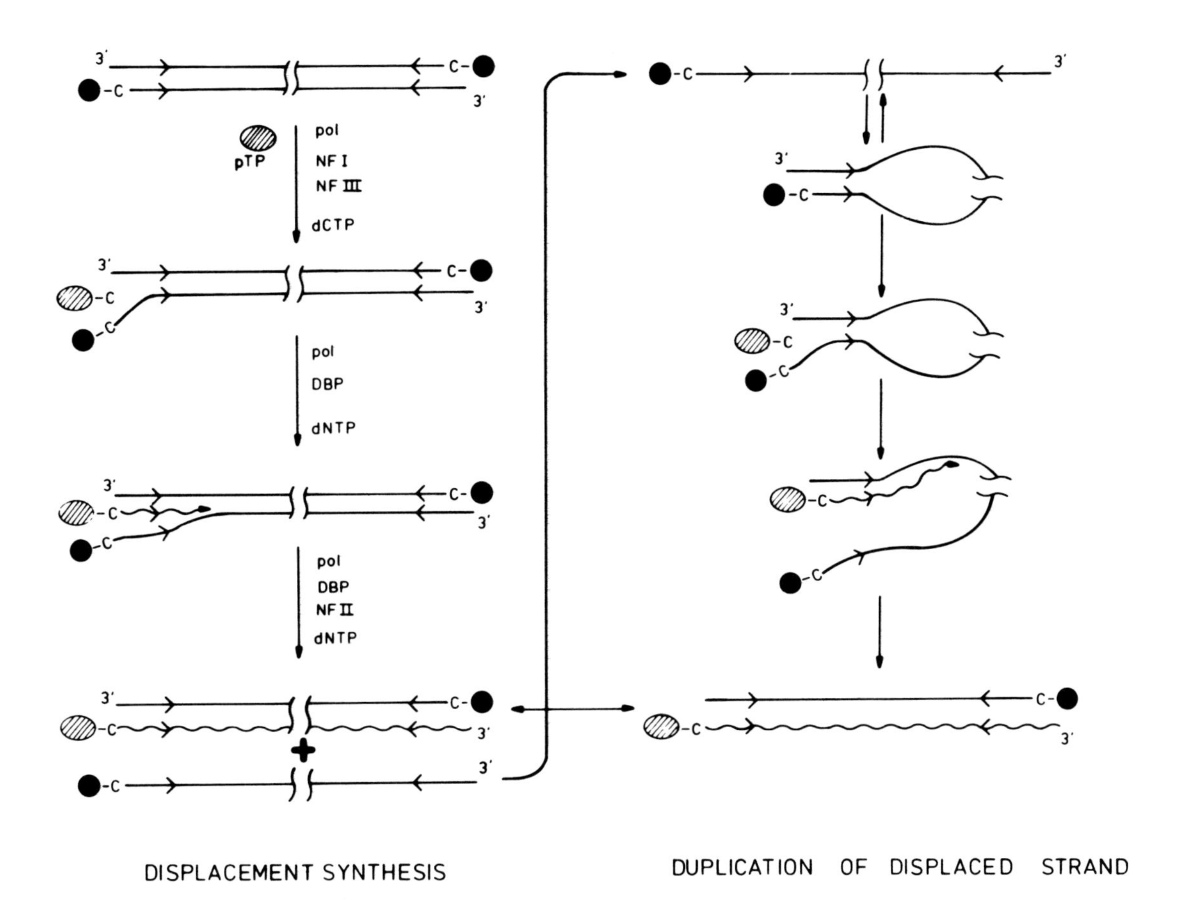

Figure 1. Model For Adenovirus DNA Replication. The model for Ad2/Ad5 is depicted. Viral DNA consists of a 36 kb linear duplex with a covalently bound terminal protein (●) and a 103 bp long inverted terminal repeat (borders: >). Initiation occurs by protein priming in which a viral precursor of the terminal protein (pTP) serves as primer. This reaction requires Ad DNA polymerase (pol) and HeLa nuclear factors I and III (NFI, NFIII). Elongation by strand displacement requires the viral DNA binding protein (DBP) and, in a later stage, nuclear factor II (NFII). The displaced strand can form a panhandle structure by intramolecular hybridization reforming an origin. Duplication of the single strand occurs after initiation at this origin, presumably employing the same protein-priming process.

An outline of the present DNA replication model, as well as the proteins involved in the different steps, is given in Figure 1. Details of the properties of the various replication proteins will be given below.

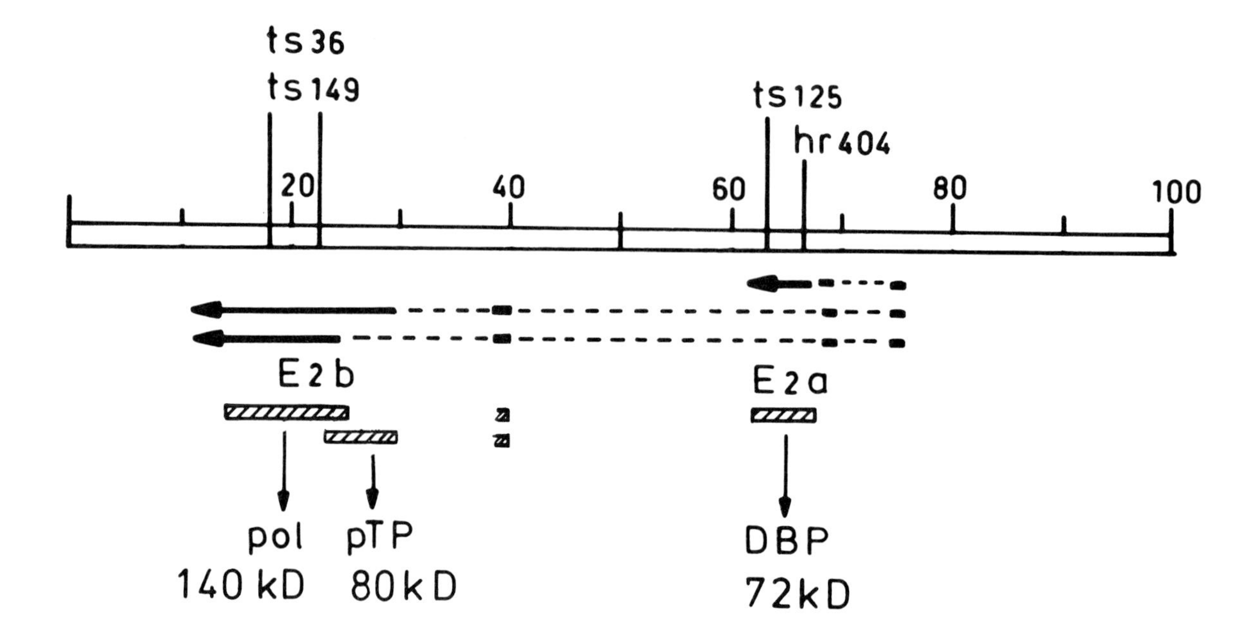

Figure 2. Adenovirus DNA Replication Proteins and Their Genes. Messenger RNA (arrows) for the three proteins, DBP, pTP and pol all originate from the E2 transcription unit starting at 75 map units and are processed by differential splicing. Continuous lines represent RNA sequences found in mature transcripts. The coding regions are indicated with shaded boxes and the location of various mutants is indicated.

VIRAL PROTEINS INVOLVED IN DNA REPLICATION

Three viral proteins are directly required for DNA replication, the DNA binding protein (DBP), the Ad DNA polymerase (pol) and the precursor terminal protein (pTP). These proteins are synthesized from one major early transcription unit, E2, with a promoter at 75 map units (see Figure 2). Although they share the same promoter, the stable transcripts show large differences in levels, the Ad DBP mRNA being at least 10-fold more abundant than the pol and pTP mRNAs (Chow *et al.*, 1980; Stillman *et al.*, 1981; Binger *et al.*, 1982). So far, the mechanisms underlying these differences are unknown but they could reflect differential splicing, well known for many other Ad transcripts.

In addition, other early viral functions are required for efficient DNA replication *in vivo* but they are dispensable *in vitro*, presumably because their effects are indirect. These include the E1A products, which transactivate the E2 promoter, possibly by modification of a cellular transcription factor (Imperiale *et al.*, 1985; Murthy *et al.*, 1985) and the E1B 19 kDa protein which prevents degradation of both cellular and viral DNA in infected cells (Stillman *et al.*, 1984).

I will confine my discussion to those proteins directly involved in DNA replication.

The DNA Binding Protein

Long before the development of cell-free replication systems the Adenovirus DNA binding protein was detected, purified and recognized as an essential component of viral DNA replication. This was accomplished thanks to (i) the abundance of the protein (approximately 2×10^7 molecules per cell) (ii), the strong binding to single stranded DNA facilitating its purification by DNA affinity chromatography and (iii) the temperature-sensitivity of DBP isolated from cells infected with the DNA-negative mutant *H5ts125* (Van der Vliet *et al.*, 1975).

From there on, studies on the phenotype of DBP mutants, in particular the prototype *H5tsl25*, have indicated that the Ad DBP is a multifunctional protein involved in several aspects of DNA replication as well as in the control of early (Carter and Blanton, 1978; Nevins and Winkler, 1980) and late (Klessig and Grodzicker, 1979) viral transcription, host range (Rice and Klessig, 1984) and possibly virus assembly (Nicolas *et al.*, 1982).

How are all these functions to be reconciled? One clue comes from recent observations that DBP binds to RNA as well as to single-stranded DNA (Cleghon and Klessig, 1986; Van Amerongen *et al.*, 1987). Binding to RNA could be instrumental in the control of gene expression. Another indication is the presence of at least two functionally separable domains in DBP.

Structure of DBP: Two Domains

Several lines of evidence indicate that DBP is a two-domain protein. These include partial proteolytic degradation, mutant studies and comparison of the DBP sequence in different serotypes. Limited proteolysis of the intact 72 kDa Ad2 DBP using trypsin or chymotrypsin resulted in preferential degradation products with C-terminal fragments in the range of 36-48 kDa, as well as a N-terminal fragment of 27 kDa (Klein *et al.*,1979; Schechter *et al.*, 1980; Linne and Philipson, 1980). In one case (Tsernoglou *et al.*, 1985) a chymotryptic site of Ad5 was analysed by amino acid analysis and located at position 174 of the 529 amino acids, *i.e.* at one-third of the molecule. Fragments of similar size have also been found within infected cells. The two domains have strikingly different amino acid compositions: the N-terminal part is devoid of tryptophan, cysteine and tyrosine and extremely low in histidine and phenylalanine while it contains the majority of proline and glutamic acid residues.

This is reflected in an aberrant mobility on SDS gels. The intact protein runs with an apparent molecular weight of 72 kDa while its actual molecular mass is 59.049 kDa (Kruijer *et al.*, 1981). This difference is caused by the N-terminal 173 amino acids which run as 27 kDa while the 39 kDa C-terminal fragment behaves normally. Since DBP is an asymmetric protein (Sugawara *et al.*, 1977; Van der Vliet *et al*, 1978), this could indicate the presence of an extended N-terminal domain coupled to a globular C-terminal domain. Such a model is at present still speculative and will await the outcome of structural studies. In contrast to the C-terminal domain, the N-terminal part is strongly phosphorylated (Klein *et al.* , 1979; Linné and Philipson, 1980).

Dephosphorylation by extensive phosphatase treatment does not impair DNA binding, in agreement with the observation that the isolated C-terminal domain can bind to DNA and is functional in DNA replication (Ariga *et al.*, 1980; Tsernoglou *et al.*, 1985). The smallest fragment tested to possess this property is 34 kDa (Friefeld *et al.*, 1983; Krevolin and Horwitz, 1987).

Analysis of the phenotype of mutants located in the two domains further strengthens the notion of two domains and, moreover, indicates that they can function independently. Several mutants located in the N-terminal domain, at positions 130 and 148, have been mapped (Kruijer *et al.*, 1981; Brough *et al.*, 1985). These mutants do not inhibit DNA replication. Instead, their host range is changed. The mutants grow well in monkey cells, in contrast to wild type. This is apparently caused by the acquired property to produce functional fiber protein, a capsid protein, and may be related to a change in the processing of the fiber mRNA, thus implicating this domain in splicing control (Anderson and Klessig, 1984). On the other hand, point mutations in the C-domain, at positions 280, 282 and 413 lead to temperature-sensitive DNA replication (Van der Vliet *et al.*, 1975; Kruijer *et al.*, 1981, 1982; Prelich and Stillman, 1986) without an effect on host range.

Comparison of the nucleotide sequences of DBP genes in several Ad serotypes confirmed and extended the model of a two-domain protein. The total lengths of the proteins vary between 529 amino acids for Ad5 and Ad2 to 473 amino acids for Ad40 (Vos *et al.*, 1988). Computer alignments show that this difference is mainly due to the N-terminal region which is most heterogeneous. Large deletions in this area in Ad2 do not diminish viral infectivity (Vos *et al.*, 1989).

In contrast, the C-terminal part is more conserved. Several highly conserved regions have been detected (Kruijer *et al.*, 1983; Kitchingman, 1985; Vos *et al.*, 1988) which may be involved in DNA binding but the actual DNA binding site has not yet been established. Structural analysis of at least the C-terminal part, which has been crystallized (Tsernoglou *et al.*, 1984), will give more information about possible DNA binding domains. A summary of the various DBP properties is shown in Table 1.

Interaction of DBP with DNA

Upon binding to single stranded nucleic acids DBP forms a regular, rigid and extended structure covering, at saturation, about 10 bases per protein molecule. The complex can be best described by a considerable tilt of the bases combined with a small rotation per base (Van Amerongen *et al.*, 1987). In these respects, the protein has many properties in common with the T4 gene 32 protein, the prototype helix destabilizing protein (for a review see Chase and Williams, 1986), but the effects of DBP on *e.g.* the circular dichroism spectra are considerably weaker while the cooperativity constant is also much lower than the values reported for the T4 gene 32 protein (Kuil *et al.*, 1989). The regular DBP-DNA structure might facilitate duplication by Ad DNA polymerase and apparently also protects against nucleolytic degradation (Nass and Frenkel, 1980).

DBP can also bind double-stranded DNA and a preference for molecular ends has been reported (Fowlkes *et al.*, 1979; Schechter *et al.*, 1980). Recent experiments in our laboratory using methidium propyl EDTA footprinting indicate that DBP can change the sensitivity to this intercalating, strand breaking reagent in such a way that a regular pattern of protection and hypersensitivity is observed (Stuiver and Van der Vliet, 1990). This behavior is reminiscent of the cellular nuclear factor IV (see below) and the p6 protein of *B. subtilis* phage ϕ29, a system in which protein priming also occurs with properties similar to Ad DNA replication (Prieto *et. al.*, 1988). This peculiar behavior is unexplained so far. This type of binding might change the dsDNA structure, thereby facilitating initiation or elongation. Another interesting possibility is that DBP dissociates nucleosomes from the parental DNA or prevents nucleosome assembly,

Table 1. Properties of the adenovirus DNA binding protein.

PROPERTIES	**FUNCTIONS**
Abundance (2.10^7 mol./cell)	DNA replication: - increases rate of chain elongation - protects displaced ssDNA
473-529 amino acids	
C-terminal 350 amino acids conserved	Early transcription control*
C-terminal part binds nucleic acids (ssDNA, dsDNA, RNA)	mRNA stability*
Forms regular nucleoprotein complex	Late gene expression (host range)
N-terminal part phosphorylated	Virus assembly

* These functions are derived from analysis of the phenotype of DBP ts mutants but are not impaired in the absence of DBP. Therefore they may relate to properties of mutated DBP rather than those of wild type DBP (Rice and Klessig, 1985).

thereby permitting a high rate of DNA chain elongation. This could explain the absence of nucleosomes in replicating Adenovirus DNA (Kedinger *et al.*, 1978; Matsuguchi *et al.*, 1979; Brown and Weber, 1980).

Role of DBP in DNA Replication

Addition of anti-DBP to isolated nuclei leads to inhibition of viral DNA synthesis (Van der Vliet *et al.*, 1977; Kedinger *et al.*, 1978) . Since these nuclei are only capable of elongating preexisting replication intermediates, a role of DBP in elongation was suspected. This has been further substantiated by the temperature-sensitive elongation in crude nuclear extracts from DBP mutants like *H5ts125* and *H5ts1O7* (Horwitz, 1978; Kaplan *et al.*, 1979; Van Bergen and Van der Vliet, 1983; Friefeld *et al.*, 1983; Ostrove *et al.*, 1983). Direct evidence for a role in DNA chain elongation comes from the observation that DBP enhances the rate of polymerization by the Ad DNA polymerase considerably and catalyzes efficient displacement synthesis, possibly by forming some kind of complex(Field *et al.*, 1984; Lindenbaum *et al.*, 1986. see also below). Also, DBP-single-stranded DNA complexes are much better templates for Ad pol than protein-free DNA, emphasizing the function of DBP in duplication of the displaced strand. This is not just a simple protection against nucleolytic breakdown since other helix destabilizing proteins like *E. coli* SSB can not substitute for DBP in this reaction (Field *et al.*, 1984). This high degree of specificity might indicate a close link between the two proteins, DBP and pol, during viral evolution, in agreement with their presence in one transcriptional unit.

In contrast to the well-established function of DBP in DNA chain elongation, its role in initiation has been more debated. Initial studies on the phenotype of *H5tsl25* (Van der Vliet and Sussenbach, 1975) provided evidence for such a function while *in vitro* DBP preparations stimulated pTP-dCMP formation about 4-fold, provided that NFI is present (Cleat and Hay, 1989; Stuiver and Van der Vliet, 1990). In the absence of NFI inhibition by DBP was observed (Nagata *et al.*, 1982; Kenny and Hurwitz, 1988). However, crude nuclear extracts from *H5tsl25* infected cells grown at the non-permissive temperature or inactivated *in vitro* by pre-incubation at 40°C supported initiation normally while elongation, even of the first 26 nucleotides, was severely inhibited. This was taken as evidence that DBP is dispensable for initiation (Challberg *et al.*, 1982; Friefeld *et al.*, 1983; Van Bergen and Van der Vliet, 1983; Prelich and Stillman, 1986). An alternative explanation is that these crude extracts contain a cellular protein that substitutes for the inactivated DBP, or otherwise stimulates initiation. Some studies using purified DBP confirmed that initiation is independent of the presence of DBP (Rosenfeld *et al.*, 1987). These authors assumed that preparations of DBP used in earlier studies still contained nuclear proteins that stimulated initiation but this could be ruled out by extensive purification of DBP (Kenny and Hurwitz, 1988). Although this discrepancy is not easily solved, one explanation could come from differences in effective concentrations of DBP used. The effect of DBP on initiation is both template and concentration dependent. For instance, with single-stranded templates stimulation is observed at low concentration while at high concentrations DBP inhibits, possibly because it prevents other proteins like pTP-pol from finding its binding site. This could indicate that the state of unwinding and the effective DBP concentration together determine the effects observed *in vitro*. Another explanation could lie in the reaction conditions used, like salt and dCTP concentration. Interestingly, stimulation of initiation of ϕ29 DNA synthesis by the related p6 protein is dependent on dATP concentration (Blanco *et. al.*, 1986) while its effect on early replication is salt dependent (Blanco *et. al.*, 1988).

The Terminal Protein and Its Precursor: Protein Priming

Isolation of DNA from viruses under conditions that keep covalent protein-DNA bonds intact led to the discovery of the Terminal protein (Robinson *et. al.*, 1973). One molecule of this 55 kDa protein (Rekosh *et. al.*, 1977) is bound to each 5′-end by a serine-dCMP phosphodiester bond (Desiderio and Kelly, 1981) and remains bound during the entire replication cycle.

How does the newly synthesized DNA attach to the TP? The TP is encoded by the E2B region (Figure 2) as a 80 kDa precursor (pTP)(Stillman *et al.*, 1981; Binger *et al.*, 1982). In Ad2 and Ad5 infected cells, the pTP is translated as a 671 amino acid long protein, presumably from a mRNA containing an exon at position 39 (Pettit *et al.*, 1988) coding for only three amino acids including the start codon (Stunnenberg *et al.*, 1988). The pTP becomes bound to the viral DNA during initiation (Challberg *et al.*, 1980; Lichy *et al.*, 1981; Tamanoi and Stillman, 1982; Challberg *et al.*, 1982; Van Bergen and Van der Vliet, 1983), in a reaction that requires at least ATP, dCTP, Adpol, pTP and the essential core origin sequences. No ATP hydrolysis is observed, suggesting that ATP acts as an effector molecule (Kenny *et al.*, 1988). A dCMP residue is linked with its 5′-phosphate group to the β-OH group of a serine residue located at position 580 in the C-terminal part of the protein (Smart and Stillman, 1982). The free 3′OH group of dCMP serves as a primer for further elongation. pTP remains bound to the DNA in subsequent replication rounds and apparently pTP containing molecules are equally efficient templates as parental TP-containing DNA, both *in vivo* and *in vitro* (Challberg and Kelly, 1981). This is exemplified by the efficient replication of viral DNA containing the 80 kDa protein due to a processing defect (Stillman *et al.*, 1981). Only late in infection is pTP converted to TP (Challberg and Kelly, 1981) by a 19 kDa viral protease, a late protein that also serves to process viral coat-protein precursors like pVI and pVII (Chatterjee and Flint, 1987). This protease presumably recognizes the sequence 344-DMTGGVF in pTP (Smart and Stillman, 1982) which has strong homology with the proteolytic cleavage sequence in pVI (NMSGG↓AF) (Akusjarvi and Persson, 1981). The trimming of pTP to TP does not change its covalent linkage to DNA.

In infected cells, the pTP is present in a strong complex with the DNA polymerase (pTP-pol complex) and can only be dissociated by denaturing agents like urea (Enomoto *et al.*, 1981). In addition to binding to pol, pTP presumably binds also to TP itself although direct binding studies have not yet been performed. This assumption is based upon the strong TP-TP interaction which occurs in virion DNA and which causes the DNA to appear circular in the electron microscope (Robinson *et al.*, 1973).

Although the position of the serine residue which is essential for priming is known, many aspects of the protein-priming mechanism are still to be resolved. One of these is the possible binding of pTP to origin DNA. Employing an indirect binding assay evidence was obtained that pTP can recognize the strongly conserved region 9-22 in the origin (see Figure 3) (Rijnders *et al.*, 1983). It has been difficult to confirm this result using direct binding studies (Nagata *et al.*, 1983b; Adhya *et al.*, 1986; Wides *et al.*, 1987) but recently specific binding of the pTP pol complex could be demonstrated using a gel retardation assay, provided that part of the 3′-strand was in the single-stranded form (Kenny and Hurwitz, 1988). Comparison of various mutated oligonucleotides indicated that pTP-pol binding correlated well with the ability of these fragments to serve as templates in initiation, indicating that recognition of this region by the pTP-pol complex is an essential event during initiation.

Role of the Parental Terminal Protein

The presence of the TP in viral DNA has a profound enhancing effect on DNA infectivity in transfection assays (Sharp *et al.*, 1976). This might simply be due to a decrease in degradation after transfection but, more likely, the TP enhances the efficiency of DNA replication. This is substantiated by comparison of TP-containing and protein-free DNA fragments as templates *in vitro*. A 10-20 fold higher efficiency was observed in the presence of TP (Tamanoi and Stillman, 1982; Van Bergen *et al.*, 1983; Challberg and Rawlins, 1984; Guggenheimer *et al.*, 1984a).

A priori, several functions in DNA replication might be envisaged for TP. The protein could stabilize initiation complexes or could enhance formation of a multi-protein initiation complex, *e.g.* by complexing the pTP. This might lead not only to an increase in the rate of initiation but also to an accurate positioning of the various initiation components like pol, serving to prevent internal starts. It is well known that with plasmid DNA devoid of TP initiation occurs with almost equal frequency employing the first and the fourth G as template (Tamanoi and Stillman, 1982; Van Bergen *et al.*, 1983; Kenny and Hurwitz, 1988). Such an aberrant start is never seen with the natural template.

Another possibility is that TP could help to unwind the origin. Indirect evidence for such a role comes from the increased S1 nuclease sensitivity of TP-containing DNA (Ariga *et al.*, 1979) and from the observation that single-stranded template DNA is effective in promoting the formation of a pTP-dCMP complex, both in Ad2 (Kenny and Hurwitz, 1988) and in Ad4 (Harris and Hay, 1988).

Unfortunately, conditional lethal mutants have not yet been described for TP. Clearly pTP can take over the role of TP in promoting the efficiency of replication since pTP-containing virus particles, obtained from *H2ts1s* mutants lacking the viral protease, replicate normally. (Challberg and Kelly, 1981). Moreover, *in vitro* second rounds of replication, with pTP-containing templates, can be easily observed (Horwitz and Ariga, 1981; Van Bergen and Van der Vliet, 1983). On the other hand, it is still questionable if TP can take over the priming role of pTP. Experimentally this question has been difficult to answer since TP, free of DNA, can not be isolated without the use of denaturing agents disrupting the covalent bond. Now that expression of functional pTP in heterologous systems, either in Cos cells (Pettit *et al.*, 1988) or employing recombinant vaccinia viruses (Stunnenberg *et al.*, 1988) has been successful, production of TP in the appropriate vector constructs or after protease cleaving of pTP will be helpful in approaching this problem.

Does the TP function in other aspects of DNA replication like DNA chain elongation? Some evidence was obtained for such a role using antisera (Rijnders *et al.*, 1983b) but this remains to be confirmed by mutant studies. Movement of the TP in the replication fork is an attractive hypothesis since it could bring the ends of the displaced strand in close vicinity, thereby facilitating the formation of a panhandle structure after termination of the displacement reaction.

The Adenovirus DNA Polymerase

During purification of pTP from infected cells a 140 kDa protein copurified with it (Enomoto *et al.*, 1981). This protein appeared to have DNA polymerase activity and differed in several aspects from known cellular DNA polymerases. Evidence that this was indeed an adenovirus-coded protein came from the observation that the protein could complement extracts obtained from cells infected with the so-called group N mutants (Van Bergen and Van der Vliet, 1983; Ostrove *et al.*, 1983; Friefeld et al., 1983; Stillman *et al.*, 1982). These mutants, exemplified by *H5ts36* and *H5ts149* are defective both in initiation and DNA chain elongation at the non-permissive temperature. They map in the large open reading frame of the E2B region, between map position 18.5 and 22.0 (Galos *et al.*, 1979) Recently, *H5ts36* was shown to possess a single point mutation (Miller and Williams, 1987).

To obtain direct proof that this region encodes the Ad DNA polymerase several attempts have been made to express the E2B ORF in heterologous systems, either intact (Rekosh *et al.*, 1985; Sasaguri *et al.*, 1987) or as a β-galactosidase fusion protein (Friefeld *et al.*, 1985). Antibodies against these proteins recognized the native DNA polymerase. However, no enzymatically active proteins were obtained. Recently, similar to pTP, expression of the pol gene into functionally active protein has been successful (Shu *et al.*, 1987; Stunnenberg *et al.*, 1988). In both cases additional sequences from an exon at 39 map units were added, in particular the coding sequence for three amino acids including the initiation codon(Stunnenberg *et al.*, 1988). Thus, it is most likely that previous failures to express the Ad DNA polymerase must be attributed to the use of the second ATG in the large open reading frame as initiation codon, thereby deleting an essential, 142 amino acids long, N-terminal part of the protein.

The Ad pol contains 5 regions of homology with many other DNA polymerases, both from prokaryotic (T4, ϕ29, PRD-1) and eukaryotic (yeast polI, human pol α, *Herpes simplex* virus, *Cytomegalovirus*, *vaccinia* virus and Epstein Barr virus) sources (Argos *et al.*, 1986; Wong *et al.*, 1988; Bernad *et al.*, 1987; Jung *et al.*, 1987; Larder *et al.*, 1987) and in particular with DNA polymerases coded by terminal protein containing genomes (Bernad *et al.*, 1987). However, it is completely different from *E. coli* DNA polymerase I and III, T7 DNA polymerase and from DNA polymerase β (Wilson *et al.*, 1988).

Interaction with Other Viral Proteins

The presence of a pTP-pol complex in infected cells immediately suggests that the DNA polymerase could be involved in the formation of a pTP-dCMP initiation complex. Indeed, this initiation reaction is completely dependent upon the DNA polymerase as was shown by dissociating the pTP-pol complex using 1.7 M urea containing glycerol gradients (Lichy *et al.*, 1982). The complex between pTP and pol is quite strong and forms immediately in abstracts obtained from cells infected separately with either pol or pTP recombinant vaccinia virus (van Miltenburg and van der Vliet, unpublished observations). Nevertheless it could well be that pTP and pol dissociate after initiation, since purified pTP inhibits polymerization on synthetic templates and since the DNA polymerase might associate with DBP during elongation (Field *et al.*, 1984; Lindenbaum *et al.*, 1986). A DBP-pol complex has been assumed based upon the increased thermostability of pol in the presence of DBP (Lindenbaum *et al.*, 1986) but direct evidence for a physical complex is still lacking. Specific interactions between single-stranded DBP's and DNA polymerases have been reported in prokaryotic systems like T4 (Huberman and Kornberg, 1971) and T7 (Reuben and Gefter, 1973).

A High Rate of Processive Polymerization and Strand Displacement

What effect has DBP on DNA polymerase? With single-stranded templates like oligo-dT primed poly-dA the Ad DBP increases the rate of chain elongation by pol as much as 100-fold (Field *et al.*, 1984) and creates processivity. DNA chains up to 30,000 nucleotides long are synthesized on much smaller templates suggesting a "slippage" mechanism. This effect is specific for the DBP-pol combination and might be explained by the regular, extended structure of single-stranded DNA-DBP complexes, by direct pol-DBP contacts or by both. Most interestingly, with primed partial duplex DNA as template the combination of pol and DBP is able to perform strand displacement synthesis (Lindenbaum *et al.*, 1986) by translocating through duplex DNA. This indicates a very efficient elongation reaction coupled to unwinding without the help of a helicase, making the Ad pol a rather unique enzyme. Moreover, this property is in complete agreement with its proposed role *in vivo*. ATP, but no ATP hydrolysis, is required for this displacement reaction. During chain elongation hydrolysis of dNTP's to dNMP's is observed, possibly related to a proofreading reaction, in agreement with the intrinsic $3' \rightarrow 5'$ exonuclease activity specific for single-stranded DNA which is present in the DNA polymerase molecule (Lindenbaum *et al.* , 1986).

Effect of Inhibitors

The sensitivity of Ad pol to inhibitors has been studied in some detail. With activated calf thymus DNA as a template, polymerisation is resistant to the DNA polymerase α-inhibitor aphidicolin and sensitive to N-ethylmaleimide, araCTP and dideoxynucleotide triphosphates at low concentrations (Field *et al.*, 1984). Ad DNA replication is also very sensitive to ddTTP (Van der Vliet and Kwant, 1978), but, in contrast to Ad pol is sensitive to high aphidicolin concentrations (Longiaru *et al.*, 1979; Kwant and Van der Vliet, 1980; Pincus *et al.*, 1981, Nagata *et al.*, 1983a). This effect is on elongation and suggests the involvement of another factor which is the target for aphidicolin. Alternatively, it reflects different sensitivities of Ad pol to aphidicolin when present in an active replication fork or when using artificial templates like activated calf thymus DNA. The latter hypothesis is supported by the observation that inhibition of DNA replication *in vitro* by aphidicolin can be overcome at high dCTP concentration, suggesting that the DNA polymerase is the prime target for the drug (Sussenbach and Van der Vliet, 1983).

CELLULAR PROTEINS STIMULATING DNA REPLICATION

In the absence of other factors, a very limited initiation can occur with the three viral proteins pTP, pol and DBP. This low level of synthesis, 1-3%, requires only the first 18 basepairs of the ITR ,thereby defining the core origin (Guggenheimer *et al.*, 1984b; Rawlins *et al.*, 1984; Wides *et al.*, 1987).

A nuclear extract from uninfected cells enhances initiation of DNA replication considerably, up to the level obtained with crude extracts from Ad infected cells. Employing straightforward protein purification two stimulating cellular proteins have been isolated, characterized and named nuclear factors I and III. These proteins proved to be sequence-specific DNA binding proteins recognizing two closely spaced sequences in the origin beyond the core sequence, between position 18 and 55 (Nagata *et al.*, 1982, 1983; Rawlins *et al.*, 1984; De Vries *et al.*, 1985; Pruijn *et al.*, 1986b; Rosenfeld *et al.*, 1987). NFI and NFIII enhance initiation, 10-30 fold and 3-7 fold respectively, and their effects are additive. Deletion of the NFI recognition sequence in Ad2 aborts DNA replication in transfected cells, thereby establishing that NFI-DNA interactions are also essential *in vivo* for viral replication (Hay, 1985b; Wang and Pearson, 1985). In addition to NFI and NFIII, several other cellular proteins have been isolated. Among these is nuclear factor II, a topoisomerase required for elongation beyond 30% of the genome (Nagata *et al.*, 1983). The various cellular proteins will be discussed in detail below, with special emphasis on NFI and NFIII.

Nuclear Factor I

NFI was originally purified as a 47 kDa HeLa nuclear protein which stimulates initiation (Nagata *et al.*, 1982). Subsequent DNase I footprint analysis showed that the protein bound between nucleotide 19 and 48, a sequence containing at least two blocks of nucleotides conserved in many Ad serotypes (Nagata *et al.*, 1983). A slightly different purification procedure identified proteins of 52-66 kDa and 160 kDa with similar stimulating properties as well as similar recognition sequences (Rosenfeld and Kelly, 1986; Diffley and Stillman, 1986) and the amount of NFI has been estimated at 2×10^5 per cell (Diffley and Stillman, 1986). Much effort has been invested in the detailed analysis of the NFI binding site. A dissociation constant of 2 × 10^{-11}M was obtained (Rosenfeld and Kelly, 1986) and the recognition sequence must be in the double-stranded form. Combined deletion analysis and extensive mutagenesis using a variety of synthetic binding sites has led to a symmetrical consensus sequence TGGC/A NNNNN GCCAA (Gronostajski *et al.*, 1984; Leegwater *et al.*, 1985; Schneider *et al.*, 1986; De Vries *et al.*, 1985; Wides *et al.*, 1987; Gronostajski, 1986, 1987). Although not as important as the two conserved blocks, the spacer sequence (N5) is not without influence on binding, but the rules governing strong binding are not yet clear (Gronostajski, 1987). Interestingly, a completely symmetric spacer region binds most tightly (Meisterernst *et al.*, 1988). The importance of the two conserved blocks has been established by mutagenesis (De Vries *et al.*, 1985; Schneider *et al.*, 1986; Gronostajski, 1986) and is also reflected in the contacts that are made between NFI and DNA. Based upon alkylation interference and methylation protection data, as well as an analysis of T contacts, a model has been proposed in which NFI binds as a dimer, with contacts almost exclusively at one side of the helix (De Vries *et al.* , 1987a). The base contacts extend between T-25 and T-39, in agreement with the minimal binding site derived from deletion analysis (Leegwater *et al.*, 1986) (see Figure 3). A dimeric binding mode is also consistent with the stoichiometry of the DNA-NFI complex (Diffley and Stillman, 1986). The way in which NFI finds its recognition sequence is still largely unexplored but modification of C-residues surrounding the binding site suggests a sliding model (De Vries *et al.*, 1987b).

Role of NFI Binding in Initiation

How important is NFI binding for stimulation of initiation? Mutations in the two binding blocks that reduce NFI binding also reduce DNA replication (De Vries *et al.*, 1985; Schneider *et al.*, 1986; Wides *et al.*, 1987). This suggests a direct relation between binding and stimulation, but the situation is more complicated. Some mutations (*e.g. pm 34A*, De Vries *et al.*, 1985) that inhibit NFI-binding 25 fold only slightly reduce DNA replication. Others, *e.g.* at positions 20, 21, 22 and 23, outside the minimal binding site, reduce initiation while retaining normal NFI binding (De Vries *et al.*, 1985; Wides *et al.*, 1987). A simple explanation for the first observation is difficult to give, but the latter results indicate that sequences just outside the NFI core binding site are important for its function. This notion is further substantiated by studying the required position of the NFI binding site within the origin. The NFI recognition site can be inverted without loss of stimulating activity, in agreement with a dimeric protein binding at one side of the helix (Adhya *et al.*, 1986; De Vries *et al.*, 1987a) but the position of the NFI site is rather critical. Addition or deletion of as little as two basepairs at position 22 already severely impairs NFI function (Adhya *et al.*, 1986; Wides *et al.*, 1987; De Vries *et al.*, unpublished). This could be explained either by a requirement for protein-protein interactions, for instance between NFI and pTP or pol, or by a need for flexibility in this region, or both. Preliminary observations indicate that insertions of AT-basepairs are better tolerated than GC basepairs (De Vries, Bloemers and Van der Vliet, unpublished).

Several Proteins Recognize the NFI Site

So far, we have considered NFI as one protein entity. The converse seems to be true. As mentioned, very different molecular weights have been reported for NFI-like proteins. Originally reported to comigrate with a 47 kDa protein (Nagata *et al.*, 1983), human NFI has since been isolated either as a 160 kDa protein (Prelich and Stillman, 1986; De Vries *et al.*, 1987b) or a mixture of related polypeptides migrating as 52-66 kDa (Rosenfeld and Kelly,

1986) also called CTF-1 (CCAAT-box transcription factor, Jones *et al.*, 1987). NFI from pig liver has a molecular weight of 36 kDa (Meisterernst *et al.*, 1988). While all these proteins bind the NFI site with high affinity, preliminary sequence comparisons have indicated large regions of totally unrelated sequences between human 55-65 kDa NFI/CTF-1 and pig liver NFI, apart from identical stretches (Meisterernst *et al.*, 1988).

Recent cloning and expression of NFI cDNAs has confirmed the presence of a family of proteins, presumably originating from alternatively spliced mRNAs (Santoro *et al.*, 1988; Paonessa *et al.*, 1988). Interestingly, the DNA binding domain, located in the N-terminus, is conserved. This explains the presence of lower molecular weight NFI proteins containing the conserved part of the protein and possibly originating from proteolysis. Unrelated sequences could then be derived from the non-conserved C-terminal domain.

Cellular NFI Sites

The cloning of cellular sequences that were isolated as DNA-protein complexes based upon their property to bind NFI led to an estimation of about 10^5 cellular NFI sites (Gronostajski *et al.*, 1984). Several of these so-called FIB (FI Binding) sites have been characterized in more detail by sequence analysis (Gronostajski *et al.*, 1985) but no obvious relation between such sites and particular cellular genes have been found, presumably because of the limited number of sites analyzed so far. Some cellular FIB sites can efficiently substitute for the Ad NFI site in DNA replication (Adhya *et al.* , 1986) .

Another approach has been to look for potential NFI binding sites in known genes and to study whether they actually bind purified NFI. Several such sites have been found upstream the myc oncogene (Siebenlist *et al.*, 1984), the chicken lysozyme gene (Borgmeyer el al., 1984; Leegwater *et al.*, 1986), IGM genes (Hennighausen *et al.*, 1985), and the (2-I)collagen promoter (Rossi *et al.* , 1988). Also, in viral regulatory regions like the MoLV and MMTV LTR (Speck and Baltimore, 1987; Nowock *et al.*, 1985), Hepatitis B virus S gene promoter (Shaul *et al.*, 1986) and cytomegalovirus immediate early gene promoter (Hennighausen and Fleckenstein, 1986; Jeang *et al.* , 1987) NFI sites are present. Most sequences are located in regulatory regions and deletion or mutation of NFI sites reduces gene expression 4-10 fold, except for the clustered high affinity NFI sites in cytomegalovirus (Jeang *et al.*, 1987). In the latter case it has been proposed that NFI sites may act indirectly to overcome the effects of repressors, thus providing constitutive transcription in different cellular backgrounds.

In conclusion, NFI sites seem to play an important role in transcription. More direct evidence comes from the 3-5 fold stimulation of RNA synthesis *in vitro* by NFI sites placed in front of the TATA box of the Ad major late promoter (Gronostajski, 1988). Interestingly, in the collagen promoter the presence of an NFI site confers inducability of collagen gene expression by transforming growth factor, suggesting that modification of the NFI binding protein by TGF-β is a mechanism for gene control (Rossi *et al.*, 1988).

Is the same protein functional in transcription and in Ad DNA replication? From the plethora of proteins recognizing the NFI consensus sequence it will be clear that this question is difficult to answer. Several attempts have been made. Leegwater *et al.*(1986) showed that a partially purified protein from chicken erythrocytes, binding to the lysozyme upstream NFI site, could functionally substitute for human NFI in Ad DNA replication, suggesting a high conservation of this protein. Similarly, CTF-I , isolated as a protein binding to the sequence CCAAT present in the α-globin gene, was able to substitute for NFI in DNA replication. Strong evidence comes also from heterologous expression of NFI cDNAs. The expressed proteins are functional in stimulation of DNA replication. Most interestingly, analysis of deletion mutants has shown that the conserved N-terminal DNA-binding domain is sufficient for stimulation (Mermud *et al.*, 1989; Gounari *et al.*, 1990). This domain is also involved in dimerization and is distinct from the C-terminal transcription activation domain. Thus, it is likely that Adenovirus DNA replication can be enhanced by all different members of the NFI family, independent of C-terminal divergence.

Nuclear Factor III

Screening of crude nuclear extracts for additional stimulating proteins, in the presence of excess NFI, led to the discovery of nuclear factor III (NFIII) (Pruijn *et al.* , 1986b). Independently, a protein with similar properties called Origin Recognition Protein C (ORP C) was identified (Rosenfeld *et al.*, 1987) which turned out to be identical to NFIII (O'Neill and Kelly, 1988).

NFIII is a 92-95 kDa monomer (O'Neill and Kelly, 1988; Pruijn *et al.*, 1988) which stimulates initiation of Ad2 DNA 3 to 7-fold by binding to a sequence close to the NFI site.

The NFIII Binding Site in Ad2

DNase I footprint analysis showed that the binding site in Ad2 lies between nucleotides 35 and 55, partially overlapping the NFI site. MPE footprinting as well as deletion analysis defined a core binding region of 9 nucleotides, TATGATAAT, between position 39 and 47 (Pruijn *et al.*, 1988). This site is not symmetric like the NFI site. Detailed contactpoint analysis showed that all bases within this core region are contacted. Modification of any of the bases, either by chemical means or by mutation ,interferes with NFIII binding. The level of interference differs, with some mutations showing only a 2-fold reduction in binding affinity and others, like the one at position 46 (A→G) a 10-fold reduction. Both major and minor groove contracts are detected, and contacts are not confined to one side of the helix (Pruijn *et al.* , 1988). This mode of binding contrasts sharply to that of NFI (De Vries *et al.*, 1987a) .

Remarkably, NFI and NFIII sites are very closely spaced and deletion of an AT basepair at 39 reduces both NFI and NFIII binding independently. In spite of this close spacing, no strong cooperativity between the proteins has been observed so far.

How important is NFIII binding for Ad2 initiation? Mutagenesis of positions 42 or 46 have indicated a direct relation between binding and stimulation. This has been confirmed by studies of other serotypes. Ad12 origins possess a low affinity NFIII site and are stimulated only 2-fold, even at high NFIII concentration while Ad4 origins, which bind NFIII with high affinity, are stimulated 10-fold, at least in the presence of Ad2 viral proteins (Pruijn *et al.*, 1988).

NFIII is an Octamer Binding Protein

The high affinity NFIII recognition site in Ad4 differs at two positions from the one in Ad2, being TATGCAAT instead of TATGATAAT. These two mutations increase the binding affinity 2.5-fold. Interestingly, the Ad4 sequence includes the octamer ATGCAAAT which is crucial for the expression of a number of cellular genes, like the immunoglobulin genes (Grosschedl and Baltimore, 1985) the histone H2B genes (Sive and Roeder, 1986) and the small nuclear RNA U1, U2 and U6 genes (Carbon *et al.*, 1987). These genes have in common that they are transcribed at high rates and in all these cases the octamer sequence is essential for transcription. Purified NFIII has been shown to bind to these cellular sequences and also to a number of viral enhancers containing the octamer (Pruijn *et al.*, 1987). Moreover, nuclear proteins have been isolated that bind to these transcription regulatory sequences and some of these are indistinguishable from NFIII. The best studied example is Octamer Transcription Factor I (OTFI) which is essential for octamer dependent H2B transcription (Fletcher *et al.*, 1987). This protein, with the same molecular weight as NFIII, is functionally identical to NFIII (O'Neill *et al.*, 1988; Pruijn *et al.*, 1989)). Although the formal possibility remains that NFIII/OTFI preparations are equal mixtures of two different proteins with identical molecular weights, it seems more likely that one and the same protein acts both in Ad DNA replication and in H2B transcription.

Is NFIII Required for DNA Replication In Vivo?

Viruses with deletions of the NFIII site in both ITR's are difficult to construct. For this reason, the effects of NFI and NFIII sites *in vivo* have been analysed in cells transfected with origin containing plasmid DNA in the presence of Ad helper DNA, and the level of plasmid DNA replication has been measured (Hay *et al.*, 1984). With Ad2 helper DNA, deletion of both NFI and NFIII sites reduces replication more than 100-fold while a 5-fold reduction is observed upon deletion of the Ad4 NFIII site (Hay, 1985b; Wang and Pearson, 1985). This result is in good agreement with all *in vitro* data. Interestingly, with Ad4 helper DNA no such effect is seen and in this case the core origin (1-18) replicates equally well as DNA containing the intact ITR (Hay, 1985b). This result was confirmed in a crude *in vitro* system (Harris and Hay, 1988). In view of the strong NFIII binding site in Ad4 this result is surprising. Apparently the Ad4 viral proteins, in particular the pTP-pol complex are able to overcome the need for additional cellular factors, in contrast to Ad2. It remains enigmatic why a strong NFIII site has nevertheless been conserved in Ad4 during evolution.

The Adenovirus Origin: Functional Domains

The ability of linearized plasmid DNA containing the origin to function as template *in vitro* (Tamanoi and Stillman, 1982; Van Bergen et al 1983) enabled, using site-directed mutagenesis, a detailed study of the essential DNA sequences. The results show that for Ad2, optimal DNA replication already occurs with templates containing the first 51 nucleotides of the ITR (Wides *et al.*, 1987). The sequence between 51 and 103, the end of the ITR, contains several potential protein binding sites for SP1 (89-GGGCGG-95), CREB (96-TGACGT-101) and E4F2 (81-TGGGAA-86). Indeed, proteins binding to the latter two sites could be easily demonstrated by DNase I footprint analysis (Van Driel *et al.*, unpublished). Nevertheless, binding of these proteins does not seem to have any effect in the systems used for *in vitro* DNA replication. Possibly these sites function in early gene expression or in the synthesis of complementary strands.

Within the Ad2 origin three functional blocks can be distinguished which coincide in part with the sequences conserved in the various human Ad serotypes (See Figure 3). The first block constitutes the sequence 9-ATAATATACC-18 which supports a basal level of initiation (Guggenheimer *et al.*, 1984b; Wides *et al.*, 1987) and is most likely involved in pTP binding (Rijnders *et al.*, 1983a; Kenny *et al.*, 1988). It is remarkably AT-rich and highly conserved. Mutations in this region greatly reduce template activity (Tamanoi and Stillman, 1983; Van Bergen *et al.*, 1983; Wides *et al.*, 1987). This conserved block forms part of the core region,

Figure 3. Functional regions in the adenovirus type 2/5 origin

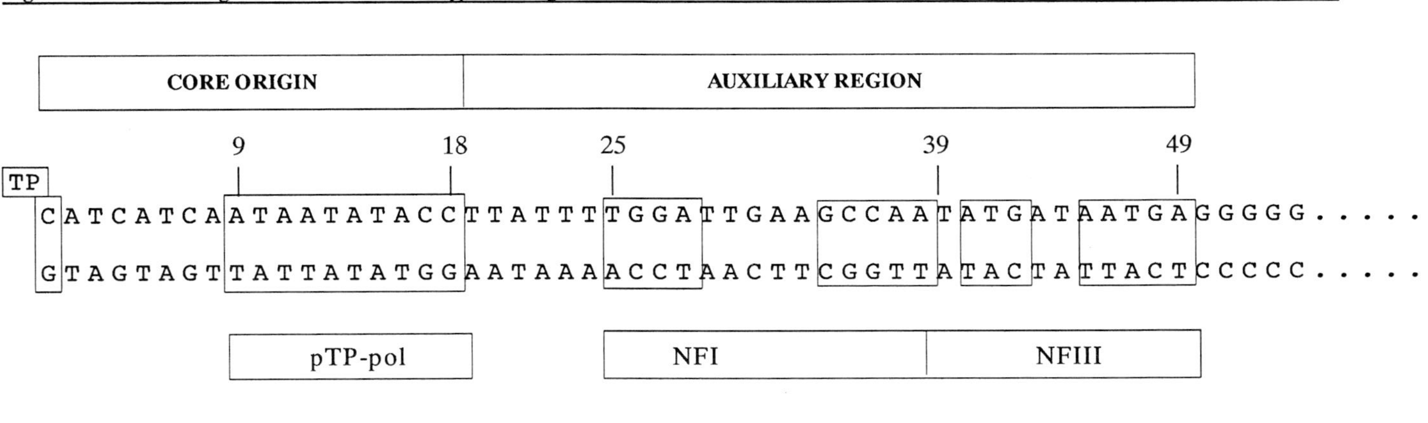

Part of the inverted terminal repetition of Ad2 is depicted. TP: terminal protein; pTP: precursor terminal protein; pol: AdDNA polymerase; NF: nuclear factor. The binding sites of NFI and NFIII are well established but the borders of the pTP-pol binding sites are not exactly known. The auxiliary region enhances DNA replication *in vitro* about 100-fold and is essential *in vivo*.

basepairs 1-18. Several basepairs in region 1-9 can be mutated without loss of activity (Challberg and Rawlins, 1984) and this region may function as a spacer (Van Bergen *et al.*,1983). Within the core region, the 3'-ended strand is most important as concluded from studies with single-stranded templates (Kenny *et al.*, 1988; Harris and Hay, 1988).

The second and third blocks constitute the NFI and NFII recognition sites, which enhance replication at least 30 fold and represent the auxiliary region (Figure3). The relative position of the functional blocks seems to be rather critical. The NFI site can be inverted or replaced by cellular sequences but its position must be fixed within 2 nucleotides (Adhya *et al.*, 1986; De Vries *et al.*, 1987b; Wides *et al.*, 1987). For NFIII such studies are still in progress (Pruijn and Van der Lugt, unpublished). With the exception of subgroups D and E all human Ad serotypes possess these three functional domains.

The origin of complementary strand synthesis has not been studied in detail yet. Synthetic panhandle constructs, devoid of the TP, have been used effectively for initiation *in vitro* and so far no difference in requirements was observed (Leegwater *et al.*, 1988), in agreement with the presence of the ITR sequences in such constructs.

RELATION BETWEEN DNA REPLICATION AND TRANSCRIPTION: ROLE OF TRANSCRIPTION FACTORS

Analysis of the adenovirus origin has clearly demonstrated that transcriptional elements act as essential components of Ad DNA replication. Transcriptional activation in the initiation of DNA is known to occur in prokaryotic systems. For instance, in *E. coli* a transcript hybridized to a region near the origin can activate replication of an otherwise inert plasmid (see Kornberg, 1988). In eukaryotic viral systems like polyomavirus, SV40 and bovine papilloma virus promoter sequences are close to origins of DNA replication (reviewed in DePamphilis, 1988). These observations indicate a close link between DNA replication and transcription, but whether these activations are due to the transcriptional process itself or just to binding of transcription factors is not always clear.

Adenovirus DNA replication presents a clear case of transcription factor activation in the absence of any transcription occurring. First of all, stimulatory effects of NFI and NFIII are seen in an *in vitro* system to which no rNTPs are added. Even more convincingly, the effects of NFI and NFIII are seen on the synthesis of a pTP-dCMP initiation complex rather than on elongation, a reaction requiring only ATP and dCTP as nucleotide triphosphates.

By what mechanism could NFI and NFIII stimulate the initiation reaction? Initial proposals for NFI concerned unwinding of the AT-rich core sequence, but experimental evidence for such unwinding is lacking. Moreover, partially unwound synthetic origins still require NFI for optimal initiation (Kenny and Hurwitz, 1988). More likely are local structural changes in the origin. Support for this comes from DNase I hypersensitivity observed adjacent to the NFI site. Such local distortions of the B-DNA structure could permit interaction of NFI with pTP-pol-bound at the core region, thereby stabilizing pTP-pol-binding or preventing DBP from binding at that position. This latter possibility finds support in experiments indicating that DBP no longer protects core DNA against nucleases in the presence of NFI (Nagata *et al.*, 1983; Kenny *et al.*, 1988). Clearly, protein-protein interactions might govern the formation of initiation complexes by well-defined three-dimensional positioning of the essential replication proteins, in agreement with the need for specific positions of the NFI binding site and possibly also the NFIII binding site in the origin. In this respect there are clear parallels with prokaryotic systems

(Kornberg, 1988). A still unsolved question is why Ad DNA replication employs NFI and NFIII sites, closely connected, rather than an array of sites for other transcription factors *e.g.* SP1. This could indicate the requirement for special structural constraints rather than just a general loosening of the DNA structure in the origin.

OTHER CELLULAR FACTORS INFLUENCING AD DNA REPLICATION

Nuclear Factor II

In contrast to NFI and NFIII, NFII is not required for initiation and is not a sequence-specific DNA binding protein. The requirement for NFII was observed when the replication of intact viral DNA containing TP was studied. DNA chain elongation is hampered after about 10,000 bp have been duplicated and NFII can relieve this block (Nagata *et al.*, 1983). Based on this property NFII was purified as a 30 kDa polypeptide with topoisomerase I activity. Interestingly, the role of NFII in DNA replication could be substituted by the intact 100 kDa HeLa topoisomerase I (review, Wang, 1987). Either NFII corresponds to a core of this protein or it is a different topoisomerase. The need for NFII indicates that torsional stress originates during movement of the replication fork, even *in vitro* in the absence of any detectable attachment of DNA to fixed superstructures. This suggests that the ends are not free to rotate, possibly due to TP-TP interactions between the termini of one molecule. This is an interesting hypothesis which might be tested using *e.g.* long viral DNA fragments devoid of TP at one end.

Factor pL

This protein from HeLa cell nuclei stimulated the replication with plasmid DNA while it had no effect on the natural TP-containing templates (Guggenheimer *et al.*, 1984 a,b). The purified protein (44k) was later shown to be a 5′→3′ exonuclease which degrades the 5′-strand, thereby creating a partially single-stranded DNA with high template activity (Kenny *et al.* , 1988). Although the effect of pL has enlarged our understanding of the template requirements for optimal initiation it is unlikely that the protein acts as such *in vivo* since natural TP-containing templates are resistant to this enzyme.

Nuclear Factor IV

A search for additional proteins binding to the Ad origin which employed an exonuclease III protection assay led to the discovery of a protein binding to the ends of the Ad genome. The purified protein, designated NFIV consists of a heterodimer of 72 and 84 kDa (Van der Vliet *et al.*, 1988; De Vries *et al.*, 1988). Its apparent specificity for the Ad origin appeared to be due mainly to the technique used for detection, exonuclease III protection. In the presence of nuclear factor I, which copurifies with NFIV during several steps, NFIV binds next to NFI which appeared to function as a blockade for translocation. Subsequent studies using other DNA fragments led to a binding model in which NFIV recognizes molecular ends and subsequently translocates on double-stranded DNA until it finds a preferred binding site or another DNA binding protein. New NFIV molecules bind and translocate to the position of the bound NFIV molecule, thus gradually filling up the DNA in a concentration dependent manner. This leads to a DNA-protein complex in which NFIV is regularly spaced with 27 bp spacing. No gross distortions of the DNA structure or its length are seen in the electron microscope. One of the preferred binding sites is the Ad origin sequence 1-18. NFI stimulates the binding of NFIV. In spite of this, the protein does not stimulate Ad DNA replication significantly. With parental

Table 2. Factors required for optimal adenovirus DNA replication *in vitro*.

FACTOR	DERIVED FROM	FUNCTIONS IN DNA REPLICATION
pTP	Virus	Initiation, protein priming
pol	Virus	Initiation, formation of pTP-dCMP complex
		Elongation, strand displacement (with DBP)
DBP	Virus	Elongation
		- binding to displaced single strands
		- improving polymerization by pol
		Initiation, enhancing NFI binding
NFI	Cell	Initiation, binding to origin
NFII	Cell	Elongation, topoisomerase I activity
NFIII	Cell	Initiation, binding to origin
Cytosol factor	Cell	Initiation and elongation, possibly as RNP
Factor PL	Cell	Initiation, only with templates devoid of TP
		5′→3′ exonuclease activity
Other factors*	Cell	Unknown

* The possible role of other proteins like NFIV and OrpA is discussed in the text.

terminal protein containing templates a 50% stimulation is observed while with protein-free templates NFIV inhibits DNA replication strongly, presumably because it prevents the binding of other essential replication components.

The role of NFIV in uninfected cells is presently unclear but may be related to processes involving molecular DNA ends like telomer replication, DNA recombination or DNA repair (De Vries *et al.*, 1988).

Origin Recognition Protein A (ORP A)

This Hela nuclear protein also binds to the very ends of the Ad2 origin. Footprint analysis shows binding between nucleotides 1-12, whether located terminally or internally (Rosenfeld *et. al.*, 1987). In contrast to NFIV, ORPA does not show affinity for other sites. Its effects on Ad DNA replication are marginal.

Cytosol Factor

A heat-stable factor from the cytosol of uninfected cells stimulates initiation 3-fold and replication of origin fragments 5 to 10-fold (Van der Vliet *et. al.*, 1984). Part of the activity is due to the presence of a phenol extractable, ribonuclease sensitive moiety, presumably low molecular weight RNA, but hybridization with origin sequences could not be detected. Inhibition by anti-trimethyl cap antibodies (P. Laquel *et. al.*, unpublished) may indicate that the stimulation is due to a small nuclear or cytoplasmic RNP particle.

Other Stimulating Factors

Two other additional factors have been reported. In HeLa nuclear extracts a protein is present that stimulates Ad4 replication but is not identical to NFI or NFIII (Harris and Hay, 1988). Similarly, during NFIII purification a stimulating side fraction, not NFI or NFIII, was detected (O'Neill and Kelly, 1988). Neither of these have been further characterized so far. The existence of other stimulating fractions is also suggested by the inhibition of DNA replication *in vitro* by auto immune antibodies (Horwitz *et. al.*, 1982; Pruijn *et. al.*, 1986a).

CONCLUSIONS AND PREVIEW

At present the adenovirus system is the most advanced DNA replication system from higher eukaryotic cells. The six most important proteins (pTP, pol, DBP, NFI, NFII and NFIII) have been purified extensively and the system can be fully reconstituted. The discovery of NFI and NFIII has enlarged our understanding of transcription activation. Moreover it has become clear that the origin is built up of several conserved protein-recognizing regions, the spacing of which is important for optimal DNA synthesis. A hypothetical scheme for the initial steps in Ad DNA replication and the role of the various proteins is given in Figure 4.

How relevant is Ad DNA replication for cellular DNA synthesizing processes? Clearly, the protein-priming mechanism differs from the initiation process in cellular origins, which most likely occurs by RNA priming comparable to initiation of *e.g.* SV40 DNA synthesis. Rather, Ad DNA synthesis may find its equivalent in gene amplification processes in replication of extrachromosomal DNA or in the duplication of telomers, serving as a prototype for replication of linear DNA. A strand displacement type of replication has been found for pea extrachromosomal DNA (Krimer and Van't Hof, 1983).

What will the direction of future research be? Now that we know most of the participants in the reaction, the field is open for a detailed study of the reaction mechanism and the various protein-protein and protein-DNA interactions. This will require large amounts of purified proteins. Presently the three viral proteins can be produced in milligram quantities, enabling a physico-chemical and structural analysis. Such studies have already begun for DBP (Tsernoglou *et al.*, 1984). Also, studies of the interactions between the three viral proteins and their relation to core DNA synthesis can start. Thus, this minimal initiation process may serve as a first model system for functional protein-protein and protein-DNA interaction in eukaryotic DNA replication. Several questions will become experimentally testable, such as:

- the interaction of pTP and pol and their binding to the origin
- the effect of DBP on the rate of polymerization by pol
- the structure of the origin during initiation
- the switch from initiation to elongation
- the role of the parental TP

Moreover, mutagenesis of the viral proteins, combined with expression of the mutated proteins, will teach us more about the various functions of DBP, the location of the primase and polymerase domains in pol and the differences in activity between pTP and TP. These studies are now underway in several laboratories.

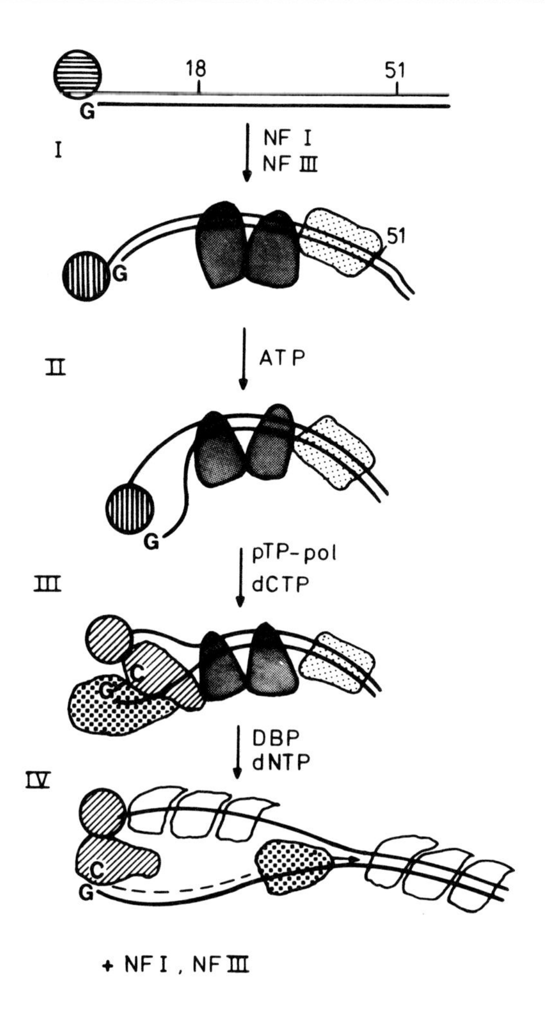

Figure 4. Hypothetical scheme for the initial steps of adenovirus DNA replication. In step I, NFI and NFIII bind to the origin leading to an unknown structural change, here arbitrarily indicated as bending. In step II, the AT-rich core region partially unwinds with ATP as effector (Kenny *et al.*, 1988), possibly involving the parental terminal protein. In step III, the pTP-pol complex binds to the 3′-strand of the partially unwound core region 9-18, presumably employing the precursor part of pTP (Rijnders *et al.*, 1983) . The binding might be stabilized by interactions with NFI as well as with the parental TP. This leads to positioning of the initiation proteins, enabling the formation of a pTP-dCMP complex. Possibly DBP is already bound to the 5′-strand (not shown). Finally, in step IV elongation starts requiring dNTP and DBP. This step may be coupled with dissociation of the pTP-pol complex and assembly of a replication fork including DBP and pol possibly as a pol-DBP complex (Lindenbaum *et al.*, 1986). pTP remains bound to TP in this model and DBP is bound to ss-DNA and ds-DNA, forming a regular DNA-protein complex. Since their binding to ss-DNA is weak, NFI and NFIII are displaced upon passing of the replication fork. I stress that many aspects of this model, in particular the order of the events, are still hypothetical.

Finally, recent cloning and expression of NFI and NFIII will enable studies of the central role of these cellular proteins in the activation of initiation. Such studies will likely employ biophysical methods and may lead to better understanding of the common features of DNA replication and transcription control. Related to this, the observed modification of NFIII, which is an important first step in *Herpes* virus infection (O'Hare and Goding, 1988) may be an attractive system for studying modification of transcription factors as a mechanism of gene regulation.

ACKNOWLEDGEMENTS

The experiments performed in the author's laboratory were supported in part by the Netherlands Foundation for Chemical Research (SON) with financial support from the Netherlands Organization for Scientific Research (NWO). I gratefully acknowledge receiving information from Drs. R. T. Hay, M. K. Kenny, J. Hurwitz, E. L. Winnacker, E. A. O'Neill and T. J. Kelly before it was published as well as other information received from R. M. Gronostajski and D. Klessig. I thank E. de Vries for critical comments.

REFERENCES

Adhya, S., P. S. Shneidman, and J. Hurwitz (1986) Reconstruction of adenovirus replication origins with a human nuclear factor I binding site. *J.Biol.Chem.* **261**: 3339-3346

Akusjarvi, G., and H. Persson (1981) Gene and mRNA for precursor polypeptide VI from adenovirus type 2. *J. Virol.* **38**: 469-482

Anderson, K. P., and D. F. Klessig (1984) Altered mRNA splicing in monkey cells abortively infected with human adenovirus may be responsible for inefficient synthesis of the virion fiber polypeptide. *Proc. Natl. Acad. Sci. USA* **81**: 4023-4027

Argos, P., A. D. Tucker, and L. Philipson (1986) Sequence comparison of proteins involved in viral and bacterial DNA replication. *Virology* **149**: 208-216

Ariga, H., H. Klein, A. J. Levine, and M. S. Horwitz (1980) A Cleavage product of the adenovirus DNA Binding Protein is Active in DNA replication *in vitro*. *Virology* **101**: 307-310

Ariga, H., H. Shimojo, S. Hidaka, and K. Miura (1979) Specific cleavage of the terminal protein from the adenovirus 5 DNA under the condition of single-strand scission by nuclease S1. *FEBS Letts* **107**: 355-358

Bernad, A., A. Zaballos, M. Salas, and L. Blanco (1987) Structural and functional relationships between prokaryotic and eukaryotic DNA polymerases. *EMBO J.* **6**: 4219-4225

Binger, M. H., S. J. Flint, and D. M. Rekosh (1982) Expression of the gene encoding the adenovirus DNA terminal protein precursor in productively infected and transformed cells. *J. Virol.* **42**: 488-501

Blanco, L., A. Bernad, and M. Salas (1988) Transition from initiation to elongation in the protein-primed phi 29 DNA replication. Salt dependent stimulation by the viral protein p6. *J. Virol* **62**: 4167-4172

Blanco, L., J. Gutierrez, J. M. Lazaro, A. Bernad, and M. Salas (1986) Replication of phage phi 29 DNA *in vitro*: role of the viral protein p6 in initiation and elongation. *Nucleic Acids Res.* **14**: 4923-4937

Bodnar, J. W., and G. D. Pearson (1980) Kinetics of adenovirus DNA replication II Initiation of adenovirus DNA replication. *Virology* **105**: 357-370

Borgmeyer, U., J. Nowock, and A. E. Sippel (1984) The TGGCA-binding protein: A eukaryotic nuclear protein recognizing a symmetrical sequence on double-stranded linear DNA. *Nucleic Acids Res* **10**: 4295-4311

Brough, D. E., S. A. Rice, S. Sell, and D. F. Klessig (1985) Restricted Changes in the Adenovirus DNA-Binding Protein That Lead to Extended Host Range of Temperature-Sensitive Phenotypes. *J. Virol.* **55**: 206-212

Brown, M., and J. Weber (1982) Discrete subgenomic DNA fragments in complete particles of adenovirus type 2. *J. Gen. Virol.* **62**: 81-89

Campbell, J. L. (1986) Eukaryotic DNA replication. *Ann. Rev. Biochem.* **55**: 736-771

Carbon, P., S. Murgo, J. P. Ebel, A. Krol, G. Tebb, and I. W. Mattaj (1987) A common octamer motif binding protein is involved in the transcription of U6 snRNA by RNA polymerase III and U2 snRNA by RNA polymerase II. *Cell* **51**: 71-79

Carter, T. B., and R. A. Blanton (1978) Possible role of the 72,000 dalton DNA binding protein in regulation of adenovirus type 5 early gene expression. *J. Virol.* **25**: 664-674

Challberg, M. D., S. V. Desiderio, and T. J. Kelly (1980) Adenovirus DNA replication *in vitro*: Characterization of a protein covalently linked to nascent DNA strands. *Proc. Natl. Acad. Sci. USA* **77**: 5105-5109

Challberg, M. D., and T. J. Kelly (1979) Adenovirus DNA replication *in vitro*. *Proc. Natl. Acad. Sci. USA* **76**: 655-659

Challberg, M. D., and T. J. Kelly (1981) Processing of the adenovirus terminal protein. *J. Virol.* **38**: 272-277

Challberg, M. D., J. M. Ostrove, and T. J. Kelly (1982) Initiation of adenovirus DNA replication: Detection of covalent complexes between nucleotide and the 80-kilodalton terminal protein. *J. Virol.* **41**: 265-270

Challberg, M. D., and D. R. Rawlins (1984) Template requirements for the initiation of adenovirus DNA replication. *Proc. Natl. Acad. Sci. USA* **81**: 100-104

Chase, J. W., and K. R. Williams (1986). *Ann. Rev. Biochem.* **55**: 103-136

Chatterjee, P. K., M. E. Vayda, and S. J. Flint (1986) Adenoviral protein VII packages intracellular viral DNA throughout the early phase of infection. *EMBO J.* **5**: 1633-1644

Chow, L. T., J. B. Lewis, and T. R. Broker (1980) RNA transcription and splicing at early and intermediate times after adenovirus-2 infection. *Cold Spring Harbor Symp. Quant. Biol.* **44**: 401-414

Cleat, P. H. and R. T. Hay (1989) Co-operative interactions between NFI and the Adenovirus DNA binding protein at the Adenovirus origin of replication. *EMBO J.* **8**: 1841-1848

Cleghon, V. G., and D. F. Klessig (1986) Association of the adenovirus DNA-binding protein with RNA both *in vitro* and *in vivo*. *Proc. Natl. Acad. Sci. USA* **83**: 8947-8951

De Vries, E., S. M. Bloemers, and P. C. Van der Vliet (1987b) Incorporation of 5-bromodeoxycytidine in the adenovirus 2 replication origin interferes with nuclear factor I binding. *Nucleic Acids Res.* **15**: 7223-7234

De Vries, E., W. Van Driel, W. G. Bergsma, A. C. Arnberg, and P. C. Van der Vliet (1989) A HeLa nuclear protein recognizing DNA termini and translocating on DNA forming a regular DNA-multiprotein complex. *J. Mol. Biol.* **208**: 65-78

De Vries, E., W. Van Driel, M. Tromp, L. Van Boom, and P. C. Van der Vliet (1985) Adenovirus DNA replication *in vitro*: site-directed mutagenesis of the nuclear factor I binding site of the Ad2 origin. *Nucleic Acids Res* **13**: 4935-4952

De Vries, E., W. Van Driel, S. J. L. Van den Heuvel, and P. C. Van der Vliet (1987a) Contactpoint analysis of the HeLa nuclear factor I recognition site reveals symmetrical binding at one side of the DNA helix. *EMBO J.* **6**: 161-168

DePamphilis, M. L. (1988) Transcriptional elements as components of eukaryotic origins of DNA replication. *Cell* **52**: 635-638

Desiderio, S. V., and T. J. Kelly (1981) Structure of the linkage between adenovirus DNA and the 55,000 molecular weight terminal protein. *J. Mol. Biol.* **145**: 319-337

Diffley, J. F. X., and B. M. Stillman (1986) Purification of a cellular, double-stranded DNA-binding protein required for initiation of adenovirus DNA replication by using a rapid filter binding assay. *Mol. Cell. Biol.* **6**: 1363-1373

Ellens, D. J., J. S. Sussenbach, and H. S. Jansz (1974) Studies on the mechanism of replication of adenovirus DNA. III Electron microscopy of replicating DNA. *Virology* **61**: 427-442

Enomoto, T., J. H. Lichy, J. E. Ikeda, and J. Hurwitz (1981) Adenovirus DNA replication *in vitro*: Purification of the terminal protein in a functional form. *Proc. Natl. Acad. Sci. USA* **78**: 6779-6783

Field, J., R. M. Gronostajski, and J. Hurwitz (1984) Properties of the Adenovirus DNA Polymerase. *J. Biol. Chem.* **259**: 9487-9495

Fletcher, C., N. Heintz, and R. G. Roeder (1987) Purification and characterization of OTF-1, a transcription factor regulating cell cycle expression of a human histone H2b gene. *Cell* **51**: 733-781

Fowlkes, D. M., S. T. Lord, T. Linne, U. Petterson, and L. Philipson (1979) Interaction between the adenovirus DNA-binding protein and double-stranded DNA. *J. Mol. Biol.* **132**: 163-180

Friefeld, B. R., R. Korn, P. J. De Jong, J. J. Sninsky, and M. S. Horwitz (1985) The 140-kDa adenovirus DNA polymerase is recognized by antibodies to *Escherichia coli*-synthesized determinants predicted from an open reading frame on the adenovirus genome. *Proc. Natl. Acad. Sci. USA* **82**: 2652-2656

Friefeld, B. R., M. D. Krevolin, and M. S. Horwitz (1983) Effects of the adenovirus *H5ts125* and *H5ts107* DNA binding proteins on DNA Replication *in vitro*. *Virology* **124**: 380-389

Galos, R. S., J. Williams, M. H. Binger, and S. J. Flint (1979) Location of additional early gene sequences in the adenoviral chromosome. *Cell* **17**: 945-956

Gounari, F., R. de Fransesco, J. Schmitt, P. C. van der Vliet, R. Cortese and H. Stunnenberg (1990) Recombinant aminoterminal domain of NFI binds DNA as a dimer and activates Adenovirus DNA replication. *EMBO J.* in press

Gronostajski, R. M. (1986) Analysis of nuclear factor I binding to DNA using degenerate oligonucleotides. *Nucleic Acids Res* **14**: 9117-9132

Gronostajski, R. M. (1987) Site-specific DNA binding of nuclear factor I: effect of the spacer region. *Nucleic Acids Res.* **15**: 5545-5559

Gronostajski, R. M., S. Adhya, K. Nagata, R. A. Guggenheimer, and J. Hurwitz (1985) Site-specific DNA binding of nuclear factor I: Analyses of cellular binding sites. *Mol. Cell. Biol.* **5**: 964-971

Gronostajski, R. M., J. Knox, D. Berry, and N. G. Miyamoto (1988) Stimulation of transcription *in vitro* by binding sites for nuclear factor I. *Nucleic Acids Res.* **16**: 2087-2098

Gronostajski, R. M., K. Nagata, and J. Hurwitz (1984) Isolation of Human DNA sequences that bind to nuclear factor I, a host protein involved in adenovirus DNA replication. *Proc. Natl. Acad. Sci. USA* **81**: 4013-4017

Grosschedl, R., and D. Baltimore (1985) Cell-type specificity of immunoglobulin gene expression is regulated by at least three DNA sequence elements. *Cell* **41**: 885-897

Guggenheimer, R. A., K. Nagata, J. Lindenbaum, and J. Hurwitz (1984a) Protein-primed Replication of Plasmids Containing the Terminus of the Adenovirus Genome. *J. Biol.Chem.* **259**: 7807-7814

Guggenheimer, R. A., B. W. Stillman, K. Nagata, F. Tamanoi, and J. Hurwitz (1984b) DNA sequences required for the *in vitro* replication of adenovirus DNA. *Proc. Natl. Acad. Sci. USA* **81**: 3069-3073

Harris, M. P. G., and R. T. Hay (1988) DNA sequences required for the initiation of adenovirus type 4 DNA replication *in vitro*. *J. Mol. Biol.* **201**: 57-68

Hay, R. T. (1985a) The origin of adenovirus DNA replication: minimal DNA sequence requirement *in vivo*. *EMBO J.* **4**: 421-426

Hay, R. T. (1985b) Origin of adenovirus DNA replication. Role of nuclear factor I binding site *in vivo*. *J. Mol. Biol.* **186**: 129-136

Hay, R. T., N. D. Stow, and I. M. McDougall (1984) Replication of adenovirus mini-chromosomes. *J. Mol. Biol.* **175**: 493-510

Hennighausen, L., and B. Fleckenstein (1986) Nuclear factor I interacts with five DNA elements in the promoter region of the human cytomegalovirus major immediate early gene. *EMBO J.* **5**: 1367-1371

Hennighausen, L., U. Siebenlist, D. Danner, P. Leder, D. Rawlins, P.J. Rosenfeld, and T.J. Kelly (1985) High affinity binding site for a specific nuclear protein in the human IgM gene. *Nature* **314**: 289-291

Horwitz, M. S. (1976) Bidirectional replication of adenovirus type 2 DNA. *Virology* **18**: 307-315

Horwitz, M. S. (1978) Temperature-sensitive replication of *H5ts125* adenovirus DNA *in vitro*. *Proc. Natl. Acad. Sci. USA* **75**: 4291-4295

Horwitz, M. S., and H. Ariga (1981) Multiple rounds of adenovirus DNA synthesis *in vitro*. *Proc. Natl. Acad. Sci. USA* **78**: 1476-1480

Horwitz, M. S., B. R. Friefeld, and H. D. Keiser (1982) Inhibition of Adenovirus DNA synthesis *in vitro* by sera from patients with systemic lupus erythematosus. *Mol. Cell. Biol.* **2**: 1492-1500

Huberman, J. A., and A. Kornberg (1971) Stimulation of T4 bacteriophage DNA polymerase by the protein product of T4 gene 32. *J. Mol. Biol.* **62**: 39-52

Imperiale, M. J., R. P. Hart, and J. R. Nevins (1985) An enhancer-like element in the adenovirus E2 promoter contains sequences essential for uninduced and EIA-induced transcription. *Proc. Natl. Acad. Sci. USA* **82**: 381-385

Jeang, K. T., D. R. Rawlins, P. J. Rosenfeld, J. H. Shero, J. T. Kelly, and G. S. Hayward (1987) Multiple tandemly repeated binding sites for cellular nuclear factor I that surround the major immediate early promoters of simian and human cytomegalovirus. *J. Virol.* **61**: 1559-1570

Jones, K. A., P. J. Kadonaga, T. J. Rosenfeld, T. J. Kelly, and R. Tjian (1987) A cellular DNA-binding protein that activates eukaryotic transcription and DNA replication. *Cell* **48**: 79-89

Jung, G., M. C. Leavitt, J.-.C. Hsieh, and J. Ito (1987) Bacteriophage PRD1 DNA polymerase: Evolution of DNA polymerases. *Proc. Natl. Acad. Sci. USA* **84**: 8287-8291

Kaplan, L. M., H. Ariga, J. Hurwitz, and M. S. Horwitz (1979) Complementation of the temperature-sensitive defect in *H5ts125* adenovirus DNA replication *in vitro*. *Proc. Natl. Acad. Sci. USA* **76**: 5534-5538

Kedinger, C., O. Brison, F. Perrin, and J. Wilhelm (1978) Structural analysis of viral replication intermediates isolated from adenovirus type-2 infected HeLa cell nuclei. *J. Virol.* **26**: 364-379

Kelly, T. J. (1984) Adenovirus DNA replication. **In**: H.S. Ginsberg (ed): *The adenoviruses*. New York: Plenum Press, pp. 271-308.

Kenny, M. K., L. A. Balogh, and J. Hurwitz (1988) Initiation of adenovirus replication: I Mechanism of action of a host protein required for replication of adenovirus DNA templates devoid of the terminal protein. *J. Biol. Chem.* **263**: 9801-9808

Kenny, M. K., and J. Hurwitz (1988) Initiation of adenovirus DNA replication: II. Structural requirements using synthetic oligonucleotide adenovirus templates. *J. Biol. Chem.* **263**: 9809-9817

Kitchingman, G. R. (1985) Sequence of the DNA binding protein of a human subgroup E adenovirus (type 4): comparisons with subgroup A (type 12), subgroup B (type 7) and subgroup C (type 5). *Virology* **146**: 90-101

Klein, H., W. Maltzman, and A. J. Levine (1979) Structure function relationships of the adenovirus DNA binding protein. *J. Biol. Chem.* **254**: 11051-11060

Klessig, D. F., and T. Grodzicker (1979) Mutations that allow human Ad2 and Ad5 to express late genes in monkey cells map in the viral gene encoding the 72K DNA binding protein. *Cell* **17**: 957-966

Kornberg, A. (1988) *DNA replication*. *J. Biol. Chem.* **263**: 1-4

Krevolin, M. D., and M. S. Horwitz (1987) Functional interactions of the domains of the adenovirus DNA binding protein. *Virology* **156**: 167-170

Krimer, D. B., and J. Van't Hof (1983) Extrachromosomal DNA of pea (*Pisum sativum*) root-tip cells replicates by strand displacement. *Proc. Natl. Acad. Sci. USA* **80**: 1933-1937

Kruijer, W., F. M. A. Van Schaik, J. G. Speijer, and J. S. Sussenbach (1983) Structure and function of adenovirus DNA binding protein: Comparison of the amino acid sequences of the Ad5 and Ad12 proteins derived from the nucleotide sequence of the corresponding genes. *Virology* **128**: 140-153

Kruijer, W., F. M. A. Van Schaik, and J. S. Sussenbach (1981) Structure and organization of the gene coding for the DNA binding protein of adenovirus type 5. *Nucleic. Acids. Res.* **9**: 4439-4457

Kruijer, W., F. M. A. Van Schaik, and J. S. Sussenbach (1982) Nucleotide sequence of the gene encoding adenovirus type 2 DNA binding protein. *Nucl. Acids Res.* **10**: 4493-4500

Kuil, M. E., H. van Amerongen, P. C. van der Vliet and R. van Grondelle (1990) Complex formation between the adenovirus DNA-binding protein and single-stranded poly (rA). Cooperativity and salt dependence. *Biochemistry* in press

Kwant, M. M., and P. C. Van der Vliet (1980) Differential effect of aphidicolin on adenovirus DNA synthesis and cellular DNA synthesis. *Nucleic Acids Res.* **8**: 3993-4007

Larder, B. A., S. D. Kemp, and G. Darby (1987) Related functional domains in virus DNA polymerases. *EMBO J.* **6**: 169-175

Lechner, R. L., and T. J. Kelly (1977) The structure of replicating adenovirus 2 DNA molecules. *Cell* **12**: 1007-1020

Leegwater, P. A. J., R. F. A. Rombouts and P. C. van der Vliet (1988) Adenovirus DNA replication *in vitro* : duplication of single-stranded DNA containing a panhandle structure. *Biochim. Biophys. Acta* **951**: 403-410

Leegwater, P. A. J., P. C. van der Vliet, R. A. W. Rupp, J. Nowock, and A. E. Sippel (1986) Functional homology between the sequence-specific DNA binding proteins nuclear factor I from HeLa cells and the TGGCA protein from chicken liver. *EMBO J.* **5**: 381-386

Leegwater, P. A. J., W. Van Driel, and P. C. Van der Vliet (1985) Recognition site of nuclear factor I, a sequence-specific DNA binding protein from HeLa cells that stimulates adenovirus DNA replication. *EMBO J.* **4**: 1515-1521

Lichy, J. H., J. Field, M. S. Horwitz, and J. Hurwitz (1982) Separation of the adenovirus terminal protein precursor from its associated DNA polymerase: Role of both proteins in the initiation of adenovirus DNA replication. *Proc. Natl. Acad. Sci. USA* **79**: 5225-5229

Lichy, J. H., M. S. Horwitz, and J. Hurwitz (1981) Formation of a covalent complex between the 80,000-dalton adenovirus terminal protein and 5'-dCMP *in vitro*. *Proc. Natl. Acad. Sci. USA* **78**: 2678-2682

Lindenbaum, J. D., J. Field, and J. Hurwitz (1986) The adenovirus DNA binding protein and adenovirus DNA polymerase interact to catalyze elongation of primed DNA templates. *J. Biol. Chem.* **261**: 10218-10227

Linne, T., and L. Philipson (1980) Further characterization of the phosphate moiety of the adenovirus type 2 DNA-binding protein. *Eur. J. Biochem.* **103**: 259-270

Longiaru, M., J. Ikeda, Z. Jarkovsky, S. B. Horwitz, and M. S. Horwitz (1979) The effect of aphidicolin on adenovirus DNA synthesis. *Nucleic Acids Res.* **6**: 3369-3386

Matsuguchi, M., F. Puvion-Dutilleul, and G. Moyne (1979) Late transcription and simultaneous replication of simian adenovirus 7 DNA as revealed by spreading lytically infected cell structures. *J. Gen. Virol.* **42**: 443-456

Meisterernst, M., I. Gander, L. Rogge, and E. L. Winnacker (1988) A quantitative analysis of nuclear factor I/DNA interactions. *Nucleic Acids Res.* **16**: 4419-4435

Mermud, N., E. A. O'Neill, T. J. Kelly and R. Tjian (1989) The proline-rich transcriptional activator of CTF/NF-1 is distinct from the replication and DNA binding domain. *Cell* **58**: 741-753

Miller, B. W., and J. Williams (1987) Cellular transformation by adenovirus type 5 is influenced by the viral DNA polymerase. *J. Virol.* **61**: 3630-3634

Murthy, S. C. S., G. P. Bhat, and B. Thimmappaya (1985) Adenovirus EIIA early promoter: Transcriptional control elements and induction by the viral pre-early EIA gene, which appears to be sequence independent. *Proc. Natl. Acad. Sci. USA* **82**: 2230-2234

Nagata, K., R. A. Guggenheimer, T. Enomoto, J. H. Lichy, and J. Hurwitz (1982) Adenovirus DNA replication *in vitro*: Identification of a host factor that stimulates synthesis of the preterminal protein dCMP complex. *Proc. Natl. Acad. Sci. USA* **79**: 6438-6442

Nagata, K., R. A. Guggenheimer, and J. Hurwitz (1983a) Adenovirus DNA replication *in vitro*: Synthesis of full-length DNA with purified proteins. *Proc. Natl. Acad. Sci. USA* **60**: 4266-4270

Nagata, K., R. A. Guggenheimer, and J. Hurwitz (1983b) Specific Binding of a Cellular DNA replication protein to the origin of replication of adenovirus DNA. *Proc. Natl. Acad. Sci. USA* **80**: 6177 6181

Nass, K., and G. D. Frenkel (1980) Adenovirus-specific DNA-binding protein inhibits the hydrolysis of DNA by DNase *in vitro*. *J. Virol.* **35**: 314-319

Nevins, J. R., and J. J. Winkler (1980) Regulation of early adenovirus transcription: A protein product of early region 2 specifically represses region 4 transcription. *Proc. Natl. Acad. Sci. USA* **77**: 1893 1897

Nicolas, J. C., D. Ingrand, P. Sarnow, and A. J. Levine (1982) A mutation in the adenovirus type 5 DNA binding protein that fails to autoregulate the production of the DNA binding protein. *Virology* **122**: 481-485

Nowock, J., U. Borgmeyer, A. W. Puschel, R. A. W. Rupp, and A. E. Sippel (1985) The TGGCA protein binds to the MMTV-LTR, the adenovirus origin of replication, and the BK virus enhancer. *Nucleic Acids Res.* **13**: 2045-2061

O'Hare, A., and C. R. Goding (1988) Herpes simplex virus regulatory elements and the immunoglobulin octamer domain bind a common factor and are both targets for virion transactivation. *Cell* **52**: 435-445

O'Neill, E. A., C. Fletcher, C. R. Burrow, N. Heintz, R. G. Roeder, and T. J. Kelly (1988) The transcription factor OTF-I is functionally identical to the adenovirus DNA replication factor NF-III. *Science* **241**: 1210-1213

O'Neill, E. A., and T. J. Kelly (1988) Purification and characterization of nuclear factor III (origin recognition protein C), a sequence specific DNA binding protein required for efficient initiation of adenovirus DNA replication. *J. Biol. Chem.* **263**: 931 937

Ostrove, J. M., P. Rosenfeld, J. Williams, and T. J. Kelly (1983) *In vitro* complementation as an assay for purification of adenovirus DNA replication proteins. *Proc. Natl. Acad. Sci. USA* **80**: 935-939

Paonessa, G., F. Gounari, R. Frank and R. Cortese (1988) Purification of a NFI-like DNA-binding protein from rat liver and cloning of the corresponding cDNA. *EMBO J.* **7**: 3115-3123

Pettit, S. C., M. S. Horwitz, and J. A. Engler (1988) Adenovirus preterminal protein synthesized in COS cells from cloned DNA is active in DNA replication *in vitro*. *J. Virol.* **62**: 496-500

Pincus, S., W. Robertson, and D. Rekosh (1981) Characterization of the effect of aphidicolin on adenovirus DNA replication: evidence in support of a protein primer model of initiation. *Nucleic Acids Res.* **9**: 4919-4938

Prelich, G., and B. W. Stillman (1986) Functional characterization of thermolabile DNA-binding proteins that affect adenovirus DNA replication. *J. Virol.* **57**: 883-892

Prieto, I., M. Serrano, J. M. Lazaro, M. Salas, and J. M. Hermoso (1988) Interaction of the bacteriophage phi 29 protein p6 with double stranded DNA. *Proc. Natl. Acad. Sci. USA* **85**: 314-318

Pruijn, G. J. M., H. G. Kusters, F. H. J. Gmelig-Meyling, and P. C. van der Vliet (1986a) Inhibition of adenovirus DNA replication *in vitro* by autoimmune sera. *Eur. J. Biochem.* **154**: 363-370

Pruijn, G. J. M., P. C. van der Vliet, N. A. Dathan and I. W. Mattaj (1989) Anti-OTF-I antibodies inhibit NFIII stimulation of *in vitro* Adenovirus DNA replication. *Nucleic Acids Research* **17**: 1845-1863

Pruijn, G. J. M., W. Van Driel, and P. C. Van der Vliet (1986b) Nuclear factor III, a novel sequence-specific DNA-binding protein from HeLa cells stimulating adenovirus DNA replication. *Nature* **322**: 656-659

Pruijn, G. J. M., W. van Driel, R. T. van Miltenburg, and P. C. van der Vliet (1987) Promoter and enhancer elements containing a conserved sequence motif are recognized by nuclear factor III, a protein stimulating adenovirus DNA replication. *EMBO J.* **6**: 3771-3778

Pruijn, G. J. M., R. T. Van Miltenburg, J. A. J. Claessens, and P. C. Van der Vliet (1988) The interaction between the octamer binding protein nuclear factor III and the adenovirus origin of DNA replication. *J. Virol.* **62**: 3092-3102

Rawlins, D. R., P. J. Rosenfeld, R. J. Wides, M. D. Challberg, and T. J. Kelly Jr. (1984) Structure and function of the adenovirus origin of replication. *Cell* **37**: 309-319

Rekosh, D., J. Lindenbaum, J. Brewster, L. M. Mertz, J. Hurwitz, and L. Prestine (1985) Expression in *Escherichia coli* of a fusion protein product containing a region of the adenovirus DNA polymerase. *Proc. Natl. Acad. Sci. USA* **82**: 2354-2358

Rekosh, D. M. K., W. C. Russell, A. J. D. Bellett, and A. J. Robinson (1977) Identification of a protein linked to the ends of adenovirus. *Cell* **11**: 283-295

Reuben, R. C., and M. C. Gefter (1973) ADNA-binding protein induced by bacteriophage T7. *Proc. Natl. Acad. Sci. USA* **70**: 1846-1850

Rice, S. A., and D. F. Klessig (1984) The function(s) provided by the adenovirus-specified DNA-binding protein required for viral late gene expression is independent of the role of the protein in viral DNA replication. *J. Virol.* **49**: 35-49

Rijnders, A. W. M., B. G. M. Van Bergen, P. C. Van der Vliet, and J. S. Sussenbach (1983a) Specific binding of the adenovirus terminal protein precursor-DNA polymerase complex to the origin of DNA replication. *Nucleic Acids Res.* **11**: 8777-8789

Rijnders, A. W. M., B. G. M. Van Bergen, P. C. Van der Vliet, and J. S. Sussenbach (1983b) Immunological characterization of the role of adenovirus terminal protein in viral DNA replication. *Virology* **131**: 287-295

Robinson, A. J., H. B. Younghusband, and A. J. D. Bellett (1973) A circular DNA-protein complex for adenoviruses. *Virology* **56**: 54-69

Rosenfeld, P. J., and T. J. Kelly (1986) Purification of nuclear factor I by DNA recognition site affinity chromatography. *J. Biol. Chem.* **261**: 1398-1408

Rosenfeld, P. J., E. A. O'Neill, R. J. Wides, and T. J. Kelly (1987) Sequence-specific interactions between cellular DNA-binding proteins and the adenovirus origin of DNA replication. *Mol. Cell. Biol.* **7**: 875-886

Rossi, P., G. Karsenty, A. B. Roberts, N. S. Roche, M. B. Sporn, and B. De Crombrugghe (1988) A nuclear factor I binding site mediates the transcriptional activation of a type I collagen promoter by transforming growth factor β. *Cell* **52**: 405-414

Santoro, C., N. Mermud, P. C. Andrews and R. Tjian (1988) A family of CCAAT-box-binding proteins active in transcription and DNA replication: cloning and expression of multiple cDNAs. *Nature* **334**: 218-224

Sasaguri, Y., T. Sanford, P. Aguirre, and R. Padmanabhan (1987) Immunological analysis of 140-kDa adenovirus-encoded DNA polymerase in adenovirus type 2-infected HeLa cells using antibodies raised against the protein expressed in Escherichia coli. *Virology* **160**: 389-399

Schechter, N. M., W. Davies, and C. W. Anderson (1980) Adenovirus coded deoxyribonucleic acid binding protein. Isolation, physical properties and effects of proteolytic digestion. *Biochemistry* **19**: 2802-2810

Schilling, R., B. Weingartner, and E. L. Winnacker (1975) Adenovirus 2 DNA replication. II. Termini of DNA replication. *Virology* **16**: 767-774

Schneider, R., I. Gander, U. Muller, R. Mertz, and E. L. Winnacker (1986) A sensitive and rapid gel retention assay for nuclear factor I and other DNA-binding proteins in crude nuclear extracts. *Nucleic Acids Res.* **14**: 1303-1317

Sharp, P. A., C. Moore, and J. L. Haverty (1976) The infectivity of adenovirus 5 DNA-protein complex. *Virology* **75**: 442-456

Shaul, Y., R. Ben-Levy, and T. De-Medina (1986) High affinity binding site for nuclear factor I next to the hepatitis B virus S gene promoter. *EMBO J.* **5**: 1967-1971

Shu, L., M. S. Horwitz, and J. A. Engler (1987) Expression of enzymatically active adenovirus DNA polymerase from cloned DNA requires sequences upstream of the main open reading frame. *Virology* **161**: 520-526

Sive, H. L., and R. G. Roeder (1986) Interaction of a common factor with conserved promoter and enhancer sequences in histone 2B, immunoglobulin and U2 snRNA genes. *Proc. Natl. Acad. Sci. USA* **83**: 6382-6386

Smart, J. E., and B. W. Stillman (1982) Adenovirus Terminal Protein Precursor. *J. Biol. Chem.* **257**: 13499-13506

Speck, N. A., and D. Baltimore (1987) Six distinct nuclear factors interact with the 75-base-pair repeat of the Moloney murine leukemia virus enhancer. *Mol. Cell. Biol.* **7**: 1101-1110

Stillman, B. W., J. B. Lewis, L. T. Chow, M. B. Mathews, and J. E. Smart (1981) Identification of the gene and mRNA for the adenovirus terminal protein precursor. *Cell* **23**: 479-508

Stillman, B. W., F. Tamanoi, and M. B. Mathews (1982) Purification of an adenovirus-coded DNA polymerase that is required for initiation of DNA replication. *Cell* **31**: 613-623

Stillman, B. W., E. White, and T. Grodzicker (1984) Independent mutations in Ad2ts111 cause degradation of cellular DNA and defective viral DNA replication. *J. Virol.* **50**: 598-605

Stuiver, M. H. and P. C. van der Vliet (1990) The Adenovirus DNA binding protein forms a multimeric protein complex with double-stranded DNA and enhances binding of nuclear fractor I. *J. Virol.* in press

Stunnenberg, H. G., H. Lange, L. Philipson, R. T. Van Miltenburg, and P. C. Van der Vliet (1988) High expression of functional adenovirus DNA polymerase and precursor terminal protein using recombinant vaccinia virus. *Nucleic Acids Res.* **16**: 2431-2444

Sugawara, K., Z. Gilead, and M. Green (1977) Purification and molecular characterization of adenovirus type 2 DNA-binding protein. *J. Virol.* **21**: 338-346

Sussenbach, J. S., and M. G. Kuijk (1978) Studies on the mechanism of replication adenovirus DNA. VI. Localization of the origins of the displacement synthesis. *Virology* **84**: 509-517

Sussenbach, J. S., and P. C. Van der Vliet (1983) The mechanism of adenovirus DNA replication and the characterization of replication proteins. *Curr. Top. Microbiol. Immunol.* **109**: 53-73

Sussenbach, J. S., P. C. van der Vliet, D. S. Ellens, and H. S. Jansz (1972) Linear intermediates in the replication of adenovirus DNA. *Nature New. Biol.* **239**: 47-49

Tamanoi, F. (1986) On the mechanism of adenovirus DNA replication. **In**: W. Doerfler (ed): *Adenovirus DNA, the viral genome and its expression. Developments in molecular virology.* Nijhoff, pp. 97 219.

Tamanoi, F., and B. W. Stillman (1982) Function of adenovirus terminal protein in the initiation of DNA replication. *Proc. Natl. Acad. Sci. USA* **79**: 2221-2225

Tamanoi, F., and B. W. Stillman (1983) Initiation of adenovirus DNA replication *in vitro* requires a specific DNA sequence. *Proc. Natl. Acad. Sci. USA* **80**: 6446-6450

Tsernoglou, D., A. Tsugita, A. D. Tucker, and P. C. Van der Vliet (1985) Characterization of the chymotryptic core of the adenovirus DNA binding protein. *FEBS Letts* **188**: 248-252

Tsernoglou, D., A. D. Tucker, and P. C. Van der Vliet (1984) Crystallization of a fragment of the adenovirus DNA binding protein. *J. Mol. Biol.* **172**: 237-239

Van Bergen, B. G. M., P. A. Van de Ley, W. Van Driel, A. D. M. Van Mansfeld, and P. C. Van der Vliet (1983) Replication of origin containing adenovirus DNA fragments that do not carry the terminal protein. *Nucleic Acids Res.* **11**: 1975-1989

Van Bergen, B. G. M., and P. C. Van der Vliet (1983) Temperature sensitive initiation and elongation of adenovirus DNA replication *in vitro* with nuclear extracts from *H5ts36-,H5ts149-* and *H5ts125* infected HeLa cells. *J. Virol.* **46**: 642-648

Van der Vliet, P. C., J. Claessens, E. De Vries, P. A. J. Leegwater, G. J. M. Pruijn, W. Van Driel, and R. T. Van Miltenburg (1988) Interaction of cellular proteins with the adenovirus origin of DNA replication. *Cancer cells* **6**: 61-70

Van der Vliet, P. C., W. Keegstra, and H. S. Jansz (1978) Complex formation between the adenovirus type 5 DNA-binding protein and single-stranded DNA. *Eur. J. Biochem.* **86**: 389-398

Van der Vliet, P. C., and M. M. Kwant (1978) Role of DNA polymerase γ in adenovirus DNA replication. *Nature* **276**: 532-534

Van der Vliet, P. C., A. J. Levine, M. Ensinger, and H. S. Ginsbergs (1975) Thermolabile DNA binding proteins from cells infected with a temperature-sensitive mutant of adenovirus defective in viral DNA synthesis. *Virology* **15**: 348-354

Van der Vliet, P. C., and J. S. Sussenbach (1975) An adenovirus type 5 gene function required for initiation of viral DNA replication. *Virology* **67**: 415-426

Van der Vliet, P. C., D. Van Dam, and M. M. Kwant (1984) Adenovirus DNA replication *in vitro* is stimulated by RNA from uninfected HeLa cells. *FEBS Letts* **171**: 5-9

Van der Vliet, P. C., J. Zandberg, and H. S. Jansz (1977) Evidence for a function of the adenovirus DNA binding protein in initiation of DNA synthesis as well as in elongation on nascent DNA chains. *Virology* **80**: 98-110

Van Amerongen, M., R. Van Grondelle, and P. C. Van der Vliet (1987) Interaction between adenovirus DNA binding protein and single stranded polynucleotides studied by circular dichroism and ultraviolet absorption. *Biochemistry* **26**: 4646-4652

Vos, H. L., D. E. Brough, F. M. van der Lee, R. C. Hoeben, G. H. Verheijden, D. Dooyes, D. F. Klessig and J. S. Sussenbach (1989) Characterization of Adenovirus type 5 insertion and deletion mutants encoding altered DNA binding proteins. *Virology* **172** 634-642

Vos, H. L., F. M. Van der Lee, A. M. C. B. Reemst, A. E. Van Loon, and J. S. Sussenbach (1988) The genes encoding the DNA binding protein and the 23 K protease of adenovirus types 40 and 41. *Virology* **163**: 1-10

Wang, J. C. (1987) Recent studies of DNA topoisomerase. *Biochim. Biophys. Acta* **909**: 1-9

Wang, K., and G. D. Pearson (1985) Adenovirus sequences required for replication *in vivo. Nucleic Acids Res.* **13**: 5173-5187

Weingartner, B., E. L. Winnacker, A. Tolun, and U. Petterson (1976) Two complementary strand specific termination sites for adenovirus DNA replication. *Cell* **9**: 259-268

Wides, R. J., M. D. Challberg, D. R. Rawlins, and T. J. Kelly (1987) Adenovirus origin of DNA replication: Sequence requirements for replication *in vitro. Mol. Cell. Biol.* **7**: 864-874

Wilson, S., J. Abbotts, and S. Widen (1988) Progress toward molecular biology of DNA polymerase β. *Biochim. Biophys. Acta* **949**: 149-157

Wong, S. W., A. F. Wahl, P.-.M. Yuan, N. Arai, B. E. Pearson, K.-.i. Arai, D. Korn, and M. W. Hunkapiller (1988) Human DNA polymerase alpha gene expression is cell proliferation dependent and its primary structure is similar to both prokaryotic and eukaryotic replicative DNA polymerases. *EMBO J.* **7**: 37-48

2

DNA REPLICATION REGULATED BY *c-myc* PROTEIN

Sanae M. M. Iguchi-Ariga and Hiroyoshi Ariga

INTRODUCTION

The replication of eukaryotic DNA occurs by initiation at many chromosomal sites which are usually clustered in a specific region characteristic of the cell type and the stage of the cell cycle (Kornberg, 1980). To understand the molecular mechanism of this process, the isolation of replication origins and the identification of proteins promoting DNA synthesis are essential. To date, three main systems have been taken as models: adenovirus, SV40 and yeast. The study of their *in vitro* replication mechanisms, supported by genetic and recombination techniques, has led to the identification of several cellular proteins and of various DNA sequences involved in DNA synthesis (Cambell, 1986).

It is little known, however, what mechanisms are carried out in higher eukaryotic animal cells. We describe here recent results concerning replication origins and initiation proteins in animal cells.

ISOLATION OF THE ORIGIN OF DNA REPLICATION

Autonomously replicating sequences (ARS) are thought to be actual origins of replication because, in yeast cells, a) they share analogous properties with *OriC* plasmids in *E. coli* which contain origins of bacterial DNA, b) there are 400 chromosomal ARSs per cell, a number that fits the average replicon size, c) in contrast to mitochondrial DNA, they replicate under the control of genes regulating the cell cycle and only during S phase, d) they replicate only once per cell cycle, and e) DNA synthesis in several *in vitro* systems initiates specifically at an ARS sequence. Analogously, it could also be true that the DNA fragments able to replicate autonomously in higher eukaryotes, including animal cells, contain a replication origin.

The cloning of ARS is usually performed as follows; chromosomal DNA is digested with restriction enzymes and cloned into bacterial vectors containing a selection marker such as *URA3* in yeast, *pgt, neo, tk* in mammalian cells. Cells defective for those marker genes are then transfected with these clones and cultured in a selection medium. Transformants resistant to the chosen selection conditions are isolated and subsequently screened for ARS candidates. Several ARSs replicating in yeast cells were easily isolated using these procedures (Cambell, 1986). The isolation of ARS from mammalian cells, however, remained long unsuccessful. This was due to the fact that, in contrast to yeast cells,mammalian cells show a lower transfection efficiency and require a more laborious selection culture technique. Recently, however, we and two other groups have succeeded in cloning ARSs from mammalian cells (Ariga *et al.*, 1987; Iguchi-Ariga *et al.*, 1987a; Frappier and Zannis-Hadjopoulos, 1987; Holst *et al.*, 1988). We briefly introduce our results concerning these significant findings.

Since it might be possible that the origin sequences have been conserved, including those of animal viruses, we first looked for an ARS clone from mouse cells that could replicate in an SV40 T antigen dependent system (Ariga *et al.*, 1985). The isolated clone, pMU65, could replicate in monkey CosI cell, which constitutively produce SV40 T antigen, the initiation protein of SV40 DNA replication, as well as in an *in vitro* SV40 DNA system (Ariga and Sugano, 1983; Ariga, 1984). pARS65, a subclone of pMU65, was further examined in cells that do not produce SV40 T antigen; it turned out that pARS65 can replicate well in various mammalian cells, including human, mouse and rat cells (Ariga *et al.*, 1987). Furthermore, when the whole plasmid was tested in an *in vitro* system developed from mouse FM3A cells, it was observed that pARS65 replication is semiconservative and that it initiates at the sequences derived from the mouse DNA (Ariga *et al.*, 1987). Other ARSs from monkey cells (Frappier and Zannis-Hadjopoulos, 1987) and from mouse cells (Holst *et al.*, 1988) were later reported to be functional. An important observation was that the replication rate of pARS65 in various cells almost parallels the expression level of the protooncogene *c-myc* in the same cells (Figure 1), which prompted us to examine the correlation between DNA replication and the *c-myc* gene.

POSSIBLE FUNCTION OF *c-myc* GENE PRODUCT

Participation of the *c-myc* Protein in Cellular DNA Replication (Iguchi-Ariga *et al.*, 1987a)

When HL-60 cells were transfected with anti-human *c-myc* antibodies by a liposome-mediated transfection technique (Itani *et al.*, 1987), DNA replication, as well as cell growth, was inhibited, suggesting that the *c-myc* protein is necessary for, or heavily involved in, cellular DNA replication. Furthermore, replication of pARS65 in transfected cells (*in vivo*) and in a cell-free system (*in vitro*) was also inhibited by the addition of anti-*c-myc* antibodies (Figure 2).

Binding of *c-myc* Protein to Mouse ARS

The role of the *c-myc* protein in DNA replication, in particular of pARS65, may imply a binding of this protein to particular DNA sequences — as is the case for SV40 T antigen promoting SV40 DNA synthesis by binding to its origin (*ori*). We examined the possibility of *c-myc* protein binding to pARS65 by an immunobinding assay. Despite previous reports that the *c-myc* product has a strong affinity for DNA but has no sequence specificity (Kornberg,

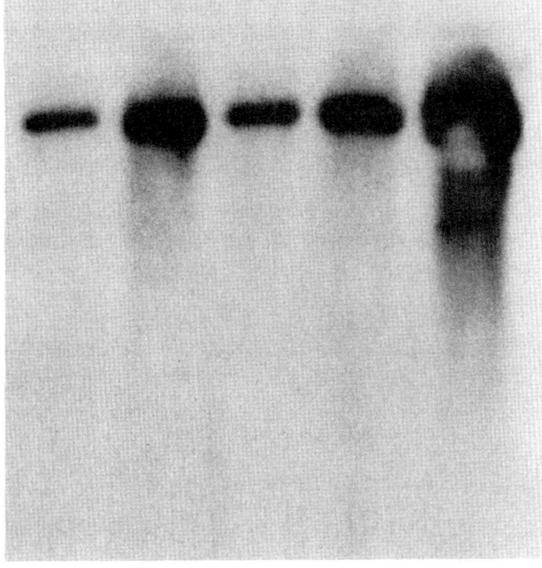

Figure 1. Southern blot analysis of the DNA from Hirt supernatants of various transfected cells. Samples of 10^6 FM3A, NS-1, and HL-60 cells were transfected with 5 μg of pARS65. 48h after transfection, the low-molecular-weight DNA was extracted from Hirt supernatants, digested with *Dpn*I and *EcoR*I, electrophoresed on 1 % agarose gel, transferred to nitrocellulose filter by the Southern method, and then hybridized with ^{32}P-labelled p64-*tk* probe. Lanes: 1, 10^2 copies per cell; 2, 10^3 copies per cell; 3, FM3A; 4, NS-1; 5, HL-60.

1982), the high sensitivity of this assay allowed us to detect specific binding over a low background of non specific binding (Iguchi-Ariga *et al.*, 1987b). pARS65 was digested with *Eco*RI, *Bgl*II, and *Bam*HI, and the resulting fragments end-labelled with α^{32}P-ATP. The fragments were then incubated with the nuclear extract of HL-60 cells. In HL-60 cells, a human promyelocytic leukemia cell line, the *c-myc* gene is amplified to more than 20 copies and is expressed vigorously, so that their nuclear extract contains a large amount of *c-myc* protein even though the protein itself is known to be labile. Upon addition of specific anti-human *c-myc* antibodies, a 2.2-2.5 kb fragment of pARS65 was precipitated. On the other hand, another portion of the plasmid, a 3.2 kb *Bam*HI-*Eco*RI fragment derived from pKSV 1, was not precipitated, which suggests that the *c-myc* protein binds specifically to sequences of the 2.2/2.5 kb fragments of pARS65. This idea is supported by the fact that no labelled material was precipitated with non-specific anti-human antibody. Since it is difficult to distinguish 2.2 from 2.5 kb fragments on a gel, the DNA fragments were recovered from the immune-complexes by phenol extraction and electrophoresed after digestion with *Hind*III (Figure 3B). A similar electrophoretic pattern with a single band of 2.5 kb was observed before and after the digestion, indicating that the specific binding site of the *c-myc* protein is present in the 2.5 kb *Eco*RI-*Bgl*II fragment of pARS65 which is derived from mouse DNA and contains the initiation site of autonomous replication (see Figure 3E). So, a conclusion from these results is that the *c-myc* protein may promote cellular DNA replication by binding to the origin of DNA replication.

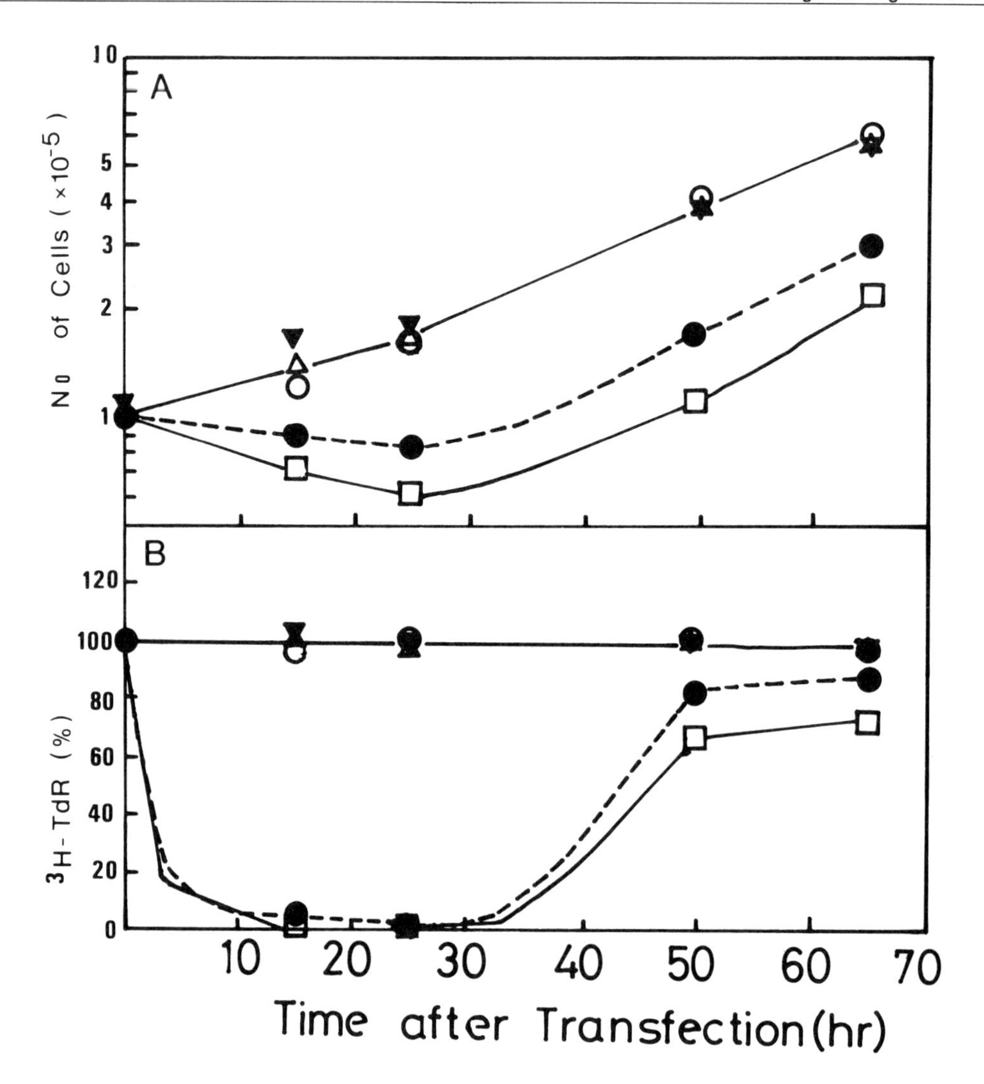

Figure 2. Inhibition of DNA replication in HL-60 cells by anti-human c-myc antibody. HL-60 cells were transfected with anti-human c-myc antibody (Oncor, Inc.) or non-specific anti-human IgG (Itani et al., 1987). At various times, the cell number was determined and the incorporation of ^{3}H-thymidine into the cells assayed after 1 h incubation with ^{3}H-thymidine at 25 μCi/ml. The amount of incorporation was expressed as the percentage of that of 10^5 cells immediately after transfection. A, Proliferation of HL-60 cells after transfection with antibodies. B Incorporation of ^{3}H-thymidine into HL-60 cells after transfection with antibodies, ○, non-transfected; ●, transfected with anti-human c-myc antibody; □, transfected with *c-myc* protein specific monoclonal antibody, IF7; △, transfected with anti-human IgG; ▼ mock-transfected with empty liposomes.

Isolation of ARS Derived from Human DNA

If our speculative conclusion above is correct then human DNA fragments selected as binding sequences for *c-myc* protein should contain ARS, in other words replication origins of human DNA. We applied the method of the immunobinding assay to the cloning of human ARS. Almost all the clones obtained by such a strategy had, indeed, ARS activity in various cells, including human, mouse, and rat cells. We also noticed in the cloned sequences several

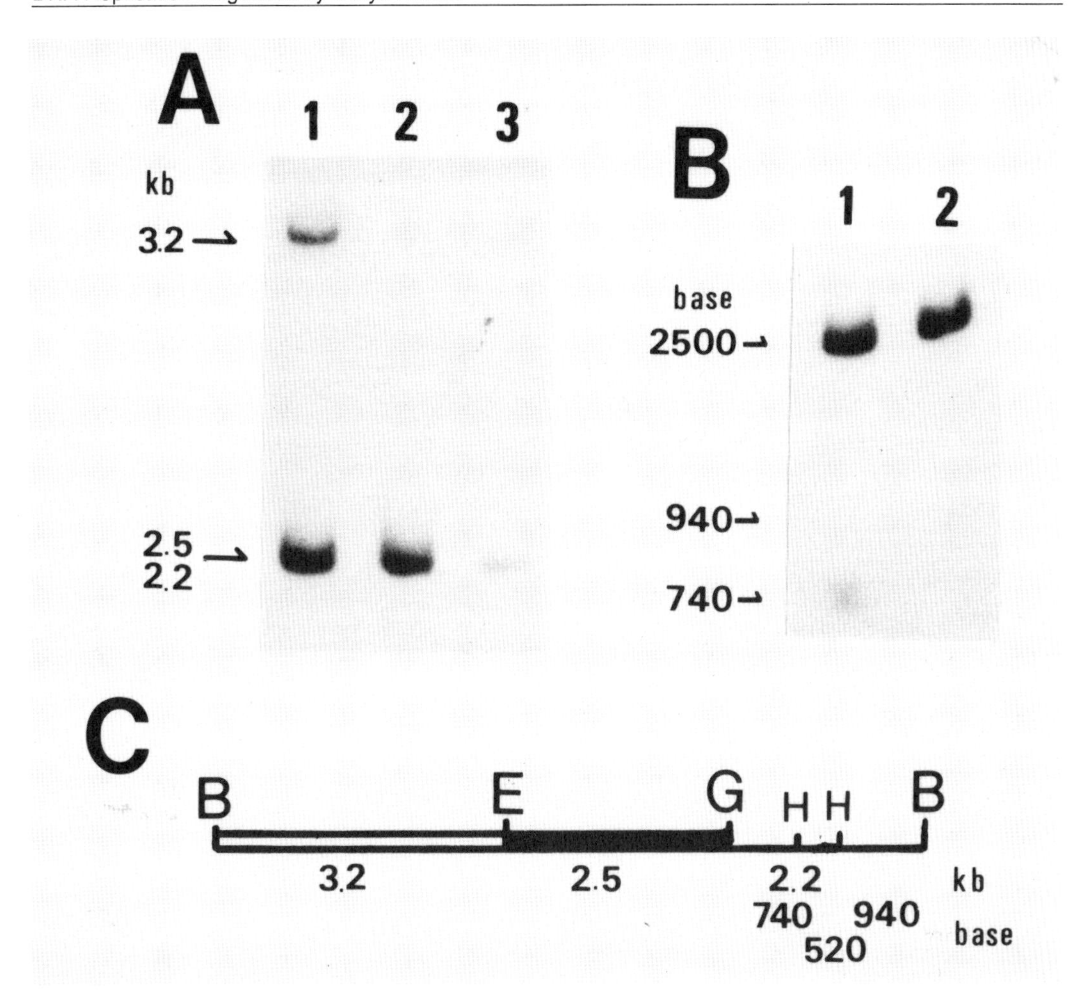

Figure 3. Binding of *c-myc* protein in HL-60 cells to sequences of pARS65. pARS65 was digested with *Eco*RI, *Bgl*II, and *Bam*HI, and each fragment then end-labelled with ^{32}P. A, The ^{32}P-labelled pARS65 fragments (lane 1) were incubated with HL-60 nuclear extract and then precipitated as immunocomplexes with 2 μg of anti-human *c-myc* antibody (lane 2) or anti-human IgG (lane 3). B, The ^{32}P-labelled pARS65 fragments recovered from immune-complexes precipitated with anti-human *c-myc* antibody (lane 1) were digested with *Hind*III (lane 2). The arrows indicate the size of the fragments. C, Structure of pARS65 cut with *Bam*HI, *Eco*RI, *Bgl*II, and *Hind*III, respectively. □, pKSV10; ■, mouse DNA; —, SV40 DNA.

inverted repeats able to form hairpin structures, two of which are shown in Figure 4. Similar structures are usually observed in oris, from bacteria to animal viruses (Kornberg, 1982).

Common Features Between SV40 T Antigen and *c-myc* Protein

The region around the origin of DNA replication in SV40 DNA consists of an enhancer, an early gene promoter, an *ori* and a T antigen coding region from 5′ to 3′ end, respectively. Only the core of *ori*, about 65 nucleotides in length, is necessary for T antigen-dependent SV40 DNA replication; cloned DNA containing the *ori* of SV40 DNA can replicate extrachromosomally in cells that produce SV40 T antigen, such as monkey CosI cells (11). Nonetheless, after comparing the common characters between T antigen and *c-myc* protein, considering that T antigen

A

```
1        10         20         30         40         50
GTATGATACA GATCGTGAGA ATACGTAGCC TCGTCACCAT TGAGCAGTAC

         60         70         80         90
GTTGTACTGG AAGAGACCAT GCTCTGACAC TGCACGACGT GACAGCATC
```

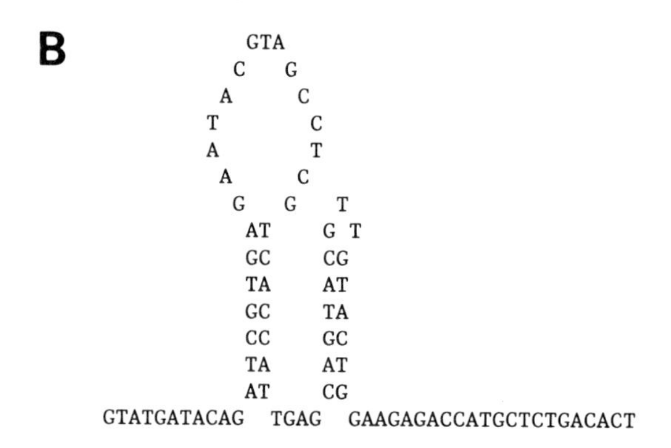

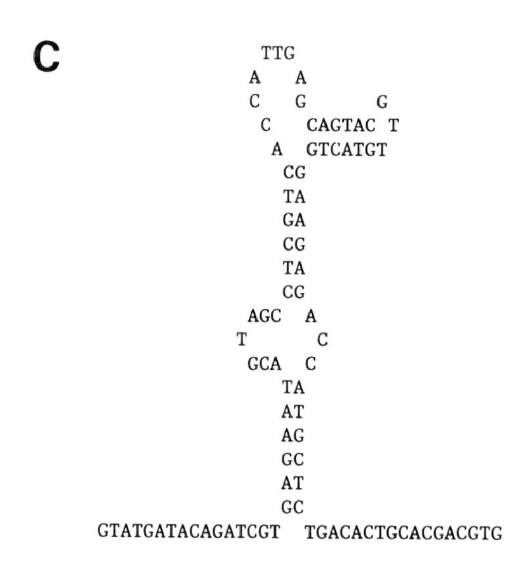

Figure 4. Sequence and possible higher structures of the cloned human ARS fragment in pHL-*myc* 1. A, The nucleotide sequence of the human DNA-derived fragment in pHL-*myc*1 was determined using the dideoxy chain termination method. B: and C, possible secondary structures of the fragment were generated by computer assisted sequence analysis.

can regulate transcription both positively and negatively and that the c-myc protein, as mentioned later, may also have transcriptional regulatory functions, we thought it possible that *c-myc* could be substituted for SV40 T antigen in a SV40 DNA replication system. In fact, when an SV40 ori containing plasmid was transfected into human cells that do produce *c-myc* actively, but do not express SV40 T antigen at all, it could replicate extrachromosomally, though dependent upon *c-myc* protein (Iguchi-Ariga *et al.*, 1987b). Results similar to ours were also reported by another group (Classon *et al.*, 1987). Moreover, the rate of replication of the SV40 ori-containing plasmid was higher in cells which were producing *c-myc* more actively than normal cells (Classon *et al.*, 1987).

REPLICATION ORIGIN AND TRANSCRIPTIONAL ENHANCER PRESENT IN *C-MYC* GENE

Considering the similarity between viral T antigens and the *c-myc* protein, it is by analogy suggested that some *c-myc* protein-binding sequences which involve an origin of replication may exist upstream of the *c-myc* gene itself, and that these sequences may also have transcriptional regulatory functions (Iguchi-Ariga *et al.*, 1988a).

A *c-myc* Protein Binding Region Exists Upstream of the *c-myc* Gene

We first examined the binding of *c-myc* protein to the *c-myc* gene itself by an immunobinding assay and found that the *c-myc* protein can bind to a *HindIII*-*Pst*I region (*myc*(H-P)) present about 2 kb upstream of the first exon of the human *c-myc* gene (Figure 5).

ARS Activity and Enhancer Activity of the *myc*(H-P) Region

We next transfected HL-60 cells with some cloned DNA containing various upstream regions of the *c-myc* gene. The results demonstrated that p*myc*(H-P) and p*myc*(H-K) (see Figure 5) replicated in HL-60 cells while neither p*myc*(K-S) nor vector pUC19 did. Replication of p*myc*(H-P) was inhibited by addition of anti-human *c-myc* antibody, but not by addition of nonspecific anti-human IgG. These data indicate that the *myc*(H-P) region contains an ARS and that its replication is dependent upon *c-myc* protein.

To examine whether this *c-myc* upstream ARS region has some function in transcription, plasmids containing this sequence joined to a bacterial chrolamphenicholacetyltransferase (CAT) gene were constructed and CAT enzyme activity was assayed after the plasmids were transfected into mouse L cells. As shown in Figure 6, the *c-myc* upstream ARS region (*myc*(H-P) region) has transcriptional enhancer activity nearly as strong as that of the SV40 enhancer, while no promoter activity was detected. The *c-myc* product, therefore, could be an enhancer binding protein as well as a replication protein.

Sequence Analysis of the *myc*(H-P) Region

The nucleotide sequence of the HindIII-PstI fragment located upstream of the human *c-myc* gene (*myc*(H-P) region) has been determined (Figure 7). This sequence can possibly fold into secondary structures with stem and loop which are commonly observed in a variety of replication origins, from bacteria to animal viruses. Two hairpin loops can be constructed in the regions from nucleotides number 80 to 105 and from 120 to 190. In the latter region, interestingly, two alternative loops can be drawn out using "TGAATAGTCAC" (nucleotides number 148 - 158) at the stem structure in both. The number of chromosomal DNA sequences homologous to that *Hind*III-*Pst*I region are limited, as determined by analysis of Southern blottings. It is quite different from the case of pARS65 and other ARSs we have cloned from human cells which present homologous sequences distributed throughout the genome (Ariga *et al.*, 1987; Iguchi-Ariga *et al.*, 1987a). Since both those genomic and *c-myc* upstream ARS replicate dependently upon *c-myc* protein, some consensus necessary for recognition by the

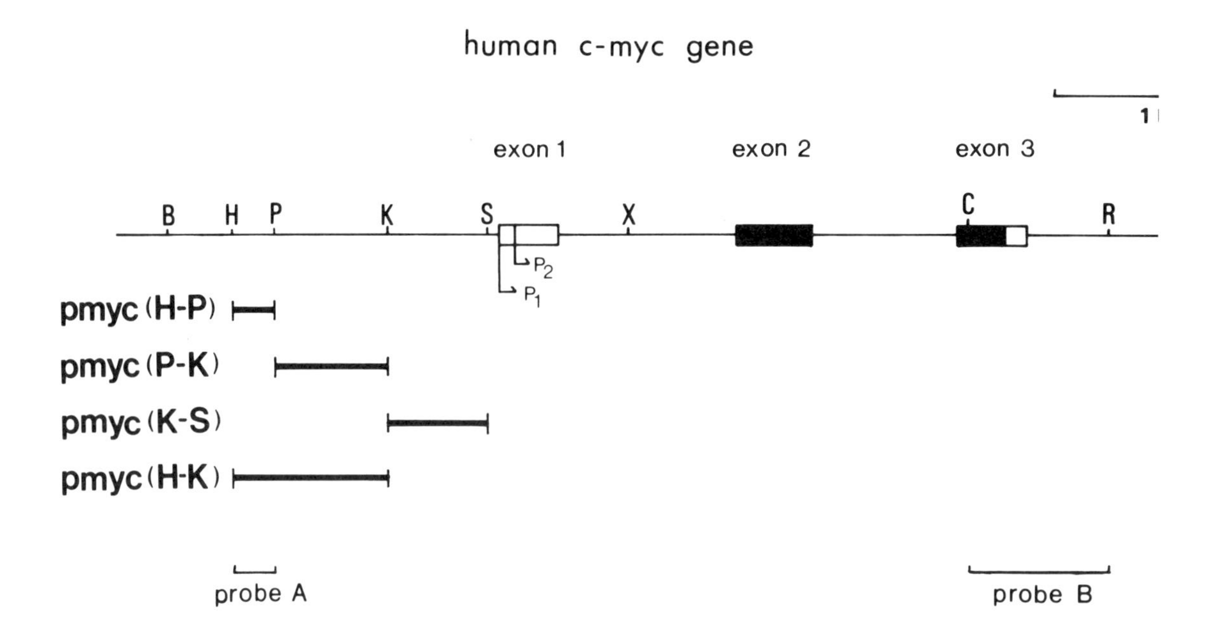

Figure 5. A physical map of the human *c-myc* gene is shown in the upper part of the figure. The upstream regions carried in the plasmids are indicated as black bars. □ , non-coding exon; ■ , coding exon; P1, P2, promoters; B, *Bam*HI; H, *Hind*III; P, *Pst*I; K, *Kpn*I; S, *Sma*I; X, *Xba*I; C, *Cla*I; R, *Eco*RI.

c-myc protein may be present in their sequences. Homologies among some of these ARSs are shown in Figure 7C. Both human pHLmyc (4) and mouse pARS65 (Ariga *et al.*, 1987) contain sequences homologous to "TGAATAGTCAC", present at the stems of possible alternative loops of p*myc*(H-P), and to "CTCTTAT" in the looped-out sequences of p*myc*(H-P) (nucleotides number 136 - 142) (Figure 7B), which might be quite important for recognition by *c-myc* protein.

To determine the sequence essential for ARS activity, several deletion mutants of p*myc*(H-P) were constructed, transfected into HL-60 cells and examined for ARS, as well as for transcriptional enhancer activity, by CAT assay in mouse L cells. It is suggested that the stem and loop structure with "TGAATAGTCAC" at the stem is necessary for ARS activity of *pmyc*(H-P) and that the sequence around nucleotides 110-121 may be responsible for its transcriptional enhancer activity. Thus, the sequence essential for initiation of DNA replication is adjacent to, but cannot be separated from, the sequence with enhancer activity. Therefore, the presence of an enhancer may be necessary for ARS activity, as is reportedly the case for polyoma virus, where DNA replication requires an enhancer in *cis* (de Villiers *et al.*, 1984). Recently, transcriptional stimulation of DNA replication or DNA replicational stimulation of

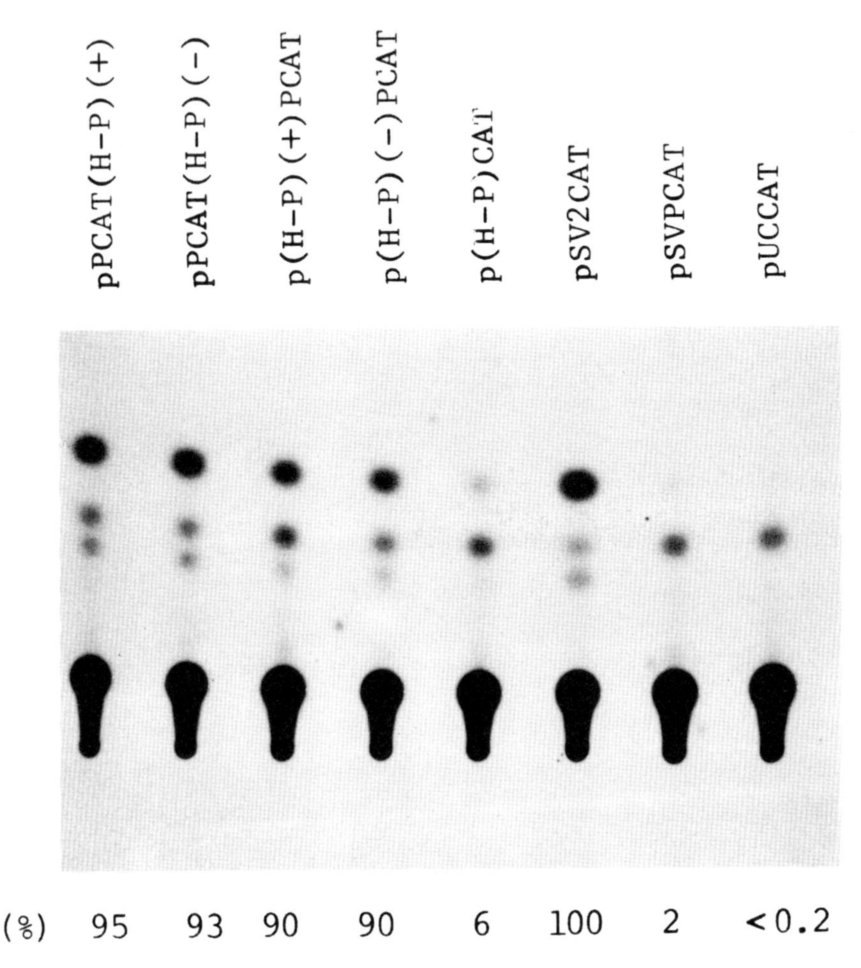

Figure 6. Transcriptional enhancer or promoter activity of the *c-myc* upstream regions with ARS activity. CAT assays were performed with lysates of mouse L cells transfected with pPCAT(H-P)(+) (pSVPCAT with a myc(H-P) fragment inserted downstream of CAT gene), pPCAT(H-P)(-) (pSVPCAT with myc(H-P) fragment inserted downstream of CAT gene in the reverse orientation of pPCAT(H-P)(+)), p(H P)(+)PCAT (pSVPCAT with a *myc*(H-P) fragment inserted upstream of the promoter), p(H-P)(-)PCAT (pSVTCAT with a *myc*(H-P) fragment inserted upstream of the promoter in the reverse orientation of p(H-P)(+)PCAT), p(H P)CAT (pUCCAT with a *myc*(H-P) fragment inserted upstream of the CAT gene), pSV2CAT (pSVPCAT with SV40 enhancer inserted upstream of the promoter), pSVPCAT (pUCCAT with SSV40 promoter (*Nsi*I-*Hind*III site of SV40 genome) inserted upstream of the CAT gene), and pUCCAT (pUC 19 carrying CAT gene inserted into a *Hind*III site), from the left, respectively. Cm, chloramphenicol; Ac-Cm, acetylated chloramphenicol. Numbers under the figure show the average number of enhancer activity after scanning the autoradiogram.

RNA transcription were reported by several groups (Grass *et al.*, 1987; Enver *et al.*, 1988; Jansen-Durr *et al.*, 1988). It was mentioned that transcription from adenovirus major late (Grass *et al.*, 1987; Enver *et al.*, 1988) and β-globin (Jansen-Durr *et al.*, 1988) promoters were stimulated by the presence of an SV40 ori sequence followed by DNA replication. Moreover, the cloned DNA fragments enriched by cellular ori sequences contained transcriptional regulatory sequences including promoter and enhancer (Tribioli *et al.*, 1987). So, it is possible

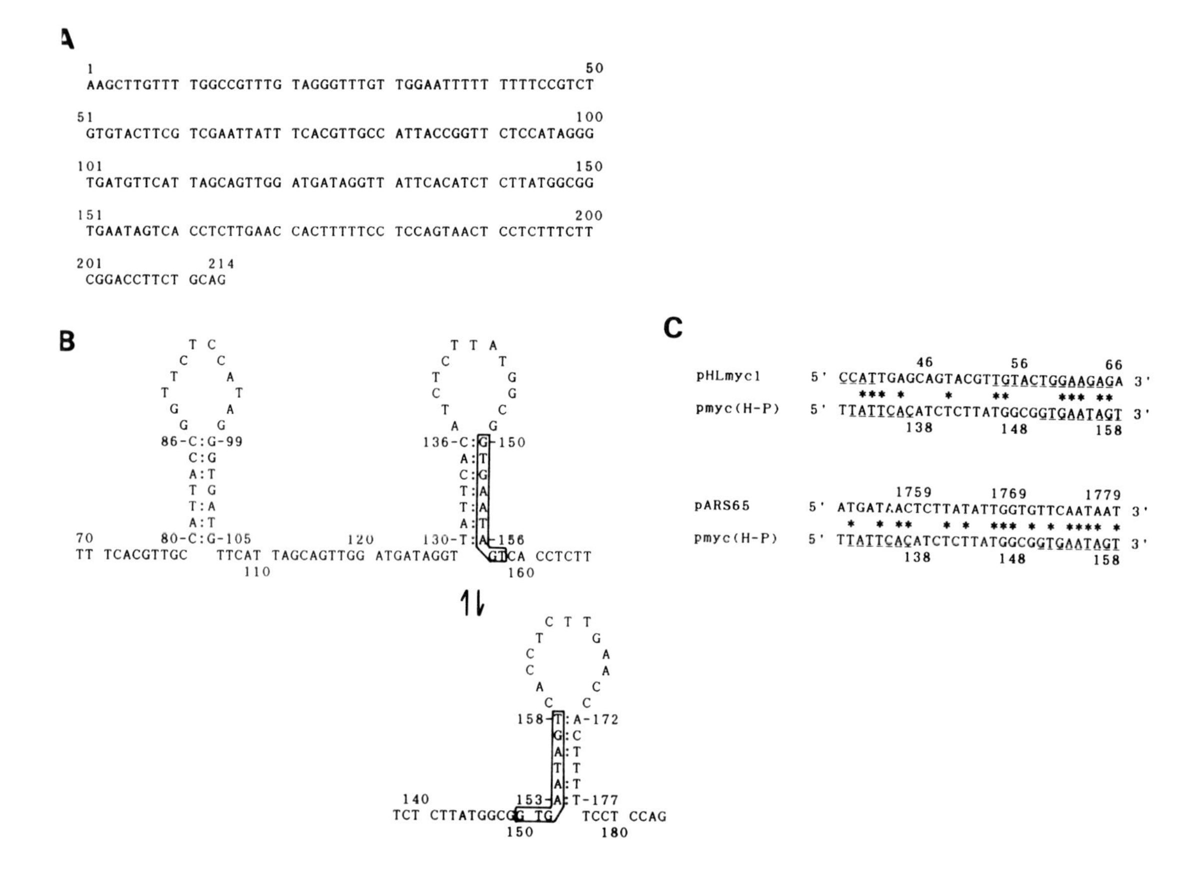

Figure 7. Sequence and possible secondary structure of the *myc*(H-P) region. A, The nucleotide the fragment from HindIII to PstI of human *c-myc* gene was determined using the dideoxy chain termination method. B, A possible secondary structure of the fragment based on computer-assisted sequence analysis. C, Comparison of the sequences among pHL*myc*1, pARS65 and p*myc*(H-P). Stars indicate matched sequences and underlined sequences are possible stems of hairpin loop structure.

that the replication ori and a transcriptional enhancer may share sequences and that each may be necessary for the function of the other.

AUTOREGULATION OF *c-myc* EXPRESSION BY THE *c-myc* PROTEIN

Our results indicate that the *c-myc* protein binds to a region upstream of the human *c-myc* gene itself. In order to know what role this protein plays in the regulation of *c-myc* expression, p*myc*(H-P), p*myc*(H-K), p*myc*(P-K), and p*myc*(KS) plasmids were introduced into HL-60 cells as binding competitors for this protein. The Northern blot hybridization presented in Figure 8 shows that the amount of *c-myc* RNA markedly decreases in the cells transfected with p*myc*(H-P) or p*myc*(H-K), while no decrease is observed in cells transfected with p*myc*(K-S) or p*myc*(P-K). It is suggested that the synthesis of *c-myc* RNA is affected by transfected plasmids

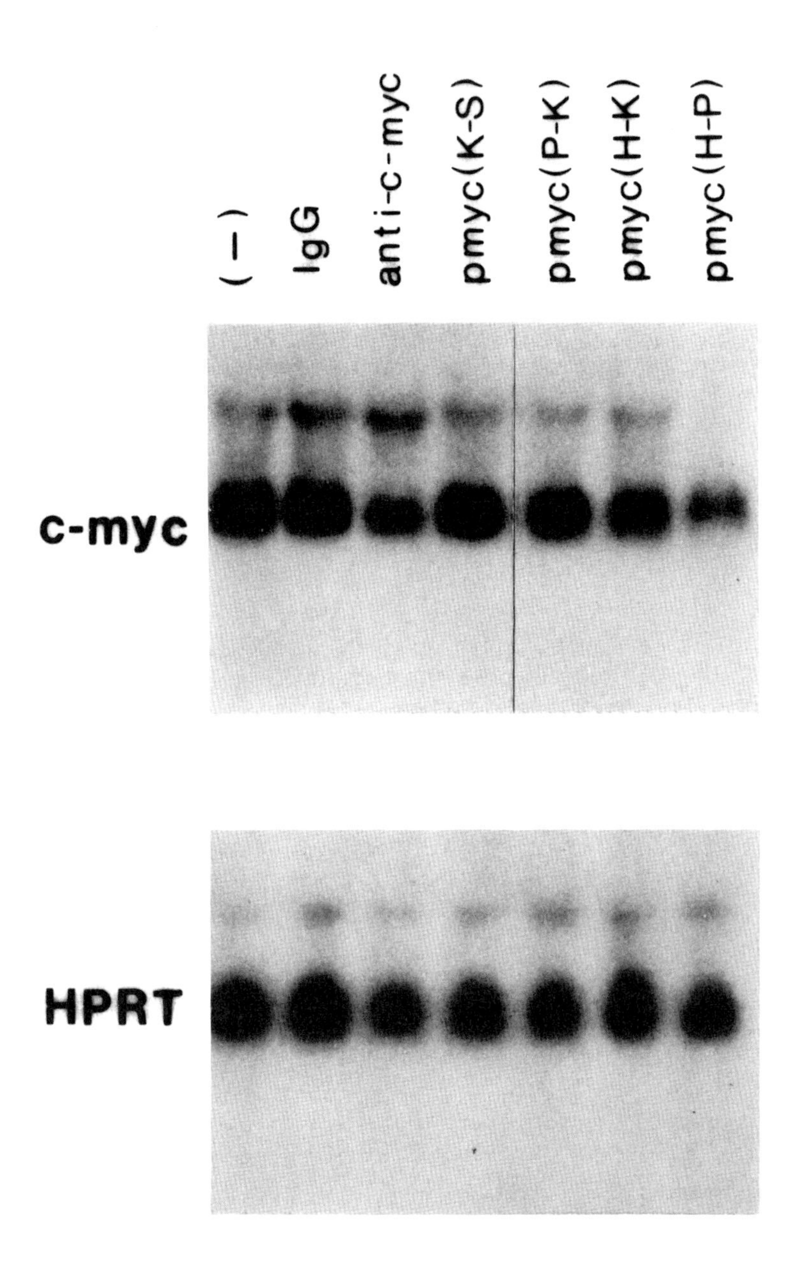

Figure 8. Expression of *c-myc* gene in HL-60 cells before and after transfection with upstream sequences. HL-60 cells (1×10^7) were transfected with 10 μg of p*myc*(H-P), p*myc*(H-K), p*myc*(P-K), p*myc*(K-S), 2 μg of *c-myc* antibody, or IgG. One day after transfection, total RNA was extracted from the cells and analysed by Northern blot hybridization. (-) shows the RNA of the HL-60 cells mock-transfected with empty liposomes. The *c-myc* fragment and human HGPRT gene were used as probes in the upper and lower figures, respectively. Numbers under the figure show the average number of *c-myc* or HGPRT expression after scanning the autoradiogram. Probe B in Figure 5 was used for *c-myc* expression.

that bind *c-myc* protein and, therefore, that the *c-myc* protein itself has some positive effect upon expression of the *c-myc* gene.

OTHER DNA REPLICATION PROTEINS

It is generally known that the cellular DNA of eukaryotes starts to replicate at many points on the chromosome, and there are approximately 10^4-10^5 DNA replication initiation sites in each mammalian cell (Cambell, 1986). So, the question arises as to whether or not only *c-myc* protein can promote all initiation of cellular DNA replication. We have searched other candidates from nuclear oncogene products, since some of the nuclear oncogene products are, like the *c-myc* protein, believed to regulate cell proliferation, especially the transition from G_1 to S phase of the cell cycle. N-*myc*, a gene of the *myc* family, is expressed in nerve cells or early developmental stage cells; amplification of the N-*myc* gene is sometimes observed in neuroblastoma cells, a situation that corresponds to that of the *c-myc* gene in leukemia cells (Alitalo *et al.*, 1987; Bishop, 1983). Further, p53, another nuclear oncogene product, is also a good candidate for a DNA replication protein. Microinjection of anti-p53 monoclonal antibody into quiescent mouse 3T3 cells inhibited the induction of DNA synthesis upon addition of fresh serum (Mercer *et al.*, 1982; Mercer *et al.*, 1984). Cell growth was also inhibited by plasmids encoding p53 antisense RNA (Shohat *et al.*, 1987). Moreover, p53 inhibits SV40 origin-dependent DNA replication by displacing DNA polymerase α from complexes with SV40 T antigen (Braithwaite *et al.*, 1987; Gannon and Lane, 1987). These experiments suggest that p53 may be involved with DNA polymerase α in cellular DNA replication.

We examined the possible functions of N-*myc* protein and p53 using the same strategies adopted in analysing the function of *c-myc* protein. We have obtained evidence suggesting that both N-*myc* protein and p53 promote cellular DNA replication by binding to the corresponding *oris*, which were later cloned from human cellular DNA (Iguchi-Ariga, 1988b). The example of a p53-dependent ori is shown in Figure 9. The interesting points were that the nucleotide sequences of the *oris* related to *c-myc* protein, N-myc protein, and p53 share no homology and that given *ori*-containing plasmids could replicate only in the corresponding cells. The p53-dependent *ori* clone, for instance, could replicate in Raji cells which produce both p53 and *c-myc* protein, but not in HL-60 cells in which the p53 coding gene is completely deleted while the *c-myc* gene is actively expressed (Iguchi-Ariga *et al.*, 1988b). It is thus concluded that in animal cells there are a variety of replication proteins that can recognize their respective *oris*.

CONCLUSIONS

The cloning of replication origins, the identification of proteins promoting DNA replication and the regulation of DNA replication, which may be correlated to RNA transcription all lead to a deeper understanding of the molecular mechanisms involved in the initiation of DNA replication in mammalian cells. Since Nuclear Factor I (NFI) and Nuclear Factor Ill (NFIII) are also functioning in both adenovirus DNA replication and transcription (Jones *et al.*, 1987; Prujin *et al.*, 1987), the question arises whether the initial step in DNA replication and RNA transcription might be carried out by the same protein and whether the same reaction would basically take place in either steps. To answer these questions, detailed biochemical analysis of the proteins and genetic analyses would be required.

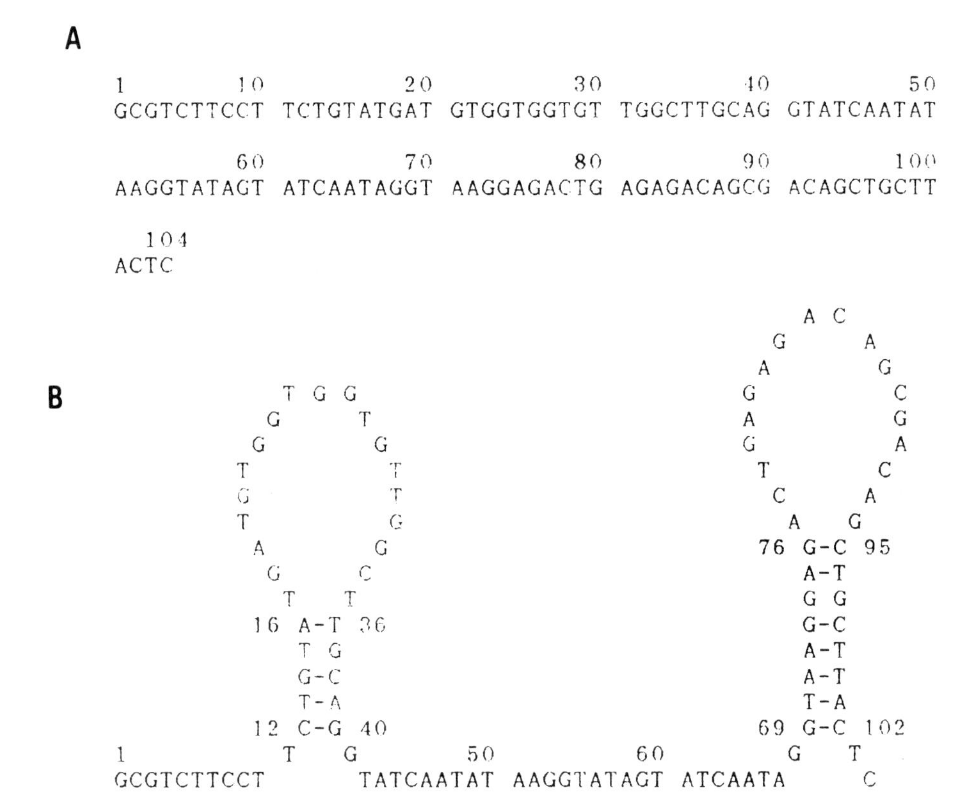

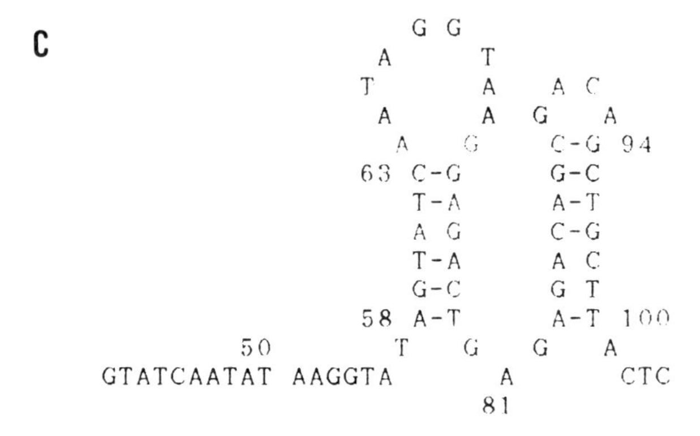

Figure 9. Nucleotide sequences of the pRJ53-1 clone and possible higher structures. A. The nucleotide sequences of the pRJ53-1 clone were determined using the dideoxy chain termination method, and the inserted human sequences are shown. B and C, possible secondary structures generated computer-assisted sequence analysis.

ACKNOWLEDGEMENTS

We gratefully acknowledge Yoshito Kaziro and Ken-ichi Arai for critical discussions and helpful suggestions. We also thank Ivo Galli for reviewing the manuscript and valuable discussions. This work was supported by grants from the Ministry for Education, Science, and Culture in Japan and by the Princess Takamatsunomiya Cancer Research Fund.

REFERENCES

Alitalo, K, Koskinen, P., Makela, T. P., Saksela, K., Sistonen, L., and Winqvist, R. (1987) *Biochem. Biophys. Acta* **907**: 1-32

Ariga, H. (1984) *Nucleic Acids Res.* **12**: 6053-6062

Ariga, H., and Sugano, S. (1983) *J. Virol.* **48**: 481-491

Ariga, H., Itani, T., and Iguchi-Ariga, S. M. M. (1987) *Mol. Cell. Biol.* **7**: 1-6

Ariga, H., Tsuchihashi, Z., Naruto, M., and Yamada, M. (1985) *Mol. Cell. Biol.* **5**: 563-568

Bishop, J. M. (1983) *Ann. Rev. Biochem.* **552**: 301-310

Braithwaite, A. W., Sturazbecher, H.-W., Addiso, C., Palmer, C., Rudge, K., and Jenkins, J. R. (1987) *Nature* **329**: 458-460

Cambell, J. L. (1986) *Ann. Rev. Biochem.* **55**: 733-771

Classon, M., Henriksson, M., Sumegi, J., Klein, G., and Hammaskjold, M.-L. (1987) *Nature* **330**: 272-274

de Villiers, J., Schaffner, W., Tyndall, C., Lupton, S., and Kamen, R. (1984) *Nature* **312**: 212-216

Enver, T., Brewer, A. C., and Patient, R. K. (1988) *Mol. Cell. Biol.* **8**: 1301-1308

Frappier, L, and Zannis-Hadjopoulos, M. (1987) *Proc. Natl. Acad. Sci. USA* **84**: 6668-6672

Gannon, J. V., and Lane, D. P. (1987) *Nature* **329**: 456-458

Grass, D. S., Read, D., Lewis, E. D., and Manley, J. L. (1987) *Genes Dev.* **1**: 1065-1074

Holst, A., Miller, F., Zastrow, G., Zentgraf, H., Schwender, S., Dinkl, E., and Grummt, F. (1988) *Cell* **52**: 355 -365

Iguchi-Ariga, S. M. M., Itani, T., Kiji, Y., and Ariga, H. (1987a) *EMBO J.* **6**: 2365-2371

Iguchi-Ariga, S. M. M., Itani, T., Yamaguchi, M., and Ariga, H. (1987b) *Nucleic Acids Res.* **15**: 4889-4899

Iguchi-Ariga, S. M. M., Okazaki, T., Itani, T., Ogata, M., Sato, Y., and Ariga, H. (1988a) *EMBO J.* **7**: 3135-3142

Iguchi-Ariga, S. M. M., Okazaki, T., Itani, T., and Ariga, H. (1988b) *Oncogene* **3**: 509-515

Itani, T., Ariga, H., Yamaguchi, N., Tadakuma, T., and Yasuda, T. (1987) *Gene* **56**: 267-276

Jansen-Durr, P., Boeuf, H., and Kedinger, C. (1988) *Nucleic Acids Res.* **16**:

Jones, K. A., Kadonaga, J. T., Rosenfeld, P. J., Kelly, T. J., and Tjian, R. T. (1987) *Cell* **48**: 79-89

Kornberg, A. (1980) *DNA replication.* W.H. Freeman and Co., San Francisco.

Kornberg, A. (1982) *Supplement to DNA replication.* W.H. Freeman and Co., San Francisco.

Mercer, W. E., Avignolo, C., and Baserga, R. (1984) *Mol. Cell. Biol.* **4**: 276-281

Mercer, W. E., Nelson, D., DeLeo, A., Old, L. J., and Baserga, R. 1982. *Proc. Natl. Acad. Sci. USA* **79**: 6309-6312

Prujin, G. J. M., van Driel, W., van Miltenburg, R. T., and van der Vilet, P. C. (1987) *EMBO J.* **6**: 3771-3778

Shohat, O., Greenberg, M., Reisman, D., Oren, M., and Rotter, V. (1987) *Oncogene* **1**: 277-283

Tooze, J. (1980) *DNA tumor viruses: molecular biology of tumor viruses.* part 2, 2nd ed., Cold Spring Harbor Lab, New York.

Tribioli, C., Biamonti, G., Giacca, M., Colonna, M., Riva, M., and Falashi, A. (1987) *Nucleic Acids Res.* **15**: 10211-10232

3

MULTIENZYME COMPLEXES FOR DNA SYNTHESIS IN EUKARYOTES: P-4 REVISITED

Linda H. Malkas, Robert J. Hickey and Earl F. Baril

FOREWORD

DNA replication in eukaryotes is a complex process with unique aspects that distinguish it from replication models that are derived from studies of DNA synthesis in the simpler prokaryotic systems. First, chromosomal DNA synthesis in eukaryotic cells is compartmentalized within the nucleus and separated from the cytoplasm which is the site of synthesis of proteins and other metabolites that function in DNA synthesis, as well as the site for mediation of extracellular stimuli that may trigger the initiation of DNA replication. Secondly, the eukaryotic chromosome is a complex nucleoprotein structure composed of both DNA, which contains multiple replication origins per DNA molecule, and protein that must be duplicated along with the DNA in order to maintain chromosome organization which in turn influences gene expression. Also, chromosomal DNA synthesis occurs in a precise, temporally and spatially regulated manner within individual replication units or replicons (DePamphilis and Wassarman, 1983). Since eukaryotic cells contain multiple chromosomes, as well as many replication origins on a single chromosome, the initiation of chromosomal DNA replication must be a highly coordinated process.

It is now apparent that intracellular metabolism does not occur by random collisions between soluble enzymes and substrates but rather through the action of organized, multienzyme complexes. The participation of multienzyme complexes in a variety of metabolic processes in both prokaryotic and eukaryotic cells is now well documented (reviewed in: Srivastava and Bernhard, 1986; Welch, 1977). Among the classic examples of thoroughly characterized multienzyme complexes that are involved in general cellular metabolism are the pyruvate dehydrogenase and the aromatic amino acid multienzyme complexes (Welch, 1977). Genome replication involves numerous protein-protein and protein-DNA interactions that must occur with high precision (Kornberg, 1980; 1982). Elegant analyses of the DNA synthesizing machinery in prokaryotes clearly indicates that the interactions occur through the action of multienzyme complexes devoted to DNA synthesis (Kornberg, 1988; Alberts, 1985). Although DNA replication in eukaryotes is more complex and less defined than in prokaryotes, a general picture is now emerging in which the DNA synthesizing apparatus in eukaryotic cells probably also involves multienzyme complexes (Baril *et al.*, 1988; Hickey *et al.*, 1988; Kaguni & Lehman, 1988; Fry & Loeb, 1986).

The purpose of this article is to present supportive evidence for the existence of multienzyme complexes for nuclear DNA replication in eukaryotes. To this end we will outline the evidence for the involvement of multienzyme complexes in genome replication and present as an "experimental case" a detailed discussion of recent evidence from our laboratory for the participation of a multienzyme system from HeLa cells in T antigen dependent replication of simian virus 40 (SV40) DNA *in vitro*.

MULTIENZYME SYSTEMS FOR DNA SYNTHESIS IN PROKARYOTES

The discovery of the involvement of multienzyme systems for DNA replication in Escherichia coli occurred in the early 1970's and was brought about by the development of *in vitro* systems for the replication of the small, circular DNA of bacteriophage such as M13, ϕX174 and G4 (Wickner *et al.*, 1973). This approach was also applied to the study of the machinery which synthesizes the large, linear DNAs of T4 (Alberts, 1985), T7 (Richardson, 1983) and λ (McMacken *et al.*, 1988) bacteriophage. For the sake of brevity, our discussion will be limited mainly to the DNA synthesizing machinery of *E. coli*.

Escherichia coli DNA Polymerase III Holoenzyme

The major DNA polymerase responsible for DNA replication (*i.e* DNA replicase) in *E. coli* is the DNA polymerase III (pol III) holoenzyme (McHenry, 1988; Kornberg, 1980). In the course of studies of the conversion primed, single-stranded DNA of M13 and ϕX174 phage into duplex, replicative forms it was discovered that pol III actually exists as a multiprotein complex that was designated DNA polymerase III holoenzyme (McHenry and Kornberg, 1977; Wickner *et al.*, 1973). DNA synthesis on single-stranded DNA templates by pol III holoenzyme is highly processive with at least 8600 nucleotides being incorporated without dissociation of the enzyme from the template (Kornberg, 1988). The pol III holoenzyme has been purified and resolved into its subunit structure and the function of many of the subunits have now been defined (McHenry, 1988; Kornberg, 1988; 1982; 1980). The holoenzyme, with a molecular mass of about 900 kDa, is composed of at least 10 separate proteins, 7 of which are now known to represent *bona fide* subunits. The catalytic core of the DNA polymerase III holoenzyme is composed of the α, ε and θ subunits and is designated pol III (McHenry, 1988; Kornberg, 1980). The 130,000 dalton α subunit is the catalytic site for polymerization (McHenry, 1988). The 27.5 kDa ε subunit is a 3′→5′ exonuclease and its association with the θ subunit provides the proofreading function of the DNA polymerase III holoenzyme (Scheuermann and Echols, 1984). The function of the 10,000 dalton θ subunit remains unknown at this time (Maki and Kornberg, 1988c). The 40 kDa β subunit is required in order for the holoenzyme to function with primed, single-stranded DNA templates (McHenry and Kornberg, 1977; Wickner *et al.*, 1973). A 200 kDa complex of auxiliary proteins, designated the γ-complex, has recently been resolved from the DNA polymerase III holoenzyme and separated from the pol III catalytic core and the β subunit (Maki and Kornberg, 1988 a-c). The γ-complex is composed of two γ subunits, one δ subunit and one each of three new holoenzyme associated polypeptides; a 33 kDa δ', a 15 kDa χ and a 12 kDa ψ protein. The highly processive synthesis that is characteristic of the pol III holoenzyme can be reconstituted *in vitro* with the isolated pol III catalytic core, the β subunit and the 200,000 dalton γ-complex (Maki and Kornberg, 1988b).

McHenry originally proposed that the functional pol III holoenzyme exists at the replication fork as an asymmetric dimer with distinguishable polymerases being involved in leading and lagging strand synthesis (reviewed in McHenry, 1988). This hypothesis is now supported by recent immunological (Hawker and McHenry, 1987) and biochemical (Maki and Kornberg, 1988) evidence.

DNA Replicase in *E. coli* Related Bacteriophage

The bacteriophage T4 and T7 induce their own DNA polymerases and also most of the other proteins that are required for their replication whereas bacteriophage λ utilizes most of the proteins of the host cell DNA synthesizing machinery (Alberts, 1985; Richardson, 1983; Kornberg, 1980). T4 DNA polymerase is a 100 kDa multifunctional protein in which the polymerase and a 3′→5′ exonuclease activity reside within the same polypeptide (Alberts, 1985). The polymerase, however, interacts with three accessory proteins encoded by T4 genes 44/62 and 45 to form a multiprotein complex that is designated the T4 holoenzyme (Cha and Alberts, 1988). The T4 polymerase holoenzyme, as in the case of the *E. coli* pol III holoenzyme, exists as a dimer at the replication fork with one half functioning in leading strand and the other in lagging strand DNA synthesis (Alberts, 1985).

Bacteriophage T7 DNA polymerase is simpler in structure than either the *E. coli* pol III holoenzyme or the T4 polymerase holoenzyme since it is composed of only two subunits (Richardson, 1983). Interestingly, one of the subunits is encoded by T7 gene 5 while the other subunit is *E. coli* thioredoxin (Marks and Richardson, 1976; Modrich and Richardson, 1975).

Multienzyme System for Replication of the *E. coli* Chromosome

The initial studies to determine which factors are required, in addition to pol III holoenzyme, for chromosome replication involved the *in vitro* analysis of the initiation of replication of small circular bacteriophage DNA such as M13, ϕX174 and G4 (Kornberg, 1980). These studies revealed that specific proteins and protein-protein interactions were required to initiate replication at a specific site on the DNA (Kornberg, 1980). This led to the isolation of the 60 kDa primase (dnaG protein) that synthesizes oligoribonucleotide primers at specific initiation sites on ϕX174 and G4 bacteriophage DNA which is coated with *E. coli* single-stranded DNA binding protein, SSB (Kornberg, 1980). It also led to the demonstration of the assembly of the 700 kDa primosome complex from the n, n′, n″, i, dnaB, dnaC and primase (dnaG) proteins and its requirement for specific initiation of ϕX174 DNA replication (Kornberg, 1982). The 76 kDa n′ protein has been shown to be a helicase with 3′ to 5′ polarity (Kornberg, 1982) while the dnaB protein is also a helicase but of 5′ to 3′ polarity (Lebowitz and McMacken, 1986). The dnaB protein signals the initiation start for the primase and unwinds the DNA at the replication fork (Kornberg, 1988).

The outcome of these studies helped to define the enzymological requirements for the specific initiation of replication of the *E. coli* chromosome. This was accomplished through the development of an *in vitro* system, reconstituted using 12 purified proteins, for the specific initiation and replication of plasmids containing the 245 bp sequence of the *E. coli* replication origin (oriC) (Baker *et al.*, 1988; Kaguni and Kornberg, 1984). The analysis has shown that the initial step in the initiation of oriC directed replication is the specific, cooperative binding of the dnaA protein to its recognition sequence within oriC (Baker *et al.*, 1988). In the presence of ATP the bound dnaA protein complex opens the duplex DNA at three repeat AT-rich 13-mer sequences within the dnaA-protein binding site. Incubation of the DNA-prepriming complex in the presence of ATP, DNA gyrase and SSB allows the dnaB helicase to unwind the DNA in both directions from oriC and primase synthesizes RNA primers at oriC which are then elongated by pol III holoenzyme for the complete replication of the template (Baker *et al.*, 1988). The primers are then excised by pol I and RNase H with the resulting gaps being filled by pol I and or pol III holoenzyme and ligation by DNA ligase. Finally, DNA gyrase decatenates and supercoils the replicated daughter molecules that are identical to the starting plasmid template (Baker *et al.*, 1988).

A multienzyme system for the specific initiation and replication of supercoiled plasmids containing the λ replication origin has been reconstituted *in vitro* using nine highly purified proteins that are involved in bacteriophage and *E. coli* DNA replication (McMacken *et al.*, 1988). The nine proteins include the λ encoded O and P initiator proteins, the *E. coli* SSB, dnaB helicase, dnaG primase, pol III holoenzyme, DNA gyrase and the dnaJ, dnaK heat shock proteins. An *in vitro* replication system that catalyzes coupled leading and lagging strand synthesis of T4 DNA has been reconstituted using seven T4 encoded proteins (Cha and Alberts, 1988). These include; T4 polymerase holoenzyme, gene 32 helix destabilizing protein and T4 primosome composed of primase and a helicase. Four proteins have been shown to function in the replication of T7 bacteriophage DNA (Huber *et al.*, 1988; Nakai *et al.*, 1988). These are; the T7 encoded DNA polymerase subunit, a multifunctional helicase-primase protein, a single-stranded DNA binding protein and *E. coli* thioredoxin, which acts as a T7 polymerase accessory protein.

MULTIENZYME COMPLEXES FOR DNA REPLICATION IN EUKARYOTES

DNA polymerase activity in eukaryotes was first discovered by Bollum (1960) in calf thymus about three decades ago. The enzyme was later designated DNA polymerase α (Weissbach *et al.*, 1975). This was followed about a decade or so later with the discovery in mammalian cells of DNA polymerases β (Baril, *et al.*, 1971; Weissbach *et al.*, 1971), γ (Fridlender *et al.*, 1972) and subsequently δ (Byrnes *et al.*, 1976). DNA polymerase β is believed to function in DNA repair, although a possible function in replication has not been ruled out, and DNA polymerase γ functions in mitochondrial DNA synthesis (Fry and Loeb, 1986). DNA polymerase δ shares many properties with DNA polymerase α and has recently been implicated in SV40 DNA replication (Downey *et al.*, 1988; Prelich and Stillman, 1988; Prelich *et al.*, 1987a,b). A δ-like polymerase, DNA polymerase III, has also been reported to be present in yeast (Burger and Bauer, 1988).

DNA polymerase α is the major polymerase in proliferating cells from a wide range of eukaryotes, from yeast to man (Fry and Loeb, 1986). In yeast and other fungi the polymerase α counterpart is designated DNA polymerase I (Campbell, 1986; Pleevani *et al.*, 1984; Chang *et al.*, 1984; Badaracco *et al.*, 1983). There is now a large body of evidence showing that DNA polymerase α functions in chromosomal DNA replication in eukaryotes (reviewed in Kaguni and Lehman, 1988; Fry and Loeb, 1986). DNA polymerase α activity was shown to increase in a variety of cells that have been stimulated to undergo proliferation (reviewed in Fry and Loeb, 1986). The results from studies using synchronized cells in culture suggested that the level of polymerase α activity is cell cycle regulated (Chiu and Baril, 1975; Spadari and Weissbach, 1974). However, recent studies using elutriated, normal cycling cells have shown that the levels of DNA polymerase α and its transcript actually remain relatively constant throughout the cell cycle (Wong *et al.*, 1988; Wahl *et al.*, 1988). Thus, the regulatory control(s) of DNA polymerase α activity is more complex than was previously thought and remains to be defined.

Multiprotein Forms of DNA Polymerase α

The elucidation of the subunit structure of DNA polymerase α over the years has been hampered by; the low amount of the enzyme present in cells, its existence in several molecular forms and its sensitivity to proteolysis (Kaguni and Lehman, 1988; Fry and Loeb, 1986; Fisher *et al.*, 1979; Wilson *et al.*, 1977; Holmes *et al.*, 1974). The recent development of improved purification procedures has helped to clarify the subunit structure of polymerase α and has led to the demonstration that DNA polymerase α from a large variety of eukaryotic cells exists as a multiprotein complex (reviewed by Kaguni *et al.*, 1988; Fry and Loeb, 1986). It is now generally observed that DNA polymerase α catalytic activity resides in a single polypeptide of about 180 kDa (Vishwanatha *et al.*, 1986a; Karawya *et al.*, 1984; Kaguni *et al.*, 1983a). Some evidence has been presented to suggest that the DNA polymerase α activity may be modulated by phosphorylation of the 180 kDa catalytic subunit (Krauss *et al.*, 1987; Donaldson and Gerner, 1987).

DNA primase activity was reported to be present in simian virus 40 (SV40) infected cell extracts (Kaufman and Falk, 1982) and it was shown almost simultaneously by three laboratories using different cell systems that primase is tightly associated with polymerase α (Conaway and Lehman, 1982; Yagura *et al.*, 1982; Tseng and Ahlem, 1983). Primase has now been shown to be tightly associated with DNA polymerase α purified from a wide variety of cells (Vishwanatha and Baril, 1986b; Chang *et al.*, 1984; Hubscher *et al.*, 1983; Yagura *et al.*, 1984; Tseng and Ahlem, 1983; Kaguni *et al.*, 1983a,b) including DNA polymerase I from yeast (Plevani *et al.*, 1984). Primase activity resides in a subunit separate from the 180 kDa polymerase α catalytic subunit and copurifies with 56 and 49 kDa polypeptides (reviewed in Kaguni and Lehman, 1988). The activity is believed to reside in a 49 kDa subunit while the role of the 56 kDa is unknown (Tseng and Ahlem, 1983; Kaguni and Lehman, 1983a,b). The association of the DNA polymerase α-primase complex with other proteins in the higher order multiprotein complex decreases the length of RNA primers synthesized relative to those observed *in vivo* (Cotterill and Lehman, 1987; Vishwanatha *et al.*, 1986c) and also increases the fidelity of replication (Reyland and Loeb, 1987).

Multiprotein forms of DNA polymerase α have now been purified to near homogeneity from *Drosophila melanogaster* embryos (Kaguni *et al.*, 1983a), HeLa cells (Vishwanatha *et al.*, 1986a) and calf thymus (Ottiger *et al.*, 1987). The DNA polymerase α complex from Drosophila melanogaster is composed of 4 subunits of 183, 73, 60 and 50 kDa (Kaguni and Lehman, 1988; 1983a, b). the 640 kDa multiprotein DNA polymerase α complex from HeLa cells is composed of 8 polypeptides, the functions of 5 of which have been defined (Vishwanatha *et al.*, 1986a), while the polymerase α complex from calf thymus is composed of 10 polypeptides with molecular masses in the range of 200 to 25 kilodaltons (Ottiger *et al.*, 1987). The three highly purified DNA polymerase α multiprotein complexes from different sources have in common the association of a 180 kDa polymerase α catalytic subunit, a primase and a 3′ → 5′ exonuclease.

The intact DNA polymerase α complex from Drosophila melanogaster embryos lacks exonuclease activity. However, the separation of the 182 kDa catalytic subunit from the 73 kDa subunit showed that the catalytic subunit has a cryptic 3′ → 5′ exonuclease activity (Cotterill and Lehman, 1987) that has proofreading activity (Reyland *et al.*, 1988). The 3′ → 5′ exonuclease that is associated with the 640 kDa polymerase α complex from HeLa cells resides with a 69 kDa polypeptide that was separated from the 183 kDa polymerase α subunit, purified to apparent homogeneity and characterized (Skarnes *et al.*, 1986).

In addition to the DNA polymerase α catalytic subunit, primase and 3′ → 5′ exonuclease, the 640 kDa multiprotein complex from HeLa cells also has associated with it a 94 kDa (dimer of 47 kDa subunits) protein of unknown function that specifically binds the unusual dinucleotide 5′, 5‴ diadenosine tetraphosphate (Baril *et al.*, 1983) and two polymerase α accessory proteins C1 (a tetramer of 24 kDa subunits) and C2 (52 kDa) (Lamothe *et al.*, 1981). Purified DNA polymerase α by itself, was found to bind nonproductively to single-stranded DNA in primed single-stranded DNA templates (Fisher *et al.*, 1979; Wilson *et al.*, 1977) and does not utilize such templates for DNA synthesis (Lamothe *et al.*, 1981). The C1,C2 accessory proteins increase the affinity of the polymerase catalytic subunit for the primer in primed single-stranded templates by about three orders of magnitude and promote synthesis by the polymerase α-C1,C2 complex on templates containing extensive single-strand regions (Pritchard and DePamphilis, 1983a; Pritchard *et al.*, 1983b; Lamothe *et al.*, 1981). Thus, the C1,C2 polymerase α accessory proteins are believed to function as primer recognition proteins (Pritchard *et al.*, 1983b).

DNA Polymerase δ

Although there is now solid evidence that DNA polymerase α functions in chromosomal DNA replication in eukaryotes (reviewed in Kaguni and Lehman, 1988; Fry and Loeb, 1986), DNA polymerase δ has also been implicated recently in eukaryotic DNA replication (Downey *et al.*, 1988; Prelich and Stillman, 1988; Prelich *et al.*, 1987a,b). It is not clear, however, that DNA polymerase δ is actually a distinct enzyme from polymerase α since they share similar properties, including the association of a 3′ → 5′ exonuclease with both (Kaguni *et al.*, 1988). The evidence for the involvement of DNA polymerase δ in chromosomal DNA replication is indirect, being based on the differences in sensitivities of polymerases α and δ to antibodies to the respective enzymes (Downey *et al.*, 1988) and butylphenyldeoxyguanosine triphosphate, BuPdGTP (Crute *et al.*, 1986; Byrnes, 1985; Lee *et al.*, 1985), as well as the association of proliferating cell nuclear antigen (PCNA), (Bravo *et al.*, 1987; Prelich *et al.*, 1987b) with DNA polymerase δ as a polymerase accessory protein (Tan *et al.*, 1986).

DNA polymerase δ has been reported to function in the elongation phase of simian virus 40 (SV40) DNA replication *in vitro* (Prelich and Stillman, 1988; Prelich *et al.*, 1988a) and is postulated to function in leading strand DNA synthesis *in vivo* (Downey *et al.*, 1988). However, DNA polymerase δ has also recently been reported to function in DNA repair (Nishida and Linn, 1988; Dressler and Frattini, 1986). Thus, it is not clear at this time if DNA polymerase δ is a distinct enzyme from polymerase α and what its actual function is in DNA metabolism.

Multienzyme Complexes for DNA Synthesis

There have been several reports over the past decade or so of the isolation of large aggregates of enzymes for DNA synthesis from extracts of eukaryotic cells (reviewed in Mathews and Slabaugh, 1986; Fry and Loeb, 1986). One of the initial reports described the cosedimentation of key enzymes for deoxyribonucleotide synthesis such as; ribonucleotide reductase, thymidine kinase, thymidylate synthetase, thymidylate kinase and nucleoside diphosphokinase with DNA polymerase α as a sedimentable subfraction (P-4) of the combined nuclear extract-postmicrosomal supernatant fraction of homogenates of regenerating rat liver and hepatomas (Baril *et al.*, 1974). In prokaryotes a complex of T4 bacteriophage ribonucleotide reductase with other enzymes for deoxyribonucleotide synthesis was also reported (Mathews and Slabaugh, 1986; Reddy *et al.*, 1978). The association of the complex was shown to regulate the activity of the respective enzymes and T4 DNA polymerase was shown to interact with the complex (Tomich *et al.*, 1974). In addition, channeling of uridine or UTP into dTTP and hydroxymethyl dCTP synthesis was demonstrated for the complex (Mathews and Slabaugh, 1986; Reddy *et al.*, 1978). The actual isolation and characterization of the complex, however, was hampered by its fragility.

Reddy and Pardee (1980) isolated a complex of enzymes for DNA synthesis from hamster fibroblast cells (CHEF) that contained key enzymes for deoxyribonucleotide synthesis such as; ribonucleotide reductase, thymidine kinase, thymidylate kinase, dihydrofolate reductase plus DNA polymerase α, topoisomerase II and nascent DNA. The complex was designated "replitase" and it was reported to effectively couple ribonucleotide reduction and channeling into deoxribonucleoside triphosphate synthesis with their subsequent incorporation into DNA *in vitro* (Reddy and Pardee, 1982). The ability of the replitase to actually channel CDP into DNA, however, has been questioned by the work of Spyron and Reichard (1983) who reported that CDP was preferentially incorporated into RNA by the "replitase". The work of Spyron and Reichard (1983) is not in conflict with the existence of multienzyme complexes for DNA synthesis but rather suggest that ribonucleotides are not channeled by such a complex for the synthesis of deoxyribonucleotides. There is now good evidence that ribonucleotide reductase in mammalian cells is actually localized in the cytoplasm in association with the endoplasmic reticulum (Engstrom *et al.*, 1984). It has been suggested that deoxynucleotide synthesis may occur at the endoplasmic reticulum and the deoxynucleotides transported into the nucleus where compartmentalization may occur within the domain of the DNA synthesizing machinery (Mathews and Slabaugh, 1986).

The actual role of the nuclear architecture in the organization of the DNA synthesizing apparatus is not very clear at this time. The nuclear matrix has been implicated in chromosomal DNA replication although the evidence for this is not compelling (Tubo and Berezney, 1987a,b; Jackson and Cook, 1986a; Berezney and Coffey, 1975). Jackson and Cook (1986b) reported the retention of a large megacomplex containing DNA polymerase α and other enzymes for DNA synthesis in agarose entrapped nuclei. The megacomplex could function in the replication of endogenous chromosomal DNA and some evidence was obtained suggesting that it may associate with the nuclear matrix (Jackson and Cook, 1986a). Tubo and Berezney (1986a,b) also reported the isolation of 100-150S megacomplexes containing DNA polymerase α-primase that were associated with the nuclear matrix from regenerating rat liver. The megacomplexes were believed to be composed of clusters of 10S and 17S complexes that underwent dissociation when the isolated megacomplexes were left at 4°C for an hour or more. Although more extensive investigation is required, it is possible that the DNA synthesizing apparatus is an association of the key enzymes for DNA synthesis into megacomplexes that then associate with components of the nuclear architecture to form a supramolecular structure for DNA synthesis.

Among the numerous problems encountered in the isolation and characterization of the putative multienzyme complexes for DNA synthesis in eukaryotes has been their instability under the usual conditions used for enzyme purification and also the lack of functional assay for the complex. Jazwinski and Edelman, (1984) reported the isolation of a two million dalton complex from yeast cells that had associated DNA polymerase I, primase, DNA ligase, topoisomerase II activities and that effectively replicated the extrachromosomal, 2 micron yeast plasmid DNA *in vitro*. As will be discussed in a later section, a multienzyme complex has now been partially purified from HeLa cells that completely replicates simian virus 40 (SV40) DNA *in vitro* (Baril *et al.*, 1988; Hickey *et al.*, 1988). Thus, the existence of functional multienzyme complexes for DNA synthesis in eukaryotes is now becoming more firmly established.

IN VITRO REPLICATION OF SV40 DNA

Advent of an *In Vitro* System for SV40 Replication

It was pointed out in an earlier section that a major contributing factor to success in defining components of the complex DNA synthesizing apparatus in *E. coli* was the development of an *in vitro* system for the replication of small bacteriophage DNAs (Kornberg, 1980,1982). This then led to the reconstitution from purified proteins of an *in vitro* system for the specific initiation of replication of plasmids having DNA inserts that contain the origin for replication of the *E. coli* chromosome (Kornberg, 1988). The recent development of an *in vitro* system for the specific initiation of replication of plasmids containing the replication origin of simian virus 40 (SV40) DNA (Stillman and Gluzman, 1985; Wobbe *et al.*, 1985; Li and Kelly, 1984; Ariga and Sugano, 1983) now permits the use of a similar approach for the analysis of the DNA synthesizing apparatus in permissive mammalian cells. This section will briefly outline the requirements for SV40 replication *in vivo* and *in vitro*.

Viral Proteins Involved in SV40 Replication *In Vitro*

The only viral encoded protein that is required for SV40 replication *in vivo* is large T antigen, a 94 kDa multifunctional phosphoprotein that specifically binds sequences in the SV40 ori, possesses ATPase activity and forms complexes with the cellular p53 protein (Tooze *et al.*, 1980). T antigen has recently been purified to homogeneity by immunoaffinity chromatography (Simanis and Lane, 1985) and this has enabled the initial characterization of the actual role of this protein in the initiation of SV40 DNA synthesis. The evidence now suggests that T antigens forms a nucleoprotein structure at the SV40 ori through an ATP-dependent reaction (Dean *et al.*, 1987; Borowiec and Hurwitz, 1988) and it has recently been shown that T antigen has associated helicase activity (Stahl *et al.*, 1986). The helicase activity is probably intrinsic to the T antigen protein since T antigen specific monoclonal antibodies inhibit the helicase and mutant T antigen proteins, defective in ATPase function, exhibit decreased helicase activity (Stahl *et al.*, 1987). DNA unwinding by T antigen, as determined by electron microscopic analysis, is initiated at the SV40 ori, proceeds bidirectionally (Dodson *et al.*, 1987) and it has been suggested than unwinding at the SV40 ori may provide a mechanism for the specific initiation of SV40 DNA replication by the DNA polymerase-primase complex (Dodson *et al.*, 1987).

Cellular Proteins Involved in SV40 Replication *In Vitro*

DNA polymerase α activity was found to account for 95% of the total DNA polymerase activity associated with replicating SV40 chromosomes (Edenberg *et al.*, 1978). Murakami and coworkers (1986) showed that an extract from the nonpermissive mouse cell line, FM3A, supported SV40 replication *in vitro* when supplemented with DNA polymerase α-primase complex isolated from the permissive HeLa cell line and SV40 encoded T antigen. These results suggest a role for DNA polymerase α as the replicative enzyme in SV40 replication. In addition, they suggest a possible role for the DNA polymerase-primase complex, and/or associated proteins, as cell permissivity factors.

The participation of topoisomerase I and II in SV40 DNA replication *in vitro* was ascertained by Yang and coworkers (1987) through reconstitution experiments using topoisomerase depleted cell extracts and purified topoisomerases I and II. Among other proteins that have been reported to be involved in the replication of SV40 DNA *in vitro* is the proliferating cell nuclear antigen, PCNA (Prelich *et al.*, 1988, 1987). PCNA was reported to be a DNA polymerase δ-associated accessory protein (Tan *et al.*, 1986) and has been implicated in the elongation phase of SV40 replication *in vitro* (Prelich *et al.*, 1988, 1987; Wold *et al.*, 1987). A single-stranded DNA binding protein that may function as a host cell SSB (Chase and Williams, 1986) in SV40 replication has recently been isolated from HeLa (Wold and Kelly, 1988) and 293 (Fairman and Stillman, 1988) cells. The HeLa cell protein, designated RP-A, is a multisubunit protein composed of polypeptides of 70, 53, 32 and 14 kDa (Wold and Kelly, 1988). RP-A binds single-stranded DNA and is required for T antigen mediated unwinding of duplex DNA molecules containing SV40 ori. The protein isolated from 293 cells, and designated RF-A, is also a multisubunit protein composed of 70, 34 and 11 kDa polypeptides, with similar binding properties to the HeLa cell RP-A protein (Fairman and Stillman, 1988). Thus, the RP-A/RF-A proteins are likely candidates for host cell SSBs that are involved in SV40 replication.

MULTIENZYME COMPLEX FOR DNA SYNTHESIS IN HELA CELLS

A 19S Multienzyme Complex from HeLa Cells that is Competent for SV40 Replication *In Vitro*

As discussed earlier, DNA polymerase α has been isolated as a multiprotein complex from a variety of eukaryotic cells (Kaguni and Lehman, 1988). The 640 kDa multiprotein DNA polymerase α -primase complex from HeLa cells is extractable from isolated nuclei by low ionic strength (*i.e.* 0.15 M KC1) buffers and is also present in the cytosol (Vishwanatha *et al.*, 1986a; Lamothe *et al.*, 1981). Interestingly, cell free extracts for the assay of T antigen dependent initiation of SV40 replication *in vitro* have also been prepared with low ionic strength buffers (Li and Kelley, 1985; Stillman and Gluzman, 1985; Wobbe *et al.*, 1985; Ariga and Sugano, 1985). We have found that the activity responsible for T antigen dependent *in vitro* initiation of SV40 replication, as in the case of the 640 kDa multiprotein polymerase α-primase complex (Vishwanatha *et al.*, 1986a), is present in the low salt nuclear extract and cytosol fractions from homogenates of HeLa cells in S-phase (Baril *et al.*, 1988). This suggests that the 640 kDa DNA polymerase α-primase multiprotein complex and the enzyme system required for origin specific initiation of SV40 replication *in vitro* are both loosely associated with intranuclear structures and readily leak from the nucleus during extractions under aqueous conditions. This nuclear leakage has commonly been observed during the isolation of DNA polymerase α from mammalian cells (Foster and Gurney, 1976; Herrick *et al.*, 1976).

We previously showed that most of the DNA polymerase α activity from mammalian cell homogenates is recovered with a sedimentable fraction during gradient centrifugation of the combined nuclear extract-postmicrosomal supernatant fractions (Baril *et al.*, 1974). This sedimentable fraction, designated P-4, contained in addition to DNA polymerase α other enzymes that function in DNA synthesis such as enzymes for the production of deoxynucleotide precursors (Baril *et al.*, 1974). It was recently shown that the P-4 fraction contains additional enzymes for DNA synthesis such as DNA primase, topoisomerase I, RNase H, DNA-dependent ATPase, DNA ligase etc. Some of these enzyme activities, such as the DNA polymerase activity with primed single-stranded DNA templates, DNA primase and DNA-dependent ATPase, are recovered almost exclusively with the sedimentable P-4 fraction whereas others, such as the DNA polymerase activity with activated (gapped) DNA templates, RNase H, topoisomerase I and DNA ligase activities are recovered with the nonsedimentable S-4 as well as the sedimentable P-4 fraction (Baril *et al.*, 1988; Hickey *et al.*, 1988). Thus, the activities associated with the P-4 fraction apparently do not arise from a general sedimentation of the individual enzymes associated with this fraction but rather is most likely attributable to the association of these activities in a complex. With the advent of an assay system for origin specific initiation of SV40 replication *in vitro*, we have reinvestigated the nature of the sedimentable complex of enzymes for DNA synthesis in the P-4 subfraction of HeLa cell homogenates.

Isolation of a 19S Multienzyme Complex for the Initiation of SV40 Replication *In Vitro*

The initial step in the isolation and partial purification of the putative multienzyme complex for DNA synthesis from the combined nuclear extract/postmicrosomal supernatant (NE/S-3) of HeLa cells involves precipitation with 5% polyethylene glycol in the presence of 2 M KCl (Figure 1). About 60-70% of the nucleic acids and 50% of the protein in the NE/S-3 fraction are precipitated by 5% PEG while the majority of the DNA polymerase α-primase and SV40 *in vitro* replication activities in this fraction are not precipitated. The distribution of enzymes for DNA synthesis in the subfractions of the NE/S-3 following the PEG precipitation step is the same as that observed previously for these enzyme activities in the sedimentable P-4 and nonsedimentable S-4 subfractions of the NE/S-3 (Baril *et al.*, 1988; Hickey *et al.*, 1988). The P-4 fraction contains the majority of the DNA primase and polymerase α activity present in the NE/S-3 that functions with primed single-stranded DNA templates. Immunoblot and chromatographic analysis showed that these activities reside with the 640 kDa multiprotein polymerase α-primase complex (Baril *et al.*, 1988; Hickey *et al.*, 1988). The nonsedimentable S-4 fraction (Figure 1), however, contains virtually no DNA primase and the polymerase α activity in this fraction functions mainly with activated (gapped) DNA templates (Baril *et al.*, 1988; Hickey *et al.*, 1988; Malkas and Baril, unpublished results).

The P-4 Fraction is Competent for SV40 Replication

Since the P-4 fraction contains a number of key enzymes for DNA synthesis, we investigated the possibility that this subcellular fraction is active in the initiation of SV40 replication *in vitro*. As shown in Figure 2 (upper panel), the low salt nuclear extract (NE) and postmicrosomal supernatant (S-3) fractions by themselves or when combined support SV40 replication *in vitro* in the presence of T antigen. Form I and form II DNAs, as well as various topological and replication intermediates are produced when the reaction is performed in the presence of T antigen (upper panel) but not when the reaction is performed in the absence of T antigen (Figure 2, middle panel). A small amount of incorporation of the labeled deoxynucleoside triphosphate into plasmid DNA occurred during incubation of the nuclear extract in the absence of T antigen (Figure 2, middle panel). This may be due to a small amount of repair synthesis that occurs with this fraction. Interestingly, the sedimentable P-4 fraction derived from gradient centrifugation of the NE/S-3 fraction supports the *in vitro* replication of SV40 DNA in the presence of T antigen (Figure 2, upper panel) whereas the nonsedimentable S-4 fraction does not.

SV40 DNA Synthesis Occurs Through a 19S Multienzyme Complex

The purified 640 kDa multiprotein DNA polymerase α-primase complex is 10S while the 183 kDa DNA polymerase α catalytic subunit is 7S (Vishwanatha *et al.*, 1986). Zonal sedimentation analysis of the P-4 and S-4 fractions on a 5 to 30% sucrose gradient in the presence of 0.5 M KCl provided evidence for the association of the 640 kDa DNA polymerase α-primase complex into a higher order complex (Figure 3 and Baril *et al.*, 1988; Hickey *et al.*, 1988). The DNA polymerase α activity with activated and primed single-stranded DNA templates and DNA primase activities sedimented coincidentally as a major 19S peak during gradient centrifugation of the P-4 fraction (Figure 3, left panel). The 19S peak, however, was absent in the sedimentation profile from gradient centrifugation of the S-4 fraction (Figure 3, right

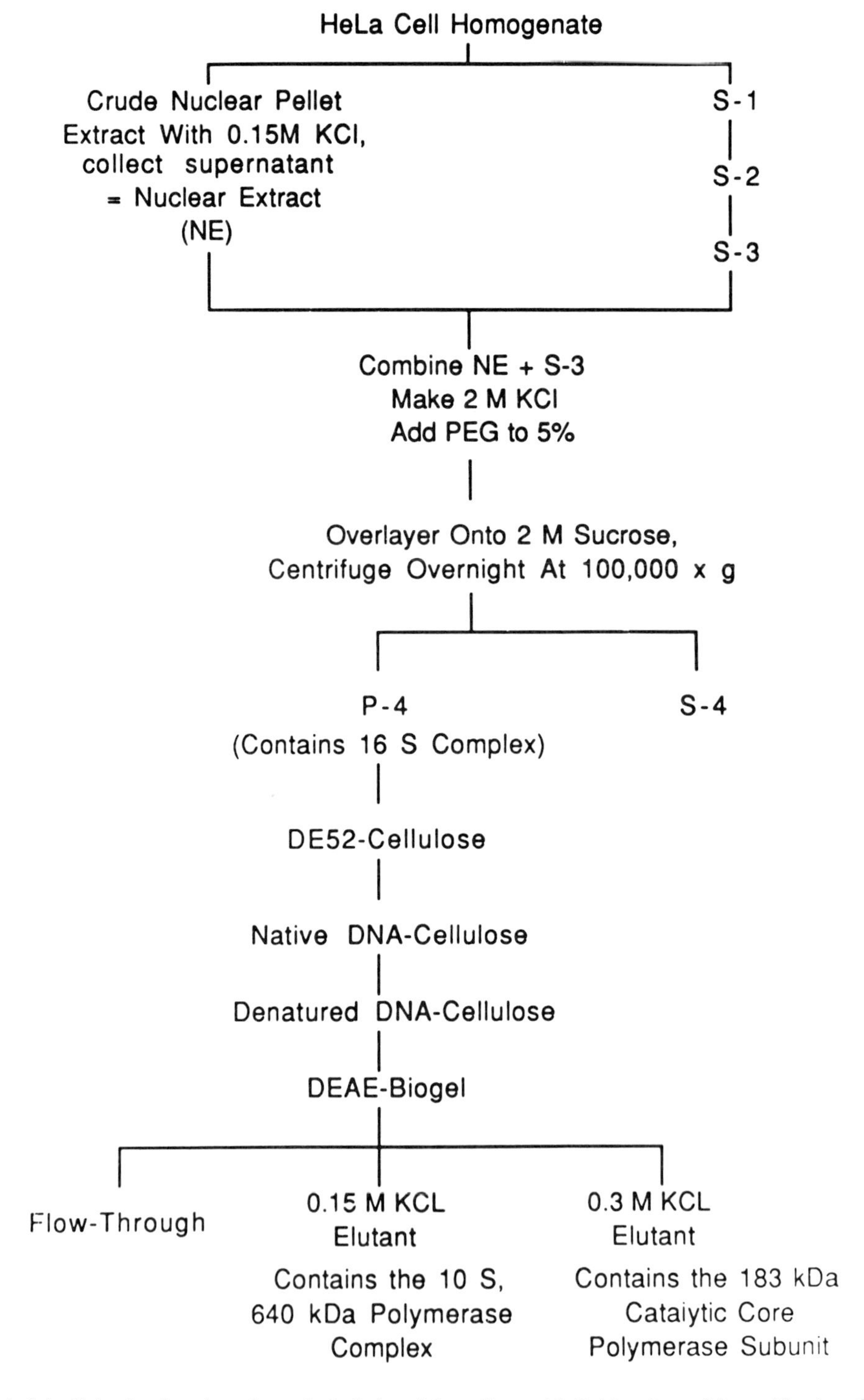

Figure 1. Subcellular fractionation scheme for isolation of the sedimentable P-4 fraction and for purification of the 640 kilodalton, 10S DNA polymerase α-primase multiprotein complex according to the procedure of Baril *et al.* (1988)

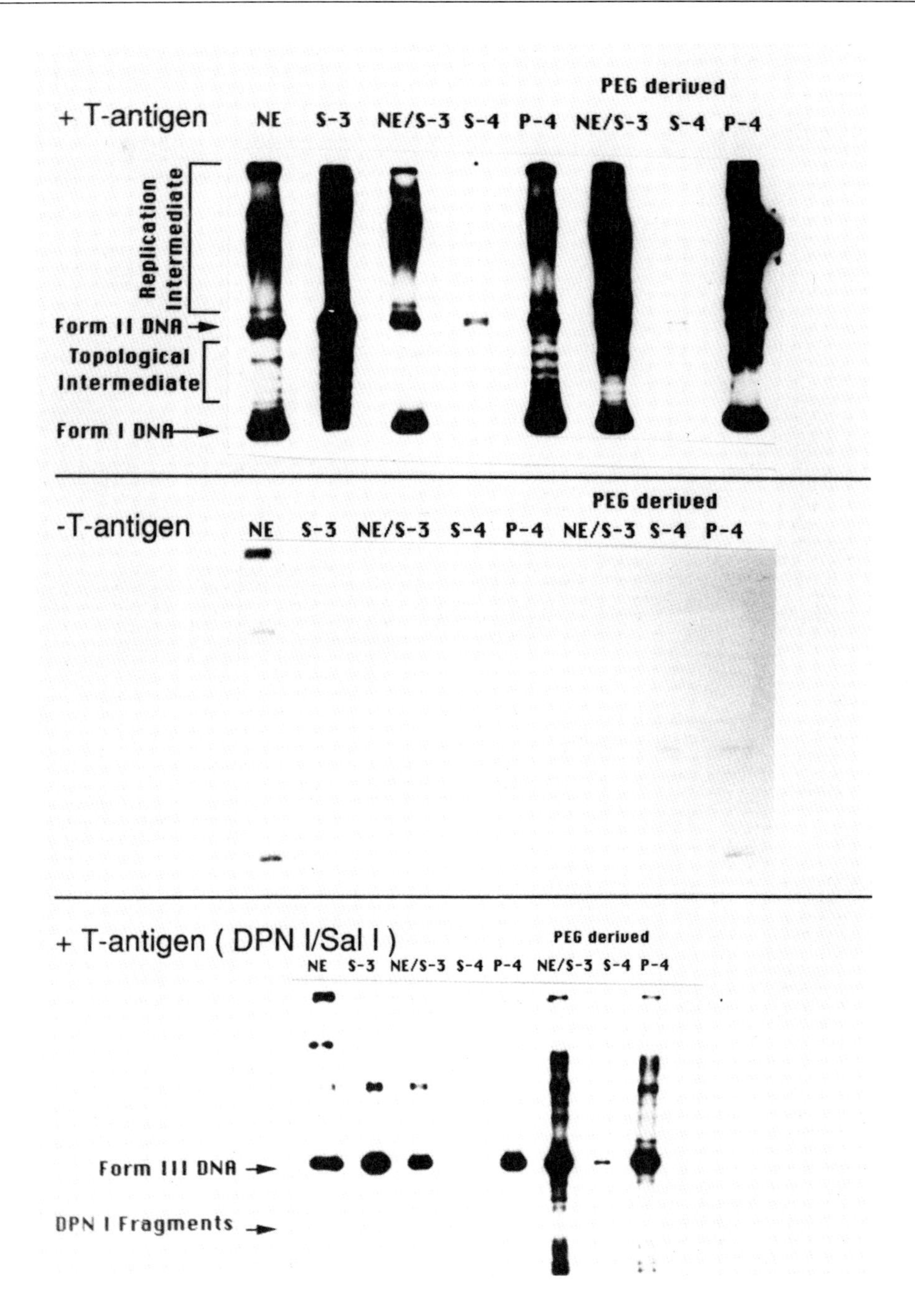

Figure 2. T antigen dependent SV40 *in vitro* replication activity in subcellular fractions isolated from a HeLa cell homogenate. SV40 *in vitro* replication activity was assayed according to the procedure of Wold and Kelly (1988). The reactions for SV40 replication were incubated at 35° for 2h, phenol extracted and the extracted DNA electrophoresed on 1% agarose gels in TBE (Malkas and Baril, 1989). The gels were dried and exposed to XAR-5 X ray film (Kodak) at -70°. The undigested product from assays performed in the presence (Upper panel) and absence (Middle panel) of T antigen. (Lower panel) Assays performed in the presence of T antigen and the DNA product digested with DPNI to remove input plasmid DNA followed by SAL I to linearize the replicated SV40 DNA.

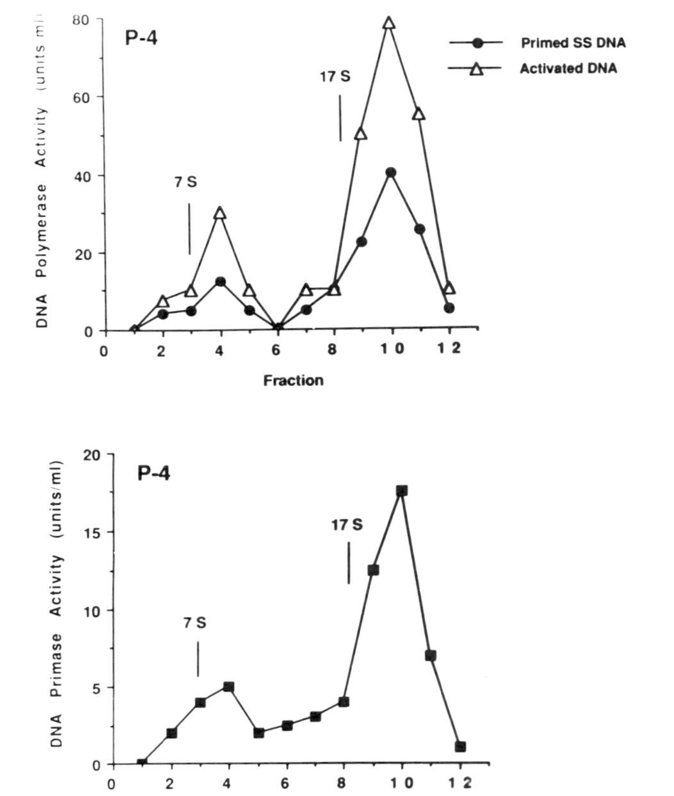

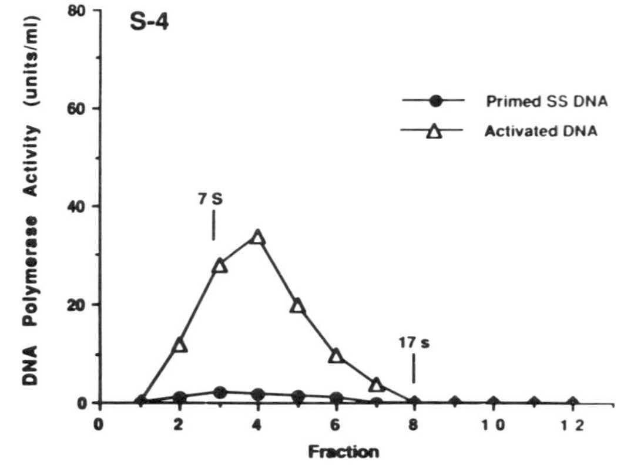

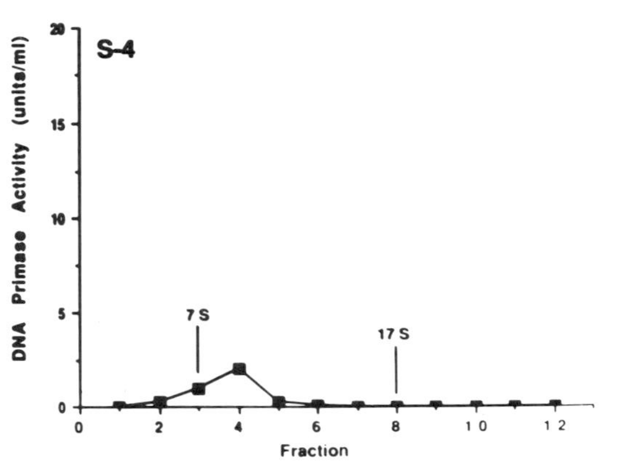

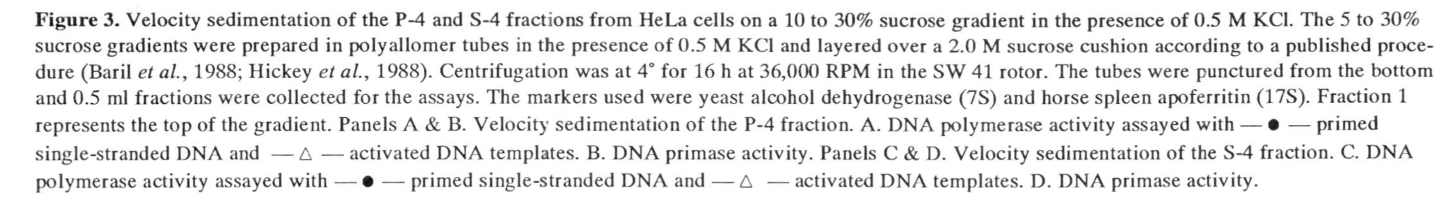

Figure 3. Velocity sedimentation of the P-4 and S-4 fractions from HeLa cells on a 10 to 30% sucrose gradient in the presence of 0.5 M KCl. The 5 to 30% sucrose gradients were prepared in polyallomer tubes in the presence of 0.5 M KCl and layered over a 2.0 M sucrose cushion according to a published procedure (Baril *et al.*, 1988; Hickey *et al.*, 1988). Centrifugation was at 4° for 16 h at 36,000 RPM in the SW 41 rotor. The tubes were punctured from the bottom and 0.5 ml fractions were collected for the assays. The markers used were yeast alcohol dehydrogenase (7S) and horse spleen apoferritin (17S). Fraction 1 represents the top of the gradient. Panels A & B. Velocity sedimentation of the P-4 fraction. A. DNA polymerase activity assayed with —●— primed single-stranded DNA and —△— activated DNA templates. B. DNA primase activity. Panels C & D. Velocity sedimentation of the S-4 fraction. C. DNA polymerase activity assayed with —●— primed single-stranded DNA and —△— activated DNA templates. D. DNA primase activity.

panel). The DNA polymerase activity in the S-4 fraction resided with an 8S peak that lacked activity with primed single-stranded DNA templates and in which primase was virtually absent (Figure 3, lower right panel). The 19S peak of DNA polymerase-primase activity from gradient centrifugation of the P-4 fraction also has associated DNA ligase and topoisomerase I activities. These activities are absent from the 8S peak of polymerase activity obtained from sucrose gradient centrifugation of the S-4 fraction (Hickey *et al.*, 1988). The T antigen dependent SV40 replication activity also cosediments with the 19S peak of enzyme activities from gradient centrifugation of the P-4 fraction (Malkas and Baril, unpublished results).

The presence of the 19S peak of associated enzymes for DNA synthesis with the P-4 fraction and its absence from the gradient profiles of the S-4 fraction probably accounts for the sedimentation of these associated activities with the P-4 fraction during the subfractionation of the NE/S-3. Also, the presence of both the associated activities of enzymes for DNA synthesis, as well as the *in vitro* SV40 replication activity with the 19S complex and the absence of the 19S complex and SV40 replication activity in the nonsedimentable S-4 fraction indicates a probable role for the 19S multienzyme complex in T antigen dependent SV40 replication *in vitro*.

The DNA Polymerase Associated with the 19S Multienzyme Complex Belongs to the α-Class Polymerases

As discussed earlier, although DNA polymerase α has clearly been shown to function in chromosomal DNA replication in eukaryotes (Fry and Loeb, 1986), there have been recent reports that DNA polymerase δ functions in the elongation phase of SV40 replication *in vitro* (Prelich and Stillman, 1988; Downey *et al.*, 1988; Wold *et al.*, 1987; Prelich *et al.*, 1987). To confirm that the DNA polymerase present in the P-4 and S-4 subfractions of the NE/S-3 of HeLa cells is of the a class DNA polymerases its activity was assessed for sensitivity to BuPdGTP (Khan *et al.*, 1984) and to the neutralizing monoclonal antibody SJK 132-20 to human polymerase α (Tanaka *et al.*, 1982). These reagents have been used to distinguish DNA polymerase α from polymerase δ. DNA polymerase δ is reported to be insensitive to the SJK 132-20 antibody (Downey *et al.*, 1988) and is 50 to 100-fold less sensitive to BuPdGTP than is polymerase α (Crute *et al.*, 1986; Byrnes *et al.*, 1985; Lee *et al.*, 1985). The SJK 132-20 antibody completely inhibited the DNA polymerase activity present in the P-4 and S-4 fractions, as well as the DNA polymerase activities that were purified from these fractions (Baril *et al.*, 1988). The DNA polymerase activity in the P-4 fraction, as well as the polymerase activity purified from both the P-4 and S-4 fractions is also inhibited by BuPdGTP at concentrations in the range of 0.5 to 10 μM (Baril *et al.*, 1988; Malkas and Baril, unpublished results). DNA polymerase δ is reported to be resistant to BuPdGTP in this concentration range (Byrnes, 1985).

The SJK 132-20 antibody to human polymerase α and a monoclonal antibody to the 640 kDa DNA polymerase α-primase complex from HeLa cells (Hickey and Baril, unpublished results) also completely inhibit the T antigen dependent *in vitro* SV40 replication activity that is associated with the P-4 fraction. In addition, BuPdGTP at a concentration of 1 μM inhibited the *in vitro* SV40 DNA replication activity associated with the P-4 fraction by over 90% (Malkas and Baril, unpublished results). This concentration of BuPdGTP is about 100-fold less than the level that is reported to be required for inhibition of DNA polymerase δ. These results indicate that the DNA polymerase associated with the sedimentable, multienzyme complex that functions in SV40 replication *in vitro* belongs to the α-class of DNA polymerases.

Chromatographic Resolution of the Components of the Multienzyme Complex for SV40 Replication

The multienzyme complex for *in vitro* replication of SV40 DNA has been partially purified by chromatography of the P-4 fraction from the PEG treated NE/S-3 on Q-Sepharose (Baril *et al.*, 1988; Hickey *et al.*, 1988). Approximately 30% of the protein in the P-4 fraction does not bind to Q-Sepharose at pH 7.5 and appears in the column flow-through fraction. There is no detectable DNA polymerase, primase or *in vitro* SV40 replication activity associated with this fraction. Elution of the column with a continuous gradient of increasing KC1 concentration from 0.05 to 0.5 M partially resolves two peaks of protein and DNA polymerase α-primase activities that elute coincidentally between 0.1 and 0.25 M KC1 concentration (Figs. 4A-C). The highest DNA polymerase and primase activities, however, reside with the peak that is eluded by KC1 concentrations between about 0.15 to 0.25 M. The DNA polymerase activity associated with the eluted fractions functions equally well with activated and primed, single-stranded DNA templates (Figure 4A). The eluted peaks of DNA polymerase α-primase also have associated topoisomerase I and DNA ligase activities (Baril *et al.*, 1988; Hickey *et al.*, 1988). In addition to containing activities for multiple enzymes for DNA synthesis the peaks also contain the activity for T antigen dependent SV40 replication *in vitro* (Malkas and Baril, unpublished results). These results suggest that the multienzyme complex remains intact during chromatography on Q-Sepharose.

In contrast to the results obtained from Q-Sepharose chromatography of the P-4 fraction, during Q-Sepharose chromatography of the nonsedimentable S-4 fraction about 50% of the protein and more than 60% of the DNA polymerase activity does not bind to Q-Sepharose and appears in the column flow-through fraction. Elution of the column with a continuous gradient of increasing KC1 concentration results in a single peak of DNA polymerase activity eluting between 0.1 to 0.25 M KC1 concentration (Figure 4D), a peak of very low primase activity coincidental with the polymerase peak (Figure 4E) and a broad protein peak eluting between 0.05 to 0.3 M KC1 concentration (Figure 4F). The eluted peak of polymerase activity from Q-Sepharose chromatography of the S-4 fraction does not contain T antigen dependent SV40 *in vitro* replication activity (Malkas and Baril, unpublished results). Although the SV40 *in vitro* replication activity that is associated with the 19S multienzyme complex for DNA synthesis survives Q-Sepharose chromatography it does not survive any of the additional steps that have been previously used for purification of the 640 kDa multiprotein DNA polymerase α-primase complex from HeLa cells (Vishwanatha *et al.*, 1986a).

Chromatography of the pooled Q-Sepharose fractions containing the polymerase α-primase and *in vitro* SV40 replication activities on a native DNA-cellulose separates the topoisomerase I and DNA-dependent ATPase activities from the remaining components of the complex and abolishes its *in vitro* SV40 replication activity (Baril *et al.*, 1988; Hickey *et al.*, 1988). Further purification of the remaining components of the multienzyme complex by successive steps of chromatography on single-stranded DNA-cellulose and DEAE-cellulose removes single-stranded DNA-binding proteins (Chase and Williams, 1986) and results in a single peak of DNA polymerase activity that is eluted from DEAE-BioGel by 0.15 M KCl. The DEAE-BioGel 0.15 M KC1 eluted fraction contains the 640 kDa multiprotein DNA polymerase α-primase

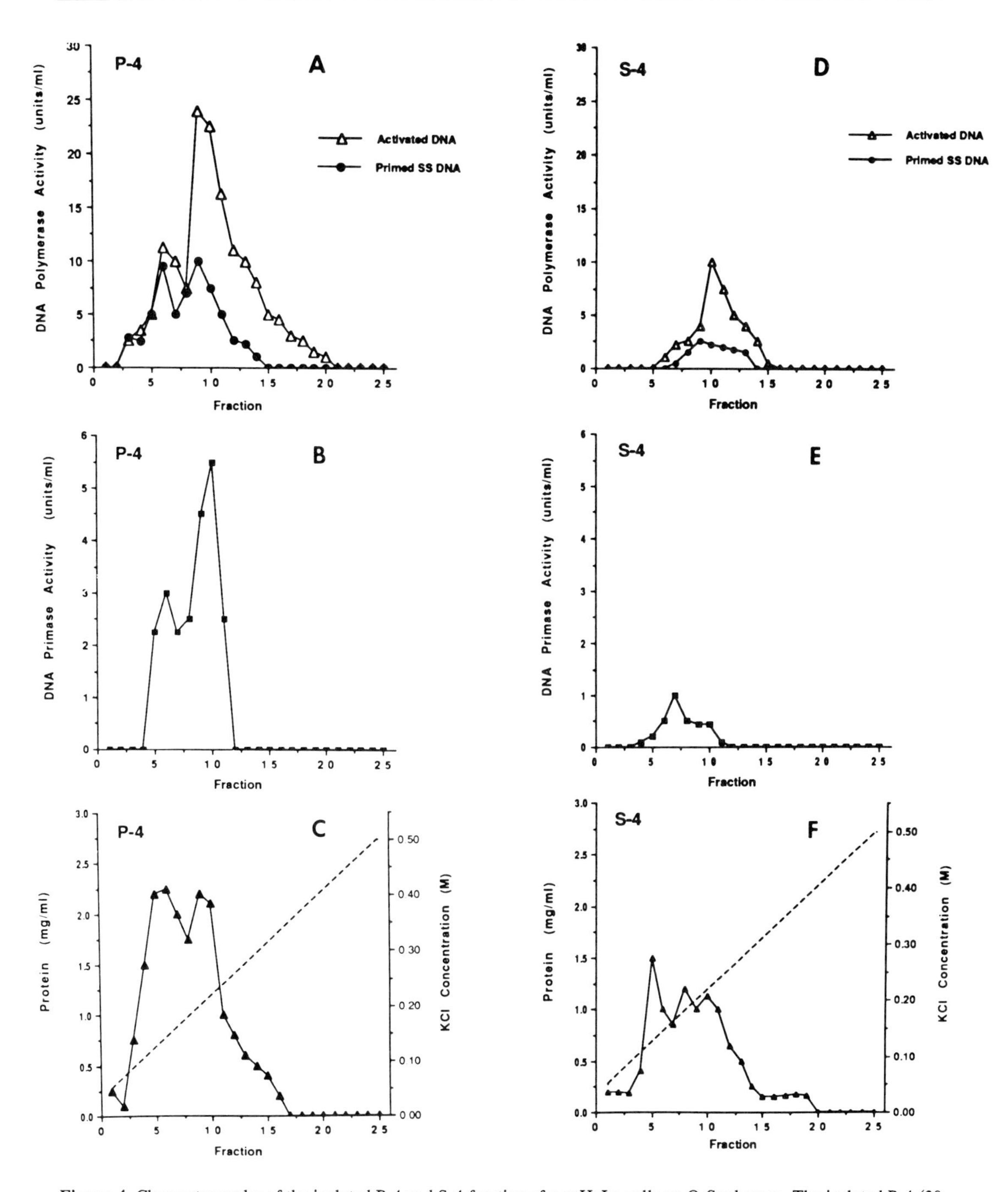

Figure 4. Chromatography of the isolated P-4 and S-4 fractions from HeLa cells on Q-Sepharose. The isolated P-4 (30 mg) and S-4 (25 mg) fractions from HeLa cells were prepared and chromatographed on 1 cm^3 columns of Q-Sepharose according to a published procedure (Baril *et al.*, 1988). A-C. Chromatography of the P-4 fraction. Fractions were assayed for, (A) DNA polymerase α activity with —●— primed single-stranded and —△ — activated DNA (B) DNA primase, and (C) protein according to published procedures (Vishwanatha *et al.*, 1986a). D-F. Chromatography of the S-4 fraction. Fractions were assayed for; (D) DNA polymerase activity with —● — primed single-stranded and —△ — activated DNA, (E) DNA primase and (F) protein as in A-C above.

complex (Vishwanatha *et al.*, 1986; Lamothe *et al.*, 1981) and additional proteins that function in DNA synthesis. Among the latter are RNase H, DNA ligase (Baril *et al.*, 1988; Hickey *et al.*, 1988) and an origin recognition, dA/dT sequence binding protein (Malkas and Baril, 1989).

SV40 Ori Recognition dA/dT Sequence Binding Protein

A dA/dT sequence binding protein that specifically recognizes the 17 base pair (bp) AT-rich tract in the core origin for SV40 replication is also associated with the 19S multienzyme complex for DNA synthesis from HeLa cell homogenates and cofractionates with the 640 kDa multiprotein DNA polymerase α-primase during the purification of the latter. The dA/dT sequence binding protein is resolved from the polymerase α-primase complex, however, by successive steps of chromatography on oligo-(dT)- cellulose and Q-Sepharose (Malkas and Baril, 1989). The binding protein has been purified to electrophoretic homogeneity and appears as a single 62 kDa protein band during polyacrylamide gel electrophoresis under denaturing conditions. The protein binds specifically to runs of A or T sequences in DNA and does not bind to G or C sequences. Gel mobility shift and competition assays showed that the protein specifically binds to the core origin for SV40 replication (Malkas and Baril, 1989). Through the use of a series of DNA fragments containing deletions within the SV40 ori the binding locus for the dA/dT sequence binding protein was shown to be the 17 bp AT-rich tract in the minimal SV40 core origin. This 17 bp AT-tract has been shown to represent a bending locus (Deb *et al.*, 1986) and to be essential for SV40 replication *in vivo* (Hertz *et al.*, 1987; Gerard and Gluzman, 1986; Stillman *et al.*, 1985) and *in vitro* (Stillman *et al.*, 1985). The A/T-rich bent DNA tracts in the replication origins of both prokaryotic and eukaryotic DNAs have recently been implicated as recognition sites for the initiation of DNA replication (Umek *et al.*, 1988). Several groups have suggested that the 17 bp AT-tract in SV40 ori is a possible site for the binding of proteins that are involved in the initiation of SV40 replication (Hertz *et al.*, 1987; Li *et al.*, 1986; Yamaguchi and DePamphilis, 1986; Deb *et al.*, 1986; Gerard and Gluzman, 1986; Stillman *et al.*, 1986). The association of a dA/dT sequence binding protein that specifically recognizes the 17 bp AT-tract in SV40 ori with the 19S multienzyme complex that is involved in the initiation of SV40 replication *in vitro* suggests a possible role for this binding protein in the initiation of SV40 replication. The dA/dT sequence binding protein may actually recognize the unusual DNA conformation in the 17 bp AT-tract and, as in the case of the *E. coli* integration host factor (IHF), induce further bending upon binding to the bent DNA (Stenzel *et al.*, 1987). It was reported recently that a partially purified 3.5S protein from HeLa cells, that actually could be related to the 62 kDa protein described here, specifically recognizes the 17 bp AT-tract in SV40 ori and produced pronounced bending in the ori containing SV40 DNA (Baur and Knippers, 1988). The dA/dT sequence binding protein, therefore, may function like the *E. coli* dnaA protein in the initiation of *E. coli* chromosome replication by specifically binding to recognition sequences in ori, producing an opening of the DNA duplex within an AT-rich region and thereby helping to direct the other proteins of the initiation complex to the opening of the duplex DNA in ori (Baker *et al.*, 1988).

In summary, there is now good evidence that the DNA synthesizing apparatus in HeLa cells that functions in origin specific initiation of SV40 replication *in vitro* exists as a loosely associated 19S, multienzyme complex for DNA synthesis.

CONCLUDING REMARKS

An understanding of the regulation of the DNA synthesizing machinery in eukaryotes is a key to our understanding the control of cell proliferation during normal growth and possible defects in this process that occur during abnormal growth such as in cancer. The past few years have seen good progress being made in the analysis of the DNA synthesizing machinery in eukaryotic cells. The development of improved purification procedures has enabled the isolation of multisubunit forms of DNA polymerase α (Kaguni and Lehman, 1988) that may prove analogous to the holoenzyme forms of prokaryotic DNA replicases (McHenry, 1988; Kornberg, 1988; Cha and Alberts, 1988). More work is required, however, to ascertain if all of the proteins associated with the multiprotein forms of DNA polymerase α are *bona fide* subunits and to define their functions in DNA synthesis. Although there is now overwhelming evidence that polymerase α functions in chromosomal DNA replication there is recent evidence, albeit indirect, suggesting the involvement of DNA polymerase δ at least in SV40 DNA replication *in vitro* (Prelich and Stillman, 1988; Prelich *et al.*, 1987a,b). DNA polymerase α and δ share many similar properties (reviewed in Kaguni and Lehman, 1988). It is imperative, therefore, in defining the DNA synthesizing machinery in eukaryotes that a rigid comparison of the physical and enzymatic properties of these enzymes be performed.

The recent development of a cell free SV40 replication system (Li and Kelly, 1984; Stillman and Gluzman, 1985) has provided a meaningful *in vitro* assay of the DNA synthesizing machinery in permissive mammalian cells. This has enabled us to demonstrate for the first time a direct interaction of the 640 kDa, multiprotein DNA polymerase α-primase complex from HeLa cells with enzymes and other proteins to form a 19S multienzyme complex for DNA synthesis (Baril *et al.*, 1988; Hickey *et al.*, 1988). Further functional analysis of this multienzyme complex, using as guidelines the approaches that were successfully used for the analysis of the DNA synthesizing machinery in prokaryotes, should help to define the enzymological machinery involved in the initiation of chromosomal DNA replication in mammalian cells. This, however, will only represent a first step toward the resolution of the complex process of chromosome replication in eukaryotes. The cell free SV40 replication system, although representing a good model for replication of single replicons, is lacking certain facets of chromosomal DNA replication that exist *in vivo*. Among these is the lack of a requirement for transcriptional activation in the initiation of SV40 replication *in vitro* that is known to exist for SV40, and probably chromosomal, DNA replication *in vivo* (DePamphilis, 1988). Also, it is likely that chromosomal DNA replication will require additional proteins and other factors that are not required for SV40 DNA replication *in vitro*. Finally, there is little information on the actual role of nuclear structures such as the nuclear matrix (Tubo and Berezney, 1987a,b; Jackson and Cook, 1986a; Berezney and Coffey, 1975) in chromosomal DNA replication. There has been some success recently in the isolation of replication origins from yeast chromosomal DNA (Huberman *et al.*, 1987; Brewer and Fangman, 1987). Thus, there is cause for optimism in thinking that the next few years will witness good progress in the enzymological analysis of the initiation of chromosomal DNA replication in eukaryotes and the possible role of the nuclear architecture in this process.

ACKNOWLEDGEMENT

We would like to thank our many colleagues who have contributed to this review. Also we apologize to all whose work we have not cited due to space limitations. The research emanating from the authors' laboratory was supported by NIH grants P30-12708, CA15187 and an NIH postdoctoral Fellowship (CA 08173) to LHM.

REFERENCES

B. M. Alberts. (1985) Protein machines mediate the basic genetic processes. *Trends In Genetics* **1**: 26-30.

H. Ariga and S. Sugano. (1983) Initiation of simian virus 40 DNA replication *in vitro*. *J. Virol.* **48**: 481-491.

G. Badaracco, L. Capucci, P. Plevani and L. M. S. Chang. (1983) Polypeptide structure of DNA polymerase I from Saccharomyces cerevisiae. *J. Biol. Chem.* **258**: 10720-10726.

T. A. Baker, L. L. Bertsch, D. Bramhill, K. Sekimizu, E. Wahle, B. Yung and Kornberg. (1988) Enzymatic mechanism of initiation of replication from the origin of the *Escherichia coli* chromosome.*Cancer Cells* **6**:19-24 Cold Spring Harbor Press,Cold Spring Harbor, NY.

E. F. Baril, L. H. Malkas, R. Hickey, C. J. Li, J. K. Vishwanatha, and S. Coughlin. (1988) A multiprotein DNA polymerase α complex from HeLa cells: Interaction with other proteins in DNA replication. *Cancer Cells* **6**: 373-384. Cold Spring Harbor Press, Cold Spring Harbor, NY.

E. F. Baril, P. Bonin, D. Burstein, K. Mara and P. Zamecnik. (1983) Resolution of the diadenosine 5′,5′ ′ ′-P^1,P^4-tetraphosphate binding subunit from a multiprotein form of HeLa cell DNA polymerase α. *Proc. Natl. Acad. Sci. USA* **80**: 4931-4935.

E. F. Baril, B. Baril, H. Elford and R. Luftig. (1974) DNA polymerases and a possible multienzyme complex for DNA biosynthesis in eukaryotes. **In**: *Mechanisms and Regulation of DNA Replication.* A. R. Kolber and M. Kohiyama (eds.), pp. 275-291. Plenum Press, New York.

E. F. Baril, O. E. Brown, M. D. Jenkins and J. Laszlo. (1971) Deoxyribonucleic acid polymerases with rat liver ribosomes and smooth membranes. Purification and properties of the enzymes. *Biochemistry* **10**: 1981-1992.

C.-P. Baur and R. Knippers, (1988) Protein-induced bending of the simian virus 40 origin of replication. *J.Mol Biol.* **203**: 1009-1019.

R. Berezney and D. S. Coffey. (1975) Nuclear protein matrix: Association with newly synthesized DNA. *Science* **189**: 291-293.

F. J. Bollum. (1960) Calf thymus polymerase. *J. Biol. Chem.* **235**: 2399-2403.

R. Bravo, R. Frank, P. A. Blundell, and H. MacDonald-Bravo. (1987) Cyclin/PCNA is the auxiliary protein of DNA polymerase-α. *Nature* (London) **326**: 515-517.

B. S. Brewer and W. L. Fangman. (1987) The localization of replication origins on ARS plasmids in *S. cerevisiae*. *Cell* **51**: 463-471.

P. M. Burger. (1988) Mammalian Cyclin/PCNA (DNA polymerase δ auxiliary protein) stimulates processive DNA synthesis by yeast DNA polymerase III. *Nucleic Acids Res.* **16**: 6297-6307.

P. M. Burger and G. A. Bauer. (1988) DNA polymerase III from *Saccharomyces cerevisiae*. *J. Biol. Chem.* **263**: 925-930.

J. J. Byrnes. (1985) Differential inhibitors of DNA polymerases alpha and delta. *Biochem. Biophys. Res. Commun.* **132**: 628-634.

J. J. Byrnes, K. M. Downey, V. L. Black, and A. So. (1976) A new mammalian DNA polymerase with 3′ to 5′ exonuclease activity: DNA polymerase δ. *Biochemistry* **15**: 2817-2823.

J. L. Campbell, M. Budd, D. Burbee, K. Sitney, K. Sweder, and F. Heffron (1988) Reverse genetics and *Saccharomyces cerevisia* DNA replication and repair. **In**: *DNA Replication and Mutagenesis.* R.E. Moses and W.C. Summers (eds.) pp. 98-102. American Society Microbiology, Washington, DC.

J. L. Campbell. (1986) Eukaryotic DNA replication. *Ann. Rev. Biochem.* 733-771.

T. A. Cha and B. M. Alberts. (1988) *in vitro* studies of the T4 bacteriophage DNA replication system. *Cancer Cells* **6**: 1-10. Cold Spring Harbor Press, Cold Spring Harbor, NY.

L. M. S. Chang, E. Rafter, C. Augl and F. J. Bollum. (1984) Purification of DNA polymerase-DNA primase complex from calf thymus glands. *J. Biol. Chem.* **259**: 14679-14687.

J. W. Chase and K. R. Williams. (1986) Single-stranded DNA binding proteins required for DNA replication. *Ann. Rev. Biochem.* **55**: 103-136.

R. C. Chiu and E. F. Baril. (1975) Nuclear DNA polymerases and the HeLa cell cycle. *J. Biol. Chem.* **250**: 7951-7957.

R. C. Conaway and I. R. Lehman. (1982a) A DNA primase activity associated with DNA polymerase α from *Drosophila melanogaster* embryos. *Proc. Natl. Sci. USA* **79**: 2523-2527.

S. M. Cotterill, M. E. Reyland, L. A. Loeb and I. R. Lehman. (1987) A cryptic proofreading 3′ → 5′ exonuclease associated with the polymerase subunit of the DNA polymerase-primase from *Drosophila melanogaster*. *Proc. Natl. Acad. USA* **84**: 5635-5639.

J. J. Crute, A. F. Wahl and R. A. Bambara. (1986) Purification and characterization of two new high molecular weight forms of DNA polymerase *Biochemistry* **25**: 26-36.

F. B. Dean, M. Dodson, H. Echols, and J. Hurwitz. (1987) ATP-dependent formation of a specialized nucleoprotein structure by simian virus 40 (SV40) large antigens at the SV40 replication origin. *Proc. Natl. Acad. Sci. USA* **84**: 8981-8985.

S. Deb, A. L. DeLucia, A. Koff, S. Tsui, and P. Tegtmeyer. (1986) The adenine-thymine domain of the simian virus 40 core origin directs DNA bending and coordinately regulates DNA replication. *Mol. Cell. Biol.* **6**: 4578-4584.

M. L. DePamphilis. (1988) Transcriptional elements as components of origins of eukaryotic DNA replication. *Cell* **52**: 635-638.

M. L. DePamphilis and P. M. Wassarman. (1980) Replication of eukaryotic chromosomes: A close-up of the replication fork. *Ann. Rev. Biochem.* **49**: 627-666.

M. Dodson, F. B. Dean, P. Bullock, H. Echols, and J. Hurwitz. (1987) Unwinding of duplex DNA from the SV40 origin of replication by T antigen. *Science* **238**: 964-967.

R. W. Donaldson and E. W. Gerner. (1987) Phosphorylation of a high molecular weight DNA polymerase α. *Proc. Natl. Acad. Sci. USA* **84**: 759-763.

S. L. Dresler and M. G. Frattini. (1986) DNA replication and U.V.-induced repair synthesis in human fibroblasts are much less sensitive than DNA polymerase α to inhibition by butylphenyl-deoxyguanosine triphosphate. *Nucleic Acids* **17**: 7093-7102.

H. J. Edenberg, S. Anderson, and M. L. DePamphilis. (1978) Involvement of DNA polymerase α in simian virus 40 DNA replication. *J. Biol. Chem.* **253**: 3273-3280.

Y. Engstrom, B. Rozell, H.-A. Hansson, S. Stemmie, and L. Thelander. (1984) Localization of ribonucleotide reductase in mammalian cells. *EMBO J.* **3**: 863-867.

M. P. Fairman and B. Stillman. (1988) Cellular factors required for multiple stages of SV40 DNA replication *in vitro*. *EMBO J.* **7**: 1211-1218.

P. A. Fisher, T. S.-F. Wang and D. Korn. (1979a) Enzymological characterization of DNA polymerase α. Basic catalytic properties, processivity and gap utilization of the homogeneous enzyme from KB cells. *J. Biol. Chem.* **254**: 6128-6137.

D. N. Foster and T. Gurney, Jr. (1976) Nuclear location of mammalian DNA polymerase activities. *J. Biol. Chem.* **251**: 7893-7898.

M. Fry and L. A. Loeb. 1986. *Animal Cell DNA Polymerases*. CRC Press, Inc. Boca Raton, FL.

V. N. Genta, D. G. Kaufman, W. K. Kaufman, and B. I. Genwin. (1976) Eukaryotic DNA replication complex. *Nature* (London) **259**: 502-503.

R. Gerard and Y. Gluzman. (1986) Functional analysis of the role of the A+T-rich region and upstream flanking sequences in simian virus 40 DNA replication. *Mol. Cell. Biol.* **40**: 4570-4577.

J. R. Hawker, Jr. and C. S. McHenry. (1987) Monoclonal antibodies specific for the τ subunit of the DNA polymerase III holoenzyme of *Escherichia coli*. *J. Biol. Chem.* **262**: 12722-12727.

G. Herrick, B. B. Spear and G. Veromett. (1976) Intracellular localization of mouse DNA polymerase-α. *Proc. Natl. Acad. Sci. USA* **73**: 1136-1139.

G. Z. Hertz, M. R. Young, and J. E. Mertz. (1987) The A+T-rich sequence of the simian virus 40 origin is essential for replication and is involved in bending of the viral DNA. *J. Virol.* **61**: 2322-2325.

R. J. Hickey, L. H. Malkas, N. Pedersen, C. Li, and E. F. Baril. 1988. Multienzyme complex for DNA replication in HeLa cells. **In**: *DNA Replication and Mutagenesis*, R. Moses and W. Summers (eds.), pp. 41-54. American Society of Microbiology Publications, Washington,DC.

A. M. Holmes, I. P. Hesslewood and I. R. Johnston. (1974) The occurrence of multiple activities in the high-molecular-weight DNA polymerase fraction of mammalian tissues. A preliminary study of some of their properties. *Eur. J. Biochem.* **43**: 487-499.

H. E. Huber, B. B. Beauchamp, J. Bernstein, H. Nakai, S. D. Rabkin, S. Tabor and C.C. Richardson. (1988) Interactions of DNA replication proteins of bacteriophage T7. *Cancer Cells* **6**: 11-17.

J. A. Huberman, L. D. Spotila, K. A. Nawotka, S. M. El-Assoli, and L. R. Davis. (1987) The *in vivo* replication origin of the yeast 2 mu plasmid. *Cell* **51**: 473-481.

U. Hubscher, P. Gerschwiller and G. K. McMaster. (1982) A mammalian DNA polymerase α holoenzyme functioning on defined *in vivo*-like templates. *EMBO J.* **1**: 1513-1519.

D. A. Jackson and P. R. Cook. (1986a) Replication occurs at a nucleoskeleton. *EMBO J.* **5**: 1403-1410.

D. A. Jackson and P. R. Cook. (1986b) Different populations of DNA polymerase α in HeLa cells. *J. Mol. Biol.* **192**: 77-86.

S. M. Jazwinski and G. M. Edelman. (1984) Evidence for participation of a multiprotein complex in yeast DNA replication *in vitro*. *J. Biol. Chem.* **259**: 6852-6857.

L. M. Johnson, M. Snyder, L. M. S. Chang, R. W. Davis and J. L. Campbell. (1985) Isolation of the gene encoding yeast DNA polymerase I. *Cell* **43**: 369-377.

J. M. Kaguni and A. Kornberg. 1984. Replication initiated at the origin (oriC) of the *E. coli* chromosome reconstituted with purified enzymes. *Cell* **38**:183-190.

L.S. Kaguni and I.R. Lehman. (1988) Eukaryotic DNA polymerase-primase: Structure, mechanism and function. *Biochem. Biophys. Acta.* **950**: 87-101.

L. S. Kaguni, J. M. Rossignol, R. C. Conaway and I. R. Lehman. (1983a) Isolation of an intact DNA polymerase-primase from embryos of *Drosophila melanogaster*. *Proc. Natl. Acad. Sci. USA* **80**: 2221-2225.

L. S. Kaguni, J. M. Rossignol, R. C. Conaway, G. R. Banks and I. R. Lehman. (1983b) Association of DNA primase with β/γ subunits of DNA polymerase α from *Drosophila melanogaster* embryos. *J. Biol. Chem.* **288**: 9037-9039.

E. Karawya, J. Swack, W. Albert, J. Fedorko, J. Minna and S. H. Wilson. (1984) Identification of a higher molecular weight DNA polymerase α catalytic polypeptide in monkey cells by monoclonal antibody. *Proc. Natl. Acad. Sci. USA* **81**: 7777-7781.

G. Kaufman and H. H. Falk. 1982. An oligoribonucleotide polymerase from SV40-infected cells with properties of a primase. *Nucleic Acids Res.* **10**: 2309-2321.

N. N. Khan, G. E. Wright, L. W. Dudycz, and N. C. Brown. (1984) Butylphenyl dGTP: A elective and potent inhibitor of mammalian DNA polymerase alpha. *Nucleic Acid. Res.* **12**: 3695-3706.

A. Kornberg. (1988) DNA replication. *J. Biol. Chem.* **263**: 1-4.

A. Kornberg. (1980) (Supplement 1982). *DNA replication.* W.H. Freeman & Co., San Francisco, CA.

S. W. Krauss, D. M. Rosen, D. E. Koshland, Jr. and S. Linn. (1987) Exposure of HeLa DNA polymerase α to protein kinase C affects its catalytic properties. *J. Biol. Chem.* **262**: 3432-3435.

P. Lamothe, B. Baril, A. Chi, L. Lee, and E. F. Baril. (1981) Accessory proteins for DNA polymerase α activity with single-strand DNA templates. *Proc. Natl. Acad. Sci. USA* **78**: 4723-4727.

J. H. Lebowitz and R. McMacken. (1986) The *Escherichia coli* DNAβ replication protein is a DNA helicase. *J. Biol. Chem.* **261**: 4738-4748.

M. Y. W. T. Lee, N. L. Toomey, and G. E. Wright. (1985) Differential inhibition of human placental DNA polymerases δ and α by BuPdGTP and BuAdATP. *Nucl. Acid. Res.* **13**: 8623-8630.

J. J. Li and T. J. Kelly. (1984) Simian virus 40 DNA replication *in vitro*. *Proc. Natl. Acad. Sci. USA* **81**: 6973-6977.

J. J. Li, K. W. C. Peden, R. A. F. Dixon and T. J. Kelly. (1986) Functional organization of the simian virus 40 origin of DNA replication. *Mol. Cell. Biol.* **6**: 1117-1128.

S. Maki and A. Kornberg. (1988a) DNA polymerase III. Holoenzyme of Escherichia coli I. Purification and distinctive functions of subunits T and γ, the dna ZX gene products. *J. Biol. Chem.* **263**: 6547-6554.

S. Maki and A. Kornberg. (1988b) DNA polymerase III. Holoenzyme of *Escherichia coli* II. A novel complex including the γ subunit essential for processive synthesis. *J. Biol. Chem.* **263**: 6555-6560.

S. Maki and A. Kornberg. (1988c) DNA polymerase III. Holoenzyme of *Escherichia coli* III. Distinctive processive polymerases reconstituted from purified proteins. *J. Biol. Chem.* **263**: 6561-6569.

L. H. Malkas and E. F. Baril. (1989) Sequence recognition protein for the 17-base-pair A+T-rich tract in the replication origin of simian virus 40 DNA. *Proc. Natl. Acad. Sci. USA* **85**: 70-74.

D. F. Mark and C. C. Richardson. (1976) *Escherichia coli* thioredoxin: A subunit of bacteriophage T7 DNA polymerase. *Proc. Natl. Acad. Sci. USA* **73**: 780-784.

C. K. Mathews and M. B. Slabaugh. (1986) Eukaryotic DNA metabolism. Are deoxynucleotides channelled to replication sites? *Exp. Cell Res.* **162**: 285-295.

C. S. McHenry. 1988. DNA polymerase III holoenzyme of *Escherichia coli*. *Ann. Rev. Biochem.* **57**: 519-550.

C. S. McHenry and A. Kornberg. (1977) DNA polymerase III holoenzyme of *Escherichia coli*. Purification and resolution into subunits. *J. Biol. Chem.* **252**: 6478-6484.

McMaken, K. Mensa-Wilmot, C. Alfano, R. Seaby, K. Carroll, B. Gomes and K. Stephens. (1988) Reconstitution of purified protein systems for the replication and regulation of bacteriophage λDNA replication. *Cancer Cells* **6**: 25-34. Cold Spring Harbor Laboratory Press, Cold Spring Harbor, NY.

P. Modrich and C. C. Richardson. (1975) Bacteriophage T7 deoxyribonucleic acid replication *in vitro*. Bacteriophage T7 DNA polymerase: An enzyme composed of phage- and host-specified subunits. *J. Biol. Chem.* **250**: 5515-5522.

Y. Murakami, C. R. Wobbe, L. Weissbach, F. B. Dean, and J. Hurwitz. (1986) Role of DNA polymerase α and DNA primase in simian virus 40 DNA replication *in vitro*. *Proc. Natl. Acad. Sci. USA* **83**: 2869-2873.

Y. Murakami, H. Yasuda, H. Miyazawa, F. Hanaoka and M. Yamada. (1985) Characterization of a temperature-sensitive mutant of mouse FM3A cells defective in DNA replication. *Proc. Natl. Acad. Sci. USA* **82**: 1761-1765.

H. Nakai, B. B. Beauchamp, J. Bernstein, H. H. Huber, S. Tabor and C. C. Richardson. (1988) Formation and propagation of the bacteriophage T7 replication fork. **In**: *DNA Replication and Mutagenesis* R. Moses and W.C. Summers (eds.) pp. 85-97. American Society Microbiology, Washington, DC.

C. Nishida and S. Linn. (1988) DNA repair synthesis in permeabilized human fibroblasts mediated by DNA polymerase δ and application for purification xeroderma pigmentosum factors. *Cancer Cells* **6**: 411-415.

H. Ottiger, P. Frei, M. Hassig, and U. Hubscher. (1987) Mammalian DNA polymerase α: A replication competent holoenzyme form from calf thymus. *Nucleic Acids Res.* **15**: 4789-4807.

P. Plevani, G. Badaracco, C. Agol and L. M. S. Chang. (1984) DNA polymerase I and DNA primase complex in yeast. *J. Biol. Chem.* **259**: 7532-7539.

G. Prelich and B. Stillman. (1988) Coordinated leading and lagging strand synthesis during SV40 DNA replication *in vitro* requires PCNA. *Cell* **53**: 117-126.

G. Prelich, M. Kostura, D. R. Marshak, M. B. Mathews, and B. Stillman. (1987a) The cell-cycle regulated proliferating cell nuclear antigen is required for SV40 DNA replication *in vitro*. *Nature* (London) **326**: 471-475.

G. Prelich, C.-K. Tan, Mikostura, M. B., Mathews, A. G. So, K. M. Downey, and B. Stillman. (1987b) Functional identity of proliferating cell nuclear antigen and a DNA polymerase-δ auxiliary protein. *Nature* (London) **326**: 517-520.

C. G. Pritchard and M. L. DePamphilis. (1983a) Preparation of DNA polymerase α C1,C2 by reconstituting DNA polymerase α with its specific stimulatory cofactors C1,C2. *J. Biol. Chem.* **258**: 9801-9809.

C. G. Pritchard, D. T. Weaver, E. F. Baril and M. L. DePamphilis. (1983b) DNA polymerase α cofactors C1,C2 function as primer recognition proteins. *J. Biol. Chem.* **258**: 9810-9819.

G. P.-V. Reddy and A. B. Pardee. (1982) Coupled ribonucleoside reduction, channeling and incorporation into DNA of mammalian cells. *J. Biol. Chem.* **257**: 12526-12531.

G. P.-V. Reddy and A. Pardee. (1980) Multienzyme complex for metabolic channeling in mammalian DNA replication. *Proc. Natl. Acad. Sci. USA.* **77**: 3312-3316.

G. P.-V. Reddy, A. Seongh, M. E. Stafford, and C. K. Mathews. (1978) Enzyme associations in T4 phage DNA precursor synthesis. *Proc. Natl. Acad. Sci. USA* **74**: 3152-3156.

M. E. Reyland, I. R. Lehman and L. A. Loeb. (1988) Specificity of proofreading by the 3′ to 5′ exonuclease of the DNA polymerase-primase of *Drosophila melanogaster*. *J. Biol. Chem.* **263**: 6518-6524.

M. E. Reyland and L. A. Loeb. (1987) On the fidelity of DNA replication. Isolation of high fidelity DNA polymerase-primase complexes by immunoaffinity chromatography. *J. Biol. Chem.* **262**: 10824-10830.

C. C. Richardson. (1983) Bacteriophage T7 minimal requirement for the replication of a complex DNA molecule. *Cell* **33**: 315-317.

R. H. Scheuermann and Echols, H. (1984) A separate editing exonuclease for DNA replication the ε subunit of *Escherichia coli* DNA polymerase III holoenzyme. *Proc. Natl. Acad. Sci. USA* **81**:7747-7751.

V. Simanis and D. P. Lane. (1985) An immunoaffinity purification procedure for SV40 large T antigen. *Virol.* **144**: 88-100.

W. Skarnes, P. Bonin and E. Baril. (1986) Exonuclease activity associated with a multiprotein form of HeLa cell DNA polymerase α. Purification and properties of the exonuclease. *J. Biol. Chem.* **261**: 6629-6636.

S. Spadari and A. Weissbach. (1974) The interrelation between DNA synthesis and various DNA polymerase activities in synchronized HeLa cells. *J. Molec. Biol.* **86**: 11-20.

G. Spyron and P. Reichard. (1983). Ribonucleotides are not channeled into DNA in permeabilized mammalian cells. *Biochem. Biophys. Res Comm.* **115: 1022-1026**

D. K. Srivastava and S. A. Bernhard. (1986) Metabolite transfer *via* enzyme-enzyme complexes. *Science* **234**: 1081-1086.

H. Stahl, P. Droge, and R. Knippers. (1986) DNA helicase activity of SV40 large tumor antigen. *EMBO J.* **5**: 1939-1944.

T. T. Steuzel, P. Patel and D. Bastia. (1987) The integration host factor of Escherichia coli binds to the bent DNA at the origin of replication of the plasmid psc101. *Cell* **49**: 709-717.

B. W. Stillman and Y. Gluzman. (1985) Replication and supercoiling of simian virus 40 DNA in cell extracts from human cells. *Mol. and Cell. Biol.* **5**:2051-2060.

B. Stillman, R. D. Gerard, R. A. Guggenheimer, and Y. Gluzman. (1985) T-antigen and template requirements for SV40 DNA replication *in vitro*, *EMBO J.* **4**: 2933-2939.

C. K. Tan, C. Castillo, A. G. So, and K. M. Downey. (1986) An auxiliary protein for DNA polymerase-δ from fetal calf thymus. *J. Biol. Chem.* **261**: 12310-12316.

S. Tanaka, S.-Z. Hu, T. S.-F. Wang and D. Korn. (1982) Preparation and preliminary characterization of monoclonal antibodies against human polymerase α. *J. Biol. Chem.* **257**: 8386-8390.

P. K. Tomick, C.-S. Chin, M. G. Novcha, and G. R. Greenberg. (1974) Evidence for a complex regulating the *in vivo* activities of early enzymes. *J. Biol. Chem.* **249**: 7613-7622.

J. Tooze. (1980) *Molecular Biology of Tumor Viruses 2nd ed.* Cold Spring Harbor Laboratory, Cold Spring Harbor, New York.

Tseng, B. Y. and C. N. Ahlem. (1983) A DNA primase from mouse cells. Purification and partial characterization. *J. Biol. Chem.* **258**: 9845-9849.

R. A. Tubo and R. Berezney. (1987a) Pre-replicative association of multiple replicative enzyme activities with the nuclear matrix during rat liver regeneration. *J. Biol. Chem.* **262**: 1148-1154. @REFERENCES = R. A. Tubo and R. Berezney. (1987b) Identification of 100 and 150S DNA polymerase α-primase megacomplexes solubilized from the nuclear matrix of regenerating rat liver. *J. Biol. Chem.* **262**: 5857-5865.

R. M. Umek, M. J. Eddy and D. Kowalski. (1988) DNA sequences required for unwinding prokaryotic and eukaryotic replication origins. *Cancer Cells* **6**: 473-478. Cold Spring Harbor Laboratory Press, Cold Spring Harbor, NY.

J. K. Vishwanatha, S. A. Coughlin, M. Wesolowski-Owen, and E. F. Baril. (1986a) A multiprotein form of DNA polymerase α from HeLa cells. Resolution of its associated catalytic activities. *J. Biol. Chem.* **261**: 6619-6628.

J. K. Vishwanatha and E. F. Baril. (1986b) Resolution and purification of free primase activity from the DNA primase-polymerase α complex of HeLa cells. *Nucleic Acids Res.* **14**: 8467-8487.

J. K. Vishwanatha, M. Yamaguchi, M. L. DePamphilis and E. F. Baril. (1986c) Selection of template initiation sites and the lengths of RNA primers synthesized by DNA primase are strongly affected by its organization in a multiprotein DNA polymerase alpha complex. *Nucleic Acids Res.* **14**: 7305-7323.

A. F. Wahl, A. M. Geis, B. H. Spain, S. W. Wang, D. Korn and T. S.-F. Wang. (1988) Gene expression of human DNA polymerase α during cell proliferation and the cell cycle. *Mol. Cell Biol.* **8**: 5016-5025.

A. Weissbach, D. Baltimore, F. Bollum, R. Gallo and D. Korn. (1975) Nomenclature of eukaryotic DNA polymerases. *Science* **190**: 401-402.

A. Weissbach, A. Schlabach, B. Fridlender and A. Bolden. (1971) DNA polymerases from human cells. *Nature* (London) New Biol. **231**:167-170.

G. R. Welch. (1977) On the role of organized multienzyme systems in cellular metabolism. A general synthesis. *Prog. Biophys. Mol. Biol.* **32**: 103-191.

W. Wickner, R. Schekman, K. Geider and A. Kornberg. (1973) A new form of DNA polymerase III and a copolymerase replicate a long single-stranded primer template. *Proc. Natl. Acad. Sci. USA* **70**: 1764-1767.

S. H. Wilson, A. Matsukage, E. W. Bohn, Y. C. Chen and M. Sivarajan. (1977) Polynucleotide recognition by DNA α polymerase. *Nucleic Acids Res.* **4**: 3981-3997.

C. R. Wobbe, L. Weissbach, J. A. Borawiec, F. B. Dean, Y. Murakami, P. Bullock and J. Hurwitz. (1987) Replication of simian virus 40 origin-containing DNA *in vitro* with purified proteins. *Proc. Natl. Acad. Sci. USA* **84**: 1834-1838.

C. R. Wobbe, F. Dean, L. Weissbach, and J. Hurwitz. (1985) *in vitro* replication of duplex circular DNA containing the simian virus 40 DNA origin site. *Proc. Natl. Acad. Sci. USA* **82**: 5710-5714.

M. S. Wold and T. Kelly. (1988) Purification and characterization of replication protein A, a cellular protein required for *in vitro* replication of simian virus DNA. *Proc. Natl. Acad. Sci. USA* **85**:2523-2527.

M. S. Wold, J. J. Li, and T. J. Kelly. (1987) Initiation of simian virus 40 DNA replication *in vitro*: Large-tumor-antigen and origin-dependent unwinding of the template. *Proc. Natl. Acad. Sci. USA* **84**: 3643-3647.

S. W. Wong, A. F. Wahl, P. M. Yuan, N. Arai, B. E Pearson, K. Arai, D. Korn, M. W. Hunkapiller and T. S.-F. Wang. (1988) Human DNA polymerase α gene expression is cell proliferation dependent and its primary structure is similar to both prokaryotic and eukaryotic replicative DNA polymerases. *EMBO J.* **7**: 37-48.

T. Yagura, T. Kozu, T. Seno, M. Saneyashi, S. Hiraga and H. Nagano. (1983) Novel form of DNA polymerase α associated with DNA primase activity of vertebrates. *J. Biol. Chem.* **258**: 13070-13075.

T. Yagura, T. Kozu and T. Seno. (1982) Mouse DNA polymerase accompanied by a novel RNA polymerase activity. Purification and partial characterization. *J. Biochem.* **91**: 607-618.

M. Yamaguchi and M. L. DePamphilis. (1986) DNA binding site for a factor(s) required to initiate simian virus 40 DNA replication. *Proc. Natl. Acad. Sci. USA* **83**: 1646-1650.

L. Yang, M. S. Wold, J. J. Li, T. J. Kelly, and L. F. Liu. (1987) Roles of DNA topoisomerases in simian virus 40 DNA replication *in vitro*. *Proc. Natl. Acad. Sci. USA* **84**: 950-954.

4

DNA POLYMERASE δ

Robert A. Bambara, Thomas W. Myers, and Ralph D. Sabatino

INTRODUCTION

Until recently it has been almost universally accepted that DNA polymerase α is the enzyme primarily responsible for mammalian chromosomal DNA replication. This conclusion was based on experiments demonstrating an increase in DNA polymerase α activity when cells were stimulated from quiescence to rapid proliferation (Thömmes *et al.*, 1986). Furthermore, aphidicolin and N-ethylmaleimide, effective inhibitors of DNA replication in living cells, were found to inhibit DNA polymerase α but not DNA polymerases β and γ (reviewed in Fry and Loeb, 1986).

DNA polymerase δ was first reported by A. G. So and colleagues as a distinct mammalian DNA polymerase (Byrnes *et al.*, 1976; Byrnes, 1984). It was purified from rabbit bone marrow, and then later from calf thymus (Lee *et al.*, 1980; 1981; 1984). The unique feature of this DNA polymerase among mammalian DNA polymerases was the presence of a nondissociable 3′→5′ exonuclease. Although this activity had been reported to also be associated with DNA polymerase α after certain approaches to purification (Chen *et al.*, 1979; Ottiger and Hübscher, 1984; Skarnes *et al.*, 1986; Cotterill *et al.*, 1987), generally, mammalian DNA polymerases have been found to lack a 3′→5′ exonuclease activity (Bollum, 1975; Falaschi and Spadari, 1978).

In prokaryotic systems, essentially all DNA polymerases have 3′→5′ exonuclease activity (Kornberg, 1982; 1984). These activities have been proposed to contribute to the fidelity of chromosomal replication. The improved fidelity is accomplished through exonucleolytic proofreading, by which misincorporated nucleotides are preferentially removed by the exonuclease. The discovery of a DNA polymerase with inherent proofreading activity could help to explain the high fidelity of chromosomal DNA replication found in mammalian cells.

During the period from 1976 to 1984, all reports concerning DNA polymerase δ came from a single laboratory. At that time, the concept of a new mammalian DNA polymerase having a tightly associated exonuclease was not widely accepted. In the last four years however, substantial evidence from a number of laboratories has supported the conclusion that DNA polymerase δ is an important component of the DNA replication and repair systems in mammalian cells. It also has been found to be sensitive to aphidicolin and present in proliferating tissue (Lee *et al.*, 1981; Byrnes, 1984; Wahl *et al.*, 1986; Marraccino *et al.*, 1987; Zhang and Lee, 1987). Furthermore, studies using specific inhibitors indicate that DNA polymerase δ is involved in chromosomal DNA replication (Hammond *et al.*, 1987; Marraccino *et al.*, 1987). Additionally, evidence suggests that DNA polymerase δ is involved in simian virus 40 (SV40) DNA replication (Decker *et al.*, 1987).

THE PROPERTIES OF DNA POLYMERASE δ

Template Specificity

The most commonly used DNA polymerase assay, which uses nuclease-activated DNA, selects for DNA polymerase α during purification protocols. DNA polymerase δ, which does not utilize nuclease-activated DNA efficiently, was isolated from rabbit bone marrow (Byrnes *et al.*, 1976) and calf thymus (Lee *et al.*, 1981; 1984) using the alternating copolymer poly(dAdT). In the early stages of purification, DNA polymerase δ also efficiently utilizes poly(dA)•oligo(dT). However, further purification of DNA polymerase δ by the above mentioned procedures demonstrated that the enzyme loses the capacity for efficient utilization of poly(dA)•oligo(dT). Recently, a 36 kilodalton (kDa) protein (Tan *et al.*, 1986), which was later identified as the well-studied proliferating cell nuclear antigen (PCNA) (Prelich *et al.*, 1987b; Bravo *et al.*, 1987; Tan, 1982, Celis and Celis, 1985b), was found to restore the capability of highly purified DNA polymerase δ to use poly(dA)•oligo(dT) efficiently (Tan *et al.*, 1987a).

In other laboratories, a different form of DNA polymerase δ has been isolated from rabbit bone marrow (Byrnes and Black, 1978; Byrnes, 1984) and calf thymus (Crute *et al.*, 1986a; Sabatino and Bambara, 1988). These enzymes, upon further purification, do not lose the capability to utilize poly(dA)•oligo(dT). It is important to note that this review discusses forms of DNA polymerase δ that may be different, depending on the purification protocol employed. However, results suggest that there is a structural relationship between these enzymes since polyclonal antibodies directed against DNA polymerase δ from calf thymus isolated using alternating poly (dAdT) (Downey *et al.*, 1988) cross-react with DNA polymerase δ isolated using poly(dA)•oligo(dT) (A.G. So, personal communication). The particular form of DNA polymerase δ discussed in the following text can be inferred from the research group that is referenced. In cases where a specific property is different in one of the purified enzymes *versus* the other, that difference will be emphasized in the text.

Physical and Structural Properties

DNA polymerase δ has generally been studied from either rabbit bone marrow or calf thymus (Byrnes, 1984; Lee *et al.*, 1981; 1984). However, the enzyme from human placenta has also been examined (Lee and Toomey, 1987). The protein has been purified to apparent or near homogeneity using conventional methods including ion exchange, gel filtration, affinity, and hydrophobic chromatography. The overall consensus for the estimated sedimentation coefficient is approximately 7 S using either sucrose or glycerol gradient ultracentrifugation containing 0.25-0.5 M KCl. At a lower salt concentration (0.05 M), the DNA polymerase apparently aggregates to form dimers or multimers having a sedimentation coefficient of approximately 11 S (Byrnes, 1984). The Stokes radius was also determined by gel filtration, at 0.25-0.5 M KCl, to be approximately 50Å. The enzyme was estimated to have a frictional ratio (F/F_0) of 1.4, a result suggesting that the enzyme has an asymmetric structure. DNA polymerase δ has also been shown to be an acidic protein (similar to DNA polymerase α) having a pI of 5.5 (Lee *et al.*, 1981; Lee and Toomey, 1987).

The proposed molecular weight of DNA polymerase δ has varied greatly since its discovery. Initially, Lee *et al.*, (1981), calculated a molecular weight for calf thymus DNA polymerase δ of 200 kDa by gel filtration. This enzyme had polypeptides corresponding to 49 and 60 kDa when analyzed by sodium dodecyl sulfate (SDS) polyacrylamide gel electrophoresis. It was suggested that these polypeptides may be similar to those of murine myeloma DNA polymerase α, having molecular weights of 48 and 50 kDa (Chen *et al.*, 1979). After further purification, using a protocol to minimize proteolysis, the isolated calf thymus enzyme had polypeptides corresponding to 48 and 125 kDa (Lee *et al.*, 1984). It was proposed that the 60 kDa band, reported in previous work, may have been derived by proteolysis of the 125 kDa protein. The molecular weight of DNA polymerase δ isolated from rabbit bone marrow was determined to be 300 kDa and it was shown to have a single polypeptide band at 122 kDa as analyzed by SDS polyacrylamide gel electrophoresis (Goscin and Byrnes, 1982). More recently, Lee and Toomey, (1987) purified DNA polymerase δ from human placenta. The apparent molecular weight, which was calculated from the sedimentation coefficient and Stokes radius, corresponded exactly with a 170 kDa polypeptide band as analyzed by SDS polyacrylamide gel electrophoresis. It was demonstrated that the 170 kDa peptide reacted, in Western blot analysis, with polyclonal sera directed against DNA polymerase δ from human placenta.

In addition, DNA polymerases δI and δII were isolated from calf thymus (Crute *et al.*, 1986a). It was determined that both of these enzymes had sedimentation coefficients of 10.1 S, and Stokes radii of 58.2 Å and 69.6 Å, respectively. The calculated molecular weights were determined to be 240 and 290 kDa for DNA polymerases δI and δII, respectively. These enzyme preparations had multiple polypeptides, including high molecular weight polypeptide bands, *e.g.*, 250 kDa, as analyzed by SDS polyacrylamide gel electrophoresis.

Sensitivity to Inhibitors

When DNA polymerase δ was initially discovered, it was found that this enzyme had similar drug sensitivities to those of DNA polymerase α (reviewed in Fry and Loeb, 1986). DNA polymerase δ is sensitive to aphidicolin, N-ethylmaleimide, and arabinofuranosyl nucleoside triphosphates, whereas it is resistant to dideoxynucleoside triphosphates (ddNTPs) (Lee *et al.*, 1981; Byrnes, 1984; Crute *et al.*, 1986a; Lee and Toomey, 1987). We have reported, however, that DNA polymerases δI and δII are slightly sensitive to ddTTP (Wahl *et al.*, 1986).

The development of monoclonal antibodies directed against DNA polymerase α isolated from human KB cells (Tanaka *et al.*, 1982) and the DNA polymerase α specific inhibitors N^2-(*p*-n-butylphenyl)-2′-deoxyguanosine 5′-triphosphate (BuPdGTP) (Khan *et al.*, 1984) and 2-(*p*-n-butylanilino)-2′-deoxyadenosine 5′-triphosphate (BuAdATP) (Khan *et al.*, 1985) have enabled investigators to distinguish DNA polymerases α and δ. Monoclonal antibodies directed against human DNA polymerase α, which inhibit calf thymus DNA polymerase α, do not cross-react with calf thymus DNA polymerase δ. An exception is calf thymus DNA polymerase δI, which is slightly sensitive to anti-DNA polymerase α antibodies SJK 132-20 IgG and SJK 287-20 IgG (Wahl *et al.*, 1986). Conversely, polyclonal sera directed against human placenta or calf thymus DNA polymerase δ, do not cross-react with the homologous DNA polymerase α (Lee and Toomey, 1987; Downey *et al.*, 1988). It is important to note, however, that polyclonal serum, directed against calf thymus DNA polymerase δ (Downey *et al.*, 1988), isolated using alternating poly(dAdT), does inhibit calf thymus DNA polymerase δII, isolated using poly(dA)•oligo(dT) as the DNA template (A.G. So, personal communication). This result strongly suggests that these two forms of DNA polymerase δ are structurally related.

DNA polymerase α, being highly sensitive to BuPdGTP, is inhibited 50% at less than 1 μM (Byrnes, 1985; Lee *et al.*, 1985; Wahl *et al.*, 1986). DNA polymerase δII, however, is not inhibited 50% until a BuPdGTP concentration of 100 μM (Wahl *et al.*, 1986). Generally, most forms of DNA polymerase δ isolated using poly(dAdT) have a similar resistance to BuPdGTP as DNA polymerase δII (Byrnes, 1985; Lee *et al.*, 1985). However, DNA polymerase δI has been shown to exhibit a sensitivity to BuPdGTP intermediate between that of DNA polymerases α and δII (Wahl *et al.*, 1986). These antibody and inhibitor studies, in conjunction with results demonstrating that DNA polymerase δI has a dissociable exonuclease (Wahl *et al.*, 1986), strongly suggest that DNA polymerase δI may not be a δ enzyme. That is, it does not fulfill the criteria of having a tightly associated exonuclease and a resistance to high concentrations of BuPdGTP. Therefore, we have not further characterized DNA polymerase δI. We have instead focused our efforts on characterization of DNA polymerase δII, which henceforth will be referred to as DNA polymerase δ.

The Elongation Reaction

An important feature of purified DNA polymerases from prokaryotes, which is not shared by those from eukaryotes, is the ability of these enzymes to perform highly processive DNA synthesis. *Escherichia coli* DNA polymerase III holoenzyme binds to DNA templates, in an ATP or dATP dependent reaction and incorporates thousands of nucleotides before dissociating (Kornberg, 1982; McHenry, 1988). Eukaryotic DNA polymerases have generally been shown to be quasi- or semi-processive, only synthesizing 5-20 nucleotides per binding event (Das and Fujimura, 1979; Fisher *et al.*, 1979; Detera *et al.*, 1981; Hockensmith and Bambara, 1981; reviewed in Fry and Loeb, 1986). Recently, we have elaborated on the findings of Tan *et al.*, (1987b) and Hohn and Grosse, (1987). These investigators were able to increase the processivity of DNA polymerase α by manipulating the pH and magnesium concentration of the DNA synthesis reaction *in vitro*. We have demonstrated that the appropriate pH and magnesium concentrations can increase the processivity values of both calf thymus DNA polymerases α and δ into the thousands on poly(dA)•oligo(dT) and singly-primed, single-stranded, circular bacteriophage M13 DNA (Sabatino *et al.*, 1988). It is important to note that conditions for optimal DNA synthesis (pH 7.0; 5 mM $MgCl_2$) do not correlate with those for optimal processivity (pH 6.0; 1-2 mM $MgCl_2$). Analysis of the products synthesized by DNA polymerase δ, by denaturing polyacrylamide gel electrophoresis (Figure 1), demonstrates the effects of pH and magnesium concentration. As can be seen, short products are synthesized at pH 8, 10 mM $MgCl_2$ (lane t) whereas very long products (200-2000 nucleotides) are made at

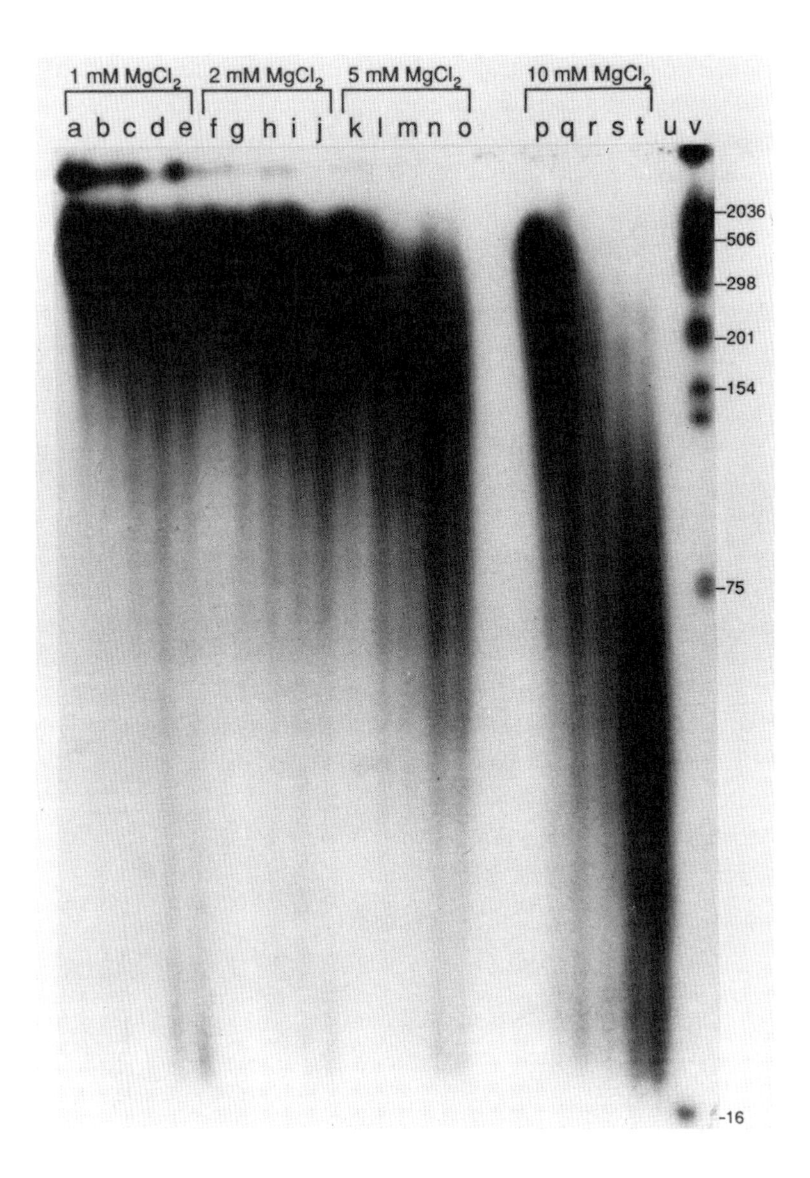

Figure 1. Analysis of the DNA products synthesized by calf thymus DNA polymerase δ. DNA products synthesized by DNA polymerase δ, on poly(dA)$_{4000\text{-}5000}$•oligo(dT)$_{16}$ (1:22), were analyzed by denaturing polyacrylamide gel electrophoresis. The $MgCl_2$ concentration was varied from 1 to 10 mM as indicated in the figure. Lanes a-e, f-g, k-o, and p-t each indicate a series of increasing pH values of buffer in half-unit increments from pH 6.0 to 8.0. Lanes u and v contain size markers. (Figure obtained from Sabatino *et al.*, 1988).

pH 6-8, 1 mM $MgCl_2$ (lanes a-e, pH is in half-unit increments from 6.0 to 8.0). Similar results have been observed for DNA polymerase α (Tan *et al.*, 1987b; Hohn and Grosse, 1987; Sabatino *et al.*, 1988) except that the measured length of processive synthesis by DNA polymerase α is always 2- to 5-fold shorter than that of DNA polymerase δ, under every tested reaction condition.

Kinetic analysis demonstrated that the K_m of the synthetic reaction with respect to substrate 3′-termini, decreases when low pH and magnesium conditions are used to raise the length of processive DNA synthesis. Furthermore, although a high concentration of oligo(dT) primers on the poly(dA) template effectively displaces the polymerase during synthesis at high pH and magnesium concentration, the polymerase can efficiently synthesize through primers at low pH and magnesium concentration. Although the physiological significance of these results is not known, it has been speculated that other factors may influence the processivity of these DNA polymerases. For example, the removal of a putative subunit or associated protein factor from partially purified calf thymus DNA polymerase α was found to decrease processive DNA synthesis (Hockensmith and Bambara, 1981). *Escherichia coli* single-stranded DNA binding protein (SSB) was found to increase processive DNA synthesis of *Drosophila melanogaster* polymerase α (Villani *et al.*, 1981) suggesting that a cellular DNA binding protein may display similar properties. ATP (Wierowski *et al.*, 1983) and spermidine (Mikhailov and Androsova, 1984) were both found to increase the processivity of DNA polymerase α, suggesting a role for nucleotide derivatives or polyamines. Additionally, Prelich *et al.*, (1987b) found that the processivity of the DNA polymerase δ, isolated using a poly(dAdT) template (Lee *et al.*, 1984; Tan *et al.*, 1986), could be increased by the addition of PCNA.

It is clear that both DNA polymerases α and δ are capable of highly processive DNA synthesis. It is not known, however, whether highly processive DNA synthesis is an important aspect of DNA replication *in vivo*, since it has been shown that eukaryotic Okazaki fragments are only 135 nucleotides in length (Anderson and DePamphilis, 1979). Additionally, DNA polymerase δ has also been shown to perform strand-displacement synthesis in the presence of PCNA (Downey *et al.*, 1988). This function may be important in concert with highly processive DNA synthesis at the replication fork (as discussed later). DNA polymerase α, on the other hand, cannot perform strand displacement synthesis.

DNA POLYMERASE δ AND ASSOCIATED ACTIVITIES

DNA Primase

DNA polymerase α has been shown to have a tightly associated DNA primase (Kaguni *et al.*, 1983; Yagura *et al.*, 1983; Hübscher, 1983; Wang *et al.*, 1984; Yamaguchi *et al.*, 1985). DNA primase synthesizes short oligoribonucleotides (or primers) which are extended by the DNA polymerase. It is hypothesized that eukaryotic primase function is similar to that of prokaryotic primase *in vivo*. In prokaryotes, the DNA primase synthesizes Okazaki fragments, required for lagging-strand DNA synthesis (Kornberg, 1980).

Initially, we reported an association of DNA primase activity with calf thymus DNA polymerase δ (Crute *et al.*, 1986a). The polymerase, exonuclease and primase activities co-sedimented through sucrose gradients, the last step in the purification protocol. After modifying the purification scheme, we have been able to simultaneously monitor polymerase, exonuclease and primase activities during each step of the purification. Using neutralizing antibodies directed against DNA polymerase α and BuPdGTP, a specific inhibitor of DNA polymerase α, we have been able to distinguish DNA polymerase δ from DNA polymerase α and related polymerase-primase complexes. Our most recent purification scheme includes a phosphocellulose column. Figure 2 indicates that both polymerase and exonuclease co-elute from this column, prior to the primase peak. It is evident from these results that DNA

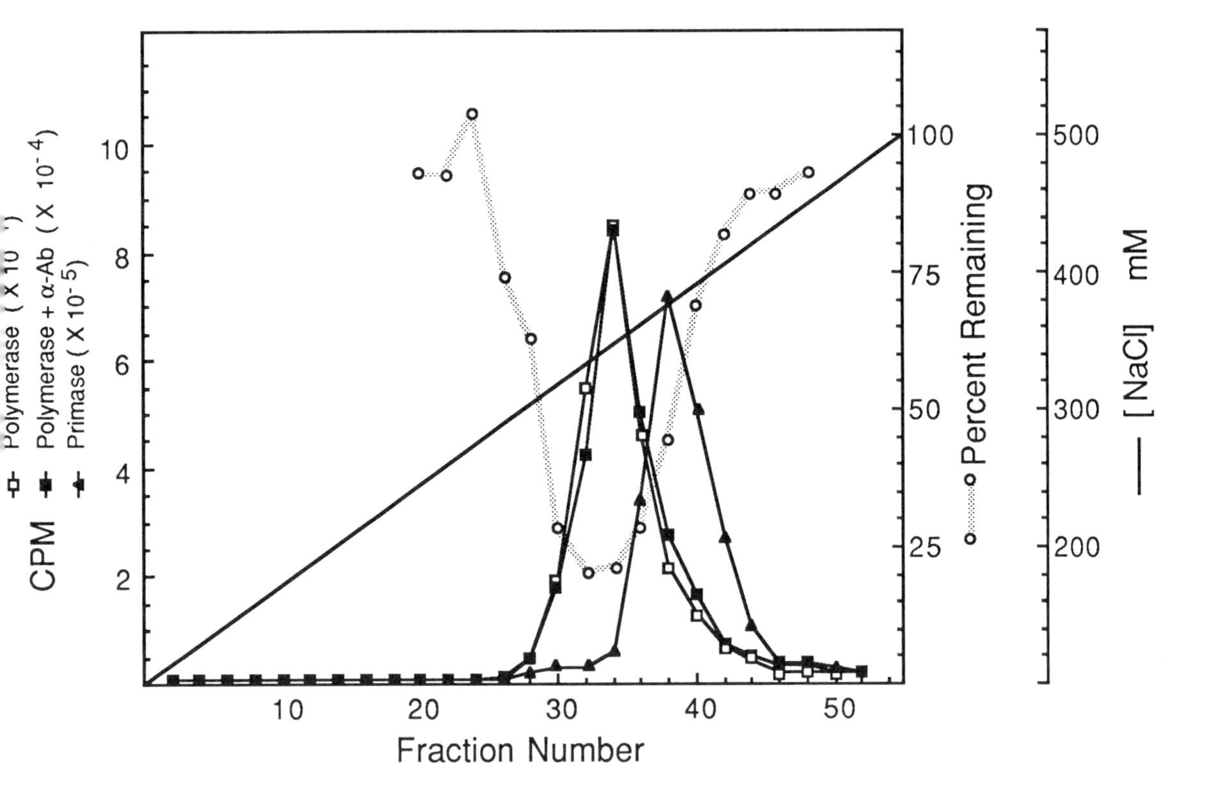

Figure 2. Phosphocellulose chromatography of calf thymus DNA polymerase δ. Assays performed on the resultant fractions are shown for DNA polymerase activity utilizing poly$(dA)_{4000\text{-}5000}$•oligo$(dT)_{16}$ in the presence (■) or absence (□) of anti-DNA polymerase α antibody (SJK 132-20 IgG), DNA primase activity utilizing poly$(dT)_{400}$ in a coupled assay with pol I (Kf) (▲), and exonuclease activity using poly $(dT)_{400}[^3H]$dTMP as substrate (○).

polymerase δ does not have a tightly associated DNA primase. Downey *et al.*, (1988) have also demonstrated that DNA polymerase δ, isolated using the template poly(dAdT), does not have an associated DNA primase.

3′→5′ Exonuclease

It has been proposed that the exonuclease associated with DNA polymerase δ removes nucleotides misincorporated during DNA synthesis, a function analogous to the exonucleolytic proofreading performed by prokaryotic DNA polymerases (Kornberg, 1980). The exonuclease digests both single-stranded and double-stranded DNA substrates in a 3′→5′ direction, with a preference for mispaired 3′-termini of double-stranded DNA substrates (Byrnes *et al.*, 1976; Lee *et al.*, 1981; Lee and Toomey, 1987; Sabatino and Bambara, 1988). We have shown that digestion of a long terminal mismatch occurs by a nonprocessive mechanism, removing one or a few nucleotides before the enzyme dissociates from the template (Sabatino and Bambara, 1988). This is evident from the transient appearance of a fractionally large population of products in which only one or two nucleotides are removed. This result is similar to those observed for the associated exonucleases of *Escherichia coli* DNA polymerase I and bacteriophage T4 DNA polymerase (Thomas and Olivera, 1978). As can be seen in Figure 3 (lanes 7-11), digestion of the long terminal mismatched DNA primer, [^{32}P]oligo(dT)$_{16}$(dCMP)$_{14\text{-}28}$ annealed to poly(dA), occurred nonprocessively. As the reaction proceeded, intermediates with mismatch lengths of 5-20 nucleotides were clearly observed. However, very few intermediates with lengths of 1-5 nucleotides were seen. This phenomenon is apparently caused by an acceleration of the exonucleolytic rate as the mismatched region becomes shorter. It does not result from an increase in the processivity during the course of digestion, since digestion of shorter mismatched molecules occurs nonprocessively (Sabatino and Bambara, 1988).

Nucleoside 5′-monophosphates are the products of prokaryotic proofreading exonucleases (Kornberg, 1980). Nucleoside monophosphates have been shown to inhibit the exonuclease activity of *Escherichia coli* DNA polymerase I (Byrnes *et al.*, 1977). 5′-AMP presumably binds to the exonuclease active site and prevents entry of the terminally mismatched base (Que *et al.*, 1978). Similarly, nucleoside 5′-monophosphates, which have been shown to be the products of exonucleolytic digestion by DNA polymerase δ, also inhibit the exonuclease activity (Byrnes *et al.*, 1977; Wahl *et al.*, 1986). Indeed, digestion of a long terminal mismatch in reactions containing 4 mM 5′-AMP was totally inhibited (Figure 3, lanes 2-6). It should also be noted that the digestion kinetics of the mismatched substrate do not reflect the sequence heterogeneity of the primer. Digestion of single-stranded [^{32}P]oligo(dT)$_{16}$(dCMP)$_{14\text{-}28}$ (Figure 3, lanes 13-17) occurred at a uniform rate. Thus, the observations made during the digestion of the mismatched primer reflect the response of DNA polymerase δ to the single- and double-stranded characteristics of the mismatched DNA molecule.

Physical and functional association of the exonuclease with DNA polymerase δ has been demonstrated by many criteria. Firstly, the polymerase and exonuclease activities co-purify and are coincident during numerous fractionation procedures, which include: ion exchange and affinity chromatography, gel filtration, sucrose and glycerol gradient centrifugation, isoelectric focusing and polyacrylamide gel electrophoresis (Byrnes *et al.*, 1976; Byrnes and Black, 1978; Goscin and Byrnes, 1982; Lee *et al.*, 1980; 1981; Crute *et al.*, 1986a; Lee and Toomey, 1987). Importantly, the ratio of polymerase to exonuclease remains constant throughout these procedures. Secondly, hemin and Rifamycin AF/013 inhibit equally both the polymerase and the associated exonuclease, by dissociating the polymerase complex from the template (Byrnes *et al.*, 1976; Byrnes, 1984). It has also been shown that both the polymerase and exonuclease activities are equally sensitive to heat inactivation (Byrnes, 1984) and inhibition by hematoporphyrin derivative (Crute *et al.*, 1986b). The exonuclease, however, is not inhibited by nucleosides, 3′-nucleotides, or cyclic nucleotides (Byrnes, 1984). Interestingly, the associated exonuclease is the only non-DNA polymerase activity inhibited by aphidicolin. DNA polymerase activity was inhibited 50% at 4 μg/ml on either poly(dA)•oligo(dT) or alternating poly(dAdT) (Byrnes, 1984; Lee *et al.*, 1984; Crute *et al*, 1986a), whereas the associated exonuclease was inhibited 50% at 10 μg/ml on poly(dAdT) (Byrnes, 1984). Lee *et al.*, (1984), however, reported that the polymerase and exonuclease had identical inhibition kinetics on poly(dAdT). In seeming contradiction, we reported that the removal of mismatched nucleotides from poly(dA)•poly(dT)([^{3}H]GMP)$_4$ was not inhibited by aphidicolin (10 μg/ml) (Sabatino and Bambara, 1988). Additionally, aphidicolin is ineffective in inhibiting the exonuclease on single-stranded substrates. These results suggest that exonuclease inhibition may be template dependent.

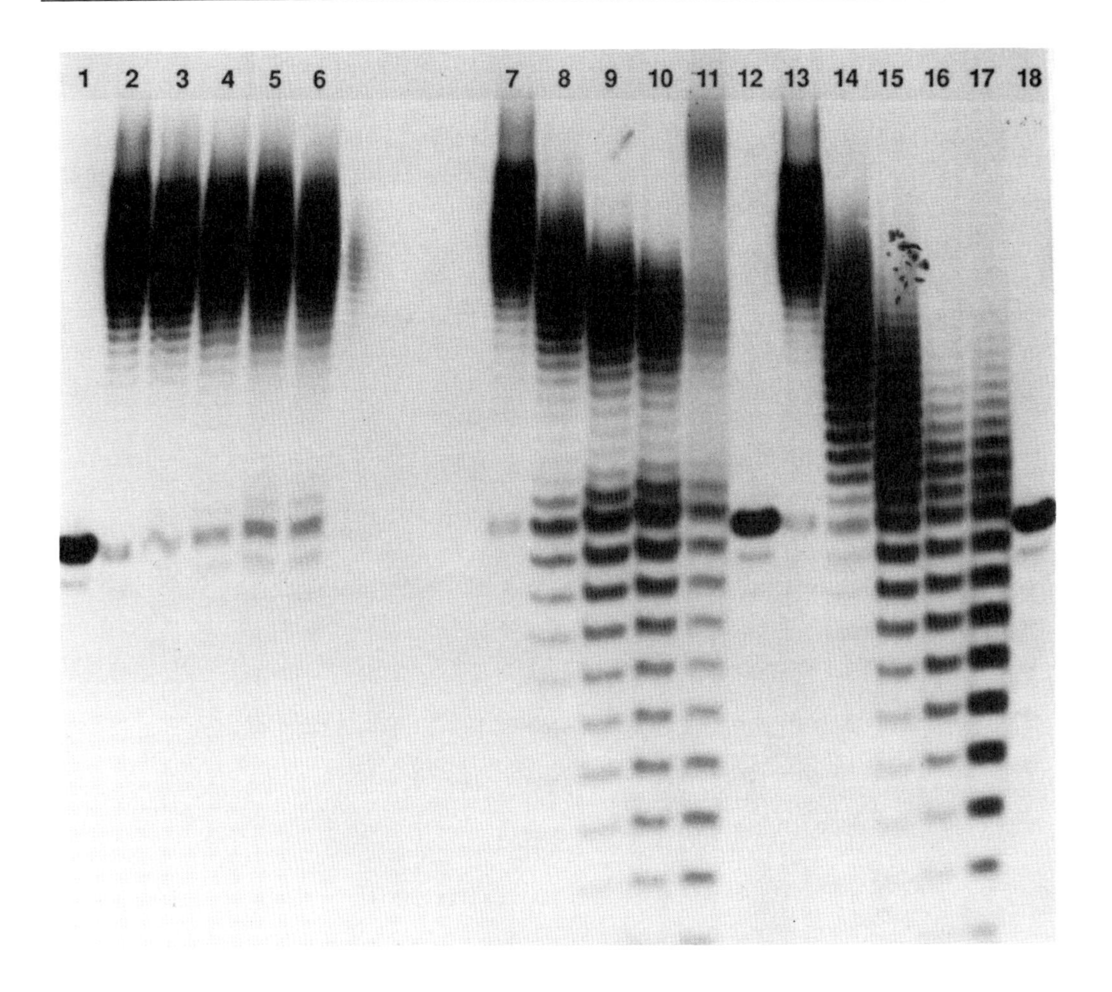

Figure 3. Digestion of single-stranded mismatched DNA by the 3'→5' exonuclease of calf thymus DNA polymerase δ. DNA products following digestion by DNA polymerase δ were analyzed by denaturing polyacrylamide gel electrophoresis. Lanes 2-6 and 7-11 show the digestion of $[^{32}P]oligo(dT)_{16}([^{3}H]dCMP)_{14\text{-}28}poly(dA)_{4000\text{-}5000}$ in the presence and absence of 4 mM AMP, respectively. Lanes 13-17 show the digestion of the single-stranded $[^{32}P]oligo(dT)_{16}([^{3}H]dCMP)_{14\text{-}28}$. For each digestion, the left lane represents zero time. Subsequent lanes moving to the right represent successive 30-min time points. Lanes 1,12, and 18 show the marker $[^{32}P]oligo(dT)_{16}$. (Figure obtained from Sabatino and Bambara, 1988)

The functional interdependence of the polymerase and exonuclease activities has been demonstrated with DNA polymerase δ. In a template dependent reaction using poly(dAdT) (Byrnes *et al.*, 1976; Lee and Toomey, 1987) or poly(dA)•oligo(dT) (Wahl *et al.*, 1986), complementary deoxynucleoside triphosphates (dNTPs) were converted to monophosphates. In the presence of non-complementary dNTPs, however, little or no conversion occurred. Thirteen percent of the nucleotides incorporated by the polymerase on poly(dAdT) were exonucleolytically removed from the 3'-termini during the synthetic reaction, demonstrating the extent of dNTP turnover by DNA polymerase δ (Byrnes *et al.*, 1976). The addition of 5'-AMP to reactions stimulated the rate of DNA synthesis 2- to 3-fold by inhibiting the exonuclease (Wahl *et al.*, 1986; Byrnes *et al.*, 1977; Lee and Toomey, 1987). In the presence of 5′-AMP, the net amount of dTMP incorporated during synthesis increased correspondingly, with a decrease in the amount of dTMP hydrolyzed.

Experiments evaluating the effects of varying DNA concentrations on stimulation of DNA synthesis by 5′-AMP were conducted in our laboratory (Wahl *et al.*, 1986). We assumed that an unassociated exonuclease would distribute itself among 3′-termini independent of the DNA polymerase. Therefore, as the DNA concentration is increased, an unassociated exonuclease should interfere progressively less with DNA synthesis, and consequently, the stimulation by 5′-AMP should decrease. In fact, we demonstrated that the addition of 5′-AMP equally stimulated the polymerase at all DNA concentrations tested. These findings, together with those described above, are all consistent with the concept that the DNA polymerase and the exonuclease activities of DNA polymerase δ are tightly associated.

THE FIDELITY OF DNA SYNTHESIS

The fidelity of eukaryotic DNA replication is an important issue because of its potential relationship to development, aging, and neoplastic transformation. Since most purified eukaryotic polymerases lack 3′→5′ exonuclease activity, alternative mechanisms have been proposed to explain the overall high fidelity of eukaryotic DNA replication (Fry and Loeb, 1986). Over the last decade, however, several polymerases, including DNA polymerase δ (Byrnes, 1984; Lee *et al.*, 1981; Crute *et al.*, 1986a), DNA polymerase γ (Kunkel and Soni, 1988), herpes virus DNA polymerase (Abbotts *et al.*, 1987; O'Donnell *et al.*, 1987), and some preparations of DNA polymerase α (Chen *et al.*, 1979; Ottiger and Hübscher, 1984; Skarnes *et al.*, 1986; Cotterill *et al.*, 1987), have been shown to have an associated 3′→5′ exonuclease activity. The exonuclease of *Drosophila melanogaster* polymerase-primase active site subunit (Cotterill *et al.*, 1987) and DNA polymerase γ (Kunkel and Soni, 1988) have been shown to contribute to the fidelity of these enzymes. These enzymes are quite accurate, producing one error for each 10^6 or 2.6×10^4 nucleotides synthesized, respectively.

Initially, Byrnes *et al.*, (1977) demonstrated that selective inhibition of the exonuclease increased the error frequency of DNA polymerase δ. Using poly(dAdT), misincorporation of dCMP was monitored in the absence and presence of 5′-GMP. In the absence of 5′-GMP, the amount of dCMP incorporated by DNA polymerase δ was very low (at the lower limits of detection in the assay). The error frequency was calculated to be less than one error for each 5.4×10^2 nucleotides synthesized. In the presence of 5′-GMP, however, the error frequency increased more than 3-fold suggesting that the exonuclease plays a role in the fidelity of DNA synthesis. It should be noted that these measurements were made using a synthetic template in the presence of Mn^{2+}, to bring the fidelity into a measurable range. This metal ion activator has been shown to decrease the fidelity of *Escherichia coli* DNA polymerase I (Kunkel and Loeb, 1979) and the DNA polymerase from avian myeloblastosis virus (AMV) (Sirover and Loeb, 1977).

To further study the accuracy of DNA polymerase δ we, in a collaborative effort, utilized an M13mp2*lacZα* nonsense codon reversion assay (Kunkel *et al.*, 1987). This assay measures the fidelity of DNA synthesis performed by a purified DNA polymerase *in vitro*. Fidelity is quantitated by monitoring the reversion of a nonsense codon, contained within a portion of the *lacZ* gene (which had been inserted into the bacteriophage), back to the wild-type gene. The reversion generates β-galactosidase activity by α-complementation after infection of an appropriate host, and is detected by plating infected cells and monitoring the hydrolysis of 5-bromo-4-chloro-3-indolyl-β-D-galactopyranoside (X-gal), which produces blue colored plaques.

DNA polymerase δ was found to be highly accurate, producing less than one error for each 10^6 nucleotides synthesized, and 10 and 500 times more accurate than immunopurified DNA polymerase α and DNA polymerase β respectively. Misincorporation can be detected not only by reversion of the *lacZ* mutant (colorless to blue change), but also by production of a mutation in the wild-type sequence other than at the nonsense codon (blue to lighter blue or colorless change). In the latter assay the exact nature of the error causing a new phenotype can be determined directly by DNA sequence analysis. This method of analysis is currently in progress.

An M13 system was also used to examine the efficiency of terminal mismatch excision by DNA polymerase δ. A mutant derivative of partially double-stranded bacteriophage M13mp2, containing a 3′-terminal cytosine in the primer strand opposite an adenine in the template strand of the *lacZ* coding region, was used as a substrate for exonucleolytic removal of the mismatched base. Efficiency of mismatch removal could be quantitated because the ratio of blue to colorless plaques depends on the percentage of mismatched termini used for synthesis compared to those removed prior to synthesis. Manipulation of the reaction conditions was used as a technique to confirm that the exonuclease functions to remove mispaired nucleotides during DNA synthesis. This approach had previously been used to demonstrate that proofreading contributes to the fidelity of several prokaryotic DNA polymerases (Clayton *et al.*, 1979; Fersht, 1979; Kunkel *et al.*, 1981). If the dNTP concentration is increased, the rate of polymerization should increase, permitting less time for the removal of the mismatched base. As shown in Table 1, the percentage of mismatch excision by DNA polymerase δ decreases with increasing dNTP concentration. For comparison, *Escherichia coli* DNA polymerase I Klenow fragment [pol I (Kf)] and AMV DNA polymerase were used in parallel experiments. As expected, pol I (Kf), which has a 3′→5′ exonuclease, responds similarly to DNA polymerase δ to increasing dNTP concentration. AMV polymerase, which lacks an exonuclease, does not excise mismatched nucleotides under any of the test conditions. In reactions containing 5′-AMP, the percentage of mismatch excision by DNA polymerases δ and pol I (Kf) also decreases (see Table 1). DNA polymerase δ, however, is least affected by increasing concentrations of 5′-AMP. As expected, AMV DNA polymerase is not affected by increasing concentrations of 5′-AMP.

INTRACELLULAR ROLE OF DNA POLYMERASE δ

Involvement in DNA Replication and Repair

Determining the roles of DNA polymerases α and δ *in vivo* is a major goal of current chromosomal replication research. An initial step in this effort has been to correlate the enzymatic activities of DNA polymerases α and δ with those required during replicative DNA synthesis. Most investigations have led to the conclusion that DNA polymerase α is the primary, or only, DNA polymerase responsible for nuclear DNA replication (Huberman, 1981; Thömmes *et al.*, 1986) based on studies with both intact cells (Ikegami, *et al.*, 1978; 1979) and subcellular systems (Wist *et al.*, 1976; Krokan *et al.*, 1979; Kwant *et al.*, 1980). However, DNA polymerase δ was either not considered in these studies, or the experiment was not designed to differentiate between DNA polymerases α and δ. The majority of the studies that have

Table 1: Deoxynucleoside triphosphate and monophosphate effects on mismatch excision[a].

dNTP (μM)	AMP(μM)	Pol δ	Pol I (Kf)	AMV Pol
1	0	-	98	1
100	0	98	76	0
1000	0	78	42	2
100	0	98	73	1[b]
100	5	75	32	3[b]
100	10	70	22	2[b]

[a] Mismatch excision is expressed as the percentage of mispaired cytosine removed from the primer terminus opposite an adenine in the template strand.
[b] Excision using AMV Pol was in the presence of 10 μM dNTP.

implicated DNA polymerase α as the predominant replicative enzyme have utilized inhibitors such as aphidicolin, N-ethylmaleimide, and arabinosyl nucleotides, that were thought to be selective inhibitors of DNA polymerase α (Kornberg, 1980; Fry and Loeb, 1986). However, these studies are also consistent with a replicative role for DNA polymerase δ (Lee *et al.*, 1984; 1985; Byrnes, 1985; Wahl *et al.*, 1986), which has also been shown to exhibit sensitivity to all three of these inhibitors (Lee *et al.*, 1981; Byrnes, 1984).

Strong evidence supporting a role of DNA polymerase α in the replication of SV40 was provided by utilizing an anti-DNA polymerase α immunoaffinity column to remove the DNA polymerase α activity from cellular extracts, resulting in a loss of SV40 replication activity (Murakami *et al.*, 1986). DNA synthetic activity could be restored by the addition of purified DNA polymerase α-primase complex of human or simian origin. These antibodies directed against DNA polymerase α have also been shown to inhibit DNA synthesis upon microinjection into human, hamster, and mouse cell lines (Kaczmarek *et al.*, 1986). In order to alleviate the technical problem of antibody traversal of the nuclear membrane, lysolecithin-permeabilized human cells have been utilized (Miller *et al.*, 1985a; 1985b). With this system, monoclonal anti-DNA polymerase α antibodies were found to inhibit the discontinuous synthesis of Okazaki DNA, and the maturation of Okazaki DNA to larger DNA. Regardless of whether the antibody directed against DNA polymerase α was microinjected into the nuclei or added to permeabilized cells, only 40-70% inhibition of replication was achieved (Kaczmarek *et al.*, 1986; Miller *et al.*, 1985a; 1985b). Although several explanations are possible for this observation, it has been suggested that the presence of DNA polymerase δ may be causing the disparity (Dresler and Frattini, 1986).

In a somewhat different approach (Dresler and Kimbro, 1987), it was shown that DNA replication and UV-induced DNA repair synthesis in permeabilized human fibroblasts were more sensitive to ddTTP inhibition than partially purified DNA polymerase α, and less sensitive than partially purified DNA polymerase β. These studies have been extended by determining the effect of BuPdGTP on DNA replication and UV-induced repair synthesis (Dresler and Frattini, 1986; Dresler and Frattini, 1988; Dresler *et al.*, 1988). DNA replication and UV-induced repair synthesis were, respectively, shown to be 500 and 3000 times more resistant to BuPdGTP than DNA polymerase α. Similarly, it was shown that the replication *in vitro* of plasmid DNA containing the SV40 *ori* using either purified DNA polymerase α or cell extracts was ten times more sensitive to inhibition by BuPdGTP than was SV40 chromatin replication in extracts (Decker *et al.*, 1987). In these experiments, control reactions indicated that both replication systems were equally sensitive to aphidicolin. Although not conclusive, these results imply that DNA polymerase δ is involved in both DNA replication and repair.

The sensitivities to the inhibitors aphidicolin, BuPdGTP, and monoclonal antibodies directed against DNA polymerase α were also determined for both partially purified DNA polymerases α and δ from CV-1 cells and for permeabilized CV-1 cells (Hammond *et al.*, 1987). Unlike most other studies, enzyme sensitivities from the same cell system could then be compared. Results of these experiments indicated that DNA polymerase δ activity accounted for 20-50% of the replication activity in the cells. Also, concentrations of BuPdGTP, which abolished DNA polymerase α activity, were shown to inhibit the synthesis and maturation of nascent DNA fragments. These findings suggest that both DNA polymerases, α and δ, have an important function in mammalian DNA replication. Similarly, a DNA polymerase δ from HeLa cells has been isolated (Nishida *et al.*, 1988) and shown to specifically promote DNA repair synthesis in permeabilized cells. DNA polymerases α and β however, were unable to complement this reaction.

The use of inhibitors specific to DNA polymerase α has allowed investigators to determine the relative abundance and variation of DNA polymerases α and δ through the cell cycle. Previous studies have suggested that DNA polymerase α activity is coordinately regulated with cellular DNA synthesis, increasing several fold during S phase (Spadari and Weissbach, 1974; Chiu and Baril, 1975; Craig *et al.*, 1975; Pedrali-Noy *et al.*, 1980). In these studies, synchronization of cells was achieved by a double thymidine block, aphidicolin block, or serum deprivation. Cells synchronized by these procedures were first growth-arrested in the G_1 phase of the cell cycle, and then the DNA polymerase activity was quantitated after the synchronizing agent was removed. In contrast, other studies have indicated that the majority of DNA polymerase α activity does not change with DNA synthesis or repair (Foster and Collins, 1985; Krauss and Linn, 1986). Taken together, the interpretation of these results has been complicated by the inability to distinguish whether the observed DNA polymerase activity increase is representative of natural entry into S phase or a result of metabolic perturbation caused by the synchronizing agent.

Potential drug induced artifactual changes in the abundance of replication factors can be avoided if DNA polymerase activities are measured in extracts from cells fractionated by centrifugal elutriation into populations representing specific cell cycle stages (Keng *et al.*, 1981). This procedure sorts cells by the frictional changes associated with increased cell volume. The DNA content of individual cells is then determined by flow cytometry in order to verify the homogeneity of cell populations. The activities of DNA polymerases α and δ, in extracts from Chinese hamster ovary (CHO) cells in G_1, S, and G_2/M phases of the cell cycle, fractionated from an exponential population, have been determined utilizing this technique (Marraccino *et al.*, 1987). The results demonstrated that only minor alterations in DNA polymerase α activity occurred throughout the cell cycle. However, the proportional amount of DNA polymerase δ increased with respect to DNA polymerase α in the G_2/M portion of the cell cycle.

Other results are also consistent with the conclusion that DNA polymerase activities do not change to a great degree with cell cycle. In a cell free SV40 replication assay, it was found that many factors required for DNA replication, including DNA polymerases α and δ, are expressed constitutively throughout the cell cycle (J.M. Roberts, personal communication). In addition, the levels of human KB cell DNA polymerase α mRNA transcript, translation product, and enzymatic activity were found to increase less than 2-fold throughout the cell cycle (Wahl *et al.*, 1988).

These studies suggest that both DNA polymerases α and δ play an essential role in DNA replication. The ultimate question remains: What is the relationship between DNA polymerases α and δ? Although experimental results are not yet available for DNA polymerase δ, the common subunit structure of DNA polymerase α appears to have an evolutionary conservation, implying that DNA polymerase α has an important function in DNA replication. A cDNA clone representing the human DNA polymerase α catalytic polypeptide has recently been isolated and the amino acid sequence deduced (Wong *et al.*, 1988). Six highly conserved regions were found between the human DNA polymerase α and the DNA polymerases of herpes virus, vaccinia virus, adenovirus, yeast (DNA polymerase I), bacteriophage T4, and *Bacillus* phage φ29. All of these DNA polymerases are replicative enzymes and, except for DNA polymerase α and yeast DNA polymerase I, all of the enzymes contain a 3′→5′ proofreading exonuclease activity.

The presence of a 3′→5′ exonuclease activity is one possible mechanism to provide for an increase in fidelity during synthesis by a DNA polymerase. A cell-free SV40 replication system has recently been utilized to measure the fidelity of bidirectional semiconservative DNA synthesis by a human DNA replication complex *in vitro* (Roberts and Kunkel, 1988). Replicative synthesis was found to be at least 40-fold more accurate than synthesis by the human DNA polymerase α-primase complex purified by immunoaffinity chromatography from HeLa cell extracts. Although several possible explanations exist, it was suggested that the increase in fidelity could result from either an altered form of DNA polymerase α, which contains an exonuclease proofreading function, or by DNA polymerase δ.

One of the predominant features distinguishing DNA polymerases α and δ is the 3′→5′ exonuclease activity of DNA polymerase δ. However, several forms of DNA polymerase α have been found to have an associated 3′→5′ exonuclease activity (Chen *et al.*, 1979; Ottiger and Hübscher, 1984; Skarnes *et al.*, 1986). Recently it was discovered that *Drosophila melanogaster* DNA polymerase-primase contains an intrinsic but cryptic 3′→5′ exonuclease activity in the catalytic 182 kDa polypeptide. Exonuclease was revealed only upon dissociation of the 182 kDa subunit from the 73 kDa subunit (Cotterill *et al.*, 1987). In addition, the fidelity of the isolated 182 kDa subunit containing the exonuclease activity was shown to be 100-fold greater than that of the multisubunit complex. The 182 kDa subunit was also found to be insensitive to BuPdGTP and BuAdATP, whereas the intact DNA polymerase-primase complex was inhibited 95% by these DNA polymerase α inhibitors (Reyland *et al.*, 1988). These results indicate that the isolation of sub- and/or superassemblies of DNA polymerase α may result in the appearance of altered enzymatic activities *in vitro*, making the distinction between DNA polymerases α and δ less clear (Kaguni and Lehman, 1988). Whether this is a general phenomenon for DNA polymerase α in animal cells, as has been suggested (Kornberg, 1988), or a specialized case for the *Drosophila melanogaster* enzyme, is yet to be determined. It is of interest to note that the primase-devoid DNA polymerase α from mouse (Prussak and Tseng, 1987) and frog (*Xenopus laevis*) (Kaiserman and Benbow, 1987), do not possess exonuclease activity and have inhibitor sensitivities characteristic of DNA polymerase α. Further evidence suggesting distinct DNA polymerase α and δ activities has been found during the study of DNA polymerases from *Saccharomyces cerevisiae*. Utilizing either protease-deficient strains and/or conditions that inhibit protease action, a new DNA polymerase, designated DNA polymerase III, was discovered that has an associated 3′→5′ exonuclease activity (Bauer *et al.*, 1988). This enzyme was found to possess inhibitor sensitivities characteristic of DNA polymerase δ (Burgers and Bauer, 1988). In addition, the yeast DNA polymerase III was shown to be immunologically distinct from the yeast DNA polymerases I and II. Taken together, these results illustrate the difficulty in determining the connection between DNA polymerases α and δ based on traditional biochemical characterizations. The availability of genetic information for both *Drosophila* and *Saccharomyces* may provide invaluable insight into the function of DNA polymerase δ *in vivo*, and its structural relationship to DNA polymerase α.

The determination of the role of both DNA polymerases α and δ in the mechanism and control of eukaryotic DNA replication has been greatly facilitated by studying the replication of SV40 in soluble extracts derived from human cells (Li and Kelly, 1984; 1985; Stillman and Gluzman, 1985; Wobbe *et al.*, 1985). The replication of SV40's double-stranded, circular, 5.2 kilobase genome is similar to the inferred mechanism of replication from a single origin within the cell's chromosome (Stillman, 1988). The replication of this virus is almost exclusively dependent on the host replication proteins. The only viral-encoded component required is the large tumor antigen (TAg) (Li and Kelly, 1984; 1985; Stillman and Gluzman, 1985; Wobbe *et al.*, 1985). Several cellular proteins have been identified as being required for SV40 DNA replication. The DNA topoisomerases I and II are well established components of the system (Yang *et al.*, 1987). Three other necessary replication factors have recently been discovered. These include an initiation factor (Wold *et al.*, 1987; Fairman and Stillman, 1988), a mammalian single-stranded DNA binding protein (Wobbe *et al.*, 1987; Fairman and Stillman, 1988), and PCNA (Prelich *et al.*, 1987a). The significance of PCNA was emphasized by the discovery that this protein is identical to a stimulatory auxiliary factor of DNA polymerase δ (Prelich *et al.*, 1987b; Bravo *et al.*, 1987), which had previously been purified (Tan *et al.*, 1986).

The determination that these two proteins are identical has been verified by diverse methodologies. Both of these proteins co-migrate on SDS polyacrylamide gel electrophoresis and on two-dimensional gel electrophoresis (Bravo *et al.*, 1987). Immunological studies demonstrated that autoantibodies to PCNA neutralize the activity of the DNA polymerase δ auxiliary protein, and Western analysis has shown that monoclonal antibodies to authentic PCNA detect the auxiliary protein (Tan *et al.*, 1987a). Complementation assays have determined that DNA polymerase δ auxiliary protein from calf thymus can substitute for human PCNA in SV40 replication systems *in vitro* and conversely, human PCNA can substitute for the auxiliary protein in its ability to allow the utilization of primed homopolymer templates by one form of DNA polymerase δ (Prelich *et al.*, 1987b).

PCNA was initially identified as a human autoantigen reactive with sera from patients with the autoimmune disease systemic lupus erythematosus (SLE) (Miyachi *et al.*, 1978; Takasaki *et al.*, 1984). Subsequently, PCNA was shown to be identical to a cell cycle-regulated protein called cyclin (Mathews *et al.*, 1984). The rate of synthesis of PCNA was found to correlate positively with the proliferation rate of normal and transformed cells from a variety of vertebrate species (Celis *et al.*, 1984). The quantity of PCNA was also found to increase coordinately with the percentage of normal or transformed cells in the S phase (Bravo, 1984a; 1984b). PCNA appears to be preferentially synthesized during S phase in cultured HeLa cells, but the protein is also present in the G_1 and G_2 phases, and stable throughout the cell cycle (Bravo and Celis, 1980; Bravo and Macdonald-Bravo, 1985; Bravo *et al.*, 1987). The induction of PCNA synthesis was shown not to be dependent upon replication because prevention of DNA synthesis by specific inhibitors that block the G_1/S transition were not effective in altering PCNA levels, implying that PCNA regulation is controlled by events prior to the initiation of S phase. (Macdonald-Bravo and Bravo, 1985; Bravo and Macdonald-Bravo, 1985; Bravo, 1986; Kurki *et al.*, 1987).

The use of indirect immunofluorescence with autoantibodies directed specifically against PCNA has shown that the expression of PCNA is observable in the nucleus, and that it precedes DNA synthesis, reaching a maximum during S phase after mitogenic stimulation (Miyachi et al, 1978; Takasaki *et al.*, 1981; Kurki *et al.*, 1986; 1987). Large changes have been observed in the nuclear distribution of PCNA during late G_1 phase and throughout S phase (Takasaki *et al.*, 1981; Chan *et al.*, 1983; Celis and Celis, 1985a; 1985b; Sadaie and Mathews 1986) and these changes have been found to be dependent upon DNA replication (Bravo and Macdonald-Bravo, 1985; 1987). It has also been shown that PCNA clusters around active centers of replication in the nucleus (Madsen and Celis, 1985; Bravo and Macdonald-Bravo, 1987).

The relationship between PCNA and DNA replication is consistent with the possibility that the synthesis of PCNA triggers the initiation of replication and is required throughout the S phase of the cell cycle. The regulation of PCNA is controlled, at least in part, at the transcriptional level (Almendral *et al.*, 1987; Celis *et al.*, 1987; Matsumoto *et al.*, 1987). In these studies, the regulation of PCNA was determined in cell populations undergoing G_0-S transitions. A recent report has shown that a human lymphoblastoid cell line treated with demecolcine, an inhibitor of cell-division, exhibited an increased PCNA signal during the G_1 phase of the cell cycle that reached its maximum in the S phase, and declined during the G_2/M phase (Kurki *et al.*, 1988). However, in order to examine the role of PCNA in the initiation and propagation of S phase DNA synthesis, it is important to study its regulation in an exponential cell population. Utilizing this approach allows the differentiation of the regulatory events accompanying the onset of S phase with changes involving the shift from quiescence to proliferation.

In an attempt to quantitate the expression of PCNA during exponential cell growth, centrifugal elutriation was used to separate an asynchronous, exponential population of CHO cells into subpopulations, representing G_1, S, and G_2/M phases of the cell cycle (Liu *et al.*, 1989). Northern hybridization analysis of the separated populations indicated that although intact PCNA mRNA is present during all phases of the cell cycle, an induction of about 3-fold occurs during S phase. Furthermore, evidence for PCNA protein induction was obtained by quantitating the PCNA and DNA content of asynchronous, exponential CHO cells by flow cytometry (Liu *et al.*, 1989). Figure 4 illustrates cells stained with propidium iodide (PI) and PCNA antiserum. The fluoroscein isothiocyanate (FITC) fluorescence results from the specific

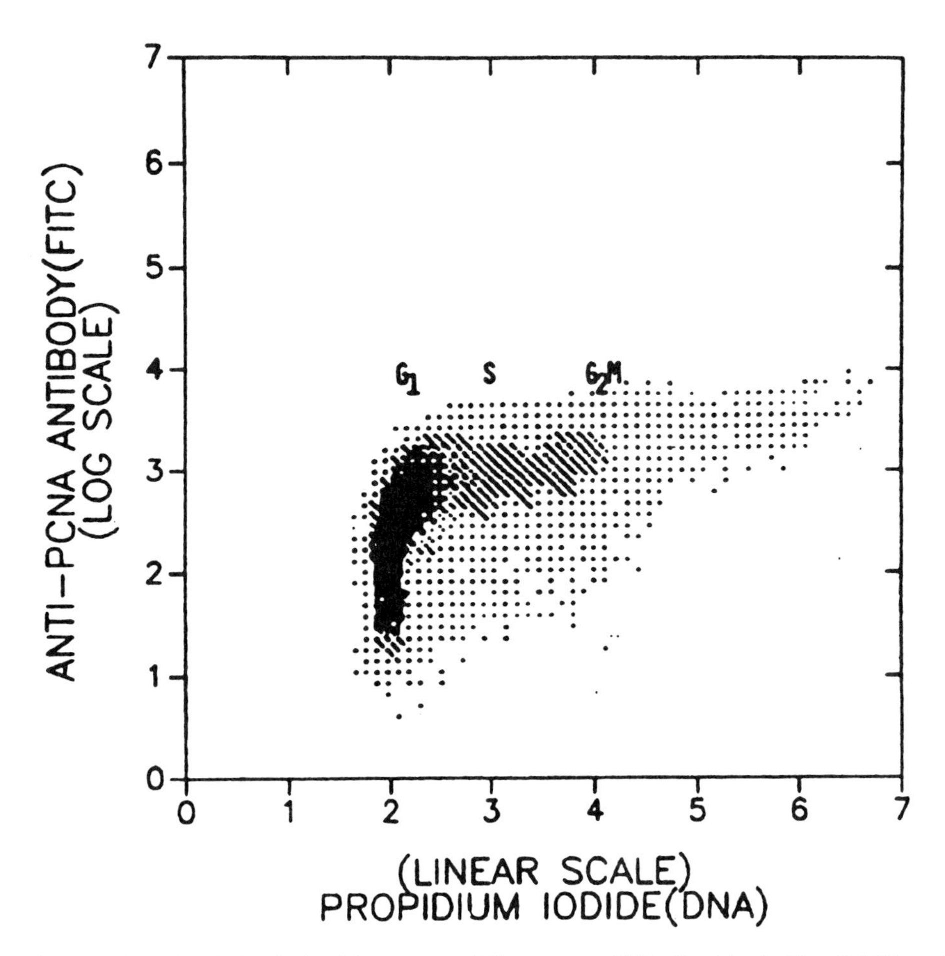

Figure 4. Three-dimensional plot obtained from exponentially growing, CHO cells stained with anti-PCNA antibody and PI. CHO cells were incubated with the murine monoclonal antibody to PCNA, 19F4 (Ogata *et al.*, 1987), followed by second step staining with a goat-antimouse antibody conjugated to FITC for PCNA content. The cells were then counterstained with PI for DNA content. The TC and PI were then quantitated by flow cytometry. The third dimension, the cell number, is represented by the density of shading. (Figure adapted from Liu *et al.*, 1989).

reaction between goat antiserum to murine antibodies, conjugated to FITC, and the murine monoclonal antibodies to PCNA, 19F4 (Ogata *et al.*, 1987). The positions of the cells in the G_1, S, and G_2 phases of the cell cycle are indicated on the chart, and were estimated by the analysis of the propidium iodide staining for DNA content. The levels of PCNA appear to increase several fold in the G_1 phase of the cell cycle, and reach a maximum and stable steady-state concentration throughout S and G_2/M.

A somewhat different approach has also been utilized in order to determine the importance of PCNA in cellular DNA replication and the cell cycle. This technique uses oligodeoxynucleotides complementary to either the sense or the antisense strand of the initial nucleotides of the protein coding sequence on the PCNA mRNA (Jaskulski *et al.*, 1988). When exponentially growing Balb/c3T3 cells were exposed to antisense oligodeoxynucleotides to PCNA mRNA, PCNA protein synthesis declined and both DNA synthesis and mitosis were completely suppressed, whereas corresponding sense oligodeoxynucleotides had no inhibitory effects. We have performed similar experiments using CHO cells (Liu *et al.*, 1989). However, our study determined the role of PCNA in the progression from quiescence (G_0/G_1) to exponential growth using early G_1 cells (G_0/G_1) isolated by centrifugal elutriation. Both PCNA and DNA content were simultaneously quantitated in each cell, as they progressed into exponential growth in the presence or absence of oligodeoxynucleotides complementary to PCNA mRNA. We have found that the antisense DNA inhibited both PCNA protein synthesis and DNA synthesis, thus blocking progression into S phase. These results suggest that some expression of PCNA may be needed throughout the cell cycle, but that an induction to a critical concentration is necessary for entry into S phase. Therefore, it appears that PCNA is needed both for proliferation competency and as part of the regulation of S phase initiation.

Mechanism in DNA Replication

The apparent requirement for PCNA in DNA replication and the cell cycle has provided a means to determine the relationship between DNA polymerases α and δ at the replication fork. Studies conducted with the SV40 replication system *in vitro* demonstrated that in reconstituted systems that contain PCNA, DNA synthesis initiates at the origin and proceeds bidirectionally on both leading and lagging strands around the template DNA to produce duplex, circular daughter molecules (Prelich and Stillman, 1988). Quite different results were obtained in the absence of PCNA. Early replicative intermediates containing short, newly synthesized strands accumulated, and replication forks continued bidirectionally from the origin. However, only products on the lagging-strand were synthesized. It was further suggested that PCNA has a function similar to that of the β subunit of *Escherichia coli* DNA polymerase III holoenzyme, indicating that PCNA either interacts with, or is a subunit of, the leading-strand DNA polymerase. The lagging-strand DNA polymerase would synthesize Okazaki fragments in a PCNA independent process, and in this way would not resemble DNA polymerase III holoenzyme.

A requirement for two types of DNA polymerase is also suggested from studies performed following the discovery of a temperature-sensitive mutant of mouse cells (Murakami *et al.*, 1985) and an aphidicolin-resistant mutant of hamster cells (Chang *et al.*, 1981; Liu *et al.*, 1983). The temperature-sensitive mutant was found to contain a heat-labile DNA polymerase α and the aphidicolin-resistant mutant was demonstrated to possess an aphidicolin-resistant DNA polymerase α. Studies carried out with the temperature-sensitive mutant demonstrated that once started, chain elongation in the nuclei of temperature sensitive cells proceeds at the normal wild type cell rate at the restrictive temperature. However, the joining of replicon-sized DNA is retarded because initiation at adjacent replicons is inhibited (Eki *et al.*, 1986). Taken together, the above data has led investigators to suggest that DNA polymerase δ may be the leading-strand DNA polymerase, while DNA polymerase α would be best suited as a lagging-strand DNA polymerase (Downey *et al.*, 1988; Focher *et al.*, 1988a; Prelich and Stillman, 1988; So and Downey, 1988). These proposals are in analogy to the proposed asymmetric dimer assembly of *Escherichia coli* DNA polymerase III holoenzyme (Maki *et al.*, 1988; McHenry, 1988) and the replication complex of bacteriophage T4 (Alberts, 1987). In fact, the analogy between the effects of PCNA on DNA polymerase δ and the effects of the β subunit of *Escherichia coli* DNA polymerase III holoenzyme on the polymerase subform that lacks this subunit has been suggested by Downey *et al.*, (1988) and So and Downey, (1988). In both cases, the addition of the subunit greatly stimulates both reaction rate and extent of processive DNA synthesis.

The properties of the eukaryotic DNA polymerases are consistent with their proposed roles at the replication fork: DNA polymerase α is suited for lagging-strand DNA synthesis because it contains a tightly associated DNA primase activity, displays processive synthesis that is less than DNA polymerase δ, is incapable of strand-displacement synthesis, dissociates from the template much more readily than that of DNA polymerase δ upon encountering the 5′-ends of primers, and it is not affected by PCNA. DNA polymerase δ conversely, is more suited for leading-strand DNA synthesis since it does not have an associated primase, is more processive than DNA polymerase α and, depending on the form of DNA polymerase δ, the synthetic activity and processivity can be stimulated by PCNA. PCNA was also shown to enable DNA polymerase δ to become active in strand-displacement synthesis, although its relevance *in vivo* is questionable since DNA helicases and topoisomerases would be expected to be present (Prelich and Stillman, 1988; Downey *et al.*, 1988; Sabatino *et al.*, 1988).

The above proposals would appear to coordinate the involvement of DNA polymerases α and δ at the replication fork. However, like most hypotheses, there are subtle problems associated with the model. One of the most perplexing issues is the inability of PCNA to stimulate all of the known forms of DNA polymerase δ. The addition of PCNA to DNA synthesis assays *in vitro* has been shown to increase the processivity of the fetal calf thymus DNA polymerase δ isolated by A.G. So and colleagues from less than 25 nucleotides to approximately 900 nucleotides on poly(dA)•oligo(dT) (Downey *et al.*, 1988). The catalytic activity of the DNA polymerase is also dramatically increased on this template (Tan *et al.*, 1986; Prelich *et al.*, 1987b). However, our calf thymus DNA polymerase δ exhibits no such stimulation by PCNA (A.G. So, personal communication and T.W. Myers, unpublished observation). There have also been two recent reports on the purification of a DNA polymerase δ, which is not stimulated by PCNA, from both calf thymus (Focher *et al.*, 1988b) and HeLa cells (Syvaoja and Linn, 1989).

In a collaborative effort with A. G. So, we have begun to address this issue. Initial experiments have shown that neither the catalytic activity nor the processivity of our DNA polymerase δ is affected by PCNA. The effect of PCNA on the processivity of the two enzymes is shown in Figure 5. DNA polymerase δ from our research group (lanes b-e) is compared with DNA polymerase δ provided by A.G. So (lanes f-i). The reactions were carried out at pH 7.5 and either 5 or 10 mM $MgCl_2$, conditions at which our DNA polymerase δ is least processive (Sabatino *et al.*, 1988). Even under these conditions our DNA polymerase δ appears to be nearly as processive, regardless of the presence of PCNA, as the other form of DNA polymerase δ in the presence of PCNA. The fact that PCNA is associated with the DNA polymerase δ isolated by A.G. So and collaborators through several initial steps of the purification (Tan *et al.*, 1986) would suggest that our DNA polymerase δ has not been purified away from the PCNA, and therefore it is always present in our reactions. However, PCNA has not been detected in our DNA polymerase δ preparations, as determined by Western analysis (A.G. So, personal communication and R.L. Marraccino, unpublished results). The structural relationship between these two forms of DNA polymerase δ remains unresolved at present, but as discussed above, the two polymerases appear to be immunologically related.

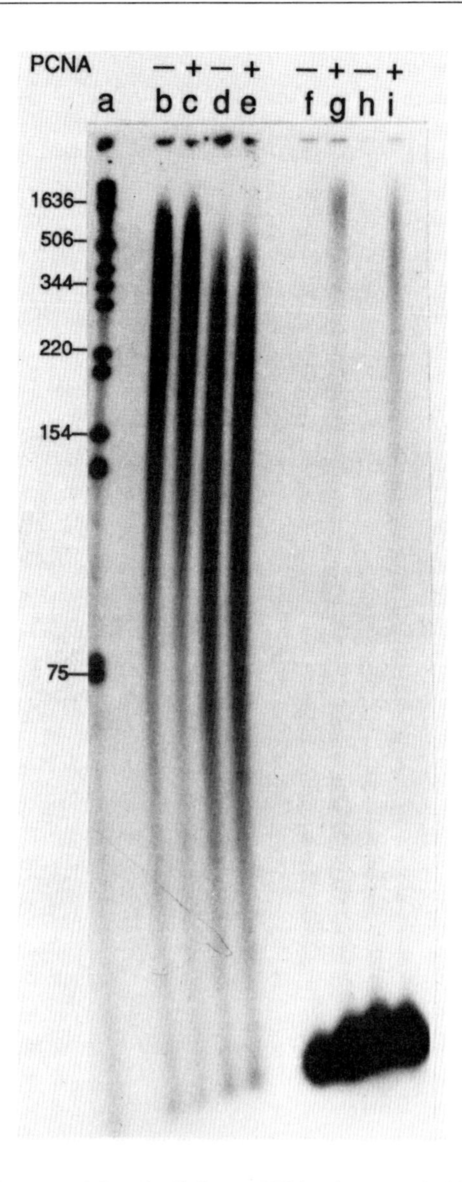

Figure 5. Effect of PCNA on the processivity of calf thymus DNA polymerase δ. DNA products synthesized by DNA polymerase δ (our laboratory) (lanes b-e) and DNA polymerase δ (laboratory of A.G. So) (lanes f-i), in the absence or presence of PCNA, were analyzed by denaturing polyacrylamide gel electrophoresis. Poly(dA)$_{800\text{-}10008}$•oligo(dT)$_{16}$ (1:1) was used as template and assays were performed using a $MgCl_2$ concentration of either 5 mM (lanes b, c, f, and g) or 10 mM (lanes d, e, h, and i) in pH 7.5 buffer.

Although much work is still required before the proposed mechanism for the involvement of both DNA polymerases α and δ at the replication fork can be proven, great strides have been made in the past several years in order to even propose the above models. Considering the vast amount of information now available and being obtained, with regard to the properties and cellular role of DNA polymerase δ, and the fact that DNA polymerases having δ-like characteristics have now been isolated from such diverse sources as rabbit bone marrow (Byrnes *et al.*, 1976), calf thymus (Lee *et al.*, 1980; Crute *et al.*, 1986a; Focher *et al.*, 1988 b), human placenta (Lee *et al.*, 1985), neonatal rat heart (Zhang and Lee, 1987), yeast (Bauer *et al.*, 1988), CV-1 cells (Hammond *et al.*, 1987), and HeLa cells (Nishida *et al.*, 1988; Syvaoja and Linn, 1989), DNA polymerase δ must now be considered to be intimately involved in the DNA metabolism of mammalian cells.

ACKNOWLEDGEMENTS

Research cited from this laboratory was supported by National Institutes of Health Grant GM24441, American Cancer Society Grant NP-566 and Cancer Center Core Grant 5-P30-CA11198. T.W.M. is a postdoctoral trainee supported by National Institutes of Health Grant T32-CA09363. We are indebted to Dr. Antero So and his research group for providing both PCNA and DNA polymerase δ for our collaborative studies. We thank Dr. Robert L. Marraccino for valuable suggestions and review of this manuscript.

REFERENCES

J. Abbotts, Y. Nishiyama, S. Yoshida, and L. A. Loeb (1987) On the fidelity of DNA replication: herpes DNA polymerase and its associated exonuclease. *Nucl. Acids. Res.* **15**: 1185-1198.

B. M. Alberts (1987) Prokaryotic DNA replication mechanisms. *Phil. Trans. R. Soc. Lond. B* **317**: 395-420.

J. M. Almendral, D. Huebsch, P. A. Blundell, H. Macdonald-Bravo, and R. Bravo. (1987) Cloning and sequence of the human nuclear protein cyclin: homology with DNA-binding proteins. *Proc. Natl. Acad. Sci. USA* **84**: 1575-1579.

S. Anderson and M. L. DePamphilis (1979) Metabolism of Okazaki fragments during simian virus 40 DNA replication. *J. Biol. Chem.* **254**: 11495-11504.

G. A. Bauer, H. M. Heller, and P. M. J. Burgers (1988) DNA polymerase III from Saccharomyces cerevisiae. I. Purification and characterization. *J. Biol. Chem.* **263**: 917-924.

F. J. Bollum (1975) Mammalian DNA polymerases. *Prog. Nucl. Acid Res. Mol. Biol.* **15**: 109-144.

R. Bravo and J. E. Celis (1980) A search for differential polypeptide synthesis throughout the cell cycle of HeLa cells. *J. Cell Biology* **84**: 795-802.

R. Bravo (1984a) Coordinated synthesis of the nuclear protein cyclin and DNA in serum-stimulated quiescent 3T3 cells. *FEBS Lett.* **169**: 185-188.

R. Bravo (1984b) Epidermal growth factor inhibits the synthesis of the nuclear protein cyclin in A431 human carcinoma cells. *Proc. Natl. Acad. Sci. USA* **81**: 4848-4850.

R. Bravo and H. Macdonald-Bravo (1985) Changes in the nuclear distribution of cyclin (PCNA) but not its synthesis depend on DNA replication. *EMBO J.* **4**: 655-661.

R. Bravo (1986) Synthesis of the nuclear protein cyclin (PCNA) and its relationship with DNA replication. *Exp. Cell Res.* **163**: 287-293.

R. Bravo, R. Frank, P. A. Blundell, and H. Macdonald-Bravo (1987) Cyclin/PCNA is the auxiliary protein of DNA polymerase-δ. *Nature* **326**: 515-517.

R. Bravo and H. Macdonald-Bravo (1987) Existence of two populations of cyclin/proliferating cell nuclear antigen during the cell cycle: association with DNA replication sites. *Cell Biology* **105**: 1549-1554.

P. M. J. Burgers and G. A. Bauer (1988) DNA polymerase III from *Sacharomyces cerevisiae*. II. Inhibitor studies and comparison with DNA polymerases I and II. *J. Biol. Chem.* **263**: 925-930.

J. J. Byrnes, K. M. Downey, V. L. Black, and A. G. So (1976) A new mammalian DNA polymerase with 3′ to 5′ exonuclease activity: DNA polymerase δ. *Biochemistry* **15**: 2817-2823.

J. J. Byrnes, K. M. Downey, B. G. Que, M. Y. W. Lee, V. L. Black, and A. G. So (1977) Selective inhibition of the 3′ to 5′ exonuclease activity associated with DNA polymerases: a mechanism of mutagenesis. *Biochemistry* **16**: 3740-3746.

J. J. Byrnes and V. L. Black. (1978) Comparison of DNA polymerases α and δ from bone marrow. *Biochemistry* **17**: 4226-4231.

J. J. Byrnes (1984) Structural and functional properties of DNA polymerase delta from rabbit bone marrow. *Mol. Cell. Biochem.* **62**: 13-24.

J. J. Byrnes (1985) Differential inhibitors of DNA polymerases alpha and delta. *Biochem. Biophys. Res. Commun.* **132**: 628-634.

J. E. Celis, R. Bravo, P. M. Larsen, and S. J. Fey (1984) Cyclin: a nuclear protein whose level correlates directly with the proliferative state of normal as well as transformed cells. *Leukemia Res.* **8**: 143-157.

J. E. Celis and A. Celis (1985a) Individual nuclei in polykaryons can control cyclin distribution and DNA synthesis. *EMBO J.* **4**: 1187-1192.

J. E. Celis and A. Celis (1985b) Cell cycle-dependent variations in the distribution of the nuclear protein cyclin proliferating cell nuclear antigen in cultured cells: subdivision of S phase, *Proc. Natl. Acad. Sci. USA* **82**: 3262-3266.

J. E. Celis, P. Madsen, A. Celis, H. V. Nielsen, and B. Gesser (1987) Cyclin (PCNA, auxiliary protein of DNA polymerase δ) is a central component of the pathway(s) leading to DNA replication and cell division. *FEBS Lett.* **220**: 1-7.

P-K. Chan, R. Frakes, E. M. Tan, M. G. Brattain, K. Smetana, and H. Busch (1983) Indirect immunofluorescence studies of proliferating cell nuclear antigen in nucleoli of human tumor and normal tissues. *Cancer Res.* **43**: 3770-3777.

C-C. Chang, J. A. Boezi, S. T. Warren, C. L. K. Sabourin, P. K. Liu, L. Glatzer, and J. E. Trosko (1981) Isolation and characterization of a UV-sensitive hypermutable aphidicolin-resistant Chinese hamster cell line. *Somat. Cell Genet.* **7**: 235-253.

Y-C. Chen, E. W. Bohn, S. R. Planck, and S. H. Wilson (1979) Mouse DNA polymerase α. Subunit structure and identification of a species with associated exonuclease. *J. Biol. Chem.* **254**: 11678-11687.

R. W. Chiu and E. F. Baril (1975) Nuclear DNA polymerases and the HeLa cell cycle. *J. Biol. Chem.* **250**: 7951-7957.

L. K. Clayton, M. F. Goodman, E. W. Branscomb, and D. J. Galas (1979) Error induction and correction by mutant and wild type T4 DNA polymerases. *J. Biol. Chem.* **254**: 1902-1912.

S. M. Cotterill, M. E. Reyland, L. A. Loeb, and I. R. Lehman (1987) A cryptic proofreading 3′→5′ exonuclease associated with the polymerase subunit of the DNA polymerase-primase from *Drosophila melanogaster*. *Proc. Natl. Acad. Sci. USA* **84**: 5635-5639.

R. K. Craig, P. A. Costello, and H. M. Keir (1975) Deoxyribonucleic Acid Polymerases of BHK-21/C13 cells. Relationship to the physiological state of the cells, and to synchronous induction of synthesis of deoxyribonucleic acid. *Biochem. J.* **145**: 233-240.

J. J. Crute, A. F. Wahl, and R. A. Bambara (1986a) Purification and characterization of two new high molecular weight forms of DNA polymerase δ. *Biochemistry* **25**: 26-36.

J. J. Crute, A. F. Wahl, R. A. Bambara, R. S. Murant, S. L. Gibson,and R. Hilf (1986b) Inhibition of mammalian DNA polymerases by hematoporphyrin derivative and photoradiation. *Cancer Res.* **46**: 153-159.

S. K. Das and R. K. Fujimura (1979) Processiveness of DNA polymerases. A comparative study using a simple procedure. *J. Biol. Chem.* **254**: 1227-1232.

R. S. Decker, M. Yamaguchi, R. Possenti, M. K. Bradley, and M. L. DePamphilis (1987) *in vitro* initiation of DNA replication in simian virus 40 chromosomes. *J. Biol. Chem.* **262**: 10863-10872.

S. D. Detera, S. P. Becerra, J. A. Swack, and S. H. Wilson (1981) Studies on the mechanism of DNA polymerase-α. Nascent chain elongation, steady state kinetics and the initiation phase of DNA synthesis. *J. Biol. Chem.* **256**: 6933-6943.

K. M. Downey, C-K. Tan, D. M. Andrews, X. Li, and A. G. So (1988) Proposed roles for DNA polymerases alpha and delta at the replication fork, In *Cancer Cells* Vol. 6, Eukaryotic DNA Replication, Cold Spring Harbor Laboratory, Cold Spring Harbor, NY, 403-410.

S. L. Dresler and M. G. Frattini (1986) DNA replication and UV-induced DNA repair synthesis in human fibroblasts are much less sensitive than DNA polymerase α to inhibition by butylphenyl-deoxyguanosine triphosphate. *Nucl.Acids Res.* **14**: 7093-7102.

S. L. Dresler and K. S. Kimbro (1987) 2′,3′-Dideoxythymidine 5′-triphosphate inhibition of DNA replication and ultraviolet-induced DNA repair synthesis in human cells: evidence for involvement of DNA polymerase δ. *Biochemistry* **26**: 2664-2668.

S. L. Dresler and M. G. Frattini (1988) Analysis of butylphenyl-guanine, butylphenyl-deoxyguanosine, and butylphenyl-deoxyguanosine triphosphate inhibition of DNA replication and ultraviolet-induced DNA repair synthesis using permeable human fibroblasts. *Biochem. Pharmacol.* **37**: 1033-1037.

S. L. Dresler, B. J. Gowans, R. M. Robinson-Hill, and D. J. Hunting (1988) Involvement of DNA polymerase δ in DNA repair synthesis in human fibroblasts at late times after UV irradiation. *Biochemistry* **27**: 6379-6383.

T. Eki, Y. Murakami, T. Enomoto, F. Hanaoka, and M-a. Yamada (1986) Characterization of DNA replication at a restrictive temperature in a mouse DNA temperature-sensitive mutant, tsFT20 strain, containing heat-labile DNA polymerase α activity. *J. Biol. Chem.* **261**: 8888-8893.

M. P. Fairman and B. Stillman (1988) Cellular factors required for multiple stages of SV40 DNA replication *in vitro*. *EMBO J.* **7**: 1211-1218.

A. Falaschi and S. Spadari (1978) The three DNA polymerases of animal cells: properties and functions, **In** *DNA Synthesis: Present and Future.* I. Molineux and M. Kohyama, Eds., Plenum Press, NY, 487-515.

A. R. Fersht (1979) Fidelity of replication of phage ϕX174 DNA by DNA polymerase III holoenzyme: spontaneous mutation by misincorporation. *Proc. Natl. Acad. Sci. USA* **76**: 4946-4950.

P. A. Fisher, T. S-F. Wang, and D. Korn (1979) Enzymological characterization of DNA polymerase α. Basic catalytic properties, processivity, and gap utilization of the homogeneous enzyme from human KB cells. *J. Biol. Chem.* **254**: 6128-6137.

F. Focher, E. Ferrari, S. Spadari, and U. Hübscher (1988a) Do DNA polymerases δ and α act coordinately as leading and lagging strand replicases? *FEBS Lett.* **229**: 6-10.

F. Focher, S. Spadari, B. Ginelli, M. Hottiger, M. Gassmann, and U. Hübscher (1988b) Calf thymus DNA polymerase δ: purification, biochemical and functional properties of the enzyme after its separation from DNA polymerase α, a DNA dependent ATPase and proliferating cell nuclear antigen. *Nucl. Acids Res.* **16**: 6279-6295.

K. A. Foster and J. M. Collins (1985) The interrelation between DNA synthesis rates and DNA polymerases bound to the nuclear matrix in synchronized HeLa cells. *J. Biol. Chem.* **260**: 4229-4235.

M. Fry and L. A. Loeb (1986) **In:** *Animal Cell DNA Polymerases*, CRC Press Inc., Boca Raton, FL.

L. P. Goscin and J. J. Byrnes (1982) DNA polymerase δ: one polypeptide, two activities. *Biochemistry* **21**: 2513-2518.

R. A. Hammond, J. J. Byrnes, and M. R. Miller (1987) Identification of DNA polymerase δ in CV-1 cells: studies implicating both DNA polymerase δ and DNA polymerase α in DNA replication. *Biochemistry* **26**: 6817-6824.

J. W. Hockensmith and R. A. Bambara (1981) Kinetic characteristics which distinguish two forms of calf thymus DNA polymerase α. *Biochemistry* **20**: 227-232.

K-T. Hohn and F. Grosse (1987) Processivity of the DNA polymerase α-primase complex from calf thymus. *Biochemistry* **26**: 2870-2878.

J. A. Huberman (1981) New views of the biochemistry of eucaryotic DNA replication revealed by aphidicolin, an unusual inhibitor of DNA polymerase α. *Cell* **23**: 647-648.

U. Hübscher (1983) The mammalian primase is part of a high molecular weight DNA polymerase α polypeptide. *EMBO J.* **2**: 133-136

S. Ikegami, T. Taguchi, M. Ohashi, M. Oguro, H. Nagano, and Y. Mano (1978) Aphidicolin prevents mitotic cell division by interfering with the activity of DNA polymerase-α. *Nature* **275**: 458-460.

S. Ikegami, S. Amemiya, M. Oguro, H. Nagano, and Y. Mano (1979) Inhibition by aphidicolin of cell cycle progression and DNA replication in sea urchin embryos. *J. Cell. Physiol.* **100**: 439-444.

D. Jaskulski, J. K. deRiel, W. E. Mercer, B. Calabretta, and R. Baserga (1988) Inhibition of cellular proliferation by antisense oligodeoxynucleotides to PCNA cyclin. *Science* **240**: 1544-1546.

L. Kaczmarek, M. R. Miller, R. A. Hammond, and W. E. Mercer. (1986) A microinjected monoclonal antibody against human DNA polymerase-α inhibits DNA replication in human, hamster, and mouse cell lines. *J. Biol. Chem.* **261**: 10802-10807.

L. S. Kaguni, J-M. Rossignol, R. C. Conaway, and I. R. Lehman (1983) Isolation of an intact DNA polymerase-primase from embryos of *Drosophila melanogaster*. *Proc. Natl. Acad. Sci. USA* **80**: 2221-2225.

L. S. Kaguni and I. R. Lehman. (1988) Eukaryotic DNA polymerase-primase: structure, mechanism and function. *Biochim. Biophys. Acta* **950**: 87-101.

H. B. Kaiserman and R. M. Benbow (1987) Characterization of a stable, major DNA polymerase α species devoid of DNA primase activity. *Nucl. Acids Res.* **15**: 10249-10265.

P. C. Keng, C. K. N. Li, and K. T. Wheeler (1981) Characterization of the separation properties of the Beckman elutriator system. *Cell Biophys.* **3**: 41-56.

N. N. Khan, G. E. Wright, L. W. Dudycz, and N. C. Brown (1984) Butylphenyl dGTP: a selective and potent inhibitor of mammalian DNA polymerase alpha. *Nucl. Acid Res.* **12**: 3695-3706.

N. N. Khan, G. E. Wright, L. W. Dudycz, and N. C. Brown (1985) Elucidation of the mechanism of selective inhibition of mammalian DNA polymerase alpha by 2-butylanilinopurines: development and characterization of 2-(*p-n*-butylanilino) adenine and its deoxyribonucleotides. *Nucl. Acid Res.* **13**: 6331-6342.

A. Kornberg (1980) *DNA Replication*, W. H. Freeman and Co., San Francisco, CA.

A. Kornberg,(1982) *Supplement to DNA Replication*, W.H. Freeman and Co., San Francisco, CA.

A. Kornberg (1984) Enzyme studies of replication of the *Escherichia coli* chromosome. *Adv.Exp. Med. Biol.* **179**: 3-16.

A. Kornberg (1988) DNA replication *J. Biol. Chem.* **263**: 1-4.

S. W. Krauss and S. Linn (1986) Studies of DNA polymerases alpha and beta from cultured human cells in various replicative states. *J. Cell. Phys.* **126**: 99-106.

H. Krokan, P. Schaffer, and M. L. DePamphilis (1979) Involvement of eucaryotic deoxyribonucleic acid polymerases α and δ in the replication of cellular and viral deoxyribonucleic acid. *Biochemistry* **18**: 4431-4443.

T. A. Kunkel and L. A. Loeb (1979) On the fidelity of DNA replication. Effect of divalent metal ion activators and deoxyribonucleoside triphosphate pools on *in vitro* mutagenesis. *J. Biol. Chem.* **254**: 5718-5725.

T. A. Kunkel, R. M. Schaaper, R. A. Beckman and L. A. Loeb (1981) On the fidelity of DNA replication. Effect of the next nucleotide on proofreading. *J. Biol. Chem.* **256**: 9883-9889.

T. A. Kunkel, R. D. Sabatino, and R. A. Bambara (1987) Exonucleolytic proofreading by calf thymus DNA polymerase δ. *Proc. Natl. Acad. Sci. USA* **84**: 4865-4869.

T. A. Kunkel and A. Soni (1988) Exonucleolytic proofreading enhances the fidelity of DNA synthesis by chick embryo DNA polymerase-γ. *J. Biol. Chem.* **263**: 4450-4459.

P. Kurki, M. Vanderlaan, F. Dolbeare, J. Gray, and E. M. Tan (1986) Expression of proliferating cell nuclear antigen (PCNA)/cyclin during the cell cycle. *Exp. Cell Res.* **166**: 209-219.

P. Kurki, M. Lotz, K. Ogata, and E. M. Tan (1987) Proliferating cell nuclear antigen (PCNA)/cyclin in activated human T lymphocytes. *J. Immunol.* **138**: 4114-4120.

P. Kurki, K. Ogata, and E. M. Tan (1988) Monoclonal antibodies to proliferating cell nuclear antigen (PCNA)/cyclin as probes for proliferating cells by immunofluorescence microscopy and flow cytometry. *J. Immunol. Methods* **109**: 49-59.

M. M. Kwant and P. C. van der Vliet (1980) Differential effect of aphidicolin on adenovirus DNA synthesis and cellular DNA synthesis. *Nucl. Acids Res.* **8**: 3993-4007.

M. Y. W. T. Lee, C-K. Tan, A. G. So, and K. M. Downey (1980) Purification of deoxyribonucleic acid polymerase δ from calf thymus: partial characterization of physical properties. *Biochemistry* **19**: 2096-2101.

M. Y. W. T. Lee, C-K. Tan, K. M. Downey, and A. G. So (1981) Structural and functional properties of calf thymus DNA polymerase δ. *Prog. Nucl. Acid Res. Mol. Biol.* **26**: 83-96.

M. Y. W. T. Lee, C-K. Tan, K. M. Downey, and A. G. So (1984) Further studies on calf thymus DNA polymerase δ purified to homogeneity by a new procedure. *Biochemistry* **23**: 1906-1913.

M. Y. W. T. Lee, N. L. Toomey, and G. E. Wright (1985) Differential inhibition of human placental DNA polymerase δ and α by BuPdGTP and BuAdATP. *Nucl. Acid Res.* **13**: 8623-8630.

M. Y. W. T. Lee and N. L. Toomey (1987) Human placental DNA polymerase δ: identification of a 170-kilodalton polypeptide by activity staining and immunoblotting. *Biochemistry* **26**: 1076-1085.

J. J. Li and T. J. Kelly (1984) Simian virus 40 DNA replication *in vitro*. *Proc. Natl. Acad. Sci. USA* **81**: 6973-6977.

J. J. Li and T. J. Kelly (1985) Simian virus 40 DNA replication *in vitro*; specificity of initiation and evidence for bidirectional replication. *Mol. Cell. Biol.* **5**: 1238-1246.

P. K. Liu, C-C. Chang, J. E. Trosko, D. K. Dube, G. M. Martin, and L. A. Loeb (1983) Mammalian mutator mutant with an aphidicolin-resistant DNA polymerase α. *Proc. Natl. Acad. Sci. USA* **80**: 797-801.

Y-C. Liu, R. L. Marraccino, P. C. Keng, R. A. Bambara, E. M. Lord, W-G. Chou, and S. B. Zain (1989) Requirement for proliferating cell nuclear antigen expression during stages of the Chinese hamster ovary cell cycle. *Biochemistry* **28**: 2967-2974.

H. Macdonald-Bravo and R. Bravo (1985) Induction of the nuclear protein cyclin in serum-stimulated quiescent 3T3 cells is independent of DNA synthesis. *Exp. Cell Res.* **156**: 455-461.

P. Madsen and J. E. Celis (1985) S-phase patterns of cyclin (PCNA) antigen staining resemble topographical patterns of DNA synthesis. A role for cyclin in DNA replication? *FEBS Lett.* **193**: 5-11.

H. Maki, S. Maki, and A. Kornberg (1988) DNA polymerase III holoenzyme of *Escherichia coli*. IV. The holoenzyme is an asymmetric dimer with twin active sites. *J. Biol. Chem.* **263**: 6570-6578.

R. L. Marraccino, A. F. Wahl, P. C. Keng, E. M. Lord, and R. A. Bambara (1987) Cell cycle dependent activities of DNA polymerases α and δ in Chinese hamster ovary cells. *Biochemistry* **26**: 7864-7870.

M. B. Mathews, R. M. Bernstein, B. R. Franza Jr., and J. I. Garrels (1984) Identity of the proliferating cell nuclear antigen and cyclin. *Nature* **309**: 374-376.

K. Matsumoto, T. Moriuchi, T. Koji, and P. K. Nakane (1987) Molecular cloning of cDNA coding for rat proliferating cell nuclear antigen (PCNA)/cyclin. *EMBO J.* **6**: 637-642.

C. S. McHenry (1988) DNA polymerase III holoenzyme of *Escherichia coli*. *Ann. Rev. Biochem.* **57**: 519-550.

V. S. Mikhailov and I. M. Androsova (1984) Effect of spermine on interaction of DNA polymerase α from the loach (*Misgurnus fossilis*) eggs with DNA. *Biochim. Biophys. Acta* **783**: 6-14.

M. R. Miller, R. G. Ulrich, T. S-F. Wang, and D. Korn (1985a) Monoclonal antibodies against human DNA polymerase-α inhibit DNA replication in permeabilized human cells. *J. Biol. Chem.* **260**: 134-138.

M. R. Miller, C. Seighman, and R. G. Ulrich (1985b) Inhibition of DNA replication and DNA polymerase α activity by monoclonal anti-(DNA polymerase α) immunoglobulin G and F(ab) fragments. *Biochemistry* **24**: 7440-7445.

K. Miyachi, M. J. Fritzler, and E. M. Tan (1978) Autoantibody to a nuclear antigen in proliferating cells. *J. Immunol.* **121**: 2228-2234.

Y. Murakami, H. Yasuda, H. Miyazawa, F. Hanaoka, and M-a. Yamada (1985) Characterization of a temperature-sensitive mutant of mouse FM3A cells defective in DNA replication. *Proc. Natl. Acad. Sci. USA* **82**: 1761-1765.

Y. Murakami, C. R. Wobbe, L. Weissbach, F.B. Dean, and J. Hurwitz (1986) Role of DNA polymerase α and DNA primase in simian virus 40 DNA replication *in vitro*. *Proc. Natl. Acad. Sci. USA* **83**: 2869-2873.

C. Nishida, P. Reinhard, and S. Linn (1988) DNA repair synthesis in human fibroblasts requires DNA polymerase δ. *J. Biol. Chem.* **263**: 501-510.

M. E. O'Donnell, P. Elias, and I. R. Lehman (1987) Processive replication of single-stranded DNA templates by the herpes simplex virus-induced DNA polymerase. *J. Biol. Chem.* **262**: 4252-4259.

K. Ogata, P. Kurki, J. E. Celis, R. M. Nakamura, and E. M. Tan (1987) Monoclonal antibodies to a nuclear protein (PCNA/cyclin) associated with DNA replication. *Exp. Cell Res.* **168**: 475-486.

H. P. Ottiger and U. Hübscher (1984) Mammalian DNA polymerase α holoenzymes with possible functions at the leading and lagging strand of the replication fork. *Proc. Natl. Acad. Sci. USA* **81**: 3993-3997.

G. Pedrali-Noy, S. Spadari, A. Miller-Faurès, A. O. A. Miller, J. Kruppa, and G. Koch (1980) Synchronization of HeLa cell cultures by inhibition of DNA polymerase α with aphidicolin. *Nucl. Acids Res.* **8**: 377-387.

G. Prelich, M. Kostura, D. R. Marshak, M. B. Mathews, and B. Stillman (1987b) The cell-cycle regulated proliferating cell nuclear antigen is required for SV40 DNA replication *in vitro*. *Nature* **326**: 471-475.

G. Prelich, C-K. Tan, M. Kostura, M. B. Mathews, A. G. So, K. M. Downey, and B. Stillman (1987b) Functional identity of proliferating cell nuclear antigen and a DNA polymerase-δ auxiliary protein. *Nature* **326**: 517-520.

G. Prelich and B. Stillman (1988) Coordinated leading and lagging strand synthesis during SV40 DNA replication *in vitro* requires PCNA *Cell* **53**: 117-126.

C. E. Prussak and B. Y. Tseng (1987) Isolation of the DNA polymerase α core enzyme from mouse cells. *J. Biol. Chem.* **262**: 6018-6022.

B. G. Que, K. M. Downey, and A. G. So (1978) Mechanism of selective inhibition of 3′ to 5′ exonuclease activity of *Escherichia coli* DNA polymerase I by nucleoside 5′-monophosphates. *Biochemistry* **17**: 1603-1606.

M. E. Reyland, I. R. Lehman, and L. A. Loeb (1988) Specificity of proofreading by the 3′→5′ exonuclease of the DNA polymerase-primase of *Drosophila melanogaster*. *J. Biol. Chem.* **263**: 6518-6524.

J. D. Roberts and T. A. Kunkel (1988) Fidelity of a human cell DNA replication complex. *Proc. Natl. Acad. Sci. USA* **85**: 7064-7068.

R. D. Sabatino and R. A. Bambara (1988) Properties of the 3′ to 5′ exonuclease associated with calf DNA polymerase δ. *Biochemistry* **27**: 2266-2271.

R. D. Sabatino, T. W. Myers, R. A. Bambara, O. Kwon-Shin, R. L. Marraccino and P. H. Frickey (1988) Calf thymus DNA polymerases α and δ are capable of highly processive DNA synthesis. *Biochemistry* **27**: 2998-3004.

M. R. Sadaie and M. B. Mathews (1986) Immunochemical and biochemical analysis of the proliferating cell nuclear antigen (PCNA) in HeLa cells. *Exp. Cell Res.* **163**: 423-433.

M. A. Sirover and L. A. Loeb (1977) On the fidelity of DNA replication. Effect of metal activators during synthesis with avian myeloblastosis virus DNA polymerase. *J. Biol. Chem.* **252**: 3605-3610.

W. Skarnes, P. Bonin, and E. Baril (1986) Exonuclease activity associated with a multiprotein form of HeLa cell DNA polymerase α. *J. Biol. Chem.* **261**: 6629-6636.

A. G. So and K. M. Downey (1988) Mammalian DNA polymerases α and δ: Current status in DNA replication. *Biochemistry* **27**: 4591-4595.

S. Spadari and A. Weissbach (1974) The interrelation between DNA synthesis and various DNA polymerase activities in synchronized HeLa cells. *J. Mol. Biol.* **86**: 11-20.

B. W. Stillman and Y. Gluzman (1985) Replication and supercoiling of simian virus 40 DNA in cell extracts from human cells. *Mol. Cell. Biol.* **5**: 2051-2060.

B. Stillman (1988) Initiation of eukaryotic DNA replication *in vitro*. *BioEssays* **9**: 56-60.

J. Syvaoja and S. Linn (1989) Characterization of a large form of DNA polymerase δ from HeLa cells that is insensitive to proliferating cell nuclear antigen. *J. Biol. Chem.* **264**: 2489-2497.

Y. Takasaki, J-S. Deng, and E. M. Tan (1981) A nuclear antigen associated with cell proliferation and blast transformation. Its distribution in synchronized cells. *J. Exp. Med.* **154**: 1899-1909.

Y. Takasaki, D. Fishwild, and E. M. Tan (1984) Characterization of proliferating cell nuclear antigen recognized by autoantibodies in lupus sera. *J. Exp. Med.* **159**: 981-992.

C-K. Tan, C. Castillo, A. G. So, and K. M. Downey (1986) An auxiliary protein for DNA polymerase-δ, from fetal calf thymus. *J. Biol. Chem.* **261**: 12310-12316.

C-K. Tan, K. Sullivan, X. Li, E. M. Tan, K. M. Downey, and A. G. So (1987a) Autoantibody to the proliferating cell nuclear antigen neutralizes the activity of the auxiliary protein for DNA polymerase delta. *Nucl. Acids Res.* **15**: 9299-9308.

C-K. Tan, M.J. So, K.M. Downey, and A.G. So (1987b) Apparent stimulation of calf thymus DNA polymerase alpha by ATP. *Nucl. Acids Res.* **15**: 2269-2278.

E. M. Tan (1982) Autoantibodies to nuclear antigens (ANA): their immunobiology and medicine. *Adv. Immunol.* **33**: 167-240.

S. Tanaka, S-Z. Hu, T. S-F. Wang, and D. Korn (1982) Preparation and preliminary characterization of monoclonal antibodies against human DNA polymerase α. J. Biol. Chem. **257**: 8386-8390.

K. R. Thomas and B. M. Olivera (1978) Processivity of DNA exonucleases. J. Biol. Chem. **253**: 424-429.

P. Thömmes, T. Reiter, and R. Knippers (1986) Synthesis of DNA polymerase α analyzed by immunoprecipitation from synchronously proliferating cells. *Biochemistry* **25**: 1308-1314.

G. Villani, P. J. Fay, R. A. Bambara, and I. R. Lehman (1981) Elongation of RNA-primed DNA templates by DNA polymerase α from *Drosophila melanogaster* embryos. *J. Biol. Chem.* **256**: 8202-8207.

A. F. Wahl, J. J. Crute, R. D. Sabatino, J. B. Bodner, R. L. Marraccino, L. W. Harwell, E. M. Lord, and R. A. Bambara (1986) Properties of two forms of DNA polymerase δ from calf thymus. *Biochemistry* **25**: 7821-7827.

A. F. Wahl, A. M. Geis, B. H. Spain, S. W. Wong, D. Korn, and T. S-F. Wang (1988) Gene expression of human DNA polymerase α during cell proliferation and the cell cycle. *Mol. Cell. Biol.* **8**: 5016-5025.

T. S-F. Wang, S-Z. Hu, and D. Korn (1984) DNA primase from KB cells. Characterization of a primase activity tightly associated with immunoaffinity-purified DNA polymerase-α. *J. Biol. Chem.* **259**: 1854-1865.

J. V. Wierowski, K. G. Lawton, J. W. Hockensmith, and R. A. Bambara (1983) Stimulation of calf thymus DNA α-polymerase by ATP. *J. Biol. Chem.* **258**: 6250-6254.

E. Wist, H. Krokan, and H. Prydz (1976) Effect of 1-β-D-arabinofuranosylcytosine triphosphate on DNA synthesis in isolated HeLa cell nuclei. *Biochemistry* **15**: 3647-3652.

C. R. Wobbe, F. Dean, L. Weissbach, and J. Hurwitz (1985) *in vitro* replication of duplex circular DNA containing the simian virus 40 DNA origin site. *Proc. Natl. Acad. Sci. USA* **82**: 5710-5714.

C. R. Wobbe, L. Weissbach, J. A. Borowiec, F. B. Dean, Y. Murakami, P. Bullock, and J. Hurwitz (1987) Replication of simian virus 40 origin-containing DNA *in vitro* with purified proteins. *Proc. Natl. Acad. Sci. USA* **84**: 1834-1838.

M. S. Wold, J. J. Li, and T. J. Kelly (1987) Initiation of simian virus 40 DNA replication *in vitro*: large-tumor-antigen- and origin-dependent unwinding of the template. *Proc. Natl. Acad. Sci. USA* **84**: 3643-3647.

S. W. Wong, A. F. Wahl, P-M. Yuan, N. Arai, B. E. Pearson, K. Arai, D. Korn, M. W. Hunkapiller, and T. S-F. Wang (1988) Human DNA polymerase α gene expression is cell proliferation dependent and its primary structure is similar to both prokaryotic and eukaryotic replicative DNA polymerases. *EMBO J.* **7**: 37-47.

T. Yagura, T. Kozu, T. Seno, M. Saneyoshi, S. Hiraga, and H. Nagano (1983) Novel form of DNA polymerase α associated with DNA primase activity of vertebrates. Detection with mouse stimulating factor. *J. Biol. Chem.* **258**: 13070-13075.

M. Yamaguchi, E. A. Hendrickson, and M. L. DePamphilis (1985) DNA primase-DNA polymerase α from simian cells. Modulation of RNA primer synthesis by ribonucleoside triphosphates. *J. Biol. Chem.* **260**: 6254-6263.

L. Yang, M. S. Wold, J. J. Li, T. J. Kelly, and L. F. Liu (1987) Roles of DNA topoisomerases in simian virus 40 DNA replication *in vitro*. *Proc. Natl. Acad. Sci. USA* **84**: 950-954.

S. J. Zhang and M. Y. W. T. Lee (1987) biochemical characterization and development of DNA polymerases α and δ in the neonatal rat heart. *Arch. Biochem. Biophys.* **252**: 24-31.

5

DNA POLYMERASE β: PHYSIOLOGICAL ROLES, MACROMOLECULAR COMPLEXES AND ACCESSORY PROTEINS*

Diane C. Rein, Anthony J. Recupero, Michael P. Reed, and Ralph R. Meyer

---

"What is the secret of life?" I asked.

"Protein," the bartender declared, "They found out something about protein."

K. Vonnegut, Jr.
Cat's Cradle

*This chapter is dedicated to Arthur Kornberg on the occasion of his 70th birthday. His work has been an inspiration and guide to us for over 20 years, and Dr. Meyer acknowledges the honor and privilege of having worked in Professor Kornberg's laboratory. Work in our laboratory has been supported by grants from the American Cancer Society and the National Cancer Institute.

INTRODUCTION AND PERSPECTIVE

In the mid 1970's we began a systematic search for regulators affecting DNA synthesis. We felt that many such regulators would enhance the ability of a DNA polymerase to synthesize DNA *in vitro*. At the outset, we realized the need for a homogeneous DNA polymerase, since impure polymerase preparations could be saturated with such factors and no stimulation of exogenous polymerase would be seen *in vitro*. Indeed, this was later confirmed. It is clear that in crude extracts, as well as in considerably more purified preparations, many such proteins (either stimulators or inhibitors) are found. Any attempt to correlate DNA polymerase activity to the actual amount of enzyme molecules present is futile. Moreover, it is very difficult to assay for stimulatory factor activity if the polymerase fraction contains any enhancer of its activity, including salt. DNA polymerase-β was purified from rat liver cytoplasm and nuclei, and later from whole cell extracts of guinea pig liver and of the Novikoff hepatoma. Most of our work has been carried out with the rat Novikoff ascites tumor, as it is easy to transplant and harvest, and we can recover large quantities of cells (10-20 g per animal) in less than one week.

As a class, β-polymerases share several properties: low molecular weight (32-40,000); alkaline pI and pH optimum; stimulation by 50-100 mM salt; inhibition by phosphate; resistance to aphidicolin; sensitivity to dideoxythymidine triphosphate (ddTTP); and general resistance to N-ethylmaleimide (NEM). All β-polymerases are devoid of any other intrinsic activity. In fact, this probably is their most distinctive feature. Unlike other DNA polymerases, β-polymerases are simple nucleotide incorporating machines, lacking the ability to carry out additional reactions, including even pyrophosphoryolysis and pyrophosphate exchange. Since the latter reactions can be considered the reverse of DNA polymerization, the β-polymerases can be construed as synthesizing DNA *in vitro* in a kinetically irreversible manner. Even at that, the enzyme can add only a single nucleotide before it dissociates from the primer.

The presence or absence of various divalent cations can significantly alter reaction mechanisms. The distributive nature of synthesis can become slightly processive in the presence of Mn^{2+} at the appropriate size gap and at reduced 3'OH primer levels. This phenomenon (called the "Mg^{2+}-Mn^{2+} effect") is restricted to β-polymerases. Optimum reaction conditions for natural and synthetic substrates can vary widely. The addition of monovalent salts can completely abolish or significantly enhance the parameter under study. Even restricted to the synthesis of natural DNA, nucleotide by nucleotide *in vitro*, polymerase-β displays the highest error rate measured for cellular DNA polymerases. In the presence of a single deoxyribonucleoside triphosphate (dNTP) substrate, the β-polymerase can continue to carry out a considerable amount of incorporation. This forgotten and potentially misunderstood phenomenon (termed "relaxed" dNTP incorporation) was initially studied in some detail to distinguish the enzyme from terminal deoxynucleotidyltransferase and is unique to β-polymerases.

The simplistic and rigid enzymatic kinetics displayed by the enzyme, the total lack of specific inhibitors and the inability to kinetically distinguish reaction intermediates have foiled these avenues of providing clues to function. Its physiological role in the cell remains unknown. Its activity is remarkably constant throughout the cell cycle and under a variety of conditions. Studies with inhibitors in permeable cells have implicated a role for polymerase-β in repair, a role it seems to share in many cell lines with polymerase-α and polymerase-δ.

In the final analysis, the one salient feature of DNA polymerase-β is that it behaves anomalously. Our research group has operated under the premise that this enzyme only appears to be simple. It is certainly well documented that the homogeneous enzyme does not behave, mechanistically, like any other characterized DNA polymerase. We feel that *in vivo*, this enzyme is modulated in a manner dramatically different than that *in vitro*. Given the complexity of protein interactions known to occur in *E. coli* DNA replication and repair, it is clear that these interactions most likely involve other proteins. To this end we have identified six stimulatory proteins for the homogeneous hepatoma polymerase-β. Some of these proteins have been shown to possess an intrinsic enzymatic activity. With one exception, no direct interaction with the polymerase has been found. However, four of the six factors examined seem to mediate their effect at the substrate: one lowers the K_m for the DNA substrate of the polymerase 40-fold; another lowers the K_m of dNTP substrates; another permits DNA synthesis to occur on long, single-stranded gapped natural and synthetic DNA's and the fourth permits synthesis on minimally gapped or nicked DNA, a normally unfavorable substrate for the polymerase. The properties of these six stimulatory proteins are listed in Table 1, and will be described in detail later.

PHYSIOLOGICAL ROLES

It has been clearly established that DNA polymerase-β is found in nuclei (Meyer and Simpson, 1968; Matsukage *et al.*, 1983; also see reviews by Fry, 1983, and Fry and Loeb, 1986). The isolation of polymerase-β in cytoplasmic fractions is, for the most part, a consequence of the extraction procedure. However, cytoplasmic localization in one tissue, namely oocytes, probably is physiologically relevant and not an artifact (Shioda *et al.*, 1977; Hobart and Infante, 1980; Fox *et al.*, 1980). (Subcellular localization is discussed in greater detail later under Macromolecular Complexes). The physiological role or roles played by DNA polymerase-β in nucleic acid metabolism have yet to be unequivocally established. There are numerous reports addressing this issue which are reviewed below. These are, for the most part, correlative studies, but nevertheless support a role for DNA polymerase-β in some aspects of DNA repair.

Inhibitor Studies

One of the approaches used to determine the function(s) of the cellular polymerases has been to use inhibitors to differentiate between DNA polymerase-α and -β. DNA polymerase-α is inhibited by aphidicolin, arabinosylcytosine triphosphate (araCTP), and sulfhydryl blocking agents such as N-ethylmaleimide (NEM). DNA polymerase-β is inhibited by dideoxythymidine triphosphate (ddTTP) and phosphate. The problem with using such agents, however, is that these enzymes are differentially rather than absolutely sensitive. Thus the data obtained must be cautiously interpreted. Using such inhibitors with isolated nuclei from HeLa cells (Wist, 1979), BHK-21/C13 cells (Burk *et al.*, 1979) and SV40-infected CV-1 cells (Edenberg *et al.*, 1978), DNA polymerase-α was implicated in DNA replication and DNA polymerase-β in DNA repair. Studies of DNA repair in human embryonic fibroblasts infected with Herpes virus (Nishiyama *et al.*, 1983) also supported a role for polymerase-β in this process. Chan and Becker (1981) examined the effects of the direct-acting carcinogens acetoxyacetylaminofluorene (AAAF), methylnitrosourea (MNU) and N-methyl-N′-nitro-N-

Table 1. DNA polymerase-β accessory proteins from the Novikoff hepatoma.

Factor Number[a]	Name[b]	Molecular Weight[c] ($\times 10^{-3}$)	Activity	Effects on DNA Polymerase-β	References
I	——	unknown	——	Stimulation	Probst *et al.*, (1975); Meyer *et al.*, unpublished
II	——	37	——	Reduces K_m of activated DNA substrate 40-fold	Probst *et al.*, (1975); Rein *et al.*, unpublished
III	SSB-48	48	Single-stranded DNA-binding protein; DNA unwinding	Increases rate and extent of DNA synthesis; allows polymerase to synthesize on long single-stranded gaps; enhances stimulation by ATPase III	Probst *et al.*, (1975); Koerner and Meyer (1983); Meyer *et al.*, (1984)
IV	DNase V	12	3′→5′ and 5′→3′ double-strand exonuclease	Complexes and stabilizes polymerase and lowers K_m for dNTP's; increases rate of reaction	Mosbaugh *et al.*, (1977) Mosbaugh and Meyer (1980) Meyer and Mosbaugh (1980)
V	DBP-29	29	Double-stranded DNA binding protein	Increases rate of reaction at nicks or short gaps	Reed *et al.*, (unpublished)
——	ATPase III	65	DNA-dependent ATPase	Increases rate of synthesis on nicked DNA; activity enhanced by SSB-48	Thomas and Meyer (1982) Meyer *et al.*, (1984)

[a] Stimulatory proteins were named in order of discovery.
[b] Factors were renamed once an activity had been found.
[c] Determined by SDS gel electrophoresis.

nitrosoguanidine (MNNG) on the cellular polymerases, finding that DNA polymerase-β was more resistant than DNA polymerase-α or -γ.

Other studies by Butt *et al.* (1978) examined *in vitro* DNA synthesis in mouse L929 cell nuclei using differential extraction of polymerases-α and -β and by the use of araCTP and NEM. Their data suggested the possibility that both polymerase-α and polymerase-β are involved in replicative DNA synthesis, while polymerase-β is clearly implicated in a DNA repair role.

Studies Using DNA Damaging Agents

Another approach has been to correlate changes in DNA polymerase activities with altered replication or repair. Pedrali-Noy and Spadari (1980) developed an assay to correlate, quantitatively, the repair of UV-damaged DNA in HeLa cells with polymerase-β activity, while Hardt *et al.* (1981) used a similar system but visualized repair synthesis by autoradiography. Both studies concluded that polymerase-β has a role in DNA repair. Further support for a role for this enzyme in the repair of UV-damaged DNA comes from earlier observations on phytohemagglutinin-stimulated lymphocytes (Bertazzoni *et al.*, 1976). One of the best studies was the work of Hübscher (Hübscher *et al.*, 1978, 1979; Waser *et al.*, 1979). Using rat neuronal nuclei, they have strongly substantiated a role for polymerase-β in DNA repair, since these cells contain 99.2% DNA polymerase-β and only 0.8% DNA polymerase-γ. These cells lack DNA polymerase-α, yet are able to carry out efficient DNA repair. Moreover, DNA repair was stimulated by UV-irradiation. The antitumor drug bleomycin (BLM) is known to induce DNA repair synthesis. Isolated rat liver nuclei (Coetzee *et al.*, 1977) and chromatin (Naryzhnyi *et al.*, 1982) from BLM-treated animals had increased DNA repair synthesis, and such repair synthesis is associated with DNA polymerase-β (Sartiano *et al.*, 1975).

There are numerous studies indicating that DNA polymerase-α also participates in DNA repair. Snyder and Regan (1981) reported that aphidicolin blocked repair synthesis in UV-irradiated human fibroblasts. Other studies involving UV-induced DNA synthesis in HeLa and other cell lines (Morita *et al.*, 1982; Seki *et al.*, 1983, 1984) have indicated that both DNA polymerase-α and -β are involved.

Another useful approach has been to use purified enzymes challenged with defined damaged DNA templates. Siedlecki *et al.* (1981) examined the abilities of polymerases-α and -β to incorporate nucleotides into a damaged synthetic template. They reported that polymerase-β apparently had the ability to synthesize DNA by a strand-displacement mode. These observations were extended (Soltyk *et al.*, 1981) using calf thymus DNA polymerase-β synthesizing on *E. coli* DNA damaged by methyl methanesulfonate or thymine starvation. DNA polymerase-β was also more efficient than polymerase-α in synthesizing on BUdR-treated *E. coli* DNA (Siedlecki *et al.*, 1980). Other studies using damaged PM2 DNA (Mosbaugh and Linn, 1983), rat liver chromatin (Verly, 1982; Goffin and Verly, 1982), and SV40 DNA (Evans and Linn, 1984) have implicated DNA polymerase-β in excision repair. Wang and Korn (1980) used duplex DNA substrates containing nicks or gaps of defined size. Their results indicated that DNA polymerase-β is most reactive (and can fill to completion) gaps of about 10 nucleotides. DNA polymerase-α, on the other hand, was most active on gaps of 30-60 nucleotides, but could not completely fill them. These observations may relate to the suggestion of Miller and Chinault (1982b) that the "patch size" of various DNA damaging agents may reflect the DNA polymerase involved in the repair process.

Permeable Cell Studies

Recent studies have attempted to mimic *in vivo* conditions by utilizing permeabilized cells in an *in situ* system. These experiments should be viewed with caution, however, since it has been reported (Patel *et al.*, 1983) that up to 25% of the total cellular protein is released during this process. This soluble fraction contains at least proteins which affect BLM-induced DNA synthesis. Initial studies by Seki and Oda (1980) used ddTTP to examine DNA synthesis in BLM-treated permeable mouse ascites sarcoma cells and indicated that polymerase-β has a role in DNA repair synthesis while polymerase-α plays a major role in replicative DNA synthesis. Seki *et al.* (1980) arrived at the same conclusions when they studied the effects of aphidicolin on the same system. A subsequent study (Seki *et al.*, 1983) concluded that both polymerases could participate in DNA repair synthesis, but their relative participation seemed to vary according to cell type and the nature and degree of damage. Recent studies which examined the completion of repair synthesis in detail in permeable HeLa cells (Seki and Oda, 1986) suggested that polymerase-β is primarily involved in the completion of repair patches in BLM-pretreated cells, and polymerase-α plays a preferential role in the repair of the intranucleosomal region of chromatin. Miller and Chinault (1982a, 1982b) reported that polymerase-β was primarily responsible for repair synthesis of DNA damage by BLM or neocarzinostatin, whereas polymerase-α played a primary role in repair of MNNG-, NMU- and UV-induced damage. Joe *et al.* (1987) demonstrated that benzo[a]pyrene-induced DNA repair synthesis was correlated with increased polymerase-α activity in permeable human fibroblasts. Data from Berger *et al.* (1979) further substantiated a role for polymerase-α in repair synthesis. Studies by Dresler and coworkers (Dresler *et al.*, 1982; Dresler, 1984) using permeable human fibroblasts have implicated polymerase-α as being involved in excision repair of UV and MNNG damage along with a possibility of accessory protein(s) interacting with the polymerase. Subsequent examination of the same system with the inhibitors N^2-(*p*-n-butylphenyl)-2′-deoxyguanosine-5′-triphosphate (BuPh-dGTP), which strongly inhibits polymerase-α but weakly inhibits polymerase-δ, and ddTTP (Dresler and Frattini, 1986; Dresler and Kimbro, 1987) suggested that polymerase-δ has a role in both DNA replication and UV-induced repair synthesis in human fibroblasts. In summary, permeable cell systems have confirmed that DNA polymerase-β is involved in DNA repair, but this function is shared by at least polymerase-α and probably polymerase-δ.

Developmental Studies

Many developmental studies have, by simple correlation, suggested a role for polymerase-β in DNA repair and DNA polymerase-α in replication. In neonatal rat heart tissue which will no longer undergo replicative synthesis, DNA polymerase-α decreases 80-fold between days 1 and 17, while DNA polymerase-β remains fairly constant (Claycomb, 1975). In regenerating rat liver (Chang and Bollum, 1972a; Lynch and Lieberman, 1973; Davis *et al.*, 1976), rat liver tumors (Baril *et al.*, 1973), and in normal liver stimulated to undergo DNA replication (Lynch *et al.*, 1976), DNA polymerase-α increases dramatically while DNA polymerase-β does not. Similar results were found in stimulated human lymphocytes (Bertazzoni *et al.*, 1976). Craddock (Craddock, 1981; Craddock and Ansley, 1979) showed that DNA polymerase-α and -β activities in carcinogen fed rats were correlated with periods of DNA replication and repair, respectively.

A role for DNA polymerase-β in rat giant trophoblast cell replication was suggested by Siegel and Kalf (1982) based upon the observation that this DNA synthesis was insensitive to aphidicolin, while being sensitive to NEM, 50 mM potassium phosphate and ddTTP to the same degree as that seen with partially purified trophoblast polymerase-β. These results were supported by Huang *et al.* (1983) using Chinese hamster cells. Other studies (Butt *et al.*, 1978; Weissbach, 1979) have also suggested a role for polymerase-β in DNA replication.

A study by Grippo *et al.* (1978) suggested that DNA polymerase-β may play a role in recombination. These authors examined DNA polymerase activities in premeiotic, meiotic and post-meiotic male mouse germ cells. DNA polymerase-β was the only polymerase found in meiotic and post-meiotic cells and was maximal during middle-late pachytene.

Cell Cycle Studies

Studies of cultured cells have addressed the question of the roles of DNA polymerase-α and -β by examining the levels of these enzymes during the cell cycle. Chang *et al.* (1973) were one of the first to show that levels of DNA polymerase-β remained relatively constant throughout the cell cycle in synchronized mouse L-cells, while polymerase-α increased up to 12-fold. Similar results were obtained with HeLa cells (Spadari and Weissbach, 1974; Chiu and Baril, 1975), BHK-21/C13 hamster cells (Craig *et al.*, 1975) and mouse thymus cells (Barr *et al.*, 1976). Pritchard *et al.* (1978) monitored DNA polymerase activities throughout the cell cycle in synchronized cells infected with bovine parvovirus. They too reported that levels of DNA polymerase-α paralleled viral DNA synthesis, while levels of DNA polymerase-β remained constant. More recently Schneider *et al.* (1985) examined two heat-sensitive (arrested in G_1 at 39.5°C) and two cold-sensitive (arrested in G_1 at 33°C) clonal cell cycle mutants for polymerase-α, -β and -γ activities. They found that polymerase-α was tightly coupled to the number of cells in S-phase, while polymerase-β and -γ activities showed no correlation to the cell cycle.

In summary, a variety of approaches, including the use of differential inhibitors, DNA damaging agents, and permeabilized cells along with developmental and cell cycle studies, have led to the conclusion that DNA polymerase-β is a DNA repair enzyme. However, repair is not carried out exclusively by this polymerase, but is shared with DNA polymerase-α and probably polymerase-δ, depending on the particular cell and the type and extent of DNA damage caused by a particular agent. There are a few studies to suggest that DNA polymerase-β may participate in some aspects of DNA replication and recombination as well.

MACROMOLECULAR COMPLEXES

Early studies of mammalian DNA polymerases quickly established the presence of a 6-8S heterogeneous cytoplasmic DNA polymerase (Chang *et al.*, 1973) and a 3-4S low molecular weight nuclear DNA polymerase (for a review, see Fry and Loeb, 1986). Considerable effort was directed towards understanding the relationship between these two enzymes. Models were proposed within the one polymerase framework established at that time from studies performed in *E. coli*. Thus, the concept that a polymerase was translated in the cytoplasm and shunted to the nucleus as a complex was suggested (Hecht and Davidson, 1973). *In vitro* studies, demonstrating that the 6-8S cytoplasmic polymerase could be converted to a 3-4S nuclear form by raising the ionic strength of cytoplasmic extracts, strengthened the precursor-product model

(Hecht, 1973a, 1973b; Lazarus and Kitron, 1973; Morioka and Terayama, 1974; Srivastava, 1974; Zunino *et al.*, 1975).

Almost simultaneously, however, confusion set in as some investigators could find a 6-8S cytoplasmic polymerase while others could not. Moreover, some investigators could convert the aggregate to a low molecular weight form, while others could not. Improved purification schemes and refinement of assays muddled the relationship further by demonstrating divergent enzymological characteristics of the two polymerases. Ultimately it became clear that mammalian cells contained two major DNA polymerases. New models were proposed that accounted for two distinct forms of nuclear DNA polymerases: a 6-8S polymerase-α that is easily leached from nuclei; and a 3-4S low molecular weight polymerase-β that is tightly associated. However, by the mid-1970's, three research groups had demonstrated that two forms of 6-8S polymerase coexisted in cytoplasmic extracts (Srivastava, 1974; Hecht, 1975; deRecondo and Abadiedebat, 1976). One of these was the 6-8S polymerase-α that could not be converted to any smaller species by salt. The second 6-8S polymerase could be reversibly converted to a 3-4S polymerase that appeared identical to nuclear polymerase-β. During this period, we had purified a homogeneous 3.3S DNA polymerase-β from Novikoff hepatoma (Stalker *et al.*, 1976). Hepatoma high salt crude extracts contain two DNA polymerases: a 12S aphidicolin-sensitive, NEM-sensitive polymerase-α (Rein, unpublished data) that was subsequently purified to near homogeneity (Thomas, 1984); and a 6-8S DNA polymerase-β. These two polymerases are completely separated from each other during ammonium sulfate fractionation. The 6-8S polymerase-β recovered from the 45-75% ammonium sulfate pellet is immediately converted to a 4.1S form upon the first chromatographic step (Mosbaugh *et al.*, 1977). This "minicomplex" persists until the final single-strand DNA-cellulose column when the 3.3S form of polymerase-β is recovered. The 4.1S minicomplex was later shown to be a highly specific, 1:1 stoichiometric interaction between DNA polymerase-β and DNase V (Mosbaugh and Meyer, 1980; Meyer and Mosbaugh, 1980). This bidirectional, double-strand exonuclease was originally identified as Factor IV and is discussed in detail below.

The 6-8S polymerase-β species can also be dissociated into a 5.8S, a 4.1S and a 3.3S subspecies by raising the ionic strength of the crude ammonium sulfate fraction. This is completely reversible by back dialysis if the dialysis tubing has an 8000 molecular weight cut-off. Partial reversion to the 5.8S species occurs following sucrose gradient sedimentation if the reconstituting fractions include the 1.9S-5.8S region of the gradient. During the process, the isoelectric point of hepatoma polymerase-β shifts from a pH of 7.5 for the 6-8S form, to a pH of 8.5 for the DNase V-polymerase-β minicomplex, and finally to a pH of 8.6 for the 3.3S homogeneous DNA polymerase-β. Other groups have shown that initial fractionation of the crude extracts on phosphocellulose results in multiple polymerase-β species exhibiting sedimentation values from 8.6 to 4.6S (Krauss and Linn, 1980). In rat liver, Morioka and Terayama (1974) have found that the fraction that does not bind to the phosphocellulose can convert their 3.3S polymerase to a 5.0S form. A change in the size of KB polymerase-β leading to a larger form has been demonstrated by similar reconstitution experiments following gel filtration (Wang *et al.*, 1975). Chiu and Sung (1973) have found a factor which is also separated from rat brain polymerase-β during gel filtration and which restores the polymerase activity lost during this procedure. All of these data collectively argue that the higher molecular weight forms are not due to self-aggregation but rather to different proteins interacting with DNA polymerase-β.

There are two physiological studies showing that the size of the polymerase-β in crude extracts can be altered during cell proliferation and differentiation. During liver regeneration the DNA polymerase-β is converted from a 3.3S form to a 10S species (deRecondo and Abadiedebat, 1976). This conversion is due to the presence of several cytoplasmic proteins appearing at approximately 24 hr following partial hepatectomy. In contrast, in normal rat adrenal gland, 80% of the total cellular DNA polymerase is a 3.3S, NEM-resistant enzyme (Nagasaka and Yoshida, 1982). This organ is the target for ACTH, which induces and maintains the differentiation of glucocorticoid-producing cells. Although ACTH does not induce cellular proliferation of adrenal gland tissue, reduction of ACTH serum levels by hypophysectomy resulted in a preferential loss of the NEM-resistant, 3.3S polymerase-β-like enzyme (Nagasaka and Yoshida, 1982). The remaining activity was found as a 5-7S cytoplasmic complex 14 days following removal of the pituitary gland. This entire process could be prevented by daily administration of ACTH immediately following hypophysectomy or reversed by the same treatment beginning 14 days later. Glucocorticoids would not substitute for ACTH in this process. In either case, a shift from a cytoplasmic 5-7S polymerase-β to a 3.3S nuclear enzyme was observed.

Many observers throughout the last decade have reported the detection of polymerase-β in the perinuclear space between the inner and outer membranes of the nucleus (Hecht, 1973b; Tsuruo *et al.*, 1974; Roufa *et al.*, 1975; deRecondo and Abadiedebat, 1976). The isolation of nuclei with both membranes intact is very difficult. Changes in ionic strength, length of time in sucrose solutions and presence of 0.5-1% Triton X-100 in buffers, will remove or damage the outer nuclear membrane. This may go unnoticed and could lead to leaching of nuclear components into the cytoplasmic fraction or to the complete loss of the perinuclear constituents during the Triton X-100 washes commonly used in preparation of nuclei prior to polymerase extraction. It is very clear that such nuclei do contain a tightly bound DNA polymerase-β which is usually recovered as a 3-4S component. It is also clear that those who have examined the Triton X-100 wash of intact nuclei have found a 6-10S polymerase complex that can be converted by salt to a 3-4S enzyme closely approximating polymerase-β (Hecht, 1973b; Tsuruo *et al.*, 1974; deRecondo and Abadiedebat, 1976). It is highly probable that the source of the so called "cytoplasmic polymerase-β" arises from a damaged outer nuclear membrane. Recently, Smith *et al.* (1984) reported that a NEM-sensitive polymerase activity is found predominantly in the interior regions of nuclei. These studies included *in vitro* activity assays of isolated nuclear matrices combined with electron microscope autoradiography of DNA synthesis in isolated nuclei and in intact cells. The authors concluded that this activity is polymerase-α. Most interesting, however, was the detection of a NEM-resistant polymerase activity that was restricted to the peripheral nuclear regions and found in isolated nuclear lamina. This polymerase activity was postulated to be DNA polymerase-β and conjectured to play a unique role in either DNA replication or DNA repair at a site restricted to a peripheral nuclear location. In an earlier study, this same group isolated nuclear matrices that contained predominantly polymerase-α (Smith and Berezney, 1982). Although polymerase-β is a minor component of this matrix, they demonstrated that its association with the matrix rose just prior (4-16 hr) and just after (24 hr) the *in vitro* rise in DNA replication during liver regeneration. This correlates well with the appearance of those proteins, reported by deRecondo and Abadiedebat (1976), which specifically shift the polymerase-β from 3.3S to 10S 24 hr post-hepatectomy. Moreover, the 10S complex was found in the 1% Triton X-100 wash of nuclei.

Gupta and Sirover (1980) originally proposed that the waves of DNA repair activity observed in WI-38 fibroblasts were, in effect, the screening of the genome for damage prior to DNA replication and a re-screening after DNA replication. Perturbations in this phenomenon were detected in Bloom's syndrome, and it was postulated as leading to the high mutation frequencies and sister chromatid exchanges that typify this disease (Gupta and Sirover, 1984). The fluctuations observed by Smith and Berezney (1982) are consistent with this model, and suggest that such a phenomenon should occur during every cell cycle. We feel that it is highly probable that, unlike DNA polymerase-α, which may be newly synthesized at every round of the cell cycle, DNA polymerase-β may be conserved. It is mobilized to perform its cellular duties mostly from a pre-existing steady-state pool of molecules. We also suggest that the 6-8S polymerase-β is heterogeneous simply because it represents a pool of *different* complexes.

The relationship between the 6-8S perinuclear polymerase-β and the tightly bound 3-4S nuclear polymerase-β needs to be clarified. It cannot be disputed that higher molecular weight forms of polymerase-β exist in cellular extracts. However, until these species can be isolated intact, resolved, and studied, we need to keep an open mind concerning their origin and possible function(s).

POLYMERASE-NUCLEIC ACID INTERACTIONS

The mechanism by which DNA polymerase-β recognizes DNA is poorly understood and when contrasted to DNA polymerase-α is, at best, unusual. The latter enzyme has at least two dNTP substrate sites (for review, see Fry and Loeb, 1986), several Mg^{2+} binding sites [some of which are involved in the catalysis reaction, some of which are involved directly in primer-template binding (Fisher and Korn, 1981a)] and at least one primer and one template binding site (Wilson *et al.*, 1977). DNA polymerase-α binds productively at 3′OH primers adjacent to single-strand template regions (Fisher and Korn, 1979a, 1979b). Fisher and Korn have postulated that these regions are recognized first by one template binding site, which in turn activates a second template site. At this point, the primer site is activated and the enzyme binds to a free 3′OH terminus. Once the template-primer sites are occupied, the dNTP is bound (Fisher and Korn, 1981b) and synthesis occurs in a quasi-processive manner (Das and Fujimara, 1979; Fisher *et al.*, 1979).

DNA polymerase-β binds to the template-primer, followed by dNTP. How it recognizes the template and primer and in what order can not be determined kinetically. Due to the lack of appropriate inhibitors to uncouple them, the events can not be separated (Wang and Korn, 1982). However, it is known that duplex DNA without a free 3′OH terminus (covalently closed DNA duplex circles) or 3′OH terminal sites at the blunt-end of restriction fragments, are completely "invisible" to the enzyme (Wang and Korn, 1982). This is in contrast to *E. coli* polymerase I which readily binds to flush-ended fragments (Englund *et al.*, 1968). This observation is important, since the polymerase-β should have bound to such fragments if it recognizes a 3′OH independently of, or prior to, template binding. This discrimination argues that polymerase-β must recognize something in addition to the primer stem for DNA binding. Blunt-ended DNA fragments can be activated for enzyme binding by creating single-strand DNA tails with exonuclease III (Wang and Korn, 1982). Productive binding (as measured by DNA synthesis) is modulated by divalent cations and will occur at a single nucleotide "tail" with Mn^{2+}. However, tails of 5-7 nucleotides are required if Mg^{2+} is used. Duplex DNA can be activated for polymerase-β by introducing internal nicks. The enzyme binds as readily to $3'PO_4$

sites as it does to 3′OH sites (although the former is a dead-end inhibitor). In fact, Chang (1973) had shown earlier that the enzyme can, under the appropriate conditions, synthesize from as much as a three nucleotide terminal mismatch at the 3′ primer stem. This lack of stringency for primer recognition by the polymerase-β only illustrates the difference between its characteristics and those of polymerase-α.

The above data collectively argue that polymerase-β recognizes 3′OH primer sites by the presence of an adjacent short, single-strand template DNA region: a characteristic shared with polymerase-α. Consequently, single-strand DNA should be an inhibitor of polymerase-β as it is for polymerase-α. However, while single-strand DNA's can inhibit, they are only very weak inhibitors, and they are uncompetitive with respect to DNA or to dNTP substrate sites on the enzyme (Korn *et al.*, 1981). It is clear from studies performed with synthetic homopolymers, that there can be a phenomenon of "too much template" strand for polymerase-β (see below). One can also "overactivate" DNA with DNase I. Whether unusual DNA structures are being created under these conditions that lead to some form of template inhibition or whether the rate of the reaction is substantially slowed without alterations in substrate affinity cannot be distinguished. The latter has been suggested by analysis of the rate of polymerization on defined gaps of increasing length (Wang and Korn, 1980).

In summary, the sequence of events and the parameters involved in template-primer recognition by polymerase-β remain elusive. Clearly, some nucleic acid structural element is required in addition to the relaxed recognition observed for the primer stem. The exact discriminatory index utilized by the enzyme to distinguish internal 3′OH nicks from 3′OH flush-end termini is also unknown. Since it binds to a variety of different primer stems, the possibility exists that the enzyme reads the phosphate backbones of the DNA helix and recognizes the absence of a phosphodiester bond opposite a stretch of single-stranded DNA. The sequence of template and primer recognition currently is refractive to kinetic analysis. Recognition could be simultaneous or independent. It could involve separate or identical sites which, in turn, could allosterically control each other.

DNA polymerase-β readily uses synthetic DNA as template primers *in vitro.* Optimum conditions can be highly variable, depending upon template, primer, pH, ionic strength and divalent cation present. There is at least one report that the absolute amount, as well as the ratio of each primer to template, varies significantly for each individual homopolymer template used (Ono *et al.*, 1979). With the exception of poly(dT) which requires Mg^{2+}, the favored divalent cation is Mn^{2+}. With poly(dA) as a template and in the presence of Mn^{2+}, significant synthesis does not occur unless the oligo(dT) primer is at least eight nucleotides in length (Chang, 1973). In the presence of Mg^{2+}, primers are not recognized and the "Mg^{2+}-Mn^{2+} effect" may be operating here (see below). Although the addition of increasing amount of primer to a given amount of template polymer demonstrates typical saturation kinetics, the addition of excess template results in a sharp optimum with rapid inhibition (Ono *et al.*, (1979) Each pair of synthetic template-primer exhibits its own optimum. It is not clear why the inhibition occurs. As discussed previously, single-strand DNA's are not significant inhibitors of DNA synthesis by polymerase-β on natural DNA. However, these studies infer that some structure is formed in the presence of excess synthetic template, resulting in an inability to continue synthesis. Alternatively, the enzyme may simply behave differently, mechanistically speaking, on synthetic homopolymers with single-strand DNA becoming an inhibitor of this reaction.

There is tantalizing evidence suggesting that the mechanism by which polymerase-β interacts with the synthetic RNA substrate, poly(rA) primed with oligo(dT), is unique and unrelated to natural DNA's. Like all synthetic and natural substrates, DNA synthesis by polymerase-β on poly(rA)•oligo(dT) substrates is stimulated by salt and, as with most synthetic substrates, requires Mn^{2+} as a cation. However, the reaction is inhibited by very low concentrations of natural DNA's (Yoshida *et al.*, 1979). The inhibition is not seen on activated DNA or on the equivalent synthetic DNA substrate, poly(dA)•oligo(dT). The inhibition cannot be explained by differential substrate affinities, since polymerase-β has a six-fold greater affinity for the synthetic RNA substrate than for natural DNA's. Moreover, the K_i of activated DNA *as an inhibitor* is 1000 times lower than its K_m *as a substrate.* Thus, a well known DNA substrate becomes a specific inhibitor for a synthetic RNA substrate that has a six-fold preference for the polymerase under the conditions of the assay.

DNA polymerase-β exhibits sigmoidal kinetics on poly(rA)•oligo(dT) for dTTP (Ono *et al.*, 1979; Yoshida *et al.*, 1979). This is manifested in upward concave curves in double-reciprocal plots, indicating positive cooperativity between dTTP molecules. Hill plots suggested 1.8 interactive sites for dTTP when poly(rA)•$(dT)_{15}$ is the substrate, 1.4 for the poly(dA)•$(dT)_{15}$ substrate and 1.0 for activated DNA. In contrast, another group has shown that a Hill plot of the hooked primer substrate, $d(A)_{57}(dT)_{13}$ has a slope of 1.0, indicating a single dTTP binding site (Chang, 1973). The discrepancy between the data for the synthetic DNA substrate may arise either from the variable reaction conditions between laboratories, or from the use of a hooked primer which could result in a different substrate structure than the traditional primer. The 1.4 interactive sites found on poly(dA)•$(dT)_{15}$ is difficult to conceptualize. The authors suggested that synthesis on this substrate may involve two different mechanisms (Yoshida *et al.*, 1979). The sigmoidal kinetics observed for dTTP by the polymerase-β has also been described for *E. coli* polymerase I on poly(rA)•oligo(dT) substrates (Cavalieri *et al.*, 1974). In the latter case, sigmoidal kinetics were also displayed when enzyme concentration was varied, a phenomenon not tested with polymerase-β. A model was proposed for *E. coli* polymerase I. Given its single dNTP binding site, this enzyme may behave cooperatively on poly(rA)•oligo(dT) substrates and act as a dimer. A similar model was proposed to explain the need for cooperation between two dTTP sites on polymerase-β (Yoshida *et al.*, 1979). Since the number of dNTP binding sites are not known for polymerase-β, it is impossible to distinguish between two molecules, each carrying a dNTP site, or one enzyme monomer carrying more than one site which affects dNTP binding. In turn, the dNTP sites could be two different substrate sites or one dNTP binding site and an allosteric site. It is also important to note that the concept of cooperativity does not require that both reactant monomers (if there are more than one) interact at the same site (presumably the primer stem in this case). One molecule can affect the cooperativity of the second at a distance.

Attempts to determine whether polymerase-β may have one or two catalytic centers on the poly(rA)•oligo(dT) substrate have thus far been unsuccessful (deRecondo *et al.*, 1978). However, it was noted that the kinetics of phosphate inhibition of the reaction catalyzed by polymerase-β on poly(rA)•oligo(dT) was different than that seen on poly(dA) templates. In addition, the incorporation observed on ribotemplates was much more sensitive to phosphate. Since phosphate probably competes for a dNTP site on the enzyme, these data argue that, mechanistically and catalytically, the enzyme has been altered. Finally, any model developed which explains the unique kinetic behavior of polymerase-β on poly(rA) templates will also have to take into account the known differences in poly(rA)•oligo(dT) conformation in solution.

POLYMERASE NTP INTERACTIONS

As with all DNA polymerases, *in vitro* DNA synthesis by polymerase-β utilizes the four deoxyribonucleoside triphosphates as substrates. However, when reactions are restricted to a single dNTP precursor, 20-50% maximal incorporation still occurs (Greene and Korn, 1970; Chang and Bollum, 1972b; de Fernandez *et al.*, 1975; Craig and Keir, 1975; Stalker, 1977). As the missing dNTP's are added back one-by-one to the assays, increased synthesis is observed. This "relaxed" dNTP requirement is seen with each of the individual dNTP's and is a unique property of polymerase-β. This was originally viewed with some consternation, and early studies focused primarily on distinguishing whether the enzyme was a true polymerase or a terminal addition enzyme. This is particularly interesting in hindsight, as we now know that the polymerase-β gene shares considerable sequence homology with the terminal deoxynucleotidyltransferase gene (Matsukage *et al.*, 1987).

The few studies performed on the relaxed dNTP phenomenon have shown that the effect is dependent upon DNA concentration and is optimal at high DNA to enzyme ratios (de Fernandez *et al.*, 1975). Activated DNA supports the relaxed dNTP effect much better than λ DNA (Greene and Korn, 1970; Chang and Bollum, 1972b). Digestion of either DNA with an exonuclease results in a shift towards more stringent dNTP utilization, indicating that the nature of the DNA substrate plays a role in determining the stringency of the dNTP requirement. These results have been interpreted to suggest that incorporation of one or a few nucleotides is occurring at many 3′OH sites and reflect a function of the distributive nature of polymerase-β (see Kornberg, 1980). However, this alone does not sufficiently explain the relaxed dNTP effect. If every other mechanistic parameter is identical, a distributive enzyme should bind as randomly (and as productively) to a given 3′OH primer as a processive enzyme does. Consequently, no difference between an enzyme exhibiting a distributive mode or a processive mode of DNA synthesis should be observed during the incorporation of the first few nucleotides. The fact that polymerase-β can still continue incorporation when other DNA polymerases can not implies that, in addition, another mechanism is operating.

It has been suggested that polymerase-β may preferentially incorporate at homopolymeric DNA sites, and that this ability is unmasked when DNA synthesis reactions are restricted to a single dNTP precursor (de Fernandez *et al.*, 1975). A very unlikely explanation would be that polymerase-β, known to avidly recognize internal 3′OH nicks (Wang and Korn, 1982), exhibits *extreme* infidelity under relaxed dNTP conditions. This model would be difficult to test. One could also postulate that the enzyme cycles rapidly at the primer site. If the appropriate dNTP is not found in a time span that is much shorter than other polymerases, polymerase-β dissociates and begins another binding cycle. Such differences in cycling time would enable polymerase-β to eventually find the appropriate nicks for incorporation, while other polymerases would carry out very limited incorporation. Whether these differences in cycling time actually exist is not known.

None of the above models adequately explain the increased stringency observed in the presence of exonuclease-digested DNA. The inability to develop models which would more thoroughly explain the relaxed dNTP effect results from our lack of knowledge concerning the mechanism of interaction of polymerase-β with its substrates. Thus far, no one has been able to determine the number of dNTP binding sites on polymerase-β. Several inhibitors appear to act competitively at such sites. These include pyrophosphate (Chang and Bollum, 1973; Wang *et al.*, 1974), pyridoxal-5-phosphate (Oguro *et al.*, 1978), and orthophosphate (Chang and Bollum, 1973; Yamaguchi *et al.*, 1980). Equilibrium dialysis studies with *E. coli* DNA polymerase I have shown that pyrophosphate does not compete at the dNTP site (Englund *et al.*, 1969). Instead, competition is with the hydroxyl ion in the pyrophosphate exchange reaction (Deutscher and Kornberg, 1969). Obviously, polymerase-β interacts with pyrophosphate in an unique manner. One is left wondering why the pyrophosphate generated from nucleotide incorporation does not bind to the dNTP site where it is generated and inhibit subsequent polymerization.

Double reciprocal plots of orthophosphate inhibition of dTTP incorporation exhibit upward-concave curves (Yamaguchi *et al.*, 1980). This suggests that high concentrations of dNTP's may relieve inhibition. It has been noted that "supersaturation" of dNTP's are required in order for polymerase-β to catalyze extensive incorporation on long stretches of single-stranded DNA (Kunkel and Loeb, 1981) and to properly gap-fill (Abbotts *et al.*, 1988). In the latter case, it was suggested that high levels of dNTP were needed to saturate a low affinity substrate site. Whether this is the same site at which orthophosphate exerts its effect remains to be tested.

It is well known that polymerase-α is relatively resistant to ddNTP's, and polymerase-γ is highly sensitive. Polymerase-β is intermediate between the two with respect to inhibition by ddNTP's. This differential sensitivity of polymerase-β to these chain terminating nucleotide analogs is commonly exploited to probe for physiological functions of the polymerase (as discussed earlier), yet little is actually known about their mode of inhibition. They can act as competitive inhibitors for dNTP incorporation by polymerase-β, supposedly at a dNTP site (Fisher *et al.*, 1979). However, it has recently been shown that both polymerase-α and polymerase-β can incorporate a dideoxy analog, suggesting that inhibition could also arise by chain termination (Reid *et al.*, 1988). These latter studies were analyzed by DNA sequencing gels using defined primers and were carried out with ddGTP in the presence of dATP, dCTP and dTTP. Since dGTP was absent in the assays, it is not clear if significant chain termination would have occurred if it were present. The differential sensitivity of polymerase-γ and polymerase-α to these analogs is still best explained by the low affinity of polymerase-α (Fisher *et al.*, 1979) and the high affinity of polymerase-γ (van der Vliet and Kwant, 1981) for ddNTP substrates. Unfortunately, the mode of action of ddNTP's in polymerase-β inhibition remains unclear. It is impossible to assess at present whether chain termination at the primer site, inhibition by interaction of the analog directly with polymerase-β, or a combination of both yields the intermediate sensitivity displayed by the enzyme. It is known, however, that suicide inactivation by irreversible binding of polymerase-β to the analog-terminated primer does not occur (Reid *et al.*, 1988).

The mode of action of another nucleotide analog (araCTP) has recently been carefully examined (Ohno *et al.*, 1988). This analog inhibits DNA synthesis. Its action as a weak competitive inhibitor for DNA polymerase-β does not fully explain its effect on this enzyme. Using a variety of different M13 DNA's as templates, araCTP incorporation was found to stall polymerase-β at sites requiring the incorporation of one, two, or three insertions of the analog. The addition of increasing amounts of dCTP to these reactions resulted in read-through at most (but not all) of these sites. However, absolute termination always occurred at a site requiring the insertion of four contiguous cytosine residues. These authors concluded that araCTP is a "pseudo chain terminator" for polymerase-β. Chain termination by this analog is dependent upon DNA sequence and the ratio of dCTP to araCTP employed in the assay. A similar finding has also been reported for the action of araATP on polymerase-β (Reid *et al.*, 1988). Although ddNTP analogs were shown not to be "pseudo chain terminators" (Ohno *et al.*, 1988), future studies with nucleotide analogs and polymerase-β should be carefully evaluated.

STIMULATORY FACTORS

Non-protein Factors

Although not requiring monovalent cations, the *in vitro* activity of DNA polymerase-β is stimulated by the presence of NaCl or KCl. This stimulation is observed on synthetic, as well as natural, DNA substrates and, in the case of the former, is higher in the presence of Mn^{2+}. Depending on the particular enzyme, 50-100 mM salt concentrations are optimal. Very high concentrations (above 300 mM NaCl) will result in decreased activity for the Novikoff hepatoma enzyme, but not below the initial activity observed in the absence of monovalent salts. In a possibly unrelated mechanism, the homogeneous hepatoma polymerase-β is thermally stabilized by the addition of 500 mM NaCl and, in combination with 0.5 mg/ml bovine serum albumin, the enzyme has survived repeated freeze-thawing after a decade of storage in liquid nitrogen.

Although salt stimulation was one of the first distinguishing characteristics ascribed to polymerase-β, very few studies have been performed to determine the underlying stimulatory mechanism. The fact that Novikoff hepatoma stimulatory proteins exert their effect best in the absence of salt (even though only one, DNase V, has actually been shown to be inhibited at salt concentrations typically employed in our factor assays) has prompted us to take a closer look at the information available on salt stimulation *in vitro*. Chang (1973) first examined this with synthetic homopolymers. Even though stimulation occurred, the rate of the reaction was not affected as dramatically as the decreased affinity of the polymerase for its substrate. It was concluded that the enzyme bound tighter to poly(dA)•oligo$(dT)_{37}$ in the absence of salt and that elongation was the parameter affected, not initiation. It is also possible that salt enhances the dissociation of the enzyme from the substrate after nucleotide insertion, a mechanism analogous to that proposed for another salt-stimulated enzyme, topoisomerase I. In another approach, we have found that the number of activated DNA binding sites does not change in the presence of salt. However, less enzyme is needed to saturate these sites, and this mechanism supersedes the ability of hepatoma factors to stimulate. Addition of NaCl will prevent the inhibition observed at high spermidine concentrations (Meyer and Rein, unpublished data) and will enhance the inhibition of pyrophosphate (Wang *et al.*, 1974), a known competitive inhibitor for a dNTP binding site. Although monovalent cations can possibly alter the conformation of nucleic acids in solution, these data above suggest that salt has altered the polymerase as well.

Spermidine, a general polymerase stimulator, appears to operate differently than salt on the hepatoma enzyme. Its major effect appears to be to reduce the number of binding sites on DNA, as well as to lower the amount of enzyme needed to saturate the sites (Meyer and Rein, unpublished data). Chiu and Sung (1972) have shown that spermidine stimulates the activity of the enzyme using a variety of substrates and stimulates better at lower enzyme concentrations, shifting the reaction from sigmoidal to linear kinetics. In a study of the mechanism of spermidine stimulation of polymerase-α (Fisher and Korn, 1980), it has been shown that spermidine exerts its effect at the primer site, presumably by preventing (or dissociating) an inefficient or unproductive binding of the polymerase. Whether such a mechanism occurs for DNA polymerase-β remains to be tested.

Protein Factors

Whole cell extracts of Novikoff hepatoma contain proteins capable of stimulating the *in vitro* DNA synthesis reaction catalyzed by DNA polymerase-β on activated DNA (Probst *et al.*, 1975). Initial studies had indicated that these proteins were very soluble, fractionating in 75-85% ammonium sulfate. However, we now know that a majority of the factors are also recoverable in the 45-75% ammonium sulfate fraction. Their presence is masked by the high levels of endogenous polymerase-β found in these preparations. Like their more soluble counterparts, they are separated into a characteristic pattern of stimulatory activity peaks on DEAE-Sephadex. Although the factors may be present in higher amounts at the lower ammonium sulfate fraction, higher levels of protein are also present, including the polymerase-β. Consequently, we have chosen to begin factor purification with the 75-85% ammonium sulfate fraction. The factors were named in order of their discovery, with the first three eluting from DEAE-Sephadex. Factor IV, which copurifies with the polymerase-β, has yet to be identified in the higher ammonium sulfate fraction.

Factor I has the lowest affinity for DEAE-Sephadex and is only retarded slightly in elution from the column in pH 8.0 buffers at low ionic strength. It is present in varying amounts, is highly unstable, and, based upon activity and elution pattern, is recovered in higher amounts from normal rat liver than from the hepatoma tissue. It is rapidly inactivated by phosphate buffers, is freeze-thaw labile and its stimulatory ability is salt-sensitive. An increase in activity occurs with storage or by treatment with NEM, indicating that the NEM-resistant Factor I is contaminated with a NEM-sensitive inhibitor. Its inherent instability upon purification has prevented recovery of activity from a variety of chromatographic steps. Consequently, little information has been obtained thus far for Factor I.

Factor II elutes as a broad stimulatory peak of activity from DEAE-Sephadex centered at 90 mM NaCl and is present in 20-fold excess in hepatoma cells compared to normal liver. On most subsequent chromatographic steps, Factor II is resolved into multiple peaks of activity. On phosphocellulose, for example, four such activities are found. Attempts to purify each of these separately eventually results in a complete loss of activity. By choosing a purification scheme which prevented such a separation, we can now purify Factor II to 90% homogeneity (Rein and Meyer, unpublished data). On denaturing polyacrylamide gels, a predominant protein of 37,000 molecular weight is present. Upon separation by two-dimensional gel electrophoresis however, four polypeptides of 37,000 molecular weight are distinguished with isoelectric points between 7 and 8. If all these species participate in polymerase-β stimulation at a minimum stoichiometry of 1:1:1:1, the native molecular weight of Factor II will exceed 100,000. Attempts to recover activity during velocity sedimentation or gel filtration have resulted in a complete loss of activity, as have further purification attempts.

As purified, Factor II contains no intrinsic polymerase, exonuclease, topoisomerase I or II, or DNA-dependent NTPase or dNTPase assayed with a variety of DNA effectors. Persistent throughout purification, however, is the presence of a Mg^{2+}-requiring endonuclease which exactly copurifies with Factor II stimulatory activity through five chromatographic steps. Upon elution from an additional CM-Sephadex column, Factor II is separated into two peaks of stimulatory activity, one of which is the endonuclease. At this stage of purification, both forms rapidly lose activity. Due to the small amounts of protein recovered, we have been unable to study these forms. Although it is tempting to speculate that each may represent a distinct isoelectric species, such analysis will have to wait until the Factor(s) II can be better stabilized.

The level of Factor II required to give optimum stimulation of polymerase-β indicates that it is needed in stoichiometric, rather than catalytic amounts. In addition, Factor II shows a definite preference for stimulation of polymerase-β on very weakly activated DNA's. Kinetic analysis has shown that the intact Factor II complex can lower the K_m of polymerase-β for activated DNA from 350 μM to 9 μM. Furthermore, the second stimulatory peak of Factor II activity recovered from the CM-Sephadex is the "K_m form" of Factor II.

The hepatoma polymerase-β displays two different divalent cation requirements, preferring Mg^{2+} over Mn^{2+} when activated DNA concentrations are above 30 μM and requiring Mn^{2+} when DNA concentrations are below this value. This "Mg^{2+}-Mn^{2+} effect" was first discovered and subsequently elaborated upon by Korn's group for human polymerase-β (Wang *et al.*, 1977; Wang and Korn, 1982) while apparently not observed for the mouse myeloma polymerase β (Tanabe *et al.*, 1982). Korn demonstrated that the difference seen in cation requirement for this enzyme was actually due to a difference in 3'OH primer concentrations. When primer concentrations are high enough (30 μM for the Novikoff hepatoma enzyme), Mg^{2+} is the preferred cation, but when primer concentrations drop (30 μM), Mn^{2+} is required. Under these latter conditions, Mn^{2+} enables the polymerase to find the primer, to productively bind to nicks and very small gaps, and to stay on the template longer, thereby increasing processivity. Magnesium ions do not have this ability. Since Mg^{2+} is the predominant intracellular metal activator and since primer sites for polymerase-β would be very low in number, we feel that proteins take the role that can be performed by Mn^{2+} *in vitro*. As indicated above, Factor II lowers the K_m of the polymerase for activated DNA to 9 μM. At low primer concentrations, the hepatoma polymerase alone *absolutely* requires Mn^{2+}. What is unique is that, in the presence of Factor II, the obligate divalent cation is switched to the physiological cation Mg^{2+}. The possibility that Factor II has replaced Mn^{2+} in the "Mg^{2+}-Mn^{2+} effect " is not unlikely. A series of similar proteins have been found for polymerases-α and -δ (Pritchard *et al.*, 1983; Tan *et al.*, 1986). Termed "primer-recognition" proteins, they have been best characterized from HeLa and CV-1 cultured cells. They were initially found interacting with polymerase-α in impure fractions (Lamothe *et al.*, 1981; Pritchard and DePamphilis, 1983). Separated from these proteins, the polymerase loses its ability to synthesize DNA on denatured substrates. Two proteins, C_1 and C_2, were eventually purified and found to lower the K_m of polymerase-α for the primer stem specifically. Synthesis on denatured DNA was apparently occurring at sparsely found, self-priming palindromic sequences which snap-back immediately following DNA denaturation. The C_1C_2 proteins do not alter polymerase pause sites, processivity or the rate of elongation. In addition, they stimulate best on substrates which contain low levels of primer, and, consequently, high levels of single-strand DNA, a potent inhibitor of polymerase-α. It has been suggested that this may be due to the ability of C_1C_2 to bind to single-strand DNA and translocate the polymerase to the primer. Perhaps the most intriguing characteristic of the C_1C_2 class of proteins is their specificity. They will not stimulate polymerase-β, nor even polymerase-α from a different tissue: C_1C_2 from HeLa cells will not stimulate polymerase-α from the monkey CV-1 cells and *vice versa*. The polymerase-δ primer recognition protein has been likened to the β-subunit of *E. coli* DNA polymerase III (Tan *et al.*, 1986) and the C_1C_2 proteins to the highly tissue specific sigma factors for RNA polymerases (Pritchard *et al.*, 1983).

Factor II, although identical in molecular weight to the polymerase-δ primer recognition protein, will not behave mechanistically in an identical manner to the C_1C_2 set of proteins. None of the primer-recognition proteins stimulate on activated DNA, as Factor II does. Similarly, polymerases-α and -δ do not display the "Mg^{2+}-Mn^{2+} effect". However, the isolation of a similar set of C_1C_2-like stimulatory proteins from *Ustilago* (Yarronton *et al.*, 1976) and *Tetrahymena* (Ganz and Pearlman, 1980) serves to illustrate the point that a class of functionally similar accessory proteins may exist for all eukaryote polymerases. Their function is to promote primer recognition through a variety of highly individualistic mechanisms.

Novikoff hepatoma Factor III is a relatively abundant protein, since we can easily obtain 1.0 mg of homogeneous protein from 300 g of cells (Koerner and Meyer, 1983). It is extremely stable, having retained stimulatory activity following a 37°C incubation for 48 hours. Its active form is a monomer and the optimum stimulation for polymerase-β occurs at very low levels of polymerase. With the use of ^{3}H-*E. coli* DNA, this factor will render a maximum of 50% of the DNA sensitive to nuclease S1, strongly implying the presence of a helix unwinding activity. This unwinding activity is sensitive to NaCl, (as is its polymerase-β stimulatory activity) and does not require, nor is it stimulated by, dNTP's or rNTP's. Its initial weak binding to the single-strand DNA-cellulose matrix seemed to rule against this factor as a classical DNA binding protein. However, as visualized by electron microscopy, its binding to DNA is highly cooperative and is prevented by 150 mM NaCl. Its interaction with DNA does not interfere with the activity of a variety of DNA-dependent ATPases, as does the strongly binding *E. coli* SSB protein. We have redesignated Factor III as SSB-48, based on its molecular weight. The salt sensitivity displayed by the binding of SSB-48 to DNA, its DNA unwinding activity and its stimulation of the polymerase suggests that these three discernable properties of this factor are interrelated.

Stoichiometric measurements indicate that 100-110 molecules of SSB-48 bind to each single-strand circular ϕX174 molecule. Although its DNA binding constant is unknown, this would represent a maximum of 45-49 nucleotides bound per SSB-48 monomer. In addition to its cooperative binding, SSB-48 condenses the ϕX174 single-strand circle 2.7-fold. Although we initially characterized SSB-48 as a single-strand DNA-binding protein, we now know, by both electron microscopy and agarose gel retardation assays, that it will bind supercoiled, double-stranded DNA molecules. This can be reversed by denaturation with detergents. Cooperativity of SSB-48 is also observed in polymerase stimulatory assays, as saturation curves of SSB-48 with activated DNA yields strongly sigmoidal kinetics.

Perhaps the most remarkable feature of SSB-48 is its effect on the extent of the DNA polymerase-β reaction. It has no effect on the K_m for activated DNA or for poly(dA)•oligo$(dT)_{15}$ DNA substrates. It increases the V_{max} of the synthetic polymer reaction two-fold, but has minimal effects on the rate of the reaction with natural DNA. In the absence of SSB-48, polymerase-β will cease incorporation after 1-2 hours. In its presence, however, the polymerase reaction is still linear at 45 hours and, if added to the assay after two hours, it reactivates DNA synthesis for at least an additional 42 hours. This stimulation will exceed 100-fold at this time and is specific for polymerase-β. Experiments longer than 45 hours have not been carried out. Approximately 20% of the template DNA is synthesized. Net synthesis might occur if reasonable levels of polymerase could be added. The very low levels required for optimum polymerase stimulation may have more to do with the presence of the 0.5 M NaCl in our polymerase sample. Stimulation by SSB-48 is very sensitive to salt, and salt may readily supersede the stimulatory reaction by this factor.

Unlike DNase V (see below), SSB-48 does not confer heat stability to DNA polymerase-β, although SSB-48 itself is not highly heat labile. No direct interaction between the polymerase and SSB-48 has been demonstrated. It does not increase the processivity of polymerase-β. In addition, the affinity of the polymerase for the 3′OH primer is unaltered. Mathematical analysis of data derived from substrate and 3′OH primer saturation kinetics have also indicated that SSB-48 does not increase the number of active polymerase molecules, suggesting that the factor is not displacing nonproductively bound enzyme. However, activated DNA's which have been additionally gapped with exonuclease III or extensively digested with DNase I, are optimum substrates for SSB-48 stimulation. By controlling the gap size with the appropriate ratio of oligo$(dT)_{15}$ to poly(dA), we have found that DNA synthesis by the hepatoma polymerase is optimal at a gap size of about five nucleotides. In the presence of SSB-48, *absolute* incorporation is optimal at a 35 nucleotide gap and is not markedly diminished at a 280 nucleotide gap. Wang and Korn (1980) have shown that the optimum gap size for KB polymerase-β is 10 nucleotides. Above that, the rate of the reaction is considerably slowed while the K_m is unaffected. It is important to reiterate that SSB-48 does not alter the K_m of the polymerase for its DNA substrate. However, it does prevent the large drop in the rate of DNA synthesis observed on gaps over 10 nucleotides. Even at that, this apparent difference in V_{max} does not fully explain the ability of the polymerase to continue the reaction for such extended periods.

We feel that SSB-48 stimulates polymerase-β by operating on the template strand. Presumably, the ability of this factor to unwind DNA plays a role in this process. Although SSB-48 may reorganize long single-strand gaps into a highly favorable conformation, processivity of the polymerase is not drastically altered, nor is there a shift in nonproductive binding of enzyme.

A DNA binding protein (HD-25) has been isolated from regenerating rat liver (Duguet and deRecondo, 1978). It also possesses helix-destabilizing activity and will stimulate homologous polymerase-β. This protein was first identified by its ability to bind to single-strand DNA-cellulose and subsequently assayed for polymerase stimulating activity. Unlike SSB-48, however, HD-25 will also stimulate polymerase-α. Moreover, HD-25 stimulates polymerase-β under high primer to template ratios. Under these conditions, nucleotide gaps are small. Consequently, HD-25 stimulates under exactly the *opposite* conditions that SSB-48 does. The HD-25 protein has subsequently been shown to be identical to chromosomal protein, HMG1 (Bonne *et al.*, 1982).

Factor IV has been purified to homogeneity. Unlike the other Novikoff stimulatory factors, it was not originally discovered in the stimulatory assay. Its presence was suspected during the final single-strand DNA-cellulose chromatographic step of Novikoff hepatoma DNA polymerase-β purification, where considerable loss of polymerase activity routinely occurs. Using the stimulatory assay, a factor was found eluting early from the column and in a position distinct from the polymerase (Mosbaugh *et al.*, 1977). We subsequently found that factor IV copurifies with DNA polymerase-β in a 1:1 stoichiometric complex. This 4.1S complex can be reconstituted from its purified components. As discussed previously, the hepatoma polymerase-β is initially extracted as a 6-8S complex and is reduced to a 4.1S form upon the first chromatographic step. We have concluded that Factor IV is one of the components of the original 6-8S complex. Further work established that Factor IV is a 12,000 molecular weight, bidirectional, double-stranded DNA exonuclease and was, therefore, renamed DNase V (Meyer and Mosbaugh, 1980; Mosbaugh and Meyer, 1980). Similar factors have been found separating from other β-polymerases on single-strand DNA-cellulose columns, and the HeLa factor has been unequivocally identified as DNase V (Mosbaugh and Linn, 1983). The specific interaction of DNase V with DNA polymerase-β should provide a 3′→5′ and 5′→3′ exonuclease activity to the polymerase, endowing the polymerase-β with *E. coli* polymerase I-like properties.

We describe DNase V as a promiscuous exonuclease. Currently, its substrate repertoire includes 3′OH, $3'PO_4$, 5′OH, $5'PO_4$, and the 3′ apyrimidinic (AP) sites and 5′ dangling dimers generated by bacteriophage T4 *denV* endonuclease action at a pyrimidine dimer (Rein, Small and Meyer, unpublished data). It will also remove a 3′ terminal mismatched nucleotide, but at a slow rate, making it improbable that this is physiologically relevant unless enhanced by other mechanisms. It will enhance the fidelity of DNA polymerase-β less than four-fold, a value considered to be minimal. However, it is the only identified protein which does enhance the highly error-prone polymerase-β.

The products produced by DNase V in a limit digest of poly(dA)•poly (dT) are 5′-dNMP's. However, 6% of the reaction products migrate with a dinucleotide marker on DEAE-cellulose. Since this technique separates according to phosphate charge, it is possible that this could be a diphosphate form of dAMP. When tested directly, DNase V cannot hydrolyze a di- or trinucleotide annealed to poly(dT) following 50 hours of digestion at 10°C [conditions under which the enzyme can hydrolyze 7.1 of the 17.8 nmoles of input poly(dAdT)]. This is perplexing, since trinucleotides are not products of the limit digests. Experiments to determine the smallest DNA fragment required for DNase V exonucleolytic digestion have yet to be performed. If dinucleotides are end products, one might expect that a tetranucleotide would be the smallest digestable piece. It is also conceivable that the mechanism of the initial products produced by DNase V are not 5′-dNMP's, but are larger oligonucleotides which are processed. At this time the origin of the dinucleotides in a limit digest of poly(dA)•poly(dT) remains unknown.

Using asymmetrically labelled synthetic DNA substrates, we have shown that DNase V is a bidirectional exonuclease, hydrolyzing DNA from the 3′ and 5′ directions at equal rates (Mosbaugh, 1979). The enzyme will also immediately reduce hydrolysis of ^{3}H-poly(dAdT) and begin degradation of ^{14}C-poly(dAdT) when the latter is added in vast excess. This is difficult to explain if DNase V possesses a highly processive mechanism of DNA degradation. This was confirmed by examination of the size of the remaining poly(dAdT) substrate following 50% hydrolysis by the enzyme under substrate excess. If the enzyme degrades large amounts of DNA before dissociating from the substrate (*i.e.* is processive), the remaining molecules would be of two distinct size classes: one very close to 50% of the initial substrate length and another very close to the original full length poly(dAdT). If, on the other hand, the enzyme degrades small amounts of DNA before dissociating and rebinding (*i.e.* the distributive mechanism of degradation), all of the remaining substrate should be uniformly one size. Two classes of products were found: the original parental length and a second size class approximately 60-80% of the full length molecules. These data, as well as data comparing the rate of 5′-terminal hydrolysis with the rate of hydrolysis of internal nucleotides, suggests that DNase V is "quasi-processive", removing about 33 nucleotides every time it binds to DNA. Since the enzyme is bidirectional, it is not clear how or in what direction the degradation occurs at each binding event.

Initially, one of the most perplexing problems with DNase V was that high molecular weight DNA's are poor substrates in comparison to poly (dAdT), even when "activated" with DNase I. We have investigated this phenomenon further using bacteriophage PM2 DNA, a double-stranded supercoiled molecule, as a substrate. In its native form, nicked once in one strand to produce a 3′OH site, or UV-irradiated and unincised, PM2 DNA is inactive as a substrate for DNase V (Rein and Meyer, unpublished data). To understand the mechanism of action of DNase V better, we have analyzed a variety of DNA substrates as inhibitors of poly(dAdT) degradation. We have found that supercoiled PM2 DNA, or relaxed, double-stranded DNA molecules have no effect on hydrolysis of poly(dAdT) by DNase V. Single-stranded DNA, blunt-ended *Hae*III restriction fragments of PM2, or PM2 DNA that has been nicked once by *Neurospora crassa* endonuclease, are extremely effective inhibitors of poly(dAdT) hydrolysis, even though they are not hydrolyzed themselves. These data explain why neither native nor activated DNA are substrates for DNase V: the former is not bound by the enzyme and the latter is probably bound unproductively, inactivating the nuclease. Digestion of the single-nicked PM2 molecule or of its *Hae*III restriction fragments with exonuclease III, results in the rapid activation of both substrates for DNase V hydrolysis. The smallest gap created by exonuclease III at a nick was 200 nucleotides, while the smallest single-strand DNA tail produced on the *Hae*III fragments was estimated to be 12-15 nucleotides. Restriction of PM2 DNA with *Hpa*II endonuclease will cleave the molecule just once, producing 3′OH ends indented by two nucleotides. This substrate was about 30% as effective for DNase V hydrolysis as was the exonuclease III-digested *Hae*III fragments. We suggest that DNase V requires a stretch of single-strand DNA to promote hydrolysis of double-strand DNA and that this "stretch" may be as little as one or two nucleotides.

DNase V shows a strong specificity for stimulation of β-polymerases. It will stimulate polymerase-α and *E. coli* polymerase I, but at only 10% of the levels seen with the β-polymerases. This small stimulation may be due to action of the nuclease on the substrate and not to a direct interaction with the other polymerases. The exonuclease will stimulate polymerase-β well on both activated calf thymus DNA and on poly(dAdT). It actually inhibits the polymerase on poly(dA)•poly(dT), even though it degrades this latter substrate almost as well as it handles poly(dAdT). Although there is no strong preference for an activated DNA substrate, it is clear the DNase V stimulates consistently better on those activated DNA substrates that the polymerase prefers. This is quite different from Factors II and V which show an almost obligate requirement for "underactivated" DNA substrates and SSB-48 which shows a strong preference for "overactivated" substrates. However, exonuclease III treatment of the optimal activated DNA does decrease stimulation by DNase V. As discussed above, activated calf thymus DNA is a poor substrate for the exonuclease activity of DNase V. Thus the paradox exists in which DNase V stimulates polymerase well on a substrate it has little affinity for and has difficulty digesting as a nuclease. It is possible that the polymerase promotes productive binding of DNase V or that stimulation of the polymerase occurs by mechanisms other than through the nuclease action of DNase V.

We have considered the possibility that DNase V stimulates DNA polymerase-β at the elongation step (Mosbaugh, 1979). Stimulation by DNase V occurs only in the presence of all four dNTP's. These interpretations are complicated by the simultaneous "relaxed dNTP" effect on polymerase-β when all nucleotides except dTTP are removed from the assays. Although DNase V does not alter the affinity of polymerase-β for activated DNA, it does increase the V_{max} five-fold. Surprisingly, it also lowers the affinity of the polymerase for all four nucleotide triphosphates: dATP, 2.2-fold; dTTP, 3.5-fold; dGTP, 4.4-fold and for dCTP, 7-fold. Double reciprocal plots show that a partial activator (DNase V) is present and two components (DNase V and polymerase-β) are competing for each dNTP. High concentrations of dNTP's do not inhibit the nuclease activity of DNase V. However, 200 μM ATP inhibits 50% and 50 μM ddTTP inhibits 40%. These data are very interesting, and suggest a nucleotide binding site on DNase V. The exonuclease could bind a dNTP and act as a nucleotide donor for the polymerase reaction, or the dNTP may participate in the reaction as a regulatory molecule, perhaps involved in an allosteric modulation of the elongation reaction. We cannot detect any dNTPase activity in homogeneous DNase V samples. We therefore doubt that dNTP's are substrates for DNase V. Studies to elucidate these roles and how they relate to the exonuclease activity of DNase V and the mechanism by which it stimulates polymerase-β are currently in progress.

Factor V elutes at 0.23 M NaCl from DEAE-Sephadex columns, and has recently been purified to homogeneity (Reed, Rein and Meyer, unpublished data). On denaturing SDS gels, a single protein band of 29,000 molecular weight is found. Preliminary results from non-denaturing polyacrylamide gels suggest a native molecular weight of approximately 100,000. This would indicate that the native form of the factor is a tetramer. Factor V has been shown to be a DNA-binding protein and has been named DBP-29 to reflect its molecular weight. The stimulatory activity is inhibited at 30-40 mM NaCl, a feature that masked its presence earlier on DEAE-Sephadex columns and which makes it the most salt-sensitive factor identified. The DBP-29 stimulates best at very low levels of polymerase. As the polymerase is maintained in 0.5 M NaCl, this could be due more to a salt inhibition than an unusual stoichiometry. In saturation experiments, a sharp optimum for DBP-29 is seen. If this is exceeded, inhibition of DNA synthesis occurs. This is unique to this stimulatory factor and remains unexplained. The DBP-29 is devoid of any exonuclease, endonuclease, topoisomerase or DNA-dependent dNTPase activities. Both the rate and extent of DNA synthesis by polymerase-β are increased for at least eight hours by DBP-29.

Like Factor II, DBP-29 clearly demonstrates a preference for "underactivated " DNA in the stimulation of polymerase-β. Unlike Factor II, however, DBP-29 retards the migration of supercoiled, double-strand linear and single-strand, circular DNA molecules in agarose gels. The binding is reversed by heating at 65°C in 1% SDS for ten minutes, with the DNA released in its original, intact state. This binding activity does not require Mg^{2+} and will occur rapidly at 4°C. In contrast to its extreme salt sensitivity in polymerase-β stimulatory assays, DNA binding to PM2 DNA occurs at 0.3 M NaCl. Since the protein elutes at 0.13 M NaCl on single-strand DNA-cellulose, it may have a much higher affinity for supercoiled DNA. It also raises the possibility that DBP-29 is a DNA binding protein that has a salt-sensitive interaction with the polymerase at the substrate. This would explain the differential effects of NaCl. DNA binding, as visualized in agarose gels, displays sigmoidal kinetics. Supercoiled PM2 DNA is abruptly shifted to a point intermediate between supercoiled DNA and linear DNA PM2 markers. Assuming a tetramer molecular weight of 100,000, 500 molecules of DBP-29 are bound to each PM2 DNA molecule. One tetramer molecule of DBP-29 is calculated to cover 35-40 nucleotides. However, addition of excess DBP-29 results in a diffuse band of slower migrating DNA species. No distinct DNA bands are discernable and migration can be significantly retarded. DNA bound by DBP-29 can still be digested with some restriction endonucleases. Thus, the interaction of DBP-29 with DNA is complicated. However, we feel that its DNA binding capability may play a significant role in its ability to stimulate DNA polymerase-β on "underprimed" DNA substrates.

Five chromatographically distinct DNA-dependent ATPases have been identified in whole cell extracts of Novikoff hepatoma (Thomas and Meyer, 1982). ATPases III and IV have been purified to homogeneity and are clearly distinct enzymes (Thomas and Meyer, 1982; Thomas *et al.*, 1988). Both ATPases II and V have been purified to near 90% homogeneity and are very similar to ATPase III (Thomas, 1984). The ATPases II, III and V will use double- and single-stranded DNA's as effectors and will hydrolyze both ATP and dATP equally well. Enzymological studies show subtle differences in their catalytic properties, and all three exist in native form as a dimer of 65,000 molecular weight subunits. Tryptic peptide maps demonstrate a marked similarity between their major peptides, although ATPase V is considerably more resistant to heat inactivation than either ATPase II or III. ATPase II can be readily phosphorylated *in vitro*. ATPase III is the only ATPase to shift sedimentation value (from 7.1S to 6.2S) in glycerol gradients in the presence of ATP or dATP suggesting that a conformational change occurs when the enzyme binds its substrate. The ability of ATPase III to stimulate hepatoma polymerase-β, however, clearly distinguishes it functionally from the other ATPases. Stimulation is specific to polymerase-β, as ATPase III will not stimulate homologous polymerase-α. In fact, inhibition of polymerase-α occurs in the presence of ATPase III. Stimulation of polymerase-β requires ATP or dATP. A potent inhibitor of ATPase III, ATPγS, is not hydrolyzed by the ATPase and abolishes stimulation of the polymerase. Other analogs, which are 10-fold less effective in inhibition of ATP hydrolysis, are, surprisingly, as efficient as ATPγS in interfering with stimulation (Meyer *et al.*, 1984). These analogs are competitive inhibitors for ATP and do not interfere with the binding of the DNA effector to ATPase III. These data suggest that ATPase activity is required for polymerase-β stimulation. Interestingly, stimulation occurs only on native DNA that is minimally nicked (a poor substrate for polymerase-β). It was expected that Factor II, the only known hepatoma factor at that time to stimulate well on such DNA substrates, would enhance the stimulation of the polymerase in the presence of ATPase III. However, this was not the case, and it was the addition of SSB-48, the factor that stimulates on overgapped DNA substrates, that provided a more than additive stimulation, to seven-fold.

No direct interaction between polymerase-β and ATPase III has been demonstrated, and neither enzyme protects the other from thermal inactivation. This ATPase contains no activity known to require ATP hydrolysis, including primase, topoisomerase, complementary DNA strand reannealing, and helicase activities. The latter studies have been exhaustively analyzed under a variety of conditions. These include: 1) altering ionic strength; 2) substitution with various cations; 3) the addition of spermidine, polymerase-β or SSB-48 to the assays; 4) under DNA synthesis conditions and 5) the inclusion of DNA substrates containing either forks , 5′ or 3′ tails.

Boxer and Korn (1980) have purified the only other DNA-dependent ATPase that is known to stimulate DNA polymerase-β. This ATPase will also stimulate only on nicked DNA substrates, but unlike Novikoff hepatoma ATPase III, it will stimulate polymerase-α. In addition, stimulation specifically requires ATP. Neither helicase nor primase activities could be demonstrated for the KB enzyme.

The ability of SSB-48 to significantly enhance stimulation by ATPase III implies that strand displacement may have occurred due to the action of the ATPase at a nick. Presumably, SSB-48 would then stabilize the template strand at the gap created by ATPase III. The relationship between ATPases II, III and V currently remain unknown. Both the KB and the hepatoma ATPases are purified in very small quantities and the inability to detect helicase activity may reflect an inability to saturate DNA substrates properly.

CONCLUSION

In this chapter we have addressed three major issues regarding DNA polymerase-β: (1) From a variety of studies a role for this polymerase in DNA repair has been established, although DNA repair is not an exclusive role for this enzyme. It is also possible that DNA polymerase-β plays a role in some aspects of replication. (2) DNA polymerase-β exists as a macromolecular complex of at least four polypeptides, and we suggest that more than one type of complex may exist. That these complexes are not artifactual is supported by the demonstration of a specific subcomplex of DNA polymerase-β and DNase V, an interaction which alters the properties of both enzymes. (3) Upon first analysis, DNA polymerase-β appears to be a simplistic enzyme, able to carry out only one reaction: polymerization. A variety of studies suggests, however, that the enzyme interacts with its dNTP and DNA substrates in a unique and potentially complex manner. *In vivo*, we suspect that it is assisted by a variety of accessory proteins which profoundly affect its activity. We have identified and characterized six such proteins from the Novikoff hepatoma that affect the enzymological properties of DNA polymerase-β. We anticipate that future studies aimed at comprehending the interaction between these accessory proteins and DNA polymerase-β will help increase our basic understanding of the mode of action of this enzyme.

REFERENCES

Abbotts, J., D. N. SenGupta, B. Zmudzka, S. G. Widen, V. Notario and S. H. Wilson (1988) Expression of human DNA polymerase β in *Escherichia coli* and characterization of the recombinant enzyme. *Biochemistry* **27**: 901-909.

Baril, E. F., M. D. Jenkins, O. E. Brown, J. Laszlo and H. P. Morris (1973) DNA polymerases I and II in regenerating rat liver and Morris hepatomas. *Cancer Res.* **33**: 1187-1193.

Barr, R. D., P. Sarin, G. Sarna and S. Perry (1976) The relationship of DNA polymerase activity to cell cycle stage. *Eur. J. Cancer* **12**: 705-709.

Berger, N. A., K. K. Kurohara, S. J. Petzold and G. W. Sikorski (1979) Aphidicolin inhibits eukaryotic DNA replication and repair - implications for involvement of DNA polymerase α in both processes. *Biochem. Biophys. Res. Commun.* **89**: 218-225.

Bertazzoni, V., M. Stefanini, G. Pedrali-Noy, E. Giulotto, F. Nuzzo, A. Falaschi and S. Spadari (1976) Variations of DNA polymerase-α and -β during prolonged stimulation of human lymphocytes. *Proc. Natl. Acad. Sci. USA* **73**: 785-789.

Bonne, C., P. Sautiere, M. Duguet and A-M. deRecondo (1982) Identification of a single-stranded DNA binding protein from rat liver with High Mobility Group Protein 1. *J. Biol. Chem.* **257**: 2722-2725.

Boxer, L. M. and D. Korn (1980) Structural and enzymological characterization of a deoxyribonucleic acid dependent adenosine triphosphate from KB cell nuclei. *Biochemistry* **19**: 2623-2633.

Burk, J. F., P. M. Duff and C. K. Pearson (1979) Effect of drugs on deoxyribonucleic acid synthesis in isolated mammalian cell nuclei. Comparison with partially purified deoxyribonucleic acid polymerases. *Biochem. J.* **178**: 621-626.

Butt, T. R., W. M. Wood, E. L. McKay and R. L. P. Adams (1978) Involvement of deoxyribonucleic acid polymerase β in nuclear deoxyribonucleic acid synthesis. *Biochem. J.* **173** :309-314.

Cavalieri, L. F., M. J. Modak and S. L. Marcus (1974) Evidence for allosterism in *in vitro* DNA synthesis on RNA templates. *Proc. Natl. Acad. Sci. USA* **71**: 858-862.

Chan, J. Y. H. and F. F. Becker (1981) Differential inhibition of rat liver DNA polymerases *in vitro* by direct-acting carcinogens and the protective effect of a thiol reducing agent. *Biochem. J.* **193**: 985-990.

Chang, L. M. S. (1973) Low molecular weight deoxyribonucleic acid polymerase from calf thymus chromatin. I. Initiation and fidelity of homopolymer replication. *J. Biol. Chem.* **248**: 6983-6992.

Chang. L. M. S. and F.J. Bollum (1972a) Variation of deoxyribonucleic acid polymerase activities during rat liver regeneration. *J. Biol. Chem.* **247**: 7948-7950.

Chang. L. M. S. and F.J. Bollum (1972b) Low molecular weight deoxyribonucleic acid polymerase from rabbit bone marrow. *Biochemistry* **11**: 1264-1272.

Chang. L. M. S. and F.J. Bollum (1973) A comparison of associated enzyme activities in various deoxyribonucleic acid polymerases. *J. Biol. Chem.* **248**: 3398-3304.

Chang, L. M. S., M. Brown and F.J. Bollum (1973) Induction of DNA polymerase in mouse L cells. *J. Mol. Biol.* **74**: 1-8.

Chiu, J-F. and S. C. Sung (1972) Effect of spermidine on the activity of DNA polymerases. *Biochem. Biophys. Acta* **281**: 535-542.

Chiu, J-F. and S. C. Sung (1973) Separation and characterization of protein factor of DNA polymerase β from developing rat brain. *Biochem. Biophys. Acta* **331**: 54-60.

Chiu, R. W. and E. F. Baril (1975) Nuclear DNA polymerases and the HeLa cell cycle. *J. Biol. Chem.* **250**: 7951-7957.

Claycomb, W. C. (1975) Biochemical aspects of cardiac muscle differentiation. Deoxyribonucleic acid synthesis and nuclear and cytoplasmic deoxyribonucleic acid polymerase activity. *J. Biol. Chem.* **250**: 3229-3235.

Coetzee, M. L., G. P. Sartiano, K. Klein and P. Ove (1977) The effect of several antitumor agents on ^{3}HTTP incorporation in host liver and hepatoma nuclei. *Oncology* **34**: 68-73.

Craddock, V. M. (1981) DNA polymerases in replication and repair of DNA during carcinogenesis induced by feeding N-acetylaminofluorene. *Carcinogenesis* **2**: 61-65.

Craddock, V. M. and C. M. Ansley (1979) Sequential changes in DNA polymerases α and β during diethylnitrosamine-induced carcinogenesis. *Biochem. Biophys. Acta* **564**: 15-22.

Craig, R. K., P. A. Costello and H. M. Keir (1975) Deoxyribonucleic acid polymerases of BHK-21/C13 cells. Relationship to the physiological state of the cells, and to synchronous induction of synthesis of deoxyribonucleic acid. *Biochem. J.* **145**: 233-240.

Craig, R. K. and H. M. Keir (1975) Deoxyribonucleic acid polymerases of BHK-21/C13 cells. Partial purification and characterization of the enzymes. *Biochem. J.* **145**: 215-224.

Das, S. K. and R. K. Fujimara (1979) Processiveness of DNA polymerases. A comparative study using a simple procedure. *J. Biol. Chem.* **254**: 1227-1232.

Davis, P. B., J. Laszlo and E. F. Baril (1976) Induction of DNA polymerase α during liver regeneration in rats on controlled feeding schedules. *Cancer Res.* **36**: 432-437.

de Fernandez, M. T. F., J. Mordoh and B. R. Fridlender (1975) New properties of the DNA polymerase-β isolated from nonstimulated normal human lymphocytes. *Biochem. Biophys. Res. Commun.* **65**: 1409-1417.

deRecondo, A-M. and J. Abadiedebat (1976) Regenerating rat liver DNA polymerases: dissimilitude or relationship between nuclear and cytoplasmic enzymes? *Nucleic Acid Res.* **3**: 1823-1837.

deRecondo, A-M., J. M. Rossignol, M. Mechali and M. Duguet (1978) Properties and interactions of DNA polymerase α, DNA polymerase β and a DNA binding protein of regenerating rat liver. **In**: *DNA Synthesis Present and Future*, eds. I. Molineux and M. Kohiyama, Plenum Press, N.Y., pp. 559-585.

Deutscher, M. P. and A. Kornberg (1969) Enzymatic synthesis of deoxyribonucleic acid. XXVIII. The pyrophosphate exchange and pyrophosphoryolysis reactions of deoxyribonucleic acid polymerase. *J. Biol. Chem.* **244**: 3019-3028.

Dresler, S. L. (1984) Comparative enzymology of ultraviolet-induced DNA repair synthesis and semiconservative DNA replication in permeable diploid human fibroblasts. *J. Biol. Chem.* **259**: 13947-13952.

Dresler, S. L. and M. G. Frattini (1986) DNA replication and UV-induced DNA repair synthesis in human fibroblasts are much less sensitive than DNA polymerase α to inhibition by butylphenyl-deoxyguanosine triphosphate. *Nucleic Acids Res.* **14**: 7093-7102.

Dresler, S. L. and K. S. Kimbro (1987) 2′,3′-Dideoxythymidine 5′-triphosphate inhibition of DNA replication and ultraviolet-induced DNA repair synthesis in human cells: evidence for involvement of DNA polymerase δ. *Biochemistry* **26**: 2664-2668.

Dresler, S. L., J. D. Roberts and M. W. Lieberman (1982) Characterization of deoxyribonucleic acid repair synthesis in permeable human fibroblasts. *Biochemistry* **21**: 2557-2564.

Duguet, M. and A-M. deRecondo (1978) A deoxyribonucleic acid unwinding protein isolated from regenerating rat liver. Physical and functional properties. *J. Biol. Chem.* **253**: 1660-1666. @REFERENCES = Edenberg, H. J., S. Anderson and M. L. DePamphilis (1978) Involvement of DNA polymerase α in simian virus 40 DNA replication. *J. Biol. Chem.* **253**: 3273-3280.

Englund, P. T., M. P. Deutscher, T. M. Jovin, R. B. Kelly, N. R. Cozzarelli and A. Kornberg (1968) Structural and functional properties of *Escherichia coli* DNA polymerase. *Cold Spring Harbor Symp. Quant. Biol.* **33**: 1-9.

Englund, P. T., J. A. Huberman, T. M. Jovin and A. Kornberg (1969) Enzymatic synthesis of deoxyribonucleic acid. XXX. Binding of triphosphates to deoxyribonucleic acid polymerases. *J. Biol. Chem.* **244**: 3038-3044.

Evans, D. H., and S. Linn (1984) Excision repair of pyrimidine dimers from simian virus 40 minichromosomes *in vitro*. *J. Biol. Chem.* **259**: 10252-10259.

Fisher, P. A. and D. Korn (1979a) Enzymological characterization of KB cell DNA polymerase α. II. Specificity of the protein-nucleic acid interaction. *J. Biol. Chem.* **254**: 11033-11039.

Fisher, P. A. and D. Korn (1979b) Enzymological characterization of KB cell DNA polymerase α. III. The polymerization reaction with single-stranded DNA. *J. Biol. Chem.* **254**: 11040-11046.

Fisher, P. A. and D. Korn (1980) KB cell DNA polymerases-α: Mechanism of primer-template recognition by the core catalytic unit. **In**: *Mechanistic Studies of DNA Replication and Genetic Recombination*, ed. B. Alberts, Academic Press, N.Y., pp. 655-664.

Fisher, P. A. and D. Korn (1981a) Properties of the primer-binding site and the role of magnesium ion in primer-template recognition by KB DNA polymerase α. *Biochemistry* **20**: 4570-4578.

Fisher, P. A. and D. Korn (1981b) Ordered sequential mechanism of substrate recognition and binding by KB cell DNA polymerase α. *Biochemistry* **20**: 4560-4569.

Fisher, P. A., T. S. Wang and D. Korn (1979) Enzymological characterization of DNA polymerase α. Basic catalytic properties, processivity and gap utilization of the homogeneous enzyme from human KB cells. *J. Biol. Chem.* **254**: 6128-6137.

Fox, A. M., C. B. Breaux and R. M. Benbow (1980) Intracellular localization of DNA polymerase activities within large oocytes of the frog, *Xenopus laevis*. *Dev. Biol.* **80**: 79-95.

Fry, M. (1983) Eukaryotic DNA polymerases. **In**: *Enzymes of Nucleic Acid Synthesis and Modification*, ed. S.T. Jacob, CRC Press, Inc., Boca Raton, FL, pp. 39-92.

Fry, M. and H. Loeb (1986) *Animal Cell DNA Polymerases*, CRC Press, Inc., Boca Raton, Fla.

Ganz, P. R. and R. E. Pearlman (1980) Purification from *Tetrahymena thermophila* of DNA polymerase and a protein which modifies its activity. *Eur. J. Biochem.* **113**: 159-173.

Goffin, C. and W. G. Verly (1982) Excision of apurinic sites from DNA with enzymes isolated from rat-liver chromatin. *Eur. J. Biochem.* **127**: 619-623.

Greene, R. and D. Korn (1970) Partial purification and characterization of deoxyribonucleic acid polymerase from KB cells. *J. Biol. Chem.* **245**: 254-261.

Grippo, P., R. Geremia, G. Locorotondo and V. Monesi (1978) DNA-dependent DNA polymerase species in male germ cells of the mouse. *Cell Differ.* **7**: 237-248.

Gupta, P. K. and M. A. Sirover (1980) Sequential stimulation of DNA repair and DNA replication in normal human cells. *Mutat. Res.* **72**: 273-284.

Gupta, P. K. and M. A. Sirover (1984) Altered temporal expression of DNA repair in hypermutable Bloom's syndrome cells. *Proc. Natl. Acad. Sci. USA* **81**: 757-761.

Hardt, N., G. Pedrali-Noy, F. Focher and S. Spadari (1981) Aphidicolin does not inhibit DNA repair synthesis in ultraviolet-irradiated HeLa cells. A radioautographic study. *Biochem. J.* **199**: 453-455.

Hecht, N. B. (1973a) Interconvertibility of mouse DNA polymerase activities derived from the nucleus and cytoplasm. *Biochem. Biophys. Acta* **312**: 471-483.

Hecht, N. B. (1973b) Enzymatically active intermediate in the conversion between the low and high molecular weight DNA polymerases. *Nature New Biology* **245**: 199-201.

Hecht, N. B. (1975) The relationship between two murine DNA-dependent DNA polymerases from the cytosol and the low molecular weight DNA polymerase. *Biochim. Biophys. Acta* **383**: 388-398.

Hecht, N. B. and D. Davidson (1973) The presence of a common active subunit in low and high molecular weight murine DNA polymerases. *Biochem. Biophys. Res. Commun.* **51**: 299-305.

Hobart, P. M., and A. A. Infante (1980) Persistent cytoplasmic location of a DNA polymerase-β in sea urchins during development. *Biochim. Biophys. Acta* **607**: 256-268.

Huang, Y., C. Chang and J. E. Trosko (1983) Aphidicolin-induced endoreduplication in Chinese hamster cells. *Cancer Res.* **43**: 1361-1364.

Hübscher, U., C. C. Kuenzle, W. Limacher, P. Scherrer and S. Spadari (1978) Functions of DNA polymerases α, β and γ in neurons during development. *Cold Spring Harbor Symp. Quant. Biol.* **43**: 625-629.

Hübscher, U., C. C. Kuenzle and S. Spadari (1979) Functional roles of DNA polymerases-β and -γ. *Proc. Natl. Acad. Sci. USA* **76**: 2316-2320.

Joe, C. O., V. L. Sylvia, J. O. Norman and D. L. Busbee (1987) Repair of benzo[a]pyrene-initiated DNA damage in human cells requires activation of DNA polymerase-α. *Mutat. Res.* **184**: 129-137.

Koerner, T. J. and R. R. Meyer (1983) A novel single-stranded DNA binding protein from the Novikoff hepatoma which stimulates DNA polymerase β. Purification and general characterization. *J. Biol. Chem.* **258**: 3126-3133.

Korn, D., P. A. Fisher and T. S-F. Wang (1981) Mechanisms of catalysis of human DNA polymerases α and β. **In**: *Prog. Nucleic Acid Res. Mol. Biol.* **26**: ed. W.E. Cohn, Academic Press, N.Y., pp. 63-81.

Kornberg, A. (1980) *DNA Replication*, W.H. Freeman and Co., San Francisco, pp. 121-125.

Krauss, S. W. and S. Linn (1980) Fidelity of fractionated deoxyribonucleic acid polymerases from human placenta. *Biochemistry* **19**: 220-228.

Kunkel, T. A. and L. A. Loeb (1981) Fidelity of mammalian DNA polymerases. *Science* **213**: 765-767.

Lamothe, P., B. Baril, A. Chi, L. Lee and E. Baril (1981) Accessory proteins for DNA polymerase-α activity with single-strand DNA template. *Proc. Natl. Acad. Sci. USA* **78**: 4723-4727.

Lazarus, L. H. and N. Kitron (1973) Cytoplasmic DNA polymerase: polymeric forms and their conversion into an active monomer resembling nuclear DNA polymerase. *J. Mol. Biol.* **81**: 529-534.

Lynch, W. E., and I. Lieberman (1973) A DNA polymerase in liver nuclei whose activity rises with DNA synthesis after partial hepatectomy. *Biochem. Biophys. Res. Commun.* **52**: 843-849.

Lynch, W. E., J. Short, and I. Lieberman (1976) The 7.1S nuclear DNA replication in intact liver. *Cancer Res.* **36**: 901-904.

Matsukage, A., K. Nishikawa, T. Ooi, Y. Seto and M. Yamaguchi (1987) Homology between mammalian DNA polymerase-β and terminal deoxynucleotidyltranferase. *J. Biol. Chem.* **262**: 8960-8962.

Matsukage, A., S. Yamamoto, M. Yamaguchi, M. Kusakabe, and T. Takahashi (1983) Immunocytochemical localization of chick DNA polymerases α and β. *J. Cell. Physiol.* **117**: 266-271.

Meyer, R. R. and D. W. Mosbaugh (1980) Characterization of mammalian DNase V, a double-strand bidirectional exonuclease which interacts with DNA polymerase-β of the Novikoff hepatoma. **In**: *Mechanistic Studies of DNA Replication and Genetic Recombination*, ed. B. Alberts, Academic Press, N.Y., pp. 639-648.

Meyer, R. R. and M. V. Simpson (1968) DNA biosynthesis in mitochondria: partial purification of a distinct DNA polymerase from isolated rat liver mitochondria. *Proc. Natl. Acad. Sci. USA*, **61**: 130-137.

Meyer, R. R., D. C. Thomas, T. J. Koerner and D. C. Rein (1984) Interaction of DNA accessory proteins with DNA polymerase-β of the Novikoff hepatoma. **In**: *Proteins Involved in DNA Replication*, eds., U. Hübscher and S. Spadari, Plenum Publ. N.Y. pp. 355-362.

Miller, M. R. and D. N. Chinault (1982a) Evidence that DNA polymerases α and β participate differently in DNA repair synthesis induced by different agents. *J. Biol. Chem.* **257**: 46-49.

Miller, M. R. and D. N. Chinault (1982b) The roles of DNA polymerases α, β, γ in DNA repair synthesis induced in hamster and human cells by different DNA damaging agents. *J. Biol. Chem.* **257**: 10204-10209.

Morioka, K. and H. Terayama (1974) A conversion factor for cytoplasmic DNA polymerase of rat liver. *Biochem. Biophys. Res. Commun.* **61**: 568-575.

Morita, T., Y. Tsutsui, Y. Nishiyama, H. Nakamura and S. Yoshida (1982) Effects of DNA polymerase inhibitors on replicative and repair DNA synthesis in ultraviolet-irradiated HeLa cells. *Int. J. Radiat. Biol.* **42**: 471-480.

Mosbaugh, D. W. (1979) *Purification and properties of stimulatory factor IV, an exonuclease which interacts with the Novikoff hepatoma DNA polymerase-*β. Ph.D. Dissertation, Univ. of Cincinnati.

Mosbaugh, D. W. and S. Linn (1983) Excision repair and DNA synthesis with a combination of HeLa DNA polymerase-β and DNase V. *J. Biol. Chem.* **258**: 108-118.

Mosbaugh, D. W. and R. R. Meyer (1980) Interaction of mammalian deoxyribonuclease V, a double strand 3′→5′ and 5′→3′ exonuclease, with deoxyribonucleic acid polymerase-β from the Novikoff hepatoma. *J. Biol. Chem.* **255**: 10239-10247.

Mosbaugh, D. W., D. M. Stalker, G. S. Probst and R. R. Meyer (1977) Novikoff hepatoma deoxyribonucleic acid polymerase. Identification of a stimulatory protein bound to the β-polymerase. *Biochemistry* **16**: 1512-1518.

Nagasaka, A. and S. Yoshida (1982) Regulation of DNA polymerase β in rat adrenal gland by adrenocorticotropic hormone. *Endocrinology* **111**: 1345-1349.

Naryzhnyi, S. N., I. V. Shevelev and V. M. Krutiakov (1982) Effect of the antitumor antibiotic bleomycin on DNA polymerase activity. *Biokhimiia* **47**: 1212-1215.

Nishiyama, Y., T. Tsurumi, H. Aoki and K. Maeno (1983) Identification of DNA polymerase(s) involved in the repair of viral and cellular DNA in herpes simplex virus type 2-infected cells. *Virology* **129**: 524-528.

Oguro, M., H. Nagano and Y. Mano (1979) The mode of inhibitory action by pyridoxal-5-phosphate on polymerase-α and -β. *Nucleic Acids Res.* **7**: 727-734.

Ohno, Y., D. Spriggs, A. Matsukage, T. Ohno and D. Kufe (1988) Effects of 1-β-D-arabinofuranoslcytosine incorporation on elongation of specific DNA sequences by DNA polymerase β. *Cancer Res.* **48**: 1494-1498.

Ono, K., A. Ohashi, K. Tanabe, A. Matsukage, M. Nishizawa and T. Takahashi (1979) Unique requirements for template primers of DNA polymerase-β from rat ascites hepatoma AH130 cells. *Nucleic Acids Res.* **7**: 715-726.

Patel, P., C. A. Myers and M. R. Miller (1983) Identification of mammalian DNA repair factors using a reconstituted subcellular system. Partial characterization and subcellular location of a DNA repair-stimulating protein in hamster cells. *Exp. Cell. Res.* **149**: 347-358.

Pedrali-Noy, G. and S. Spadari (1980) Aphidicolin allows a rapid and simple evaluation of DNA repair synthesis in damaged human cells. *Mutat. Res.* **70**: 389-394.

Pritchard, C., R. C. Bates, and E. R. Stout (1978) Levels of cellular DNA polymerases in synchronized bovine parvovirus-infected cells. *J. Virol.* **27**: 258-261.

Pritchard, C. G. and M. L. DePamphilis (1983) Preparation of DNA polymerase $\alpha \bullet C_1C_2$ by reconstituting DNA polymerase α with its specific stimulatory cofactors C_1C_2. *J. Biol. Chem.* **258**: 9801-9809.

Pritchard, C. G., D. T. Weaver, E. F. Baril and M. L. DePamphilis (1983) DNA polymerase α cofactors C_1C_2 function as primer recognition proteins. *J. Biol. Chem.* **258**: 9810-9819.

Probst, G. S., D. M. Stalker, D. W. Mosbaugh and R. R. Meyer (1975) Stimulation of DNA polymerase by factors isolated from Novikoff hepatoma. *Proc. Natl. Acad. Sci. USA* **72**: 1171-1174.

Reid, R., E-C. Mar, E-S. Huang, and M. D. Topol (1988) Insertion and extension of acyclic, dideoxy and aranucleotides by Herpesvirdae, human α and human β polymerases. A unique inhibition mechanism for 9-(1,-3-dihydroxy-2-propoxymethyl)guanine triphosphate. *J. Biol. Chem.* **263**: 3898-3904.

Roufa, D. J., R. E. Moses and S. J. Reed (1975) The DNA polymerases of chinese hamster cells. Subcellular distribution and properties of two DNA polymerases. *Arch. Biochem. Biophys.* **167**: 547-559.

Sartiano, G. P., W. Lynch, S. S. Boggs and G. L. Neil (1975) The demonstration of separate DNA polymerase activities in intact isolated rat liver nuclei by means of response to bleomycin and arabinosyl cytosine 5-triphosphate. *Proc. Soc. Exp. Biol. Med.* **150**: 718-722.

Schneider, E., B. Muller and R. Schindler (1985) Control of DNA polymerase α, β and γ activities in heat- and cold-sensitive mammalian cell-cycle mutants. *Biochim. Biophys. Acta* **825**: 375-383.

Seki, S., N. Hosogi and T. Oda (1983) Participation of DNA polymerases α and β in unscheduled DNA synthesis in mammalian cells. *Acta Med. Okayama* **37**: 213-225.

Seki, S., N. Hosogi and T. Oda (1984) Differential sensitivity to aphidicolin of replicative DNA synthesis and ultraviolet-induced unscheduled DNA synthesis *in vivo* in mammalian cells. *Acta. Med. Okayama* **38**: 227-237.

Seki, S. and T. Oda (1980) Effects of 2′, 3′-dideoxythymidine triphosphate on replicative DNA synthesis and unscheduled DNA synthesis in permeable mouse sarcoma cells. *Biochim. Biophys. Acta* **606**: 246-250.

Seki, S. and T. Oda (1986) DNA repair synthesis in bleomycin-pretreated permeable HeLa cells. *Carcinogenesis* **7**: 77-82.

Seki, S., T. Oda and M. Ohashi (1980) Differential effects of aphidicolin on replicative DNA synthesis and unscheduled DNA synthesis in permeable mouse sarcoma cells. *Biochim. Biophys. Acta* **610**: 413-420.

Shioda, M., H. Nagano and Y. Mano (1977) Cytoplasmic location of DNA polymerase-α and -β of sea urchin eggs. *Biochem. Biophys. Res. Commun.* **78**: 1362-1368.

Siedlecki, J. A., R. Nowak, A. Soltyk and B. Zmudzka (1981) Net DNA synthesis catalysed by calf thymus DNA polymerase-β. *Acta. Biochim. Pol.* **28**: 157-173.

Siedlecki, J.A., J. Szyszko, I. Pietrzykowska and B. Zmudzka (1980) Evidence implying DNA polymerase-β function in excision repair. *Nucleic Acids Res.* **8**: 361-375.

Siegel, R. L. and G. F. Kalf (1982) DNA polymerase β involvement in DNA endoreduplication in rat giant trophoblast cells. *J. Biol. Chem.* **257**: 1785-1790.

Smith, H. C. and R. Berezney (1982) Nuclear matrix-bound deoxyribonucleic acid synthesis: an *in vitro* system. *Biochemistry* **21**: 6751-6761.

Smith, H. C., E. Puvion, L. A. Buchholtz and R. Berezney (1984) Spatial distribution of DNA loop attachment and replicational sites in the nuclear matrix. *J. Cell Biol.* **99**: 1794-1802.

Snyder, R. D. and J. D. Regan (1981) Aphidicolin inhibits repair of DNA in UV irradiated human fibroblasts. *Biochem. Biophys. Res. Commun.* **99**: 1088-1094.

Soltyk, A., J. A. Siedlecki, I. Pietrzykowska and B. Zmudzka (1981) Reactions of calf thymus DNA polymerases α and β with native DNA damaged by thymine starvation or by methyl methanesulphonate treatment of *Escherichia coli* cells. *FEBS Lett.* **125**: 227-230.

Spadari, S. and A. Weissbach (1974) The interrelation between DNA synthesis and various DNA polymerase activities in synchronized HeLa cells. *J. Mol. Biol.* **86**: 11-20.

Srivastava, B. I. S. (1974) Deoxynucleotide-polymerizing enzymes in normal and malignant human cells. *Cancer Res.* **34**: 1015-1026.

Stalker, D. M. (1977) *Purification and properties of a homogeneous DNA β polymerase from the Novikoff hepatoma.* Ph.D. Dissertation, Univ. of Cincinnati.

Stalker, D. M., D. W. Mosbaugh and R. R. Meyer (1976) Novikoff hepatoma deoxyribonucleic acid polymerase. Purification and properties of a homogeneous β polymerase. *Biochemistry* **15**: 3114-3121.

Tan, C-K., C. Castillo, A. G. So and K. M. Downey (1986) An auxiliary protein for DNA polymerase-δ from fetal calf thymus. *J. Biol. Chem.* **261**: 12310-12316.

Tanabe, K., E. W. Bohn and S. H. Wilson (1979) Steady state kinetics of mouse DNA polymerase β. *Biochemistry* **18**: 3401-3406.

Thomas, D. (1984) *Isolation and characterization of DNA-dependent ATPases from the Novikoff hepatoma*. Ph.D. Dissertation, Univ. of Cincinnati.

Thomas, D. C., and R. R. Meyer (1982) Deoxyribonucleic acid dependent adenosine-triphosphatases from the Novikoff hepatoma. Characterization of a homogeneous adenosine-triphosphatase that stimulates DNA polymerase β. *Biochemistry* **21**: 5060-5068.

Thomas, D. C., D. C. Rein and R. R. Meyer (1988) Purification and enzymological characterization of DNA-dependent ATPase IV from the Novikoff hepatoma. *Nucleic Acids Res.* **16**: 6447-6464.

Tsuruo, T., K. Hirayama, M. Kawaguchi, H. Satoh and T. Ukita (1974) Low molecular weight DNA polymerase of rat ascites hepatoma cells. *Biochim. Biophys. Acta* **366**: 270-278.

van der Vliet, P. C. and M. M. Kwant (1981) Role of DNA polymerase γ in adenovirus DNA replication. Mechanism of inhibition by 2′,3′-dideoxynucleoside 5′-triphosphates. *Biochemistry* **20**: 2628-2632.

Verly, W. G. (1982) Repair of AP sites in DNA. *Biochimie* **64**: 603-605.

Wang, T. S-F., D.C. Eichler and D. Korn (1977) Effect of Mn^{2+} on the *in vitro* activity of human deoxyribonucleic acid polymerase β. *Biochemistry* **16**: 4927-4934.

Wang, T. S-F. and D. Korn (1980) Reactivity of KB cell deoxyribonucleic acid polymerases α and β with nicked and gapped deoxyribonucleic acid. *Biochemistry* **19**: 1782-1790.

Wang, T. S-F. and D. Korn (1982) Specificity of the catalytic interaction of human DNA polymerase β with nucleic acid substrates. *Biochemistry* **21**: 1597-1608.

Wang, T. S-F., W.D. Sedwick and D. Korn (1974) Nuclear deoxyribonucleic acid polymerase. Purification and properties of the homogeneous enzyme from human KB cells. *J. Biol. Chem.* **249**: 841-850.

Wang, T. S-F., W.D. Sedwick and D. Korn (1975) Nuclear deoxyribonucleic acid polymerase. Further observations on the structure and properties of the enzyme from human KB cells. *J. Biol. Chem.* **250**: 7040-7044.

Waser, J., U. Hübscher, C. C. Kuenzle and S. Spadari (1979) DNA polymerase-β from brain neurons is a repair enzyme. *Eur. J. Biochem.* **97**: 361-368.

Weissbach, A., (1979) The functional roles of mammalian DNA polymerases. *Arch. Biochem. Biophys.* **198**: 386-396.

Wilson, S. H., A. Matsukage, E. W. Bohn, Y. C. Chen and M. Sivarjan (1977) Polynucleotide recognition by DNA α-polymerase. *Nucleic Acids Res.* **4**: 3981-3996.

Wist, E. (1979) The role of DNA polymerases α, β, and γ in nuclear DNA synthesis. *Biochim. Biophys. Acta* **562**: 62-69.

Yamaguchi, M., K. Tanabe, Y. N. Taguchi, M. Nishizawa, T. Takahashi and A. Matsukage (1980) Chick embryo DNA polymerase β. Purified enzyme consists of a single M_r = 40,000 polypeptide. *J. Biol. Chem.* **255**: 9942-9948.

Yarronton, G. T., P. D. Moore and A. Sparos (1976) The influence of DNA binding protein on the substrate affinities of DNA polymerase from *Ustilago maydis*: one polymerase implicated in both DNA replication and repair. *Mol. Gen. Genet.* **145**: 215-218.

Yoshida, S., M. Yamada and S. Masaki (1979) Novel properties of DNA polymerase β with poly(rA)•oligo(dT) template-primer. *J. Biochem.* **85**: 1387-1395.

Zunino, F., R. Gambetta, A. Colombo, G. Luoni and A. Zaccara (1975) DNA polymerases of rat liver. Partial characterization and effect of various inhibitors. *Eur. J. Biochem.* **60**: 495-504.

6

THE YEAST DNA REPLICATION APPARATUS: GENETIC AND BIOCHEMICAL ANALYSIS

K. C. Sitney and Judith L. Campbell

INTRODUCTION

The yeast *Saccharomyces cerevisiae* is an attractive organism for studying eukaryotic DNA replication. First, the ease with which both biochemical and classical and modern genetic studies can be carried out in yeast is currently unparalleled in any eukaryote. Second, the organization of replication units within yeast chromosomes and their activation are typical of all eukaryotes. Third, yeast is the only eukaryote to date in which it has been possible to unequivocally identify and isolate chromosomal origins of replication.

Countering these advantages, however, has been the difficulty of devising appropriate *in vitro* replication systems that allow efficient initiation of DNA replication at chromosomal origins of replication. This initially slowed the pace of identifying the yeast analogs of proteins known to be essential for replication in other systems. The challenge of identifying *bona fide* replication proteins without an *in vitro* system has been met by combining genetics and biochemistry in novel ways, such as by the use of "reverse genetics". In yeast, the latter term implies that a purified protein is used to generate appropriate reagents for cloning the gene encoding the protein. *In vitro* mutagenesis of the gene followed by homologous gene replacement and *in vivo* analysis of the phenotype of mutants reveals the biological function of the protein. Furthermore, such methods provide a powerful tool for systematically correlating the effects of changes in structure on catalytic function *in vitro* and biological function *in vivo*.

In this review, several classes of replication protein to which reverse genetics has been successfully applied will be discussed. Knowledge gained from mutants derived by both reverse and traditional genetics will be summarized. Where enough is known, comparisons will be made to metazoan analogs of the yeast proteins. One excellent recent review covers the DNA polymerases (Burgers, 1989) in more detail and another (Newlon, 1989) is a comprehensive review of yeast DNA replication.

DNA POLYMERASES

yDNA polymerase α

Eukaryotic cells contain a variety of DNA polymerases called α, β, γ, δ_1 and δ_2. There is still lively discussion of how these enzymes are related to each other and what roles each of these enzymes plays in replication, repair and recombination. A major goal of reverse genetics in yeast has been to identify the yeast analogs of eukaryotic DNA polymerases biochemically and to genetically define the division of labor among the various cellular DNA polymerases.

DNA polymerase I was first described by Wintersberger and Wintersberger (1970). Recent isolation of monoclonal antibodies to polymerase I has allowed the purification of a holoenzyme by immunoaffinity chromatography (Plevani *et al.*, 1985). This yeast polymerase I holoenzyme contains the conserved DNA polymerase α subunit structure: a catalytic polymerase polypeptide of 180 kDa, a 70-75 kDa subunit of unknown function, and two subunits of 58 and 48 kDa, that together constitute a primase activity (Plevani *et al.*, 1985). There is no tightly associated $3' \rightarrow 5'$ exonuclease.

The gene encoding the catalytic subunit of yeast DNA polymerase I was isolated by screening a λgt11 genomic DNA library using polyclonal antibodies to DNA polymerase I (Johnson *et al.*, 1985; Lucchini *et al.*, 1985). λgt11 is a phage expression vector that can synthesize β-galactosidase fusion proteins from the inserted DNA (Young and Davis, 1983). In a protease-deficient *E. coli* strain (Lon^-), these fusion proteins are generally stable enough to be detected with antibody probes. The original library has been used to clone numerous yeast genes besides replication proteins (Young and Davis, 1983; Young and Davis, 1984; Snyder and Davis, 1985). The most important predictor of success in the use of the λgt11 library is the availability of high titer and specific antibodies. Both polyclonal and monoclonal antibody can be used effectively.

Verification of clones obtained by λgt11 can be carried out by demonstrating that the RNA is of the expected size using Northern analysis, by expression and functional analysis in *E. coli*, by overproduction of a particular activity in yeast and by various immunological tests (Johnson *et al.*, 1985). Mapping the coding sequence is facilitated by two different methods of transposon mutagenesis — one using Tn10 which can be used on the λgt11 clones directly and one using a hybrid Tn3 transposon after subcloning into a plasmid vector.

The first mutation usually made in an isolated gene is called a gene disruption. There are many ways to do this now (Rothstein, 1983; Seifert *et al.*, 1986; Snyder *et al.*, 1986). A large deletion or insertion is made in the cloned gene and the altered gene is introduced into the chromosome by transformation and homologous gene replacement. Such mutations should yield an altered phenotype provided that the gene exists as a single copy. If the gene is likely to be essential, the disruption is carried out in a diploid strain and after sporulation, the observation of two viable and two inviable haploid spores confirms that the gene is required.

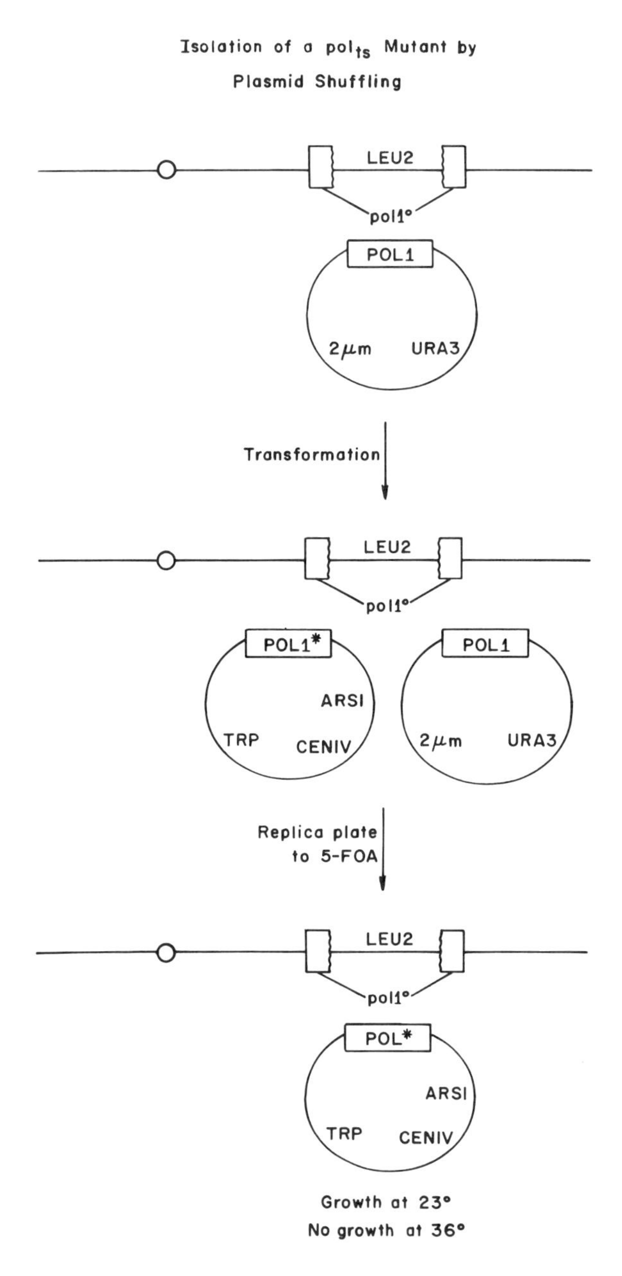

Figure 1. A plasmid shuffling strategy for isolation of mutants in cloned genes. Details are given in the text.

Gene disruption showed that *POL1* was a single copy, essential gene in yeast. No other DNA polymerase in the cell could compensate for loss of polymerase I (Johnson *et al.*, 1985).

In order to investigate the basis for the lethality of *pol1* deletions and to further characterize the *in vivo* roles of DNA polymerase I, reverse genetics was carried one step further to create conditional lethal mutations. This was accomplished by a plasmid exchange strategy shown in Figure 1 (Budd and Campbell, 1987). The original experiments designed in yeast to create conditional lethal mutants from an isolated gene used an integrative replacement technique involving integration of a mutagenized fragment of the desired gene into the chromosome (Shortle *et al.*, 1982). While this worked, there were some disadvantages. This approach required the presence of a unique restriction site at the 3′ or 5′ end of the gene and thus a physical map of the gene. Since a physical map is easy to obtain this was a minor drawback. A second problem that arose in applying this strategy was the appearance of unlinked mutations at high frequency. The reasons for this are unclear but may stem from the fact that large amounts of carrier DNA had to be used in the transformations. Finally, the mutagenized gene was obtained in the chromosome and had to be recloned before it could be studied further.

The plasmid shuffling strategy overcomes some of these problems. A plasmid containing the *POL1* gene and the selectable marker *TRP1* was mutagenized with hydroxylamine. The mutagenized plasmid was introduced into a haploid strain carrying a chromosomal disruption of the *pol1* gene, *pol1°:LEU2*, with the necessary wild-type copy of the *POL1* gene supplied on a plasmid. Since *pol1* mutations are recessive, the plasmid containing the intact copy of the gene had to be evicted to reveal the ts phenotype of the mutagenized plasmids. This was facilitated by including *URA3* as the selectable marker on the *POL1*$^+$ plasmid. *URA*$^+$ strains do not grow on medium containing 5-fluoroorotic acid (Boeke *et al.*, 1984). Thus, after selection on tryptophan-less medium for transformants containing the mutagenized plasmid, the non-mutagenized plasmid was eliminated by counterselection with 5-FOA. Resulting transformants were screened by replica plating for their inability to grow at 36°C, which indicated a ts mutation in the *POL1* gene on the plasmid. This protocol resulted in isolation of ts mutants at a frequency of about 1% of transformants, a much higher frequency than by the integrative technique. Unlinked mutations were eliminated using an intermediate screen for the inability to grow at 37°C with both plasmids present, circumventing the problem encountered in the integrative method. Plasmid shuffling is described in detail in Boeke *et al.* (1987). Another variation of plasmid shuffling has been used for yeast RNA polymerase II (Wittekind *et al.*, 1988) and may be useful for other genes as well.

The phenotypes of the various *pol1* mutants obtained by this method provide a good example of why it is important to be able to isolate many different mutant alleles of a gene. The ts mutants were screened for tight alleles. Only one of the twelve alleles, *pol1-17*, showed no residual chromosomal synthesis at the non-permissive temperature. All the rest showed considerable residual synthesis, making them inappropriate for studying the role of polymerase I in other aspects of DNA metabolism, such as repair, which requires much less extensive DNA synthesis for complete biological function. *pol1-17* mutants cease DNA synthesis as soon as they are shifted to the non-permissive temperature and thus fail to complete ongoing rounds of replication. The "quick-stop" phenotype of *pol1-17* mutants suggests a defect in the elongation stage of DNA replication (Budd *et al.*, 1989a).

pol1-17 mutants have been examined for defects in other aspects of DNA metabolism. One question was whether *POL1* was required for meiosis as well as mitosis. Some yeast genes, for instance *CDC7*, are required for mitotic replication but not for premeiotic DNA synthesis (Schild and Byers, 1978). Homozygous *pol1-17* diploids were unable to sporulate at the non-permissive temperature. Premeiotic DNA synthesis did not occur and commitment to recombination was reduced to 4% of that of wild-type (Budd *et al.*, 1989a). Thus, none of the other DNA polymerases can compensate for loss of polymerase I in meiosis.

Several studies in higher cells show that DNA repair is aphidicolin sensitive, indicating a role for polymerase α and/or δ. The resistance of DNA repair synthesis to 2′, 3′ dideoxythymidine 5′ triphosphate (ddTTP) (Dresler and Kimbro, 1987), and the ability of purified DNA polymerase δ to complement the repair synthesis deficiency exhibited by permeabilized HeLa cells (Nishida *et al.*, 1988) suggest that DNA polymerase δ is important for repair, and that DNA polymerase α may not even be involved in repair. We examined the ability of *pol1-17* mutants to repair X-ray induced single and double strand breaks at the non-permissive temperature. Both the rate and extent of repair of this type of damage was normal (Budd *et al.*, 1989a). Thus, either other polymerases can compensate for the absence of polymerase I, or polymerase I is not involved in this type of repair.

Recently, the *CDC17* gene has been shown to be the same as *POL1* by restriction and genetic mapping (M. J. Carson, Ph.D. Thesis, University of Washington, Seattle, 1987). *cdc17* mutants carry out extensive residual DNA synthesis at the non-permissive temperature, but have two interesting phenotypes. They are hyper-rec, and the telemeres are elongated (M. J. Carson, 1987; Carson and Hartwell, 1985). Hyper-rec phenotypes are also found in *Escherichia coli* DNA replication mutants and are thought to occur because DNA contains recombinogenic damage (Konrad, 1977). The fact that the *cdc17* mutant has been available for several years, but that its important role in DNA replication was only deduced quite recently, underscores the utility of the reverse genetics approach to assigning functions to particular genes.

Reverse Genetics and Subunits of yDNA Polymerase α

Although the subunit structure of polymerase α from all eukaryotes is conserved, problems of proteolysis during purification have raised the legitimate question of whether the lower molecular weight polypeptides are structurally related to the larger ones. Plevani *et al.* (1985) were able to separate the primase from the polymerase activity by immunoaffinity chromatography and to show that the active primase fraction contained the 48 and 58 kDa subunits, suggesting that one or both constitute the primase. This did not eliminate the possibility that the 58 or 48 kDa species was a breakdown product of the 70 kDa protein for which no catalytic function is known. Lucchini *et al.* (1987) isolated antibody to each of the 180, 70, 58 and 48 kDa subunits and showed no antibody prepared against a specific subunit cross reacted with any other subunit. Since proteolysis can result in deletion of important epitopes, however, it was still important to show that all polypeptides were encoded by different genes. This was accomplished by using reverse genetics.

The 70 kDa subunit has now been cloned and sequenced. The intact protein migrates with an apparent molecular weight of 85kDa on SDS gels but is easily proteolyzed to the 70kDa species previously reported in polymerase I preparations. The DNA sequence predicts a 78.7 kDa protein. Hybridization to an OFAGE blot indicates that the gene is located on chromosome II, and thus is not allelic to POLI (chromosome XIV) or any other known replication gene. The gene is essential for growth. (D. Hinkle, personal communication)

Lucchini *et al.* (1987) cloned and sequenced the 48 kDa subunit of the primase. Gene disruptions show that the gene, *PRI*1, is essential and maps to chromosome IX. Foiani *et al.* (1989) have recently cloned the 50 kDa subunit and it is also essential. Thus, all the subunits of yDNA polymerase α are genetically and structurally independent.

Structure Function Studies of yDNA Polymerase α

DNA sequence analysis of the *Eco*RI-*Bam*HI fragment of *POL1* (Johnson *et al.*, 1985; Wong *et al.*, 1988) and of the whole gene including 5′ flanking regions (Lucchini *et al.*, 1988) has been very informative. The predicted protein contains 1496 amino acids. Comparison of the primary sequence with that derived from cDNA clones of human DNA polymerase α, 4 viral and 2 prokaryotic polymerases and two other yeast DNA polymerases (*CDC2* and *REV3*, see below) reveals six regions of extensive conservation in a 400-500 amino acid domain in the C-terminal portion of the protein [amino acids 609-1085 in human DNA polymerase α; amino acids 610-1079 in DNA polymerase I (Wong *et al.*, 1988; Pizzagali *et al.*, 1988)]. While nine of the polymerases that share homology are replicative polymerases, one of them, yeast REV3, is not essential for growth and therefore is not likely to be involved in replication (A. Morrison, C. Lawrence, personal communication). Interestingly, not all DNA polymerases share these homologies. DNA polymerase β (which is homologous to terminal transferase), reverse transcriptase, and *E. coli* DNA polymerase I are all very different. Even the replicative polymerase of *E. coli*, DNA polymerase III, lacks the conserved regions, except for some similarity in one conserved subregion.

Within the region of conservation there are six stretches of extensive homology, designated regions I-VI, with I being the most homologous and VI being the least. These regions have an invariant order and similar, though not identical, spacing in all of the polymerases. The most highly conserved amino acids occur in regions first identified in viral and in prokaryotic polymerases, such as phage T4, ϕ29, and PRD1 polymerases (for refs. see Lardy, *et al.*, 1987; Jung *et al.*, 1987). They span only 150 amino acids. Based on the fact that mutations of *Herpes simplex* virus falling in regions II, III and V confer altered sensitivity to antiviral drugs, to the pyrophosphate analog phosphonoacetic acid and to aphidicolin (see Wong *et al.*, 1988 for refs.) and the fact that *Drosophila* DNA polymerase α is labeled at a lysine in the active site by pyridoxal phosphate (Diffley, 1988), regions II, III and V are proposed to be nucleotide and PPi binding domains. There is no homology to the GXXXXGKS motif conserved in other nucleotide binding proteins such as helicases, ATPases, and kinases (Möoller and Amons, 1985). No mutations have as yet been found in region I, the most highly conserved domain (90% overall). This region may be essential for catalysis (Burgers, 1989).

Based on this conservation and the close spacing between regions I and III, Wong *et al.* have proposed that the 500 nucleotides spanning domains I-VI fold into the active catalytic site of the polymerase. In this regard it is interesting that all of the ts mutants isolated by plasmid shuffling map in regions II and III (Budd *et al.*, 1989a), further suggesting that this region may be critical for protein folding or stability. Although all of the ts mutations fall in the same region as the drug resistant mutations of *Herpes*, none of the ts mutations confers on yeast resistance to aphidicolin, acyclovir or arabinosyl thymidine (D. Ma and J.L. Campbell, unpublished results). This is as might be expected, since DNA polymerase III is also essential for growth and is also sensitive to these drugs (see below). The ts mutant polymerases have not been tested for drug resistance *in vitro*, because the mutant proteins are not stable to lysis. Isolation of yeast mutants resistant to substrate analogs will require either site-directed mutagenesis or use of a differential inhibitor of polymerases I and III. Polymerase I is 100 times more sensitive to butylphenyl triphosphates than is polymerase III, thus, mutations conferring BuPdGTP resistance might be expected to map within the polymerase I gene (Khan *et al.*, 1984; Burgers and Bauer, 1988; Bauer *et al.*, 1988; Sitney *et al.*, 1989).

Since the non-conserved regions of the protein are concentrated in the N-terminal region and since the C-terminal region of polymerase I seems to be dedicated to substrate binding and catalysis, the N-terminal portion is a possible region for specific interactions with other proteins of the replication apparatus. Preliminary data suggest that an arginine to glycine change at amino acid 493 weakens the interaction of the core catalytic subunit and the primase (Lucchini *et al.*, 1988), although it is difficult to distinguish a specific effect from general effects due to denaturation of the protein.

Deletion analysis of the protein has also been informative. A 219 amino acid deletion of the C-terminus renders the polymerase unable to function *in vivo* (Johnson *et al.*, 1985). Fusion of a DNA fragment, which lacks 339 nucleotides of coding sequence to the yeast *GAL10* promoter results in a protein with a deletion of 113 amino acids from the N-terminus. This polymerase, translated from an internal, in-frame ATG, still complements a *pol1-17* mutant and retains all of the *in vitro* properties of the intact polymerase (C. Gordon and J. L. Campbell, unpublished results). All of the above results correlate with studies that have shown that a 70 kDa protein derived from DNA polymerase I by proteolysis is catalytically active (Chang, 1977). From these studies a picture of DNA polymerase α as an array of functional domains organized in linear fashion from the N- to C-terminus is emerging.

yDNA Polymerase δ — DNA Polymerase III

DNA polymerase III was discovered by Bauer *et al.* (1988) and judged to be a distinct DNA polymerase on immunological grounds. Polymerase III was not precipitated by monoclonal antibodies to polymerase I and polyclonal antibodies to polymerase III did not immunoprecipitate either DNA polymerases I or II. The enzyme was of immediate interest because it so closely resembled the DNA polymerase δ enzyme described by Downey *et al.* (1988). Highly purified DNA polymerase III preparations contain two polypeptides — one of 120 kDa and one of 55 kDa (Bauer and Burgers, 1988). The enzyme is sensitive to aphidicolin but resistant to 100 μM butylphenyl dGTP. It copurifies with a tightly associated 3′→5′ exonuclease. In particular, it is a moderately processive enzyme and is specifically stimulated by mammalian PCNA/cyclin (Burgers, 1988). PCNA/cyclin is a protein that stimulates polymerase δ by increasing the processivity of the polymerase (Tan *et al.*, 1988), and that has been shown to be required for SV40 DNA replication *in vitro* (Prelich *et al.*, 1987a, 1987b; Prelich and Stillman, 1988).

As with DNA polymerase δ, however, it was difficult to be sure that DNA polymerase III did not represent an altered or modified form of DNA polymerase I. For instance, the levels of polymerase III were apparently inversely proportional to the levels of polymerase I (Bauer and Burgers, 1988) and polymerase III was extremely sensitive to proteolysis, partly explaining why it had escaped detection for so many years. Genetic evidence now shows that polymerase III is distinct from DNA polymerase I. DNA polymerase III is present in normal amounts and is thermostable in *pol1-17* mutants (Sitney *et al.*, 1989).

Among the collection of temperature sensitive mutants (*cdc*) of Hartwell and colleagues, many are blocked in DNA synthesis. The alleles most likely to affect DNA replication directly are *cdc2*, *cdc6*, *cdc40* (Hartwell, 1976), and *cdc16* (Hartwell, 1967; Kuo *et al.*, 1983) and a series of six mutants isolated by screening for defects in *in vitro* synthesis (Kuo *et al.*, 1983).

The *CDC2* gene has been cloned and has been proposed to encode a DNA polymerase based on sequence similarity with DNA polymerase I and the other polymerases mentioned above (Boulet *et al.*, 1989). Fractionation of extracts of *CDC2* and *cdc2-1* mutant cells, shows that DNA polymerase III is deficient in *cdc2-1* cells. These biochemical data considered alone could mean either that *CDC2* regulates polymerase III or that *CDC2* is itself the polymerase III structural gene. The fact that the DNA sequence of *CDC2* shows that it encodes a DNA polymerase argues strongly that *CDC2* encodes DNA polymerase III (Boulet *et al.*, 1989; Sitney *et al.*, 1989). This is especially interesting in view of recent data implicating mammalian DNA polymerase δ in replication (Dressler and Frattini, 1986; Dressler and Kimbro, 1987; Hammond *et al.*, 1987; Decker *et al.*, 1987).

The phenotype of existing *cdc2* mutants is complex. Hartwell has shown that it is essential for DNA synthesis and that DNA synthesis is also essential for its function; that is, *cdc2* cannot complete its function during arrest of cells by hydroxyurea, a drug that inhibits DNA replication, most likely through its inhibition of ribonucleotide reductase (Hartwell, 1976). The mutants are able to initiate DNA synthesis at the restrictive temperature, but are unable to synthesize a full genome equivalent of DNA. About 30% of the DNA remains unreplicated (Conrad and Newlon, 1983). *cdc2* mutants are completely replication defective in permeabilized yeast cells. When yeast cells are permeabilized with Brij in the presence of sucrose they can continue propagation of replication forks initiated before treatment but they cannot initiate new rounds of replication. DNA synthesis in the Brij-sucrose treated *cdc2* mutants ts 328, ts 370 and ts 346 is normal at 23°C but completely defective at 37°C, consistent with a defect in elongation in *cdc2* mutants (Kuo *et al.*, 1983). The extensive *in vivo* synthesis described by Newlon and Conrad may be due to partial function of the mutant protein *in vivo* or to synthesis by DNA polymerase I in the absence of polymerase III. Isolation of new alleles of *cdc2* might reveal a mutant that, like the *pol1-17* mutant, shows no DNA synthesis at the restrictive temperature. Use of the cloned *CDC2* gene and plasmid shuffling should allow suitable new alleles to be identified.

With respect to the DNA sequence and structure/function considerations, it is interesting to note that while *CDC2* contains the conserved IV-II-VI-III-I-V domains, yeast DNA polymerase I is much more closely related to human polymerase α than it is to the yeast DNA polymerase III. For example, in region II, there are 24/37 identical amino acids between the yeast and human polymerase α's, but only 19 identities between *CDC2* and *POL1*. In region VI there are 10 identities between the α polymerases and only 8 between *CDC2* and *POL1*. In region III, 36/42 amino acids are identical between the α polymerases, but only 15 out of 42 are the same in *CDC2* and *POL1*. While region I contains a core of 6 amino acids conserved in all polymerases, the polymerase α's are nearly identical over an additional 6 amino acids on the N-terminal side and 3 out of 5 amino acids on the C-terminal side of that core. Yeast polymerases I and III diverge outside of the core. When the sequence for mammalian polymerase δ becomes available, it might then be expected to show more similarity to *CDC2* than to human polymerase α.

PCNA — A Polymerase δ Accessory Protein

PCNA (proliferating cell nuclear antigen, or cyclin) was first observed as a protein synthesized preferentially in S phase (Bravo and Celis, 1980; Bravo and Macdonald-Bravo, 1985). It has recently been demonstrated that PCNA is required for SV40 *in vitro* replication

(Prelich *et al.*, 1987b), and that it is actually an accessory protein of mammalian DNA polymerase δ (Tan *et al.*, 1986; Prelich *et al.*, 1987a; Bravo *et al.*, 1987). Though PCNA itself doesn't bind to DNA, PCNA increases the processivity of DNA polymerase δ. Interestingly, mammalian PCNA is able to interact with yeast DNA polymerase III, resulting in increased processivity of the yeast enzyme (Burgers, 1988). This finding led to a search for a yeast PCNA analog. Such a protein has recently been purified from yeast using stimulation of calf thymus polymerase δ as an assay (Bauer and Burgers, 1988). This cross-species complementation is especially striking in light of the fact that yeast PCNA is immunologically distinct from mammalian PCNA (Bauer and Burgers, 1988).

Yeast PCNA has additional properties similar to those of mammalian PCNA. It is an acidic protein with a low molecular weight, 26,000 Da, and the purification properties of the enzymes are the same. yPCNA stimulates polymerase III by increasing the processivity of the polymerase. In mammalian cells, PCNA is cell cycle regulated. In yeast, PCNA levels increase 4 to 5-fold in cells blocked in S phase by treatment with hydroxyurea. Despite these similarities, yPCNA fails to stimulate DNA synthesis in an SV40 *in vitro* replication system depleted of mammalian PCNA (Bauer and Burgers, 1988).

OTHER DNA POLYMERASE ACCESSORY PROTEINS

SSBs

The term SSB (single-stranded DNA binding protein) has been used traditionally to describe proteins whose *in vivo* roles depend on their binding single-stranded DNA. In prokaryotes, a single SSB is involved in DNA unwinding during replication, recombination and repair, however multiple species of SSB have been found in eukaryotic cells (for reviews see Chase and Williams, 1986; Kowalczykowski *et al.*, 1981). While specific protein-protein interactions mediate the stimulation of prokaryotic polymerases by their cognate SSBs, mammalian DNA polymerase α is inhibited by naked DNA, but not by SSB-coated DNA, accounting for enhanced polymerase activity.

Because any protein that binds with high affinity to single-stranded DNA will be retained during ssDNA affinity chromatography, it is not surprising that eukaryotic SSBs with functions quite different from those described above have been isolated. For example, the prototypical eukaryotic SSBs, UP1 and UP2 from calf thymus, show extensive homology with heterogeneous nuclear ribonuclear proteins (hnRNPs), as does mouse myeloma HDP1 (Herrick and Alberts, 1976; Herrick *et al.*, 1976; Williams *et al.*, 1985; see also Jong *et al.*, 1987). These and related proteins may be involved in replication as well as in RNA processing. A reverse genetics approach was taken to address this issue in yeast.

Yeast contains multiple species of SSB. Several groups have studied SSBs that stimulate DNA polymerase I. The gene encoding one of these, SSB1, was cloned using SSB1 antibodies and λgt11 (Jong and Campbell, 1986). Gene disruptions showed that the gene is not essential and strains from which the gene has been entirely deleted are viable, although no SSB1 can be detected immunologically in extracts (Jong *et al.*, 1987; Clark *et al.*, in preparation). SSB1 has been localized to the nucleolus by immunofluorescence microscopy (Jong *et al.*, 1987). Its predicted amino acid sequence shows some homology to proteins involved in RNA binding. A role for this protein in replication is unlikely.

Two other SSBs have been purified from yeast, SSB2, and a mitochondrial SSB (SSBm), however, these proteins have not been further characterized. Thus, the only SSB known to be involved in eukaryotic DNA replication is the three subunit protein required for SV40 DNA replication *in vitro*, RF-A or RP-A (Fairman and Stillman, 1988; Wold and Kelly, 1987). Isolation of a yeast replicative SSB may require a functional *in vitro* replication assay.

Helicases

While not all of the proteins discussed above are actually required for DNA replication, it is clear that the reverse genetics approach has been essential in obtaining a genetic handle on replication. One class of proteins missing from this set is helicases. The *RAD3* gene of yeast has recently been shown to encode a helicase (Sung *et al.*, 1987). The *RAD3* gene, cloned by classical means (Naumovski and Friedberg, 1982; Higgens *et al.*, 1983), encodes an essential function (Naumovski and Friedberg, 1983; Higgens *et al.*, 1983). However, a lysine to arginine substitution at position 48 (see Naumovski *et al.*, 1985 or Reynolds *et al.*, 1985 for the nucleotide and predicted amino acid sequences) abolishes helicase activity *in vitro* without affecting cell viability (Sung *et al.*, 1988), while a temperature sensitive *rad3* mutant does not arrest with the terminal phenotype of a replication mutant (dumbbell) but continues for two to four cell divisions before arresting with no uniform morphology (Naumovski and Friedberg, 1987). This *rad3* allele also has no specific defect in DNA synthesis at the restrictive temperature (Naumovski and Friedberg, 1987). While further studies are necessary to determine whether *RAD3* might encode a replicative helicase, these studies make this possibility seem unlikely. A second yeast helicase (ATPase III) has been described which, although absent in *rad3* mutants, does not appear to be the *RAD3* gene product (Sugino *et al.*, 1986). While it is odd that a structural gene for one helicase (*RAD3*) should also regulate a second helicase (ATPase III), polyclonal antiserum to the *RAD3* protein does not react with ATPase III, and ATPase III activity is not overproduced in strains which overproduce the *RAD3* protein (Sugino *et al.*, 1986). ATPase III stimulates synthesis by DNA polymerase I and increases the processivity of this polymerase, thus it will be of interest to determine, by reverse genetics, whether this helicase might be essential for replication.

Initiation of Replication — *ARS* Binding Proteins

For many years, yeast autonomously replicating sequences (ARSs; Struhl *et al.*, 1979) have been studied as the presumed chromosomal origins of replication. While much physical data supported this hypothesis (for example, Chan and Tye, 1980; Petes and Williamson, 1975; Rivin and Fangman, 1980; reviewed in Newlon, 1989), the precise physical mapping of plasmid and chromosomal origins of replication to these sequences has only recently been possible (Brewer and Fangman, 1987; Huberman *et al.*, 1987; Huberman *et al.*, 1989).

Mutational analysis of various *ARS* sequences [*ARS1* (Celniker *et al.*, 1984), the 2 μM *ARS* (Broach *et al.*, 1983), the HO *ARS* (Kearsey, 1984), the *ARS* at the *HMRE* locus (Abraham *et al.*, 1984; Brand *et al.*, 1985), the H4 *ARS* (Bouton and Smith, 1986) and *ARSs* from *Drosophila* and human DNA (Marunouchi *et al.*, 1987; Palzkill and Newlon, 1988; Montiel *et al.*, 1984)] has allowed the division of yeast *ARS* sequences into three functional domains illustrated in Figure 2. Domain A is a short stretch of nucleotides, 11-19 bp which contains a conserved 11-bp element, A/T TTTATPuTTT A/T, usually called the core consensus (Broach *et al.*, 1983; Stinchomb *et al.*, 1981). ("Domain" is used here to denote a general region defined by deletion

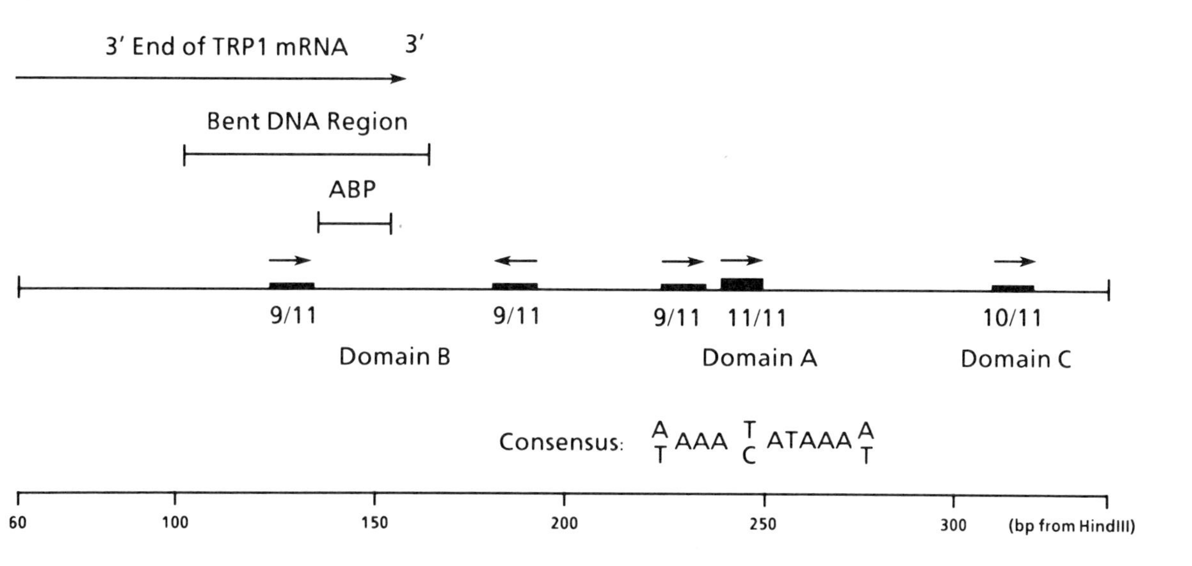

Figure 2. Architecture of autonomously replicating sequence 1 (*ARS1*). The core consensus is indicated by a box labeled 11/11. Arrows over other boxes represent the orientation of close matches to the core consensus. Other features and references are given in the text. ABP refers to a binding site for an *ARS* binding protein now designated ABFI.

mutations, while "element" is used to define a specific sequence of nucleotides defined by point mutations, sequence conservation, or protein binding.) Element A has been shown to be essential for *ARS* function, as point mutations within the consensus of *ARS1* abolish the autonomous replication property of colinear DNA (Celniker *et al.*, 1984; Kearsey *et al.*, 1984). Flanking DNA, in particular a less highly conserved 50-100 bp region, Domain B, is also critical for efficient replication (Srienc *et al.*, 1985). Domain B contains multiple copies of close matches to the core consensus, which are important for the *ARS* activity of at least at one *ARS* (Palzkill and Newlon, 1988). Domain B may also contain a sequence conserved at some but not all *ARSs* (Palzkill *et al.*, 1986). Finally, Domain B at *ARS1* contains a region of bent DNA that may contribute to *ARS* function or serve as a recognition signal (Snyder *et al.*, 1986). The third functional domain of *ARSs*, domain C, also contributes to efficient plasmid replication (Celniker *et al.*, 1984; Koshland *et al.*, 1985; Srienc *et al.*, 1985). It is in this region that replication bubbles have been observed in *in vitro* measurement of *ARS1* replication.

Since proteins involved in the initiation of replication would be expected to interact with *ARS* sequences, several laboratories have attempted to identify proteins which bind to DNA fragments containing portions of various *ARS* sequences. Recently, Shore *et al.* (1987) and Buchman *et al.* (1988) have described an activity in crude fractions of yeast extracts that binds to both *HMR-E* and to element B of *ARS1* and have suggested that binding is due to a single protein species, which they called *ARS* binding factor I (ABF-I) or silencer binding factor B

(SBF-B). Using oligonucleotide affinity chromatography, ABF-I has been purified to homogeneity (Sweder *et al.*, 1988; Diffley and Stillman, 1988). This protein has a molecular weight of 135 kDa, and binds with sequence specificity to *ARS1* and the *HMR-E ARS*. DNase I footprinting using the purified protein shows that ABF-I protects the same regions as does crude SBF-B, confirming that the proteins are identical (Sweder *et al.*, 1988; Shore *et al.*, 1987; Buchman *et al.*, 1988; Diffley and Stillman, 1988).

While Domain B itself is required for *ARS* function, the importance of the actual ABF-I binding site is less clear. A 4-bp deletion of the ABF-I site at *ARS1* decreases *ARS* activity only 10%. A deletion covering one-half of the site reduces activity by 52%. Thus, ABF-I may not be essential for *ARS1* function, but it does appear to be important. The ABF-I site has also been shown to fall in a region important for *ARS* function at *HMR-E* (Brand *et al.*, 1987).

It is interesting that the ABF-I binding sites at *ARS1* and *HMR-E ARS* show little DNA sequence homology. It has been suggested that the recognition is for a DNA structure rather than for a particular sequence. Domain B contains a bent structure, a feature shared by various origins of replication (Snyder *et al.*, 1986).

In order to asses the importance of ABF-I in *ARS* function, "reverse genetics" has been used. Antibodies against ABF-I have been used to clone the ABFI gene from a λgt11 library (Rhode *et al.*, 1989). Gene disruptions show that ABFI encodes an essential function. Although ABFI is essential for growth, since ABFI does not bind to the essential *ARS* core consensus sequence, we do not think that it is the yeast initiator protein, that is, the analog of Dna A of *E. coli* or of large T antigen of SV40. As there is some data suggesting a role for element B in transcriptional activation (Buchman *et al.*, 1988; Brand *et al.*, 1987), ABF-I may function in both transcription and replication. ABF-I in such a scenario may act to keep *ARS* DNA nucleosome free. As termination of TRP1 transcription occurs very near element B of *ARS1*, it has also been suggested that the protein acts as a transcriptional terminator, and thus prevents transcription from interfering with replication (Snyder *et al.*, 1988).

Diffley and Stillman (1988) have described a second *ARS* binding protein, ABF-II. ABF-II is a 21 kDa protein that binds all DNA with equal affinity but gives a specific binding pattern at *ARS1*. DNase I footprinting of ABF-II binding at *ARS1* occurs at multiple but specific sites spanning the *ARS1* region, leaving the core consensus sequence exposed (Diffley and Stillman, 1988).

In addition to ABF-I and II, a third, much less abundant, species that also recognized the ABF-I site was detected during oligonucleotide affinity purification of ABF-I. This factor is called ABF-III. It is at this point less strongly implicated in *ARS* function than ABF-I, since the ABF-III recognition sequence can be deleted with little effect on replication *in vivo*. Furthermore, there appears to be no interaction between ABF-I and ABF-III. ABF-III may prove to be an ancillary protein for optimal replication or a transcriptional termination factor at the end of the *TRP1* mRNA. Such studies await functional *in vitro* assays for *ARS* function.

Topoisomerases

The genes for two topoisomerases have been identified in yeast. *TOP1* encodes a type I topoisomerase, which *via* transient single strand breaks can change the linking number of either positively or negatively supercoiled DNA. *TOP2* encodes a type II topoisomerase, which makes transient double-strand breaks to relax either positively or negatively supercoiled DNA (for review see Gellert, 1981). Different approaches were used by two different groups in cloning these genes, one a reverse genetics approach, and one a more classical scheme.

Both a type I and a type II topoisomerase had been partially purified from yeast at the time the cloning was undertaken (Durnford and Champoux, 1978; Badaracco *et al.*, 1983; Goto and Wang, 1982), thus, it was of interest to identify mutants defective in either of these enzymes. By assaying extracts of individual isolates of the Hartwell collection of conditional lethal mutants (Hartwell, 1967) for topoisomerase activity, two appropriate mutants were identified. The first was defective in topoisomerase I activity, and was shown to be allelic with a gene, *MAK1*, required for the maintenance of the yeast double stranded RNA virus, killer (Thrash *et al.*, 1985). The *ts* mutation did not segregate with the topoisomerase defect. The *TOP1* gene was therefore cloned by simultaneous complementation of both the *in vitro* topoisomerase defect and the *mak1* phenotype. A genomic disruption of *TOP1* confirmed that this gene is not essential (Thrash *et al.*, 1985).

The second topoisomerase mutant from the Hartwell collection proved to be defective in topoisomerase II (DiNardo *et al.*, 1984). This mutation was mapped to chromosome XIV, and the gene was later cloned by complementation of the *ts* defect (Voelkel-Meiman *et al.*, 1986). In another case, reverse genetics was applied to obtain the cloned *TOP2* gene, using antibody against topoisomerase II to screen a λgt11 library. Unlike *TOP1*, *TOP2* does encode an essential function, as evidenced by gene disruption (Goto and Wang, 1984). Identity between the *TOP2* gene identified by reverse genetics and by screening ts mutants was established by genetic mapping. Both the original *top2* mutant (DiNardo *et al.*, 1984) and a set of temperature sensitive *top2* mutants generated by *in vitro* mutagenesis of the cloned *TOP2* gene (Holm *et al.*, 1985) arrest with large buds upon shifting to the restrictive temperature, although this terminal phenotype is only uniformly observed when a synchronous culture is examined (DiNardo *et al.*, 1984; Holm *et al.*, 1985). DNA synthesis is normal at the non-permissive temperature — both in extent and rate. The *TOP2* execution point determined by both groups is between the α-factor arrest point in the G1 phase of the cell cycle, and medial nuclear division. This enzyme is probably required for decatenation of newly replicated DNA before nuclear division can proceed (DiNardo, *et al.*, 1984).

When a mutant defective in both topoisomerases, (*top1°* *top2 ts*) is used in shift up experiments, a "quick-stop" phenotype is observed with respect to DNA synthesis, indicating that at least one topoisomerase is required for replication (Brill *et al.*, 1987). Presumably, the viable phenotype of *top1°* is due to *TOP2* partially carrying out the usual *TOP1* function.

An interesting phenotype recently attributed to both *top1* and *top2* mutants is a high frequency of mitotic recombination between ribosomal DNA repeats (Christman *et al.*, 1988). Normally, the rate of mitotic recombination in this region is extremely low, however, introduction of either a *top1* or *top2* mutation causes an increase in this rate of over 50-fold. It appears that both *TOP1* and *TOP2* act to suppress mitotic recombination within the ribosomal cistrons. The rate of mitotic recombination at other loci is not affected by either topoisomerase mutation. Interestingly, in a *top1 top2(ts)* double mutant, the rate of transcription is 3 to 6-fold lower than in wild-type cells (Brill *et al.*, 1987). Whether this somehow mediates the effect on rDNA recombination is not known.

Ribonuclease H

Enzymes that degrade the RNA component of RNA-DNA hybrids, designated RNaseH activity, may play a role in replication, for instance, by removing RNA primers or in providing specificity to initiation. An RNase H activity was discovered during a search among DNA binding proteins for proteins that specifically stimulated yeast DNA polymerase α. One partially purified polymerase stimulatory protein was shown to contain an RNase H activity, and this protein was then extensively purified using the comparatively simple RNase H assay (Karwan *et al.*, 1983). The highly purified RNase H, as well as all intermediate fractions, retained the ability to stimulate polymerase alpha on homopolymer primer-templates. The enzyme degrades RNA in RNA/DNA hybrids to oligoribonucleotides, suggesting an endonucleotyotic mode, and will hydrolyze a phosphodiester bond joining a ribonucleotide and deoxyribonucleotide, as would be required during processing of Okazaki pieces. There seems to be some specificity in the stimulation, since the protein does not stimulate DNA polymerase II. RNase H activities that specifically stimulate DNA polymerase α have also been purified from other organisms (e.g., De Francesco and Lehman, 1986).

The major polypeptide found on a Coomassie stained SDS gel was a 70 kDa species, and therefore the enzyme has been called RNaseH(70) (Karwan *et al.*, 1986). However, antiserum was prepared to the most highly purified preparations and immunoblots indicated two immunoreactive polypeptides of 70 and 160 kDa (Karwan *et al.*, 1986). Antibody prepared by immunopurification of the antiserum using the 160 kDa protein as a ligand reacts strongly with the 160 kDa protein but only very weakly with the 70 kDa band, and a partial deletion of the gene encoding the 160 kDa protein (vide infra) still produces the 70 kDa protein, suggesting that they are probably different gene products (Wintersberger *et al.*, 1988).

Under conditions where the RNaseH activity was inhibited, RNase H(70) was shown to possess an RNA-dependent, DNA synthesizing activity [reverse transcriptase, (Karwan *et al.*, 1986)]. Activity gels demonstrated that both the 160 and 70 kDa proteins have reverse transcriptase activity, and the p160 has been provisionally named SRT(160) (Karwan *et al.*, 1986; Wintersberger *et al.*, 1988). Antisera containing antibodies that recognize epitopes in both the 70 and 160 kDa polypeptides have been used to screen a λgt11 library and two sets of non-overlapping clones have been identified. One set encodes SRT(160). Gene disruptions are not lethal, but the mutant cells have an abnormal morphology and possess at least two times the normal amount of DNA. Interestingly, this is also the phenotype of one mutant *pol1* allele, *pol1-14*. RNase H(70) has been purified from the SRT(160) deletion mutant and has been shown to retain RNaseH activity and ability to stimulate DNA polymerase I. The purified protein stimulates the extent of synthesis but not the length of the product and hence is not a processivity factor.

The association of a reverse transcriptase with the 70 and 160 kDa polypeptides suggested that one or the other might be encoded by the transposable element Ty, which transposes by a retrovirus-like mechanism, or sigma. However, the cloned SRT(160) gene does not hybridize to Ty DNA (Karwan *et al.*, 1986; Wintersberger *et al.*, 1988). Sequences related to reverse transcriptases have also been found in yeast mitochondrial DNA. However, strains lacking the mitochondrial genome contain normal amounts of RNase H(70) activity (Karwan *et al.*, 1986).

Nucleases

Various exonucleases have been identified in yeast, although none of their genes has been cloned. As in prokaryotes, 3′→5′ exonuclease activity is likely to be essential in eukaryotes for high fidelity DNA replication. While polymerase I (yDNA polymerase α) has no intrinsic exonuclease activity, both yeast DNA polymerase II (Wintersberger, 1978; Chang, 1977; Budd *et al.*, 1989b) and DNA polymerase III (yDNA polymerase δ) (Bauer *et al.*, 1988) have associated 3′→5′ exonuclease activities. Although the cloned yDNA polymerase δ gene (*CDC2*) is available (see above discussion), it is not yet clear whether this presumed proofreading activity is contained in the same polypeptide. An interesting result has recently been obtained for *Drosophila* polymerase α, which contains a cryptic 3′→5′ exonuclease, active only upon dissociation of a polymerase α subunit (Cotterill *et al.*, 1987). No other eukaryotic α polymerase has revealed a similar activity.

A yeast endonuclease has been purified to near homogeneity and its gene identified through reverse genetics (Burbee, *et al.*, 1988). This protein, the *NUC3* gene product, appears to play a role in DNA repair. [*NUC3* was designated *NUC1* in Burbee, *et al.* (1988)]. Genomic disruption of the *NUC3* gene, obtained by screening λgt11 with antibody to a 140 kDa endonuclease, results in cells with an apparent inability to repair double-stranded DNA breaks (Burbee *et al.*, 1988). Mating-type switching in *nuc3*⁻ cells is a lethal event. *nuc3* mutant strains carrying a plasmid in which the *HO* endonuclease gene is under the control of the galactose-inducible promoter *GAL1,10* are unable to grow on galactose medium. Under non-inducing conditions (glucose medium), these cells are viable. The effects of the *nuc3* mutation are pleiotropic; these mutants also sporulate poorly, the products of meiosis being inviable, and are sensitive to the radiomimetic chemical MMS. Most aspects of this phenotype are shared by a group of yeast mutants defective in repair of x-ray induced double stand breaks (*rad51, rad52, rad54, rad55,* and *rad57*; for reviews see Haynes and Kunz, 1981; Game, 1983). Although *NUC3* has not been mapped, complementation analysis indicates that it is not allelic to any of the *rad52* group mutants. The cloned *RAD52* gene has recently been shown to complement the growth defects of T4 gene 46 and gene 47 endonuclease mutants, indicating that *RAD52* may also encode an endonuclease (Chen and Bernstein, 1988).

Activity gel analysis during the course of the *NUC3* purification revealed six active endonuclease polypeptides in addition to *NUC3* (Burbee *et al.*, 1988). Although some of these may be active proteolytic fragments of *NUC3*, it will be of interest to determine the *in vivo* roles of other endo and exonucleases.

Cell Cycle Regulation

The availability, through reverse and classical genetics, of genes encoding replication proteins, offers a tremendous opportunity for investigating how DNA replication is integrated into the eukaryotic cell cycle. Extensive collections of yeast mutants that affect progression through the cycle have been isolated, and many genes encoding DNA replication functions have been identified among them (*polymerase I* (α) = *CDC17*; *polymerase III* (δ) = *cdc2*; DNA ligase = *CDC9*, topoisomerase II = ts 14-16 [*TOP2*]). While cell cycle control of mammalian replication can be studied at the transcriptional level, the ease with which yeast cultures can be synchronized and with which yeast DNA can be manipulated, and in particular, the amenability of yeast to genetic study has allowed the execution points and temporal relationships of several replication genes to be assessed.

One of the crucial points in the cell cycle occurs during G_1 when the cells commit to entry into S phase and progression through mitosis. In yeast, several genes have been shown to function sequentially in the pathway to S phase. *cdc28*, *cdc4*, *cdc7*, for instance, define a temporal order. *cdc28* acts at the point of commitment, called "start", and has been shown to encode a protein kinase. These genes may regulate expression of gene products needed for proliferation. It is rather striking, in this context, that while most genes in yeast are constitutively transcribed, including *CDC28* and *CDC7*, many genes whose products function in S phase have been shown to be cyclically expressed. Using synchronized cell cultures, *POL1*, *PRI1*, *CDC21*, CDC8, CDC9, RNR2, HO, SWI5, and histones have been shown to be cell cycle regulated.

To investigate whether this pattern of expression is dependent upon *CDC28* or 7 function, the ability of mutants blocked at these points in G_1 to express the replication genes was assessed. Cells blocked at the *cdc28* step, "start", failed to express *POL1*, *PRI1*, *CDC21*, *CDC8*, *CDC9*, *RNR2*, *HO*, *SWI5*, and histones. In cells blocked subsequent to "start", at the *cdc4* step, all of these genes except histones were expressed. Cells blocked at the *cdc7* stage, which falls at the G_1/S transition expressed all genes including histones. Thus, passage through start is sufficient to turn on all of the replication genes and this transcriptional event may be one of the results of signals set in motion at "start" (see Breeden, 1988).

CONCLUSIONS

One of the major purposes of developing an *in vitro* replication system is to provide functional assays for replication proteins, either through complementation assays or through fractionation/reconstitution regimes. It is somewhat ironic that many replication proteins will be in hand before a useful yeast *in vitro* replication system is available. The search for some of the proteins described above was spurred by studies with prokaryotic *in vitro* replication systems, namely, helicases, topoisomerases and RNase H's. Other proteins were suggested by results obtained using the SV40 *in vitro* system, namely polymerase δ and PCNA. Reverse genetics applied to these proteins has greatly accelerated the characterization of yeast replication by obviating the need for the identification of mutants with a defined, predicted phenotype and the long subsequent search for proteins affected by the mutations.

Probably the most important insight gained to date from the work on yeast is the remarkable similarity between the yeast and mammalian DNA polymerases - from the subunit structure of DNA polymerase α to the specific interaction between PCNA and DNA polymerase δ. The similarity between yeast and mammalian polymerases at the DNA sequence level and the fact that yeast PCNA has similar effects on both yeast and mammalian DNA polymerase δ raise the hope that such homologies exist between other components of the two replication complexes. It may even be possible, given the specific stimulation of mammalian polymerases by a yeast protein, to directly combine existing mammalian *in vitro* replication systems with yeast proteins and thereby use classical and reverse genetics to gain important insights into the mechanism of eukaryotic DNA replication. At the least, the work of the last two years increases confidence that results obtained using yeast can be extended directly to mammalian systems. For instance, the participation of PCNA in SV 40 DNA replication *in vitro* suggests that DNA polymerase δ plays a role in SV40 DNA replication. However, this result was inherently incomplete. The demonstration that yeast DNA polymerase III (δ) is encoded by a gene essential for chromosomal DNA replication makes an essential role for polymerase δ in mammalian systems seem inescapable. The ability to extrapolate makes the utility of the yeast genetic system more profound than ever.

REFERENCES

Abraham, J., K. A. Nasmyth, J. N. Strathern, A. J. S. Klar, and J. B. Hicks. (1984) Regulation of mating type information in yeast. *J. Mol. Biol.* **176**: 307-331.

Badaracco, G., P. Plevani, W. T. Ruyechen, and L. M. S. Chang. (1983) Purification and characterization of yeast topoisomerase I. *J. Biol. Chem.* **258**: 2022-2026.

Bauer, G. A., H. M. Heller, and P. M. J. Burgers. (1988) DNA polymerase III from *Saccharomyces cerevisiae*. I. Purification and characterization. *J. Biol. Chem.* **263**: 917-924.

Bauer, G. A., and P. M. J. Burgers. (1988) The yeast analog of mammalian cyclin/proliferating-cell nuclear antigen interacts with mammalian DNA polymerase delta. *Proc. Natl. Acad. Sci. USA* **85**: 7506-7510.

Boeke, J. D., F. LaCroute, and G. R. Fink. (1984) A positive selection for mutants lacking orotidine-5′-phosphate decarboxylase activity in yeast: 5-fluoroorotic acid resistance. *Mol. Gen. Genet.* **197**: 345-346.

Boeke, J. D., J. Truehart, G. Natsoulis, and G. Fink. (1987) 5-Fluoroorotic acid as a selective agent in yeast molecular genetics. *Meth. Enzymol.* **154**: 164-175.

Boulet, A., Simon, M., Faye G., Bauer, G. A. and Burgers, P. M. J. (1989) Structure and function of the *Saccharomyces cerevisiae CDC2* gene encoding the large subunit of DNA polymerase III. *EMBO J.* **8**: 1849-1854

Bouton, A. H., and M. M. Smith. (1986) Fine-structure analysis of the DNA sequence requirements for autonomous replication of *Saccharomyces cerevisiae* plasmids. *Mol. Cell. Biol.* **6**: 2354-2363.

Brand, A. H., L. Breeden, J. Abraham, R. Sternglanz, and K. Nasmyth. (1985) Characterization of a "silencer" in yeast: a DNA sequence with properties opposite to those of a transcriptional enhancer. *Cell* **41**: 41-48.

Brand, A. H., G. Micklem, and K. Nasmyth. (1987) A yeast silencer contains sequences that can promote autonomous plasmid replication and transcriptional activation. *Cell* **51**: 709-719.

Bravo, R., and J. E. Celis. (1980) A search for differential polypeptide synthesis throughout the cell cycle of Hela cells. *J. Cell Biol.* **84**: 795-802.

Bravo, R., and H. MacDonald-Bravo. (1985) Changes in the nuclear distribution of cyclin (PCNA) but not its synthesis depend on DNA replication. *EMBO J.* **4**: 655-661.

Bravo, R., R. Frank, P. A. Blundell, and H. MacDonald-Bravo. (1987) Cyclin/PCNA is the auxiliary protein of DNA polymerase-delta. *Nature* **326**: 515-517.

Breeden, L. (1988) Cell cycle-regulated promoters in budding yeast. *Trends in Genetics*. **4**: 249-253.

Brewer, B. J., and W. L. Fangman. (1987) The localization of replication origins on *ARS* plasmids in *S. cerevisiae*. *Cell* **51**: 463-471.

Brill, S. J., S. DiNardo, K. Voelkel-Meiman, and R. Sternglanz. (1987) Need for DNA topoisomerase activity as a swivel for DNA replication and transcription of ribosomal DNA. *Nature* **326**: 414-416.

Broach, J. R., Y-.Y. Li, J. Feldman, M. Jayaram, J. Abraham, K. A. Nasmyth, and J. B. Hicks. (1983) Localization and sequence analysis of yeast origins of DNA replication. *Cold Spring Harbor Symp. on Quant. Biology* **47**: 1165-1173.

Buchman, A. R., W. J. Kimmerly, J. Rine, and R. D. Kornberg. (1988) Two DNA-binding factors recognize specific sequences at silencers, upstream activating sequences, autonomously replicating sequences, and telomeres in *Saccharomyces cerevisiae*. *Mol. Cell. Biol.* **8**: 210-225.

Budd, M. E. and Campbell, J. L. (1987) Temperature sensitive mutations in the yeast DNA polymerase I gene. *Proc, Natl. Acad. Sci. USA* **84**: 2838-2842.

Budd, M. E., K. D. Wittrup, J. E. Bailey, and J. L. Campbell. (1989a) DNA polymerase I is required for DNA replication but not for repair in *Saccharomyces cerevisiae*. *Mol. Cell. Biol.* **9**: 365-376.

Budd, M. E., K. C. Sitney, and J. L. Campbell. (1989b) Purification of DNA polymerase II from *Saccharomyces cerevisiae*. *J. Biol. Chem.* **264**: 6557-6565.

Burbee, D., J. L. Campbell, and F. Heffron. (1988) The purification and characterization of the NUCl gene product, a yeast endodeoxyribonuclease required for double-strand break repair, **In**: E. Friedberg, and P. Hanawalt (eds.), *Mechanisms and Consequences of DNA Damage Processing*, UCLA Symposia on Molecular and Cellular Biology, **83**: 223-230.

Burgers, P. M. J.(1988) Mammalian cyclin/PCNA (DNA polymerase delta auxiliary protein) stimulates processive DNA synthesis by yeast DNA polymerase III. *Nucl. Acids Res.* **16**: 6297-6307.

Burgers, P. M. J., and G. A. Bauer. (1988) DNA polymerase III from *Saccharomyces cerevisiae*: Inhibitor studies and comparison with DNA polymerases I and II. *J. Biol. Chem.* **263**: 925-930.

Burgers, P. M. J. (1989) Eukaryotic DNA polymerases α and δ: Conserved Properties and Interactions from Yeast to Mammalian Cells. Prog. *Nucleic Acids Res. Mol. Biol.*; in press.

Carson, M. J. (1987) Ph.D. Thesis, University of Washington, Seattle.

Carson, M. J., and L. Hartwell. (1985) *CDC17*: An essential gene that prevents telomere elongation in yeast. *Cell* **42**: 249-257.

Celniker, S. E., K. Sweder, F. Srienc, J. E. Bailey, and J. L. Campbell. (1984) Deletion mutations affecting autonomously replicating sequence *ARS1* of *Saccharomyces cerevisiae*. *Mol. Cell. Biol.* **4**: 2455-2466.

Chan, B. K., and B-.K. Tye. (1980) Autonomously replicating sequences in *Saccharomyces cerevisiae*. *Proc. Natl. Acad. Sci. USA* **77**: 6329-6333.

Chang, L. M. S. (1977) DNA polymerases from baker's yeast. *J. Biol. Chem.* **252**: 1873-1880.

Chase, J. W., and K. R. Williams. (1986) Single-stranded DNA binding proteins required for DNA replication. *Ann. Rev. Biochem.* **55**: 103-136.

Chen, D. S., and H. Bernstein. (1988) Yeast Gene *RAD52* can substitute for phage T4 gene *46* or *47* in carrying out recombination and DNA repair. *Proc. Natl. Acad. Sci. USA*, **85**: 6821-6825.

Christman, M. F., F. S. Dietrich, and G. R. Fink. (1988) Mitotic recombination in the rDNA of *S. cerevisiae* is suppressed by the combined action of DNA topoisomerases I and II. *Cell* **55**: 413-425.

Conrad, M. N., and C. S. Newlon. (1983) *cdc2* mutants of yeast fail to replicate approximately one-third of their nuclear genome. *Mol. Cell. Biol.* **3** :1000-1012.

Cotterill, S. M., M. E. Reylard, L. A. Loeb, and I. R. Lehman. (1987) A cryptic proofreading 3′-5′ exonuclease associated with the polymerase subunit of the DNA polymerase-primase from *Drosophila melanogaster*. *Proc. Natl. Acad. Sci. USA* **84**: 5635-5639.

Decker, R. S., R. Possanti, M. K. Bradley, and M. L. DePamphilis. (1987) *in vitro* initiation of DNA replication in Simian virus. *Biochem. Biophys. Res. Commun.* **262**: 10863-10872.

Diffley, J. F. X. (1988) Affinity labelling the DNA polymerase α complex. *J. Biol. Chem.* **263**: 14669-14677.

Diffley, J. F. X., and B. Stillman. (1988) Purification of a yeast protein that binds to origins of DNA replication and a transcriptional silencer. *Proc. Natl. Acad. Sci. USA* **85**: 2120-2124.

DiFrancesco, R. A., and I. R. Lehman. (1985) Intercation of ribonuclease H from *Drosophila melanogaster* embryos with DNA polymerase-primase. *J. Biol. Chem.* **260**: 14764-14770.

DiNardo, S., K. Voelkel, and R. Sternglanz. (1984) DNA topoisomerase II mutant of *Saccharomyces cerevisiae*: topoisomerase II is required for segregation of daughter molecules at the termination of DNA replication. *Proc. Natl. Acad. Sci. USA* **81**: 2616-2620.

Downey, K. M., C. K. Tan, D. M. Andrews, X. Li, and A. G. So. (1988) p. 403-410. *Cancer Cells: Eukaryotic DNA Replication*, vol. 6. Cold Spring Harbor, New York.

Dresler, S. L., and K. S. Kimbro. (1987) 2′,3′-Dideoxythymidine-5′-triphosphate inhibition of DNA replication and ultraviolet-induced DNA repair synthesis in human cells, evidence for involvement of DNA polymerase delta. *Biochemistry* **26**: 2664-2668.

Dresler, S. L., and M. G. Frattini. (1987) Inhibition of DNA replication, DNA repair synthesis, and DNA polymerase alpha and delta by butylphenyl deoxy-guanosine triphosphate. *Fed. Proc.* **46**: 2208.

Durnford, J. M., and J. J. Champoux. (1978) DNA untwisting enzyme. *J. Biol. Chem.* **253**: 1086-1089.

Fairman, M. P., and B. Stillman. (1988) Cellular factors required for multiple stages of SV40 DNA replication *in vitro*. *EMBO J.*, **7**: 1211-1218.

Foiani, M., Santocanale, C., Plevani, P. and Lucchini, G. (1989) A single essential gene, *PR12*, encodes the large subunit of DNA primase in *Saccharomyces cerevisiae*. *Mol. Cell. Biol.* **9**: 3081-3087.

Game, J. L. (1983) Radiation sensitive mutants and repair in yeast. **In**: J.F.T. Spencer, D.M. Spencer and A. R. W. Smith (eds.), *Yeast Genetics, Fundamental and Applied Aspects.* Springer-Verlag, New York: pp. 109-137.

Gellert, M. (1981) DNA Topoisomerases. *Ann. Rev. Bioch.* **50**: 879-910.

Goto, T., and J. C. Wang. (1982) An ATP-dependent type II topoisomerase that catalyzes the catenation, decatenation, unknotting, and relaxation of double-stranded DNA rings. *J. Biol. Chem.* **257**: 5866-5872.

Goto, T., and J. C. Wang. (1984) Yeast DNA topoisomerase II is encoded by a single copy, essential gene. *Cell* **36**: 1073-1080.

Hammond, R. A., J. J. Byrnes, and M. R. Miller. (1987) Identification of DNA polymerase delta in CV-1 cells: studies implicating both DNA polymerase delta and DNA polymerase alpha in DNA replication. *Biochemistry* **26**: 6817-6824.

Hartwell, L. H. (1967) Macromolecule synthesis in temperature-sensitive mutants of yeast. *J. Bacteriol.* **93**: 1662-1670.

Hartwell, L. H. (1976) Sequential function of gene products relative to DNA synthesis in the yeast cell cycle. *J. Mol. Biol.* **104**: 803-814.

Haynes, R. H., and B. A. Kunz. (1981) DNA repair and mutagenesis in yeast. **In**: J. N. Strathern, E. W. Jones, J. R. Broach (eds.), *The Molecular Biology of the Yeast Saccharomyces.* Cold Spring Harbor Laboratory, New York: pp. 371-414.

Herrick, G., and B. Alberts. (1976a) Purification and physical characterization of nucleic acid helix-unwinding proteins from calf thymus. *J. Biol. Chem.* **251**: 2124-2132.

Herrick, G., and B. Alberts. (1976b) Nucleic acid helix-coil transitions mediated by helix-unwinding proteins from calf thymus. *J. Biol. Chem.* **251**: 2133-2141.

Herrick, G., H. Delius, and B. Alberts. (1976) Single stranded DNA structure and DNA polymerase activity in the presence of nucleic acid helix-unwinding proteins from calf thymus. *J. Biol. Chem.* **251**: 2142-2146.

Higgens, D. R., S. Prakash, P. Reynolds, R. Polakowska, S. Weber, and L. Prakash. (1983) Isolation and characterization of the *RAD3* gene of *S. cerevisiae* and inviability of *rad3* deletion mutants. *Proc. Natl. Acad. Sci. USA* **80**: 5680-5684.

Holm, C., T. Goto, J. C. Wang, and D. Botstein. (1985) DNA topoisomerase II is required at the time of mitosis in yeast. *Cell* **41**: 553-563.

Huberman, J. A., L. D. Spotila, K. A. Nawotka, S. M. El-Assouli, and L.R. Davis. (1987) The *in vivo* replication origin of the yeast 2 micron plasmid. *Cell* **51**: 473-481.

Huberman, J. A., Zhu, J., Davis, L. R. and Newlon, C. S. (1988) Close association of a DNA replication origin and an *ARS* element on chromosome III of the yeast *Saccharomyces cerevisiae*. *Nucl. Acids Res.* **16**: 6373-6384.

Johnson, L. M., M. Snyder, L. M. S. Chang, R. W. Davis, and J. L. Campbell. (1985) Isolation of the gene encoding yeast DNA polymerase I. *Cell* **43**: 369-377.

Jong, A. Y. S., and J. L. Campbell. (1986) Isolation of the gene encoding yeast single-stranded nucleic acid binding protein I. *Proc. Natl. Acad. Sci. USA* **83**: 877-881.

Jong, A., M. W. Clark, M. Gilbert, and J. L. Campbell. (1987) Yeast SSB-1 and its relationship to nucleolar RNA binding proteins. *Mol. Cell. Biol. 7*: 2947-2955.

Jung, G., M. C. Leavitt, J. C. Hsieh, and J. Ito. (1987) Bacteriophage PRDI DNA polymerase: Evolution of DNA polymerases. *Proc. Natl. Acad. Sci. USA* **84**: 8287-8291.

Karwan, R., H. Blutsch, and U. Wintersberger. (1983) Physical association of a DNA polymerase stimulating activity with a ribonuclease H purified from yeast. *Biochemistry* **22**: 5500-5507.

Karwan, R., H. Blutsch, and U. Wintersberger. (1984) **In**: *Advances in Experimental Medicine and Biology*, eds. Hüubscher, U. and. Spadari, S. (Plenum, N.Y.) Vol. 179, pp. 513-318.

Karwan, R., C. Kuhne, and U. Wintersberger. (1986) Ribonuclease H(70) from *Saccharomyces cerevisiae* possesses cryptic reverse transcriptase) *Proc. Natl. Acad. Sci. USA* **83**: 5919-5923.

Karwan, R., and U. Wintersberger. (1988) In addition to RNase H(70) two other proteins of *Saccharomyces cerevisiae* exhibit ribonuclease H activity. *J. Biol. Chem.* **263**: 14970-14977.

Kearsey, S. (1984) Structural requirements for the function of a yeast chromosomal replicator. *Cell* **37**: 299-307.

Khan, N. N., G. E. Wright, L. W. Dudycz, and N. C. Brown. (1984) Butylphenyl dGTP: a selective and potent inhibitor of mammalian DNA polymerase alpha. *Nucleic Acids Res.* **12**: 3695-3706.

Konrad, E. B. (1977) Method for Isolation of *Escherichia coli* mutants with enhanced recombination between chromosomal duplications. *J. Bacteriol.* **130**: 167-172

Koshland, D., J. C. Kent, and L. H. Hartwell. (1985) Genetic analysis of the mitotic transmission of minichromosomes. *Cell* **40**: 393-403.

Kowalczykowski, S. T., D. G. Bear, and P. H. vonHippel. (1981) p. 373-444. *The Enzymes*, vol. **XIV**. Academic Press, Inc.,

Kuo, C-.L., and J. L. Campbell. (1983) Cloning of *Saccharomyces cerevisiae* DNA replication genes: isolation of the *CDC8* gene and two genes that compensate for the *cdc8-1* mutation. *Mol. Cell. Biol.* **3**: 1730-1737.

Kuo, C-.L., N-.H. Huang, and J. L. Campbell. (1983) Isolation of yeast DNA replication mutants using permeabilized cells. *Proc. Natl. Acad. Sci. USA* **80**: 6465-6469.

Larder, B. A., S. D. Kemp, and G. Darby. (1987) Related functional domains in virus DNA polymerases. *EMBO J.* 6: 169-175.

Lucchini, G., A. Brandazza, G. Badaracco, M. Bianchi, and P. Plevani. (1985) Identification of the yeast DNA polymerase I gene with antibody probes. *Curr. Genet.* **10**: 245-252.

Lucchini, G., S. Francesconi, M. Foiani, G. Badaracco, and P. Plevani. (1987) Yeast DNA polymerase-DNA primase complex: cloning of PRI1, a single essential gene related to DNA primase activity. *EMBO J.* **6**: 737-742.

Lucchini, G., C. Mazza, E. Scacheri, and P. Plevani. (1988) Genetic mapping of the *Saccharomyces cerevisiae* DNA polymerase I gene and characterization of a *pol1* temperature-sensitive mutant altered in DNA primase-polymerase complex stability. *Mol. Gen. Genet.* **212**: 459-465.

Marunouchi, T., Y-.I. Matsumoto, H. Hosoya, and K. Okabayashi. (1987) In addition to the *ARS* core, the *ARS* box is necessary for autonomously replicating sequences. *Mol. Gen. Genet.* **206**: 60-65.

Möoller, W. and R. Amons. (1985) Phosphate-binding sequences in nucleotide-binding proteins. *FEBS Letters* **186**: 1-7.

Montiel, J. F., M. F. Norbury, M. F. Tuite, M. J. Dobson, J. S. Mills, A. J. Kingsman, and S. M. Kingsman. (1984) Characterization of human chromosomal DNA sequences which replicate autonomously in *Saccharomyces cerevisiae*. *Nucl. Acids Res.* **12**: 1049-1067.

Naumovski, L., and E. C. Friedberg. (1982) Molecular cloning of eucaryotic genes required for excision repair of UV-irradiated DNA: Isolation and partial characterization of the *RAD3* gene of *Saccharomyces cerevisiae*. *J. Bacteriol.* **152**: 323-331.

Naumovski, L., and E. C. Friedberg. (1983) A DNA repair gene required for the incision of damaged DNA is essential for viability in *Saccharomyces cerevisiae*. *Proc. Natl. Acad. Sci. USA* **80**: 4818-4821.

Naumovski, L., G. Chu, P. Berg, and E. C. Friedberg. (1985) *RAD3* gene of *Saccharomyces cerevisiae*: Nucleotide sequence of wild type and mutant alleles, transcript mapping and aspects of gene regulation. *Mol. Cell. Biol.* **5**: 17-26.

Naumovski, L., and E. C. Friedberg. (1987) The *RAD3* gene of *Saccharomyces cerevisiae*: Isolation and characterization of a temperature-sensitive mutant in the essential function and of extragenic suppressors of this mutant. *Mol. Gen. Genet.* **209**: 458-466.

Newlon, C. S. (1989) Yeast Chromosome Replication and Segregation. *Microbiological Revs.* **52**: 568-510.

Nishida, C., P. Reinhard, and S. Linn. (1988) DNA repair synthesis in human fibroblasts requires DNA polymerase delta. *J. Biol. Chem.* **263**: 501-510.

Palzkill, T. G., S. G. Oliver, and C. S. Newlon. (1986) DNA sequence analysis of *ARS* elements from chromosome III of *Saccharomyces cerevisiae*: identification of a new conserved sequence. *Nucl. Acids Res.* **14**: 6247-6264.

Palzkill, T. G., and C. S. Newlon. (1988) A yeast replication origin consists of multiple copies of a small conserved sequence. *Cell* **43**: 441-450.

Petes, T. D., and D. H. Williamson. (1975) Fiber autoradiography of replicating yeast DNA. *Exp. Cell Res.* **95**: 103-110.

Pizzagalli, A., P. Valsasnini, P. Plevani, and G. Lucchini. (1988) DNA polymerase I gene of *Saccharomyces cerevisiae*: nucleotide sequence, mapping of a temperature-sensitive mutation, and protein homology with other DNA polymerases. *Proc. Natl. Acad. Sci. USA* **85**: 3772-3776.

Plevani, P., M. Foiani, P. Valsasni, G. Badaracco, E. Cheriathundam, and L.M.S. Chang. (1985) Polypeptide structure of DNA primase from a yeast DNA polymerase-Primase complex. *J. Biol. Chem.* **260**: 7102-7107.

Prelich, G., C-.K. Tan, M. Kostura, M. B. Mathews, A. G. So, K. M. Downey, and B. Stillman. (1987) Functional identity of proliferating cell nuclear antigen and a DNA polymerase-delta auxiliary protein. *Nature* **326**: 517-520.

Prelich, G., M. Kostura, D. R. Marshak, M. B. Mathews, and B. Stillman. (1987) The cell-cycle regulated proliferating cell nuclear antigen is required for SV40 DNA replication *in vitro*. *Nature* **326**: 471-475.

Prelich, G., and B. Stillman. (1988) Coordinated leading and lagging strand synthesis during SV40 DNA replication *in vitro* requires PCNA. *Cell* **53**: 117-126.

Reynolds, P., D. R. Higgins, L. Prakash, and S. Prakash. (1985) The nucleotide sequence of the *RAD3* gene of *Saccharomyces cerevisiae*: A potential adenine nucleotide binding amino acid sequence and a non-essential acidic carboxy terminal region. *Nucl. Acids Res.* **13**: 2357-2372.

Rhode, P. R., Sweder, K. S., Oegema, K. F. and Campbell, J. L. (1989) The gene encoding ARS binding factor I is essential for the viability of yeast. *Genes & Devel.* **3**: 1926-1939.

Rivin, C. J., and W. L. Fangman. (1980) Replication fork rate and origin activation during the S phase of *Saccharomyces cerevisiae*. *J. Cell Biol.* **85**: 108-115.

Rothstein, R. J. (1983) One-step gene disruption in yeast. *Meth. Enzymol.* **101**: 202-209.

Schild, D., and B. Byers. (1978) Meiotic effects of DNA defective cell division cycle mutations of *Saccharomyces cerevisiae*. *Chromosoma (Berl.)* **70**: 109-130.

Seifert, H. S., E. Chen, M. So, and F. Heffron. (1986) Transposon mutagenesis for *Saccharomyces cerevisiae*. *Proc. Natl. Acad. Sci. USA* **83**: 735-739.

Shore, D., D. J. Stillman, A. H. Brand, and K. A. Nasmyth. (1987) Identification of silencer binding proteins from yeast: possible roles in SIR control and DNA replication. *EMBO J.* **6**: 461-467.

Shortle, D., P. Grifasi, S. J. Benkovic, and D. Botstein. (1982) Gap misrepair mutagenesis: efficient site-directed induction of transition, transversion, and frameshift mutations *in vitro*. *Proc. Natl. Acad. Sci. USA* **79**: 1588-1592.

Sitney, K. C., M. E. Budd, and J. L. Campbell. (1989) DNA polymerase III, a second essential DNA polymerase, is encoded by the *Saccharomyces cerevisiae CDC2* gene. *Cell* **56**: 599-605.

Snyder, M., and R. W. Davis. (1985) *Hybridomas in Biotechnology and Medicine*, New York. Plenum

Snyder, M., A. R. Buchman, and R. W. Davis. (1986) Bent DNA at a yeast autonomously replicating sequence. *Nature* **324**: 87-89.

Snyder, M., R. J. Sapolsky, and R. W. Davis. (1988) Transcription interferes with elements important for chromosome maintenance in *Saccharomyces cerevisiae*. *Mol. Cell. Biol.* **8**: 2184-2194.

Srienc, F., J. E. Bailey, and J. L. Campbell. (1985) Effect of *ARS1* mutations on chromosome stability in *Saccharomyces cerevisiae*. *Mol. Cell. Biol.* **5**: 1676-1684.

Stinchcomb, D. T., C. Mann, E. Selker, and R. W. Davis. (1981) DNA sequences that allow the replication and segregation of yeast chromosomes, p. 473-488. **In**: D.S. Ray(ed.), *The Initiation of DNA Replication*, ICN-UCLA Symposia on Molecular and Cellular Biology, vol. 22. Academic Press, New York.

Struhl, K., D. T. Stinchcomb, S. Scherer, and R. W. Davis. (1979) High-frequency transformation of yeast: autonomous replication of hybrid DNA molecules. *Proc. Natl. Acad. Sci. USA* **76**: 1035-1039.

Sugino, A., B. H. Ryu, T. Sugino, L. Naumovski, and E. C. Friedberg. (1986) A new DNA dependent ATPase which stimulates yeast DNA polymerase I and has DNA unwinding activity. *J. Biol. Chem.* **261**: 11744-11750.

Sung, P., L. Prakash, S. W. Matson, and S. Prakash. (1987) *RAD3* protein of *Saccharomyces cervisiae* is a DNA helicase. *Proc. Natl. Acad. Sci.USA* **84**: 8951-8955.

Sung, P., D. Higgins, L. Prakash, and S. Prakash. (1988) Mutation of lysine-48 to arginine in the yeast *RAD3* protein abolishes its ATPase and DNA helicase activities but not the ability to bind ATP. *EMBO J.* **7**: 3263-3269.

Sweder, K. S., P. R. Rhode, and J. L. Campbell. (1988) Purification and characterization of proteins that bind to yeast *ARSs*. *J. Biol. Chem.* **263**: 17270-17277.

Tan, C-.K., C. Castillo, A. G. So, and K. M. Downey. (1986) An auxiliary protein for DNA polymerase-delta from fetal calf thymus. *J. Biol. Chem.* **261**: 12310-12316.

Thrash, C., K. Voelkel, S. DiNardo, and R. Sternglanz. (1984) Identification of *Saccharomyces cerevisiae* mutants deficient in DNA topoisomerase I activity. *J. Biol. Chem.* **259**: 1375-1377.

Thrash, C., A. T. Bankier, B. G. Barrell, and R. Sternglanz. (1985) Cloning, characterization and sequence of the yeast DNA topoisomerase I gene. *Proc. Natl. Acad. Sci. USA* **82**: 4374-4378.

Voelkel-Meiman, K., S. DiNardo, and R. Sternglanz. (1986) Molecular cloning and genetic mapping of the DNA topoisomerase II gene of *Saccharomyces cerevisiae*. *Gene* **42**: 193-199.

Williams, K. R., K. L. Stone, M. B. LoPresti, B. M. Merrill, and S. R. Planck. (1985) Amino acid sequence of the UP1 calf thymus helix-destabilizing protein and its homology to an analogous protein from mouse myeloma. *Proc. Natl. Acad. Sci. USA* **82**: 5666-5670.

Wintersberger, U., and E. Wintersberger. (1970) Studies on deoxyribonucleic acid polymerase from yeast. I. Partial purification and properties of two DNA polymerases from mitochondria-free cell extracts. *Eur. J. Biochem.* **13**: 11-19.

Wintersberger, E. (1978) Yeast DNA polymerases: antigenic relationship, use of RNA primer and associated exonuclease activity. *Eur. J. Biochem.* **84**: 167-172.

Wintersberger, U., C. Kuhne, and R. Karwan. (1988) Three ribonucleases H and a reverse transcriptase from the yeast, *Saccharomyces cerevisiae*. *Biochim. Biophys. Acta* **951**: 322-329.

Wittekind, M., J. Dodd, L. Vu, J. M. Kolb, J.-M. Buhler, A. Sentenac, and M. Nomura. (1988) Isolation and characterization of temperature-sensitive mutations in RPA190, the gene encoding the largest subunit of RNA polymerase I from *Saccharomyces cerevisiae*. *Mol. Cell. Biol.* **8**: 3997-4008.

Wold, M. S., J. J. Li, and T. J. Kelly. (1987) Initiation of Simian virus 40 DNA replication *in vitro*: large-tumor-antigen- and origin-dependent unwinding of the template. *Proc. Natl. Acad. Sci. USA* **84**: 3643-3647.

Wong, S. W., L. R. Paborsky, P. A. Fisher, T. S-.F. Wang, and D. Korn. (1986) Structural and enzymatic and characterization of immunoaffinity-purified DNA polymerase alpha - DNA primase complex from KB cells. *J. Biol. Chem.* **261**: 7958-7968.

Wyers, F., A. Sentenac, and P. Fromageot. (1976) Role of DNA-RNA hybrids in eukaryotes: purification of two ribonucleases H from yeast cells. *Eur. J. Biochem.* **69**: 377-383.

Wyers, F., J. Huet, A. Sentenac, and P. Fromageot. (1976) Role of DNA-RNA hybrids in eukaryotes: characterization of yeast ribonucleases H1 and H2. Eur. *J. Biochem.* **69**: 385-395.

Young, R. A., and R. W. Davis. (1983) Efficient isolation of genes using antibody probes. *Proc. Natl. Acad. Sci. USA* **80**: 1194-1198.

7

HUMAN DNA POLYMERASE α: A RETROSPECTIVE AND PROSPECTIVE

Teresa S.-F. Wang

PERSPECTIVES

Eukaryotic genomic DNA replication is a complex and tightly regulated process. Recent studies suggest that the eukaryotic DNA replication process, like that in prokaryotes, initially requires the formation of a nucleoprotein complex between a specific initiator protein and a specific DNA sequence to serve as the nucleus of an initiation complex. This is followed by a series of specific and ordered but coordinated protein-protein/protein-DNA interactions that lead to DNA priming and elongation by DNA primase and DNA polymerase (Kornberg, 1980;

1982; 1988; Campbell, 1986). A key component in this complex eukaryotic chromosomal DNA replication process is the replicative DNA polymerase. Based on studies with inhibitors (Ikegami *et al.*,1978), the characterization of a temperature sensitive mutant (Murakami *et al.*, 1985), the observation of enzymatic activity variation during cell proliferation (Fry and Loeb, 1986; Thommes *et al.*, 1986; Wahl *et al.*, 1988), immunological investigation with specific monoclonal antibodies (Tanaka *et al.*, 1982; Miller *et al.*, 1985 and 1986) and *in vitro* viral DNA replication data (Li and Kelly, 1984; Li and Kelly, 1985; Murakami *et al.*, 1986; Stillman and Gluzman, 1985; Wobbe *et al.*, 1985), DNA polymerase α is generally accepted as the principal eukaryotic chromosomal DNA replicative enzyme.

DNA polymerase α was originally identified from calf thymus thirty years ago (Bollum, 1960). Despite intense biochemical investigation by various laboratories, the very low abundance of the DNA polymerase α protein, its apparent structural complexity, and extreme lability during conventional biochemical purification have plagued and confounded investigators for decades and distinguished this enzyme as one of the more problematic in the literature (Fry and Loeb, 1986). Thus, until the mid nineteen eighties, little was known about this essential enzyme's protein structure, biochemical properties, substrate interaction mechanisms, genetic structures, structure-function relationships or about the elements that regulate its expression.

During the period from 1981 to 1988, several landmark achievements have been accomplished. A set of extensive mechanistic studies of polymerase α and its substrate interaction was reported with biochemically purified polymerase α that led to the model of a rigidly sequential order of substrate recognition and binding (Fisher and Korn, 1981 a and b; Fisher *et al.*, 1981; Korn *et al.*, 1983). A panel of stable murine hybridomas that produce monoclonal antibodies specifically against DNA polymerase α was established (Tanaka *et al.*, 1982). Using this panel of monoclonal antibodies, a DNA polymerase α/DNA primase complex was isolated and thoroughly characterized (Wang *et al.*, 1984; Hu *et al.*, 1984). The intracellular localization as well as the chromosomal locus of this enzyme were also identified (Bensch *et al.*, 1982; Wang *et al.*, 1985). Recently, a full length cDNA clone of human DNA polymerase α was isolated and characterized (Wong *et al.*, 1988). The deduced primary protein sequence of this polymerase has provided suggestions for the functional domains and begun to shed light on the structure-function question (Wong *et al.*, 1988; Wang *et al.*, 1989). The available cDNA also has made it possible to address the question of the regulation of this gene during the cell cycle and activation of cell proliferation (Wahl *et al.*, 1988) as well as aiding in the identification of genomic regulatory elements responsible for the expression of this gene. These accomplishments are major advances in understanding this replicative DNA polymerase.

A review of the last eight years progress in our understanding of DNA polymerase α is presented here.

MONOCLONAL ANTIBODIES

A near homogeneous preparation of DNA polymerase α purified from human KB cells (Fisher and Korn, 1977) was used as antigen to immunize a set of mice. The resulting panel of 16 stable murine hybridoma monoclones produce homogeneous antibodies that specifically recognize DNA polymerase α and show no cross-reactivity with DNA polymerase β and γ (Tanaka *et al.*, 1982). All 16 hybridoma clones are capable of binding DNA polymerase α. Five of the antibodies have been used to perform quantitative binding studies, the results of which generate the theoretically predicted linear Scatchard plots with estimated individual binding titers varying over a 10-fold range from 3.4×10^{-10}M to 3.2×10^{-9}M . Three of these hybridoma clones also exhibit potent neutralizing activity against DNA polymerase α. The general immunological properties of these hybridoma clones are summarized in Table I.

Five of the 16 hybridoma monoclones, three neutralizing clones SJK132-20, SJK211-14, and SJK287-38, and two non-neutralizing clones STK1 and SJK237-71, were selected for evaluation of epitope specificities. The epitope specificities of these five clones were assessed by solid phase competitive radioimmunoassay. In general, the results demonstrate that the binding site for each of the five antibodies on DNA polymerase α appears to be unique and independent (Table II; Wong *et al.*, 1986). One anomalous exception is that the binding of SJK211-14 (as first antibody) prevents the subsequent binding of STK1 (as second antibody): the blocking effect is as potent as self versus self control. However, the binding of STK1 as first antibody does not significantly prevent the subsequent binding of SJK211-14 as second antibody. Thus, three of these five well characterized monoclones, SJK237-71, SJK132-20 and SJK287-38, were later selected and extensively used in various investigations.

INTRACELLULAR LOCALIZATION

The intracellular localization of DNA polymerase α had been a controversial issue in the literature (Bollum, 1975; Brown *et al.*, 1981). It was reported from various laboratories that the majority of the polymerase α activity could be recovered in the post-organellar cytoplasmic fraction of aqueously extracted tissues or cells. However, it has been observed by several investigators that a small but significant fraction of the polymerase α activity is reproducibly associated with detergent purified nuclei and could only be released by high salt extraction (Filpula, 1982; Sedwick *et al.*, 1972; Weissbach *et al.*, 1971; Chiu and Baril, 1975; Eichler *et al.*, 1977). Several laboratories reported using various non-conventional cell fractionation methods and claimed the sole nuclear localization of this enzyme, supporting the argument that the cytoplasmic recovery of polymerase α is an artifact of leakage of the aqueous cell fractionation method (Lynch *et al.*, 1975; Foster and Gurney, 1976; Herrick *et al.*, 1976). However, these novel but non-conventional methods have themselves been subjected to criticism, it being suggested that they result in methodological artifacts and are the result of a cross contaminated nuclear fraction with the cytoplasmic fraction, or *vice versa*.

Using four of the monoclonal antibodies specific for human DNA polymerase α and immunoperoxidase detection methods, exclusive intranuclear localization of polymerase α was demonstrated in three cultured human cell lines (Bensch *et al.*, 1982; Fry and Loeb, 1986). The antigens distribute in a diffuse pattern within the nucleoplasm, but the nucleoli are clearly negative. In cultures of transformed cell lines KB and BeWo, almost every cell in the un-synchronized culture is antigen positive. In contrast, in normal diploid fibroblasts, WI-38, only a small fraction of the cells are antigen positive, and there is no detectable antigen in the closely apposed cells of microcolonies that are presumed to be contact-arrested and non-mitotically cycling cells. A detailed report of this study is described in Bensch *et al.*, 1982.

TABLE I

Immunological Properties of Monoclonal Anti-DNA Polymerase α Antibodies

Monoclone	Binding activity		Neutralizing activity		Scatchard plot[c]	Binding affinity K_d
	(+/-)	Titer[a]	(+/-)	Titer[b]		
		ng		*ng*		*M*
STK 1	+	20	-	25	Linear	$2.0x10^{-9}$
SJK 104-4	+	20	-			
SJK 132-20	+	4	+	25	Linear	$3.4x10^{-10}$
SJK 135-45	+	20	-			
SJK 160-4	+	30	-			
SJK 164-35	+	30	-			
SJK 186-7	+	30	-			
SJK 210-33	+	30	-			
SJK 211-14	+	5	-	250	Linear	$1.4x10^{-9}$
SJK 216-13	+	30	-			
SJK 230-16	+	20	-			
SJK 237-71	+	7	-		Linear	$3.2x10^{-9}$
SJK 253-7	+	30	-			
SJK 258-22	+	30	-			
SJK 276-35	+	20	-			
SJK 287-38	+	5	+	60	Linear	$1.4x10^{-9}$

a Binding activity titer is expressed as the nanograms of monoclonal IgG required to bin 50% of 1 unit of DNA polymerase α activity under standard binding assay conditions (Tan *et al* , 1982)

b Neutralizing activity titer is expressed as the nanograms of monoclonal IgG required tc neutralize 50% of 1 unit of DNA polymerase α activity under standard binding assay conditions (Tanaka *et al* , 1982). A negative entry means that less than 10% of the polymerase a activity was neutralized in assays that contained up to 400 ng of monclonal IgG.

c Quantitative binding assays have been performed to date only with 5 monoclones show

Reprinted, with permission, from Tanaka *et al* , 1982.

TABLE II
Epitope Specificity of Monoclonal Anti-DNA Polymerase α Antibodies
Measured by Solid-Phase Competitive Radioimmunoassay

Solid-phase competitive radioimmunoassay was performed as described under Methods. The specific activities of the [^{125}I] IgGs were: STK1, 2117 cpm/ng; SJK 132-20, 1664 cpm/ng; SJK 211-14, 1442 cpm/ng; SJK 237-71, 852 cpm/ng; SJK 187-38, 846 cpm/ng. (Reprinted, with permission fron Wong *et al* ., 1986).

Second antibody (radioiodinated)	First antibody (unlabeled)					
	P3	STK 1	SJK 132-20	SJK 211-14	SJK 127-71	SJK 187-28
STK 1	250	300	2,100	270	1,400	960
SJK 132-20	100	800	90	11,000	7,200	4,600
SJK 211-14	700	4,300	18,000	700	26,500	10,300
SJK 237-71	300	1,100	5,000	6,400	300	5,000
SJK 287-38	140	1,900	5,000	5,300	3,200	300

PROTEIN STRUCTURE

Polypeptide Composition of the Immunoaffinity Purified DNA Polymerase α

Two independent immunoaffinity protocols were developed and used to define the polypeptide composition of human DNA polymerase α. One immunoaffinity protocol was designed to obtain active enzyme fractions and another was designed to identify antigen polypeptide(s). In obtaining maximum yields of active enzyme, SJK237-71 is the most useful antibody. The procedure which yields the active enzyme with binding antibody SJK237-71 is summarized in Table III and has been described in detail (Wang *et al.*, 1984; Wong *et al.*, 1986). Three groups of polypeptides are reproducibly obtained by this protocol: (a) a group of large polypeptides, predominantly of 180 and 165 kDa, with the minor presence of 140 kDa and 125 kDa polypeptides and an occasionally observed 200 kDa polypeptide; (b) a polypeptide of 77 kDa; and (c) two polypeptides of 55 and 49 kDa, Figure 1A. The enzyme fraction obtained from this protocol is an immunocomplex, therefore, the presence of 52 kDa heavy chain of monoclonal antibody is also presented in the enzyme fraction (Wang *et al.*, 1984).

TABLE III

Immunoaffinity Purification of DNA Polymerase-α with Monoclonal Antibody SJK 237-71

Step	Fraction	Volume	Protein	Polymerase-α activity	Specific activity	Yiel
		ml	*mg*[a]	*units*[b]	*units/mg*	%
Crude extract	I	139	293	9,560	33	100
Phosphocellulose	IIA	17	25	5,800	230	61
Protein A-IgG	IIIA	17	0.06	3,000	50,000	31
DNA-Cellulose	IVA	2.5	0.025	2,400	96,000	25

[a] Protein was determined by the method of Schaffner and Weissman (1973) with bovine serum albumin as the standard.

[b] The standard unit of DNA polymerase-α activity is the amount that incorporates 1 nmol of labeled dNMP into acid-insoluble product in 1 h at 37oC (Fisher and Korn, 1977).

Reprinted, with permission from Wang *et al*., 1984.

Another protocol designed to identify antigen polypeptide(s) utilizes covalently linked antibody-Sepharose resin as an immunoadsorbent. This protocol allows more stringent washing and uses a harsh elution condition to recover only the pertinent antigen polypeptide(s) but not enzymatic activity. The method has been described in detail (Wong *et al.*, 1986) and the antigen polypeptide(s) is shown in Figure 1B. The antigen polypeptide(s) recovered by this protocol are primarily comprised of the group of large polypeptides from 180 to 125 kDa and the 77 kDa polypeptide

To investigate which of these polypeptides contain the antigenic epitope and catalytic function for DNA polymerization, an immunoblot with 12 of the monoclonal antibodies including the three potent neutralizing antibodies and non-immune P3 IgG was performed, Figure 2. All monoclonal antibodies tested recognized an antigenic epitope on the polypeptide of 180 kDa in this enzyme preparation. This finding corroborates the results of renaturing gels which indicate that the large polypeptides, 180 to 125 kDa, are the principal polypeptides that contain or are required for the catalytic function of DNA polymerase α (Wong *et al.*, 1986).

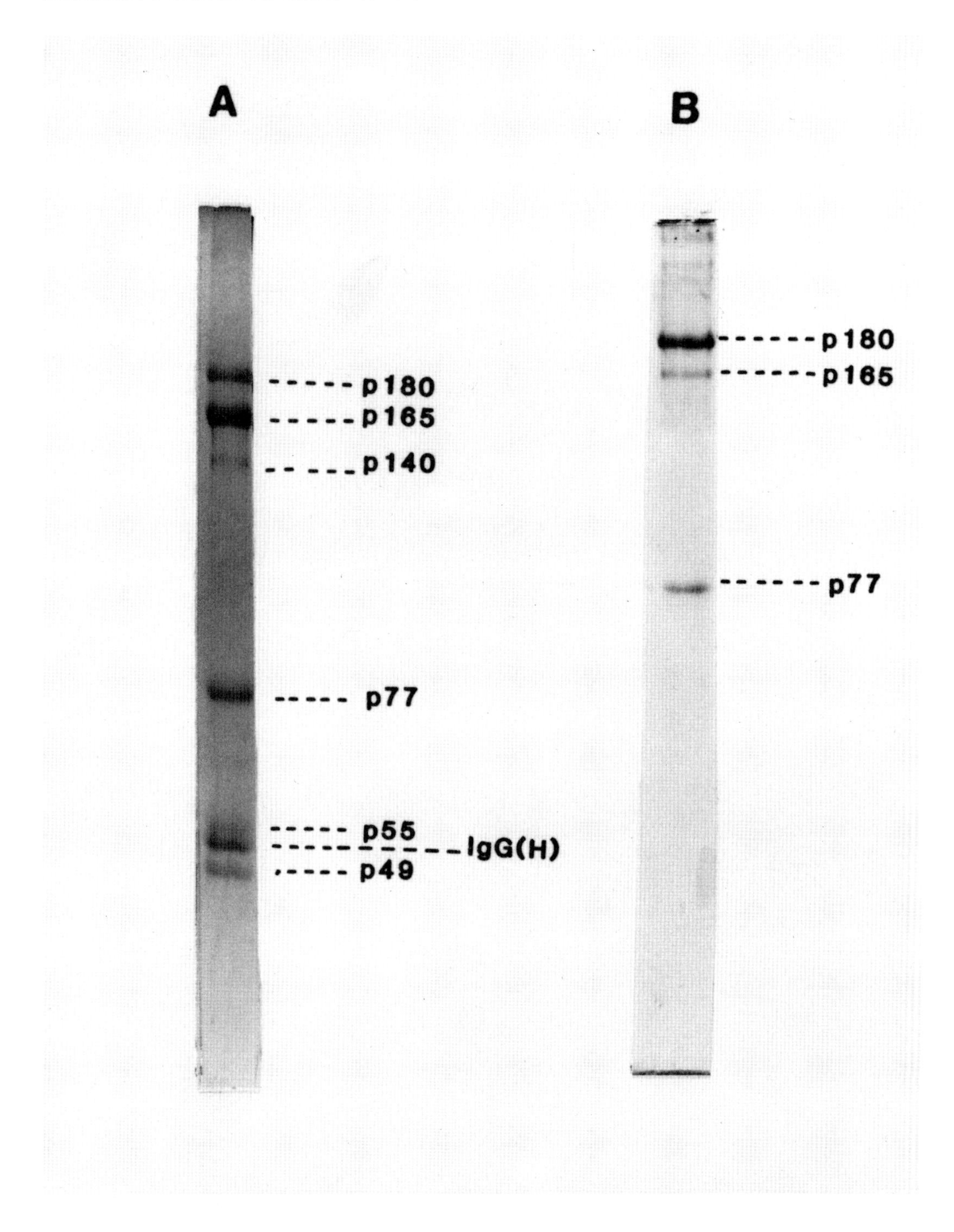

Figure 1. Immunoaffinity purified human DNA polymerase α. A. Immunoaffinity purified DNA polymerase α enzyme fraction from monoclonal IgG-protein A Sepharose 4B column. B. Immunoaffinity purified DNA polymerase α antigen from covalently linked monoclonal IgG-Sepharose 4B column. (Adapted, with permission, from Wang *et al.*, 1984 and Wong *et al.*, 1986).

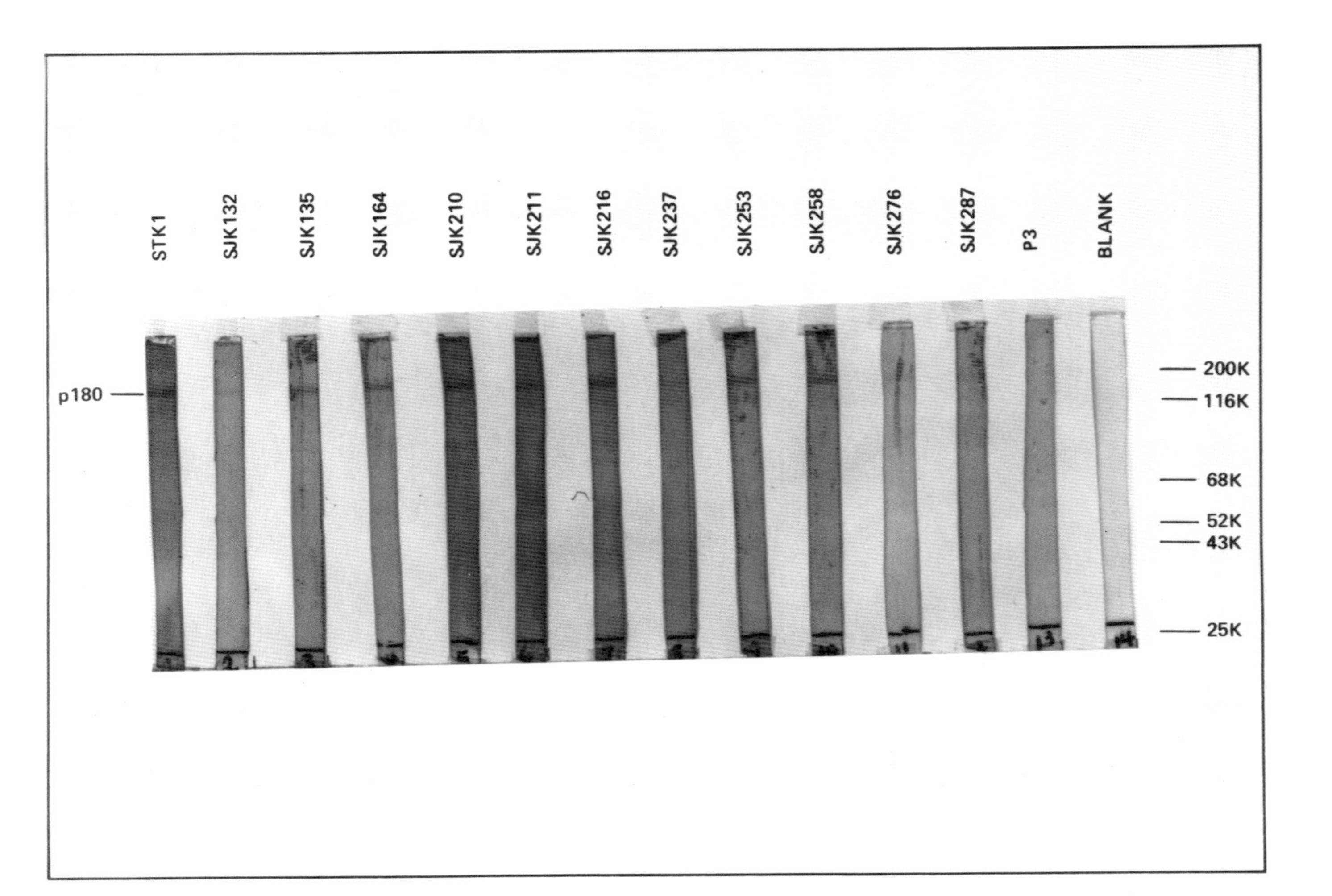

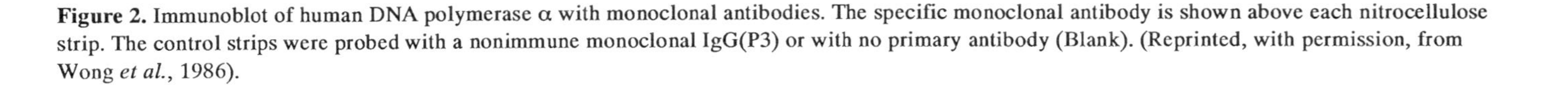
Figure 2. Immunoblot of human DNA polymerase α with monoclonal antibodies. The specific monoclonal antibody is shown above each nitrocellulose strip. The control strips were probed with a nonimmune monoclonal IgG(P3) or with no primary antibody (Blank). (Reprinted, with permission, from Wong *et al.*, 1986).

Inter-Relationship of the Polymerase α Components

To assess the possible structural relationship among these three groups of polymerase α polypeptide components, the polypeptides were subjected to two extensive peptide mapping analyses (Wong *et al.*, 1986). By two-dimensional tryptic peptide mapping (Elder *et al.*, 1977), the two predominant species of the large polypeptides of 180 and 165 kDa appear to have similar if not identical peptide maps, Figure 3. The other polypeptides of 77 kDa, 55 kDa and 49 kDa are found to be unique, distinct and unrelated to the large polypeptides of 180 to 165 kDa, and to each other. The minor species of the large polypeptides, such as the 140 and 125 kDa polypeptides, or the occasionally observed 200 kDa polypeptide, were also found to have identical peptide maps to the 180 and 165 kDa polypeptides (data not shown). The results of these peptide analyses were further confirmed by partial chemical cleavage mapping with N-chlorosuccinimide which specifically cleaves tryptophanyl bonds in the protein (Lischwe and Ochs, 1982). The partial chemical cleavage data again demonstrate similar if not identical cleavage patterns among the large polypeptides and unique, distinct and unrelated patterns for the 77, 55 and 49 kDa polypeptides.

The results from these analyses clearly indicate there are basically four distinct polypeptides in this immunoaffinity purified DNA polymerase α protein component: (i) a group of large polypeptides demonstrated by peptide mapping to be derivatives of the same primary structure constituting the catalytic polypeptide of DNA polymerase α; (ii) a medium size polypeptide of 77 kDa of unknown function; (iii) a smaller polypeptide of 55 kDa; and (iv) another small protein of 49 kDa. The two smaller polypeptides (*i.e.* 55 and 49) have been reported by others to be related to DNA primase activity (Kaguni *et al.*, 1983a; Tseng and Ahlem, 1983; Nasheuer and Grosse, 1988).

The relative molar ratio of each of these polypeptide components was quantitated by densitometric analysis of the Coomassie blue stained gels of DNA polymerase α polypeptides from both immunoaffinity protocols. The results summarized in Figure 4, in general, indicate that the 77 kDa polypeptide and the group of large polypeptides exist in an equal molar ratio when isolated by both immunoaffinity protocols, while the DNA primase-related 55/49 kDa polypeptides are variable.

Phosphorylation of DNA Polymerase α

To assess the possible post-translational modification of DNA polymerase α, human KB cells were labeled *in vivo* with ^{32}P orthophosphate and immunoprecipitated as described (Wong *et al.*, 1986). Both the 180 and 165 kDa catalytic polypeptides and the 77 kDa polypeptide were found to be phosphoproteins (Figure 5A). The phospho-amino acids of these polypeptides are phosphoserine and phosphothreonine, Figure 5B.

The polymerase α polypeptides can be dephosphorylated by acid phosphatase with no apparent effect on either the polymerase or the primase activity. These polypeptides can also be phosphorylated *in vitro* by cyclic AMP dependent protein kinase and the resulting enzyme fraction again demonstrates no apparent change in either enzymatic activity. The biological effect of this post-translational phosphorylation is currently unknown.

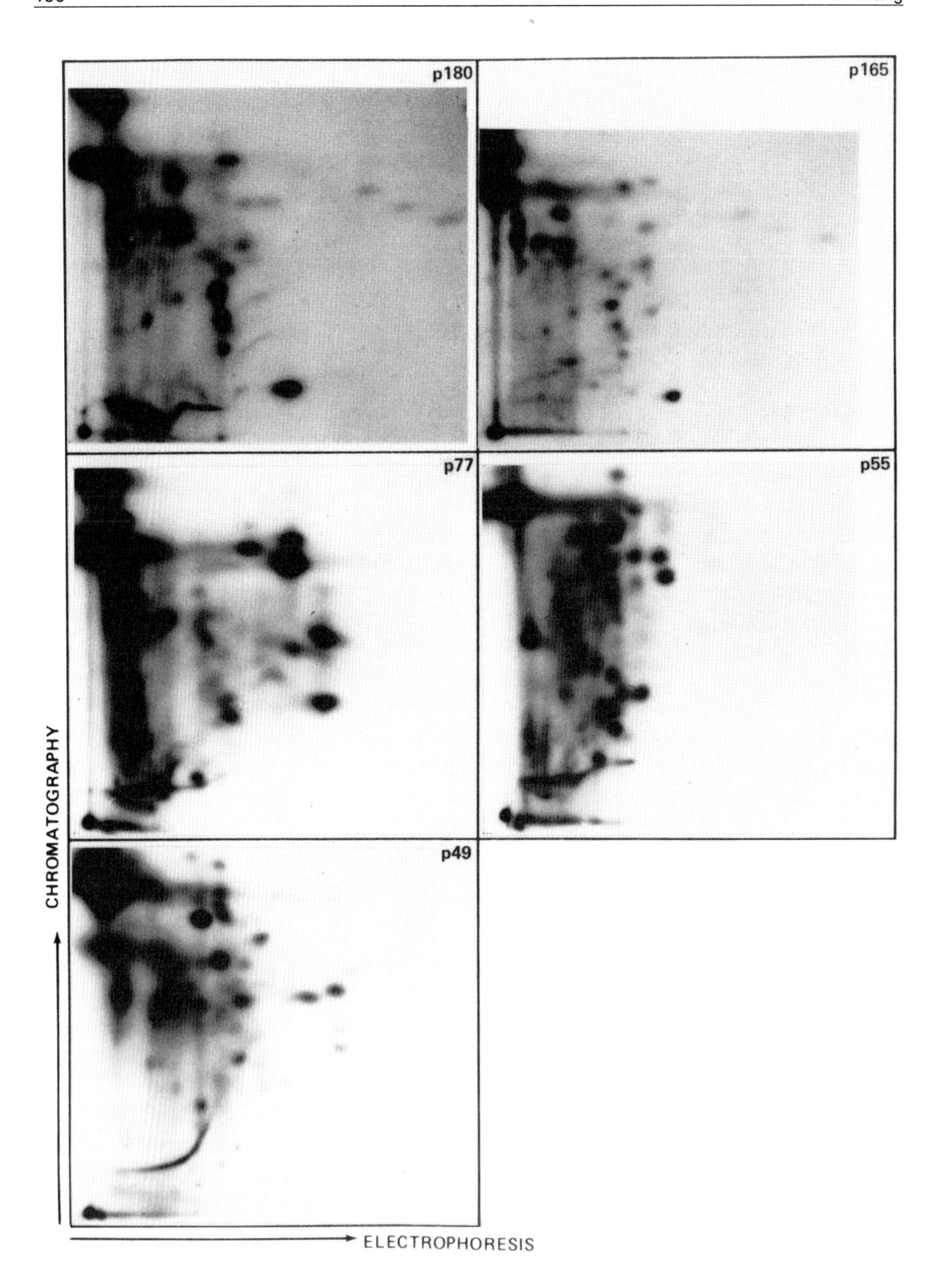

Figure 3. Primary structure comparison by tryptic mapping of DNA polymerase α polypeptides purified by immunoaffinity protocol. (Reprinted, with permission, from Wong *et al.*, 1986)

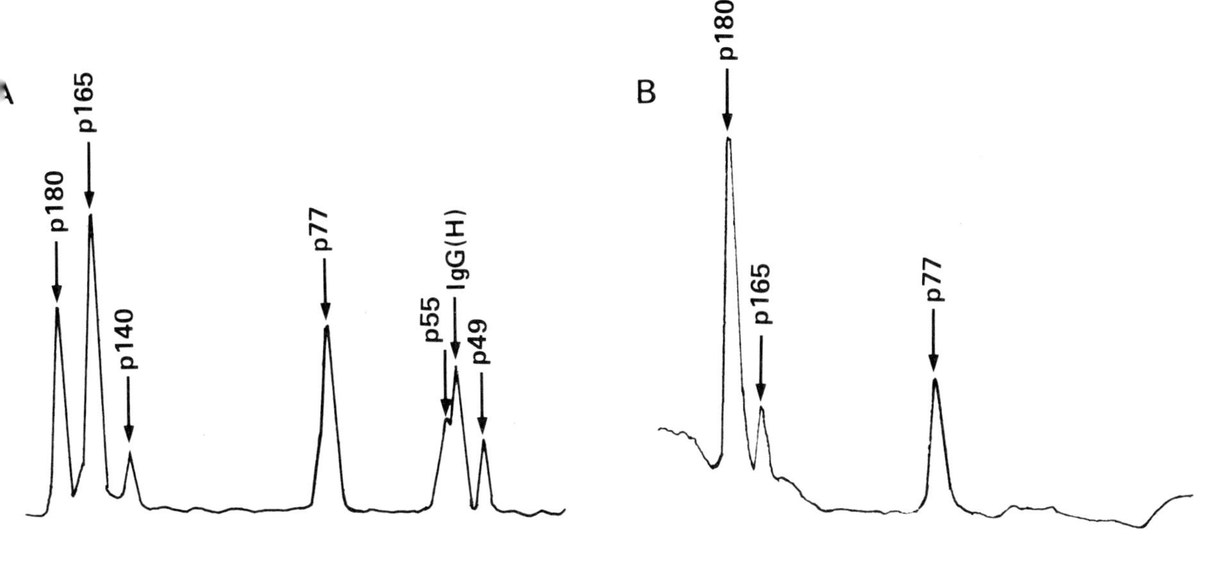

Polypeptide Species	A IgG • Protein A-Sepharose Chromatography		B IgG-Sepharose Chromatography	
	Molar Quantity	Molar Ratio	Molar Quantity	Molar Ratio
p180, p165, p140	0.21	1.0	0.09	1.0
p77	0.17	0.81	0.07	0.80
p55	0.07	0.33	—	—
p49	0.09	0.43	—	—

Figure 4. Stoichiometry of polypeptides in immunoaffinity purified DNA polymerase α. A. Densitometric scan of Coomassie Brilliant Blue stained SDS gel of DNA polymerase α purified by monoclonal IgG-protein A-Sepharose 4B column. B. Densitometric scan of Coomassie Brilliant Blue stained SDS gel of covalently linked monoclonal IgG-Sepharose 4B purified antigen polypeptides. C. Quantitative summary of the immunoaffinity purified DNA polymerase α polypeptides from A. and B.

ENZYMOLOGICAL PROPERTIES

Mechanisms of Substrate Recognition and Binding

Steady-state kinetic analyses, together with sedimentation binding assays, have demonstrated that the interaction of the catalytic core protomer of human DNA polymerase α with its substrates obeys a rigidly ordered sequential ter mechanism (Fisher and Korn, 1981 a and b; Korn *et al.*, 1983). The polymerase α catalytic core protomer binds template (single-stranded DNA) first, followed by primer stem, and finally, by dNTP. Kinetically significant binding of dNTP is absolutely dependent on antecedent primer binding and specification of

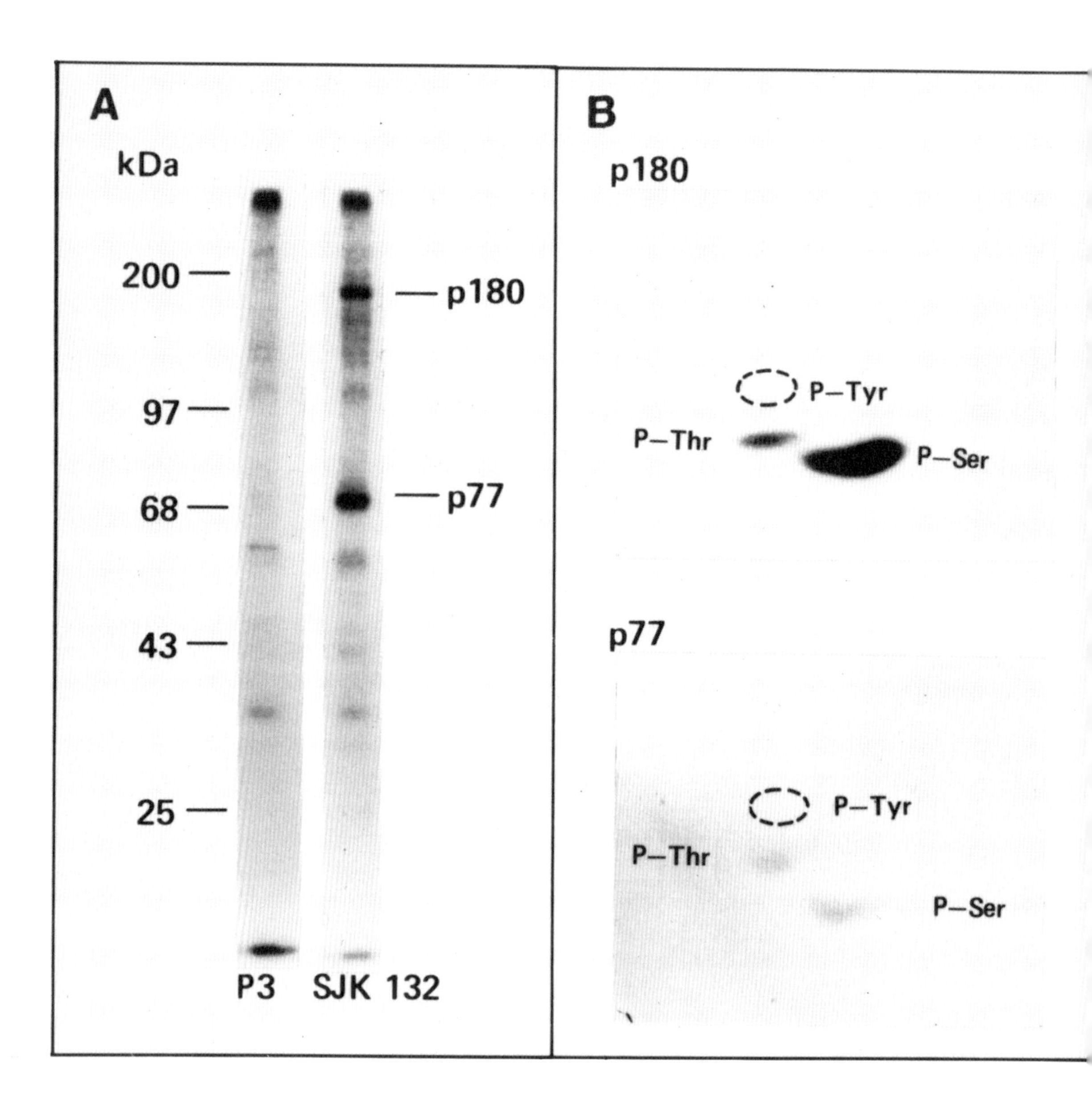

Figure 5. Phosphorylation of DNA polymerase α. A. Immunoprecipitation of DNA polymerase α antigens from crude lysate of [^{32}P]-orthophosphate-labeled KB cells. B. Phosphoamino acids from DNA polymerase α phosphoprotein. (Adapted, with permission, from Wong *et al.*, 1986.)

which dNTP can be added to the polymerase active site is strictly determined by template sequence. Primer binding requires a minimum length of 8 nucleotides, of which the terminal 3-5 nucleotides must be template-complementary. A single mispaired terminal primer nucleotide prevents correct primer binding as well as the subsequent step of dNTP addition to the enzyme. Primer binding, which is satisfied by 3′-terminal H or OH, but blocked by PO_4, occurs through the coordinated participation of 4 Mg^{2+}-primer binding subsites that may serve as a "Mg^{2+}-shuttle" and thereby facilitate not only phosphodiester bond formation but polymerase translocation as well. Each catalytically active polymerase α molecule appears to possess 2 positively cooperative single-stranded DNA binding sites, and very possibly 2 complete active centers. The results of these studies suggest polymerase α is a conformationally active protein that responds to signals generated by template sequence and transduced *via* template binding site(s) interactions.

The immunoaffinity purified enzyme fraction, which contains 4 distinct species of polypeptides, an associated DNA primase activity and an approximately equimolar quantity of monoclonal IgG, was also investigated to determine its mechanisms of substrate recognition and binding (Wong *et al.*, 1986). The results of the mechanistic analyses of catalytic core protomer and the multi-protein immunoaffinity purified polymerase α are compared and summarized in Table IV. The basic mechanisms of substrate recognition and binding with either polymerase α preparations are qualitatively indistinguishable in all respects tested, in regard to the precisely ordered, sequential ter reactant mechanism by which it recognizes and binds its substrates: template, primer and dNTP, respectively.

Associated DNA Primase Activity

In both prokaryotes and eukaryotes, the mechanism of initiation of nascent chain DNA synthesis involves the formation of short oligoribonucleotide primers (Kornberg, 1980; 1982 and 1988; Bramhill and Kornberg, 1988; DePamphillis and Wasserman, 1980). It has been found in the past that a significant fraction of the nascent DNA chains recoverable from a variety of normal and virus-infected mammalian cells contain transient, covalently attached 5′-terminal oligoribonucleotide moieties approximately 10 nucleotides in length and of variable sequence (Reichard *et al.*, 1974; Eliasson and Reichard, 1979; Tseng and Goulian, 1977; Tseng *et al.*, 1979; Waqar and Huberman, 1975). From 1982 to 1984, a DNA primase activity was identified in various phylogenetically distant eukaryotic organisms (Yagura *et al.*, 1982; Kaufmann and Falk, 1982; Tseng and Ahlem, 1982 and 1983; Riedel *et al.*, 1982; Shioda *et al.*, 1982; Conaway and Lehman, 1982 a and b; Kaguni *et al.*, 1983 a; Wang *et al.*, 1984). The immunoaffinity purified human DNA polymerase α contains a DNA primase activity (Wang *et al.*, 1984). The associated human primase synthesizes oligoribonucleotide chains initiated with ATP in a reaction that is strictly dependent on added template and ribonucleotide triphosphates (rNTPs). In the presence of added dNTPs and single stranded DNA template, the human DNA primase produces a short oligoribonucleotide primer, predominantly of 6 to 10 nucleotides, which is extended by polymerase α into long DNA chains up to several thousand nucleotides. There is no evidence for nucleotide preferences at RNA/DNA junctions. The primase and polymerase α activities are distinguishable by several physical and chemical criteria, and the primase reaction is partially sensitive to two potent neutralizing monoclonal antibodies specific for DNA polymerase α. The two enzyme activities are functionally and structurally distinct. A detailed report of the human DNA polymerase α and associated primase activity has been described (Wang *et al.*, 1984). In the absence of added dNTPs, the primers synthesized by the primase are heterogeneous in size and composition, and susceptible to cleavage by pancreatic DNase I due to the content of short oligodeoxynucleotide tracts synthesized by primase from trace contaminant dNTPs in the rNTPs. A working model for DNA primase catalysis has been proposed (Hu *et al.*, 1984). The model predicts that the physiologically significant primer for

TABLE IV

Enzymological comparison of the biochemically purified catalytically active core component of DNA polymerase-α with the immunoaffinity purified DNA polymerase-α/DNA primase holoenzyme

Property	Core Component	Holoenzyme
Nucleic Acid binding affinity:		
a. Single-stranded circular DNA	K_i=20-40 uM	K_i=20-40 uM
b. Duplex circular DNA	None	None
c. Duplex linear DNA	None	None
d. "Nicked" duplex DNA	None	None
Template binding specificity		
a. Homopolymeric templates	$(dT)_n$>dC$)_n$>>$(dA)_n$	$(dT)_n$>$(dC)_n$>>)dA$)_n$
b. Heteropolymeric templates	$(dABT)_n$>>$(dACT)_n$>$(dGCT)_n$	$(dACT)_n$>>$(dACT)_n$≅$(dGCT)_n$
Cooperativity of single-stranded DNA binding (Hill coefficienc) (Template binding site)	1.7 + 0.1	1.6 + 0.1
Cooperativity of Mg++ binding (Hill coefficient) (Primer binding site)	3.9 + 0.2	3.8 + 0.3
Substrate binding order	template—> primer—>dNTP	template—>primer—>dNTP

Reprinted, with permission from Wong *et al*, 1986.

DNA polymerase α is a mixed 5′-oligoribo-3′-oligodeoxynucleotide and the signal which governs the switch from RNA to DNA synthesis is intrinsic to the primase mechanism and is generated by ambient dNTPs. A detailed discussion of the model has been published (Hu *et al.*, 1984).

CONSERVATION OF PROTEIN EPITOPES AMONG EUKARYOTIC SPECIES

The constituent polypeptides of DNA polymerase α preparations purified from a variety of eukaryotic sources display remarkable similarities (Table V). From unicellular fungi such as yeast (Plevani *et al.*, 1985) or invertebrates such as *Drosophila* (Kaguni *et al.*, 1983 a and b) to mammals such as rodent (Mechali *et al.*, 1980), calf (Wahl *et al.*, 1984; Chang *et al.*, 1984; Nasheuer and Grosse, 1987), simian (Ymaguchi *et al.*, 1985) and human (Wang *et al.*, 1984; Wong *et al.*, 1986), each comprises a cluster of large polypeptides of 180 to 120 kDa identified as the catalytic polypeptide, a medium size polypeptide of ≈70 kDa of unknown function and two small polypeptides in the size range of 46 to 56 kDa reported to be primase-related polypeptides. The finding that this essential DNA replication enzyme maintains similar polypeptide composition across such phylogenetically diverse organisms, strongly suggests the possibility of an evolutionary pressure to conserve critical functional domains on the DNA polymerase α molecule.

This possibility was tested immunologically by four of the DNA polymerase α specific monoclonal antibodies (Tanaka *et al.*, 1982; Miller *et al.*, 1988). The study was performed with three neutralizing monoclonal antibodies and a single binding monoclonal antibody. The results are summarized in Figure 6 and Table VI. The data in general indicate that while the single non-neutralizing antibody, SJK237-71, recognizes higher mammalian DNA polymerase only (Figure 6D) the three neutralizing monoclonal antibodies, SJK132-20, 211-14 and 287-38, cross react to variable extents with polymerase α from vertebrate species (Figure 6 A, B and C). This result corroborates the finding presented in Table II that each of these antibodies recognizes a unique and independent epitope on the polymerase α molecule. The most cross-reactive antibody, SJK132-20, exhibits identical reactivity with all mammalian polymerase α tested and strongly neutralizes polymerases from lower vertebrates down to the amphibian level (Figure 6A) but not fish. The other two neutralizing antibodies, SJK211-14 and SJK287-38, are able to neutralize the mammalian polymerases tested, but are weakly cross-reactive with avian polymerase, and unable to react with polymerases from reptilian, amphibian and fish cell extract (Figure 6B and C). None of the antibodies tested cross-reacts with DNA polymerase α from insect (*Drosophila*) or fungus (yeast) (Table VI).

The protein epitope conservation is further demonstrated with the most cross-reactive monoclonal antibody, SJK132-20, which recognizes a polypeptide of 165-180 kDa from species as diverse as human and amphibian (Figure 7). These immunological data strongly support the conclusion that there is an evolutionary pressure to conserve critical functional domains on the DNA polymerase α catalytic polypeptide. This result underscores the evolutionary conservation of this specific epitope over a period of 350 million years and indicates the critical functional role of this particular epitope in DNA polymerase α catalysis.

Analysis of conservation at the gene level, described later, indicates the conservation of several nucleotide regions which extend beyond vertebrates into invertebrates and unicellular fungi. Given this striking similarity in polypeptide composition, mechanisms of catalysis and nucleotide sequence, it is surprising that neither SJK132 antibody nor any of our monoclonal

TABLE V

Cell Source	Subunit Composition and Function		References
	Mass (kDa)		
Human: KB cells	180-125	Catalytic core	Wang *et a*l, 1984 and Wong *et al*,1986
	77		
	55	> primase	
	40		
Simian : CV-1	176-118		Yamaguchi et al, 1985
	62		
	57		
	53		
	30		
Calf thymus	160	Catalytic core	Chang *et al*, 1984; Nasheuer and Grosse, 1987
	68		
	55	> primase	
	48		
Murine	185		Mechali *et al*, 1987; Tseng and Ahlem, 1983
	68		
	56	> primase	
	46		
Insect : D.Melanogaster	182	Catalytic core	Kaguni *et al*, 1983b
	73		
	60	> primase	
	50		
Yeast	180-140	Catalytic core	Plevani *et al*, 1985
	74		
	58	> primase	
	48		

antibodies cross-reacts with DNA polymerases of invertebrate organisms (*Drosophila*) or unicellular organism (yeast). Conservation at the genetic level will be discussed further in a later section. A detailed discussion of the evolutionary conservation of DNA polymerase α is described in Miller *et al.*, 1988.

CHROMOSOME MAPPING

Mapping of Human DNA Polymerase α Gene on the X Chromosome

The finding that the non-neutralizing monoclonal antibody, SJK237-71, recognizes polymerase α from higher mammals (human, simian, canine, and bovine), but not lower mammals such as rodents (Figure 6D and Table VI; Miller *et al.*, 1988), provides a convenient tool to map the expression of the human DNA polymerase α gene on a specific chromosome by solid phase immunoassay and simultaneously karyotyping a set of rodent-human somatic hybrids (Wang *et al.*, 1985). Three sets of experiments were performed to map the expression of human polymerase α on a specific chromosome. Initially seven rodent human somatic hybrid clones derived from five independent hybridizations and containing different complementations of human chromosomes were screened. The results suggested a correlation between the presence of the X chromosome and the expression of human DNA polymerase α. This observation was supported by cell sorting of two rodent-human somatic hybrids, each containing X but not Y chromosomes in a proportion of their cells, for the presence of a cell surface marker encoded by the human X and Y chromosomes (Levy *et al.*, 1979; Goodfellow *et al.*, 1980). The data clearly confirm a correlation between the presence of the X chromosome and the expression of human DNA polymerase α enzymatic activity. This conclusion was further confirmed by medium selection and the counter-selection of three X chromosome-containing somatic hybrid clones from HAT (hypoxanthine/aminopterin/thymidine) medium to 6TG (6-thioguanine) medium. A set of experiments was performed based on the fact that those hybrid clones which survived in HAT medium must contain the HPRT (hypoxanthine phosphoribosyltransferase) gene encoded on the X chromosome, and after counter-selection in 6TG medium, those clones containing human X chromosome would be eliminated. The three somatic hybrid clones were first maintained in HAT medium, and then propagated in 6TG medium; the clones were simultaneously karyotyped, assayed for human polymerase α activity and antigen protein by solid phase immunoassays, and also assayed for X chromosome marker enzyme G6PD and HPRT as references. The results clearly indicate an absolute correlation between the expression of human DNA polymerase α and the presence of the human X chromosome. Thus, the genetic locus can be assigned to the X chromosome (Wang *et al.*, 1985).

Assignment of Human Polymerase α Gene Expression to the Short Arm of the X Chromosome

Further refined mapping of the expression of the human DNA polymerase α gene was accomplished using a set of rodent-human somatic hybrid clones, which contain X and autosome translocations with defined X chromosome breakpoints or interstitial deletions (Figure 8). Initial analysis indicated that the expression of human DNA polymerase α was positive in a hybrid clone (CF25-8) containing an X/13 chromosome translocation with a breakpoint at Xp22.1 but negative in a clone (CF37-6) containing an X/11 chromosome translocation with a breakpoint at Xp21.2. The shortest region of overlap is between Xp21.2 to Xp22.1 on the short arm of the X chromosome. Final, precise localization of the expression of human polymerase α was accomplished by analysis of a hybrid clone, XVIII-54A-2a, containing an interstitial deletion of the entire Xp21 band (Francke, 1984; Lindgren *et al.*, 1984). This hybrid clone was found to be strongly positive for expression of human polymerase α activity. Two other cell lines, one with an intact inactivated X chromosome, another with a translocated

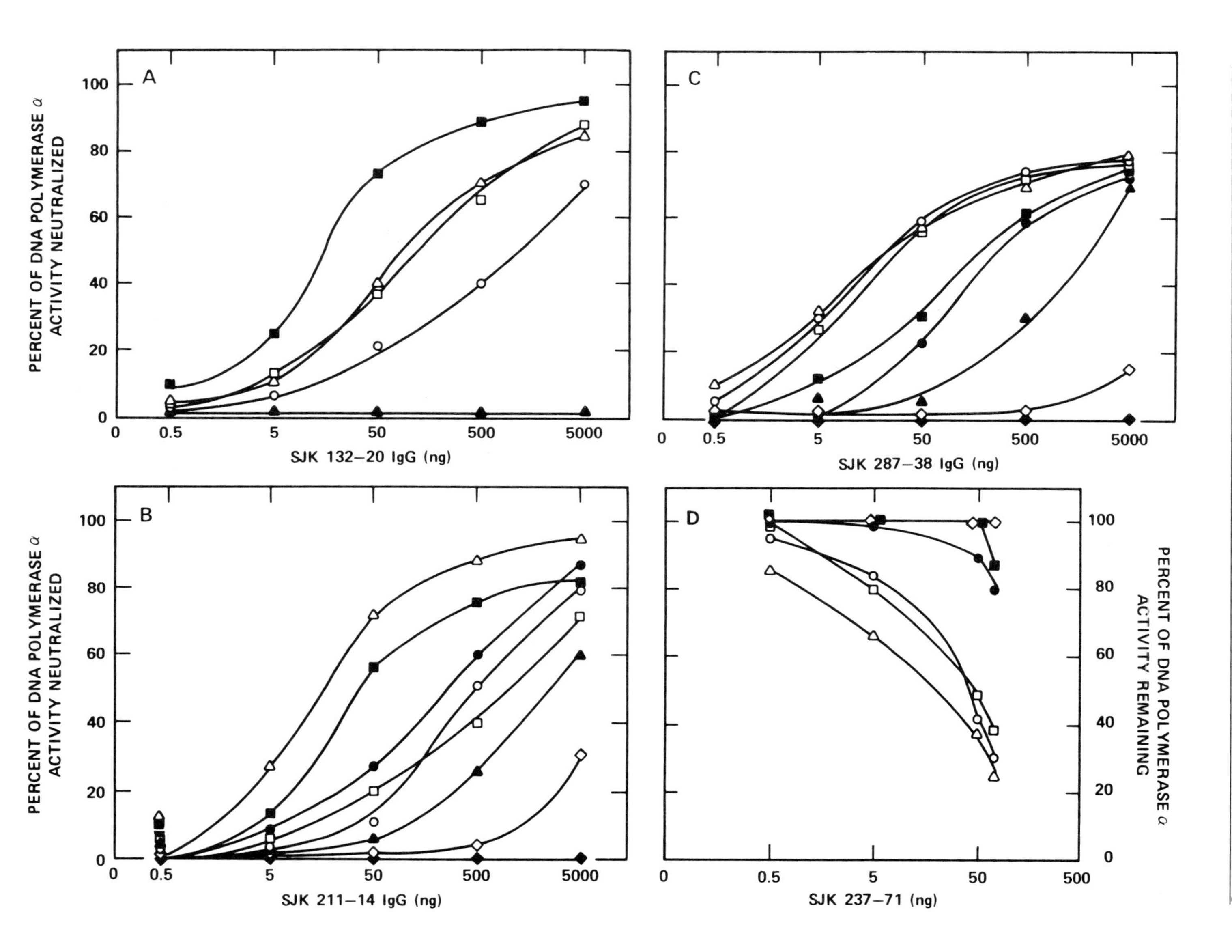
A
PERCENT OF DNA POLYMERASE α ACTIVITY NEUTRALIZED
SJK 132–20 IgG (ng)
C
SJK 287–38 IgG (ng)
B
PERCENT OF DNA POLYMERASE α ACTIVITY NEUTRALIZED
SJK 211–14 IgG (ng)
D
PERCENT OF DNA POLYMERASE α ACTIVITY REMAINING
SJK 237–71 (ng)

inactivated X chromosome with breakpoint at Xp22.1, both resulted in negative expression of human polymerase α activity. Thus, the expression of human polymerase α gene can be precisely assigned to the junctional region of Xp21.3 to Xp22.1 on the short arm of the X chromosome and the gene does not escape X chromosome inactivation (Wang *et al.*, 1985).

cDNA OF HUMAN DNA POLYMERASE α

cDNA and Deduced Primary Protein Structure

Despite substantial biochemical investigation of DNA polymerase α, nothing is known of its structure-function relationship or the regulation of DNA polymerase α gene expression. In an attempt to investigate these two important issues, a full length cDNA of human DNA polymerase α was isolated (Wong *et al.*, 1988). A total of five overlapping cDNA clones were isolated. After sequencing and assembling, a cDNA of 5433 bp encoding a single open reading frame of 1462 amino acids was obtained. An in-frame initiator ATG codon flanked by nucleotides matching Kozak's criteria for a translation initiation site was identified. The derived molecular weight from this recombinant DNA polymerase α cDNA is 165 kDa. A restriction map of the full-length cDNA and the 5 overlapping cDNA clones originally isolated are illustrated in Figure 9 and the entire 5433 bp cDNA sequence and the deduced protein sequence are reported in Wong *et al.*, 1988.

Structural Gene Mapping

The cDNA was proven to be X chromosome linked by genomic Southern hybridization of normal male DNA (46XY) and DNA from cell line GM1202A of 4X, karyotyped 49, XXXXY. As shown in Figure 10A, two EcoR1 digested genomic DNA bands are observed by hybridization of the Pst1 restriction fragment from cDNA insert with perfect signal intensity ratio of 1:4. The precise structural gene localization was mapped by genomic Southern hybridization using the same Pst1 fragment with EcoR1 digested DNA samples from a panel of human-rodent hybrids which were previously used to map the polymerase gene by expression, Figure 8 (Wang *et al.*, 1985). Under conditions that exclude cross-hybridization to rodent DNA, the structural gene of human DNA polymerase α was mapped precisely at the previously determined

Figure 6. (Left) Reactivities of monoclonal anti-human DNA polymerase α antibodies with eukaryotic DNA polymerase α from various phylogenetic species. 0.5 unit of DNA polymerase α (fraction IIA) from human (KB), simian (Baboon lymphoblast and monkey CV-1), canine (dog thymus), bovine (calf thymus), murine (LA-9), avian (Pekin duck), reptilian (Gekko), amphibian (*Xenopus laevis*), or piscine (CHSE) cell extracts was incubated with neutralizing antibody SJK 132-20, SJK 211-14, or SJK 287-38 and surviving polymerase α activity was measured as described (13,15). For the non-neutralizing antibody assay, 0.5 unit of DNA polymerase α (fraction IIA) from each cell line was incubated with non-neutralizing antibody SJK 237-71, and polymerase activity in the supernatant was measured as described (13). (A) Reactivity of antibody SJK 132-20 with vertebrate polymerase α activities. Percent of mammalian (human, simian, canine, bovine and murine) (■), avian (△), reptilian (○), amphibian (□), or piscine (▲) cell polymerase α activity neutralized. (B) Reactivity of neutralizing antibody SJK 211-14 with vertebrate polymerases. Percent of human (△), canine (■), bovine (●), baboon (□), monkey (○), murine (▲), or avian (◇) cell polymerase α activity neutralized; (◆), percent of either reptilian, amphibian, or piscine cell polymerase α activity neutralized. (C) Reactivity of neutralizing antibody SJK 287-38 with vertebrate polymerases. Symbols for each species are the same as in (B). Values of 2.2±5.8% (mean ±2SD) polymerase activity neutralized are defined as background. (D) Reactivity of non-neutralizing antibody SJK 237-71 with a panel of eukaryotic DNA polymerases α. Percent of human (△), baboon(□), monkey (○), bovine (●), canine (■) cell polymerase α activity remaining; (◇), percent of all other lower vertebrate (murine, avian, reptilian, amphibian, and piscine) cell polymerase α activity remaining. Values of 98.2% ± 8.0% (mean ±2SD) polymerase activity remaining in supernatant are defined as background. (Reprinted, with permission, from Miller *et al.*, 1988)

TABLE VI
Summary of the Conservation of DNA Polymerase α Epitopes Recognized by Anti-KB Cell DNA Polymerase α Monoclonal Antibodies

Cell Source	Representative Phylogenetic Class	Neutralization Titers			Binding Titers
		132-20 (ng)	211-14 (ng)	287-38 (ng)	237-71 (ng)
KB	Mammalian (human)	25	25	30	27.5
Baboon	Mammalian (simian)	25	2000	30	37.5
CV-1	Mammalian (simian)	25	500	30	37.5
Dog Thymus	Mammalian (canine)	25	45	250	NB
Calf Thymus	Mammalian (bovine)	25	350	350	NB
BW	Mammalian (murine)	25	3500	3000	NB
LA-9	Mammalian (murine)	25	3500	3000	NB
Pekin Duck	Avian	125	NN	NN	NB
Gekko	Reptilian	2000	NN	NN	NB
Xenopus	Amphibian	275	NN	NN	NB
CHSE	Piscine	NN	NN	NN	NB
D. melanogaster	Insect	NN	ND	ND	ND
Yeast	Fungus	NN	ND	ND	ND

Titer of binding and neutralizing activity is defined as in Table I.

NN = not neutralize; NB = not bind; ND = not determine.

expression locus on the short arm of X chromosome at the junction of Xp21.3 to Xp22.1, Figure 10B, (Wang *et al.*, 1985; Wong *et al.*, 1988).

PRIMARY PROTEIN SEQUENCE SIMILARITY TO BOTH PROKARYOTIC AND EUKARYOTIC REPLICATIVE DNA POLYMERASES

Comparison of the deduced primary sequence of human DNA polymerase α with known protein sequences of both prokaryotic and eukaryotic replicative DNA polymerases reveal several regions of striking similarity (Figure 11). From amino acid number 609 to 1081 of human DNA polymerase α, there are six regions of similarity with viral DNA polymerases from herpes simplex I, cytomegalovirus, Epstein-Barr and vaccinia (Earl *et al.*, 1986; Larder *et al.*, 1987; Kouzardies *et al.*, 1987; Wong *et al.*, 1988; Wang *et al.*, 1989) as well as with yeast DNA polymerase 1 (Pizzagalli *et al.*, 1988; Wong *et al.*, 1988; Wang *et al.*, 1989). Four of these six regions are also found in the DNA polymerases of *E. coli* phage T_4, *Bacillus* phage φ29 and

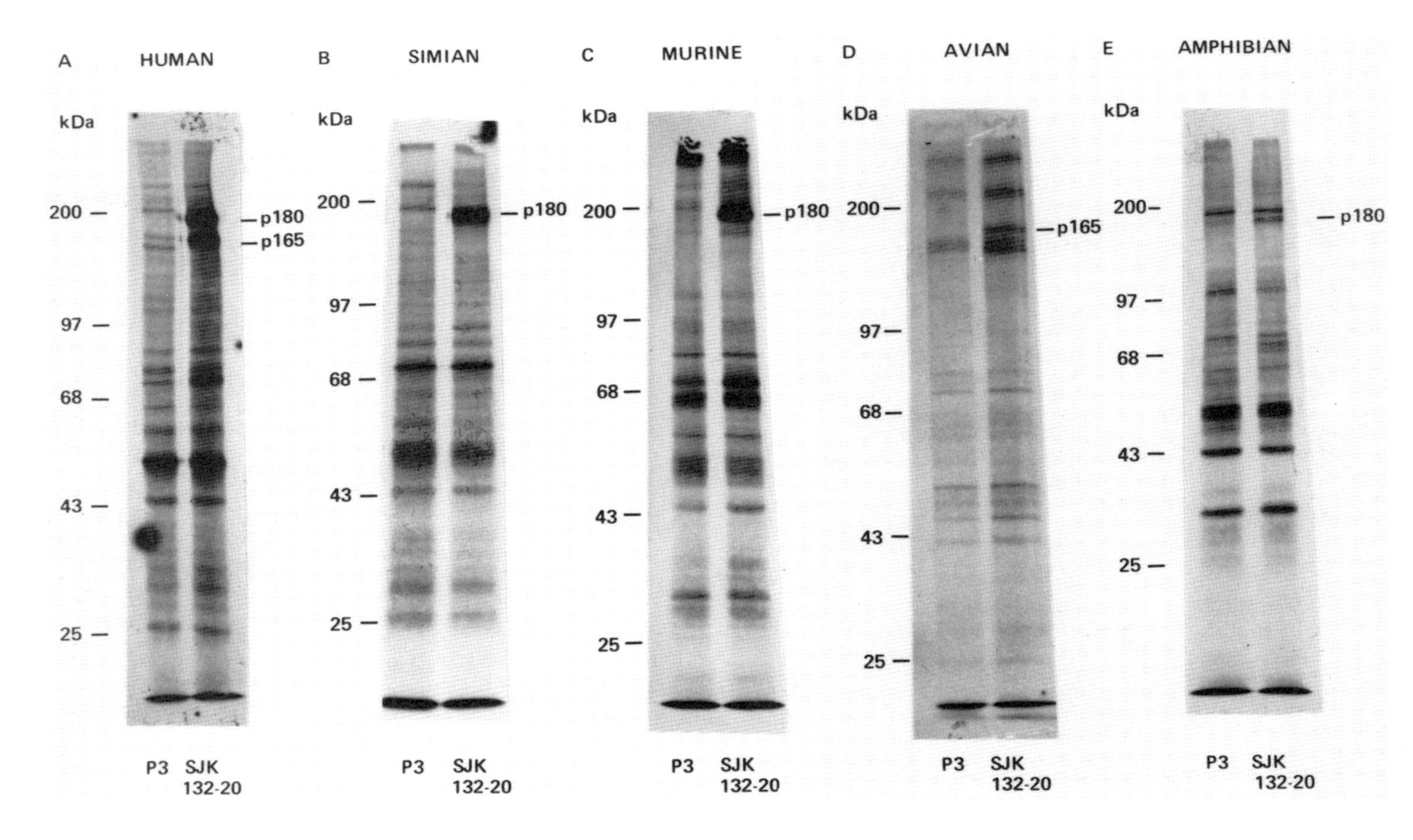

Figure 7. Immunoprecipitation of DNA polymerase α antigens from crude lysates of [^{35}S]-methionine-labeled vertebrate cells with neutralizing antibody SJK 132-20. Lysates were prepared and radiolabeled polypeptides were immunoadsorbedon SJK 132-20 IgG-sepharose or non-immune P3 IgG sepharose and analyzed by denaturing slab gel electrophorosis. (Reprinted, with permission, from Miller, *et al.*, 1988).

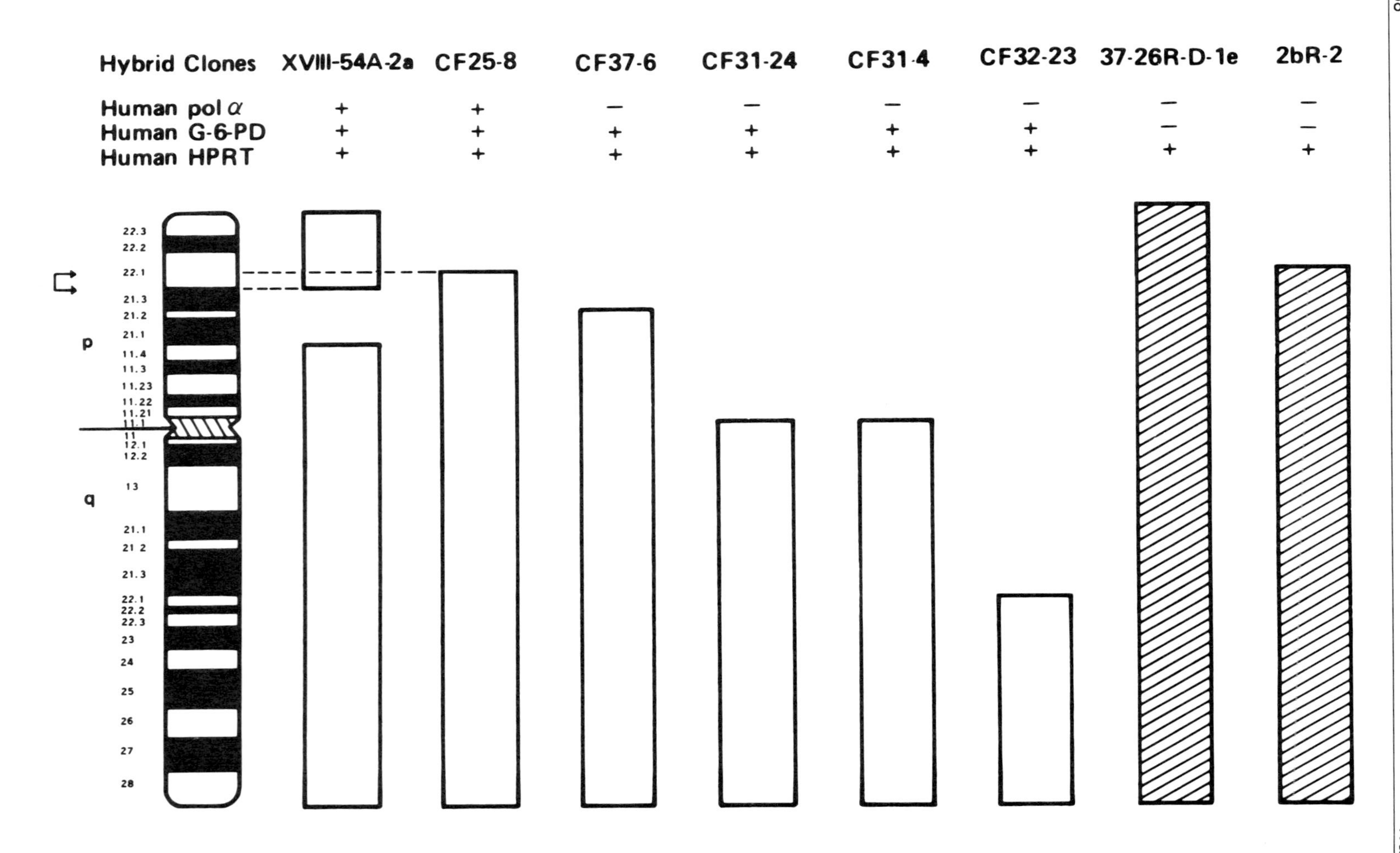
Hybrid Clones
XVIII-54A-2a
CF25-8
CF37-6
CF31-24
CF31-4
CF32-23
37-26R-D-1e
2bR-2
Human pol α
Human G-6-PD
Human HPRT
+ + + + + + + + − + + − + + − + + − + + − − + − − +
22.3
22.2
22.1
21.3
21.2
21.1
11.4
11.3
11.23
11.22
11.21
11.1
11
12.1
12.2
13
21.1
21.2
21.3
22.1
22.2
22.3
23
24
25
26
27
28
p
q

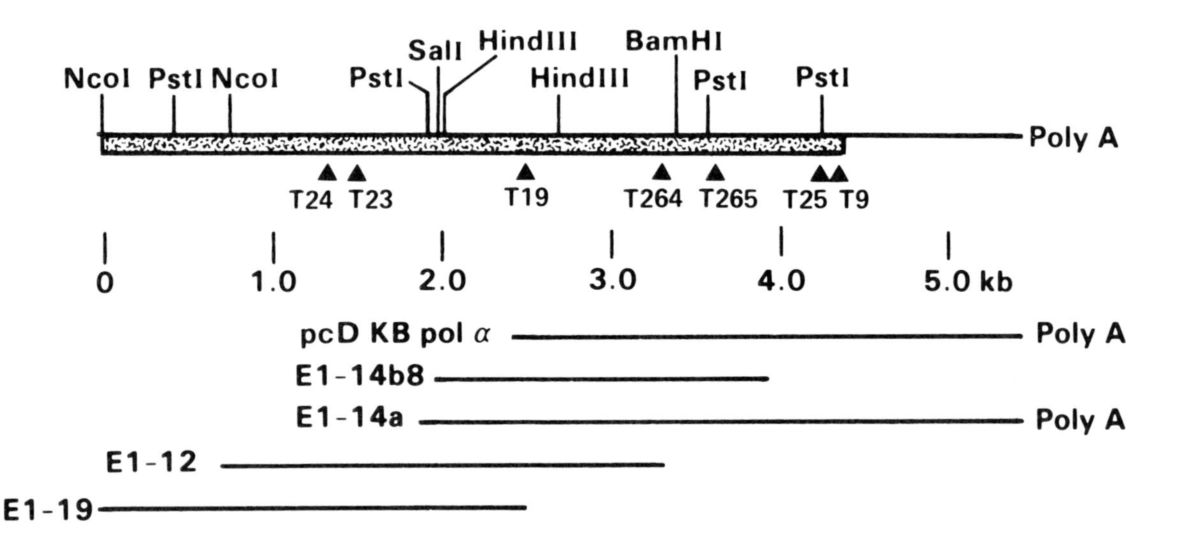

Figure 9. Human DNA polymerase α cDNA. Restriction map of human DNA polymerase α and overlapping cDNA clones. The stippled box represents the coding region of human DNA polymerase α and the solid line indicates the 5′ and 3′ non-coding region. ▲ indicates the locations of each of the previously determined amino acid sequences. The five overlapping cDNA clones are pcD-KBpola, E1-14b8, E1-14a, E1-12 and E1-19. (Reproduced, with permission, from Wong *et al.*, 1988).

Adenovirus (Wang *et al.*, 1989; Wong *et al.*, 1988; Spicer *et al.*, 1988; Bernard *et al.*, 1987; Gingeras *et al.*, 1982). Three of these regions are similar to sequences of *E. coli*PRD1 phage (Jung *et al.*, 1987; Wang *et al.*, 1989), and two potential DNA polymerases, S1 mitochondrial DNA from maize (Paillard *et al.*, 1985) and linear plasmid-like DNAs from yeast plasmid, pGKL1 (Stark *et al.*, 1984; Kikuchi *et al.*, 1985). These regions are designated regions I to VI according to the extent of similarity, with region I being the most similar and region VI the least

Figure 8. (Left) Assignment of the genetic locus of DNA polymerase α to the short arm of the X chromosome near the junction of bands Xp21.3 and Xp22.1. The idiogram of the trypsin-Giesma banding pattern of the human X chromosome is presented at the left. All hybrid clones were analyzed for human DNA polymerase α expression by SJK237-71 binding assays and for the presence of human G6PD and HPRT activities as described (Wang *et al.*, 1985). A + represents 14% of total polymerase activity bound by monoclonal antibody SJK237-71 in CF25-8 cells and 11% in XVIII-54A-2a cells, presence of human G6PD or HPRT activity of 16nmol of IMP produced per hr per 10^7 cells; a - represents <2.6% of total DNA polymerase α activity bound, absence of human G6PD, or HPRT activity of <0.27 nmol of IMP produced per hr per 10^7 cells. The arrows and the broken lines indicate the region of the X chromosome where the gene for DNA polymerase α expression is assigned. Hybrid clone XVIII-54A-2a contains *de novo* interstitial deletion of most of the Xp21 band [46,X del(X) (pter—21.3::p21.1—qter)]; and the other five CF series hybrid clones contain X/autosome translocations: CF25-8 containing an X/13 translocation, retained der(X), Xqter—Xp22.1; CF37-6 containing an X/11 translocation, retained der(X), Xqter—Xp21.2; CF31-24 and CF31-4 containing an X/20 translocation with der(X), Xqter—Xcent. The hatched pattern designates two hybrid clones, 37-26R-D-1e and 2bR-2, containing an inactive X or der(X) chromosome, respectively. (Reproduced, with permission, from Wang *et al.*, 1985).

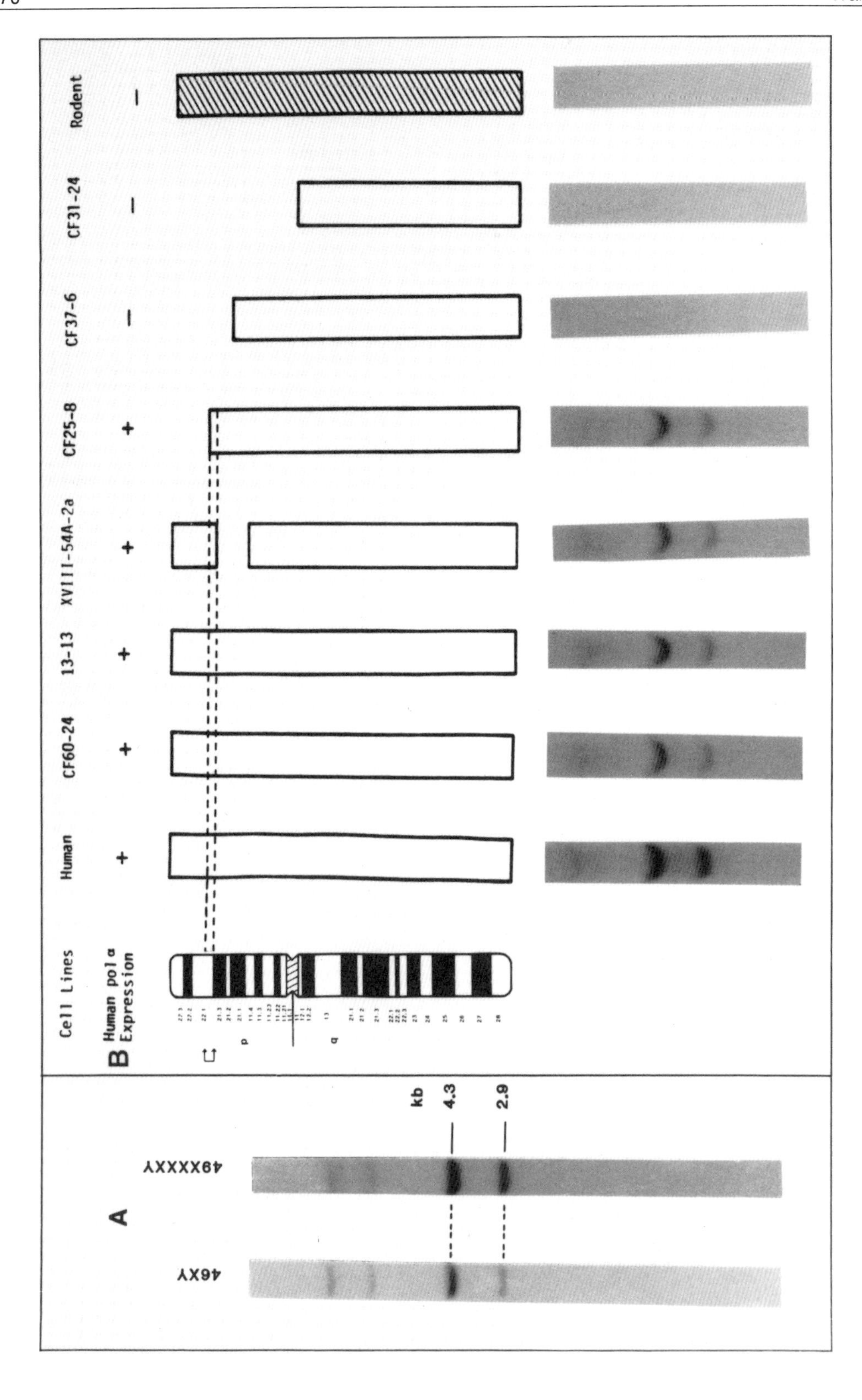
A
46XY
49XXXXY
kb
4.3
2.9
B
Cell Lines
Human pol α Expression
Human
CF60-24
13-13
XVIII-54A-2a
CF25-8
CF37-6
CF31-24
Rodent
p
q

(Figure 11; Wong *et al.*, 1988; Wang *et al.*, 1989). The significance of these regions is further underscored by the relative location of these regions on the respective polymerase polypeptides. The linear spatial arrangement of these six regions on each polymerase polypeptide has a similar order of IV-II-VI-III-I-V (Figure 12). However, the distances between each of these regions are variable.

Regions I, II and III are presented in all twelve DNA polymerases. Region V is present in all eukaryotic polymerases except the two potential eukaryotic polymerases, maize S1 mitochondrial DNA and yeast plasmid pGKL1. The three prokaryotic DNA polymerases (phage T_4, ϕ29 and PRD1) do not contain region V. Adenovirus 2 DNA polymerase and the two potential DNA polymerases (yeast plasmid pGKL1 and maize S1 mitochondrial DNA) are the only eukaryotic DNA polymerases which lack region VI, while T_4 DNA polymerase is the only prokaryotic polymerase containing both regions IV and VI. It is interesting to note that, in this class of replicative DNA polymerases, human DNA polymerase α and yeast polymerase 1 are the largest polymerases (165 kDa) containing the consensus sequences, while *E. coli* phage PRD1 is the smallest DNA polymerase, of 63 kDa molecular weight, containing only regions I, II and III (Wang *et al.*, 1989). This suggests that regions I, II and III might comprise the basic core sequences required for the DNA polymerase function. Comparison of protein sequences to other eukaryotic polymerases such as DNA polymerase β, terminal transferase (Zmudzka *et al.*, 1986; Matsukage *et al.*, 1987) and retroviral reverse transcriptase or prokaryotic *E. coli* polymerase 1 (Joyce *et al.*, 1982) reveal no significant similarity to the regions described above. However, a region II-like sequence is identified in *E. coli* polymerase III (Tomasiewicz and McHenry, 1987).

CONSERVATION OF GENETIC STRUCTURE AMONG EUKARYOTIC SPECIES

The demonstration of the presence of six consensus sequences in the deduced amino acid sequence of human DNA polymerase α to bacteriophage, yeast and human viral DNA polymerases suggests the existence of these sequences in polymerase α genes from phylogenetically diverse eukaryotic organisms. To evaluate this possibility, five cDNA fragments which represent 90% of the coding region of human polymerase α cDNA (Figure 13A), were used in a set of genomic Southern hybridizations with DNA samples from phylogenetically representative species (Figure 13B). The extent of conservation at the nucleotide level was evaluated by washing the genomic Southern blots in a buffer of defined ionic strength but with increasing temperature. Starting from the 3′-end of the coding region of the cDNA, the Pst1/Pst1 fragment hybridizes weakly with calf, rodent, *Drosophila*, and yeast genomic DNA

Figure 10. (Left) Physical Mapping of human DNA Polymerase α cDNA on X chromosome. (A) Genomic Southern blot. Ten micrograms of human genomic DNA isolated from 46XY (1X normal male) and 49,XXXXY (4X, from cell line GM1202A) were digested with EcoR1 and hybridized as described (Wong *et al.*, 1988). (B) Human DNA polymerase α structural gene mapping. The idiogram of trypsin-Giemsa banding pattern of human X chromosome is presented at the left. EcoR1 digested genomic DNA, 10 μg each from human KB cells, rodent AKR thymoma cell line BW5147 and somatic hybrids (Wang, *et al.*, 1985): CF60-24 containing human chromosome 4, 13, 17, 19, 21 and X; hybrid 13-13 containing only human X chromosome; hybrid clone XVIII-54A-2a containing *de novo* interstitial deletion of most of the Xp21 band [46,X del(X) (pter—21.3::p21.1—qter)]; and the other three CF series hybrid clones containing X/autosome translocations: CF25-8 containing an X/13 translocation, retained der(X), Xqter—Xp22.1; CF37-6 containing an X/11 translocation, retained der(X), Xqter—Xp21.2; CF31-24 containing an X/20 translocation with der(X), Xqter—Xcent, were hybridized at 42°C with 30 ηg ^{32}P-labeled (10^9 cpm/μg) Pst1 700 bp fragment of pcD-KBpola. Hybridization and washing conditions were as described (Wong *et al.*, 1988). (Reproduced, with permission, from Wong *et al.*, 1988)

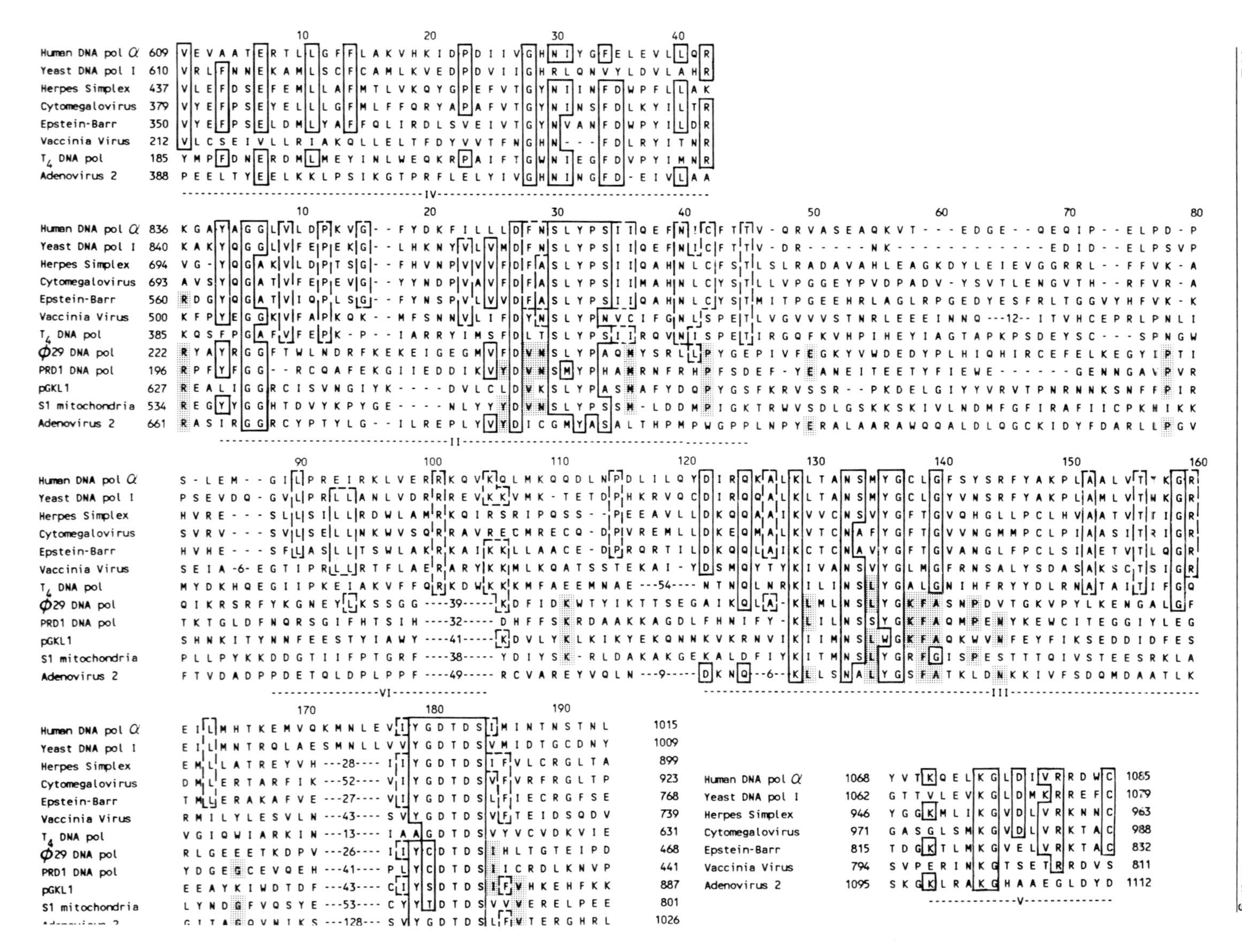
Human DNA pol α
Yeast DNA pol I
Herpes Simplex
Cytomegalovirus
Epstein-Barr
Vaccinia Virus
T4 DNA pol
Φ29 DNA pol
PRD1 DNA pol
pGKL1
S1 mitochondria
Adenovirus 2
IV
II
VI
III
V

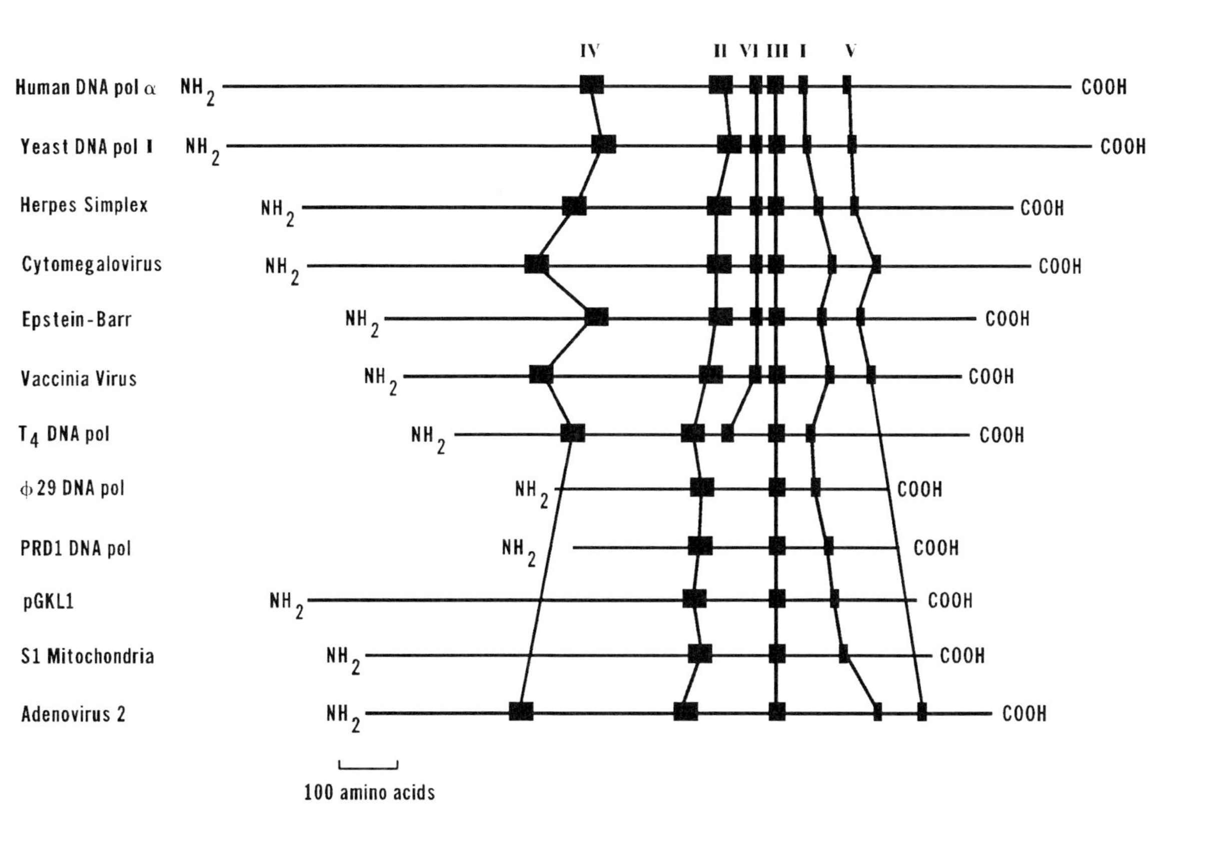

Figure 12. Relative linear spatial arrangements of the conserved regions of DNA polymerases. Each DNA polymerase polypeptide is represented by a straight line with NH_2 and COOH denoting the amino and carboxyl termini, respectively. The heavy black bars represent the consensus sequences of each region. Similar regions of each DNA polymerase polypeptide are aligned by vertical lines. (Reproduced, with permission, from Wang *et al.*, 1989 and a portion of this figure is adapted, with permission, from Wong *et al.*, 1988)

Figure 11. (Left) Conserved regions of human DNA polymerase α and other selected DNA polymerases. Amino acid residues 609-1015 from human DNA polymerase α were aligned with amino acid residues derived from other DNA polymerases; identical residues of five or more sequences between human DNA polymerase α and other DNA polymerases are boxed. Gaps are indicated by dashes and extensive gaps are marked by the number of amino acids contained within the gap. The designated conserved regions are marked by dashed lines under the amino acid residues. Amino acids 998-1005 of human DNA polymerase α are defined as region I; amino acids 839-878 are region II; amino acids 943-984 are region III; amino acids 609-650 are defined as region IV; regions V and VI are amino acids 1075-1081 and 909-926, respectively. Amino acid residues which are uniquely similar among DNA polymerases that use terminal protein priming mechanisms are outlined by stippling; amino acid residues which are uniquely similar among polymerases that use or are predicted to use DNA primase mechanisms are boxed in dashed lines. (Reproduced, with permission, from Wang *et al.*, 1989 and portions of this figure are adapted, with permission, from Wong *et al.*, 1988).

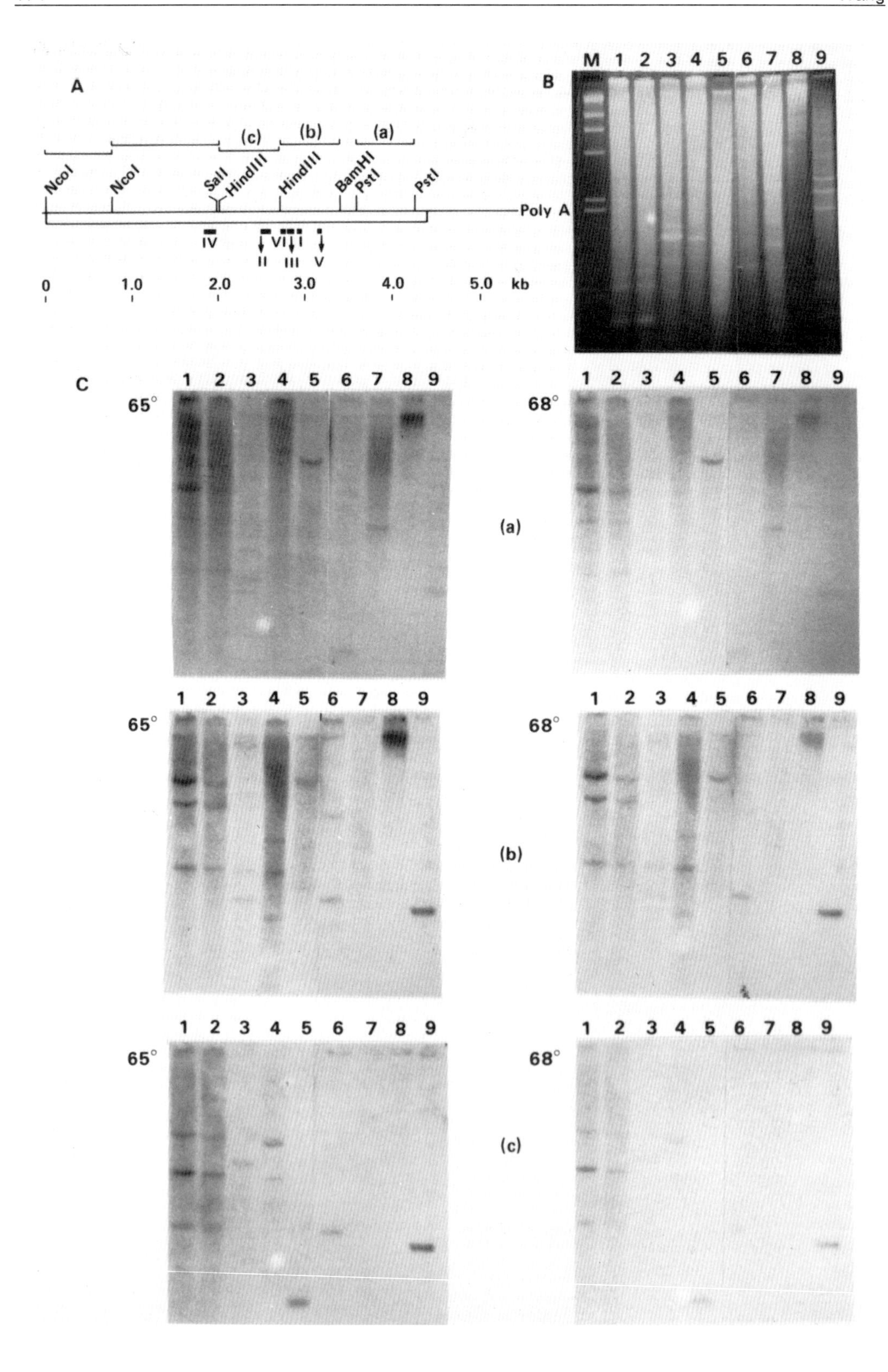
A
(c)
(b)
(a)
NcoI
NcoI
SalI
HindIII
HindIII
BamHI
PstI
PstI
Poly A
IV
II
VI
III
I
V
0
1.0
2.0
3.0
4.0
5.0
kb
B
M 1 2 3 4 5 6 7 8 9
C
65°
68°
1 2 3 4 5 6 7 8 9
(a)
(b)
(c)

but hybridizes relatively strongly with *Xenopus* and tobacco genomic DNA (Figure 13C(a)). The further upstream fragment, BamH1/Hind III, hybridizes to genomic DNA samples from calf, rodent, *Xenopus*, *Drosophila* and yeast after washing at 68°C but does not hybridize to tobacco or algae genomic DNA (Figure 13C(b)). Similarly, the Hind III/Hind III fragment hybridizes to genomic DNA of calf, rodent, *Xenopus*, *Drosophila* and yeast but not with plant DNA (Figure 13C(c)). Both the Pst1/Pst1 and BamH1/Hind III fragments produce a non-discrete large size smear hybridization signal with algae genomic DNA (Figure 13C(a) and (b)). This is probably due to the non-specific hybridization between these cDNA fragments and the undigested portion of the algae DNA and is not representative of true hybridization. Additional experiments with restriction fragments further 5′-upstream, Sal1/Nco1 and Nco1/Nco1, resulted in no significant hybridization with genomic DNA from lower mammal, vertebrate, invertebrate, plant or yeast. These two fragments also hybridize poorly to human genomic DNA at 55°C. The finding that the 5′ end of the human polymerase gene comprises several small exons of less than 100 base pairs separated by large introns of greater than 4000 base pairs may explain the weak hybridization of these two fragments. The strong hybridization of the two human cDNA fragments, BamH1/Hind III and Hind III/Hind III, with yeast genomic DNA is most likely due to the absence of introns in the yeast polymerase I gene.

Five of the six consensus protein sequences are localized in these two most cross-hybridized fragments, BamH1/Hind III and Hind III/Hind III (Figure 13A; Wong *et al.*, 1988; Wang *et al.*, 1989). The ability of these two cDNA fragments to hybridize with genomic DNA from a variety of eukaryotic species indicates the presence of extensive nucleotide sequence conservation in these regions of the polymerase genes. This observation further supports the conclusion of an evolutionary pressure to conserve these regions of DNA sequence which correspond to critical functional domains required for DNA polymerase catalysis. The results also suggest that there is a class of DNA polymerases from vertebrate, invertebrate, unicellular fungi, and from human DNA virus and bacteriophage T_4, PRD1 and ϕ29 all containing these consensus sequences, which are all replicative DNA polymerases and may all have evolved from a single primordial gene. A detail of this study is reported in Miller *et al.*, 1988.

PREDICTED FUNCTIONAL DOMAINS

During the DNA replication process, the catalytic polypeptide of DNA polymerase α is involved in a series of complex and coordinated reactions with DNA, dNTPs and other DNA replication required proteins. A major focus of DNA enzymology is to understand the

Figure 13. (Left) Southern hybridization of human cDNA fragments with genomic DNA samples from various phylogenetic species. (A) Restriction map of human DNA polymerase α cDNA. The open box represents the coding region of human DNA polymerase α and the solid line indicates the 5′ and 3′ non-coding regions. The solid bars on the top labeled (a), (b) and (c) represent the corresponding cDNA fragments used for hybridization. The black bars underneath represent the designated consensus sequences (14). (B). Ethidium bromide stain of the digested genomic DNA samples from representative species. The hybridization and wash conditions are described in (Miller *et al.*, 1988). Lane M, Hind III digested lambda DNA marker; lane 1, human genomic DNA of the 4 X chromosome (49, XXXXY) cell line GM1202A; lane 2, human DNA from KB cells; lane 3, Calf thymus DNA; lane 4, murine DNA; lane 5, *Xenopus* DNA; lane 6, *Drosophila* DNA; lane 7, tobacco DNA; lane 8, green algae DNA and lane 9, yeast DNA. (C) Genomic Southern hybridization: (a) Hybridization with the Pst1/Pst1 fragment of human DNA polymerase α cDNA. (b) Hybridization with the Bam H1/Hind III cDNA fragment. (c) Hybridization with the Hind III/ Hind III cDNA fragment. The washing temperature of each blot is shown at the top left corner of the respective panel. Blots were washed under conditions of increasing stringency from 55°C to 68°C. Only the results obtained at the two highest temperatures are shown. DNA samples of each lane are as described in (B). (Reproduced, with permission, from Miller *et al.*, 1988).

mechanism by which DNA polymerase α acts within these three parameters. Based on the deduced primary protein sequence of the polymerase α catalytic polypeptide, the predicted functional domains which are postulated to be responsible for these interactions are discussed below.

DNA Interacting Domains

In eukaryotic cells, protein structures identified as potential DNA interacting motifs are found to contain cysteine (C)- and histidine (H)- rich sequences. These sequences have the potential to chelate metal ions and form DNA interacting domains designated as "zinc fingers". This was originally documented in *Xenopus laevis* transcription factor TFIIIA (Miller *et al.*, 1985) and later found to be a surprisingly common structure in a variety of nucleic acid binding proteins (Berg, 1986). Generally, there are two classes of "fingers". The C_2H_2 type which is typified by TFIIIA and SP1 transcription factors and usually exists in tandom repeats with 7 to 8 amino acid linkers (designated as X) separating the C/H units. Another C_n type exists as a cluster of 4, 5 or 6 cysteines identified in yeast GAL4 and steroid hormone receptors (Evans and Hollenberg, 1988). All these finger structural motifs identified are transcription regulation proteins. Thus far, no DNA replication proteins have been reported to contain this kind of motif with the proven function of interacting with DNA. Analysis of the deduced primary protein sequence of human DNA polymerase α reveals a cysteine-rich region (Figure 14). One region from amino acid 650 to 715 contains $C\text{-}X_3\text{-}H\text{-}X_{27}\text{-}C\text{-}X_4\text{-}C\text{-}X_{11}\text{-}C\text{-}X_3\text{-}H$, where X represents amino acids other than cysteine and histidine capable of forming an extended protein loop containing many amino acids with side chains capable of interacting with the DNA backbone (Wong *et al.*, 1988). Towards the carboxyl-terminus of the human DNA polymerase α sequence, from amino acid 1244 to 1391, there is another C/H rich region. This region contains sequences which have extensive similarity with yeast DNA polymerase 1 and cysteine residues with the potential to form a DNA interacting finger motif (Figure 14). This C-terminal region is proposed to be the potential DNA interacting domain. Whether the motif observed in DNA polymerase α represents the actual DNA interacting domains of DNA replicative protein or these cysteine/histidine rich sequences are only "gilt by association" (quoted from Evens and Hollenberg, 1988), is a question to be challenged by site specific mutagenesis and functional studies of these mutants in the future.

Deoxynucleotide Interacting Domains

The finding that human DNA polymerase α contains a consensus amino acid sequences with DNA polymerases from bacteriophage and DNA viruses implies an evolutionary pressure to maintain these sequences for essential catalytic function. By marker rescue and marker transfer experiments, many herpes simplex virus mutants conferring altered anti-viral drug sensitivity have been mapped to herpes DNA polymerase gene (HSV pol) within several of these consensus regions (Coen *et al.*, 1983; Quinn and McGeoch, 1985). Most mutants either demonstrate altered sensitivity to the pyrophosphate analog, phosphonoacetic acid (PAA) or the nucleotide competitive inhibitor such as aphidicolin, or nucleotide analog-like, anti-viral drugs such as acyclovir (ACV), bromovinyldeoxyuridine (BVdU), ganciclovir (DHPG), or vidarabine (araA) (Coen *et al.*, 1983; Larder *et al.*, 1987). Recent sequence analyses of several of these mutants derived from three viral strains indicate that a majority of the mutants analyzed contain single amino acid substitutions in regions II and III (Larder *et al.*, 1987; Tsurumi *et al.*,

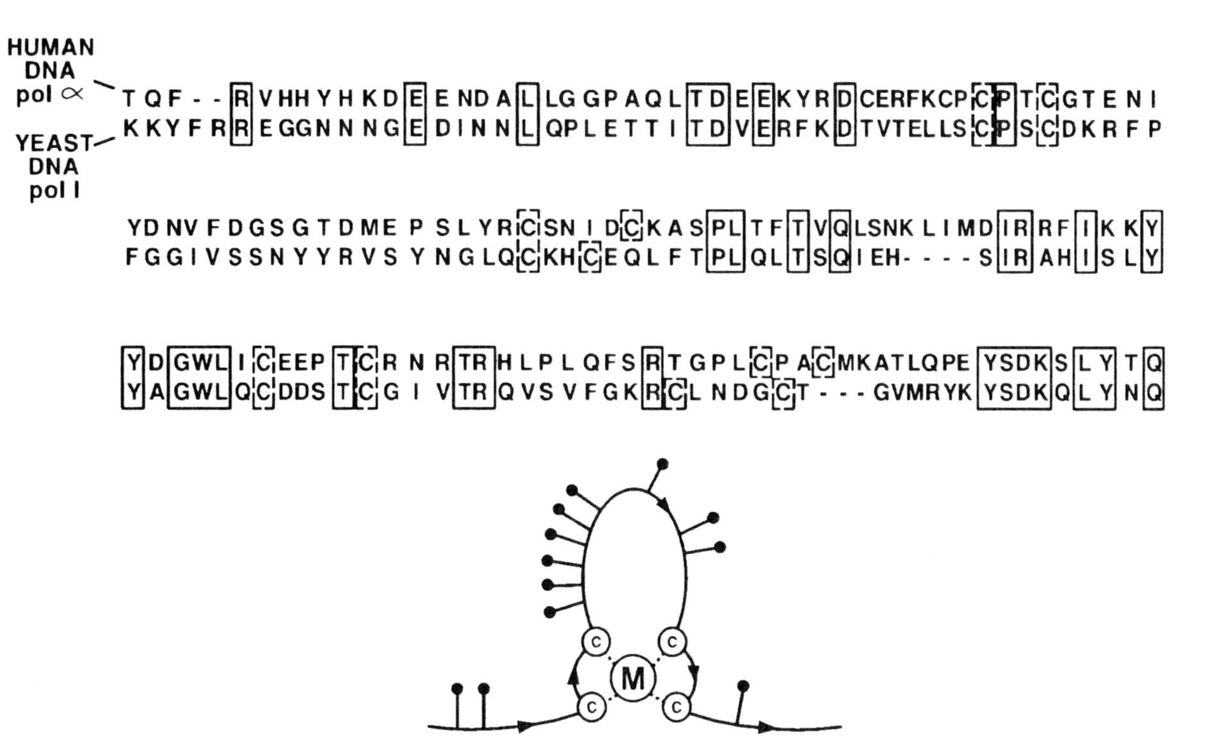

Figure 14. Amino acid sequence of possible DNA interaction region of human DNA polymerase α and yeast DNA polymerase I. Human DNA polymerase α amino acids 1244 to 1391 and yeast pol I amino acids 1246 to 1389 are depicted. Amino acid sequence of possible DNA interaction regions of human DNA polymerase α and yeast DNA polymerase I. Human DNA polymerase α amino acids 1244 to 1391 and yeast pol I amino acids 1246 to 1389 are depicted. Identical amino acids are boxed in solid lines, and cysteines are boxed in dashed lines. A schematic potential DNA-binding loop is illustrated with most probable DNA-binding side chains marked as black circles. The cysteine residues potentially involved in metal ion coordination are boxed in dash line. Human DNA polymerase α sequence is derived from (Wong *et al.*, 1988) and yeast DNA polymerase I sequence from (Pizzagalli *et al.*, 1988, with permission) (Reproduced, with permission, from Wang *et al.*, 1989).

1987; Gibbs *et al.*, 1988). Mutations conferring altered sensitivity to drugs which are analogs of either pyrophosphate or deoxynucleotides are inferred to be dNTP and PP_i interacting domains. Sequence comparison between human DNA polymerase α and herpes simplex I viral DNA polymerase reveals extensive sequence similarity in all six regions. Region I has the highest, 87.5%, followed by region II, 60%; region V, 57%; region III, 47%; region IV, 26%; and region VI, 10.5% (Figure 15)(Wong *et al.*, 1988; Wang *et al.*, 1989). Based on the data from the herpes DNA polymerase mutant, regions II and III of human DNA polymerase α may be directly involved in deoxynucleotide binding or pyrophosphate hydrolysis.

Most recently, two mutations of herpes simplex virus have been identified that do not reside within regions II and III. One mutation in region V has a single amino acid substitution of $asparagine_{961}$ to lysine which confers altered sensitivity to aphidicolin and ganciclovir (DHPG) (Gibbs *et al.*, 1988); the corresponding residue in human DNA polymerase α is aspartate(Figure

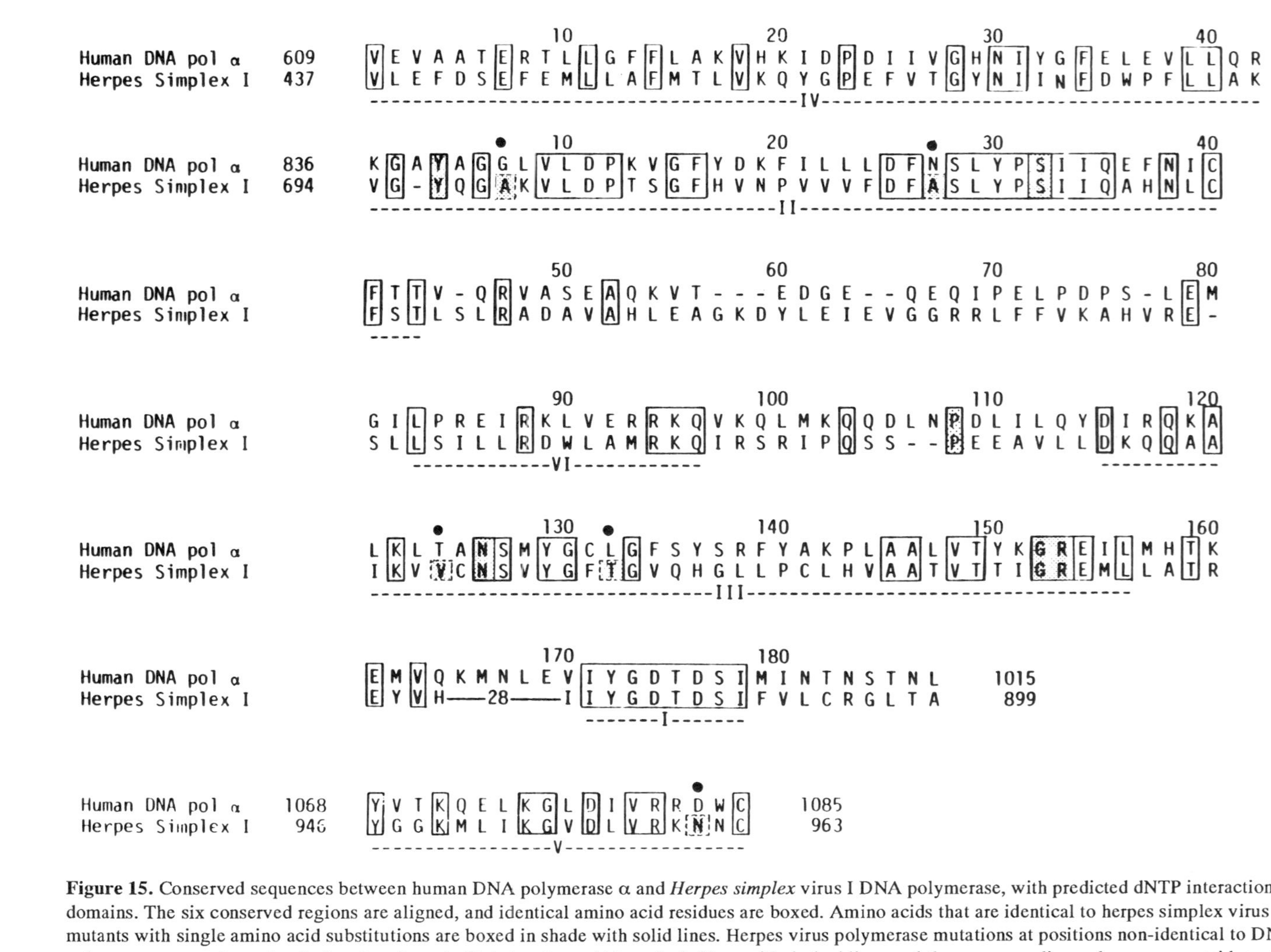

Figure 15. Conserved sequences between human DNA polymerase α and *Herpes simplex* virus I DNA polymerase, with predicted dNTP interaction domains. The six conserved regions are aligned, and identical amino acid residues are boxed. Amino acids that are identical to herpes simplex virus mutants with single amino acid substitutions are boxed in shade with solid lines. Herpes virus polymerase mutations at positions non-identical to DNA polymerase α are shown by designating the substituted amino acid in a shaded box with dashed lines and the corresponding polymerase α residue with a black dot. (Reprinted, with permission, from Wang *et al.*, 1989 and Wong *et al.*, 1988).

15). Another herpes virus mutant, PAA^{r}C, contains a single amino acid substitution of proline$_{797}$ to threonine between regions VI and III, conferring resistance to phosphonoacetic acid (PAA). Interestingly, human DNA polymerase α has a proline residue at the same position. Recently, two additional mutants of herpes simplex virus polymerase with altered drug sensitivity phenotypes were identified from wild-type HSV type I strain KOS. The mutations were mapped to the HSV *pol* gene designated as region A, from amino acid residues 565 to 637. These two mutants show sequence similarity only to other viral DNA polymerases that exhibit sensitivity to certain anti-viral drugs, but this region A is not present in the human DNA polymerase α sequence (Gibbs *et al.*, 1988). Furthermore, four other mutations of HSV *pol* gene within regions II and III have substitutions at amino acid residues that are not identical to the human DNA polymerase α sequence.

Data from herpes simplex DNA polymerase mutants with altered sensitivity to anti-viral drugs have failed to define any specific region on the HSV polymerase molecule as the site solely responsible for dNTP or anti-viral drug binding. These findings suggest that the dNTP interacting site(s) may require the interaction of several separate regions on the polymerase molecule. Thus far, there are no herpes DNA polymerase mutants identified in regions I, IV or VI. The expression of enzymatically active recombinant adenovirus 2 DNA polymerase from cloned cDNA was recently accomplished (Shu *et al.*, 1987; Stunnenberg, *et al.*, 1988). Attempts to identify functional mutants within region I have been unsuccessful (J. A. Engler, personal communication). This suggests that regions I, IV and VI could be critical for polymerase function and that mutations in these regions may be extremely damaging to DNA polymerization function and viral viability.

Protein Interacting Domains

The protein domain(s) required for interaction with other DNA replicative proteins are essentially unknown at present. Most recently, the gene of yeast polymerase I was sequenced (Pizzagalli *et al.*, 1988). A temperature sensitive mutant of yeast polymerase I which is defective in DNA primase association was isolated by *in vitro* mutagenesis (Lucchini *et al.*, 1988). At the non-permissive temperature, this mutant can complete one round of cell division and DNA replication before it arrests. Sequence analysis of this mutant reveals a point mutation of G to A, resulting in a single amino acid substitution of glycine at codon 493 by positively charged arginine (Lucchini *et al.*, 1988). The point mutation is mapped to the consensus region IV of yeast DNA polymerase I gene within a region which is perfectly homologous to human DNA polymerase α, Figure 16. The change of glycine$_{493}$ to arginine is a drastic charge change and the defect in primase association could be caused by indirect charge effect on the peptide folding. Nevertheless, this temperature sensitive mutant of yeast polymerase I might just define a protein domain that is responsible for primase interaction.

Further examination of the primary sequence of this class of DNA polymerases also reveals the presence of several sequences unique to polymerases which utilize a protein priming mechanism instead of oligoribonucleotide primers synthesized by DNA primase(Figure 11). The implication of these unique sequences in polymerases such as PRD1, ϕ29 and adenovirus or even in the two potential DNA polymerases, yeast pGKL1 and maize S1 mitochondrial DNA remains to be verified by future mutagenesis studies.

Based on the available data thus far, the predicted functional domains of human DNA polymerase α are illustrated in Figure 17.

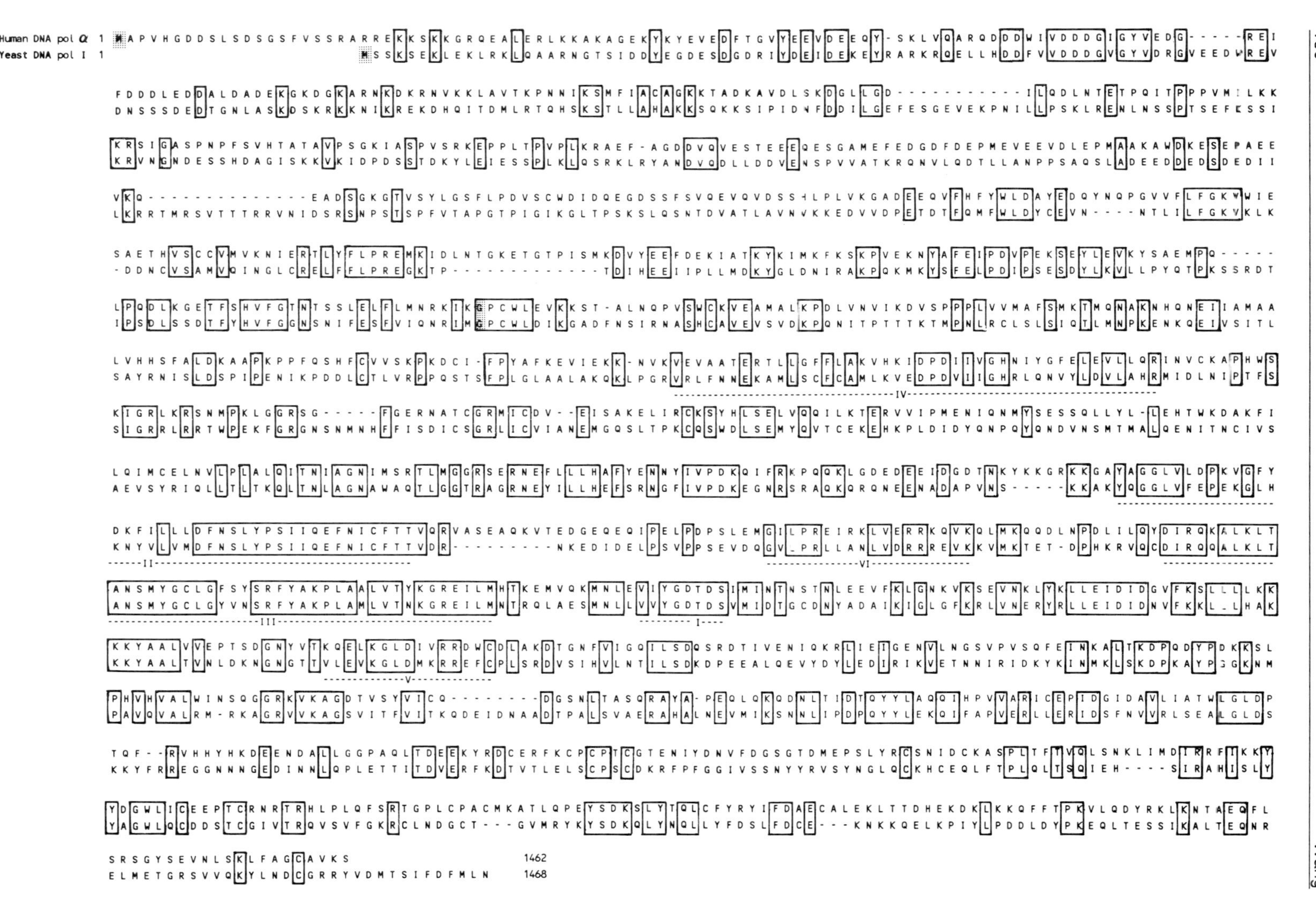
Human DNA pol α 1
Yeast DNA pol I 1
1462
1468

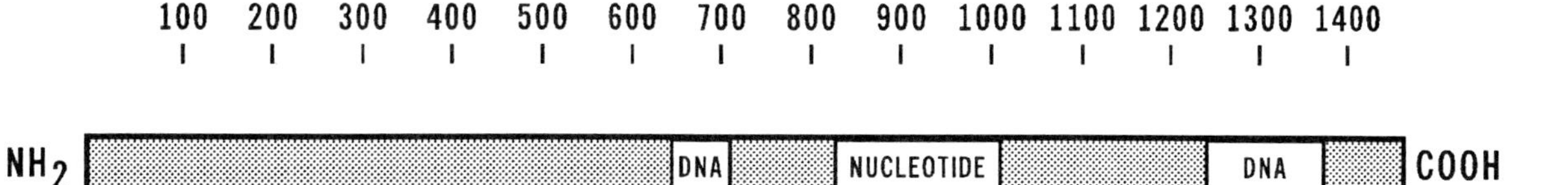

Figure 17. Schematic representation of the putative functional domains of human DNA polymerase α. (Reprinted, with permission, from Wang *et al.*, 1989; adapted and modified from Wong *et al.*, 1988).

GENE EXPRESSION

Cell proliferation and the cell cycle are tightly regulated processes. Increasing evidence indicates that these processes are controlled by environmental signals such as the spatial restriction of cell to cell contact or by extracellular growth factors and hormones. Exactly how these growth factors and mitogens transduce and mediate signals within the cell to influence cell proliferation is a subject of intense interest. Recent studies suggest that certain proto-oncogenes play a regulatory role in cellular proliferation. To understand how the cell cycle and cellular proliferation are regulated in eukaryotes and how certain mitogens and growth factors affect these events, it is necessary to elucidate the regulation of the expression of genes involved directly in DNA replication. Several genes involved indirectly in DNA synthesis, such as dihydrofolate reductase (*dhfr*) (Farnham and Schimke, 1985), thymidine kinase (*tk*) (Groudine and Casimir, 1984; Coppock and Pardee, 1987; Sherley and Kelly, 1988) and thymidylate synthetase (*ts*) (Storm *et al.*, 1984; Ayusawa *et al.*, 1986), demonstrate transient expression in the cell cycle or during cell proliferation. DNA polymerase α plays a direct role in DNA synthesis and its enzymatic activity and enzyme protein were previously reported to correlate positively with cellular proliferation and transformation (Thommes *et al.*, 1986; Bensch *et al.*, 1982; Wong *et al.*, 1988; Wahl *et al.*, 1988). A gene that exhibits transient expression during cell proliferation or during the cell cycle could either be the regulator responsible for cells entering proliferation and the cell cycle, or could itself be the target of a proliferative or cell cycle specific regulatory signal. With the available cDNA of human DNA polymerase α, it is possible to address the question of the regulation of the expression of this gene, which is directly involved in DNA synthesis during the cell cycle and cellular proliferation, and to begin to ask whether the expression of this gene is a direct target of the cascade of biochemical events induced by growth factors, proto-oncogenes and various mitogens.

Figure 16. (Left) Primary protein sequence comparison of human DNA polymerase α and yeast DNA polymerase I. Identical amino acid residues are boxed, and the six consensus regions are underlined by dashed lines. The first methionine residues are stippled, as is the $glycine_{493}$ of yeast polymerase I, the site of a yeast pol I temperature sensitive mutation with altered stability of DNA primase association. (Human DNA polymerase α sequence is reprinted, with permission, from Wong *et al.*, 1988 and yeast DNA polymerase I sequence, with permission, from Pizzagalli *et al.*, 1988).

An important consideration in studying gene expression in various cell biological events is to distinguish the cell cycle from cell proliferation. Regulatory mechanisms in actively cycling cells could be entirely different from those in growth stimulated proliferating cells. Thus, it is absolutely necessary to discriminate between the methods used to obtain cell populations which reflect cycling cells and those which stimulate quiescent cells to proliferate. Many investigations studying the "cell cycle" regulation of gene expression have used either serum or amino acid deprivation, thymidine blockage or growth to high cell density to achieve cell synchronization. These methods severely perturb normal cell growth and artificially bring the cells into G_0 phase from which, upon stimulation, the cells have to enter G_1 to S phase. It is problematic whether the time required for the cells to enter the cell cycle is a factor of the type of metabolic block applied or, in addition, whether the passage through an artificial G_0 phase will affect the normal gene expression events occurring later in the cell cycle. Thus, it is important that the study of gene expression during the cell cycle be done only in cells that are actively cycling, not in cells activated to proceed from quiescence to proliferation.

Human DNA polymerase α gene expression is reviewed here in three biological events: (i) in a comparison of transformed cells versus normal cells; (ii) in the activation of quiescent cells to proliferate; and (iii) in different stages of the cell cycle in actively cycling cells.

Gene Expression in Transformed Cells

Human DNA polymerase α as analyzed by Northern blot reveals a single 5.8 kb signal (Figure 18A). Comparison of message level in normal proliferating human cells and transformed cells is presented in Figure 18B. With equal quantities of poly$(A)^+$ mRNA from human breast carcinoma cell line (MDA4) and 20 week actively growing normal human placenta, the transformed cell line yields a readily detectable polymerase α message while, with the same length of exposure of the autoradiograph, there is no detectable signal with normal human placenta mRNA. This experiment gives a semi-quantitative evaluation of the abundance of polymerase α steady-state message in transformed cells as compared to normal cells.

A careful and refined quantitative comparison of DNA polymerase α transcript level and enzymatic activity in normal and transformed cells was also performed and is shown in Table VII. To quantitate polymerase α mRNA, known quantities of total RNA from transformed cell lines, KB and 293, and non-transformed but proliferating cell lines, TNHF and GM1604, are compared with known amounts of standard sense human DNA polymerase α specific RNA corresponding to a cDNA restriction fragment, by RNase protection analysis using a corresponding anti-sense RNA probe. Based on a value of 26 picograms of total RNA per mid-logarithmic growth cultured human fibroblast (Alberts *et al.*, 1983), normal human cell lines GM1604 and TNHF contain 15 and 4 molecules of human polymerase α specific message, respectively. The transformed cell lines KB and 293 contain 150 and 170 molecules per cell, respectively. Comparative assays of DNA polymerase α enzymatic activities in these cell lines also indicate about a 10 fold increase in polymerase enzymatic activity in transformed cells as compared to normal proliferating cells. Thus, in transformed cells there is a greater than 10-fold amplification of human DNA polymerase α gene expression at both transcriptional and post-translational levels as compared to normal cells (Wahl *et al.*, 1988). Comparison of the polymerase α gene dosage from transformed and non-transformed cells by genomic Southern blot analysis indicate that the greater than 10 fold amplification of polymerase α gene expression in transformed cells is not due to polymerase α gene amplification. These data suggest

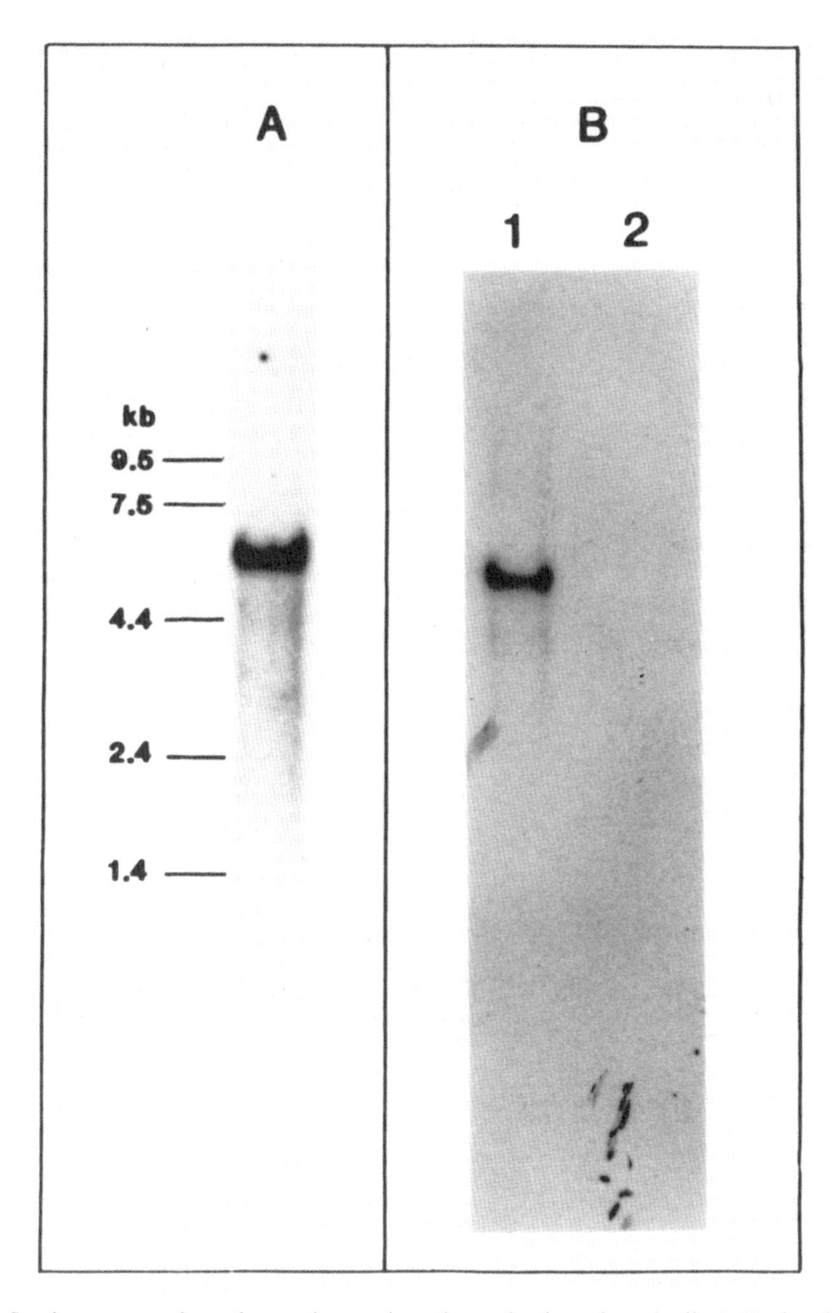

Figure 18. Steady-state transcript analyses and comparison of normal and transformed cells. A. Northern hybridization analysis of human DNA polymerase α mRNA. Ten micrograms of polyadenylated mRNA from early mid log human KB cells was hybridized with 50 μg of ^{32}P-labeled Hind III/BamH1 700 bp restriction fragment as described in Wong *et al.*, 1988. RNA sizes are given in kilobases (kb) and were determined by staining of parallel blot with RNA standard ladder marker from Bethesda Research Laboratory (BRL). B. Steady-state messenger RNA comparison of transformed and normal human cells. Five μg of poly(A)+mRNA from transformed cell line, MDA4, (lane 1) and from normal human proliferating tissue isolated from 20 week placenta (lane 2) were compared for relative abundance of steady state DNA polymerase α mRNA. (Adapted, with permission, from Wong *et al.*, 1988)

TABLE VII

DNA Polymerase α Activity and Steady State Transcript Levels in Transformed and Non-Transformed Human Cells

Transformed Cells	Activity Units per 10^7 Cells[a]	mRNA Copies per cell
293	19.0	170
KB	14.0	150
Non-Transfomed Cells		
TNHF	2.1	4
GM1604	1.7	15

[a] One unit of DNA polymerase incorporates 1 nmol of dAMP hr^{-1} at 37°. Reprinted, with permission from Wahl *et al* , 1988.

that,in transformed cells, there is either higher transcriptional rate of this gene or greater stability of the polymerase α message.

Gene Expression During Activation of Cell Proliferation

Gene expression of human DNA polymerase α at transcriptional and post-transcriptional levels was investigated in quiescent cells stimulated to proliferate. Normal human lung fibroblasts, IMR90, of 28 population doubling passages were serum deprived by maintaining cells in 0.1% serum and were processed by two approaches: (i) various times of serum deprivation but harvesting of all cell samples at a defined time as depicted in Figure 19A; (ii) equal length of time of serum deprivation but harvesting at times designated in Figure 19B. In both approaches, cellular DNA synthesis starts 18 hours after serum stimulation, peaks at 24 hours, and then declines after 30 hours as monitored by $[H^3]$-thymidine incorporation. Using equal amounts of total RNA (verified by spectrophotometry and ethidium bromide stained ribosomal RNA bands on agarose gel electrophoresis, (Figure 19A and B), steady-state mRNA was analyzed by either Northern blot hybridization (Figure 19A), or the RNase protection method (Figure 19B), at different times after serum stimulation. Two control gene probes, histone H3, known for its transient expression prior to DNA synthesis, and actin, known for its

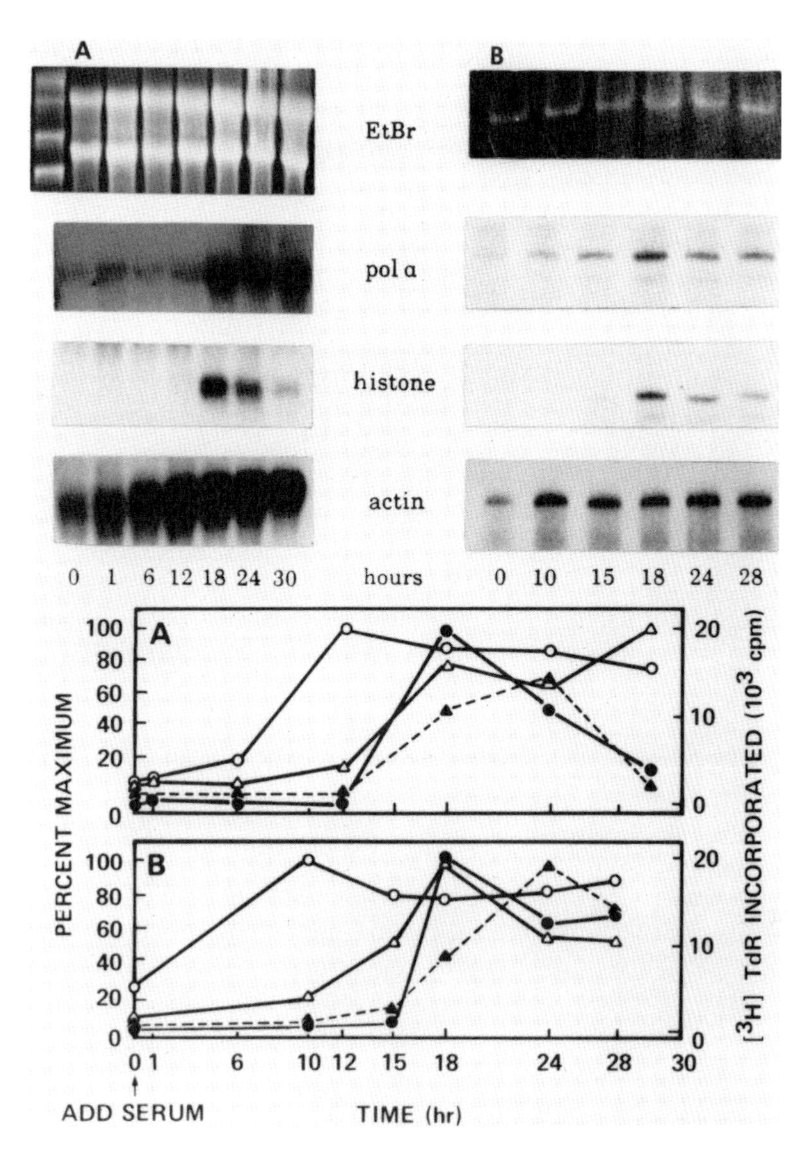

Figure 19. Steady state message level of human DNA polymerase α in quiescent IMR90 cells stimulated to proliferate. Northern blot hybridization and RNase protection analysis of total RNA from IMR90 cells from quiescence to progressive proliferation were performed to measure relative levels of human DNA polymerase α, histone H3 and actin steady-state mRNA. Equal amounts of total RNA were used for each experiment and verified by ethidium bromide staining of rRNA bands (EtBr). Shown are RNA size markers (BRL Inc.) and 18S and 28S rRNA staining of the 30 μg samples prior to transfer and Northern analysis (A), and 28S rRNA staining of 10 μg aliquots of RNA used for RNase protection analysis (B). The relative intensity of each resultant autoradiogram was measured by densitometry and presented as percent maximum intensity for each experiment; DNA polymerase α (△), histone H3 (●), actin (○), versus time after serum stimulation. The level of *in vivo* DNA synthesis determined by [^{3}H]-thymidine incorporation into parallel cultured cells for each time point is also shown (—▲ —). Column A shows results from 30 μg total RNA samples from IMR90 cells starved and restimulated by method (A) described in Wahl *et al.*, 1988 and analyzed by Northern blot hybridization. Column B shows results of 7 μg (for histone H3 or actin) or 10 μg (for DNA polymerase α) total RNA from IMR90 cells starved and restimulated by method (B) and analyzed by RNase protection as described in Wahl *et al.*, 1988. (Reprinted, with permission, from Wahl *et al.*, 1988)

constitutive expression during cell proliferation, were used as references. By either methods of serum deprivation, cells harvested for steady-state mRNA analysis demonstrated a low but detectable basal level of human DNA polymerase α steady-state message which appears in the first 12 hours after stimulation of quiescent cells (G_0) by serum. Eighteen hours after serum stimulation, the steady-state mRNA increases 8-fold but is constitutively expressed even after DNA synthesis (thymidine incorporation) declines. For histone H3 gene, no detectable steady-state message was found 12 to 15 hours after serum stimulation, while a burst of mRNA induction was readily detectable prior to the peak of DNA synthesis at 18 hours, followed by a 7 fold decrease 30 hours after serum addition, when *in vivo* DNA synthesis declines. Conversely, actin steady-state message is present throughout serum induction, peaks 10 to 12 hours after stimulation and is constitutively expressed even after DNA synthesis declines. The results of this study of human polymerase α gene expression at the transcriptional level relative to histone H3 and actin are summarized in Figure 19.

The post-transcriptional expression of human DNA polymerase α, *i.e.* the rate of nascent polymerase protein synthesis and enzymatic activity, was also studied at progressive stages of cell proliferation. At different stages of proliferation, normal human lung fibroblasts were pulse-labeled with $[S^{35}]$-methionine, DNA polymerase α protein was immunoprecipitated by a neutralizing monoclonal antibody SJK132 (Wong *et al.*, 1986) and subsequently analyzed on SDS gel (Figure 20). Like the transcriptional expression result, a low but detectable basal level of polymerase *de novo* protein synthesis is observed in all early fractions up to 12 hours after serum stimulation, followed by an 11-fold induction of polymerase protein at 18 hours, just prior to the peak of DNA synthesis. After the decline of DNA synthesis, there is an observed slight decrease in the rate of polymerase protein synthesis, which is still about 7-fold above the basal level. The enzymatic activity was also studied during the activation of cell proliferation. In order to distinguish polymerase α specific enzymatic activity, DNA polymerase activity was measured by assaying dNMP incorporation into acid insoluble DNA (Fisher and Korn, 1977), in the absence and presence of neutralizing anti-polymerase α monoclonal antibody (Figure 20; Tanaka *et al.*, 1982). *In vivo* DNA synthesis was monitored by thymidine incorporation in parallel with the enzyme assay. Polymerase α enzymatic activity increases in parallel with *in vivo* DNA synthesis. However, like the steady-state message and nascent polymerase α protein, the enzymatic activity does not decline as does DNA synthesis *in vivo* 30 hours after serum stimulation (Figure 20; Wahl *et al.*, 1988).

Reports of similar cell proliferation experiments but addressed as "the cell cycle" (Baril *et al.*, 1973; Chiu and Baril, 1975; Spardari *et al.*, 1978; Thommes *et al.*, 1986), have demonstrated a decline of enzymatic activity at later time points after serum stimulation of cells or the release of a metabolic block. In cell proliferation experiments, regardless of the method used the only biologically significant time frame is the transition from G_0 to G_1/S phase. The biological events after S phase may reflect cell cycle changes or the recovery of induction to proliferate. Therefore, the expression of human DNA polymerase α in this biological event can only be interpreted meaningfully within the narrow window of G_0 to G_1/S transition. In conclusion, the expression of human DNA polymerase α at both the transcriptional and the post-transcriptional levels (*i.e.* both nascent protein synthesis and enzymatic activity) is positively correlated with cell proliferation. Once cells are in an active proliferating state, however, the expression of this gene appears to be constitutive.

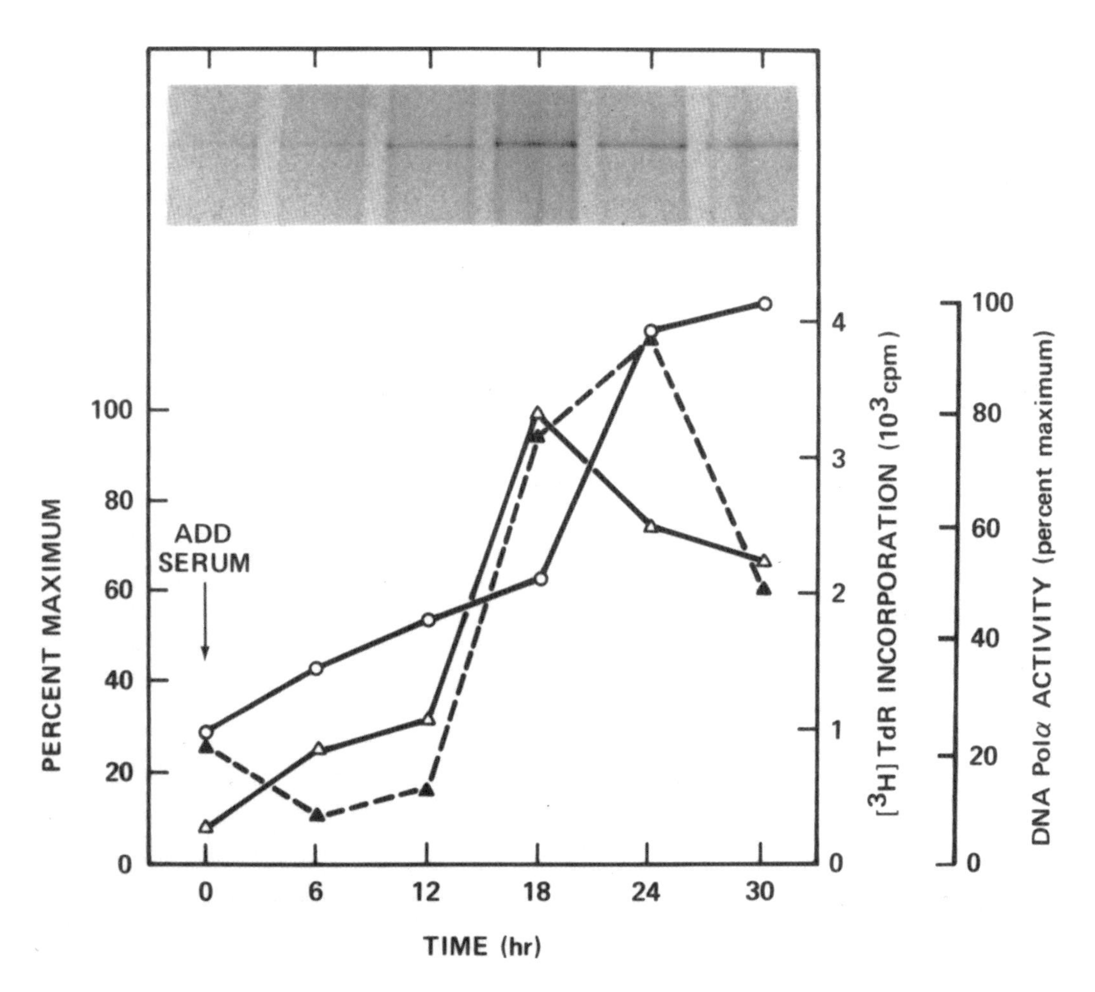

Figure 20. Post-transcriptional expression of human DNA polymerase α in quiescent IMR90 cells stimulated to proliferate. Cells were starved and restimulated by method (A), pulse labeled with [^{35}S]-methionine/ cysteine, and lysed. Polymerase α antigens were immunoprecipitated from the normalized lysates, separated by SDS-polyacrylamide gel and subjected to autoradiography. The relative intensity of each 165 kDa (antigen) protein band was determined by integration of scanning densitometry, and the percent maximum intensity was plotted versus time after serum stimulation (Δ). DNA polymerase α activity (O) was measured from normalized cell lysates of the same IMR90 cells progressing from quiescence to proliferation as described in Wahl *et al.*, 1988. The 100 % value was 100 pmol of dAMP hr^{-1} 2.5 × 10^{-5} $cells^{-1}$. Cellular DNA synthesis was determined by parallel [^{3}H]-thymidine incorporation assay and plotted against each time point (—▲ —). (Reprinted, with permission, from Wahl *et al.*, 1988).

Gene Expression in the Cell Cycle

To distinguish gene expression events occurring during the activation of quiescent cells to proliferate from events regulated within the cell cycle, cells in exponentially growing cultures were separated by counterflow centrifugal elutriation, which fractionates cell populations on the basis of cell size. Cells undergo a linear increase in volume as they progress through the cell cycle, thus non-synchronized and actively growing cells can be separated by size into subpopulations representing progressive stages of the cell cycle. This method, unlike cell proliferation experiments, does not artificially perturb cellular metabolism as do metabolic block and growth factor deprivation experiments.

Exponentially growing cells, 293 or K562 (Figure 21), were separated into subpopulations of increasing average cell volume, which in turn represent different stages of the cell cycle. To verify the effective separation of cell cycle populations, aliquots of each cell fraction were analyzed for their relative concentration of DNA and RNA per cell by microfluorometry (Figure 21 A,B and D,E). The profile of average fluorescence per cell of each fraction was estimated and assigned a ploidy value. Fractions 1-3 are of 2n DNA and correspond to G_1, fractions 4-6 represent the progressive stages of S phase, and fractions 7-9 represent the progressive stages of G_2 of the cell cycle. Fractions 10 and 11 contain a high population of aneuploid cells, *i.e.* 4n DNA and cells with more than 2 nuclei. Ribosomal RNA increases in direct proportion to cell volume as cells progress from G_1 to G_2/M. To avoid variation in the quantitation of polymerase α mRNA attributed to the increase of total RNA, equal amounts of total RNA samples were used for analysis, instead of equal numbers of cells, in the measurement of steady-state DNA polymerase α messages. By the RNase protection method, the steady-state mRNA levels of human DNA polymerase α from both cell lines 293 and K562 are found to be constitutively expressed throughout the cell cycle, with a nominal increase during S phase. Similarly, the levels of actin are constitutively present throughout the cell cycle, with a slight increase during S phase in both 293 and K562 cell lines. Conversely, histone H3 steady-state message level transiently and dramatically increases during S phase and declines in G_2 phase (Figure 21). As measured by scanning densitometry, the mRNA levels of human DNA polymerase α vary less than 2-fold between any two fractions of cell-cycle dependent subpopulations in several experiments with both cell lines (Figure 21).

To ensure the generality of this finding, synchronous cell populations were also obtained by the second method of mitotic shake-off. Unlike centrifugal elutriation, which separates cells by size and where each cell fraction does not contain a pure, discrete cell population representing distinct stages of the cell cycle, the mitotic shake-off method provides more discrete fractions of synchronous cell populations. Synchronization of Hela cell fractions obtained by mitotic shake-off was monitored by [H^3]-thymidine incorporation and mitotic index (Milarsky and Morimoto, 1986). By RNase protection analysis, the steady-state mRNA of human DNA polymerase α again is found to be constitutively expressed during the cell cycle with a slight increase from G_1 to S. However, there is a more than 2-fold decrease of polymerase mRNA levels from S_2 to S/G_2 and G_2/M (Figure 22). In contrast, histone H3 again demonstrates no detectable expression in G_1, a transient and dramatic expression in S phase and a decline when cells enter G_2 (Figure 22). This finding of a slight decline of human DNA polymerase α steady-state message in S/G_2 and G_2/M of Hela cells could either be an intrinsic property of Hela cells different from K562 and 293 cell lines, or is due to experimental variation.

To analyze post-transcriptional events of human DNA polymerase α expression in various stages of the cell cycle, cells were pulsed-labeled with [H^3]-leucine for 30 minutes prior to elutriation. Since cell mass is in direct proportion to cell volume during progression through the cell cycle, the labeled cells from each elutriated fraction were lysed and lysates from different cell fractions were adjusted to normalize for equal amounts of incorporated label. This allows an analysis of changes in polymerase α protein levels that are not due to changes in cell mass during the cell cycle. The polymerase protein was immunoprecipitated from these normalized lysates by monoclonal antibody SJK132-Sepharose, and enzymatic activity was assayed directly from these normalized lysates. Like steady-state message, both polymerase gene-encoded protein synthesis and enzymatic activity were constitutively expressed throughout the cell cycle in either 293 or K562 cells (Figure 23). No striking variation in the synthesis of an immunoprecipitated 180 and 165 kDa protein was observed and a less than 20% variation in enzymatic activity was also found throughout the cell cycle.

Taken together, the results from three different cell lines derived from three independent modes of transformation and from populations of cells at substages of the cell cycle derived by two different methods of cell synchronization, indicate that the gene of human DNA polymerase α is constitutively expressed in actively cycling cells (when cells are not in G_0 state). The results of this set of experiments strongly suggest that the regulatory mechanisms of expression of this gene during the entrance of cells from a quiescent state into the mitotic cycle are entirely different from those that exist in continuously cycling cells. The concordant expression of polymerase α gene at transcriptional and post-transcriptional levels in cell transformation, in activation of cellular proliferation and throughout the cell cycle supports the conclusion that the regulation of human DNA polymerase α gene expression is at the transcriptional level.

CONCLUSION AND FUTURE QUESTIONS

During the last few years, substantial progress has been made in the understanding of eukaryotic DNA replication, mainly due to numerous breakthroughs and the advancement of three model systems for eukaryotic DNA replication: yeast, adenovirus and SV40. In parallel with the general progress in the study of these model systems, a major step forward in the understanding of several key proteins involved in DNA replication has been accomplished. As reviewed here, several significant questions about DNA polymerase α from human cells have been answered. The initial step was the establishment of a panel of hybridomas producing stable monoclonal antibodies specific for DNA polymerase α. The development of immunoaffinity purification with these monoclonal antibodies has resolved the much debated structural problem. The question of intracellular localization, which was a controversial issue for years, has been resolved by immunocytochemical detection methods where DNA polymerase α has been proven to be a nuclear enzyme. By the species specificity of a monoclonal antibody, the gene of human DNA polymerase α was mapped to a specific locus on the short arm of the X chromosome. A full length cDNA was subsequently isolated based on the knowledge of the immunoaffinity purified antigen protein sequence. Primary protein structure deduced from cDNA sequence reveals the genealogy between human DNA polymerase α and DNA polymerases from several lower eukaryotes, human viruses and bacteriophage. The evolutionary conservation of DNA polymerase α is defined both at the protein epitope level by differential monoclonal antibody reactivity and at the genetic structural level by the cDNA sequence homology. Using the genetic mutation data of herpes virus polymerase as a reference, protein domains on the polymerase α molecule required for substrate binding and interaction have been identified. With the cDNA probe, the study of the regulation of this gene's expression has begun. Most recently a genomic clone of human DNA polymerase α was isolated (Pearson, Nasheuer and Wang, submitted). This genomic clone contains the first three exons and the first two introns and a portion of the third intron as well as 1.6 kb upstream sequence from the transcriptional start site. This makes it possible to begin to investigate regulatory elements for the expression of this essential DNA replication gene.

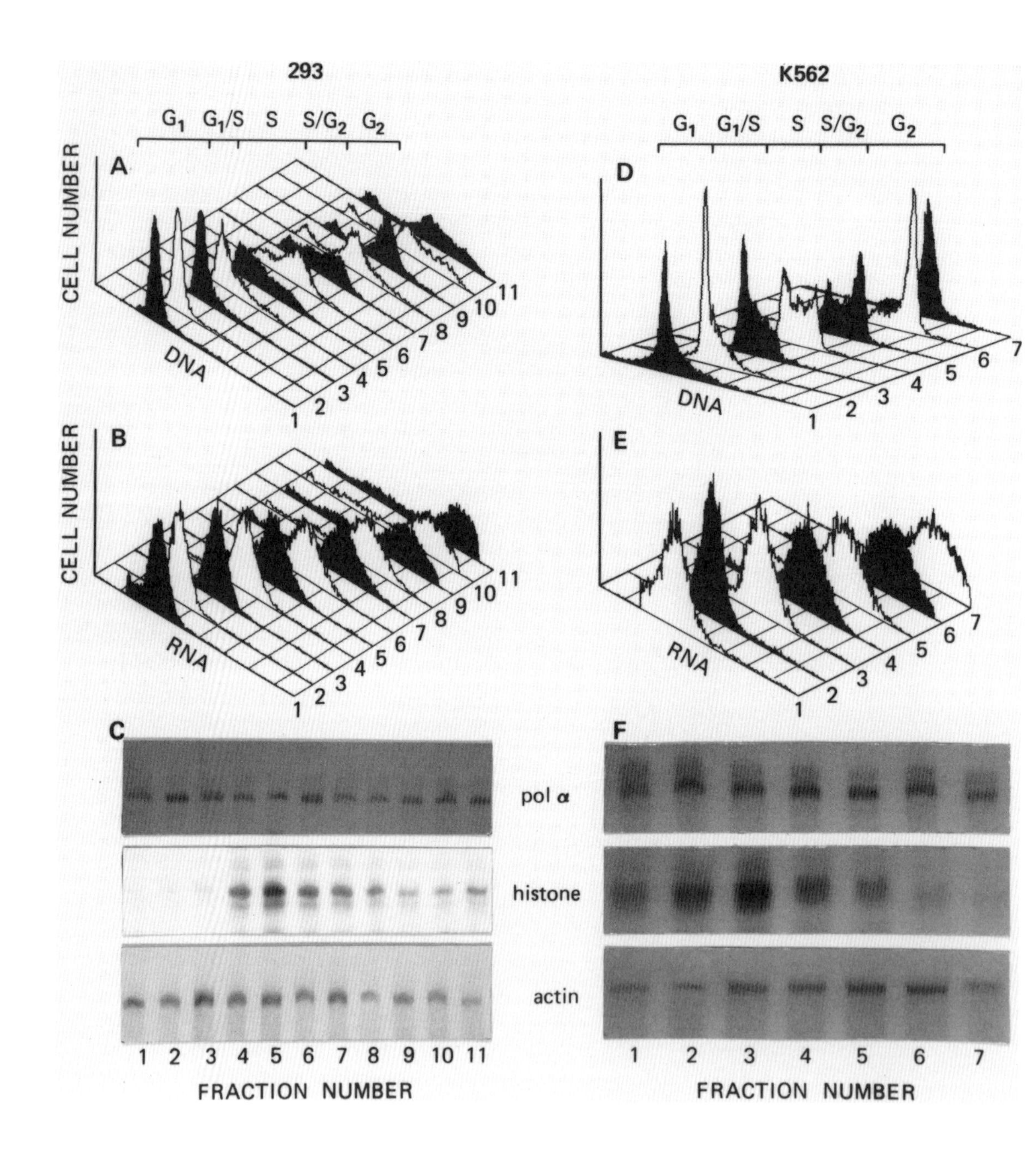
293
K562
G1 G1/S S S/G2 G2
CELL NUMBER
A
B
C
D
E
F
DNA
RNA
pol α
histone
actin
1 2 3 4 5 6 7 8 9 10 11
1 2 3 4 5 6 7
FRACTION NUMBER

In the future, the cDNA can also be utilized to elucidate the detailed structure-function relationship of DNA polymerase α. DNA polymerase α, being the principal player in DNA replication, is the key enzyme for maintenance and transmission of error free genetic information from one generation to the next. Mutation of this enzyme could induce an erroneous DNA polymerization mechanism which subsequently results in serious errors in the genome and having detrimental consequences for the organism. Therefore, which protein domains or amino acids of this enzyme are critical for the catalytic mechanism and the maintenance of replication fidelity, and how this is accomplished, are important biochemical issues. From a practical point of view, drugs may be designed to target this enzyme responsible for chromosomal DNA replication in order to arrest cell proliferation and retard (or halt) tumor growth. An understanding of the structure-function relationship of the human DNA polymerase α protein domains required for substrate recognition would be extremely valuable in the design of such agents.

To achieve this objective, it is necessary to functionally express the recombinant polymerase, followed by identification of the amino acids or domains responsible for substrate (dNTP/DNA) binding or protein-protein interaction. Site specific mutagenesis can be performed on the expressed recombinant human DNA polymerase α molecule. The isolation of catalytic mutants and comparison of kinetic parameters of these functionally active mutants versus the well-characterized wild-type polymerase can identify specific amino acids responsible for substrate interaction. Those conserved regions, especially the predicted dNTP interacting regions, are the obvious target regions for mutagenesis. Eventually, all six conserved regions should be definable by this approach. These experiments should shed light on how this large and complicated protein molecule recognizes and interacts with its several substrates to achieve accurate DNA replication. Furthermore, the identification of domains that interact with other accessory DNA replication proteins should greatly advance the understanding of the details of eukaryotic DNA replication *in vitro* and *in vivo*.

The studies of polymerase gene expression at the transcriptional and post-transcriptional levels, *i.e.* at the steady-state message level, *de novo* protein synthesis rate and enzymatic activity, in three defined biological events have left us with many interesting thoughts. It is also necessary to investigate these parameters during down-regulation such as in terminal cell differentiation and senescence. After thoroughly understanding the polymerase gene expres-

Figure 21. (Left) Steady state message levels of human DNA polymerase α in actively cycling cells. Logarithmically growing cultures of 293 or K562 cells were separated by counterflow centrifugal elutriation into fractions of increasing cell mass. Cell volume increases were verified by a Coulter Channelizer (data not shown). An aliquot of each fraction was fixed, stained with Acridine Orange and the relative DNA and RNA content determined by flow microfluorometry. Unelutriated cells of both 293 and K562 gave a bimodal distribution of DNA typical of unsynchronized exponentially growing cells (data not shown). Assignment of cell cycle stages was based on conversion of the 2N (haploid number) DNA complement of early fractions to the 4N DNA complement of later elutriation fractions. Fractions 10 and 11 of the elutriated 293 cells contain many anneuploid cells (multinucleated with >4N DNA content) typical of transformed populations and do not represent a specific stage of cell cycle. The final fraction of K562 (not shown) also contains cells with >2 nuclei and was therefore not used for subsequent analysis. Steady state message levels from these cell fractions were determined by RNase protection. (A) DNA content of 293 cell fractions; (B) RNA content of 293 cell fractions; (C) steady state message levels of human DNA polymerase α, histone H3 and actin from elutriated 293 cell fractions. (D) DNA content of elutriated K562 cell fractions; (E) RNA content of K562 cell fractions; (F) steady state message levels of DNA polymerase α, histone H3 and actin from elutriated K562 cell fractions. Histone and actin results are from RNase protection of 7 μg of total RNA per fraction and film exposures of 5 and 2 hr, respectively. Polymerase α results are from 10 μg total RNA per fraction and 30 hr film exposure. (Reprinted, with permission, from Wahl *et al.*, 1988).

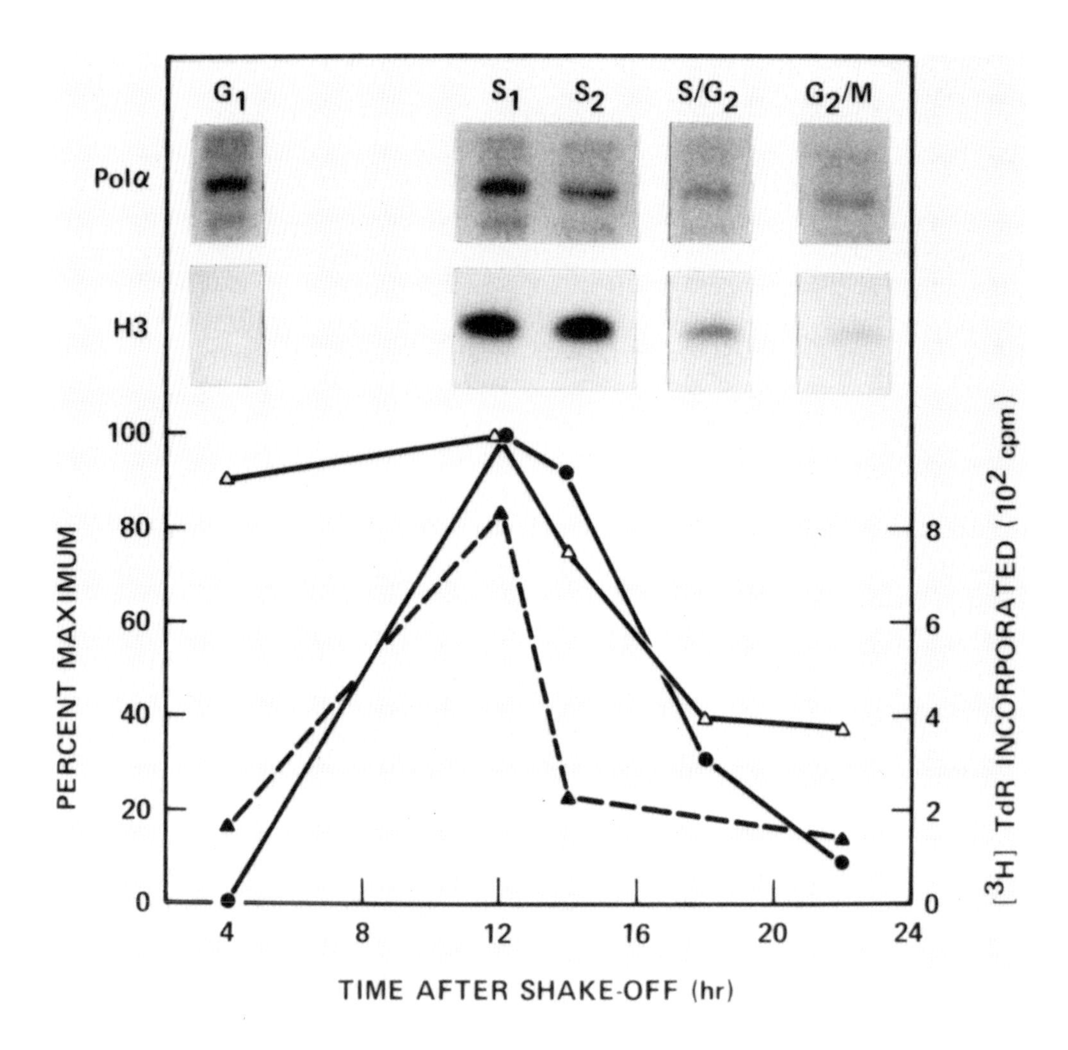

Figure 22. DNA polymerase α and histone H3 steady state message levels from HeLa cells synchronized by the mitotic shake-off method. Mitotic shake-off, preparation of total RNA and [^{3}H]-thymidine incorporation for cellular DNA synthesis were as described in Wahl *et al.*, 1988. Assignment of cell cycle stages were based on mitotic index. RNase protection analysis was used to measure steady state mRNA. Results of RNase protection were measured by scanning densitometry and presented as percent maximum intensity for DNA polymerase α message (△) and histone H3 message (●) and plotted against the time after shake-off. A parallel culture was used to monitor cellular DNA synthesis by [^{3}H]-thymidine incorporation (— ▲ —) and plotted versus time. (Reprinted, with permission, from Wahl *et al.*, 1988.)

sion in these biological events, the molecular mechanisms that regulate cellular proliferation and differentiation should be explored. Cellular proliferation and differentiation probably involve signal transduction from membrane receptors to biochemical processes mediated by nuclear proteins which presumably induce up-or down-regulation of the expression of DNA polymerase α. Several such regulatory nuclear proteins, including those of the proto-oncogene family, are possible candidates. In view of the reported time-frame of these proto-oncogenes' expression during cellular proliferation, the induction of DNA polymerase α gene expression

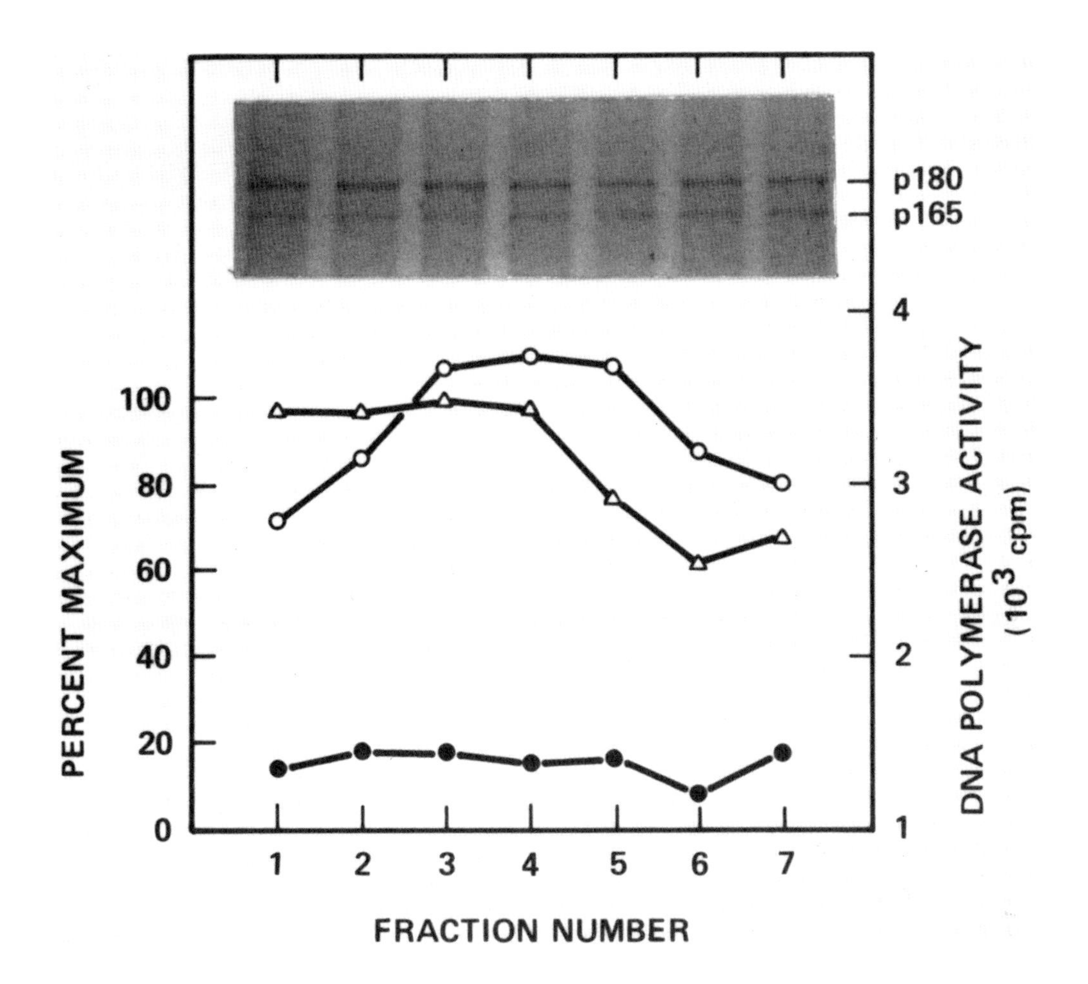

Figure 23. Post-transcriptional expression of DNA polymerase α from K562 cells. Nascent protein and enzymatic activity are presented from normalized cell lysates of K562 cells separated into fractions representing progressive stages of the cell cycle. K562 cells were pulse labeled with [^{3}H]-leucine for 30 min, elutriated by counterflow centrifugation, lysed, and immunoprecipitated by anti-polymerase monoclonal IgG. Immunoprecipitated nascent human DNA polymerase α catalytic polypeptides of 180 and 165 kDa were separated on a SDS-polyacrylamide gel and subjected to autoradiography. Fractions 1 through 7 are results from each elutriation separated population. The intensity of each 180 and 165 kDa band was measured by scanning densitometry, summed for each fraction and expressed as a percent of the maximum level for the experiment (△). DNA polymerase α activity (○) and other polymerase activities (●) in normalized cell lysates were determined by standard polymerase assay in the presence and absence of neutralizing SJK132-20 IgG. (Reprinted, with permission, from Wahl *et al.*, 1988).

is apparently one of the later targets of the biochemical cascade events in the activation of cellular proliferation.

It is reported in the literature that several enzymes involved in DNA substrate synthesis are transiently expressed during the cell cycle, with a maximum expression near G_1/S phase boundary (Ayusawa *et al.*, 1986; Coppock and Pardee, 1987; Farham and Schimke, 1985; Tricoli *et al.*, 1985; Shereley and Kelly, 1988). A coordinated production of these DNA replication related enzymes should be temporally coupled to DNA replication during S phase. The constitutive expression of DNA polymerase α throughout the cell cycle is rather unexpected and raises interesting questions. What mechanism(s) or which protein factor(s) are responsible for the precise timing of onset and termination of DNA replication during a discrete period of the cell cycle? Nuclear transcription rate measurement by run-on assays indicates that transcription, like the steady-state transcript, is also constitutive throughout the cell cycle. It is possible that post-transcriptional modification might play a role in the cell-cycle dependent activation of the polymerase activity during a discrete period of the cell cycle. The constitutive expression of polymerase α *activity* throughout the cell cycle was derived from *in vitro* assays of polymerase α with gapped DNA primer template, which is not a realistic reflection of the cellular events *in vitro* where replication takes place on chromosomes with a possible genomic replication origin. Transient production of protein factor(s) may be required to activate polymerase α for chromosomal DNA replication at a precisely defined time. When cells traverse the S phase, these transient factor(s) may decay, and additional factor(s) such as the cis-acting negative control factor (CAN factor) may then participate in preventing the reinitiation of replicons that have already been replicated (Roberts and Weintraub, 1988). The data from these sets of experiments suggest that although DNA polymerase α is a key element in this complex chromosomal replication machinery, regulation of the onset of S phase is not caused by the transient expression of this enzyme (Roberts and D'urso, 1988). Other cell cycle phase-specific factors which interact with and modulate DNA polymerase α activity *in vivo* are most likely to play a critical role in the precise timing and order of DNA replication in the cell cycle.

To further investigate the regulation of DNA polymerase α expression at the molecular level, nucleotide elements such as promoter and enhancer sequences should further advance our knowledge of the effect of biological events on the regulation of the expression of the polymerase α gene. Once these nucleotide elements are defined, regulatory protein factors that interact with these elements and which could produce either positive or negative effects on polymerase α gene expression in cellular proliferation and terminal differentiation or senescence will be identified. This will shed light on the molecular events involved in DNA replication.

In view of more than a quarter of a century of intense hard work on DNA polymerase α by many investigators, much progress has been accomplished in the last ten years. The new found knowledge of this enzyme as reviewed in this chapter provides new tools to explore and means to approach the unknowns of this key eukaryotic chromosomal DNA replication enzyme.

ACKNOWLEDGEMENT

The author would like to express her special thanks to David Korn, who, in past years, provided her with inspiration, encouragement and many insightful thoughts during the course of pursuing the challenging study of eukaryotic DNA polymerases. The author also would like to thank many colleagues who have passed through her laboratory in the last eight years and whose extensive contributions have made this study possible. Research described in this article was supported by National Institute of Health Grant CA14835 and gifts from the George C. Smith Fund and the Donald and Delia Baxter Fund.

REFERENCES

B. Alberts, D. Bray, J. Lewis, M. Raff, K. Roberts and J. D. Watson (1983) *Molecular Biology of The Cell*, New York and London, Garland Publishing, Inc. pp. 412.

D. Ayusawa, K. Schimizu, H. Koyama, S. Kaneda, K. Takeishi and T. Seno (1986) Cell-cycle-directed regulation of thymidylate synthetase messenger RNA in human diploid fibroblasts stimulated to proliferate. *J. Mol. Chem.* **190**: 559-567.

E. F. Baril, M. D. Jenkins, O. E. Brown, J. Laszlo and H. P. Morris (1973) DNA polymerase I and II in regenerating rat liver and Morris hepatoma. *Cancer Res.* **33**: 1187-1193.

K. G. Bensch, S. Tanaka, S.-Z. Hu, T. S.-F. Wang and D. Korn (1982) Intracellular localization of human DNA polymerase α with monoclonal antibodies. *J. Biol. Chem.* **257**: 8391-8396.

J. M. Berg (1986) Potential metal-binding domains in nucleic acid binding proteins. *Science* **232**: 485-487.

A. Bernard, A. Zaballos, M. Salas and L. Blanco (1987) Structural and functional relationship between prokaryotic and eukaryotic DNA polymerases. *EMBO J.* **6**: 4219-4225.

F. J. Bollum (1960) Calf thymus polymerase. *J. Biol. Chem.* **235**: 2399-2404.

F. J. Bollum (1975) Mammalian DNA polymerase. *Prog. Nucleic Acid Res. Mol. Biol.* **15**: 109-144.

D. Bramhill and A. Kornberg (1988) A model for initiation at origin of DNA replication. *Cell* **54**: 915-918.

M. Brown, F. J. Bollum and L. M. S. Chang (1981) Intracellular localization of DNA polymerase -α. *Proc. Natl. Acad. Sci. USA* **78**: 3049-3052.

J. L. Campbell (1986) Eukaryotic DNA Replication. *Ann. Rev. Biochem.* **55: 733-771.**

L. M.-S. Chang, E. Rafter, C. Augl and F. J. Bollum (1984) Purification of a DNA polymerase-DNA primase complex from Calf thymus glands. *J. Biol. Chem.* **259**: 14679-14687.

R. W. Chiu and E. F. Baril (1975) Nuclear DNA polymerases and the Hela cell cycle. *J. Biol. Chem.* **250**: 7951-7957.

D. M. Coen, P. A. Fuman, D. Aschman and P. A. Schaffer (1983) Mutations in herpes simplex virus DNA polymerase gene conferring hypersensitivity to aphidicolin. *Nucleic Acids Res.* **11**: 5287-5297.

R. C. Conaway and I. R. Lehman (1982a) A DNA primase activity associated with DNA polymerase α from *Drosophila melanogaster* embryos. *Proc. Natl. Acad. Sci. USA.* **79**: 2523-2527.

R. C. Conaway and I. R. Lehman (1982) Synthesis by the DNA primase of *Drosophila melanogaster* of a primer with a unique chain length. *Proc. Natl. Acad. Sci. USA.* **79**: 4585-4588.

D. Coppock and A. B. Pardee (1987) Control of thymidine kinase mRNA during the cell cycle. *Mol. Cell. Biol.* **7**: 2925-2932.

M. L. DePamphilis and P. M. Wassarman (1980) Replication of eukargotic chromosomes: a close-up of the replication fork. *Ann. Rev. Biochem.* **49**: 627-666.

P. L. Earl, E. P. Jones and B. Moss (1986) Homology between DNA polymerases of poxiviruses, herpes viruses, and adenoviruses: Nucleotide sequence of the vaccinia virus DNA polymerase gene. *Proc. Natl. Acad. Sci. USA* **83**: 3659-3663.

D. C. Eichler, P. A. Fisher and D. Korn (1977) Effect of calcium on the recovery and distribution of DNA polymerase α from cultured human cells. *J. Biol. Chem..* **252**: 4011-4014.

J. H. Elder, R. A. Pickett, J. Hampton and R. A. Lerner (1977) Radioiodination of proteins in single polyacrylamide gel slices. Tryptic peptide analysis of all the major members of complex multicomponent systems using microgram quantities of total protein. *J. Biol. Chem.* **252**: 6510-6515.

R. Eliasson and P. Reichard (1979) Replication of polyoma DNA in isolated nuclei, VII. Initiator RNA synthesis during nucleotide depletion. *J. Mol. Biol.* **129**: 393-409.

R. M. Evans and S. M. Hollenberg (1988) Zinc fingers: Gilt by association. *Cell* **52**: 1-3.

P. J. Farnham and R. T. Schimke (1985) Transcriptional regulation of mouse dihydrofolate reductase in the cell cycle. *J. Biol. Chem.* **260**: 7675-7670.

D. Filpula, P. A. Fisher and D. Korn (1982) DNA polymerase α, common polypeptide core structure of three enzyme forms from human KB cells. *J. Biol. Chem.* **257: 2029-2040.**

P. A. Fisher and D. Korn (1977) DNA polymerase α. Purification and structural characterization of the near homogeneous enzyme from human KB cells. *J. Biol. Chem.* **252: 6528-6535.**

P. A. Fisher and D. Korn (1981a) Ordered sequential mechanism of substrate recognition and binding by KB cell DNA polymerase α. *Biochemistry* **20**: 4560-4569.

P. A. Fisher and D. Korn (1981b) Properties of the primer-binding site and the role of magnesium ion in primer-template recognition by KB cell DNA polymerase α. *Biochemistry* **20**: 4570-4578.

P. A. Fisher, J. T. Chen and D. Korn (1981) Enzymological characterization of KB cell DNA polymerase α. Regulation of template binding by nucleic acid base composition. *J. Biol. Chem.* **256**: 133-141.

D. N. Foster and T. Gurney, Jr. (1976) Nuclear location of mammalian DNA polymerase activities. *J. Biol. Chem.* **251**: 7893-7898.

U. Francke (1984) Random X inactivation resulting in Mosaic nullisome of region Xp21.1-p21.3 associated with heterozygosity for ornithininetranscarbamylase deficiency and for chronic granulomatous disease. *Cytogenet. Cell Genet.* **38**: 298-307.

M. Fry and L. A. Loeb (1986) **In**: *Animal Cell DNA Polymerases*. Florida, CRC Press, pp. 13-60.

J. S. Gibbs, H. C. Chiou, K. F. Bastow, Y-C. Cheng and D. M. Coen (1988) Identification of amino acids in herpes simplex virus DNA polymerase involved in substrate and drug recognition. *Proc. Natl. Acad. Sci USA* **85**: 6672-6676.

J. S. Gibbs, H. C. Chiou, J. D. Hall, D. W. Mount, M. J. Retondo, S. K. Weller and D. M. Coen (1985) Sequence and mapping analyses of the herpes simplex virus DNA polymerase gene predicted a C-terminal substrate binding domain. *Proc. Natl. Acad. Sci. USA* **82**: 7969-7973.

T. R. Gingeras, D. Sciaky, R. E. Gelinas, J. Bing-Dong, B. L. Parsons, M. M. Kelly, P. A. Bullock, C. E. Yen, K. E. O'Neil and R. J. Roberts (1982) Nucleotide sequences from the Adenovirus-2 genome. *J. Biol. Chem.* **257**: 13475-13491.

P. Goodfellow, G. Banting, R. Levy, S. Povey and A. McMichael (1980) A human X-linked antigen defined by a monoclonal antibody. *Somatic Cell Genet.*. **2**: 777-787.

M. Groudine and C. Casimir (1984) Post-transcriptional regulation of the chicken thymidine Kinase gene. *Nucleic Acids Res.* **12**: 1427-1446.

G. Herrick, B. B. Spear and G. Veomett (1976) Intracellular localization of mouse polymerase α. *Proc. Natl. Acad. Sci. USA* **73**: 1136-1139.

S. Z. Hu, T. S.-F. Wang and D. Korn (1984) DNA primase from KB cells. Evidence for a novel model of primase catalysis by a highly purified primase/polymerase α complex. *J. Biol. Chem.* **259**: 2602-2609.

S. Ikegami, T. Toguchi, M. Ohasi, M. Oguro, H. Nagano and Y. Mano (1978) Aphidicolin prevents mitotic cell division by interfering with the activity of DNA polymerase α. *Nature* **275**: 458-459.

C. M. Joyce, W. S. Kelley and N. D. F. Grindley (1982) Nucleotide sequence of *Escherichia coli* pol A gene and primary structure of DNA polymerase I. *J. Biol. Chem. 257*: 1958-1964.

G. Jung, M. C. Leavitt, J.-C. Hsieh and J. Ito, Bacteriophage PRDI DNA polymerase: Evolution of DNA polymerases. *Proc. Natl. Acad. Sci. USA* **84**: 8287-8291.

L. S. Kaguni, J.-M. Rossignol, R. C. Conaway, G. R. Banks and I. R. Lehman (1983a) Association of DNA primase with the subunits of DNA polymerase α from *Drosophila melanogaster* embryos. *J. Biol. Chem. 288*: 9037-9039.

S. Kaguni, J.-M. Rossignol, R. C. Conaway and I. R. Lehman 91983b) Isolation of an intact DNA polymerase-primase from embryos of *Drosophila melanogaster*. *Proc. Natl. Acad. Sci. USA* **80**: 2221-2225.

G. Kaufmann and H. H. Falk (1982) An oligoribonucleotide polymerase from SV40-infected cells with properties of a primase. *Nucleic Acid Res.* **10**: 2309-2321.

Y. Kikuchi, K. Hirai, N. Gunge and F. Nishinuma (1985) Hairpin plasmid - a novel linear DNA of perfect hairpin structure. *EMBO J.* **4**: 1881-1886.

D. Korn, P. A. Fisher and T. S.-F. Wang (1983) Enzymological characterization of human DNA polymerases α and β. **In**: *New Approaches in Eukaryotic DNA Replication.* Plenum Publishing Corp., pp. 17-55.

A. Kornberg (1980) **In**: *DNA Replication*, San Francisco, W. H.Freeman,

A. Kornberg (1982) *DNA Replication*, San Francisco, W. H. Freeman.

A. Kornberg (1980) DNA replication, *J. Biol. Chem.* **263**: 1.

T. Kouzardies, A. T. Banker, S. C. Satchwell, K. Weston, P. Tomlinson and B. C. Barrell (1987) Sequence and transcription analysis of the human cytomegalovirus DNA polymerase gene. *J. Virol.* **61**: 125-133.

B. A. Larder, S. D. Kemp and G. Darby (1987) Related functional domains in virus DNA polymerases. *EMBO J.* **6**: 169-175.

R. Levy, J. Dilley, R. I. Fox and R. Warnke (1979) A human thymus-leukemia antigen defined by hybridoma monoclonal antibodies. *Proc. Natl. Acad. Sci. USA* **76**: 6552-6556.

J. J. Li and T. J. Kelly (1984) Simian virus 40 DNA replication *in vitro*. *Proc. Natl. Acad. Sci. USA* **81**: 6973-6977.

J. J. Li and T. Kelly (1985) SV40 DNA replication *in vitro*: Specificity of initiation and evidence for bidirectional replication. *Mol. Cell. Biol.* **5**: 1238-1246.

V. Lindgren, B. DeMartinvile, L. E. Horwich and U. Francke (1984) Human Ornithine transcarbamylase locus mapped to band Xp22.1 near the Duchenne Muscular Dystrophy locus. *Science* **226**: 698-700.

M.A. Lischwe and D. Ochs (1982) A new method for partial peptide mapping using N-chorosuccinimide/urea and peptide silver staining in sodium dodecyl sulfate polyacrylamide gels. *Anal. Biochem.* **127**: 453-457.

G. Lucchini, C. Mazza, E. Scacheri and P. Plevani (1988) Genetic mapping of the Saccharomyces cerevisiae DNA polymerase I gene and characterization of a pol I temperature-sensitive mutant altered in DNA primase-polymerase complex stability. *Mol. Gen. Genet.* **212**: 459-465.

W. E. Lynch, S. Surry and I. Lieberman (1975) Nuclear deoxyribonucleic acid polymerases of liver. *J. Biol. Chem.* **250**: 8179-8183.

A. Matsukage, K. Nishikawa, T. Ooi, Y. Seto and M. Yamaguchi, Homology between mammalian DNA polymerase β and terminal deoxynucleotidyl-transferase. *J. Biol. Chem.* **262**: 8960-8962.

M. Mechali, J. Abadiedebat and A.-M. de Recondo (1980) Eukaryotic DNA polymerase α. Structural analysis of the enzyme from regenerating rat liver. *J. Biol. Chem.* **255**: 2114-2122

J. Miller, A. D. McLachlan and A. Klug (1985) Repetitive zinc-binding domains in the protein transcription factor III A from *Xenopus Oocytes*. *EMBO.J.* **4**: 1609-1614.

M. A. Miller, D. Korn and T. S.-F. Wang (1988) The evolutionary conservation of DNA polymerase α. *Nucleic Acid Res.* **16**: 7961-7973.

M. R. Miller, C. Seighman and R. G. Ulrich (1986) Inhibition of DNA replication and DNA polymerase α activity by monoclonal anti-(DNA polymerase α) immunoglobulin G and F(ab) fragments. *Biochemistry* **24**: 7440-7445.

M. R. Miller, R. G. Ulrich, T. S.-F. Wang and D. Korn (1985) Monoclonal antibodies against human DNA polymerase α inhibit DNA replication in permeabilized human cells. *J. Biol. Chem.* **260**: 134-138.

Y. Murakami, C. R. Wobbe, L. Weissbach, F. B. Dean and J. Hurwitz (1986) Role of DNA polymerase α and DNA primase in simian virus 40 DNA replication *in vitro*. *Proc. Natl. Acad. Sci. USA*. **83**: 2869-2873.

Y. Murakami, H. Yasuda, H Miyazawa, F. Hanaoka and M. Yamada (1985) Characterization of a temperature-sensitive mutant of mouse FM3A cell defective in DNA replication. *Proc. Natl. Acad. Sci. USA* **83**: 2869-2873.

H.-P. Nasheuer and F. Grosse (1988) DNA polymerase α-primase from calf thymus. Determination of the polypeptide responsible for primase activity. *J. Biol. Chem.* **263**: 8981-8988.

H.-P. Nasheuer and F. Grosse (1987) Immunoaffinity-purified DNA polymerase α displays novel properties. *Biochemistry* **26**: 8458-8466.

M. Paillard, R. R. Sederiff and C. S. Levings (1985) Nucleotide sequence of the S-1 mitochondrial DNA from the cytoplasm of Maize. *EMBO J.* **4**: 1125-1985.

V. Pigiet, R. Eliasson and P. Reichard, Replication of polyoma DNA in isolated nuclei III. The nucleotide sequence at the RNA-DNA junction of nascent strands. *J. Mol. Biol.* **84**: 187-216.

A. Pizzagalli, P. Valsasnini, P. Plevani and G. Lucchini (1988) DNA polymerase I gene of Saccharomyces cerevisiae: Nucleotide sequence, mapping of temperature-sensitive mutation and protein homology with other DNA polymerases. *Proc. Natl. Acad. Sci. USA* **85**: 3772-3776.

P. Plevani, M. Foiani, P. Valsasnini, G. Bodaracco, E. Cheriathundam and L.M.S. Chang (1985) Polypeptide structure of DNA primase from a yeast DNA polymerase primase complex. *J. Biol. Chem.* **260**: 7120-7107.

J. P. Quinn and D. J. McGeoch (1985) DNA sequence of the region in the genome of herpes simplex virus type I containing the genes for DNA polymerase and major DNA binding protein. *Nucleic Acids Res.* **13**: 8143-8163.

P. Reichard, R. Eliasson and G. Soderman (1974) Initiator RNA in discontinuous polyoma DNA synthesis. *Proc. Natl. Acad. Sci. USA* **61**: 4901-4905.

H.-D. Riedel, H. Konig, H. Stahl and R. Knipper (1982) Circular single stranded phage M13-DNA as a template for DNA synthesis in protein extracts from *Xenopus laevis* eggs: evidence for a eukaryotic DNA priming activity. *Nucleic Acids Res.* **10**: 5621-5635.

J. M. Roberts and H. Weintraub (1988) Cis-acting negative control of DNA replication in eukaryotic cells. *Cell* **52**: 397-404.

J. M. Roberts and G. D'urso (1988) An origin unwinding activity regulates initiation of DNA replication during mammalian cell cycle. *Science* **241**: 1486-1489.

W. D. Sedwick, T. S.-F Wang and D. Korn (1972) Purification and properties of nuclear and cytoplasmic deoxyribonucleic acid polymerases from human KB cells. *J. Biol. Chem.* **247**: 5026-5033.

J. L. Sherley and T. J. Kelly (1988) Regulation of human thymidine kinase during the cell cycle. *J. Biol. Chem.* **263**: 8350-8358.

M. Shioda, E. M. Nelson, M. L. Bayne and R. W. Benbow (1982) DNA primase activity associated with DNA polymerase α from *Xenopus laevis* ovaries. *Proc. Natl. Acad. Sci. USA* **79**: 7209-7213.

L. Shu, M. S. Horwitz and J. A. Engler (1987) Expression of enzymatically active adenovirus DNA polymerase from cloned DNA requires sequences upstream of the main open reading frame. *J. Virol.* **161**: 520-526.

S. Spardari and A. Weissback (1974) The interrelation between DNA synthesis and various DNA polymerase activities in synchronized Hela cells. *J. Mol. Biol.* **86**: 11-20.

S. Spadari, G. Villani and N. Hardt (1978) *Exp. Cell. Res.* **113**: 57-62.

E. K. Spicer, J. Rush, C. Fung, L. J. Reha-Krantz, J. D. Karam and W. H. Konigsberg (1988) Primary Structure of T_4 DNA Polymerase. Evolutionary relatedness to eukaryotic and other procaryotic DNA polymerases. *J. Biol. Chem.*. **263**: 7478-7486.

M. J. Stark, A. J. Milehem, M. A. Romanos and A. Boyd (1984) Nucleotide sequence and transcription analysis of a linear DNA plasmid associated with the killer character of the yeast *Klyveromyes lactis*. *Nucleic Acid Res.* **12**: 6011-6030.

B. W. Stillman and Y. Gluzman (1985) Replication and supercoiling of simian virus 40 DNA in cell extracts from human cells. *Mol. Cell. Biol.* **5**: 2051-2060.

R. K. Storm, R. W. Ord, M. T. Greenwood, B. Mirdamadi, F. K. Chu and M. Belfort (1984) Cell cycle-dependent expression of thymidylate synthetase in *Saccharomyces cerevisiae*. *Mol. Cell. Biol.* **4**: 2858-2864.

H. G. Stunnenberg, H. Large, L. Philipson, R. T. van Miltenberg and P. C. van der Vliet (1988) High expression of functional adenovirus DNA polymerase and precursor terminal protein using recombinant vaccinia virus. *Nucleic Acid Res.* **16**: 2431-2444.

S. Tanaka, S.-Z. Hu, T. S.-F. Wang and D. Korn (1982) Preparation and preliminary characterization of monoclonal antibodies against human DNA polymerase α. *J. Biol. Chem.* **257**: 8386-8390.

P. Thommes, T. Reiter and R. Knippers (1986) Synthesis of DNA polymerase α analyzed by imminoprecipitation from synchronously proliferating cells. *Biochemistry* **25**: 1308-1314.

H. G. Tomasiewicz and C. S. McHenry (1987) Sequence analysis of the dnaE gene of *Escherichia coli*. *J. Bacteriol.* **169**: 5735-5744.

J. V. Tricoli, B. M. Shai, P. J. McCormick, S. J. Jarlinski, J. S. Bertram, D. Kowalski (1985) DNA topoisomerases I and II activities during cell proliferation and the cell cycle in mouse embryo fibroblast (C3H 10T1/2) cells. *Exp. Cell. Res.* **158**: 1-14.

B. Y. Tseng and C. N. Ahlem (1982) DNA primase activity from human lymphocytes. *J. Biol. Chem.* **257**: 7280-7283.

B. Y. Tseng and C. N. Ahlem (1983) A DNA primase from mouse cells. Purification and partial characterization. *J. Biol. Chem.* **258**: 9845-9849.

B. Y. Tseng, J. M. Erickson and Goulian (1979) Initiator RNA of nascent DNA from animal cells. *J. Mol. Biol.* **129**: 531-545.

B. Y. Tseng and M. Goulian (1977) Initiator RNA of discontinuous DNA synthesis in human lymphocytes. *Cell* **12**: 483-489.

T. Tsurumi, K. Maeno and Y. Nishiyama (1987) A single-base change within the DNA polymerase locus of herpes simplex virus type 2 can confer resistance to aphidicolin. *J. Virol.* **61**: 388-394.

M. A. Wagar and J. A. Huberman (1975) Covalent attachment of RNA to nascent DNA in mammalian cells. *Cell* **6**: 551-557.

A. F. Wahl, A. M. Geis, B. H. Spain, S. W. Wong, D. Korn and T. S.-F. Wang (1988) Gene expression of human DNA polymerase α during cell proliferation and the cell cycle. *Mol. Cell. Biol.* **8**: 5016-5025.

A. F. Wahl, S. P. Kowalski, L. W. Harwell, E. M. Lord and R. A. Bambara (1984) Immunoaffinity purification and properties of a high molecular weight calf thymus DNA α-polymerase. *Biochemistry* **23**: 1894-1899.

T. S.-F. Wang, S.-Z. Hu and D. Korn (1984) DNA primase from KB cells. Characterization of a primase activity tightly associated with immunoaffinity-purified DNA polymerase α. *J. Biol. Chem.* **259**: 1854-1865.

T. S.-F. Wang, B. E. Pearson, H. A. Suomalainen, T. Mohandas, L. J. Shapiro, J. Schroder and D. Korn (1985) Assignment of the gene for human DNA polymerase α to the X chromosome. *Proc. Natl. Acad. Sci. USA* **82**: 5270-5274.

T. S.-F. Wang, S. W. Wong and D. Korn (1989) Human DNA polymerase α Predicted Functional Domains and Relationships with Viral DNA Polymerases. *FASEB J.* **3**: 14-21.

A. Weissbach, A. Schlabach, B. Fridlender and A. Bolden (1971) DNA polymerase from human cells. *Nature New Biol.* **231**: 167-170.

C. R. Wobbe, L. J. Weissback, J. Hurwitz (1985) *In vitro* replication of duplex circular DNA containing the simian virus 40 DNA origin site. *Proc. Natl. Acad. Sci. USA* **82**: 5710-5714.

C. R. Wobbe, L. J. Weissbach, J. A. Borowiec, F. B. Dean, Y. Mura Kami, P. Bullock and J. Hurwitz (1987) Replication of simian virus 40 origin-containing DNA *in vitro* with purified proteins. *Proc. Natl. Acad. Sci. USA* **84**: 1834-1838.

S. W. Wong, L. R. Paborsky, P. A. Fisher, T. S.-F. Wang and D. Korn (1986) Structural and enzymological characterization of immunoaffinity-purified DNA polymerase α. DNA primase complex from KB cells. *J. Biol. Chem.* **261**: 7958-7968.

S. W. Wong, A. F. Wahl, P.-M. Yuan, N. Arai, B. E. Pearson, K-I. Arai, D. Korn, M. W. Hunkapiller and T. S.-F. Wang (1988), Human DNA polymerase α gene expression is cell proliferation dependent and its primary structure is similar to both prokaryotic and eukaryotic replicative DNA polymerases. *EMBO J.* **7**: 37-47.

T. Yagura, T. Koza and T. Seno (1982) Mouse DNA replicase. DNA polymerase sociated with a novel RNA polymerase activity to synthesize initiator RNA of strict size. *J. Biol. Chem.* **257**: 11121-11127.

M. Yamaguchi, E. A. Henderickson, and M. L. Depamphilio (1985) DNA primase-DNA polymerase α from simian cells. Modulation of RNA primer synthesis by ribonucleotide triphosphates. *J. Biol. Chem.* **260**: 6254-6263.

B. Z. Zmudzka, D. SenGupta, A. Matsukage, F. Cobianti, P. Kumar and S. H. Wilson (1986) Structure of rat DNA polymerase β revealed by partial amino acid sequencing and cDNA cloning. *Proc. Natl. Acad. Sci. USA* **83**: 5106-5110.

8

GENE REGULATION AND STRUCTURE-FUNCTION STUDIES OF MAMMALIAN DNA POLYMERASE β

Samuel H. Wilson

INTRODUCTION

DNA polymerase beta (β-pol) is found in all vertebrate species as a 40 kDa DNA polymerase lacking intrinsic accessory activities, such as 3′ to 5′ exonuclease, endonuclease, dNMP turnover, or even the reverse of the DNA synthesis reaction, pyrophosphorolysis. In light of its size and limited catalytic repertoire, this enzyme is considered the simplest naturally occurring DNA polymerase and is an ideal model for studies of the DNA synthesis nucleotidyltransferase mechanism. The usual 40 kDa β-pol of vertebrates is not present in other eukaryotes, and β-pol does not yet have recognized homologues in yeast or in lower eukaryotes. The enzyme conducts "gap-filling" synthesis on single-stranded DNA templates (Chang, 1973a, 1975; Fry, 1983; Fry and Loeb, 1986) at about the same catalytic rate as the *E. coli* DNA repair polymerase *Pol*I (*i.e.*, 1 to 2 nucleotides/sec), which incidently is 100 to 1000 times slower than the rate of most replicative DNA polymerases. This gap-filling reaction is consistent with repair patch synthesis during DNA repair (Mosbaugh and Linn, 1983; Siedlecki *et al.*, 1980; Wang and Korn, 1980). Inhibitor studies with mammalian cell systems have implicated β-pol in some types of mammalian DNA repair (Dresler and Lieberman, 1983; Miller and Chinault, 1982; Smith and Okumoto, 1984), and a good example of this is repair synthesis following bleomycin damage (DiGiuseppe and Dresler, 1987). β-pol, however, clearly is not the only repair DNA polymerase activity of mammalian cells (Dresler and Frattini, 1986; Dresler and Kimbro, 1987; Smith and Okumoto, 1984). It seems plausible also that β-pol could have a required role in some vertebrate cell-specific process, such as the recombination involved in meiosis or terminal differentiation. Yet, at the moment, such a role for the enzyme remains obscure, and most experimental work (for review see Wilson *et al.*, 1988) focuses on a DNA repair role for the enzyme and the use of the enzyme as a model for both structure-function studies of DNA polymerase mechanism and gene expression studies with a DNA metabolism enzyme. This chapter will summarize recent work in my laboratory and others in each of these latter areas, namely properties of the β-pol gene and its control and properties of the β-pol protein. The organization of the chapter, as for the "central dogma," of molecular biology is from the gene to the protein. In keeping with the "central dogma", the chapter begins with the gene.

THE HUMAN β-POL GENE

McBride *et al.* (1987) and Matsukage *et al.* (1986) initially made use of Southern blot analysis of DNA from somatic cell hybrids segregating human chromosomes to localize the β-pol gene. The Southern blot patterns obtained were inconsistent with a multicopy gene and permitted unambiguous localization of the gene to human chromosome 8. Sublocalization of the gene on human chromosome 8 was accomplished by analysis of DNAs from cells with known chromosome 8 translocations (McBride *et al.*, 1987). Additional analysis of β-pol gene localization in the mouse (McBride *et al.*, 1989) and *in situ* hybridization with human chromosomes (Cannizzaro *et al.*, 1988) indicate that the human β-pol gene is on the proximal short arm of chromosome 8 (Matsukage *et al.*, 1986). Further sublocalization studies of the gene, as yet unpublished, using DNA from individual families indicate that the gene is near (within approximately 100 CM) the tissue plasminogen activator gene on the proximal short arm of chromosome 8 (Olson and McBride, 1989). One polymorphism of the β-pol gene was noted during our initial chromosome localization study (McBride *et al.*, 1987). The frequency of this polymorphism in the human population is 10 to 20%. No additional polymorphism has been noted since, and in surveys of various cell types associated with DNA repair deficiencies, no rearrangements of the β-pol gene have been detected as yet. Likewise, examination of the β-pol gene in various human tumors, such as sarcomas and lung carcinomas, have failed to reveal rearrangement of the β-pol gene.

The Southern blot patterns with human DNA also permitted estimates of the overall size of the β-pol gene. These estimates indicate a gene of about 35 kb. Genomic cloning by Padmini Kedar in our laboratory using a human fetal liver library resulted in the isolation of a 14.6-kb DNA segment containing sequences from the 5′ end of the cDNA, together with associated introns (Widen *et al.*, 1988). This DNA segment is illustrated in Figure 1. Several of the restriction fragments identified in the gene localization studies (McBride *et al.*, 1987) are contained within this 14.6-kb segment and, in addition, the polymorphism noted above has been localized to the 5′ half of this DNA segment (Olson, McBride and Wilson, unpublished observations). Sequencing of portions of the 14.6-kb segment revealed splice sites for exons 1 and 2 and also sequence characteristics of the region just 5′ of exon 1. This region has been identified as the functional promoter for the β-pol gene and is described in detail below. The first intron is only 56 nucleotides long, whereas the second intron is over 2.5 kb. The sequence of the first intron has characteristics similar to intron sequences with transcriptional control activity, but this property does not appear to be shared by β-pol intron 1, as will be described below.

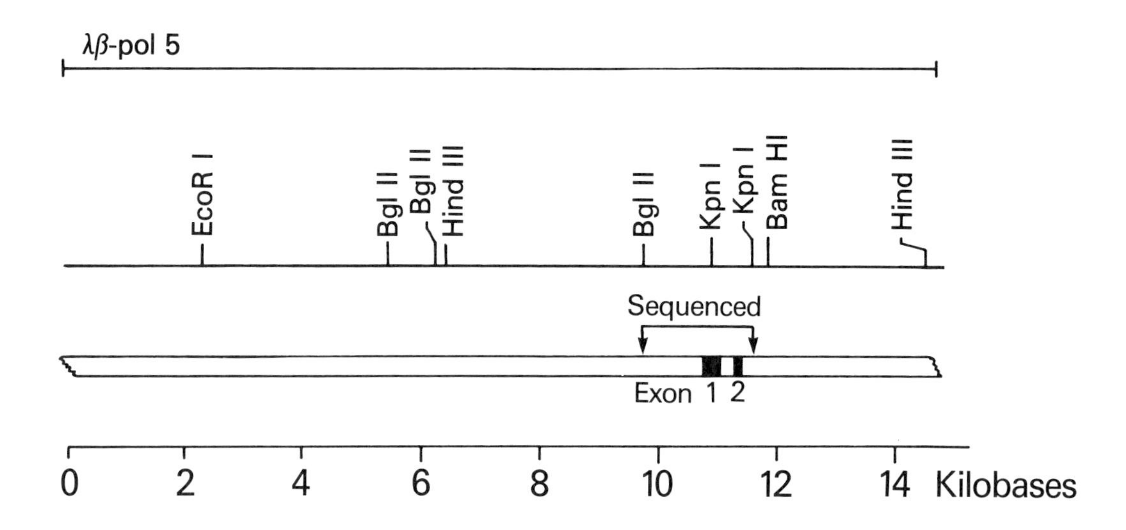

Figure 1. Map of genomic clone β-pol 5. A restriction map of the 5′ end of the human β-pol gene is shown. The first two exons are indicated by the solid blocks, and the portion that was sequenced is indicated. Taken from Widen *et al.*, 1988, with permission..

THE HUMAN β-POL PROMOTER

To characterize transcription control of β-pol mRNA, the region of the β-pol gene just 5′ of exon 1 was isolated. This DNA fragment extends about 11 kb 5′ of the start of exon 1. The region around this exon was sequenced, as indicated in Figure 1, and the transcription start site was characterized by Sl analysis and by primer extension analysis (Widen *et al.*, 1988). These results, which were obtained by Steven Widen, are summarized in Figure 2. The DNA sequence spanning 150 nucleotides 5′ of the major start site does not contain an obvious TATAA sequence element or CAAT sequence element. It does contain three GC boxes, as indicated in Figure 2, and a 10 nucleotide perfect palindrome ending 49 residues upstream from the major start site.

Promoter activity of the DNA sequence 5′ of exon 1 was examined in detail by Steven Widen, who fused the sequence to the bacterial reporter gene for chloramphenicol acetyltransferase (CAT) and then conducted transient expression assays in HeLa cells. The results indicated that the 4.8-kb sequence noted in Figure 1 is indeed capable of promoter activity. Surprisingly, however, the sequence just 114 nucleotides 5′ of the major start site has equal promoter activity with that of the sequence measuring 4.8 kb 5′ of the start site. These results are summarized in Table 1. Therefore, activity of this 5′ segment of the human gene does not appear to contain enhancers or negative control elements, such as the negative elements found near the rodent β-pol core promoter (Yamaguchi *et al.*, 1987) and major transcription start site by Yamaguchi *et al.* (1988).

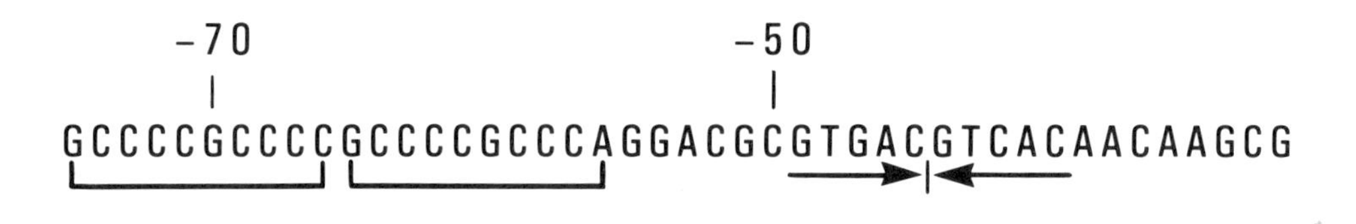

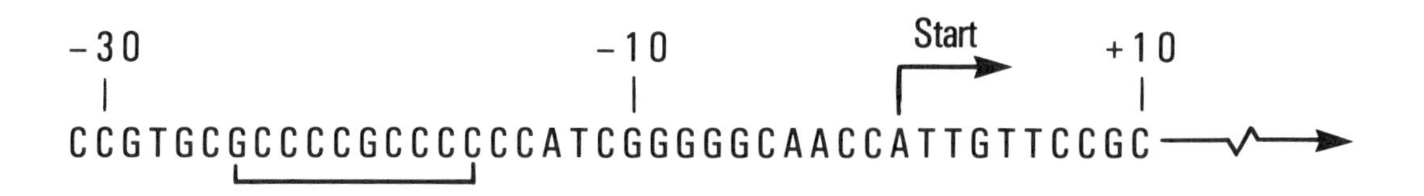

Figure 2. Sequence illustrating features of the β-pol gene around the major transcription start site. Sequences matching Sp1 factor binding sites (⌴) and a decanucleotide element with dyad symmetry (→←) are noted. Transcription start site is noted. Taken from Widen *et al.*, 1988, with permission.

One of the GC boxes, termed the proximal GC box, is located between the palindrome and the major start site. Note that a minor start site is present near position -30. The other two GC boxes are contiguous and are positioned just 5′ of the palindrome. The deletion analysis summarized in Table 1 indicated that removal of DNA with the tandom GC boxes lowered activity of the promoter by about 50%. GC box elements have, of course, been implicated as binding sites for the RNA polymerase II transcription factor termed SPl. We also noted in the analysis shown in Table 1 that deletion to residue -41, removing 7 bp of the decanucleotide palindrome, substantially reduced activity of the promoter. These results indicated that both the decanucleotide element and sequences further upstream (possibly the GC boxes) are required for full promoter activity in this transient expression assay. Overall, the sequence several hundred nucleotides 5′ of exon 1 is relatively CpG rich compared with genomic DNA, suggesting hypomethylation in this region, as found with a number of mammalian promoters.

Several additional transient expression plasmids were constructed to further examine the idea of a promoter activity requirement for the decanucleotide palindromic sequence. Starting with plasmid pβP8, which contains 114 bp of flanking sequence and shows full promoter activity in our assay, the palindrome was disrupted by deletions and insertions (Table 2). When the core ACGT sequence was replaced with an 8-bp *Sal*I linker (GGTCGACC) to make pβP8*, expression was reduced to about one-tenth that seen with intact pβP8 (Table 2). In a second construct, 4 bp were removed from the *Sal*I linker insertion, resulting in a plasmid with the ACGT core replaced with GGCC (pβP8*A). The expression of this plasmid was higher than that of pβP8*, but still only about one-third the level of pβP8. Two plasmids with deletions of 3 and 44 bp, respectively, also showed reduced activity (Table 2). These results demonstrate that the decanucleotide palindromic element is required for full promoter activity. In addition, plasmid pβP8*A has the same number of base pairs as pβP8 and also is palindromic, showing that at least part of the effect seen in Table 2 is due to changes in the sequence rather than to

Table 1.

Deletion Analysis of the 5′-Flanking Region of the β-pol Gene

The β-pol-CAT deletion clones are listed with the 5′ endpoint of the clone as determined by DNA sequencing or restriction mapping (pβP1). The activity of each plasmid is compared to pβP8 set to an arbitrary value of 100. The number of independent transfections (n) and the standard deviation (S.D.) are listed. At least two different preparations of DNA were used for each plasmid.

CAT Construct		Activity	S.D.	n
pβP1	-4800	106	20	6
pβP2	-1153	94	30	4
pβP3	-715	91	9	6
pβP4	-577	92	17	4
pβP5	-402	99	7	4
pβP6	-270	95	46	4
pβP7	-126	120	63	4
pβP8	-114	100	18	12
pβP9	-51	54	21	4
pβP10	-41	12	4	8

changes in the number of bases inserted or deleted or the apparent integrity of the palindrome.

Finally, it should be noted that the palindromic element of the β-pol promoter is similar to promoter elements in several viral and cellular genes (Table 3). All of the elements listed share the ACGT central motif; similarity is seen in each case and is striking for several of the elements such as the IAP LTR and the somatostatin promter, for example. Some of the promoters listed, but not all, are responsive to cAMP regulation.

The relative transient expression activity of the human β-pol promoter is low as a fusion gene with CAT. This fact was illustrated by an experiment where the SV40 promoter (Figure 3) in pSV2 CAT is compared with a construct containing the β-pol promoter. pSV2 CAT is approximately 10 times more active per microgram of DNA than pbP2. In searching for effectors or activators of the β-pol promoter, we noted by the computer search summarized in Table 3 that sequences similar to the β-pol promoter palindromic element are present in the adenovirus major late promoter, a promoter strongly activated in *trans* by adenovirus ElA-ElB proteins. We tested for this effect by cotransfecting an adenovirus early region expression plasmid with two different β-pol-CAT constructs, one with 1.2 kb (pbP2) and the other with 114

Table 2.

Internal Changes in the β-pol Promoter

Activities of several promoter mutants are listed with the number of independent transfections (n) and the standard deviation (S.D.) for each. The specific changes in each plasmid are listed with the changed nucleotides being underlined. The dots for pβP*-3 indicate deleted bases. Activity is relative to the activity of the intact promoter (pβP8 = 100).

Construct	Activity	S.D.	n	Alterations from GTGACGTCAC
pβP8*	13	10	6	GTGGGTGCACCCAC
pβP8*A	29	13	8	GTGGGCCCAC
pβP8*-3	35	15	4	GTGGG . . . AC
pβP8*-44	17	4	4	Deletion (-60 to -15)

bp of 5′ flanking sequence (pβP8). Both plasmids were strongly activated by cotransfection with the early gene expression plasmid (Figure 4). In the experiment shown, pbP2 expression in HeLa cells was increased 200-fold, and pβP8 expression was increased about 70-fold. We also tested for expression of the β-pol-CAT plasmids in 293 cells, a line derived from human kidney cells that constitutively expresses adenovirus ElA and ElB proteins. In this experiment, a plasmid containing 4.8 kb of 5′ flanking sequence was included (pbPl). We found that the β-pol-CAT plasmids are expressed at much higher levels in 293 cells than in HeLa cells (Figure 3). Because of this higher level of expression, we chose to characterize the β-pol promoter in more detail using 293 cells. We analyzed the RNA products from the β-pol-CAT genes to determine if initiation of transcription occurs at the same place as it does with the cellular β-pol gene. RNA from 293 cells that were transfected with pβP2 or with pβP8 was isolated and analyzed by Sl nuclease protection mapping. The probe was derived from pβP2 by labeling at the *Eco*RI site present in the CAT gene 264 bp from the junction with the β-pol fragment. A doublet of about 326-330 nucleotides is predicted for correctly initiated transcripts. This pattern, indeed, is observed with both plasmids (Figure 4), indicating correct initiation of transcription. Also, these results show that both plasmids express similar levels of RNA, and this is consistent with our observation of similar CAT activity with each plasmid (Figure 4).

Table 3.

Comparison of Viral and Cellular Promoter Sequences with β-pol

Sequences showing similarities to the β-pol promoter are listed with identical nucleotides underlined. The number of matches (n), the position of the 5′ most nucleotide relative to the transcription start site, and the gene name and reference are noted. A consensus sequence is derived at the bottom of the table and is compared to the β-pol sequence. The AP-1 binding site is also listed.

Sequence	n	Gene	5′ Nucleotide
Viral			
ATGACGTAGT	6	Ad2 E2	-77
CTTTCGTCAC	7	Ad2 E3	-62
GTTACGTCAT	8	Ad5 E4	-47
GTGACGTAAC	8	Ad5 E4	-164
GTGACGTAGG	7	Ad5 E4	-230
GTGACGTGGC	8	Ad5 E4	-262
CTGACGTGTC	7	HTLV I LTR	-197
CTGACGTCTC	8	HTLV I LTR	-228
CTGACGTCCC	8	HTLV II LTR	-179
CTGACGTCTC	8	HTLV II LTR	-229
GAGACGTCAG	8	BLV LTR	-161
GTGACGTCAA	9	IAP LTR, Mouse M1A14	-101
TTGACGTCAA	8	CMV enhancer	
Cellular			
GTGACGTAGG	7	c-*fos*	-67
TTGACGTCAT	8	α-Chorionic gonadotropin	-124
GTGACGTCAT	9	Parathyroid hormone	-76
GTGACGTCTT	8	Vasoactive intestinal peptide	-77
CTTACGTCAG	7	Phosphoenolpyruvate carboxykinase	-91
CTGACGTCAG	8	Somatostatin	-49
Summary			
NTGACGTCAN		Consensus of above	
GTGACGTCAC		β—Polymerase	-49
TGA-GTCAG		TPA[a]-inducible element	
TGAC-TCA		AP-1 binding site	

[a]TPA, 12-*O*-tetradecanoylphorbol-13-acetate

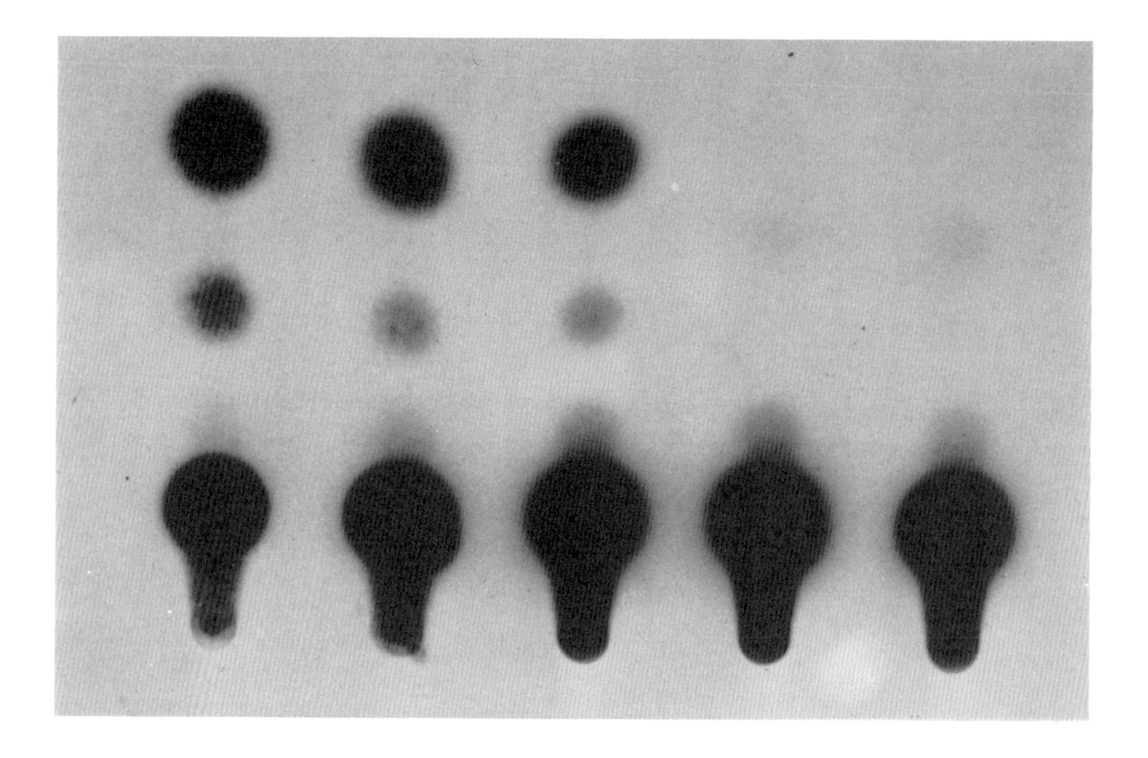

Figure 3. Autoradiogram showing transient expression of a β-pol-CAT fusion gene in HeLa cells. HeLa cells were transfected with the indicated plasmids as described (Widen *et al.*, 1988). The upper two rows of spots are the acetylated derivatives of chloramphenicol, and the lower row of spots is unacetylated chloramphenicol. The amount of added DNA in micrograms is shown above each lane. Taken from Widen *et al.*, 1988, with permission.

When untransfected 293 cell RNA was tested with the same probe, no signal was detected (not shown).

The β-pol promoter also is activated in *trans* by the activated *ras* protein (Kedar *et al.*, 1989a). A cotransfection experiment with a Harvey *ras* expression plasmid, pJCS-1 in 3T3 cells is shown in Figure 5. Expression of pβP8 is enhanced approximately 5-fold by *ras* plasmid cotransfection. Further, the activity of the β-pol promoter construct is approximately 5-fold higher in 3T3 cells (133B) stably transformed and expressing the Harvey *ras* protein (Figure 6). These results with adenovirus E1A-E1B proteins and with activated *ras* protein indicate that the β-pol promoter is responsive to activation by some types of growth control proteins. We found in both cases that the activation requires the presence of the β-pol palindrome. Thus,

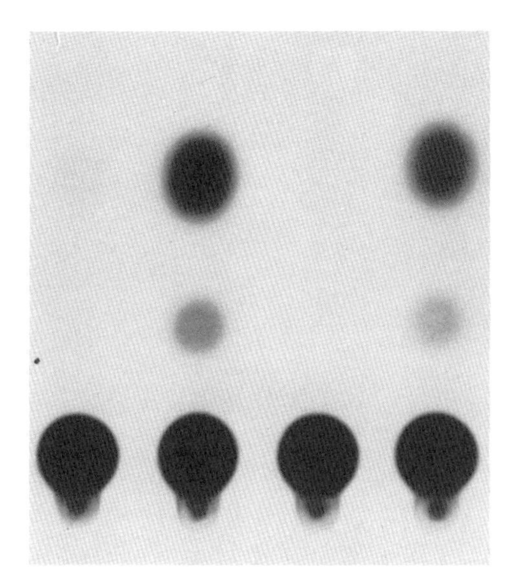

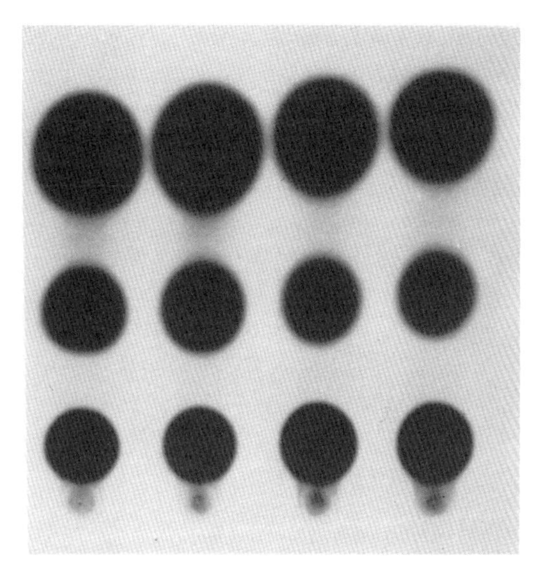

Figure 4. Autoradiograms showing transient expression of β-pol-CAT genes in the presence of adenovirus early region gene products. **A**: HeLa cells were transfected with 10 μg of pβP2 or pβP8 either with (lanes 2 and 4) or without (lanes 1 and 3) 10 μg of pGC212, an early region expression plasmid. **B**: human 293 cells, constitutively expressing E1A and E1B products, were transfected with 20 μg of pβP1 (lane 1), pβP2 (lane 2), or pβP8 (lane 3) or 2 μg of pSV2-CAT (lane 4). Taken from Widen *et al.*, 1988, with permission.

modification of the four central residues of the palindrome, ACGT - GGCC, resulted in substantial loss of β-pol promoter activity, whether the experiment was conducted in HeLa, 293, 3T3, or 133 B4 cells. Since *ras* and ElA-ElB proteins are not considered to be DNA binding proteins themselves, the mechanism of their stimulation may be through activation of a cellular protein that binds directly to the β-pol palindrome and increases transcription by RNA polymerase II. The obvious suspicion is that this activation is mediated through phosphoprotein metabolism, since *ras* and ElA/ElB proteins are involved in the signal transduction system. In view of results to be described below, showing that DNA damage of CHO cells produces a rapid increase in β-pol mRNA level, we were interested in examining the activity of the β-pol promoter constructs as a function of DNA damage. Typical results from these experiments are shown in Figure 7. Cells were transfected with a β-pol CAT construct for 20 hr, then were treated with MNNG (30 μM), and eventually examined for CAT activity as a function of time after MNNG treatment. At 16 hr after treatment, CAT activity was approximately 4-fold higher in treated cells than in untreated cells (Kedar *et al.*, 1989b). RNA start site analysis with these cells confirmed a corresponding increase in correctly initiated

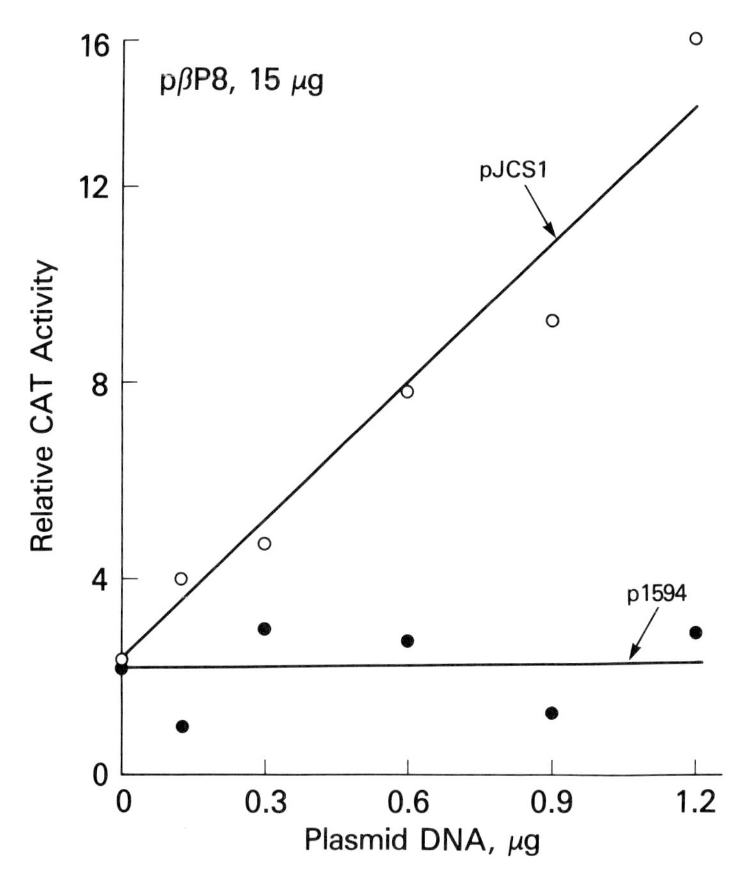

Figure 5. CAT activity after cotransfection of normal mouse 3T3 cells with 15 μg pβP8 and different amounts of β-pol fusion gene, v-*ras*^{H} protein expression plasmid pJCS1, or its vector alone, p1594. Taken from Kedar *et al.*, 1989a, with permission.

mRNA. Thus, the response to DNA damage extends to the level of β-pol promoter activity as well as to mRNA abundance (see below). As in the case of induction or stimulation by growth control proteins, the promoter stimulation by DNA damage was not seen with any of the β-pol promoter constructs containing the modifications of the decanucleotide palindrome shown in Table 2.

SURVEY OF PROTEINS BINDING TO THE β-POL PROMOTER

A tissue and cell survey was conducted by Ella Englander to identify proteins in crude nuclear extracts that are capable of binding to the cloned human β-pol promoter. This was done by DNase I footprinting analysis with the minimal promoter sequence shown in Figure 2 as probe (Englander and Wilson, submitted). Typical results with human fibroblast AG-1522, 293 cells, and HeLa cells are shown in Figure 8. The predominant footprint pattern observed with each cell extract corresponds to the decanucleotide palindrome plus 5 and 7 nucleotides on the 5′ and 3′ side, respectively. Protection with the 293 cell extract and HeLa extract is slightly better than with the 1522 extract. Slight, but significant, protection is seen over the double GC box region. These results of DNase I footprinting do not parallel the results of mRNA level

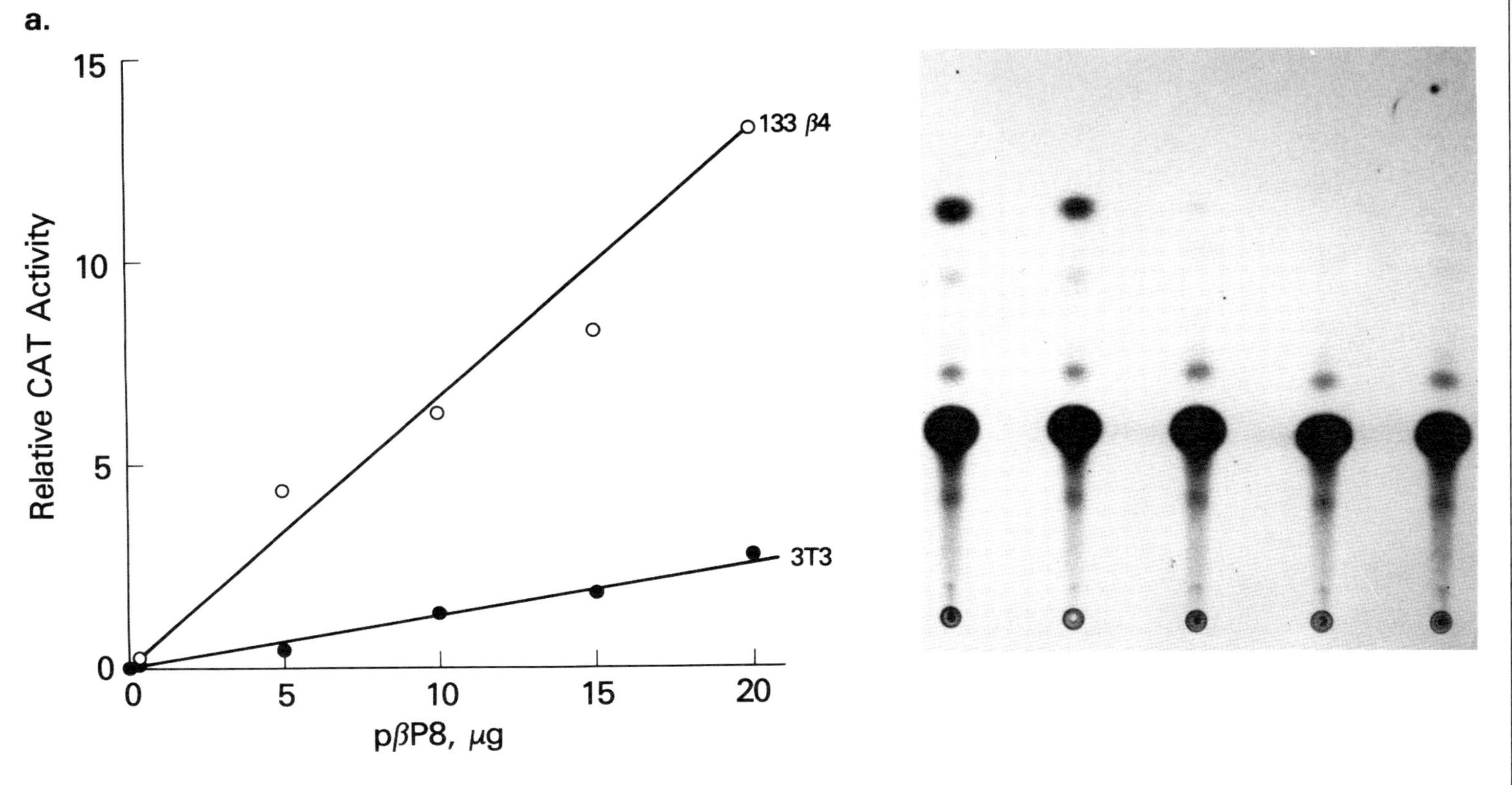

Figure 6. CAT activity after transfection of β-pol fusion gene pβP8 into various cell lines. **A**: different amounts of pβP8 were transfected into 3T3 or 133 B4 (v-*ras*^{H} expressing) cells; equal amounts of cell extract were used for measurement of CAT activity. **B**: autoradiogram showing results of expression with 15 μg pβP8 in different mouse 3T3-derived lines, as indicated in parentheses. EJ expresses an independently derived, yet identical, version of v-*ras*^{H} protein; SRD expresses v-*src* protein, and SPONT is a spontaneously transformed line. Taken from Kedar *et al.*, 1989a, with permission.

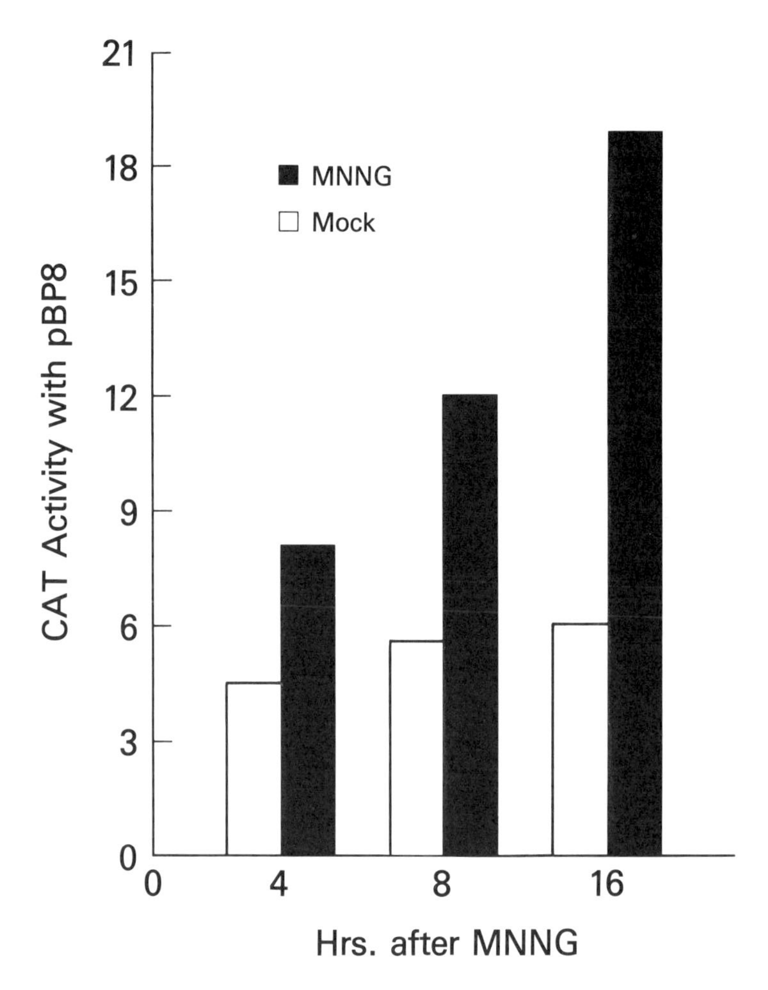

Figure 7. CAT activity in Chinese hamster cell extracts as a function of time after treatment of cells with the DNA damaging agent MNNG (30 μM). Cells had been transfected with pβP8 20 hr before MNNG treatment. Taken from Kedar *et al.*, 1989b, with permission.

analysis, since 293 cells have about five times more mRNA than AG-1522 cells. Examination of CHO cells treated with MNNG along with untreated CHO cells also is shown in Figure 8. The treated cell extract has the same palindrome binding activity as the untreated cell extract. Neither extract shows significant binding to the GC box regions or to other regions of the promoter. Again, no striking differences in the footprints were found to correlate with the 4-fold elevation in mRNA level observed after MNNG treatment.

Examination of proteins in extracts of rat liver and testes is shown in Figure 9. The testes extract contains strong palindrome binding activity and some weak binding activity to the proximal GC box. Binding to the double GC box is weak. Likewise, the liver extract shows binding to the palindrome region, but strong binding to both proximal and distal GC boxes is shown. In addition, there is footprinting in the region just downstream of the major start site; this protection clearly is not shown with the testes extract and could be due to a tissue-specific binding element and protein.

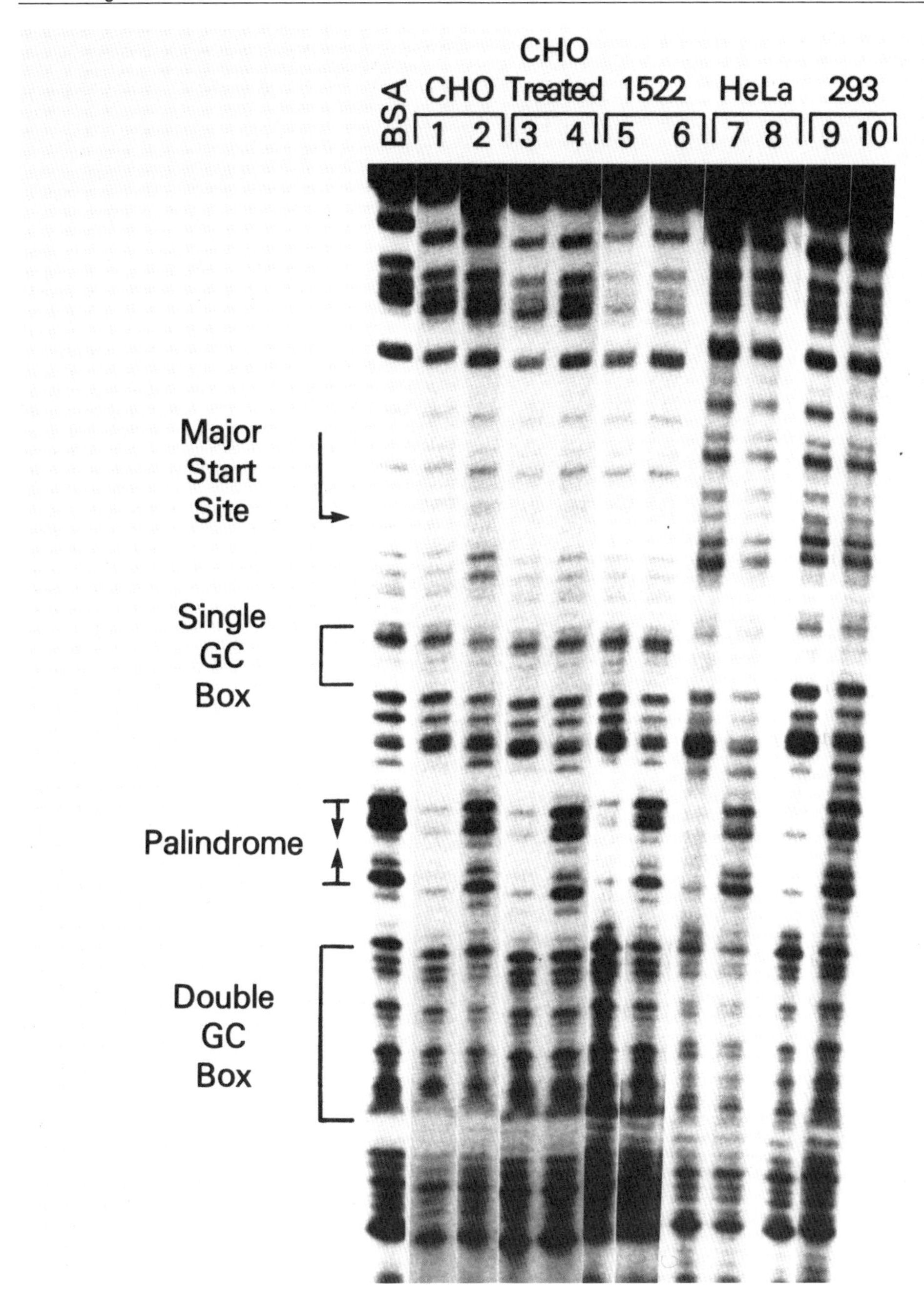

Figure 8. A composite of autoradiograms showing survey of DNase I footprinting of the β-pol minimal promoter sequence (-114 to +62) by nuclear extracts from the cell lines indicated. In the case of each extract, 200 μg of protein was mixed with 0.1 μg of DNA probe 5′ end labeled; this level of protein was the optimal footprinting level for the palindrome sequence. After DNase I digestion, products were resolved on a 6% polyacrylamide sequencing gel. Sequencing lanes were run in parallel. The control digestion with BSA as the added protein is shown on the left. Odd-numbered lanes are without and even-numbered lanes are with a 100-fold molar excess of the 24-residue palindrome oligonucleotide. CHO-treated refers to extract from cells 4 hr after treatment with 30 μM MNNG. Taken from Englander and Wilson, submitted, with permission.

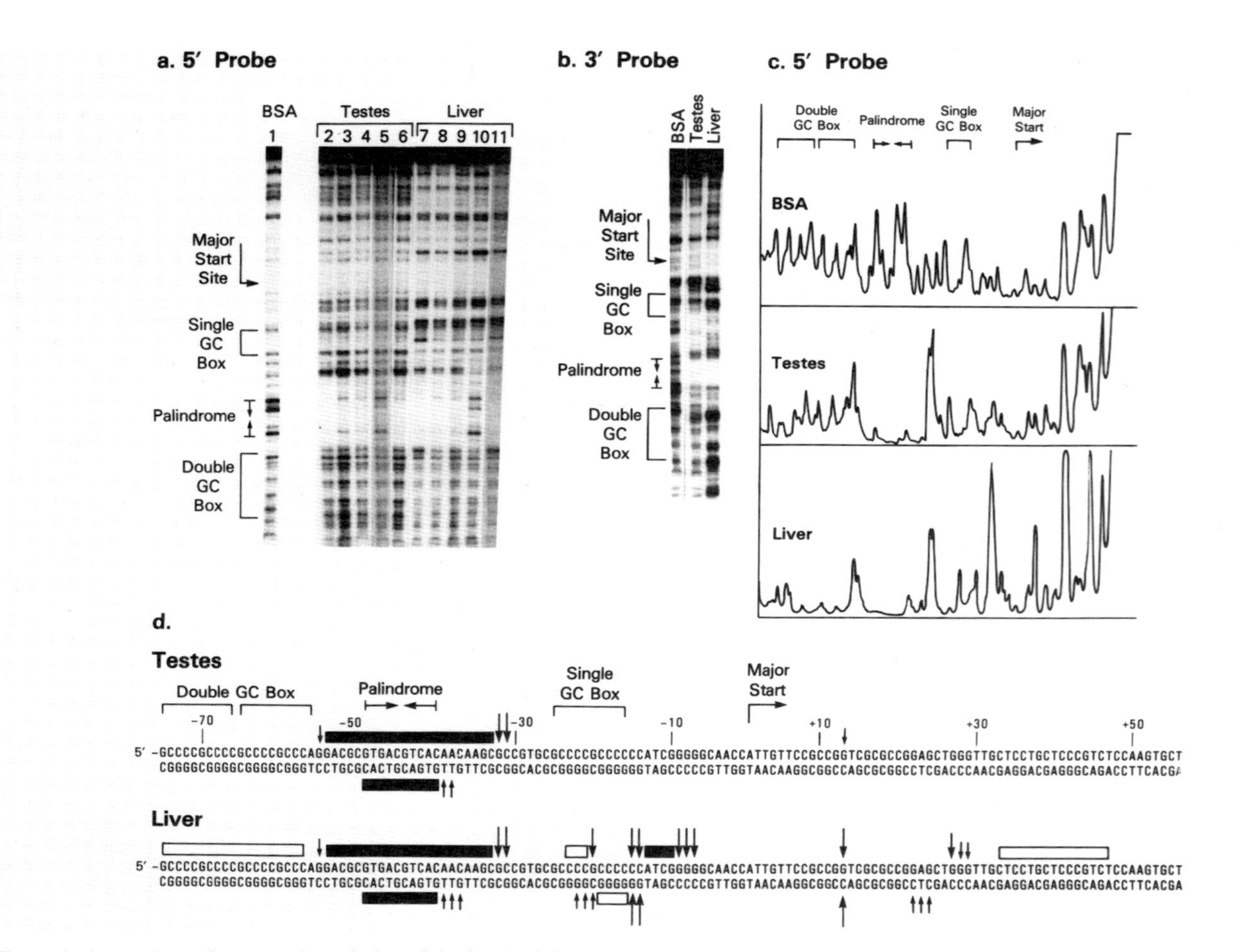

Figure 9. Comparison of DNase I footprinting of the β-pol minimal promoter sequence by nuclear extract from rodent testes and liver. Autoradiograms showing results with 5′ end labeled (**A**) and 3′ end labeled (**B**) probes were scanned by densitometry (**C**). Also shown in **A** and **B** are results of competition with an excess of the palindrome oligonucleotide. **D** shows a summary of footprinting results in the other panels. Arrows indicate hypersensitive sites and are proportional in size to band strength of the site. Regions protected by footprinting are indicated as blocks, solid representing stronger footprinting and open representing weaker footprinting. Taken from Englander and Wilson, submitted, with permission.

These results, taken together, illustrate the binding capacity of the palindrome element (about 22 residues in length) and the presence of palindrome region (22 nucleotides) binding protein(s) in each nuclear extract examined. Also, the apparent level of this binding activity fails to correlate with the level of transcriptional activity evident from steady-state mRNA levels (Kedar *et al.*, 1989a). An exception to this picture may be the liver extract; this extract contains a high level of palindrome binding activity, yet much stronger binding to the GC box regions is shown and novel downstream binding activity is shown also. As will be described below, liver contains much less (approximately 20-fold) mRNA than testes.

PURIFICATION AND CHARACTERIZATION OF A β-POL PROMOTER PALINDROME BINDING PROTEIN

Palindrome binding protein has been purified by Steven Widen from a nuclear extract prepared from thawed bovine testes (Widen and Wilson, 1989). This tissue was chosen as the source because it appears to be enriched for palindrome binding protein activity and, of course, because of the availability of large amounts of starting material. The purification of the protein was conducted using a now-routine oligonucleotide gel mobility shift or "bandshift" assay. Our implementation of this assay, illustrated with a testes extract, is shown in Figure 11. First, we found that DNase I footprint over the palindrome region afforded by the testes extract is completely reversed by competition with an oligonucleotide spanning the palindrome. This oligonucleotide is 24 residues long and is illustrated in the bottom right-hand portion of Figure 10. Bandshift assays with this oligonucleotide, 5′ end-labeled, as probe yield a linear dose-response curve with crude nuclear extract of bovine testes (Figure 11) and also with various purified fractions. Therefore, the bandshift assay could be used to follow purification of palindrome binding protein.

The results of the purification of palindrome binding protein using the bandshift assay are shown in Table 4. The purification started with 120 g of tissue and yielded approximately 3 μg of purified protein. Major purification was obtained with the FPLC MonoQ step and with the final step in the purification, oligonucleotide affinity chromatography. Examination of these fractions by SDS-polyacrylamide gel analysis, using Coomassie blue staining according to the usual procedures, revealed that the final fraction contains multiple polypeptide species with a predominant species of approximately 50,000 molecular weight; the other, less abundant polypeptides are of lower M_r, ranging from 45,000 to 25,000. The 50,000 M_r protein appears as a relatively minor component of the MonoQ fraction and is probably masked in the other two earlier fractions.

We found that the gel filtration behavior of the binding activity in each of the fractions shown in Table 4 corresponds to an average globular protein of approximately 105,000 M_r. Therefore, the native molecular weight of the binding activity did not change from step to step during the purification, and the results are consistent with the interpretation that the native palindrome binding protein is a dimer of the approximately 50,000 M_r polypeptide or a heterodimer of this polypeptide and other minor polypeptides in the final fraction. The actual binding complex, of course, could contain multimers of this 50,000 M_r polypeptide or other lower M_r polypeptides not readily detected in our SDS-PAGE analysis. In studies to confirm the identity of the 50,000 M_r polypeptide as a component of the binding species, we found that bandshift activity could be renatured from the region of SDS-PAGE containing the 50,000 M_r polypeptide. However, recovery of activity is very low (about 1%), and we could not conclude

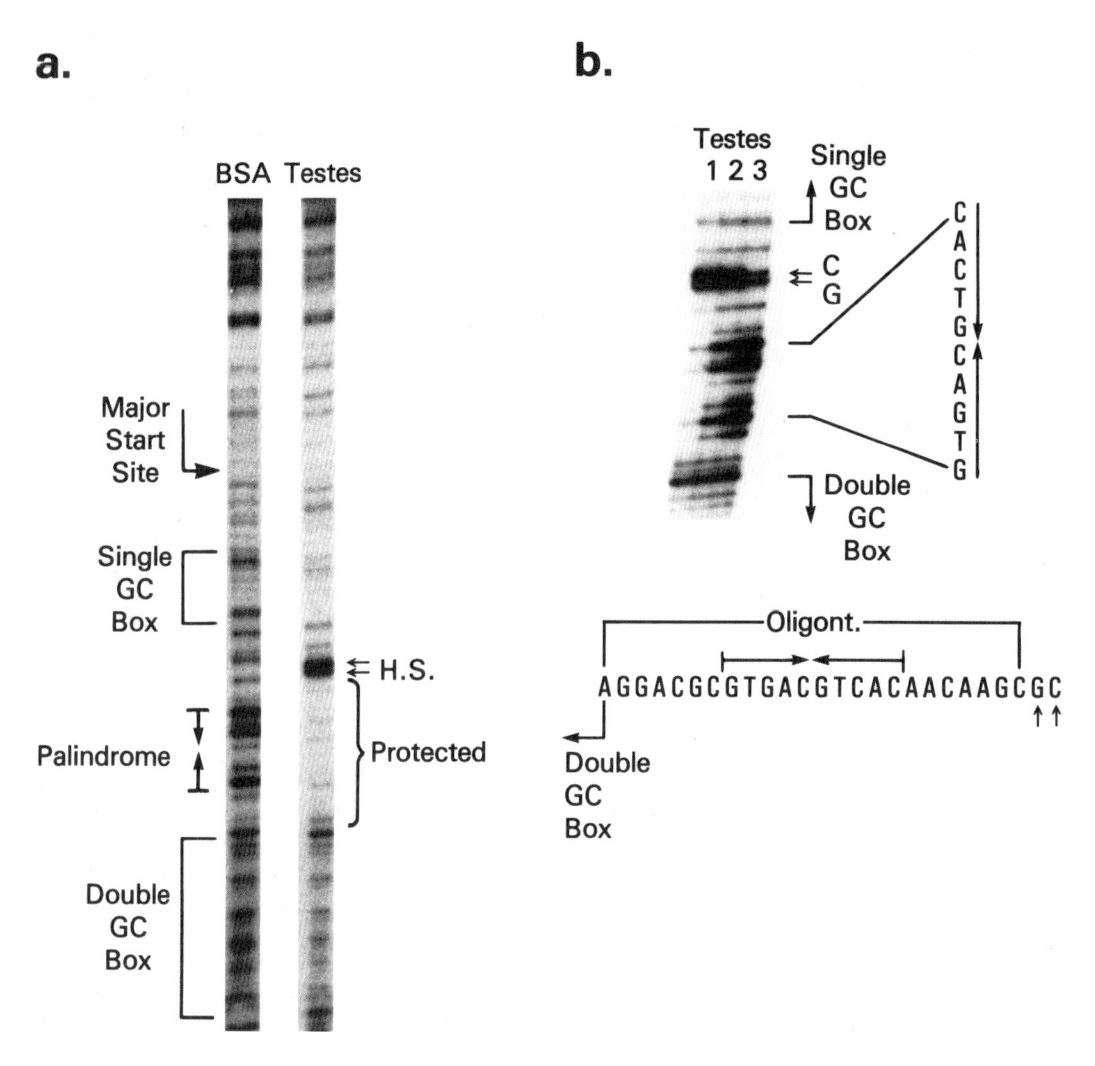

Figure 10. Autoradiograms showing results of DNase I footprinting of the human β-pol promoter (-114 to +62) by 200 μg of a crude nuclear extract from testes. **A**: the DNase I footprint around the palindrome at -49 to -40 in the promoter is noted. (See Fig. 2 for sequence.) **B**: competition with the oligonucleotide indicated is shown; lane 1, no competition DNA; lane 2, 10-fold molar excess over probe of competition DNA; lane 3, 100-fold molar excess of competition DNA. Take from Englander and Wilson, submitted, with permission.

from the experiment that all of the binding activity of the final fraction is due to the 50,000 M_r polypeptide. Next, we found that this polypeptide and the other (lower M_r) polypeptides all co-fractionate precisely with the binding activity through several chromatographic procedures, including gel filtration and single-stranded DNA column chromatography. Therefore, all of these proteins appear to participate in the binding activity, even though the 50,000 polypeptide is quantitatively the most abundant and is capable of binding activity alone after renaturation.

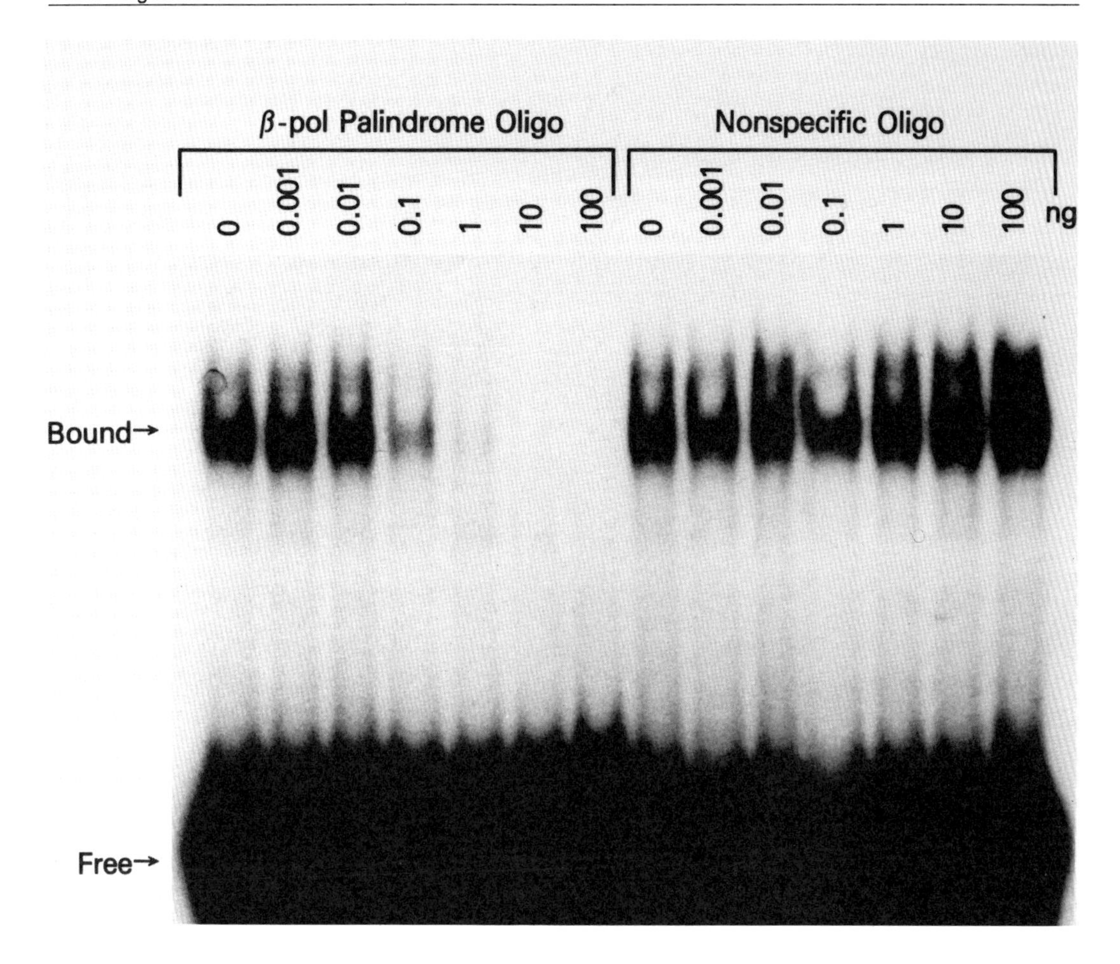

Figure 11. Autoradiogram showing results of typical gel mobility shift (bandshift) experiment with crude nuclear extract of bovine testes. The probe at 0.1 ng was ^{32}P-labeled oligonucleotide shown in Fig. 10. Competition was with this oligonucleotide on a random oligonucleotide, as indicated. Taken from Widen and Wilson, 1989, with permission.

IN VITRO ACTIVITY OF THE PURIFIED PALINDROME BINDING PROTEIN

The purified palindrome binding protein is well behaved (Widen and Wilson, 1989), in the sense that it gives linear dose-response curves for protein in the bandshift assay and that binding is fully competed by the specific oligonucleotide, but not by a random (nonspecific) oligonucleotide (Figure 12). The protein also is capable of transcriptional activity using the 1.2-kb β-pol promoter fragment cloned in a plasmid as template (Widen and Wilson, 1989). Typical results illustrating this are shown in Figure 13. We prepared a crude extract from 293 cell nuclei using ammonium sulfate for protein extraction; this extract itself is found to have transcriptional activity with a start site corresponding to a minor start site (near -30) found by analysis of endogenous cellular RNA; note that this extract contains some palindrome binding protein, as well as other transcription proteins. This minor start site (near position -30) is approximately 15 nucleotides downstream from the 3′ end of the decanucleotide palindrome

Table 4.

Summary of Purification of Palindrome Binding Protein

Step	Activity %	Protein mg
Nuclear Extract	—	1,200
0-40% Amm. Sulfate ppt.	100	300
Heparin-Sepharose	30	15
Mono Q	25	0.15
Oligo-Sepharose	20	0.003

Experiment with 120g bovine testes.

(Figure 2). In this experiment, however, no initiation at the major start site is detected (Figure 13). In the presence of the purified palindrome binding protein, such that a stoichiometric ratio of added binding protein to promoter sequence was present, much a higher level of initiation at this minor start site is observed. The palindrome binding protein, therefore, stimulates the activity of the 293 cell transcription extract, and on this basis, the protein appears to have transcriptional activity.

β-POL MESSENGER RNA

The β-pol cDNAs (SenGupta *et al.*, 1986; Zmudzka *et al.*, 1986) have been used to characterize tissue levels of β-pol mRNA as well as levels of the mRNA as a function of the cell cycle and stages of cultured cell growth (Zmudzka *et al.*, 1988). Two predominant size transcripts are shown in most rodent tissue at 1.4 and 4.0 kb, respectively (Nowak *et al.*, 1989; Zmudzka *et al.*, 1986). The 1.4-kb species is the functional mRNA for the 40,000 M_r protein. This species is present in all cell and tissue types, whereas the 4-kb species is absent in many cases and is more abundant in some species than in others (*i.e.*, rat >> human). The precise identity of the 4-kb transcript is unknown, as is its relationship to the 1.4-kb transcript.

Cell cycle experiments by Barbara Zmudzka with synchronized HeLa cells revealed that the 1.4-kb mRNA is the only species detected, and the level of this species is essentially constant over the course of the cell cycle and as cells traverse into a second cell cycle after release from double-thymidine block (Figure 14) (Zmudzka *et al.*, 1988). An examination of β-pol mRNA steady-state levels as cells are undergoing a change in the amount of replicative DNA synthesis is provided by the regenerating rat liver system. Northern blot analysis of poly(A) RNA from regenerating liver (Nowak *et al.*, 1989), as a function of hours after

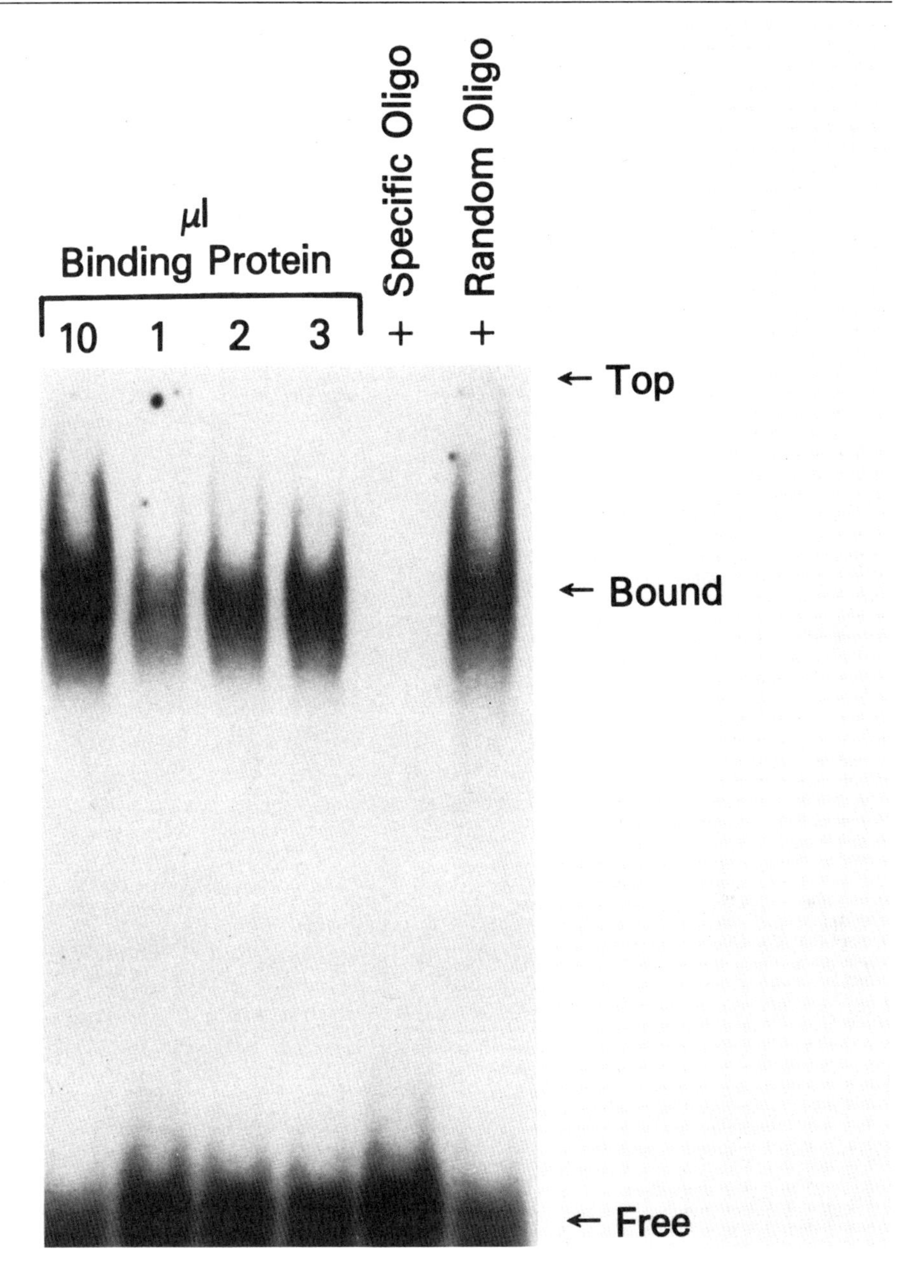

Figure 12. Autoradiogram showing results of dose-response and specificity tests with the purified palindrome-binding protein from bovine testes. A gel mobility shift assay is shown.

hepatectomy is shown in Figure 15. This tissue contains a relatively large amount of the 4-kb transcript. Note that little change in the level of the 1.4-kb transcript is detected and, similarly, the 4.0-kb transcript does not change significantly.

There is, however, differential regulation of the level of the β-pol transcript in some human cell types; for example, adenovirus ElA-ElB transformed cells termed (293) and teratocarcinoma cells have about 5-fold higher levels than fibroblast AG-1522 cells. Similarly, a tissue

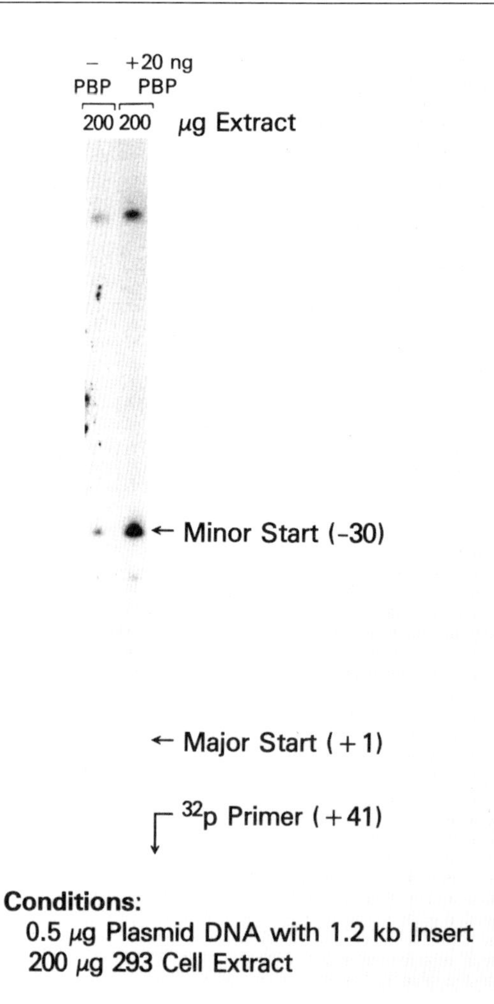

Figure 13. Autoradiogram showing primer extension analysis of products of *in vitro* transcription with and without the purified palindrome-binding protein (PBP). The start site observed with the mixture containing the PBP corresponds to a minor start site located at about -30 in the human β-pol promoter, as seen in Fig. 2.

survey of various rat organs reveals that the level of β-pol mRNA is much higher in testes than in other tissues (Nowak *et al.*, 1989). These results and all the mRNA level data available are summarized in Table 5.

Regulation of the steady-state level of β-pol mRNA also has been observed as a function of DNA damage of cells (Fornace *et al.*, 1989). When CHO cells are treated with a variety of DNA damaging agents, an increase of β-pol mRNA level is observed with some damaging agents, but not with others. Treatment with MNNG, MMS, and AAAF, for example, resulted in an approximately 4-fold higher mRNA level after 4 hr. This is illustrated in Figure 16, where the level of the 1.4-kb β-pol mRNA was examined by dot blot analysis. Similar results were obtained for MMS or AAAF treatment, but not for treatment with a number of other damaging agents (Table 6). The induction, therefore, is specific for the type of damaging agent used; a common feature of the inducing agents appears to be that they produce lesions in single base residues, as opposed to cross-links between residues or strand breaks. As already described above, the human β-pol minimal promoter in a fusion gene construct, *i.e.*, pβP8, is itself

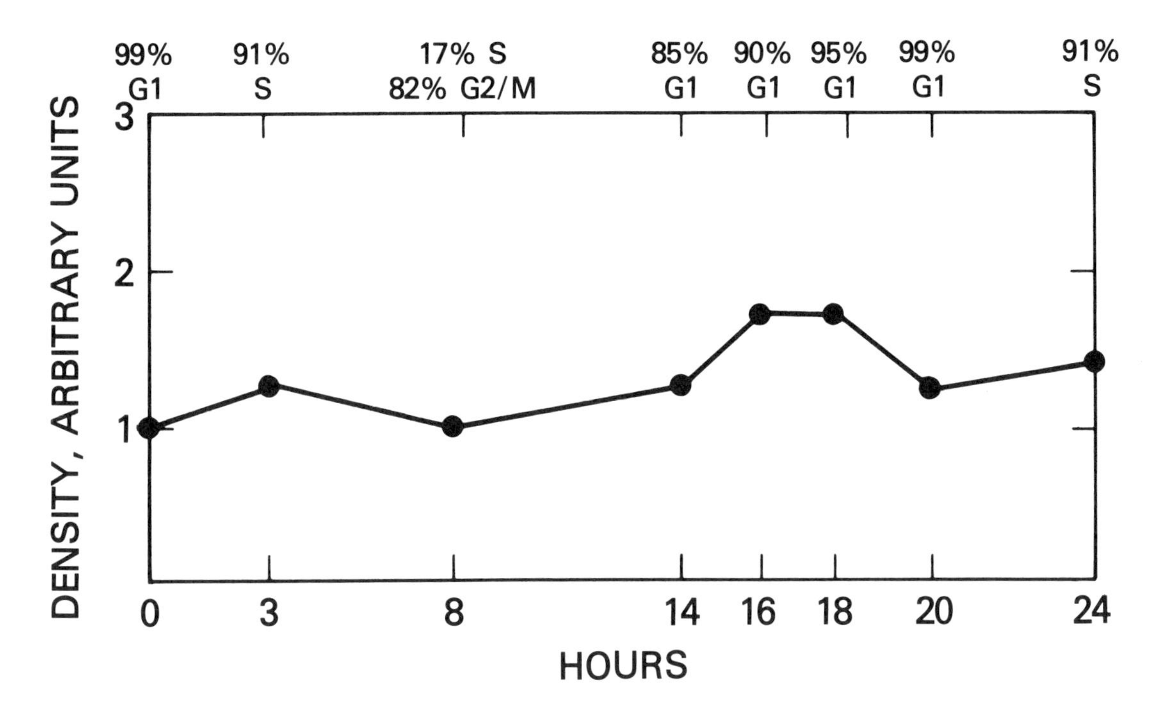

Figure 14. Quantitative analysis of β-pol mRNA levels in synchronized HeLa cells at different phases of the cell cycle. Poly(A^+) RNA was prepared from HeLa S3 cells collected 0, 3, 8, 14, 16, 18, 20, and 24 hr after release from thymidine block. The plot shows relative β-pol poly(A^+) RNA levels as a function of time after release from block (lower abscissa) or percentage of cells in phases of the cell cycle (upper abscissa). Taken from Zmudzka *et al.*, 1988, with permission.

responsive to stimulation by MNNG treatment. The sequence elements in the promoter that are responsible for the response to MNNG have not yet been identified, although the palindrome appears to be involved, since constructs with the modified palindrome do not show substantial stimulation of their basal activity when transfected into MNNG-treated cells.

β-POL PROTEIN: HISTORICAL ASPECTS

The discovery of β-pol (Baril *et al.*, 1971; Chang and Bollum, 1972; Spadari and Weissbach, 1974), like that of γ-polymerase, was a secondary product of the search for cellular reverse transcriptases in the early and mid-1970s. Among various laboratories working in the field, β-pol was universally recognized as the smallest DNA polymerase in the cell extract (Matsukage *et al.*, 1974) with a native M_r of about 40,000. Indeed, this fact was one of the very few points in this field that enjoyed general agreement. Like reverse transcriptases and γ-polymerase, β-pol is active with several widely used RNA homopolymer templates. Thus, when size measurements were omitted in studies, β-pol was misidentified as a cellular reverse transcriptase and, in some cases, as DNA γ-polymerase. In any event, the agreement on the size

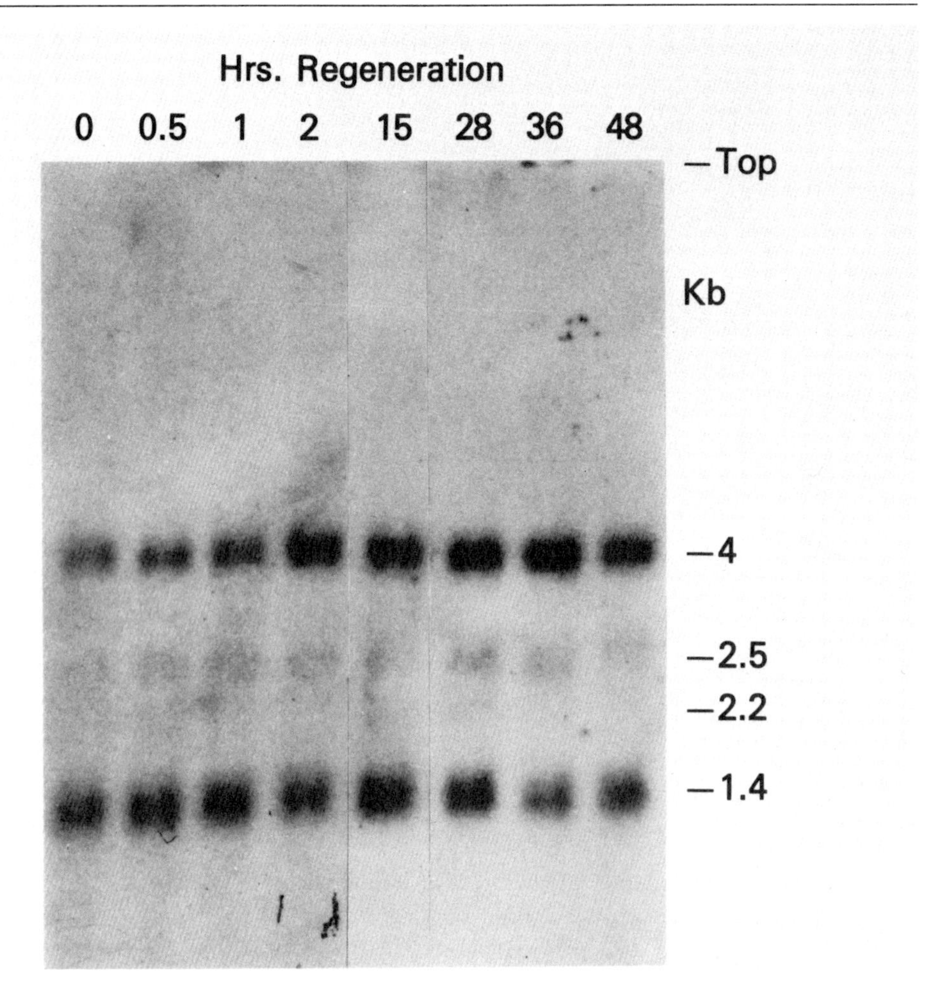

Figure 15. Northern blot analysis of rat liver poly(A^+) RNA as a function of time after partial hepatectomy. Taken from Nowak *et al.*, 1989, with permission.

of native β-pol did not extend to precise properties of the polypeptide structure of the enzyme. For example, Meyer and associates (Stalker *et al.*, 1976) consistently obtained highly purified preparations from rat tissue where the polypeptide mass was 32 kDa; Chang (1973b), on the other hand, obtained a polypeptide mass of 45 kDa for a near-homogeneous preparation from calf tissue. Wilson and his associates (Tanabe *et al.*, 1979) and Korn and his associates (Wang *et al.*, 1974) obtained near-homogeneous preparations from mouse and human cells, respectively, with polypeptides of about 40 kDa. To resolve this discrepancy in polypeptide chain size, Planck *et al.* (1980) went on to examine the possibility that β-pol activity in their mouse enzyme preparation could be associated with some other very minor polypeptide rather than with the major 40 kDa polypeptide. They found that the 40 kDa polypeptide in the purified enzyme preparation could be resolved from all other proteins by isoelectric focusing in native polyacrylamide gels and that this protein contained all of the β-pol activity of the preparation. Thus, the results clearly indicated that the 40 kDa polypeptide is β-pol.

Table 5.

Cell Type/Tissue	Relative mRNA/Cell
Human Normal Fibroblast (AG1522)	1
Human Adenovirus E1A/E1B Transf. (293)	4
Rodent or Bovine Testes	>10
Rodent Liver	0.5
CHO	1
CHO, MNNG Treated	4

The discrepancy between the sizes of the calf, rat, and mouse β-polymerases was further examined by Tanabe *et al.* (1981) and was found to be due to differences in calibration of SDS-polyacrylamide gels. For example, Tanabe *et al.* (1981) and Zmudzka and her associates (unpublished observations) independently found that purified mouse, rat, and calf β-pol all precisely co-migrate during electrophoresis in SDS-polyacrylamide gels. The rat β-pol purified by Tanabe *et al.* (1981) represented good recovery of total activity and had no polypeptide components at 32 kDa on SDS-polyacrylamide gels, and its 40 kDa polypeptide was homologous with the 40 kDa mouse β-pol polypeptide as revealed by tryptic peptide mapping. Therefore, the idea that rat β-pol is smaller is incorrect and the idea that calf β-pol is 5 kDa larger than mouse, rat, or human β-pol is erroneous also, even though this point has appeared in the literature as recently as 1988 (Cannizzaro *et al.*, 1988).

Continuing beyond these details about β-pol polypeptide structure, the early work of Chang (1973b) contributed greatly to the body of information on the enzyme. This was particularly so in the areas of the discovery of the enzyme as a novel cellular DNA polymerase, discovery of an effective purification protocol, and documentation of the constitutive nature of β-pol activity levels as a function of cell growth (Chang *et al.*, 1973). Likewise, studies by Meyer and his associates established the idea that β-pol in the cell is associated with accessory proteins

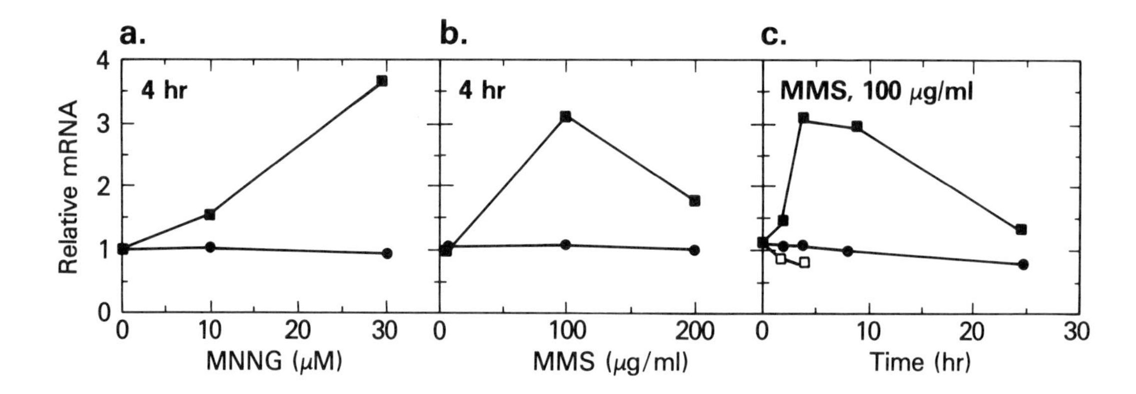

Figure 16. Dose-response and time course for β-pol induction by alkylating agents. The relative RNA abundance of β-pol (■) and β-actin (●) transcripts was determined. **A** and **B**: Dose-response experiments. CHO cells were treated with the indicated concentrations of alkylating agents for 4 hr. **C**: Time study. MMS (100 μg/ml) was added at the start of the experiment and removed after 2 hr for the 2-hr time point or after 4 hr for the other samples; values on the ordinate are relative to untreated controls. β-pol mRNA in cells also treated with 5 μg of actinomycin D per ml added 10 min before MMS is also indicated (□). Taken from Fornace *et al.*, 1989, with permission.

conferring various key activities, such as exonuclease and endonuclease (Meyer *et al.*, 1984). Such β-pol-containing macromolecular complexes are described in another chapter in this volume.

cDNA cloning of a rat β-pol coding sequence (Zmudzka *et al.*, 1986), and then later the cloning of a human β-pol cDNA (SenGupta *et al.*, 1986), resulted in the identification of the complete sequence of β-pol from these species. The deduced protein is 335 amino acids from these two mammalian species. The assignment of the start codon for translation of this mRNA was based on identification of the RNA cap site by using a genomic clone spanning the β-pol promoter and the first two exons of the β-pol coding sequence (Widen *et al.*, 1988). The translation start is the first in-frame Met codon in the mRNA, and this codon is preceded upstream by an in-frame termination codon, TGA (Abbotts *et al.*, 1988). The coding sequence cannot extend 5′ of this stop codon, because the reading frame is closed by virtue of frequent stop codons and because longer RNA transcripts could not be detected by either SI or primer extension analysis (Abbotts *et al.*, 1988; Widen *et al.*, 1988).

The only remaining obvious possibility that the β-pol mRNA could extend 5′ of the start site noted in Figure 2 is that β-pol mRNA could be produced from a gene not yet characterized. This seems very unlikely in view of Southern blot patterns of human genomic DNA, which indicate a single copy gene for β-polymerase (McBride *et al.*, 1987). The possibility of alternate splicing and of production of different forms of β-pol and its mRNA should not be overlooked, because many mammalian species contain multiple species of β-pol mRNA on Northern blot analysis, and in the case of rat tissues, the Northern blot pattern is tissue specific and age specific for brain. Adult brain contains predominantly a 4-kb mRNA species, instead of the

Table 6.

Comparison of β-Polymerase and Other mRNA Levels in CHO Cells after Treatment with DNA-Damaging Agents

DNA Damaging Agent	Dose	Relative mRNA Level				Poly(A)$^+$ RNA (μg)
		β-Polymerase	β-Actin	A1	HSP70	
MNNG	30 μM	3.6	0.9	0.9	0.5	71.3
MMS	100 μg/ml	3.5	0.8	1.0	0.8	75.9
AAAF	20 μM	2.9	0.8	1.0	0.9	77.9
H_2O_2	400 μM	2.3	1.1	1.3	0.7	77.1
UV Light	14 J/m^2	0.6	1.0	1.1	0.4	66.9
Near-UV Light	300 J/m^2	0.8	1.2	1.1	0.4	65.4
HN_2	40 μM	1.2	1.0	1.0	1.0	71.7
cis-Pt	45 μg/ml	0.9	0.8	0.8	1.1	71.4
X-ray	40 Gy	1.1	1.1	—	0.9	75.0
Adriamycin	400 ng/ml	1.0	1.1	1.1	1.0	43.0
Bleomycin	50 μg/ml	1.1	0.9	—	1.0	70.0
Heat shock	9 min at 45.5°C	0.9	0.7	1.2	26.2	82.9
Control		1	1	1	1	73.8

1.4-kb species normally seen in newborn rat brain and in various human cell lines (Nowak *et al.*, 1989; Zmudzka *et al.*, 1986).

β-POL PROTEIN: THEORETICAL CONSIDERATIONS

The deduced and measured structure of rat β-pol is shown in Figure 17. This amino acid sequence is listed in single letter code and extends 334 residues after the NH_2-terminal Met. The sequence contains one Try and three Cys residues. There is a possible nuclear localization signal in the extreme NH_2-terminal portion of the sequence. There is no obvious homology with *E. coli* DNA pol I, with reverse transcriptases, or with the DNA polymerase conserved sequences described by Salas and her associates (Bernad *et al.*, 1987) and by Wang and her associates (1989). The sequence contains some homology with terminal transferase, as revealed by computer-based matching techniques (Anderson *et al.*, 1987; Matsukage *et al.*, 1987). The significance of this homology with terminal transferase on a biological level is completely obscure; however, the two regions exhibiting the highest match between β-pol and terminal

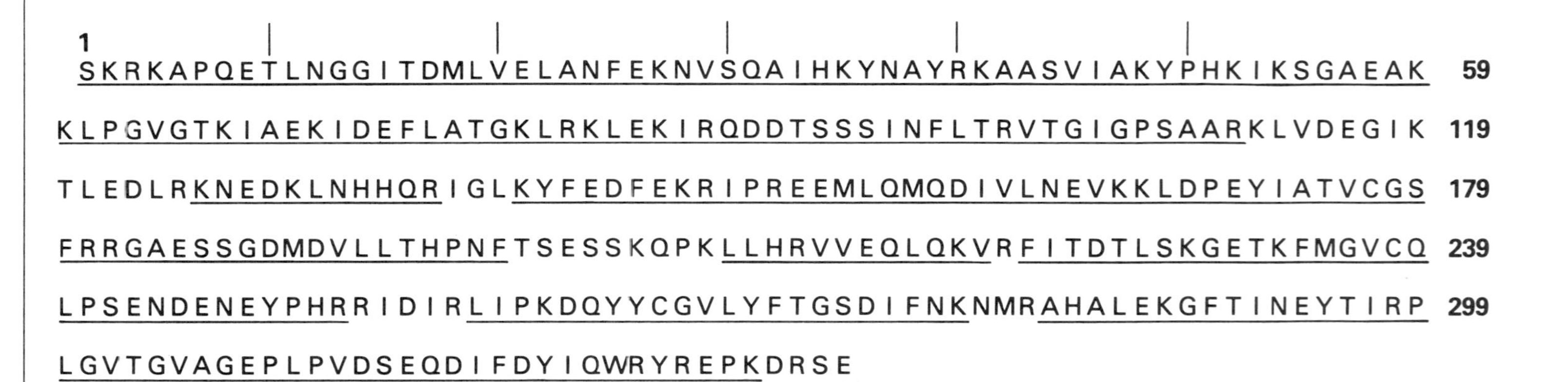

Figure 17. The sequence of rat β-pol deduced from the cDNA sequence. The sequence begins with Ser 2 of the coding sequence. Residues underlined have been confirmed by sequence analysis of recombinant rat β-pol. Taken from Kumar *et al.*, in press, with permission.

transferase may be important in polynucleotide substrate recognition and in nucleotide recognition, as described below.

The β-pol sequence does not lend itself to simple zinc finger motif identification or to the identification of other structures well known in nucleotide enzymes, such as the nucleotide binding fold of dehydrogenase or the so-called ATP binding domain. Computer-derived secondary structure analysis of rat β-pol is shown in Figure 18. The protein is predicted to be organized in seven almost equally spaced regions of α-helix. These regions are separated by turns and some segments of random coil structure. Two of the segments of random coil structure are hydrophilic and could potentially be preferred sites for protease cleavage of the protein. These sites are shown in Figure 18 around residues 75-80. As will be described below, residues around 75-80 correspond to a protease hypersensitive hinge in the protein. Overall, the ordered spacing of predicted α-helix bundles and the number of these structures suggests that the β-pol protein is tightly folded in solution. Comparison of the sequence of rat β-pol with the sequence of human β-pol indicates that the enzymes match in over 95% of the individual residues; yet, surprisingly, several of the amino acid differences appear to involve a strong chemical difference (SenGupta *et al.*, 1986).

β-POL PROTEIN: EXPERIMENTAL CONSIDERATIONS

To examine structure-function relationships of β-pol, we overexpressed both the rat and human proteins in *E. coli* and purified the recombinant proteins to homogeneity as fully active enzymes. Yield of the rat protein is greater than the human protein due to a suppressive effect within the 3′ one-third of the human coding sequence; therefore, the rat protein was selected for detailed study. Amino acid sequencing revealed that the purified rat enzyme begins with residue Ser 2 (Figure 17), rather than residue Met l, suggesting that Met l had been removed (Kumar *et al.*, in press). The enzymatic specific activity of the recombinant protein and its general catalytic properties are indistinguishable from those of comparably purified proteins isolated from mammalian tissues, mouse myeloma, and rat liver. Thus, direct comparisons between the recombinant rat protein, the recombinant human protein, and the natural mouse β-pol have failed to reveal significant differences thus far. The pure recombinant rat protein is free of exonuclease activities and DNA endonuclease activities. The enzyme exists as a monomer of approximately spherical dimensions in solutions of physiological salt concentrations (Kumar *et al.*, in press).

Domain mapping of the enzyme was undertaken by Amal Kumar (in press) using controlled digestion with various proteases. The results of these experiments indicate that the enzyme is organized as a two domain protein with a protease hypersensitive hinge-like region joining the two domains. The domains are ≈ 8 and 31 kDa, respectively. Neither domain alone, in purified form, has significant DNA polymerase activity, and an artificial mixture of the two purified domains likewise is devoid of DNA polymerase activity. On the other hand, the 8 kDa domain, which is derived from the NH_2-terminal region of the protein, has strong single-stranded DNA binding activity and is similar in this property to the intact protein. This is illustrated by the experiment shown in Figure 19. Chromatography over single-stranded DNA agarose was conducted with the intact enzyme, the 8 kDa domain, and the 31 kDa domain. It is apparent from these results that the 8 kDa domain and the intact enzyme bind to the single-stranded DNA-agarose column and are eluted at essentially identical salt concentrations. Therefore, the 8 kDa domain itself is a single-stranded DNA binding protein and may function in the enzyme

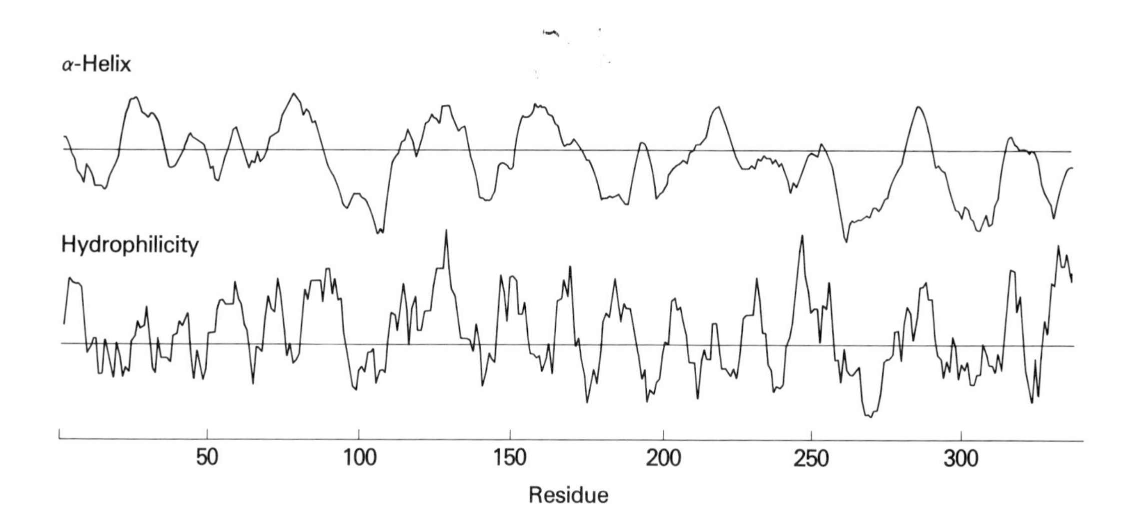

Figure 18. Secondary structure prediction for rat β-pol. For the α-helix prediction, values above the line represent 5-fold higher probability for α-helix than any other structure. For hydrophilicity prediction, values above the line are hydrophilic and values below the line are hydrophobic. Taken from Kumar *et al.*, in press, with permission.

in a modular fashion, conferring single-stranded DNA binding capacity. Binding capacity to single-stranded DNA was further examined using a gel retardation assay, such as that recently employed for the analysis of the Al hnRNP complex protein, a prototype single-stranded nucleic acid binding protein (Cobianchi *et al.*, 1988; Sapp *et al.*, 1986). Mixture of the intact β-pol protein with single-stranded M13 DNA results in an all or none pattern of complex formation. This is consistent with protein-protein cooperativity for the binding reaction; the results were very similar to results previously obtained with the Al protein (Cobianchi *et al.*, 1988). The intact enzyme, therefore, is a cooperative single-stranded nucleic acid binding protein. The 8 kDa domain, on the other hand, binds to single-stranded DNA by the gel retardation assay, but is noncooperative in its behavior. These results suggest the 31 kDa domain of the intact protein is responsible for, or at least is required for, protein-protein interaction resulting in positive binding cooperativity. This observation of positive cooperativity in DNA binding for β-pol may explain some of the peculiar behavior of this enzyme reported previously with synthetic homopolymer template primers. For example, it generally has been very difficult to obtain a range of linearity with enzyme concentration, and indeed, sigmoidal-shaped enzyme concentration curves for β-pol enzymatic activity with homopolymer template primer systems have been reported in this laboratory and by others (Ono *et al.*, 1979).

Binding to double-stranded DNA agarose by the β-pol proteolytic fragments and by the intact protein was studied also. Neither the 8 kDa domain fragment nor the intact protein nor the 31 kDa domain fragment is capable of binding to double-stranded DNA under the conditions used in Figure 19 (with single-stranded DNA). The results of these domain mapping

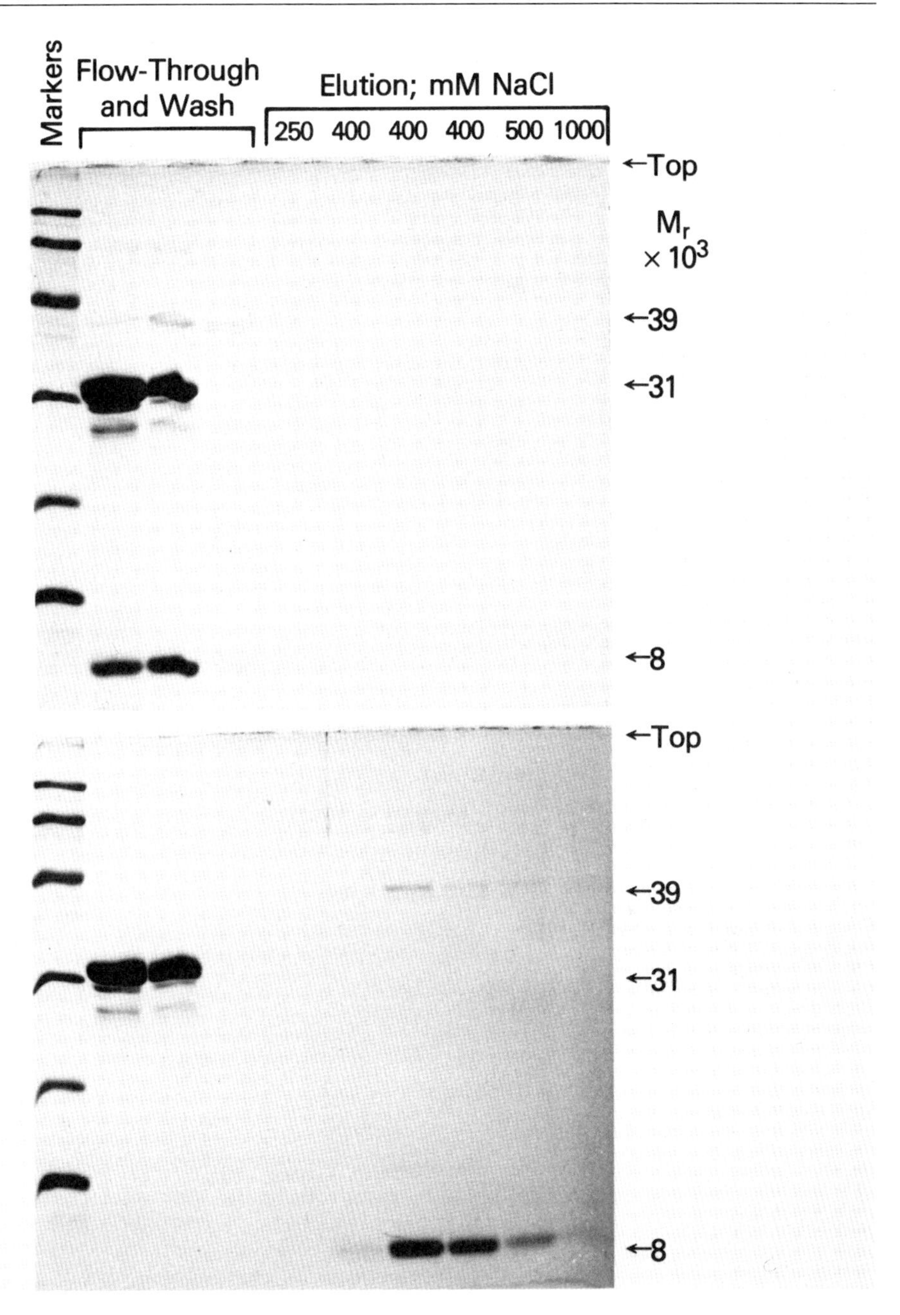

Figure 19. DNA agarose column chromatography of a mixture of the 31 kDa and 8 kDA fragments. Photographs of stained SDS-polyacrylamide gels are shown. In the upper panel, the proteins were chromatographed on double-stranded DNA-agarose. In the lower panel, chromatography was on single-stranded DNA-agarose. The flow-through and wash fractions are indicated, as are the fractions obtained by NaCl elution. The buffer was Tris-HCl at pH 7.6, and each fraction corresponded to 1.5 column bed volumes. The *Mr* values on the right were calculated from positions of marker proteins, as shown. A small amount of intact β-pol was present in the protein mixture also, and it behaved similarly to the 8 kDa fragment, as seen in the lower panel. Taken from Kumar *et al.*, in press, with permission.

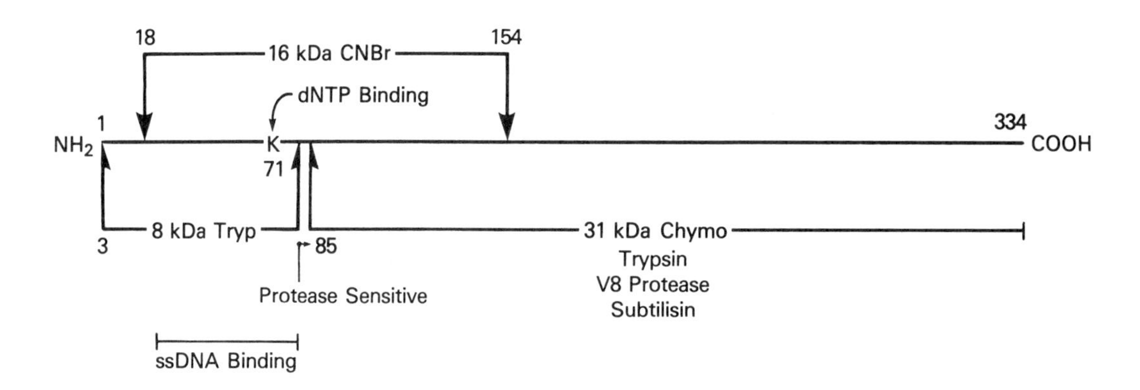

Figure 20. Scheme illustrating positions of the 8 kDa domain, 31 kDa domain, and PLP Lys target residue in the primary structure of rat β-pol. Proteases used for fragment isolation are indicated. The cyanogen bromide fragment of 16 kDa has single-stranded DNA binding activity. The protease-sensitive region is indicated.

studies are summarized in Figure 20. The peptide fragments were aligned along the sequence of the intact protein by direct NH_2-terminal sequencing (Kumar *et al.*, in press). Equilibrium binding studies with poly(ethenoadenylate) as fluorescent probe (Cobianchi *et al.*, 1988) confirmed single-stranded nucleic acid binding capacity by the 8 kDa fragment and the complete absence of binding by the 31 kDa fragment. However, these studies revealed differences in binding properties of the intact protein and the 8 kDa fragment. The occluded site size for binding is 12 nucleotide residues for β-pol and 6 nucleotides for the 8 kDa fragment; in 0.05 M NaCl, the association constant for β-pol is 3×10^7 M^{-1}, whereas the value for the 8 kDa fragment is only 2×10^5 M^{-1} (Kumar *et al.*, in press). Thus, as noted above, the 31 kDa domain clearly contributes to the binding mechanism of the intact protein.

NUCLEOTIDE BINDING POCKET

Identification of a residue in the vicinity of the nucleotide binding pocket was performed using the pyridoxal 5′ phosphate-labeling approach. In the absence of nucleotide substrate, pyridoxal phosphate is incorporated at just four Lys residues of the enzyme (Basu *et al.*, 1989). Only one of these Lys residues is protected from labeling in the presence of a substrate dNTP. Tryptic peptide preparation and sequencing revealed that this Lys residue is Lys 71 in the primary structure of the recombinant rat protein. This residue is within the 8 kDa domain near its COOH-terminal end, as illustrated in Figure 20. Another region of β-polymerase also may be involved in forming the dNTP binding pocket. This idea comes from the computer-based homology between β-pol and terminal transferase. One of the two regions where the enzymes show close similarity corresponds to a dNTP binding pocket of terminal transferase proposed by Evans *et al.* (1989). This region in β-pol is around residue 270 in the 31 kDa domain.

β-POL ENZYMATIC ACTIVITY

The proposed kinetic reaction scheme for DNA synthesis by β-pol is ordered BiBi (Tanabe *et al.*, 1979). This scheme depicts an ordered sequential reaction where the enzyme combines first with template primer and then with dNTP substrate. The products are depicted as ordered release of pyrophosphate first and then newly formed DNA second. Evidence for the ordered addition of substrates to the enzyme was first obtained by Tanabe *et al.* (1979), who used a steady-state kinetic approach with product inhibition by pyrophosphate. They found that pyrophosphate produced a slope effect in the primary reciprocal plots with dNTP as variable substrate, but did not produce a slope effect in primary reciprocal plots with DNA as variable substrate. Thus it was proposed that addition of pyrophosphate and dNTP to the enzyme is reversibly connected in the kinetic scheme. In contrast, addition of DNA to the enzyme is separated from and not reversibly connected to addition of pyrophospate, since pyrophosphate does not cause a slope effect for DNA primary reciprocal plots. The simplest and most straightforward interpretation of these results was that DNA adds to the enzyme before dNTP, and hence, the addition of dNTP isolates DNA from the slope effect by pyrophosphate. The observation of direct single-stranded DNA binding by the intact enzyme is, of course, consistent with this ordered addition of DNA to the enzyme, yet this type of physical binding is not necessarily evidence in favor of the proposed kinetic scheme, since binding to single-stranded DNA in a column experiment or equilibrium binding experiment may not be equivalent to the binding step observed during the enzymatic reaction under steady-state conditions.

The results of pyridoxal 5′ phosphate modification experiments described in the previous section do, however, corroborate the kinetic scheme. Thus, pyridoxal 5′ phosphate is not displaced from the dNTP binding pocket by dNTP alone (without template primer). Instead, both a template primer and the template complementary dNTP are required; this result is consistent with the kinetic scheme because binding to the template primer is required before the dNTP binding event can occur.

The recombinant β-pol, like natural β-pol, can gap fill on a single-stranded template (Abbotts *et al.*, 1988). The enzymatic turnover number for this activity is about the same as that of *E. coli Pol*I and the natural β-pols. However, the enzyme does not completely fill a gap when the reactions are conducted under usual steady-state conditions of limiting enzyme. The enzyme begins having difficulty when the gap closes to ~11 nucleotide residues and is refractory to synthesis when the gap is closed to 6 residues. This interpretation is based on experiments such as that shown in Figure 21, where a defined ϕX174 DNA template primer system was used (Abbotts *et al.*, 1988). Note that the size of this 11-residue gap matches well with the binding site size of β-pol on the poly(ethenoadenylate) probe (Kumar *et al.*, in press). The failure of pure β-pol to completely fill small gaps *in vitro* is interesting, since the predominant activity of the enzyme in repairing bleomycin damage of cells is synthesis to fill short gaps (DiGiuseppe and Dresler, 1989). Therefore, the expectation is that the enzyme is somehow recruited into the short bleomycin-induced gaps by other proteins.

Strand-displacement activity of β-pol was evaluated with a singly nicked plasmid DNA template primer made from pML2 DNA by controlled *Eco*RI digestion. These experiments were conducted by Barbara Zmudzka. As expected from the gap-filling studies described above, the enzyme is completely inactive with this nicked double-stranded DNA as substrate under normal steady-state conditions of limiting enzyme and substrate excess. However, when

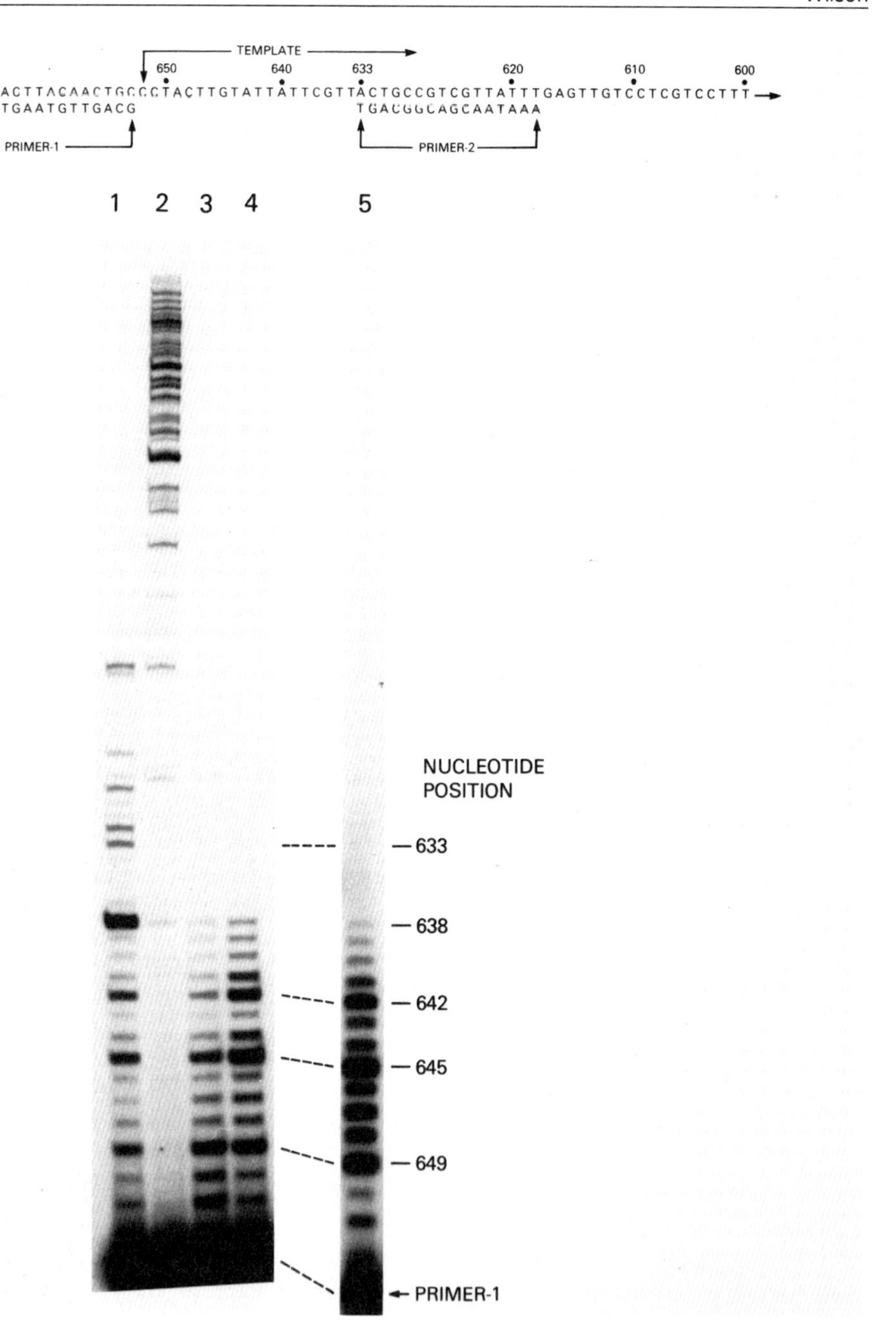

Figure 21. Gap filling analysis of recombinant β-pol with φX174 DNA as template primer. Photographs of autoradiograms are shown. At the top is shown synthetic primer 1 5′ end labeled and hybridized to φX174 DNA. DNA synthesis reactions were carried out as usual, except in some cases both primer 1 and unlabeled primer 2 were hybridized to φX174 DNA. Conditions were as follows: lane 1, only primer 1 hybridized, 20-min incubation; lane 2, identical with lane 1, 60-min incubation; lane 3, both primers hybridized, 20-min incubation; lane 4, both primers hybridized, 60-min incubation; lane 5, both primers hybridized, reaction contained 100 μM NaCl, 60-min incubation. Position indicated on the right refers to template nucleotide opposite the last nucleotide of the product. Taken from Abbotts *et al.*, 1988, with permission.

the enzyme is added to reaction mixtures in molar excess over primer termini, abundant DNA synthesis is observed. This activity corresponds to addition of an average of 700 nucleotide residues per nick after a 10-min incubation. Agarose gel analysis of products of this reaction revealed that the distribution of products is unimodal, and this is consistent with strand-displacement synthesis from each of the nicked substrate molecules, such that the distribution of products in the gel is roughly the same as expected from the average dNMP incorporation per nicked template molecule. Further characterization of these product molecules revealed the surprising finding that they are digested by *Eco*RI endonuclease to molecules of ≈30 residues in length. Nucleotide sequence analysis of these digested products revealed the sequence to be a repeat. We conclude that synthesis proceeds by strand displacement for ≈12 template residues before the primer misaligns and anneals to itself, thus providing a substrate for "snap-back synthesis." True strand-displacement synthesis and copying of the template DNA was not detected.

CONCLUSIONS

I have summarized work aimed at establishing the β-pol promoter as a system for detailed studies of transcriptional control of the gene and the β-pol protein as a system for studies of DNA polymerase mechanism. Research during the next several years on promoter function will focus on the nature of the palindrome-binding protein and its relationship to proteins that bind to similar sequences in other promoters. Experiments on the β-pol protein will focus on physical biochemical approaches, such as crystallography and NMR, to implicate certain amino acid residues in the mechanism. Another interesting line of experiments on β-pol will focus on the *in vivo* activity of the enzyme, as revealed by techniques used to modulate the level of the enzyme in living cells.

REFERENCES

Abbotts, J., D. N. SenGupta, B. Zmudzka, S. Widen, V. Notario, and S. H. Wilson.(1988) Expression of human DNA polymerase beta in *E. coli* and characterization of the recombinant enzyme. *Biochemistry* **27**: 901-909

Anderson, R. S., C. B. Lawrence, S. H. Wilson, and K. L. Beattie. (1987) Genetic relatedness of DNA polymerase beta and terminal deoxynucleotidyl-transferase. *Gene* **60**: 163-173

Baril, E. F., O. E. Brown, M. D. Jenkins, and J. Laszlo. (1971) Deoxyribonucleic acid polymerases with rat liver ribosomes and smooth membranes. Purification and properties of the enzymes. *Biochemistry* **10**: 1981-1992

Basu, A., P. Kedar, S. H. Wilson, and M. J. Modak. (1989) Active site modification of mammalian DNA polymerase b with pyridoxal 5′ phosphate: Mechanism of inhibition and identification of a deoxynucleoside 5′-triphosphate binding site residue. *Biochemistry* **28**: 6305-6309.

Bernad, A., A. Zaballos, M. Salas, and L. Blanco. (1987) Structural and functional relationship between prokaryotic and eukaryotic DNA polymerases. *EMBO J.* **6**: 4219-4225

Cannizzaro, L. A., F. J. Bollum, K. Huebner, C. M. Croce, L. C. Cheung, X. Xu, B. K. Hecht, F. Hecht, and L. M. S. Chang. (1988) Chromosome sublocalization of a cDNA for human DNA polymerase-β to 8p11-p12. *Cytogenet. Cell Genet.* **47**: 121-124

Chang, L. M. S. (1973a) Low molecular weight deoxyribonucleic acid polymerase from calf thymus chromatin. I. Initiation and fidelity of homopolymer replication. *J. Biol. Chem.* **248**: 6983-6992

Chang, L. M. S. (1973b) Low molecular weight deoxyribonucleic acid polymerase from calf thymus chromatin. I. Preparation of homogeneous enzyme. *J. Biol. Chem.* **248**: 3789-3795

Chang, L. M. S. (1975) The distributive nature of enzymatic DNA synthesis. *J. Mol. Biol.* **93**: 219-235

Chang, L. M. S., and F. J. Bollum. (1972) Low molecular weight deoxyribonucleic acid polymerase from rabbit bone marrow. *Biochemistry* **11**: 1264-1272

Chang, L. M. S., M. Brown, and F. J. Bollum. (1973) Induction of DNA polymerase in mouse L cells. *J. Mol. Biol.* **74**: 1-8

Cobianchi, F., R. L. Karpel, K. L. Williams, V. Notario, and S. H. Wilson. (1988) Mammalian hnRNP complex protein Al: Large-scale overproduction in *E. coli* and cooperative binding to single-stranded nucleic acids. *J. Biol. Chem.* **263**: 1063-1071

DiGiuseppe, J. A., and S. L. Dresler. (1989) Repair of Bleomycin-induced DNA repair synthesis in permeabilized human cells. *Biochemistry*, in press.

Dresler, S. L., and M. G. Frattini. (1986) DNA replication and UV-induced DNA repair synthesis in human fibroblasts are much less sensitive than DNA polymerase-α to inhibition by butylphenyl-deoxyguanosine triphosphate. *Nucleic Acids Res.* **14**: 7093-7102

Dresler, S. L., and K. S. Kimbro. (1987) 2′,3′-Dideoxythymidine 5′-triphosphate inhibition of DNA replication and ultraviolet-induced DNA repair synthesis in human cells: Evidence for involvement of DNA polymerase. *Biochemistry* **26**: 2664-2668

Dresler, S. L., and M. W. Lieberman. (1983) Identification of DNA polymerases involved in DNA excision repair in diploid human fibroblasts. *J. Biol. Chem.* **258**: 9990-9994

Englander, E. W., and S. H. Wilson. ( ) Protein binding elements in the human β-polymerase promoter. *J. Biol. Chem.* Submitted.

Evans, R. K., C. M. Beach, and M. S. Coleman. (1989) Photoaffinity labeling of terminal deoxynucleotidyl transferase. 2. Identification of peptides in the nucleotide binding domain. *Biochemistry* **28**: 713-720

Fornace, A., B. Zmudzka, M. C. Hollander, and S. H. Wilson. (1989) Induction of DNA polymerase b mRNA by DNA damaging agents in Chinese hamster ovary cells. *Mol. Cell. Biol.* **9**: 851-853

Fry, M. (1983) Eukaryotic DNA polymerases. In *Enzymes of nucleic acid synthesis and modification*, (Ed. S. T. Jacob), pp. 39-92. CRC Press, Inc., Boca Raton, FL.

Fry, M., and L. Loeb. (1986) *Animal Cell DNA Polymerases*, CRC Press, Inc., Boca Raton, FL.

Kedar, P., S. G. Widen, A. Fornace, and S. H. Wilson. (1989a) Activation of a β-polymerase promoter fusion gene in CHO cells by DNA damaging agents. *Nature* Submitted.

Kedar, P., S. G. Widen, D. R. Lowy, and S. H. Wilson. (1989b) Stimulation of a β-polymerase promoter fusion gene in mouse 3T3 fibroblast by Harvey *ras* protein. *Mol. Cell. Biol.* Submitted.

Kumar, A., S. G. Widen, K. R. Williams, R. L. Karpel, P. Kedar, and S. H. Wilson. ( ) Studies of the domain structure of mammalian DNA polymerase β: Identification of a discrete template binding domain. *J. Biol. Chem.* In press.

Matsukage, A., E. W. Bohn, and S. H. Wilson. (1974) Multiple forms of DNA polymerase in mouse myeloma. *Proc. Natl. Acad. Sci. USA* **71**: 578-582

Matsukage, A., K. Nishikawa, T. Ooi, Y. Seto, and M. Yamaguchi. (1987) Homology between mammalian DNA polymerase b and terminal deoxynucleotidyl-transferase. *J. Biol. Chem.* **262**: 8960-8962

Matsukage, A., M. Yamaguchi, K. R. Utsumi, Y. Hayashi, R. Ueda, and M. C. Yoshida. (1986) Assignment of the gene for human DNA polymerase beta (POLB) to chromosome 8. *Jpn. J. Cancer Res.* **77**: 330-333

McBride, O. W., C. A. Kozak, and S. H. Wilson. (1989) Mapping of DNA polymerase-β to mouse chromosome 8. *Cytogenet. Cell Genet.* In press.

McBride, O. W., B. Z. Zmudzka, and S. H. Wilson. (1987) Chromosomal location of the human gene for DNA polymerase-β. *Proc. Natl. Acad. Sci. USA* **84**: 503-507

Meyer, R. R., D. C. Thomas, T. J. Koerner, and D. C. Rein. (1984) Interaction of DNA accessory proteins with DNA polymerase-β of the Novikoff hepatoma. In: *Proteins involved in DNA replication*, (Ed. U. Hubscher and S. Spadari), pp. 355-362. Plenum Publ., New York.

Miller, M. R., and D. N. Chinault. (1982) The roles of dNA polymerases-α, -β and -γ in DNA repair synthesis induced hamster and human cells by different DNA damaging agents. *J. Biol. Chem.* **257**: 10204-10209

Mosbaugh, D. W., and S. Linn. (1983) Excision repair and DNA synthesis with a combination of HeLa DNA polymerase-β and DNase V. *J. Biol. Chem.* **258**: 108-118

Nowak, R., J. A. Siedlecki, L. Kaczmarek, B. Z. Zmudzka, and S. H. Wilson. (1989) Levels and size complexity of DNA polymerase b mRNA in rat regenerating liver and other organs. Biochim. *Biophys. Acta* **1008**: 203-207

Olson, S., and O. W. McBride. (1989) Mapping of β-polymerase on human chromosome 8. *Cytogenet. Cell Genet.* Submitted.

Ono, K., A. Ohashi, K. Tanabe, A. Matsukage, M. Nishizewa, and T. Takahashi. (1979) Unique requirements for template primers of DNA polymerase-β from rat ascites hepatoma AH130 cells. *Nucleic Acids Res.* **7**: 715-726

Planck, S. R., K. Tanabe, and S. H. Wilson. (1980) Distinction between mouse DNA polymerases a and b by tryptic peptide mapping. *Nucleic Acids Res.* **8**: 2771-2782

Sapp, M., R. Knippers, and A. Richter. (1986) DNA binding properties of a 110 kDA nucleolar protein. *Nucleic Acids Res.* **14**: 6803-6820

SenGupta, D. N., B. Z. Zmudzka, P. Kumar, F. Cobianchi, J. Skowronski, and S. H. Wilson. (1986) Sequence of human DNA polymerase b mRNA obtained through cDNA cloning. *Biochem. Biophys. Res. Commun.* **136**: 341-347

Siedlecki, J. A., J. Szysko, I. Pietrzykowska, and B. Zmudzka. (1980) Evidence implying DNA polymerase-β function in excision repair. *Nucleic Acids Res.* **8**: 361-375

Smith, C. A., and D. S. Okumoto. (1984) Nature of DNA repair synthesis resistant to inhibitors of polymerase a in human cells. *Biochemistry* **23**: 1383-1390

Spadari, S., and A. Weissbach. (1974) The interrelation between DNA synthesis and various DNA polymerase activities in synchronized HeLa cells. *J. Mol. Biol.* **86**: 11-20

Stalker, D. M., D. W. Mosbaugh, and R. R. Meyer. (1976) Novikoff hepatoma deoxyribonucleic acid polymerase. Purification and properties of a homogeneous β-polymerase. *Biochemistry* **15**: 3114-3121

Tanabe, K., E. W. Bohn, and S. H. Wilson. (1979) Steady-state kinetics of mouse DNA polymerase beta. *Biochemistry* **18**: 3401-3407

Wang, T. S.-F., and D. Korn. (1980) Reactivity of KB cell deoxyribonucleic acid polymerases-α and -β with nicked and gapped deoxyribonucleic acid. *Biochemistry* **19**: 1782-1790

Wang, T. S.-F., D. C. Eichler, and D. Korn. (1977) Effect of Mn^{2+} on the *in vitro* activity of human deoxyribonuclease acid polymerase-β. *Biochemistry* **16**: 4927-4933

Wang, T. S.-F., W. D. Sedwick, and D. Korn. (1974) Nuclear deoxyribonucleic acid polymerase. Purification and properties of the homogeneous enzyme from human KB cells. *J. Biol. Chem.* **249**: 841-850

Wang, T. S.F., S. W. Wong, and D. Korn. (1989) Human DNA polymerase α: Predicted functional domains and relationships with viral DNA polymerases. *FASEB J.* **3**: 14-21

Widen, S., and S. H. Wilson. (1989) Purification and characterization of palindrome binding proteins for the β-polymerase promoter. *Biochemistry*. Submitted.

Widen, S., P. Kedar, and S. H. Wilson. (1988) Human β-polymerase gene: Structure of the 5′-flanking region and active promoter. *J. Biol. Chem.* **263**: 16992-16998

Wilson, S. H., J. Abbotts, and S. Widen. (1988) Progress toward the molecular biology of DNA polymerase β. *Biochim. Biophys. Acta* **949**: 149-157

Yamaguchi, M., Y. Hayashi, and A. Matsukage. (1988) Mouse DNA polymerase b gene promoter: Fine mapping and involvement of Sp1-like mouse transcription factor in its function. *Nucleic Acids Res.* **16**: 8773-8787

Yamaguchi, M., F. Hirose, Y. Hayashi, Y. Nishimoto, and A. Matsukage. (1987) Murine DNA polymerase b gene: Mapping of transcription initiation sites and the nucleotide sequence of the putative promoter region. *Mol. Cell Biol.* **7**: 2012-2018

Yamaguchi, M., K. Tanabe, Y. N. Taguchi, M. Nishizawa, T. Takahashi, and A. Matsukage. (1980) Chick embryo DNA polymerase-β. Purified enzyme consists of a single *M*r = 40,000 polypeptide. *J. Biol. Chem.* **255**: 9942-9948

Zmudzka, B. A., A. Fornace, J. Collins, and S. H. Wilson. (1988) DNA polymerase b mRNA: Cell-cycle and growth response in cultured human cells. *Nucleic. Acids Res.* **16**: 9587-9596

Zmudzka, B. Z., D. SenGupta, A. Matsukage, F. Cobianchi, P. Kedar, and S. H. Wilson. (1986) Structure of rat DNA polymerase b revealed by partial amino acid sequencing and cDNA cloning. *Proc. Natl. Acad. Sci. USA* **83**: 5106-5110

9

PRIMASE FROM EUKARYOTIC CELLS

Ben Y. Tseng and Charles E. Prussak

INTRODUCTION

The inability of DNA polymerases to initiate DNA synthesis *de novo* led to the proposal that RNA priming was a mechanism for initiation. The demonstration of this mechanism in *E. coli* was based on the requirement for a rifampicin sensitive activity, RNA polymerase, in the initial conversion of infecting M13 phage single stranded DNA to double stranded form (Kornberg, 1980). However, for ϕX174 SS to RF synthesis, a new rifampicin-resistant RNA polymerase was necessary and purification of this specialized RNA polymerase, termed primase, was carried out using a bacteriophage G4 DNA template which had simpler component requirements (Bouche *et al.*, 1978). *E. coli* primase was identified as the product of the dnaG gene which had been previously found to be essential for the synthesis of Okazaki fragments in cells (Lark, 1972). Several *E. coli* bacteriophages (Scherzinger *et al.*, 1977; Romano and Richardson, 1979; Hinton and Nossal, 1985; Cha and Alberts, 1986) and plasmids (Boulnois and Wilkins, 1979; Lanka and Barth, 1981) were also found to encode distinct primases. *E. coli* primase was found to have a specific requirement for *E. coli* single stranded DNA binding protein (Kornberg, 1980) or dnaB protein (Arai and Kornberg, 1979) for its synthesis. This led to a general belief that eukaryotic DNA primase would only be detected by specific replication systems and/or specific interactions with a DNA binding protein which turned out not to be the case.

Early evidence of an RNA priming mechanism for eukaryotic DNA synthesis came from the characterization of nascent DNA. Two types of analysis indicated an RNA primer was used for initiating DNA synthesis. One, transfer of ^{32}P-label from incorporated α-^{32}P-dNTP to rNMP after alkaline hydrolysis of nascent DNA, was indicative of the presence of ribonucleotides at a position 5′ to the labeled DNA. These label-transfer experiments demonstrated that the RNA-DNA junctions were on most nascent DNA fragments, were transiently present on nascent DNA, and had the same dinucleotide frequencies as genomic DNA (Pigiet *et al.*, 1974; Hunter and Francke, 1974; Tseng and Goulian, 1975; Anderson *et al.*, 1977). The latter results indicated general rather than specific sites of initiation of primer synthesis for Okazaki fragments. In another type of analysis, the RNA primer itself was studied by labeling of the RNA and isolation from the nascent DNA population. Nuclei isolated from papovavirus infected or uninfected cells could be used to label the RNA primer by incorporation of α-^{32}P-rNTPs. These studies indicated an RNA primer of 10 nucleotides in length was present on the nascent DNA that was started with a 5′ ribonucleoside triphosphate, indicating *de novo* initiation (Reichard *et al.*, 1974; Tseng and Goulian, 1977). The distinctive size of the primer was suggested to be a unique characteristic of primer synthesis that signaled the transition between RNA and DNA elongation (Reichard *et al.*, 1974) but whether this size was due to an intrinsic mechanism of RNA primer synthesis or to the termination and initiation of DNA synthesis could not be determined. Examination of continuously growing cells for RNA primers on nascent DNA has also been carried out by isolating the nascent DNA and labeling the 5′ end of the nascent DNA fraction after alkaline phosphatase treatment followed by treatment with γ-^{32}P-ATP and polynucleotide kinase. DNase digestion was used to remove the DNA portion of nascent fragments as well as to digest the labeled free DNA. A labeled DNase I resistant nonanucleotide was demonstrated to be present on nascent DNA. The transient presence of the oligoribonucleotides in the nascent DNA fraction was demonstrated by their absence after a short 10 min inhibition of DNA synthesis with hydroxyurea, consistent with the presence of the oligoribonucleotides only on newly synthesized DNA (Tseng *et al.*, 1979).

Deoxynucleotide incorporation is a characteristic of the *E. coli* primase, the dnaG product, and utilizes dNTP with facility (Rowen and Kornberg, 1978). In mammalian cells, incorporation of cognate dNTPs for rNTPs was observed in the synthesis of papovavirus primers in isolated nuclei when rNTPs were omitted (Eliasson and Reichard, 1978), similar to the facile incorporation described for *E. coli* primase. In addition, synthesis of primer appeared to have rather relaxed properties of base pairing when rNTPs were omitted in that, in addition to a low level of cognate dNTP incorporation, the nearest neighbor frequencies of the RNA primer were altered in rather specific ways which suggested the misincorporation of rNTPs (Tseng and Goulian, 1980). These various types of studies provided evidence for a primer RNA that initiates nascent DNA synthesis with features that could be used to distinguish the product from other types of RNA transcripts.

PROPERTIES OF PRIMASE

Enzyme Activity

In eukaryotes, primase activity was first reported by several laboratories, independently, and from different cell sources. These were from *Drosophila* embryos (Conaway and Lehman, 1982a), mouse Ehrlich ascites cells (Yagura *et al.*, 1982), human lymphocytes (Tseng and Ahlem, 1982), and SV40 infected monkey cells (Kaufmann and Falk, 1982). A key characteristic of the activities detected was the synthesis of a RNA primer of 10 nucleotides in length that primed DNA synthesis. In the case of the *Drosophila* activity, the highly purified and near homogeneous DNA polymerase α enzyme, which had four subunits, contained the priming activity and indicated a tight association between polymerase α and primase activities (Conaway and Lehman, 1982a). Reports of primase activity have subsequently been made from a diverse variety of sources including *Xenopus* (Shioda *et al.*, 1982; Riedel *et al.*, 1982), chick fibroblasts (Hirose *et al.*, 1988), mouse cells (Enomoto *et al.*, 1985), rat liver (Philippe *et al.*, 1986), human cells (Wang *et al.*, 1984; Gronostajski *et al.*, 1984), calf thymus (Yoshida *et al.*, 1983; Chang *et al.*, 1984; Grosse and Krauss, 1985), yeast (Singh and Dumas, 1984; Plevani *et al.*, 1984), salmon (Izuta *et al.*, 1985), and in most cases the association of primase and polymerase α has been noted.

Assay for Primase Activity

Two methods are commonly used to detect primase activity. Direct measurement of RNA synthesis and analysis of the product oligoribonucleotides in polyacrylamide gels indicated that primases characteristically synthesize oligoribonucleotides in multiples of approximately 10 nucleotides (Tseng and Ahlem, 1982; Tseng and Ahlem, 1983a). Synthetic DNA polymers that do not form double stranded structures are found to be the most active templates. Single stranded DNA, from biologic sources, is much less active as a template due possibly to secondary structure formation or to the site specificity of the enzyme (Conaway and Lehman, 1982b; Konig *et al.*, 1983; Grosse and Krauss, 1984; Tseng and Ahlem, 1984), described below.

Assay with a synthetic polymer consisting of two non-base-pairing nucleotides, *e.g.* poly $(dGdT)_n$, allows a more sensitive assay of oligoRNA synthesis than the use of a homopolymer, such as poly dT. The K_m for rATP is higher (about 0.3 mM) than for rCTP (about 0.03 mM) and use of an alternating single stranded polymer allows a more sensitive assay by reducing the level of radiolabeled rNTP required whereas the initiating nucleotide can be maintained at a high concentration thus reducing background levels (Tseng and Ahlem, 1983a). Another assay is the coupled DNA synthesis reaction where a non-self priming template, such as poly dT or circular single stranded DNA (M13 DNA), is used as a template for the DNA polymerase reaction, either by the associated DNA polymerase α or by addition of DNA polymerase such as *E. coli* polymerase I (Conaway and Lehman, 1982a; Tseng and Ahlem, 1982). Since the DNA polymerases are unable to initiate on such templates, DNA synthesis depends upon production of a primer. The ribonucleotide dependence of the reaction is indicative of RNA priming and analysis of the primer as a decaribonucleotide gives a definitive characterization of primase activity. This would also demonstrate that priming was not the consequence of other cellular RNA polymerases or nuclease activation of circular templates. The coupled reaction is a more sensitive assay as the RNA primer is extended by labeled DNA giving a many-fold amplification of the signal. However, it is also more prone to interference or mis-identification due to other RNA polymerase activities. Characterization of the RNA primer size is essential when this assay is used.

Enzyme Structure

Eukaryotic DNA primase was first isolated to homogeneity from mouse hybridoma cells. The biochemically isolated enzyme has an apparent native molecular weight of approximately 110 kDa and a sedimentation coefficient of 5.5 S. The enzyme activity co-purified with two polypeptides of 58 and 49 kDa, on SDS-PAGE, that were always present in a 1:1 stoichiometry (Tseng and Ahlem, 1983a). The native size indicated the enzyme to be a heterodimer of p58 and p49. Polyclonal antibodies have been produced against the individual p58 and p49 subunits of mouse primase, which were separated by SDS-PAGE, and it has been found that these subunits are immunologically distinct although both antibodies neutralize primase activity (Prussak *et al.*, 1989). Furthermore, the p58 and p49 polyclonal seras do not react with either component of the biochemically purified core mouse DNA polymerase α (p185 and p68) (Prussak and Tseng, 1987). Whereas *Drosophila* embryo polymerase α and primase were shown to be tightly associated and copurified (Conaway and Lehman, 1982a), the mouse cell primase (from mouse hybridoma cells) separated easily from DNA polymerase α (Tseng and Ahlem, 1983a). This may be due to the solvents (10% ethylene glycol and 20% dimethyl sulfoxide) used to effect purification. Mouse polymerase α and primase activities are associated, though. Using a monoclonal antibody produced against human DNA polymerase α that cross-reacts with mouse DNA polymerase α (SJK 132-20, Tanaka *et al.*, 1982), immunoprecipitation of DNA polymerase α from a mouse cell extract also precipitated all the mouse primase detectable in the extract by anti-primase antibodies on Western blots and primase antigen was recovered in the immunoprecipitate (Prussak *et al.*, 1986). Therefore, the intact mouse DNA polymerase α and primase activities consists of 185 and 68 (Prussak and Tseng, 1987), and 58 and 49 kDa (Tseng and Ahlem, 1983a) subunits, respectively. The catalytic subunit of DNA polymerase α from animal cells has been shown to be 185 kDa and other high molecular weight polypeptides of 160-140 kDa show related tryptic peptides (Wong *et al.*, 1986) and are probably degradation products of p185. The subunit structure of mouse cell DNA polymerase α/primase is similar to that described for the *Drosophila* complex (Kaguni *et al.*, 1983b). Additional complexes isolated from other mammalian sources and yeast cells, as described below, also yield similar polypeptide sizes and indicates a conservation of polypeptide sizes and, probably, amino acid sequences. (Kaguni and Lehman, 1988).

Primase was found to copurify along with DNA polymerase α from *Drosophila melanogaster* embryos (Conaway and Lehman, 1982a). This purified complex contained polypeptides of 185, 70, 60 and 50 kDa. The two activities could be dissociated in the presence of moderate concentrations of urea, which inactivated more than 90% of both the polymerase α and primase activities. Sedimentation analysis with urea treatment demonstrated the primase activity was associated with one or both of the 60 and 50 kDa polypeptides, which could not resolved by the sedimentation analysis (Kaguni *et al.*, 1983a). Similar results were also observed by treatment with 50% ethylene glycol that resulted in higher recovery of primase activity from the complex (Cotterill *et al.*, 1987).

The development of specific monoclonal antibodies to DNA polymerase α and their use for immunoaffinity purification (Chang *et al.*, 1984; Hu *et al.*, 1984; Wahl *et al.*, 1984) has allowed the rapid isolation of the relatively low abundance DNA polymerase α/primase enzymes from a wide variety of sources. The human DNA polymerase α/primase complex, isolated by a monoclonal antibody directed against the polymerase, contains a set of polypeptides with predominant members having molecular weights of 185, 165, 77, 55 and 49 kDa. This complex also contained a DNA primase activity as evidenced by the ability of this complex to produce ribo-oligonucleotide chains in the presence of ribonucleotide triphosphates and a single stranded template. Like the *Drosophila* polymerase α/primase complex, the primase activity was bound quite tightly to the complex and could not be dissociated without destroying substantial amounts of either the polymerase or primase activities. Since the monoclonal antibodies used for these studies interacted with only the large molecular weight polypeptides and did not affect primase activity, the primase enzyme activity was apparently contained on the lower molecular weight polypeptides (Wang *et al.*, 1984).

Likewise, the DNA polymerase α/primase complex was isolated from calf thymus using monoclonal antibodies (Wahl *et al.*, 1984; Chang *et al.*, 1984; Nasheuer and Grosse, 1987). Immunoaffinity purified calf thymus DNA polymerase α/primase complex contained five major polypeptides of 160, 68, 55, and 48 kDa (Chang *et al.*, 1984; Nasheuer and Grosse, 1987). The immunoaffinity purified polypeptides were used to produce polyclonal antibodies to aid in the determination of the polypeptides required for the polymerase and primase activities of the complex (Holmes *et al.*, 1986). Free calf thymus DNA primase could be eluted from the immunoaffinity column-bound DNA polymerase α/primase complex at 1M NaCl and had a molecular weight of 110 kDa and consisted of polypeptides with a molecular mass of 55 and 48 kDa. The 68 and 48 kDa polypeptides from the complex were found to be immunologically related and antiserum produced against the 68, 55 and 48 kDa polypeptides all inhibited primase activity (Holmes *et al.*, 1986). It was suggested, based on the immunoreactivity of anti-p68 antibodies with the p48 subunit, that the primase activity resides in the 68 kDa polypeptide and p48 was a degradation product (Holmes *et al.*, 1986). Another study of calf thymus polymerase α/primase led to a different conclusion. Polymerase α/primase was purified by elution from an immunoaffinity column at pH 12.5 and immediately neutralized. Primase activity could then be separated from polymerase α by sucrose gradient centrifugation and was associated with p59 and p48 polypeptides (Nasheuer and Grosse, 1987). Their analysis confirmed the immunological cross-reactivity of p68 and p48 subunits with a different set of antibodies but found the absence of p68 in free primase fractions and it was concluded that the antigenic reactivity may not indicate that the p48 is a proteolyzed product of p68 (Nasheuer and Grosse, 1988). Also, in studies with human DNA polymerase α/primase (Yagura *et al.*, 1987), a monoclonal antibody (E4) which reacted with the 77 kDa polypeptide (probably the same as p68 in other gel systems), neutralized primase activity but not polymerase α activity. However, p77 did not separate with the primase activity but remained associated with DNA polymerase α, indicating that the inhibitory effect of E4 antibody on primase may be steric in nature. This may be similar to the neutralization of primase activity in the calf thymus DNA polymerase α/primase complex with anti-p68 antibody (Holmes *et al.*, 1986).

An immunopurified yeast DNA polymerase α/primase complex contained five major polypeptides with molecular masses of 180, 140, 74, 58 and 48 kDa. The yeast primase activity was eluted from immunoaffinity column bound complex with a gradient of $MgCl_2$. The isolated primase contained protein components of 58 and 48 kDa, a molecular mass of 110 kDa and a sedimentation coefficient of 5.5 S. The primase activity isolated from the complex was labile although it was stable before separation from DNA polymerase α (Plevani *et al.*, 1985).

As evidenced by the studies detailed above, DNA primase activity has been associated with a heterodimer of approximately 110 kDa with subunits of 60 and 50 kDa and a sedimentation coefficient of approximately 5.5 S. An apparent difference in the isolated eukaryotic polymerase α/primase complexes is the interaction between the two enzymatic activities. While the *Drosophila* (Conaway and Lehman, 1982a) and human (Wang *et al.*, 1984) polymerase α/primase are tightly associated in the complex and the yeast (Plevani *et al.*, 1984) and calf thymus enzymes (Nasheuer and Grosse, 1987) apparently less so, the mouse primase (Tseng and Ahlem, 1983a) can be dissociated from the complex without a substantial loss of enzymatic activity.

The identification of the primase catalytic polypeptide has proven difficult. Separation of mouse primase activity from the replicase activity (polymerase α/primase complex) indicated only a single polypeptide of 58 kDa was present (Yagura *et al.*, 1986). Recently, two groups attributed primase activity to individual polypeptides of p60 or p48 which retained a small portion of the catalytic activity. Hirose and co-workers (Hirose *et al.*, 1988) developed a monoclonal antibody which specifically inhibited primase activity of the DNA polymerase α/primase complex from chick embryos. An immuno-affinity column produced from this antibody bound the chick polymerase-primase complex and upon elution with alkaline solutions primase activity was eluted, after polymerase α, along with a single polypeptide of 60 kDa; the purified activity was very labile. In contrast, Nasheuer and Grosse have separated the calf thymus primase from DNA polymerase α after immunoaffinity purification by DEAE chromatography (Nasheuer and Grosse, 1988). Approximately 10% of the total primase activity applied to a DEAE cellulose column was detected in the column flow-through and this material, when analyzed by SDS-PAGE, was composed of greater than 95% p48 and only a trace of p59 protein. This p49 primase activity isolated from the column was extremely unstable with a half-life of only 60 min at 0°C. Additionally, polyclonal antibodies were produced against p59 and p48. Anti-p48 antibody inhibited primase activity while the anti-p59 antibody only had a marginal affect. Lastly, UV cross-linking studies of the polymerase-primase complex revealed that only the p48 subunit bound ^{32}P-rGTP. These results with calf thymus primase suggest that the p48 subunit contains the catalytic activity (Nasheuer and Grosse, 1988).

These conflicting results suggest either that the activities from the different species sources reside on different polypeptides or that both subunits are required for stable primase activity. Since the subunit structure of DNA polymerase α/primase appears to be highly conserved between species (Kaguni and Lehman, 1988), the former possibility is less likely. The identification of the catalytic polypeptide(s) will require additional evidence and may be determined when the individual polypeptides are cloned and primase activity reconstituted and expressed.

Several reports indicate that DNA primase activity may be purified as a single polypeptide, perhaps different from those mentioned above with heterodimeric structure. Activities from calf thymus (Hübscher, 1983) and *Xenopus laevis* (Konig *et al.*, 1983) have been reported that demonstrated a rNTP stimulated DNA polymerase activity on M13 single stranded circular DNA upon renaturation and assay in the gel of polypeptides separated by SDS-PAGE. These studies suggested a co-migration of both DNA polymerase α and primase activities in a polypeptide of ≈100-125 kDa. However, purification of these activities to homogeneity has not yet been described. Isolation of free primase activity from HeLa cell polymerase α/primase indicated that a single polypeptide of 70 kDa was present in the primase fraction that synthesized oligoribonucleotides of characteristic size (Vishwanatha and Baril, 1986). It has not yet been determined if these activities constitute different primase activities in the cell.

In yeast, primase activity differing, apparently, in structure from the heterodimer described before has been purified to near homogeneity by several laboratories and has been correlated with a single polypeptide of 59 - 65 kDa depending upon the laboratory (Jazwinski and Edelman, 1985; Wilson and Sugino, 1985; Biswas *et al.*, 1987). The relation of these activities to each other or to the heterodimer primase, possibly the p60 subunit, from yeast described above (Plevani *et al.*, 1985) has yet to be resolved. These single polypeptide activities exhibit characteristics described for primase in the synthesis of discrete sized oligoribonucleotides

(Jazwinski and Edelman, 1985; Wilson and Sugino, 1985; Biswas *et al.*, 1987) and, in the case examined, the ability to incorporate dNTP at 10% the rate for rNTP (Wilson and Sugino, 1985). Recent molecular cloning of the p48 subunit of heterodimer yeast primase should help resolve the issue. The possibility that the p48 polypeptide is a degradation product of a larger p60 primase is made unlikely by the finding of a single open reading frame in the yeast p48 gene encoding a 47 kDa protein (Plevani *et al.*, 1987). The gene for the similar-sized subunit of mouse primase has also recently been isolated and comparison of the amino acid sequence with the yeast protein indicates that the two polypeptides are highly conserved in the amino-terminal halves of the molecules (Prussak *et al.*, 1989), as described below, consistent with their identification as intact primase subunits.

Organelle and Viral Primase

Cellular organelles also have specific priming activities as primase activity has been described from human (Wong and Clayton, 1985) and rat liver mitochondria (Ledwith *et al.*, 1986) cells. In studies of human mitochondrial DNA synthesis, the H-strand or D-loop origin is primed from a mitochondrial RNA polymerase transcript initiated from an upstream promoter and cleaved by an endoribonuclease to generate a primer. The L-strand origin is initiated by a specific primase which is capable of synthesizing short oligoribonucleotides of 9-12 nucleotides in length. A highly conserved stem loop structure, found in mitochondria from different species, is the site of initiation for the L-strand origin (Wong and Clayton, 1986). The mitochondrial primase activity sediments at 30 and 70 S, a consequence of an association with an RNA component the major species of which was shown to be cytosolic 5.8 S rRNA. The RNA component appears essential for the function of the mitochondrial priming activity since specific degradation of the co-fractionating 5.8 S RNA component with RNase H and 5.8 S specific oligodeoxynucleotides substantially reduced primase activity (Wong and Clayton, 1986). Although nuclear primase activities do not appear to require an RNA component, the finding of a small RNA species required for mitochondrial L-strand priming activity points out the necessity to examine other priming reactions for a possible role of associated RNA species in priming reactions.

A herpes simplex virus-specific primase was reported that fractionated with a polypeptide of 40 kDa and is suggested to be a viral encoded activity (Holmes *et al.*, 1988). Additional study should resolve whether this is a distinct primase but since HSV encodes a DNA polymerase it is likely the primase may also be viral in order to provide specific enzyme-enzyme interactions.

Mechanisms of Synthesis

Early characterizations of the primer found on nascent DNA in cells revealed that primer remained relatively invariant even upon perturbations of rNTP pools and suggested that a canonical primer size was important (Reichard *et al.*, 1974; Eliasson and Reichard, 1978; Tseng and Goulian, 1980). It was not evident from those studies if the characteristic primer size was an intrinsic property of the enzymatic synthesis itself or a consequence of the coupling of RNA primer to DNA synthesis. Upon purification of the enzymatic activity, one of the characteristic features of primase synthesis is the chain length of the products. There is a discontinuous distribution of RNA products in multiples of approximately 10 nucleotides in the absence of DNA polymerase α (Tseng and Ahlem, 1983a; Badaracco *et al.*, 1985; Singh *et al.*, 1986). Processive synthesis may be approximately 20 nucleotides in length with additional lengths a consequence of rebinding (Tseng and Ahlem, 1983a; Singh *et al.*, 1986). However, the longest products are not thousands of nucleotides long, as synthesized by the RNA polymerases, but rather less than several hundred nucleotides. Synthesis in steps of 10 may be a consequence of a decrease in the chain elongation rate after 10 nucleotides and offers the rationale that this step allows access of DNA polymerase to the ribonucleotide primer. In the absence of DNA chain elongation, primase can continue elongation but again in steps of 10 ribonucleotides. When DNA polymerase α is present, either added exogenously (Tseng and Ahlem, 1983b) or in complex (Badaracco *et al.*, 1985; Singh *et al.*, 1986), the primer length is limited to a 10-mer due to DNA chain elongation at the unit primer size.

The isolated mouse primase also has the capacity to incorporate dNTPs at approximately one-tenth the rate of rNTPs (Tseng and Prussak, 1989b), consistent with the facile incorporation of deoxynucleotides into RNA primers of nascent DNA seen in isolated nuclei (Eliasson and Reichard, 1978; Tseng and Goulian, 1980). Surprisingly, *Drosophila* polymerase α/primase can synthesize and prime DNA elongation on M13 single stranded template when only rATP is included in the reaction by synthesis of an oligo-rA primer (Cotterill *et al.*, 1987). The suggested mechanisms are either a reiterative synthesis or a misincorporation of rNTP by primase but a based-paired primer is required as primase does not synthesize on poly dT with rGTP alone. This may be similar to the misincorporation of ribonucleotides observed in the synthesis of primer RNAs in isolated nuclei when synthesis was carried out upon omission of individual ribonucleotides (Tseng and Goulian, 1980). These observations point out the apparent need of the enzyme to synthesize a primer of a particular size, although smaller primers may also prime synthesis (Hay and DePamphilis, 1982; Yamaguchi *et al.*, 1985a; Tseng and Prussak, 1989b). In coupling the priming and DNA elongation reaction, the enzyme may not leave the template but primer synthesis and DNA polymerization may occur in a coordinated fashion. The stability of a partially mismatched primer, such as those synthesized in the absence of individual ribonucleotides, would be very poor and would suggest that primase may help stabilize the primer for DNA chain elongation by DNA polymerase α. The mechanism of the transition between RNA and DNA synthesis has not yet been examined but reconstitutiton experiments do indicate that separated polymerase α and primase can physically reassociate, as seen by changes in sedimentation after mixing the separated activities, and provides a means for studying the reaction mechanisms (Suzuki *et al.*, 1985).

Removal of RNA Primer

The absence of ribonucleotides in genomic DNA indicates that removal of the RNA primer occurred before ligation of the nascent DNA. RNaseH activities in the cell presumably carry out the removal. The stimulation by an RNaseH activity of coupled DNA synthesis by *Drosophila* primase/polymerase α suggests that this is a plausible function for this activity (DiFrancesco and Lehman, 1985). In addition, an exonuclease activity has been found to be required for the covalent closure of the coupled reaction products in the presence of RNase H and ligase. In a reaction with mouse cell enzymes and M13 circular single stranded DNA, an activity was required for covalent closure of complementary strand synthesis in addition to DNA polymerase α/primase, RNAse H, and ligase (Goulian *et al.*, 1988). The activity is a 5′ to 3′ exonuclease with properties similar to those reported previously for DNase IV. Its suggested function is to remove a residual 1 or 2 ribonucleotides that apparently remain after RNaseH action and allow ligation of the nascent DNA and suggests that two activities are required for proper removal of the RNA primer in order for ligation to occur (Goulian *et al.*, 1988).

Site Specificity of Origin Initiations

Primase requires a single stranded DNA template for activity. This has been found for the various primases reported and appears to be a consistent property. That the enzyme initiates with a purine nucleotide is suggested by the fact that only poly dT or poly dC templates are effective whereas poly dG or poly dA templates are ineffective as templates (Holmes *et al.*, 1985; Parker and Cheng, 1987; Badaracco *et al.*, 1986; Philippe *et al.*, 1986). Also, the labeling of the 5′ ends of the RNA by incorporation of γ-^{32}P rPuTP or with vaccinia mRNA capping enzyme indicated purine nucleotides initiate synthesis of the primers (Tseng and Ahlem, 1983a; Tseng and Ahlem, 1984; Yamaguchi *et al.*, 1985a). Primase showed some template sequence preference in the use of single stranded circular DNA of M13. Non-random initiations have been reported for polymerase α/primase from *Drosophila*, *Xenopus*, calf thymus, mouse, and monkey cells on M13 single stranded DNA (Conaway and Lehman, 1982b; Konig *et al.*, 1983; Grosse and Krauss, 1984; Tseng and Ahlem, 1984; Yamaguchi *et al.*, 1985b). Mapping of initiation sites has been carried out with the single stranded linear DNA of minute virus of mice and indicated a consensus sequence of $C_2A_{1\text{-}2}(C_{2\text{-}3}/T_2)$ for initiation but this occurred only at some of the consensus sites in the 5 kb genome (Faust *et al.*, 1985).

Using the SV40 origin of replication as a model origin, site specificity of initiation by purified mouse primase (Tseng and Ahlem, 1983a) was investigated using the separated strands of DNA that spanned the SV40 origin (Tseng and Ahlem, 1984). Specific initiation sites were observed and mapped to only two regions when the early (E) strand was used as template, at np 5145-5147 and at np 5210-5220 (Tseng and Ahlem, 1984) (Figure. 1). On the late (L) strand template, sites of initiation were found at each of the six 3′-CCCGCC-5′ repeats which are the G/C boxes or SP1 binding sites of the early promoter. The 27-bp inverted repeat sequence (T-ag binding site II) promotes initiation at the proximal initiation sites at np 5210-5220 on the E-strand but is not required for initiation at distal sites (np 5145-5147) or at the G/C boxes on the L-strand. This was shown by a 6-bp deletion within the 27-bp inverted repeat that substantially decreased initiations at the proximal sites (np 5210-5220) but had no effect on initiations at distal sites (Tseng and Ahlem, 1984). This mutant had been shown to inactivate the SV40 DNA ori function when introduced into cells (Myers and Tjian, 1980). A second site on the L-strand similarly situated as the proximal sites of the E-strand has been subsequently identified on the L-strand at np 22 and initiation from this site also required an intact 27-bp inverted repeat for promoting initiation (Figure 1) (Tseng and Prussak, 1989b). However, the inverted repeat structure is not sufficient as a promoter since a mutant that deletes the sequences at the E strand proximal initiation sites, but does not affect the 27-bp inverted repeat sequence, is unable to initiate synthesis on the E-strand at the new proximal sequences (Tseng and Prussak, 1989b). The 27-bp sequence is a perfect inverted repeat and from the calculation of the free energy of formation, it would be expected to form a stem-loop structure in these reactions since it is present as single stranded DNA. These results suggest that there are two modes of recognition by mouse primase, one that requires a template sequence and a secondary structure (as found for the sites proximal to the inverted repeat on both E (np 5210-5220) and L strands (np22)), and another that has sequence recognition alone (as for the distal sites (np 5140, G/C boxes) adjacent to the SV40 minimal origin and possibly those on MVM DNA (Faust *et al.*,

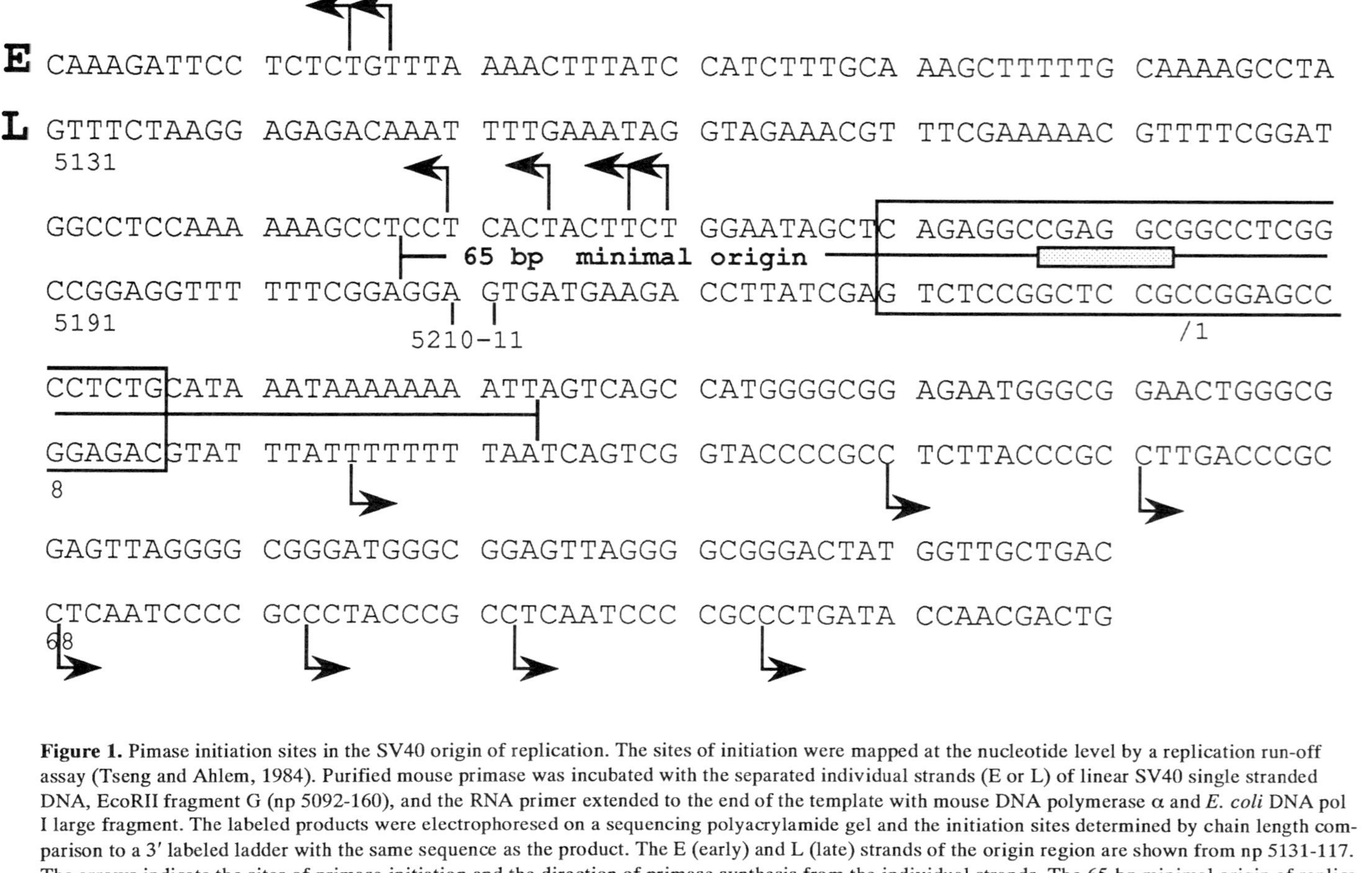

Figure 1. Pimase initiation sites in the SV40 origin of replication. The sites of initiation were mapped at the nucleotide level by a replication run-off assay (Tseng and Ahlem, 1984). Purified mouse primase was incubated with the separated individual strands (E or L) of linear SV40 single stranded DNA, EcoRII fragment G (np 5092-160), and the RNA primer extended to the end of the template with mouse DNA polymerase α and *E. coli* DNA pol I large fragment. The labeled products were electrophoresed on a sequencing polyacrylamide gel and the initiation sites determined by chain length comparison to a 3′ labeled ladder with the same sequence as the product. The E (early) and L (late) strands of the origin region are shown from np 5131-117. The arrows indicate the sites of primase initiation and the direction of primase synthesis from the individual strands. The 65-bp minimal origin of replication (DiMaio and Nathans, 1980; Myers and Tjian, 1980) is indicated by the bar between the E and L strands; the 27-bp inverted repeat is within the boxed sequence; and the 6-bp deletion that inactivates origin activity (Myers and Tjian, 1980) and greatly reduces the proximal primase starts is indicated by the stipled box within the 27-bp inverted repeat. Figure is modified from (Tseng and Ahlem, 1984; Tseng and Prussak, 1989b).

1985)). The G/C sequences resemble the consensus sequence of the MVM sites in that both specify a primer of high purine content and the template sequence has several pyrimidine residues interrupted by one or two purine residues.

Similar examination of primase initiation sites within the SV40 origin with monkey cell polymerase α/primase has indicated additional sites of initiation (Yamaguchi *et al.*, 1985a). In this case, the DNA polymerase α/primase was not as highly purified as the mouse primase and additional factors may relax the specificity of the initiation reaction, as required for lagging strand initiations. Interestingly, a more complex enzyme fraction from HeLa cells that contains DNA polymerase α/primase activities, which fractionates by gel filtration as a high molecular weight complex and includes other factors for DNA replication, has even more sites of initiation on SV40 DNA (Vishwanatha *et al.*, 1986). In the case of purified mouse primase, additional regions of SV40 DNA (Hind III D, E, F fragments) have been examined for primase initiation sites and of about 2000 nucleotides of single stranded DNA examined, only one additional region was observed at np 2412 on the E-strand (Tseng and Prussak, 1989b). This demonstrates the high specificity of initiations by purified mouse primase for the SV40 origin of replication region.

Studies of the RNA primers found on nascent DNA isolated from SV40 replicative intermediates indicate that RNA primers of approximately 10 nucleotides long are found on nascent DNA synthesized from within the minimal origin of replication, suggesting that primase also initiates leading strand synthesis from the origin (Hay and DePamphilis, 1982). However, all of these sites do not have a one to one correspondence with sites mapped by the *in vitro* reactions (Tseng and Ahlem, 1984; Yamaguchi *et al.*, 1985a). The *in vivo* initiation sites have been found to have a bidirectional transition point at np 5210-5211 where Okazaki fragments are no longer detected beyond this point in the leading strand directions (Hay and DePamphilis, 1982). The *in vitro* initiation sites of mouse primase proximal to the 27-bp repeat on the E-strand are the expected sites of RNA initiation (10 nucleotides upstream) for the E-strand at the bidirectional transition point (np 5210-11). The corresponding site on the L-strand, however, has not been observed with *in vitro* initiation with purified enzymes. The evidence from *in vivo* examination of RNA-DNA junctions in the origin region of replicative intermediates suggests that primase is active in leading strand synthesis from the SV40 origin of replication (Hay and DePamphilis, 1982), although the precise sites are not all the same as those observed *in vitro* with purified enzymes. The highly specific initiation with purified mouse primase within the SV40 minimal origin of replication and their position in the origin (Tseng and Ahlem, 1984; Tseng and Prussak, 1989b) suggests that primase may be active for site specific leading strand initiation at an origin of replication as well as for Okazaki fragment initiation.

Recent studies indicate that DNA polymerase δ may also be involved in the replication process (Prelich *et al.*, 1987) providing leading strand DNA synthesis at a replication fork (Prelich and Stillman, 1988). Primase activity is not tightly associated with DNA polymerase δ and raises the question of whether the same or a different mechanism initiates synthesis for DNA polymerase δ. Two possibilities can be considered; either primase initiates synthesis for DNA polymerase δ or polymerase α/primase synthesizes an initial nascent DNA that can be used as a primer for DNA synthesis by DNA polymerase δ. The latter mechanism may be suggested by the aphidicolin resistant DNA synthesis described for SV40 replicating intermediates and for *in vitro* primase/DNA polymerase α synthesis (Decker *et al.*, 1986). The role of primase in the initiation of DNA polymerase δ synthesis will be of importance in understanding the mechanism of DNA replication.

Stimulatory Factors

Several reports of the stimulation of coupled primase, polymerase α DNA synthesis have demonstrated the existence of factors which facilitate primase initiation. DNA polymerase α from Ehrlich ascites mouse cells was found to fractionate into two activities, one with associated primase and another without primase. The activity with associated primase was termed replicase (Yagura *et al.*, 1982). A stimulatory factor increased the replicase activity on M13 single stranded DNA as measured by the extent of DNA synthesis. The factor was shown to increase the affinity of replicase for template but did not increase the size of either the RNA primer or the DNA product (Yagura *et al.*, 1983a). Fractionation indicated a molecular size of 130 kDa, a dimer of 63 kDa subunits (Yagura *et al.*, 1983b). This stimulatory factor was reported to be found only from mouse cell extracts but replicase activity could be detected from a wide number of species (Yagura *et al.*, 1983b).

Another, and possibly similar, stimulatory factor was reported from calf thymus. This also stimulated primase activity specifically but in this case the primary effect appeared to be on the RNA synthesis rate and the rATP requirement for RNA synthesis. In the absence of stimulatory factor, the K_m for rATP was increased whereas the V_{max} of the reaction was decreased by 3-4 fold (Itaya *et al.*, 1988). The reported protein structure is however different from the mouse factor (Yagura *et al.*, 1983b) and consists of two subunits of 146 kDa and 47 kDa (Itaya *et al.*, 1988). The mechanism by which these factors exert their effect is unknown.

A characterized activity that stimulates RNA primed DNA synthesis by *Drosophila* DNA polymerase α/primase is an RNase H activity that increased the frequency of primer synthesis, suggesting that nuclease activity resulted in recycling of the DNA polymerase α/primase. The activity also is different in size from those described above in that the native structure is tetramer of 2 subunits of 92 kDa and 2 subunits of 39 kDa (DiFrancesco and Lehman, 1985).

Nucleotide Inhibitors

Some nucleotide analogues are potent antimetabolites whose target of action is frequently the process of DNA polymerization itself. The arabinosyl nucleosides are a group of such analogues and their mechanism of action was initially thought to be as chain terminators for DNA polymerization. Studies with isolated mouse primase indicated that primer synthesis was at least as sensitive to araCTP as was reported for DNA synthesis (Tseng and Ahlem, 1983a) and must also be considered a potential target of these antimetabolites. Additional studies with calf thymus polymerase α/primase showed that the K_i for araATP for primer RNA synthesis was 20 μM and that the ratio of K_i/K_m was 0.15 (Yoshida *et al.*, 1985). Inhibition occurred by slowing chain elongation of the RNA and not initiation of primers. It was also indicated that chain termination did not occur since priming by araNMP containing RNA was observed (Yoshida *et al.*, 1985), although incorporation of araNMP into internal sites was not directly demonstrated. Study with human polymerase α/primase from acute lymphocytic leukemia cells also demonstrated that araNTP were potent competitive inhibitors of primase RNA synthesis, however the inhibitor did not have to be complementary to the template suggesting binding was to a single nucleotide incorporation site and did not have a strong preference for complementary nucleotides (Reiter *et al.*, 1987).

Studies with HeLa cell polymerase α/primase have indicated that the analogues GTPγS, GMP-PNP, or GMP-PCP do not inhibit the RNA priming measured by incorporation of dGTP into DNA with poly dC templates, although higher concentrations are required than for GTP. In contrast, the analogues ATPγS, AMP-PNP, or AMP-PCP do inhibit the incorporation of dATP into DNA with poly dT template (Gronostajski *et al.*, 1984). This suggests that ATP hydrolysis of the γ phosphate may be required for initiating DNA synthesis although ATPase activity could not be detected in this or other primase preparations (Conaway and Lehman, 1982a; Tseng and Ahlem, 1983a; Gronostajski *et al.*, 1984). ATP analogues do bind since they inhibit the incorporation of dGTP with poly dC, again indicating a single nucleotide binding site (Gronostajski *et al.*, 1984).

The antitrypanocidal agent suramin has been found to be a potent inhibitor of primase/DNA polymerase α from human KB cells and had different modes of action for the two activities. With primase and poly dC template, suramin was a competitive inhibitor of GTP. With DNA polymerase activity, measured on activated DNA, the drug was a competitive inhibitor with respect to template-primer concentration (Ono *et al.*, 1985). The inhibition constant for primase nucleotide binding was 2.6 μM (Ono *et al.*, 1988). This indicates a potentially new mechanism of trypanocidal action of the drug.

MOLECULAR STRUCTURE

Recent molecular genetic analysis of the p49 subunit from yeast has allowed isolation of the gene and its characterization (Plevani *et al.*, 1987). The gene encodes a 47,623 Da protein that expresses an essential function demonstrated by gene disruption experiments (Lucchini *et al.*, 1987). The full cDNA of the similar sized subunit of mouse primase was independently isolated and the identity of the clone confirmed by the presence of tryptic peptides of mouse p49 in the deduced amino acid sequence of the cDNA clone (Prussak *et al.*, 1989). The mouse p49 subunit encodes a protein of 49,295 Da. Comparison of the mouse and yeast amino acid sequences indicated a homology between the two subunits that extended over the N-terminal halves of the proteins (Figure 2). There are four regions (I-IV) within the N-terminal halves that are highly conserved and may represent exon domains. The yeast gene is a single exon whereas the mouse gene is spread over 17 kb of genomic DNA with multiple exons, not yet fully mapped. With a 2 residue shift, the two sequences are aligned and show the amino acid homologies to extend in phase, without insertions or deletions, over approximately the first 190 amino acid residues (Figure 2). This indicates that not only are the four highly similar regions conserved but that the amino acid spacing between them is conserved as well. The C-terminal halves of the proteins are not similar between yeast and mouse except for one region (V, Figure 2). To examine the nucleotide sequence divergence in the C-terminal region, a 3′ portion of the cDNA with predominantly protein coding sequences was used to probe polyA+ RNA from different mammalian sources. The 3′ mouse probe detected a similar sized poly A+ RNA (about 1.6 kb) from rat but not from hamster or human cells, even with low stringency hybridization conditions. Whereas, the mouse full cDNA probe detected a similar sized and amount of polyA+ RNA from all the cell lines (Prussak *et al.*, 1989). This indicates that the 3′ portion of the coding sequence has rapidly diverged and suggests the C-terminal amino acid sequences have also diverged.

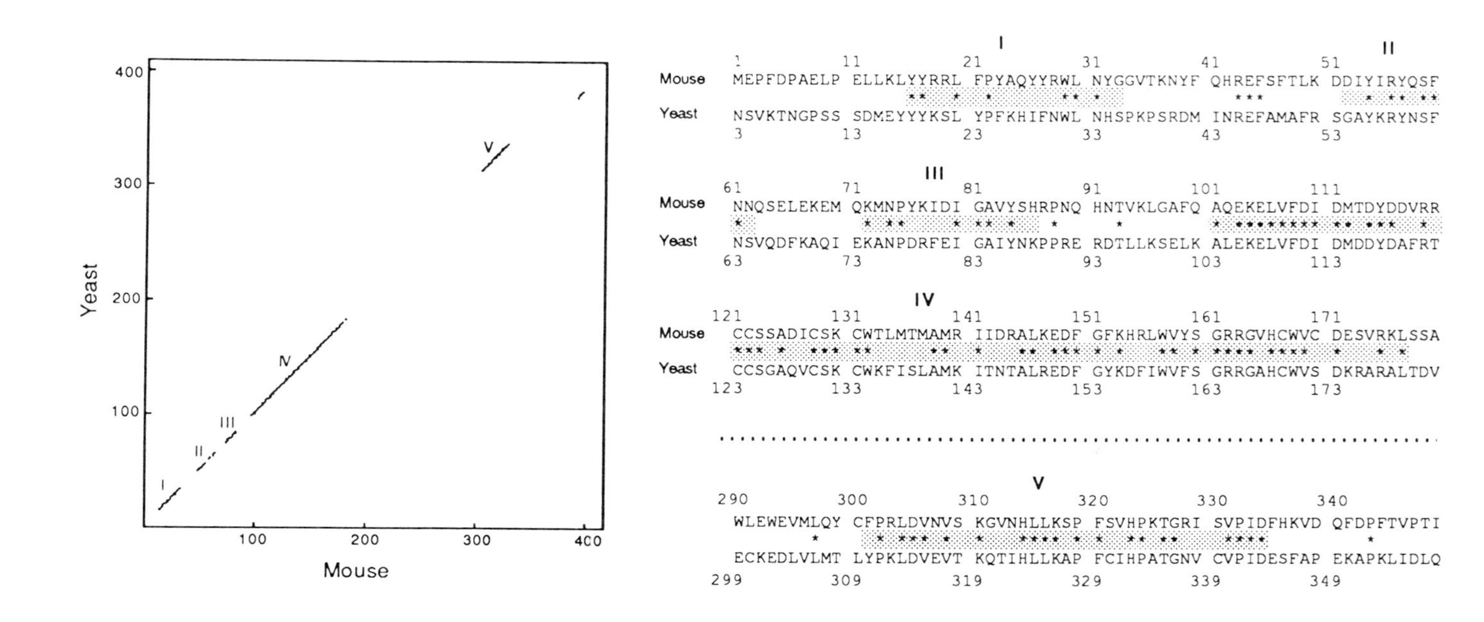

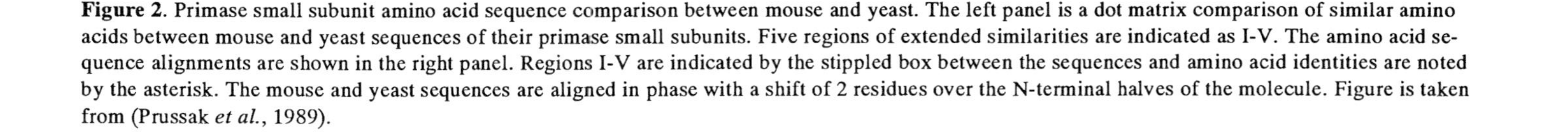
Figure 2. Primase small subunit amino acid sequence comparison between mouse and yeast. The left panel is a dot matrix comparison of similar amino acids between mouse and yeast sequences of their primase small subunits. Five regions of extended similarities are indicated as I-V. The amino acid sequence alignments are shown in the right panel. Regions I-V are indicated by the stippled box between the sequences and amino acid identities are noted by the asterisk. The mouse and yeast sequences are aligned in phase with a shift of 2 residues over the N-terminal halves of the molecule. Figure is taken from (Prussak *et al.*, 1989).

A possible consequence of this rapid divergence is the host range specificity seen with papovaviruses. The basis for the host range has been examined with *in vitro* DNA viral replication systems (Murakami *et al.*, 1986). The replication of polyoma viral DNA with polyoma T-ag and non-permissive cell extracts from HeLa cells, was not supported unless polymerase α/primase from mouse cells (immunoaffinity column purified) was added. Addition of SV40 T-ag to the HeLa cell extract did not support polyoma DNA replication *in vitro*. These studies and the converse with SV40 DNA and mouse cell extracts indicated that the host range of viral replication could be accounted for by the presence of appropriate viral T-ag and permissive host cell polymerase α/primase (Murakami *et al.*, 1986). An interaction between SV40 T-antigen and HeLa cell DNA polymerase α/primase has been demonstrated by the co-immunoprecipitation of the proteins and may be the basis for the species specific requirement (Smale and Tjian, 1986). As described above, it was found that there was a rapid divergence in nucleotide sequence in the 3′ portion of the primase p49 gene from mouse to hamster or human nucleotide sequences which implies a rapid divergence of amino acid sequences in this region which may possibly be a determinant domain of the host range specificity exhibited to the viral T-antigens.

CELL EXPRESSION OF PRIMASE mRNA

In quiescent or resting cells, produced by serum starvation or obtained from mature rat liver, the levels of primase enzymatic activity is low but increases after serum repletion (Reiter *et al.*, 1987) or partial hepatectomy (Philippe *et al.*, 1986) at the beginning of S-phase, together with polymerase α activity. As cells enter the cell cycle, the nuclear location of primase has been examined by measuring the activity of the enzyme in cells stimulated to proliferate with serum. The nuclear matrix fraction has been found to be the cellular fraction in which primase is localized as cells enter S-phase (Tubo and Berezney, 1987). There also appears to be a differential retention of primase in the matrix fraction, as 70% of the activity is found in the matrix fraction whereas only 30% of the DNA polymerase α is in the fraction despite the fact that they are isolated as a complex (Wood and Collins, 1986). The nuclear localization of replication enzymes to specific nuclear structures remains an interesting area of current studies.

Recent molecular cloning of the p49 subunit of mouse primase provided the opportunity to examine the transcriptional expression of the gene. Primase enzymatic activity during the transition between G_0 and S phase indicated a low level of activity in G_0 cells and an increase, approximately 10-fold, that starts at the beginning of S phase and reaches maximal levels in mid to late S (Reiter *et al.*, 1987). Examination of p49 mRNA steady state levels in mouse 3T6 cells, starved and then repleted with serum, showed an increase of about 10-fold in the levels of p49 mRNA that preceded both S phase and histone H3 mRNA appearance by several hours (Tseng *et al.*, 1989a). A control point in the mammalian cell cycle, termed the R point, has been characterized by Pardee (Pardee, 1974) and is postulated to regulate cells in G_1 to progress through the cell cycle. The serum induced p49 mRNA expression was examined to determine if this occurred before or after the R point. Prior to the R-point, inhibition of protein synthesis by as little as 50% prevents cell progression into S phase whereas after the R-point this level of inhibition has no effect on cell progression (Pardee, 1974; Campisi *et al.*, 1982; Zetterberg and Larsson, 1985). If cycloheximide is added to cells at low concentrations which reduce protein synthesis by about 50%, after serum repletion of 3T6 cells the induction of p49 mRNA levels is affected minimally whereas these cells are blocked from entering S-phase. Higher concentrations of cycloheximide did block p49 mRNA induction. These results indicated that the serum induction of p49 mRNA is not a primary event in that protein synthesis was required and also that this occurred prior to the R-point control of cell progression (Tseng *et al.*, 1989a). It had been commonly assumed, until recently, that the events observed upon serum stimulation of quiescent cells reflect the events in the G_1 stage of the cell cycle. Studies of the expression of c-myc gene indicated that the induction of c-myc mRNA and protein seen upon serum stimulation do not reflect the G_1 phase of continuously dividing cells separated according to cell cycle stage by size. c-myc mRNA and protein were constitutively expressed in dividing cells at all cell cycle stages whereas they are not expressed in G_0 cells (Thompson *et al.*, 1985; Hann *et al.*, 1985). A similar situation has been demonstrated recently for several genes involved in nucleotide metabolism or required for DNA synthesis. Human thymidine kinase (Sherley and Kelly, 1988), human DNA polymerase α (Wahl *et al.*, 1988), and mouse primase p49 (Tseng *et al.*, 1989a) mRNA levels are found not to vary greatly during the cell cycle but are down regulated in G_0 cells and increase substantially after serum stimulation of G_0 cells. The mRNA induction in these cases appears to be more a consequence of the low level of mRNA in G_0 cells than cell cycle regulated events. The replication linked histone genes, in contrast, are cell cycle regulated at both transcriptional and post-transcriptional levels and are expressed only during S-phase (Alterman *et al.*, 1984; Sittman *et al.*, 1983; Heintz *et al.*, 1983).

Understanding the functions and gene expression of primase and DNA polymerase α can now be approached with the isolation of the mouse p49 (Prussak *et al.*, 1989) and human DNA polymerase α (Wong *et al.*, 1988) cDNAs. Isolation of the genes of the other polypeptides of the DNA polymerase α/primase complex is expected in the near future. Study of the structure/function relation of these enzymes and the regulation of their expression will yield additional important insights into the growth regulation of cells as the activation of these genes is essential for genetic continuity and proliferation of cells.

ACKNOWLEDGMENTS

The work carried out in the author's laboratory was supported by grants from USPHS GM29091 and from the American Cancer Society NP-594.

REFERENCES

R. M. Alterman, S. Ganguly, D. H. Schulze, W. F. Marzluff, C. L. Schildkraut and A. I. Skoultchi. (1984) Cell cycle regulation of mouse H3 histone mRNA metabolism. *Mol. Cell. Biol.* **4**: 123-132.

S. Anderson, G. Kaufmann and M. L. DePamphilis. (1977) RNA primers in SV40 DNA replication: identification of transient RNA-DNA covalent linkages in replicating DNA. *Biochemistry* **16**: 4990-4998.

K. Arai and A. Kornberg. (1979) A general priming system employing only dnaB protein and primase for DNA replication. *Proc. Natl. Acad. Sci. USA* **76**: 4308-4312.

G. Badaracco, M. Bianchi, P. Valsasnini, G. Magni and P. Plevani. (1985) Initiation, elongation and pausing of *in vitro* DNA synthesis catalyzed by immunopurified yeast DNA primase: DNA polymerase complex. *EMBO. J.* **4**: 1313-1317.

G. Badaracco, P. Valsasnini, M. Foiani, R. Benfante, G. Lucchini and P. Plevani. (1986) Mechanism of initiation of *in vitro* DNA synthesis by the immunopurified complex between yeast DNA polymerase I and DNA primase. *Eur. J. Biochem.* **161**: 435-440.

E. E. Biswas, P. E. Joseph and S. B. Biswas. (1987) Yeast DNA primase is encoded by a 59-kilodalton polypeptide: purification and immunochemical characterization. *Biochemistry* **26**: 5377-5382.

J. P. Bouche, L. Rowen and A. Kornberg. (1978) The RNA primer synthesized by primase to initiate phage G4 DNA replication. *J. Biol. Chem.* **253**: 765-769.

G. J. Boulnois and B. M. Wilkins. (1979) A novel priming system for conjugal synthesis of an Inc1 alpha plasmid in recipients. *Mol. Gen. Genetics* **175**: 275-279.

J. Campisi, E. E. Medrano, G. Morreo and A. B. Pardee. (1982) Restriction point control of cell growth by a labile protein: evidence for increased stability in transformed cells. *Proc. Natl. Acad. Sci. USA* **79**: 436-440.

T. A. Cha and B. M. Alberts. (1986) Studies of the DNA helicase-RNA primase unit from bacteriophage T4. A trinucleotide sequence on the DNA template starts RNA primer synthesis. *J. Biol. Chem.* **261**: 7001-7010.

L. M. Chang, E. Rafter, C. Augl and F. J. Bollum. (1984) Purification of a DNA polymerase-DNA primase complex from calf thymus glands. *J. Biol. Chem.* **259**: 14679-14687.

R. C. Conaway and I. R. Lehman. (1982a) A DNA primase activity associated with DNA polymerase alpha from *Drosophila melanogaster* embryos. *Proc. Natl. Acad. Sci. USA* **79**: 2523-2527.

R. C. Conaway and I. R. Lehman. (1982b) Synthesis by the DNA primase of *Drosophila melanogaster* of a primer with a unique chain length. *Proc. Natl. Acad. Sci. USA* **79**: 4585-4588.

S. M. Cotterill, G. Chui and I. R. Lehman. (1987) DNA polymerase-primase from embryos of *Drosophila melanogaster* DNA primase subunits. *J. Biol. Chem.* **262**: 16105-16108.

R. S. Decker, M. Yamaguchi, R. Possenti and M. L. DePamphilis. (1986) Initiation of simian virus 40 DNA replication *in vitro*: aphidicolin causes accumulation of early-replicating intermediates and allows determination of the initial direction of DNA synthesis. *Mol. Cell. Biol.* **6**: 3815-3825.

R. A. DiFrancesco and I. R. Lehman. (1985) Interaction of ribonuclease H from *Drosophila melanogaster* embryos with DNA polymerase-primase. *J. Biol. Chem.* **260**: 14764-14770.

D. DiMaio and D. Nathans. (1980) Cold-sensitive regulatory mutants of simian virus 40. *J. Mol. Biol.* **140**: 129-142.

R. Eliasson and P. Reichard. (1978) Primase initiates Okazaki pieces during polyoma DNA synthesis. *Nature* **272**: 184-185.

T. Enomoto, M. Suzuki, M. Takahashi, K. Kawasaki, Y. Watanabe, K. Nagata, F. Hanaoka and M. Yamada. (1985) Purification and characterization of two forms of DNA polymerase alpha from mouse FM3A cells: a DNA polymerase alpha-primase complex and a free DNA polymerase alpha. *Cell Struct. Funct.* **10**: 161-171.

E. A. Faust, R. Nagy and S. K. Davey. (1985) Mouse DNA polymerase alpha-primase terminates and reinitiates DNA synthesis 2-14 nucleotides upstream of C2A1-2(C2-3/T2) sequences on a minute virus of mice DNA template. *Proc. Natl. Acad. Sci. USA* **82**: 4023-4027.

M. Goulian, C. Carton, L. DeGrandpre, C. Heard, B. Olinger and S. Richards. (1988, Discontinuous DNA synthesis by purified proteins from mammalian cells) **In**: *Cancer Cells.* Vol. **6** Eukaryotic DNA Replication, edited by T. Kelly and B. Stillman. Cold Spring Harbor Laboratory, Cold Spring Harbor: p. 393-396.

R. M. Gronostajski, J. Field and J. Hurwitz. (1984) Purification of a primase activity associated with DNA polymerase alpha from HeLa cells. *J. Biol. Chem.* **259**: 9479-9486.

F. Grosse and G. Krauss. (1984) Replication of M13mp7 single-stranded DNA *in vitro* by the 9-S DNA polymerase alpha from calf thymus. *Eur. J. Biochem.* **141**: 109-114.

F. Grosse and G. Krauss. (1985) The primase activity of DNA polymerase alpha from calf thymus. *J. Biol. Chem.* **260**: 1881-1888.

S. R. Hann, C. B. Thompson and R. N. Eisenman. (1985) c-myc oncogene protein synthesis is independent of the cell cycle in human and avian cells. *Nature* **314**: 366-369.

R. T. Hay and M. L. DePamphilis. (1982) Initiation of simian virus 40 DNA replication *in vivo*: location and structure of 5'-ends of DNA synthesized in the *ori* region. *Cell* **28**: 767-779.

N. Heintz, H. L. Sive and R. G. Roeder. (1983) Regulation of human histone gene expression: kinetics of accumulation and changes in the rate of synthesis and in the half-lives of individual histone mRNAs during the HeLa cell cycle. *Mol. Cell. Biol.* **3**: 539-550.

D. M. Hinton and N. G. Nossal. (1985) Bacteriophage T4 DNA replication protein 61. Cloning of the gene and purification of the expressed protein. *J. Biol. Chem.* **260**: 12858-12865.

F. Hirose, S. Yamamoto, M. Yamaguchi and A. Matsukage. (1988) Identification and subcellular localization of the polypeptide for chick DNA primase with a specific monoclonal antibody. *J. Biol. Chem.* **263**: 2925-2933.

A. M. Holmes, E. Cheriathundam, F. J. Bollum and L. M. Chang. (1985) Initiation of DNA synthesis by the calf thymus DNA polymerase-primase complex. *J. Biol. Chem.* **260**: 10840-10846.

A. M. Holmes, E. Cheriathundam, F. J. Bollum and L. M. Chang. (1986) Immunological analysis of the polypeptide structure of calf thymus DNA polymerase-primase complex. *J. Biol. Chem.* **261**: 11924-11930.

A. M. Holmes, S. M. Wietstock and W. T. Ruyechan. (1988) Identification and characterization of a DNA primase activity present in herpes simplex virus type 1-infected HeLa cells. *J. Virol.* **62**: 1038-1045.

S. Z. Hu, T. S. Wang and D. Korn. (1984) DNA primase from KB cells. Evidence for a novel model of primase catalysis by a highly purified primase/polymerase-alpha complex. *J. Biol. Chem.* **259**: 2602-2609.

U. Hübscher. (1983) The mammalian primase is part of a high molecular weight DNA polymerase alpha polypeptide. *EMBO. J.* **2**: 133-136.

T. Hunter and B. Francke. (1974) *In vitro* polyoma DNA synthesis: involvement of RNA in discontinuous chain growth. *J. Mol. Biol.* **83**: 123-130.

A. Itaya, T. Hironaka, Y. Tanaka, K. Yoshihara and T. Kamiya. (1988) Purification and properties of a specific primase-stimulating factor of bovine thymus. *Eur. J. Biochem.* **174**: 261-266.

S. Izuta, M. Kohsaka and M. Sancyoshi. (1985) Inhibitory effects of various 3'-deoxyribonucleotides on DNA polymerase alpha 2-primase from developing cherry salmon (*Oncorhynchus masou*) testes. *Nucleic. Acids. Symp. Ser.* 241-244.

S. M. Jazwinski and G. M. Edelman. (1985) A DNA primase from yeast. Purification and partial characterization. *J. Biol. Chem.* **260**: 4995-5002.

L. S. Kaguni, J. M. Rossignol, R. C. Conaway, G. R. Banks and I. R. Lehman. (1983a) Association of DNA primase with the beta/gamma subunits of DNA polymerase alpha from *Drosophila melanogaster* embryos. *J. Biol. Chem.* **258**: 9037-9039.

L. S. Kaguni, J. M. Rossignol, R. C. Conaway and I. R. Lehman. (1983b) Isolation of an intact DNA polymerase-primase from embryos of *Drosophila melanogaster. Proc. Natl. Acad. Sci. USA* **80**: 2221-2225.

L. S. Kaguni and I. R. Lehman. (1988) Eukaryotic DNA polymerase-primase: structure, mechanism and function. *Biochim. Biophys. Acta* **950**: 87-101.

G. Kaufmann and H. H. Falk. (1982) An oligoribonucleotide polymerase from SV40-infected cells with properties of a primase. *Nucleic. Acids. Res.* **10**: 2309-2321.

H. Konig, H. D. Riedel and R. Knippers. (1983) Reactions *in vitro* of the DNA polymerase-primase from *Xenopus laevis* eggs. A role for ATP in chain elongation. *Eur. J. Biochem.* **135**: 435-442.

A. Kornberg. (1980) *DNA Replication* W.H.Freeman, San Francisco:

E. Lanka and P. T. Barth. (1981) Plasmid RP4 specifies a deoxyribonucleic acid primase involved in its conjugal transfer and maintenance. *J. Bacteriol.* **148**: 769-781.

K. G. Lark. (1972) Genetic control over the initiation of synthesis of the short deoxynucleotide chains in *E. coli. Nature New Biol.* **240**: 237-240.

B. J. Ledwith, S. Manam and G. C. Van Tuyle. (1986) Characterization of a DNA primase from rat liver mitochondria. *J. Biol. Chem.* **261**: 6571-6577.

G. Lucchini, S. Francesconi, M. Foiani, G. Badaracco and P. Plevani. (1987) Yeast DNA polymerase-DNA primase complex; cloning of PRI 1, a single essential gene related to DNA primase activity. *EMBO. J.* **6**: 737-742.

Y. Murakami, T. Eki, M. Yamada, C. Prives and J. Hurwitz. (1986) Species-specific *in vitro* synthesis of DNA containing the polyoma virus origin of replication. *Proc. Natl. Acad. Sci. USA* **83**: 6347-6351.

R. M. Myers and R. Tjian. (1980) Construction and analysis of simian virus 40 origins defective in tumor antigen binding and DNA replication. *Proc. Natl. Acad. Sci. USA* **77**: 6491-6495.

H-P. Nasheuer and F. Grosse. (1987) Immunoaffinity-purified DNA polymerase alpha displays novel properties. *Biochemistry* **26**: 8458-8466.

H-P. Nasheuer and F. Grosse. (1988) DNA polymerase alpha-primase from calf thymus. Determination of the polypeptide responsible for primase activity. *J. Biol. Chem.* **263**: 8981-8988.

K. Ono, H. Nakane and M. Fukushima. (1985) Inhibition of the activities of DNA primase-polymerase alpha complex from KB cells by hexasodium sym-bis(m-aminobenzoyl-m-amino-p-methylbenzoyl-1-naphthylamino-4 ,6,8-trisulfonate)carbamide. *Nucleic. Acids. Symp. Ser.* 249-252.

K. Ono, H. Nakane and M. Fukushima. (1988) Differential inhibition of various deoxyribonucleic and ribonucleic acid polymerases by suramin. *Eur. J. Biochem.* **172**: 349-353.

A. B. Pardee. (1974) A restriction point for control of normal animal cell proliferation. *Proc. Natl. Acad. Sci. USA* **71**: 1286-1290.

W. B. Parker and Y. C. Cheng. (1987) Inhibition of DNA primase by nucleoside triphosphates and their arabinofuranosyl analogs. *Mol. Pharmacol.* **31**: 146-151.

M. Philippe, J. M. Rossignol and A. M. De Recondo. (1986) DNA polymerase alpha associated primase from rat liver: physiological variations. *Biochemistry* **25**: 1611-1615.

V. Pigiet, R. Eliasson and P. Reichard. (1974) Replication of polyoma DNA in isolated nuclei III. The nucleotide sequence at the RNA DNA junction of nascent strands. *J. Mol. Biol.* **84**: 197-216.

P. Plevani, G. Badaracco, C. Augl and L.M. Chang. (1984) DNA polymerase I and DNA primase complex in yeast. *J. Biol. Chem.* **259**: 7532-7539.

P. Plevani, M. Foiani, P. Valsasnini, G. Badaracco, E. Cheriathundam and L.M. Chang. (1985) Polypeptide structure of DNA primase from a yeast DNA polymerase-primase complex. *J. Biol. Chem.* **260**: 7102-7107.

P. Plevani, S. Francesconi and G. Lucchini. (1987) The nucleotide sequence of the PRI1 gene related to DNA primase in *Saccharomyces cerevisiae. Nucleic. Acids. Res.* **15**: 7975-7989.

G. Prelich, M. Kostura, D. R. Marchak, M. B. Mathews and B. Stillman. (1987) The cell-cycle regulated proliferating cell nuclear antigen is required for SV40 DNA replication *in vitro. Nature* **326**: 471-475.

G. Prelich and B. Stillman. (1988) Coordinated leading and lagging strand synthesis during SV40 DNA replication *in vitro* requires PCNA. *Cell* **53**: 117-126.

C. E. Prussak, M. T. Almazan and B. Y. Tseng. (1986) Mouse primase and association with DNA polymerase alpha. *J. Cell. Biochem. supp.* **10B**: I333.

C. E. Prussak, M. T. Almazan and B. Y. Tseng. (1989) Mouse primase p49 subunit, molecular cloning indicates conserved and divergent regions. *J. Biol. Chem.* (In Press)

C. E. Prussak and B. Y. Tseng. (1987) Isolation of the DNA polymerase alpha core enzyme from mouse cells. *J. Biol. Chem.* **262**: 6018-6022.

P. Reichard, R. Eliasson and G. Soderman. (1974) Initiator RNA in discontinuous polyoma DNA synthesis. *Proc. Natl. Acad. Sci. USA* **71**: 4901-4905.

T. Reiter, R. Fett and R. Knippers. (1987) Cell-cycle-dependent expression of DNA primase activity. *Eur. J. Biochem.* **164**: 59-63.

H. D. Riedel, H. Konig, H. Stahl and R. Knippers. (1982) Circular single stranded phage M13-DNA as a template for DNA synthesis in protein extracts from *Xenopus laevis* eggs: evidence for a eukaryotic DNA priming activity. *Nucleic. Acids. Res.* **10**: 5621-5635.

L. J. Romano and C. C. Richardson. (1979) Characterization of the ribonucleic acid primers and the deoxynucleic acid products synthesized by DNA polymerase and gene 4 protein of bacteriophage T7. *J. Biol. Chem.* **254**: 10483-10489.

L. Rowen and A. Kornberg. (1978) A ribo-deoxyribonucleotide primer synthesized by primase. *J. Biol. Chem.* **253**: 770-774.

E. Scherzinger, E. Lanka, G. Morelli, D. Seiffert and A. Yuki. (1977) Bacteriophage-T7-induced DNA-priming protein. A novel enzyme involved in DNA replication. *Eur. J. Biochem.* **72**: 543-558.

J. L. Sherley and T. J. Kelly. (1988) Regulation of human thymidine kinase during the cell cycle. *J. Biol. Chem.* **263**: 8350-8358.

M. Shioda, E. M. Nelson, M. L. Bayne and R. M. Benbow. (1982) DNA primase activity associated with DNA polymerase alpha from *Xenopus laevis* ovaries. *Proc. Natl. Acad. Sci. USA* **79**: 7209-7213.

H. Singh, R. G. Brooke, M. H. Pausch, G. T. Williams, C. Trainor and L. B. Dumas. (1986) Yeast DNA primase and DNA polymerase activities. An analysis of RNA priming and its coupling to DNA synthesis. *J. Biol. Chem.* **261**: 8564-8569.

H. Singh and L. B. Dumas. (1984. A DNA primase that copurifies with the major DNA polymerase from the yeast Saccharomyces cerevisiae. *J. Biol. Chem.* **259**: 7936-7940.

D. B. Sittman, R. A. Graves and W. F. Marzluff. (1983) Histone mRNA concentrations are regulated at the level of transcription and mRNA degradation. *Proc. Natl. Acad. Sci. USA* **80**: 1849-1853.

S. T. Smale and R. Tjian. (1986) T-antigen-DNA polymerase alpha complex implicated in simian virus 40 DNA replication. *Mol. Cell. Biol.* **6**: 4077-4087.

M. Suzuki, T. Enomoto, F. Hanaoka and M. Yamada. (1985) Dissociation and reconstitution of a DNA polymerase alpha-primase complex. *J. Biochem. (Tokyo)* **98**: 581-584.

S. Tanaka, S. Hu, T. S. Wang and D. Korn. (1982) Preparation and preliminary characterization of monoclonal antibodies against human DNA polymerase alpha. *J. Biol. Chem.* **257**: 8386.

C. B. Thompson, P. B. Challoner, P. E. Neiman and M. Groudine. (1985) Levels of c-myc oncogene mRNA are invariant throughout the cell cycle. *Nature* **314**: 363-369.

B. Y. Tseng, J. M. Erickson and M. Goulian. (1979) Initiator RNA of nascent DNA from animal cells. *J. Mol. Biol.* **129**: 531-545. B.Y.

Tseng and C. N. Ahlem. (1982) DNA primase activity from human lymphocytes. Synthesis of oligoribonucleotides that prime DNA synthesis. *J. Biol. Chem.* **257**: 7280-7283.

B. Y. Tseng and C. N. Ahlem. (1983a) A DNA primase from mouse cells. Purification and partial characterization. *J. Biol. Chem.* **258**: 9845-9849.

B. Y. Tseng and C. N. Ahlem. (1983b, DNA primase from mouse cells) **In**: *Mechanism of DNA Replication and Recombination*, Vol.**10**, edited by N. Cozzarelli. A.R.Liss,Inc., New York: p. 511-516.

B. Y. Tseng and C. N. Ahlem. (1984) Mouse primase initiation sites in the origin region of simian virus 40. *Proc. Natl. Acad. Sci. USA* **81**: 2342-2346.

B. Y. Tseng and M. Goulian. (1975) Evidence for covalent association of RNA with nascent DNA in human lymphocytes. *J. Mol. Biol.* **99**: 339-346.

B. Y. Tseng and M. Goulian. (1977) Initiator RNA of discontinuous DNA synthesis in human lymphocytes. *Cell* **12**: 483-489.

B. Y. Tseng and M. Goulian. (1980) Initiator RNA synthesis upon ribonucleotide depletion: evidence for base substitutions. *J. Biol. Chem.* **255**: 2062-2067.

B. Y. Tseng, C. E. Prussak and M. T. Almazan. (1989a) Primase p49 mRNA expression is serum stimulated but does not vary with cell cycle. *Mol. Cell. Biol.* (In Press)

B. Y. Tseng and C. E. Prussak. (1989b) Sequence and structural requirements for primase initiation in the SV40 origin of replication. *Nucleic Acids Res.* (In Press)

R. A. Tubo and R. Berezney. (1987) Pre-replicative association of multiple replicative enzyme activities with the nuclear matrix during rat liver regeneration. *J. Biol. Chem.* **262**: 1148-1154.

J. K. Vishwanatha and E. F. Baril. (1986) Resolution and purification of free primase activity from the DNA primase-polymerase alpha complex of HeLa cells. *Nucleic. Acids. Res.* **14**: 8467-8487.

J. K. Vishwanatha, M. Yamaguchi, M. L. DePamphilis and E. F. Baril. (1986) Selection of template initiation sites and the lengths of RNA primers synthesized by DNA primase are strongly affected by its organization in a multiprotein DNA polymerase alpha complex. *Nucleic. Acids. Res.* **14**: 7305-7323.

A. F. Wahl, S. P. Kowalski, L. W. Harwell, E. M. Lord and R. A. Bambara. (1984) Immunoaffinity purification and properties of a high molecular weight calf thymus DNA alpha polymerase. *Biochemistry* **23**: 1895-1899.

A. F. Wahl, A. M. Geis, B. H. Spain, S. W. Wong, D. Korn and T. S. Wang. (1988) Gene expression of human DNA polymerase alpha during cell proliferation and the cell cycle. *Mol. Cell. Biol.* **8**: 5016-5025.

T. S. Wang, S. Z. Hu and D. Korn. (1984) DNA primase from KB cells. Characterization of a primase activity tightly associated with immunoaffinity-purified DNA polymerase-alpha. *J. Biol. Chem.* **259**: 1854-1865.

F. E. Wilson and A. Sugino. (1985) Purification of a DNA primase activity from the yeast Saccharomyces cerevisiae. Primase can be separated from DNA polymerase I. *J. Biol. Chem.* **260**: 8173-8181.

S. W. Wong, L. R. Paborsky, P. A. Fisher, T. S. Wang and D. Korn. (1986) Structural and enzymological characterization of immunoaffinity-purified DNA polymerase alpha.DNA primase complex from KB cells. *J. Biol. Chem.* **261**: 7958-7968.

S. W. Wong, A. F. Wahl, P. Yuan, N. Arai, B. E. Pearson, K. Arai, D. Korn, M.W. Hunkapiller and T.S. Wang. (1988) Human DNA polymerase alpha gene expression is cell cycle proliferation dependent and its primary structure is similar to both prokaryotic and eukaryotic replicative DNA polymerases. *EMBO. J.* **7**: 37-47.

T. W. Wong and D. A. Clayton. (1985) Isolation and characterization of a DNA primase from human mitochondria. *J. Biol. Chem.* **260**: 11530-11535.

T. W. Wong and D. A. Clayton. (1986) DNA primase of human mitochondria is associated with structural RNA that is essential for enzymatic activity. *Cell* **45**: 817-825.

S. H. Wood and J. M. Collins. (1986) Preferential binding of DNA primase to the nuclear matrix in HeLa cells. *J. Biol. Chem.* **261**: 7119-7122.

T. Yagura, T. Kozu and T. Seno. (1982) Mouse DNA polymerase accompanied by a novel RNA polymerase activity: purification and partial characterization. *J. Biochem. (Tokyo)* **91**: 607-618.

T. Yagura, T. Kozu and T. Seno. (1983a) Mechanism of stimulation by a specific protein factor of *de novo* DNA synthesis by mouse DNA replicase with fd phage single stranded circular DNA. *Nucleic. Acids. Res.* **11**: 6369-6380.

T. Yagura, T. Kozu, T. Seno, M. Saneyoshi, S. Hiraga and H. Nagano. (1983b) Novel form of DNA polymerase alpha associated with DNA primase activity of vertebrates. Detection with mouse stimulating factor. *J. Biol. Chem.* **258**: 13070-13075.

T. Yagura, T. Kozu and T. Seno. (1986) Size difference in catalytic polypeptides of two active forms of mouse DNA polymerase alpha and separation of the primase subunit from one form, DNA replicase. *Biochim. Biophys. Acta* **870**: 1-11.

T. Yagura, T. Kozu, T. Seno and S. Tanaka. (1987) Immunochemical detection of a primase activity related subunit of DNA polymerase alpha from human and mouse cells using the monoclonal antibody. *Biochemistry* **26**: 7749-7754.

M. Yamaguchi, E. A. Hendrickson and M. L. DePamphilis. (1985a) DNA primase-DNA polymerase alpha from simian cells: sequence specificity of initiation sites on simian virus 40 DNA. *Mol. Cell. Biol.* **5**: 1170-1183.

M. Yamaguchi, E. A. Hendrickson and M. L. DePamphilis. (1985b. DNA primase-DNA polymerase alpha from simian cells. Modulation of RNA primer synthesis by ribonucleoside triphosphates. *J. Biol. Chem.* **260**: 6254-6263.

S. Yoshida, R. Suzuki, S. Masaki and O. Koiwai. (1983) DNA primase associated with 10 S DNA polymerase alpha from calf thymus. *Biochim. Biophys. Acta* **741**: 348-357.

S. Yoshida, R. Suzuki, S. Masaki and O. Koiwai. (1985) Arabinosylnucleoside 5'-triphosphate inhibits DNA primase of calf thymus. *J. Biochem. (Tokyo)* **98**: 427-433.

A. Zetterberg and O. Larsson. (1985) Kinetic analysis of regulatory event in G_1 leading to proliferation of quiescence of Swiss 3T3 cells. *Proc. Natl. Acad. Sci. USA* **82**: 5365-5369.

10

THE GENETIC AND BIOCHEMICAL COMPLEXITY OF DNA REPAIR IN EUKARYOTES: THE YEAST *SACCHAROMYCES CEREVISIAE* AS A PARADIGM

Errol C. Friedberg

INTRODUCTION

Nucleotide excision repair defines processes by which damaged nucleotides are enzymatically removed from the genome (Friedberg, 1985a). This process is distinguished from base excision repair, in which damaged bases are enzymatically removed as free bases (Friedberg, 1985a). In the latter process the specificity of damage recognition and of base excision is determined by a class of repair enzymes called DNA glycosylases (see Friedberg, 1985a). Both prokaryotic and eukaryotic cells contain multiple such enzymes. All known DNA glycosylases are relatively small (about 18-30kDa) single polypeptides which specifically recognize non-bulky purine or pyrimidine derivatives such as uracil, hypoxanthine or 3-methyladenine, in either single- or double-stranded DNA. A few DNA glycosylases recognize related forms of simple base damage, *e.g.*, N^3-methylated purines and O^2-methylated pyrimidines, or multiple forms of ring-opened pyrimidines. The molecular basis for this relatively relaxed substrate specificity is not fully understood.

One might expect that in the interests of genetic economy, cells could not afford to encode specific DNA glycosylases for every type of environmental or spontaneous base damage that accompanied biological evolution. Hence, it is not surprising that most organisms evolved a more general mode of excision repair that recognizes a conformational change in DNA common to many forms of base damage, and which results in the removal of oligonucleotide tracts rather than single bases.

In contrast to the elegant simplicity of base excision repair based on a motif of one base adduct one enzyme, the generalization of excision repair poses special problems for the specificity of damage recognition. It has been frequently suggested that the substrate specificity of nucleotide excision repair is determined by DNA conformation. However, to the extent that similar conformations may be generated by other metabolic transactions such as replication, transcription or recombination, or by spontaneous folding, bending or kinking of DNA, the genome is potentially vulnerable to spurious degradation in the absence of DNA damage.

This problem is not a unique one for living cells. As pointed out by Echols (1986), the basis for the exceptional locational precision of initiation of DNA replication, for site-specific integration of prophages and possibly for other DNA reactions, is not immediately evident from a simple consideration of the DNA binding properties of the proteins involved in these events. Site-specific DNA binding proteins also bind to DNA non-specifically (albeit with lower affinity), and bind very specifically to incorrect sites with sequences which are identical to or closely resemble those of the correct sites. Echols (1986) has suggested that the DNA binding proteins that localize high precision events do so through the elaboration of multiprotein-DNA complexes which constitute specialized nucleoprotein structures in which DNA sequences are folded or wound in a highly specific fashion, thereby generating unique substrates for particular site-specific catalytic events .

Specialized nucleoprotein structures could provide the essential basis for substrate specificity during nucleotide excision repair. Simply stated, the substrate for damage-specific incision of DNA might depend on the ordered assembly of a series of proteins at or near sites of damage that alter DNA conformation in a highly specific way, thereby generating a distinctive nucleo-protein complex (Figure 1). This complex could include an endonuclease activity whose catalytic sites are placed in immediate proximity with the damaged strand at appropriate distances from the site of base damage (Figure 1). Two fundamental predictions stem from this model.

1. The model predicts that nucleotide excision repair is biochemically complex since it depends on multiple, specific protein-protein and protein-DNA interactions. Biochemical complexity does not necessarily imply corresponding genetic complexity if the protein complexes are multimers of a small number of different polypeptides. However, biochemical transactions that utilize multiprotein complexes (including nucleotide excision repair in the context of the present discussion) may have a complex genetic basis if all or most of the polypeptide components are encoded by different genes.

2. The ordered assembly of a progressively more complex nucleo-protein structure provides a satisfactory heuristic explanation for biochemical specificity during nucleotide excision repair. But what determines the recognition specificity for the initiation of a specific multi-protein-DNA complex ? If nucleation of the complex requires the binding of only a single polypeptide to DNA one must assume that this initiating protein is endowed with sufficient specificity to distinguish sites of conformational distortion caused by DNA damage from those that arise spuriously. [Perhaps in the living cell this problem is resolved kinetically, *i.e.*, spurious conformational distortions are transient while those generated by base damage are permanent.] Alternatively, the specificity of the initial protein-DNA interaction might be altered by the rapid binding of subsequent proteins in the complex, or specificity for binding at

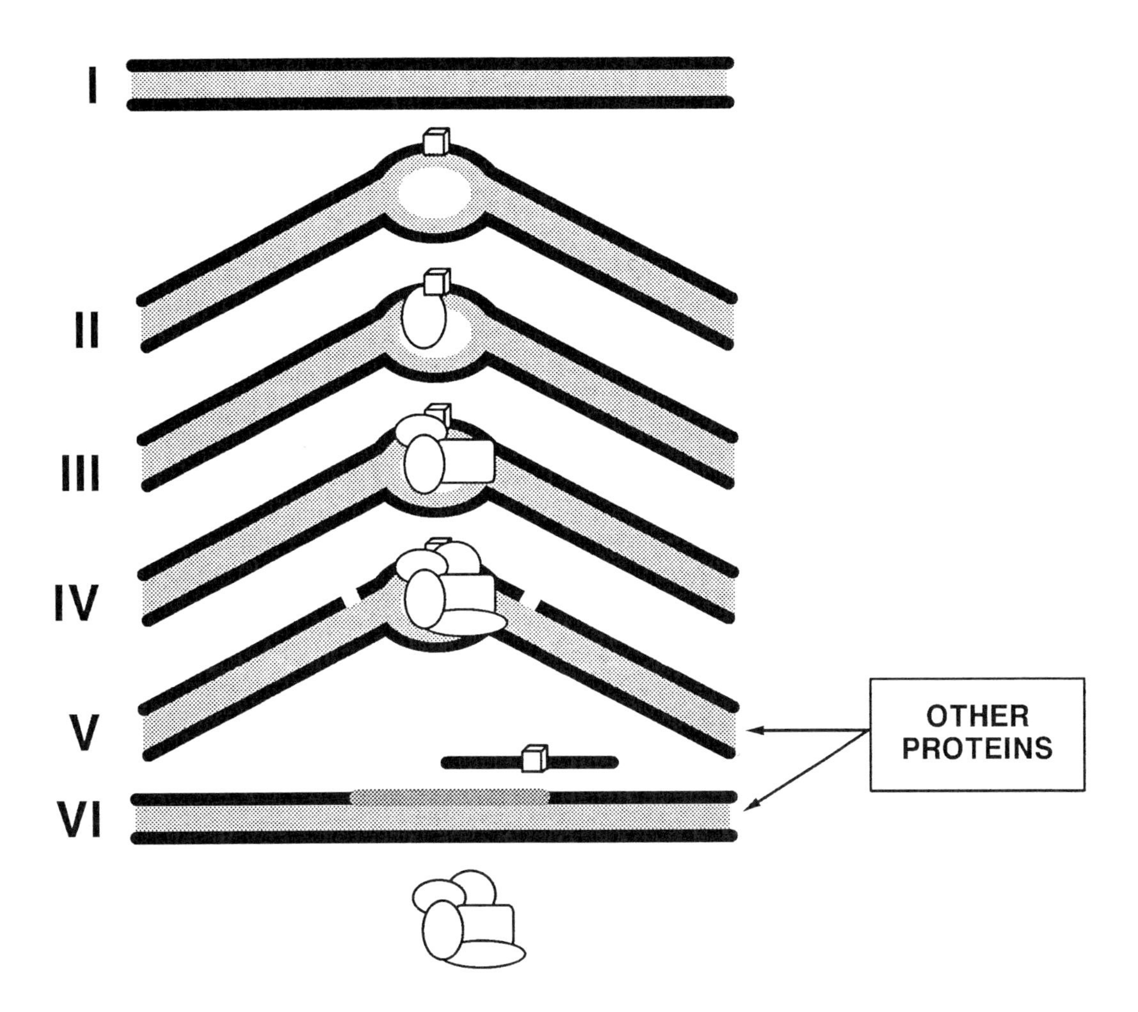

Figure 1. Diagrammatic representation of a hypothetical damage-specific multiprotein complex for nucleotide excision repair in yeast. The presence of a bulky adduct in DNA is shown to result in kinking or bending (II). This conformational distortion is recognized by a protein, which binds at the site of base damage (III). A multiprotein complex is progressively assembled consisting of 5 proteins encoded by the *RAD1*, *RAD2*, *RAD3*, *RAD4* and *RAD10* genes (IV and V). When this complex is assembled it constitutes a catalytically active endonuclease that cuts the damaged strand specifically (V). As suggested in the figure, other proteins may be involved in this process *in vivo*. Other proteins may also be required to displace the protein complex thereby facilitating turnover, and for release (excision) of an oligonucleotide containing the damaged base, for repair synthesis and for DNA ligation (VI).

sites of DNA damage may depend on the simultaneous interaction of more than one protein with DNA.

In summary, the assembly of a specialized nucleoprotein complex offers a plausible explanation for the biochemical mechanism of both damage-specific recognition in DNA and of site-specific and strand-specific incision of DNA during nucleotide excision repair in living cells.

THE COMPLEXITY OF NUCLEOTIDE EXCISION REPAIR IN EUKARYOTES

The problems inherent in substrate specificity discussed above are expected to be common to all levels of biological organization. Based on several considerations one might anticipate that additional elements of complexity are unique to eukaryotic organisms. Firstly, multi-protein structures for nucleotide excision repair might have acquired increased complexity as a consequence of evolutionary progression and refinement. Such evolutionary progression could lead to complete loss of homology between prokaryotic and eukaryotic proteins when viewed at the level of primary structure, even though fundamental mechanistic elements may be conserved. At the extreme, such evolutionary divergence could eventually result in distinct biochemical mechanisms for the excision of oligonucleotides with damaged bases. Secondly, since the earliest theoretical considerations of nucleosome conformation (Kornberg, 1974; Kornberg, 1977; McGhee and Felsenfeld, 1980) it was intuitively obvious to students of DNA repair that the structural association of DNA with histones and non-histone proteins, as well as the coiling of nucleosomes to yield higher levels of chromatin compaction, might pose special problems for the access of proteins and/or enzymes involved in the processing of base damage. One hypothetical solution to this problem posits the evolution of components of repair complexes whose specific function is to provide access to sites of damage by modifying chromatin structure (Friedberg, 1985a).

A third element of potential complexity derives from the notion of metabolic channeling or functional compartmentalization, *i.e.*, that enzymes of like function may be compartmentalized or otherwise physically juxtaposed, to facilitate the channeling of metabolic intermediates and/or the maintenance of concentration gradients of the precursors for macromolecular synthesis, with the highest concentrations being at the actual sites of utilization (Mathews *et al.*, 1979). DNA synthesis is an integral component of many forms of repair, including nucleotide excision repair. Hence, it would not be totally unexpected if enzymes for deoxynucleoside triphosphate biosynthesis were components of multi-protein complexes for cellular processing of genomic injury.

A fourth element of potential complexity stems from a consideration of the possible overlapping association of selected components of multiprotein complexes for DNA repair with complexes for other cellular functions, *i.e.*, proteins required for different aspects of DNA metabolism may assemble into alternative complexes. The participation of individual polypeptides in other cellular functions does not contribute biochemical complexity to a multiprotein structure for DNA repair. However, in theory, optimal stability of these polypeptides might necessitate their physical association with a specific subset of proteins which do not have a specific mechanistic function during DNA repair. Such a situation would have obvious potential for increased structural complexity.

Metabolic transactions that incorporate selected DNA repair proteins could be unrelated to DNA metabolism. However, it is obviously more interesting to consider that such transactions do involve DNA, and that the participation of proteins in both the repair of, and in other aspects of, DNA metabolism is physiologically relevant. In this context one must bear in mind the possibilities that overlapping polypeptides may have a *single* function operative both in DNA repair and in some other cellular process(es), or that these polypeptides have *dual*, non-overlapping functions.

Finally, one must consider the possibility that regulated expression of individual genes or of entire biochemical pathways could contribute importantly to the genetic as well as biochemical complexity of DNA repair in eukaryotes. The SOS and *ada* regulatory systems in *E. coli* and in other bacteria constitute well documented examples of complex negative and positive regulatory mechanisms respectively, governing selected aspects of DNA repair in prokaryotes (Witkin, 1976; Walker *et al.*, 1985). The literature contains numerous examples of regulated responses to DNA damage in eukaryotes (Radman, 1980; Friedberg, 1985a; Moustacchi, 1987; Friedberg, 1988). However, the search for complex regulons embracing defined DNA repair mechanisms in higher organisms remains elusive .

The aim of this article is to examine evidence for the biochemical and genetic complexity of DNA repair in the context of the general concepts discussed above. The term "DNA repair" is used in a generic sense since there are indications that both genetic and biochemical complexity are characteristic of diverse cellular responses to DNA damage. However, the focus of these discussions will be on the process of nucleotide excision repair, using the bacterium *E. coli* as a prokaryotic biochemical paradigm, and the yeast *Saccharomyces cerevisiae* as a eukaryotic genetic model.

THE BIOCHEMICAL COMPLEXITY OF NUCLEOTIDE EXCISION REPAIR IN *E. COLI*

E. coli is currently the only organism for which fairly detailed information on the biochemistry of nucleotide excision repair is available. It is well established that mutant strains designated *uvrA*, *uvrB* and *uvrC* are abnormally sensitive to killing by a variety of physical and chemical DNA-damaging agents (Friedberg, 1985a). These agents share the capacity to produce bulky base adducts that result in significant distortion of secondary structure in DNA. While the term "conformational distortion" is banded about in the literature rather freely, its precise physical basis is not well defined.

Evidence that UV irradiation alters the structure of DNA comes from numerous indirect physical studies, including sedimentation (Denhart and Kato, 1973), gel electrophoresis (Ciarrocchi and Pedrini, 1982), sensitivity to nucleases (Legerski *et al.*, 1977) and sensitivity to chemical modification (Shafranovskaya *et al.*, 1973). Perlman *et al.* (1985) constructed structural models of B-DNA containing a psoralen cross-link or a pyrimidine (thymine) dimer by applying energy minimization techniques and model-building procedures to data obtained from X-ray crystallographic studies. The models so derived showed substantial kinking (Figure 2) and unwinding at the sites of damage, more so in the molecule with a psoralen cross-link than with a pyrimidine dimer. These authors noted numerous changes in the torsion angles of the DNA structure containing a pyrimidine dimer. Interestingly, they suggested the presence of greater conformational derangement in the non dimer-containing strand because the dimer bases open at an angle to this strand.

The *E. coli uvr* mutants are specifically defective in excision of bulky base damage, and the potential for such damage to arrest transcription and generate mutations in essential genes provides a reasonable explanation for their sensitivity to killing. For many years attempts to isolate the proteins encoded by the *uvr* genes were hampered by the fact that these proteins are present at extremely low intracellular concentrations. Indeed, it appears to be generally true that living cells constitutively express very small amounts of excision repair proteins. This may be

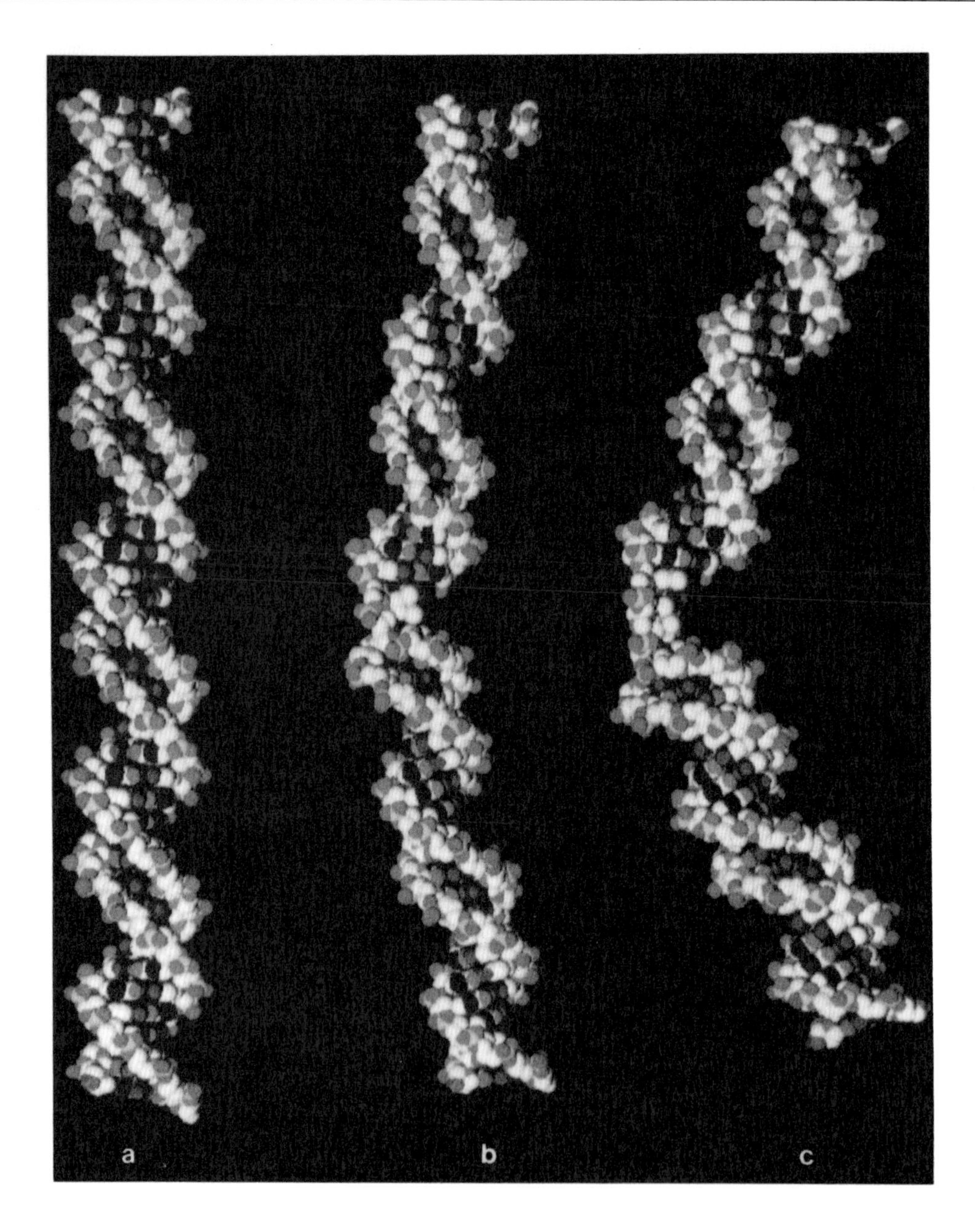

Figure 2. Space-fillings model of DNA. **a**, B-form native DNA; **b**, DNA containing a thymine dimer; **c**, DNA containing a psoralen cross-link. (From Pearlman *et al.*, 1985. Reproduced with permission).

an important mechanism for limiting the potential for spurious degradation of native DNA or of DNA that assumes conformations resembling those produced by damage.

With the advent of molecular cloning the marked UV sensitivity of the *uvr* mutants provided a powerful selectable phenotype for screening genomic libraries, and the *uvrA*, *uvrB* and *uvrC* genes were cloned and overexpressed to yield large amounts of soluble proteins, thereby facilitating biochemical studies (Grossman *et al.*, 1988; Van Houten *et al.*, 1988 and references therein). The general mechanism of nucleotide excision repair in *E. coli* is now understood in considerable detail, and there is substantial agreement on the results of independent studies in different laboratories. However, much remains to be established about the details of damage-specific recognition by proteins, protein-protein and protein-DNA interactions, the precise substrate for damage-specific incision, and the exact mechanism of oligonucleotide excision and repair synthesis of DNA.

Grossman *et al.* (1988) have proposed that UvrA protein (a protein of molecular weight about 114,000) binds to DNA as a dimer with 2 moles of ATP bound per mole of protein. The binding of ATP is postulated to change the conformation of the protein and to facilitate its dimerization. The subsequent binding of UvrB protein (molecular weight about 84,000) generates a $(UvrA/ATP_2)_2$/UvrB/DNA nucleoprotein complex. These authors propose that the protein complex translocates to sites of bulky base damage by a helicase action catalyzed by the $UvrA_2$/UvrB protein complex, but not by either protein alone. A unique feature postulated for this helicase is that it is arrested at sites of bulky base damage. This allows UvrC protein (molecular weight about 67,000) to bind to the $UvrA_2$/UvrB/DNA complex and to mediate incision exclusively on the strand with the damaged base, at precise distances 5′ and 3′ to the damage.

Grossman *et al.* (1988) have pointed out that at equilibrium the affinity of UvrA protein for sites of bulky base damage is only 10^4 greater than for native DNA. They suggest that this difference is too low to satisfactorily account for the discrimination of damage-specific incision of DNA in the *E. coli* genome, and postulate that kinetic determinants play an important role in offsetting this relatively indiscriminate specificity, as evidenced by the observation that off-rates of UvrA protein from damaged sites are considerably slower than for undamaged sites. This kinetic distinction, together with the postulated arrest of DNA unwinding by the UvrAB helicase at sites of damage allows time for UvrC protein to complete a catalytically productive complex.

This model does not specifically delineate the maximal distance of the site of initial assembly of the $UvrA_2$/UvrB complex from a site of bulky base damage. Hence as stated, an implication of the model is that this complex could translocate by helicase action for considerable distances along the genome before arresting at sites of damage, an energetically expensive process.

Grossman *et al.* (1988) have detected a proteolytically degraded form of UvrB protein of molecular weight about 70,000 designated UvrB*. The amino acid sequence of the site of proteolytic cleavage is very similar to that of Ada protein, a protein involved in the repair of O^6 alkylguanine and O^4 alkylthymine in *E. coli* (Lindahl *et al.*, 1988). UvrB* protein can participate in a $UvrA_2$/UvrB*/DNA complex. However, this complex does not support the helicase activity of the $UvrA_2$/UvrB complex, nor the formation of a catalytically active endonuclease. It is suggested that proteolysis of UvrB protein may be a mechanism for aborting UvrABC complexes formed at inappropriate sites in the genome, which could potentially result in spurious nicking of DNA.

The model of Sancar and his colleagues (Van Houten *et al.*, 1988) differs in some respects. These investigators have suggested that UvrA protein recognizes sites of conformational distortion and binds at such sites as a monomer in the absence of ATP (Figure 3). The subsequent interaction of UvrB protein with the UvrA/DNA complex generates a $UvrA_1/UvrB_1$/DNA complex which induces a specific change in the secondary structure of the

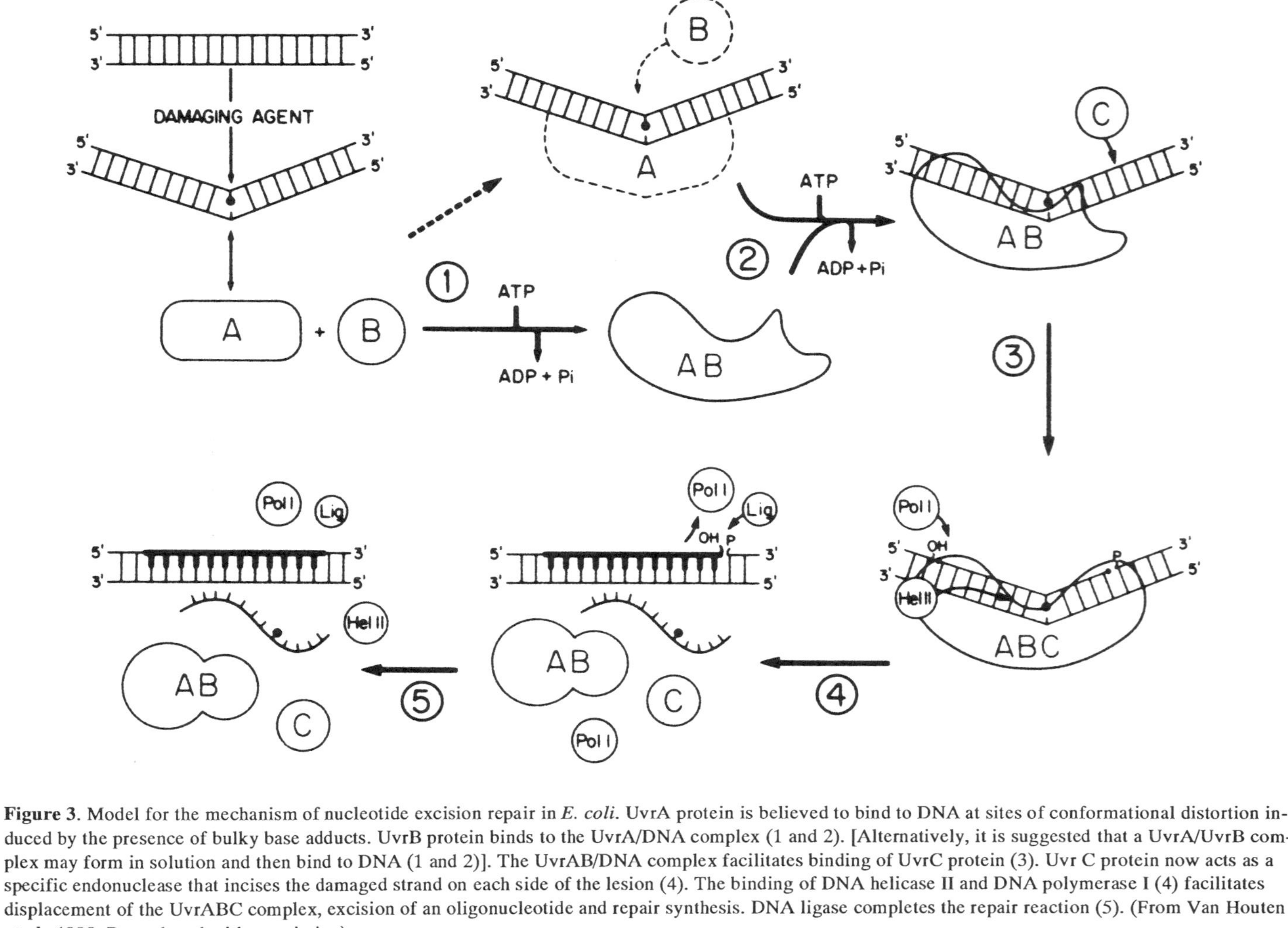

Figure 3. Model for the mechanism of nucleotide excision repair in *E. coli*. UvrA protein is believed to bind to DNA at sites of conformational distortion induced by the presence of bulky base adducts. UvrB protein binds to the UvrA/DNA complex (1 and 2). [Alternatively, it is suggested that a UvrA/UvrB complex may form in solution and then bind to DNA (1 and 2)]. The UvrAB/DNA complex facilitates binding of UvrC protein (3). Uvr C protein now acts as a specific endonuclease that incises the damaged strand on each side of the lesion (4). The binding of DNA helicase II and DNA polymerase I (4) facilitates displacement of the UvrABC complex, excision of an oligonucleotide and repair synthesis. DNA ligase completes the repair reaction (5). (From Van Houten *et al.*, 1988. Reproduced with permission).

DNA, thereby increasing the stability of the complex. It is suggested that this conformational change requires the energy of ATP hydrolysis and also favors the binding of UvrC protein (Figure 3). The UvrABC complex so generated has a unique three-dimensional conformation that determines the site-specificity of incision.

Sancar and his colleagues (Van Houten *et al.*, 1988; Orren and Sancar, 1988) have observed dimerization of UvrA protein in the absence of DNA, and the formation of $UvrA_2/UvrB1$ complexes in solution. They suggest that such complexes might constitute an alternative intermediate that could bind damaged DNA (Figure 3). However such a mechanism requires the loss of one mole of UvrA protein to reconcile their stoichiometric observations that Uvr proteins bound to DNA are in the ratio $UvrA_1/UvrB_1/UvrC_1$.

Both groups of investigators agree that the UvrABC/DNA complex is stable and turns over very slowly *in vitro*. Dissociation of the complex and turnover of these proteins requires the addition of UvrD protein and of DNA polymerase I. Sancar and his colleagues have suggested that these two proteins may bind on opposite strands at the nick 5′ to the site of base damage (Van Houten *et al.*, 1988). DNA polymerase I could utilize the nicked strand to catalyze repair synthesis in the 5′→3′ direction. [There is no obvious role for the 5′→3′ exonuclease function of the polymerase in this reaction, since the limit of the synthesis reaction is determined by the 3′ nick generated during the initial incison reaction. Indeed, in their hands, about 90% of the repair patches generated *in vitro* are 12 nucleotides in length (Van Houten *et al.*, 1988)]. The UvrD helicase (a 3′→5′ DNA helicase) could translocate in the same direction on the opposite (intact) strand. The net effect of the movement of both the DNA polymerase and DNA helicase could displace the UvrABC complex (which would then dissociate), as well as the oligonucleotide generated by the bimodal incision (Figure 3).

In summary, the study of nucleotide excision repair in *E. coli* confirms the predictions of biochemical complexity for damage-specific recognition and site-specific incision of DNA during nucleotide excision repair. At least 5 distinct proteins are required. *In vitro*, these proteins participate in a series of reactions that can be conveniently broken down into 3 major steps: (i) damage recognition/complex assembly, (ii) DNA incison, (iii) excision/resynthesis/ligation/protein turnover. *In vivo* these and possibly other proteins (*e.g.*, DNA ligase) may associate to form a large multiprotein repairosome for nucleotide excision repair. Recent studies in *M. luteus* have resulted in the molecular cloning and sequencing of two genes homologous to the *E. coli uvrA* and *uvrB* genes (Nakayama and Shiota, 1988), suggesting that genes for nucleotide excision repair are highly conserved in prokaryotes.

THE GENETIC COMPLEXITY OF NUCLEOTIDE EXCISION REPAIR IN YEAST

Detailed genetic analysis (Friedberg, 1988 for a recent review) has defined a surprisingly large number of loci for excision repair of bulky base damage in the yeast *S. cerevisiae*. Strains defective in at least 5 genes (designated *RAD1*, *RAD2*, *RAD3*, *RAD4* and *RAD10*) are extremely sensitive to killing by UV radiation and by UV-mimetic agents such as 4-nitroquinoline-1-oxide (4NQO). These mutant strains are also defective in damage-specific incision of DNA and in excision of bulky base adducts during post-irradiation incubation *in vivo*.

Other loci have been identified which render cells less sensitive to killing by UV radiation when mutated. These include the *RAD7*, *RAD14*, *RAD16*, *RAD23*, *RAD24* and *MMS19* genes. Consistent with the phenotype of reduced sensitivity to killing, mutations in these genes also result in a partial rather than a complete defect in nucleotide excision repair (Friedberg, 1988).

The reason for the phenotypic differences between these two groups of mutants is not understood. However, the following possibilities merit special consideration.

(i) The proteins encoded by the *RAD1*, *RAD2*, *RAD3*, *RAD4* and *RAD10* genes are likely to constitute essential components for "building" a specialized nucleoprotein structure for damage-specific incision (Figure 1). In keeping with the *E. coli* paradigm, efficient excision of oligonucleotides and rapid turnover of this multi-protein complex may require other proteins (Figure 1). The observation that mutants in the *RAD7*, *RAD14*, *RAD16*, *RAD23* and *MMS19* genes have phenotypic characteristics similar to those of *uvrD* and *polA* mutants of *E. coli* (Friedberg, 1988), is certainly consistent with this notion.

(ii) The proteins encoded by the secondary group of genes may be specifically involved in providing access of the incision complex to sites of base damage in chromatin.

(iii) The secondary group of genes may regulate, or modulate expression of, some or all of the primary group.

(iv) As suggested earlier, the proteins encoded by these genes may have a primary role in some other aspect of DNA metabolism, but may confer optimal structural stability to specific polypeptides required for nucleotide excision repair. Hence they are components of the multi-protein repair complex.

A number of genes in the yeast nucleotide excision repair pathway have recently been cloned and characterized. Similar progress has been made with genes from other epistasis groups for DNA repair in yeast (Friedberg, 1988). Efforts in several laboratories are directed towards the overexpression of these genes and towards the purification and biochemical characterization of the proteins they encode (see the article in this volume by S. Prakash and L. Prakash). Until this considerable task is completed there is essentially no information on the biochemistry and molecular mechanism of any DNA repair pathway in yeast; a sobering conclusion that suggests an obvious point at which to terminate this review. However, I suggested earlier a number of theoretical attributes that might account for the evolution of complex repair functions in eukaryotes. It might be instructive to explore the extent to which these and/or other explanations for genetic complexity are supported by information gained from the characterization of cloned DNA repair genes of *S. cerevisiae*.

The Eukaryotic Repairosome for Nucleotide Excision Repair is Fundamentally Distinct from that in *E. coli*

The yeast Rad1, Rad2, Rad3, Rad4 and Rad10 proteins and the *E. coli* UvrA, UvrB, UvrC, UvrD and PolI proteins for nucleotide excision repair do not share extensive similarities in amino acid sequences. Additionally, functional complementation of yeast or *E. coli* mutants has not been observed following the introduction of cloned genes from one species into cells of the other (Friedberg, 1985b). The amino acid sequences of DNA ligase from yeast and phage T4 are similar (Barker *et al.*, 1985), but such evolutionary conservation presumably reflects the essential role of DNA ligase in cellular metabolism rather than its role in DNA repair.

The one Rad protein about which some biochemical information is currently to hand is Rad3. Examination of the amino acid sequence of the yeast *RAD3* gene reveals a nucleotide binding domain which is conserved in the *E. coli uvrA*, *uvrB* and *uvrD* genes (Friedberg, 1988). Additionally, Rad3 protein has been extensively purified and shown to possess DNA-dependent ATPase and 5′→3′ DNA helicase activities (Sung *et al.*, 1987a; Sung *et al.*, 1987b; see article in this volume by S. Prakash and L. Prakash). To what extent can these functions be extrapolated to those associated with excision repair proteins from *E. coli*? UvrA protein and the UvrAB complex have ATPase activity, however, neither are DNA-dependent. As indicated earlier, a UvrAB complex has been reported to catalyze DNA helicase activity with the same polarity (5′→3′) as that of Rad3 protein (Oh and Grossman, 1987). Rad3 protein also bears some resemblance to *E. coli* UvrD protein, which is a DNA-dependent ATPase and DNA helicase. Indeed, the amino acid sequence of this protein and that of Rad3 protein share a number of similarities (Foury and Lahaye, 1987). However, the DNA helicase of UvrD protein has 3′→5′ polarity, the reverse of that of Rad3 protein (Matson, 1986). Furthermore, UvrD protein is not absolutely required for damage-specific incision of DNA, whereas Rad3 protein is.

Mutagenesis of the nucleotide binding domain in a plasmid-borne *RAD3* allele results in its inability to complement the UV sensitivity of excision-defective *rad3* mutant strains (Naumovski and Friedberg, 1986). This observation is consistent with a requirement for binding of ATP to effect a conformational change of Rad3 protein during damage-specific incision of DNA (as suggested for the UvrA protein of *E. coli*). However, this observation is equally consistent with a requirement for ATPase activity (and perhaps for DNA helicase activity) for damage-specific incision.

Referring to the *E. coli* model proposed by Grossman and his colleagues (Grossman *et al.*, 1988) one could accomodate the need for a DNA helicase if the incision mechanism in yeast required translocation of a protein complex (of which Rad3 is a component) from a site of initial binding to a site at which damage-specific incision was catalyzed, *i.e.*, Rad3 protein may be functionally homologous to the UvrAB complex of *E. coli*. An alternative explanation for the requirement of a DNA helicase for damage-specific incision of DNA in yeast is that inactivation of any single component of the multi-protein complex for incision and post-incision events might arrest the entire repair pathway. Hence, even though the DNA helicase function may be directly involved in a post-incision event (*e.g.*, displacement of oligonucleotide fragments), inactivation of Rad3 protein could shut down incision of DNA.

In the absence of detailed biochemical information from any eukaryotic system, speculation about the relationship between the mechanism of nucleotide excision repair in *E. coli* and in yeast constitutes nothing more than an intellectual exercise. Perhaps such a relationship can be reasonably defended by arguing that, as is true of other parameters of DNA metabolism such as replication and transcription, the *fundamental* mechanism of nucleotide excision repair is likely to be conserved in all cells. Nonetheless, one must consider the possibility that the biochemistry of damage-specific incision in yeast is fundamentally distinct from that in *E. coli* and that nucleotide binding, DNA-dependent ATPase activity and/or DNA helicase activity may be required for biochemical events in excision repair that are distinct from those identified in *E. coli*.

The arguments just presented are predicated on the assumption that the ATPase and DNA helicase activities of Rad3 protein are functionally linked. However, it has been established that the *RAD3* gene is involved in at least two other cellular processes (see later discussion). Thus, yet another viewpoint to be considered is that the known biochemical functions of Rad3 protein are involved in other aspects of yeast metabolism. As a first step toward unravelling these many unresolved questions it is obviously important to isolate several mutant Rad3 proteins that correlate with known phenotypes of *rad3* mutants and to determine whether individual mutations selectively inactivate nucleotide binding, ATPase activity and DNA helicase activity.

Yeast (and Other Eukaryotic) DNA Repair Genes Are Involved in Chromatin Modelling

The most compelling evidence in support of the idea that some of the complexity of cellular responses to DNA damage in eukaryotes may relate to perturbations of chromatin structure, comes from recent studies on the product of the *RAD6* gene of *S. cerevisiae*. As evidenced by the pleiotropic phenotype of *rad6* mutants, the *RAD6* gene is apparently involved in diverse aspects of DNA metabolism, including post-replicative DNA repair, mutagenesis induced by a variety of DNA-damaging agents, and sporulation (Friedberg, 1988). The *RAD6* gene was cloned and sequenced by L. Prakash and her colleagues (see accompanying article in this volume). The presence of an extremely acidic polyaspartate carboxy terminus identified in the predicted amino acid sequence suggested that this protein might interact with histones (Reynolds *et al.*, 1985). More recently, the gene was independently isolated as the *UBC2* gene and it was shown that Rad6 protein is a ubiquitin-conjugating enzyme with apparent specificity for histones H2A and H2B (Jentsch *et al.*, 1987).

While this interesting observation invites obvious speculation about the role of chromatin remodelling during DNA repair, the details of such a role are not established. Furthermore, there are compelling experimental indications that the *RAD6* gene does not overlap with the *RAD3* epistasis group, *i.e.*, *RAD6* is not involved in nucleotide excision repair. Hence, if chromatin remodelling is a general phenomenon during cellular responses to DNA damage, multiprotein complexes for each of these responses must have different associated remodelling enzymes. At present no such function has been identified for the genes associated with nucleotide excision repair, *i.e.*, the *RAD3* epistasis group.

Components of Eukaryotic Repair Complexes are Involved in Other Cellular Functions

The *RAD6* gene product is required for mutagenesis, post-replication repair and sporulation. Other examples of multifunctional genes come from recent studies on members of the *RAD3* epistasis group.

The RAD3 *gene of* S. cerevisiae

As indicated earlier, a number of *rad3* mutants are highly UV sensitive and are known to be defective in nucleotide excision repair. To the extent that phenotypic characterization of these mutants has been pursued, these mutants have no other known defect. Other *rad3* mutants carrying specific alleles designated *rem* (Malone and Hoekstra, 1984; Hoekstra and Malone, 1987; Montelone *et al.*, 1988) are not particularly UV sensitive but are hypermutable and hyperecombinogenic, thereby defining a second function of the *RAD3* gene. Additionally, chain-terminating nonsense mutations, insertion mutations or extensive deletions of one or other end of the *RAD3* gene result in lethality, defining a third, essential function of the gene.

The relationships between the excision repair, recombination/mutation and essential functions of *RAD3* are not known. The ability to differentially inactivate these functions suggest that they are distinct, *i.e.*, that Rad3 protein is multifunctional and that different functions are required for repair, recombination/mutation and viability. An alternative possibility is that different amounts of a single Rad3 activity (*e.g.*, ATPase-associated DNA helicase) are required for these different aspects of DNA metabolism.

As is true of the excision repair function of Rad3 protein, it is not yet established which (if any) of the known biochemical attributes of this protein are associated with the essential function. I mentioned earlier that mutational inactivation of the putative nucleotide binding domain of the *RAD3* gene correlates with inactivation of excision repair and not with the function required for viability of yeast cells. If the ATPase and DNA helicase activities are biochemically linked, these results suggest that the DNA helicase is not required for viability.

The only *rad3* mutant thus far characterized with a defect exclusively in the essential function is a conditional-lethal mutant designated *rad3*-ts1 (Naumovski and Friedberg, 1987). This temperature-sensitive mutant has extremely limited growth capacity at elevated temperatures, and shows only slight UV sensitivity under these conditions. A single mutation has been identified in this strain that maps to codon position 73 in the Rad3 polypeptide (Naumovski and Friedberg, 1987). However, extensive phenotypic characterization of this mutant has not been particularly informative. Clearly, it would be interesting to examine the thermostability of the ATPase and DNA helicase activities of Rad3 protein purified from this strain. Additionally, there is an urgent need to isolate and characterize other conditional-lethal *rad3* mutants and their gene products.

A second mutant *rad3* allele defective in the essential function contains tandem missense mutations in codons 595 and 596 of the *RAD3* open reading frame (778 codons) (Naumovski and Friedberg, 1986). Plasmids carrying this allele are also unable to complement the UV sensitivity of excision-defective haploid cells, indicating a defect in the excision repair function as well. Nonetheless, if it were determined that mutants defective in just the excision repair function have normal DNA helicase activity, it would be informative to examine the helicase expressed from this allele.

One cannot eliminate the possibility that the essential function of *RAD3* is completely unrelated to the repair function. However, in the context of the present discussion one would guess that a likely candidate for this function is some aspect of DNA replication. A persuasive argument for such a relationship stems from the observation that DNA replication slows down or is arrested by bulky adducts in many organisms exposed to DNA damaging agents (Friedberg, 1985b). Such replicative arrest may result from the inability of the replication machinery to negotiate template strands containing non-instructional bulky base adducts. Alternatively (or additionally), replicative arrest may be a physiological response to DNA damage, that allows time for excision repair to be completed before replication resumes at normal rates. One way that such a response could be effected is by the overlapping function of one or more proteins in both repair and replication. Thus, if the amount of Rad3 protein were limiting for both functions, sequestration of the protein by a repair complex could limit DNA replication, thereby mitigating the potential for mutations associated with error-prone, trans-lesion DNA synthesis.

The acquisition of extensive nucleotide and amino acid sequence data bases, and the increasing sophistication of algorithms for seeking sequence relationships has fostered a growing interest in "computer biochemistry". Functional extrapolations from sequence information can be revealing, but are to be interpreted with caution in the absence of corroborating genetic and/or biochemical evidence. The admonition "don't expect your computer to tell you the truth" (Staden, 1988) must be taken seriously. With these reservations in mind I offer the following recent observations by Hodgman (1988). This author reported a set of 21 related proteins involved in nucleic acid replication (RNA or DNA) and/or recombination. All include the consensus nucleotide binding domain in the Rad3 sequence referred to earlier and in the sequences of the *E. coli* UvrA and UvrB excision repair proteins. However, all 21 polypeptides also include other conserved sequence domains not present in the Rad3, UvrA or UvrB proteins. The omission of the latter proteins from this set suggests that they are not involved in DNA replication!

A final speculation about the multifunctional *RAD3* gene concerns the Rem phenotype described above. The phenotype of hyperecombinogenicity and hypermutability is very reminiscent of that displayed by *E. coli mut* mutants defective in post-replicative mismatch repair of DNA (Radman and Wagner, 1986). In *E. coli* this process requires DNA helicase II, the product of the *uvrD* gene. Hence the DNA helicase function of Rad3 protein may be specifically required for mismatch correction in yeast.

In summary, the *RAD3* gene is extremely complex. It is absolutely required for damage-specific incision during nucleotide excision repair, for the viability of haploid yeast cells in the absence of DNA damage and for normal levels of recombination and mutagenesis. Rad 3 protein is a DNA-dependent ATPase and a DNA helicase, and likely binds ATP, and possibly other, nucleotides. Considerable further work is required to definitively establish the precise relationships between the functions of the gene and those of the protein.

The RAD2 *gene of* S. cerevisiae

Among the set of 5 *RAD* genes required for damage-specific incision of DNA during nucleotide excision repair, only the *RAD2* gene is inducible by exposure of cells to DNA-damaging agents (Robinson *et al.*, 1986; Madura and Prakash, 1986). Induction of the gene results in a 3-6 fold increase in the steady-state levels of *RAD2* mRNA. Available assays for Rad2 protein are not sensitive enough to directly determine whether induction results in a corresponding increase in the level of native protein. However, there is a corresponding increase in the level of β-galactosidase in cells transformed with *RAD2-lacZ* fusion genes.

The physiological significance of this inducible response is unclear. A UV sensitive strain carrying a mutation in the coding region of the *RAD2* gene was transformed with a centromeric plasmid carrying a deletion in the regulatory region required for induction, or with a wild-type control plasmid. Following exposure to UV radiation the latter cells showed a very modest enhancement of survival exclusively in the late logarithmic phase of growth (W. Siede and E. C. Friedberg, unpublished observations). Hence, induction of *RAD2* is not obviously required for nucleotide excision repair of DNA.

These results prompted an exploration of other phenotypic consequences of the inducible response. Expression of β-galactosidase from a *RAD2-lacZ* fusion plasmid increases when diploid cells are incubated under conditions that promote sporulation (W. Siede and E. C. Friedberg, unpublished observations). More direct experiments have demonstrated a significant increase in the steady-state levels of *RAD2* mRNA during meiosis; quantitatively greater than that detected after exposure of cells to DNA-damaging agents (W. Siede, T. Malley and E. C. Friedberg, unpublished observations). These experiments suggest a role for Rad2 protein in meiosis. However, the nature of this role is unclear. Diploid cells homozygous for *rad2* mutations sporulate normally. A more subtle defect such as reduced spore viability has not yet been extensively investigated (W. Siede, T. Malley and E. C. Friedberg, unpublished observations).

The RAD1 *gene of* S. cerevisiae

Like the *RAD3* and *RAD2* genes, *RAD1* is required absolutely for damage-specific incision of DNA in yeast cells (Friedberg, 1988). There are intriguing indications that this gene also plays a role in transcription. Keil and Roeder (1984) discovered a recombinational hot spot in the yeast genome designated *HOT1*. This region is located in the repeated ribosomal gene cluster and stimulates mitotic interchromosomal and intrachromosomal recombination about 7- and 50-fold, respectively, by a transcription-dependent mechanism, even when inserted into novel locations in the genome. The transcriptional dependence of this phenomenon has provoked models in which recombination is stimulated by localized unwinding of the helix associated with transcription, leading to regions of exposed single-stranded DNA (Volkel-Meiman *et al.*, 1987).

In an effort to determine which genes might influence *HOT1*-mediated recombination Keil and his associates (personal communication) examined a number of mutants defective in cellular responses to DNA damage. The *rad1*-2 allele had no effect on mitotic intrachromosomal recombination in the absence of *HOT1*. However, recombination was reduced 10-fold in its presence. Mutations in the *RAD3* or *RAD4* genes did not produce this effect, suggesting that this is not a general response to defective nucleotide excision repair.

The observation that the *RAD1* gene is specifically involved in *HOT1*-mediated mitotic recombination, and that the latter process is transcription-dependent, suggests that *RAD1* may play a role in transcription. However, disruption or deletion of *RAD1* is not lethal to haploid cells. Thus, such a putative function is presumably not essential. This particular role for a DNA repair protein is especially interesting in light of recent observations indicating that some component(s) of nucleotide excision repair in mammalian cells is transcriptionally linked. Specifically, it has been demonstrated that the rate and extent of excision repair in actively transcribed genes in Chinese hamster cells is significantly greater than in the genome overall or in silent genes (Bohr *et al.*, 1985; Madhani *et al.*, 1985), and in human cells the rate of loss of bulky adducts from the genome is greater in transcribed genes (Mellon *et al.*, 1986). Hanawalt and his collaborators have suggested that in mammalian cells an excision repair complex may be directly coupled to the transcription machinery (Hanawalt, 1988).

Preferential repair of transcriptionally-active genes has not been examined in yeast. However, the suggestion that *RAD1* may be involved in transcription leads to speculation that the protein encoded by this gene executes its function during repair by facilitating unfolding of the genome and that this function also stimulates transcription. The role of *RAD1* in preferential repair of actively transcribed yeast genes warrants close examination.

YEAST AS A EUKARYOTIC PARADIGM FOR EXCISION REPAIR OF DNA

Comparisons of the genetic complexity of nucleotide excision repair in yeast and higher eukaryotes (including human cells from the hereditary repair-defective disease xeroderma pigmentosum) have been discussed in detail elsewhere (Friedberg, 1985b), and I will not retread this ground here. Nonetheless, in the context of the present discussions it is relevant to explore experimental observations that support or detract from the utility of yeast as a general model for understanding the biochemistry of DNA repair in eukaryotes.

Structural and Functional Homology Between Yeast *RAD* and Human *ERCC* Genes

ERCC1 was the first human repair gene to be isolated by molecular cloning (Westerveld *et al.*, 1984). Its name derives from the ability to phenotypically complement excision repair-defective Chinese hamster ovary (CHO) cells with human DNA [hence, *e*xcision repair gene *c*ross *c*omplementing (Thompson *et al.*, 1988)]. A comparison of the amino acid sequence of the *ERCC1* cDNA revealed similarity with the yeast gene *RAD10* (van Duin *et al.*, 1986). There is an extended region of sequence similarity between the carboxy terminal two-thirds of Rad10 protein (amino acids ≈70-210) and a region of the Ercc1 polypeptide between amino acids ≈80-210. In this 138 amino acid overlap, 28.3 % of the residues are identical and this value increases to 39.4% in a region defined by a stretch of 73 amino acids that includes the carboxy terminal one-third of Rad10.

The observation of structural similarity between these two proteins prompted studies in my laboratory aimed at the demonstration of possible functional relatedness. The *RAD10* gene was expressed in CHO cells of genetic complementation group 1 (the particular genetic group phenotypically complemented by *ERCC1*). Increased resistance to killing of the CHO cells by UV radiation or by other DNA damaging agents was observed under conditions in which Rad10 protein was detectable in these cells by immunoblotting (Lambert *et al.*, 1988). Additionally, *RAD10*-transfectants demonstrated increased excision repair of UV irradiated plasmids containing the *E. coli cat* gene. This partial phenotypic complementation of defective repair was specific to CHO cells from genetic complementation group 1 (Lambert *et al.*, 1988).

More recently, the amino acid sequence derived from the cloned cDNA of a second *ERCC* gene [*ERCC2* (Weber *et al.*, 1988)] has been shown to share extensive homology with that of the yeast *RAD3* gene (C. Weber and L. H. Thompson, personal communication). The Ercc2 polypeptide is expected to be 760 amino acids long, just 18 amino acids shorter than the Rad3 polypeptide. The two proteins share 52.3% identity in amino acid sequence (C. Weber and L. H. Thompson, personal communication) and the overall similarity is about 84% (E. C. Friedberg, unpublished observations). This extensive evolutionary conservation suggests that like *RAD3*, *ERCC2* is an essential gene. Experiments designed to demonstrate functional complementation of various *rad3* mutants with the cloned *ERCC2* cDNA are in progress.

The demonstration that both of the sequenced human *ERCC* genes are homologous to yeast DNA repair genes suggests that many if not all yeast and mammalian genes for nucleotide excision repair are conserved, and that yeast is not only an experimentally convenient model system, but is in fact a biologically relevant eukaryotic paradigm. The generality of these observations awaits sequence comparisons between other cloned mammalian and yeast genes and attempts to complement the phenotype of CHO mutants from other genetic complementation groups with other yeast *RAD* genes.

CONCLUSIONS

The molecular mechanism and biochemistry of nucleotide excision repair in eukaryotes is not yet understood. The genetic framework established by study of the yeast *S. cerevisiae* indicates a profound complexity that poses a formidable challenge to students of nucleic acid enzymology. Indications are that at least 5 proteins must be purified to reconstitute a damage-specific DNA incising activity *in vitro*. At least 5 other gene products may be required to catalyze incision (and possibly excision) reactions at rates approximating those in living cells.

The molecular cloning of some of these genes (*RAD1*, *RAD2*, *RAD3*, *RAD4* and *RAD10*) has been achieved by complementation of the phenotype of marked UV sensitivity. A strategy for cloning the remaining genes of interest is less obvious, since mutants in these genes are considerably less sensitive to killing by genotoxic agents. Unfortunately, the amino acid sequences of the cloned genes have been relatively uninformative. With the exception of *RAD3*, in which the recognition of a nucleotide binding domain suggested possible ATPase and helicase functions, there are no compelling clues as to the function of the products of the other cloned *RAD* genes. Hence, further understanding of their functions must await purification and detailed characterization of these proteins.

The nature of the biochemical complexity of nucleotide excision repair, as well as other cellular responses to DNA damage in *S. cerevisiae* predicted by the extensive genetic characterization of this organism can only be guessed at. As far as nucleotide excision repair is concerned, the *E. coli* paradigm suggests that biochemical complexity derives principally from the building of a specific nucleoprotein substrate for damage-specific incision of DNA. I would predict that this fundamental principle obtains in eukaryotes as well. In the course of this review I have attempted to identify additional and alternative contributions to the biochemical complexity of DNA repair in eukaryotes, including mechanisms for remodelling chromatin to facilitate accessibility of repair complexes to nucleotides in DNA, and the organization of enzyme complexes for metabolic channelling during DNA repair. Finally, I have provided specific examples of multifunctional repair proteins in yeast.

It is likely that the biochemistry of DNA repair in eukaryotes is as complex as that of DNA replication and transcription. The availability of cloned genes from yeast and human cells and the recent establishment of a cell-free system for measuring repair in human cells is expected to facilitate progress in this area in the coming years.

ACKNOWLEDGEMENTS

Studies in the author's laboratory are supported by research grant CA-12428 from the U.S.P.H.S. I wish to acknowledge past and present collaborations with numerous graduate students and post-doctoral fellows whose enthusiastic interest in DNA repair in yeast has provided constant stimulation and challenge. Special thanks are due to Helmut Burtscher, Jane Cooper, Linda Couto, Louie Naumovski, Wolfram Siede and Bill Weiss for critical reading of the manuscript and for helpful discussions.

REFERENCES

D. G. Barker, J. H. M. White and L. H. Johnston. (1985) The nucleotide sequence of the DNA ligase gene (*CDC9*) from *Saccharomyces cerevisiae*: a gene which is cell-cycle regulated and induced in response to DNA damage. *Nucleic Acids Res.* **13**: 8323-8337

V. A. Bohr, C. A. Smith, D. S. Okomuto and P. C. Hanawalt. (1985) DNA repair in an active gene: removal of pyrimidine dimers from the *DHFR* gene of CHO cells is much more efficient than in the genome overall. *Cell* **40**: 359-369 .

G. Ciarrocchi and A. M. Pedrini. (1982) Determination of pyrimidime dimer unwinding angle by measurement of DNA electrophoretic mobility. *J. Mol. Biol.* **155**: 177-183

D. T. Denhart and A. C. Kato. (1973) Comparison of the effect of ultraviolet irradiation and ethidium bromide intercalation on the conformation of superhelical ϕX174 replicative form DNA. *J. Mol. Biol.* **77**: 479-494

H. Echols. (1986) Multiple DNA-protein interactions governing high-precision DNA transactions. *Science* **233**: 1050-1056

F. Foury and A. Lahaye. (1987) Cloning and sequencing of the PIF gene involved in repair and recombination of yeast mitochondrial DNA. *EMBO J.* **6**: 1441-1449

E. C. Friedberg. (1985a) *DNA Repair*. W. H. Freeman, New York

E. C. Friedberg. (1985b) Nucleotide excision repair of DNA in eukaryotes: comparisons between human cells and yeast. *Cancer Surv.* **4**: 529-555

E. C. Friedberg. (1988) Deoxyribonucleic acid repair in the yeast *Saccharomyces cerevisiae. Microbiol. Rev.* **52**: 70-102

L. Grossman, S. J. Mazur, P. R. Caron and E. Y. Oh. (1988) Dynamics of the *E. coli uvr* DNA repair system. **In**: *Mechanisms and Consequences of DNA Damage Processing.* E. C. Friedberg and P. C. Hanawalt (eds.) Alan R. Liss, New York. pp. 73-78.

P. C. Hanawalt. (1988) Retrospective perspectives on DNA repair. **In**: *Mechanisms and Consequences of DNA Damage Processing.* E. C. Friedberg and P. C. Hanawalt (eds.) Alan R. Liss, New York. pp. 1-6.

T. C. Hodgman. (1988) A new superfamily of replicative proteins. *Nature* **333**: 22-23

M. F. Hoekstra and R. E. Malone. (1987) Hyper-mutation caused by the *rem1* mutation in yeast is not dependent on error-prone or excision repair. *Mutation Res.* **178**: 201-210

S. Jentsch, J. P. McGrath and A. Varshavsky. (1987) The yeast DNA repair gene *RAD6* encodes a ubiquitin-conjugating enzyme. *Nature* **329**: 131-134

R. L. Keil and G. S. Roeder. (1984) Cis-acting, recombination stimulating activity in a fragment of the ribosomal DNA of *S. cerevisiae. Cell* **39**: 377-386

R. D. Kornberg. (1974) Chromatin structure: a repeating unit of histones and DNA. *Science* **184**: 868-871

R. D. Kornberg. (1977) Structure of chromatin. *Ann. Rev. Biochem.* **46**: 931-954

C. Lambert, L. B. Couto, W. A. Weiss, R. A. Schultz, L. A. Thompson and E. C. Friedberg. (1988) A yeast DNA repair gene partially complements defective excision repair in mammalian cells. *EMBO J.* **10**: 3245-3253.

R. J. Legerski, H. B. Gray, Jr. and D. L. Robberson. (1977) A sensitive endonuclease probe for lesions in deoxyribonucleic acid helix structure produced by carcinogenic and mutagenic agents. *J. Biol. Chem.* **252**: 8740-8746

T. Lindahl, B. Sedgwick, M. Sekiguchi and Y. Nakabeppu. (1988) Regulation and expression of the adaptive response to alkylating agents. *Ann. Rev. Biochem.* **57**: 133-158

H. D. Madhani, V. A. Bohr and P. C. Hanawalt. (1986) Differential DNA repair in transcriptionally active and inactive proto-oncogenes: *c-abl* and *c-mos. Cell* **45**: 417-423

K. Madura and S. Prakash. (1986) Nucleotide sequence, transcript mapping, and regulation of the *RAD2* gene of *Saccharomyces cerevisiae. J. Bacteriol.* **166**: 914-923

R. E. Malone and M. F. Hoekstra. (1984) Relationships between a hyper-rec mutation (*REM1*) and other recombination and repair genes in yeast. *Genetics* **107**: 33-48

C. K. Mathews, T. W. North and G. P. V. Reddy. (1979) Multienzyme complexes in DNA precursor biosynthesis. *Adv. Enz. Reg.* **17**: 133-156

S. W. Matson. (1986) *Escherichia coli* helicase II (uvrD gene product) translocates unidirectionally in a 3'→5' direction. *J. Biol. Chem.* **261**: 10169-10175

I. M. Mellon, V. A. Bohr, C. A. Smith and P. C. Hanawalt. (1986) Preferential DNA repair of an active gene in human cells. *Proc. Natl. Acad. Sci. USA.* **83**: 8878-8882

J. D. McGhee and G. Felsenfeld. (1980) Nucleosome structure. *Ann. Rev. Biochem.* **49**: 1115-1156

B. A. Montelone, M. F. Hoekstra, and R. F. Malone. (1988) Spontaneous mitotic recombination in yeast: the hyper-recombinational *rem1* mutations are alleles of the *RAD3* gene. *Genetics* **119**: 289-301

E. Moustacchi. (1987) DNA repair in yeast: genetic control and biological consequences. *Adv. Radiat. Biol.* **13**: 1-30

H. Nakayama and S. Shiota. (1988) Excision repair in *Micrococcus luteus*: evidence for a UvrABC homolog. **In**: *Mechanisms and Consequences of DNA Damage Processing.* E. C. Friedberg and P. C. Hanawalt (eds.) Alan R. Liss, New York. pp. 115-120.

L. Naumovski and E. C. Friedberg. (1986) Analysis of the essential and excision repair functions of the *RAD3* gene of *Saccharomyces cerevisiae* by mutagenesis. *Mol. Cell. Biol.* **6**: 1218-1227

L. Naumovski and E. C. Friedberg. (1987) The *RAD3* gene of *Saccharomyces cerevisiae*: isolation and characterization of a temperature-sensitive mutant in the essential function and of extragenic suppressors of this mutant. *Mol. Gen. Genet.* **209**: 458-466

E. Y. Oh and L. Grossman. (1987) Helicase properties of the *E. coli* UvrAB protein complex. *Proc. Natl. Acad. Sci. USA.* **84**: 3638-3642

D. K. Orren and A. Sancar. (1988) Subunits of ABC excinuclease interact in solution in the absence of DNA. **In**: *Mechanisms and Consequences of DNA Damage Processing.* E. C. Friedberg and P. C. Hanawalt (eds.) Alan R. Liss, New York. pp. 87-94.

D.A. Perlman, S. R. Holbrook, D. H. Pirkle and S-H Kim. (1985) Molecular models for DNA damaged by photoreaction. *Science* **227**: 1304-1308

M. Radman. (1980) Is there SOS induction in mammalian cells? *Photochem. Photobiol.* **32**: 823-830

M. Radman and R. Wagner. (1986) Mismatch repair in *Escherichia coli. Ann. Rev. Genet.* **20**: 523-538

P. Reynolds, S. Weber, and L. Prakash. *RAD6* gene of *Saccharomyces cerevisiae* encodes a protein containing a tract of 13 consecutive aspartates. *Proc. Natl. Acad. Sci. USA.* **82**: 168-172

G. W. Robinson, C. M. Nicolet, D. Kalainov and E. C. Friedberg. A yeast excision-repair gene is induced by DNA damaging agents. *Proc. Natl. Acad. Sci. USA.* **83**: 1842-1846

N. N. Shafranovskaya, E. N. Trifonov, Y. S. Lazurkin and M. D. Frank-Kamenetskii. (1973) Clustering of thymine dimers in ultraviolet irradiated DNA and the long-range transfer of electronic excitation along the molecule. *Nature New Biol.* **241**: 58-60

R. Staden. (1988) Genetic packages. *Nature* **333**: 27

P. Sung, L. Prakash, S. Weber and S. Prakash. The *RAD3* gene of *Saccharomyces cerevisiae* encodes a DNA-dependent ATPase. *Proc. Natl. Acad. Sci USA.* **84**: 6045-6049

P. Sung, L. Prakash, S. W. Matson and S. Prakash. *RAD3* protein of *Saccharomyces cerevisiae* is a DNA helicase. *Proc. Natl. Acad. Sci. USA.* **84**: 8951-8955

L. H. Thompson, T. Shiomi, E. P. Salazar and S. A. Stewart. An eighth complementation group of rodent cells sensitive to ultraviolet radiation. *Somatic Cell and Mol. Genet.* **14**: 605-612.

M. van Duin, J. deWit, H. Odijk, A. Westerveld, A. Yasui, M. H. M. Koken, J. H. J. Hoeijmakers and D. Bootsma. (1986) Molecular characterization of the human excision repair gene *ERCC1*: cDNA cloning and amino acid homology with the yeast DNA repair gene *RAD10*. *Cell* **44**: 913-923

B. Van Houten, H. Gamper, J. Hearst and A. Sancar. (1988) Factors influencing the assembly and turnover of ABC excinuclease. **In**: *Mechanisms and Consequences of DNA Damage Processing.* E. C. Friedberg and P. C. Hanawalt (eds.) Alan R. Liss, New York. pp. 79-86.

Voelkel-Meiman, K., Keil, R. L. and Roeder, G. S. (1987) Recombination stimulating sequences in yeast ribosomal DNA correspond to sequences regulating transcription by RNA polymerase I. *Cell* **48**: 1071-1079.

G. C. Walker. (1985) Inducible DNA repair systems. *Ann. Rev. Biochem.* **54**: 425-457

C. A. Weber, E. P. Salaz, S. A. Stewart and L. H. Thompson. (1988) Molecular cloning and biological characterization of a human gene, *ERCC2*, that corrects the nucleotide excision repair defect in CHO UV5 cells. *Mol. Cell. Biol.* **8**: 1137-1146

A. Westerveld, J. H. J. Hoeijmakers, M. van Duin, J. deWit, H. Odijk, A. Pastink and D. Bootsma. (1984) Molecular cloning of a human DNA repair gene. *Nature* **310**: 425-429

E. M. Witkin. (1976) Ultraviolet mutagenesis and inducible DNA repair in *Escherichia coli*. *Bacteriol. Rev.* **40**: 869-907

11

STRUCTURE AND FUNCTION OF *RAD3, RAD6* AND OTHER DNA REPAIR GENES OF *SACCHAROMYCES CEREVISIAE*

Satya Prakash, Patrick Sung, and Louise Prakash

PERSPECTIVES AND SUMMARY

In *Escherichia coli*, the role of various genes in the repair of ultraviolet (UV) light damaged DNA is now fairly well understood. The *uvrA*, *uvrB*, and *uvrC* gene products are required for incision of damaged DNA strands (Sancar and Rupp, 1983; Yeung *et al.*, 1983) and uvrD and DNA polymerase I effect the dissociation of the uvrABC incision complex from the damage site in DNA (Caron *et al.*, 1985; Husain *et al.*, 1985). The RecA protein acts as a positive regulator of the SOS regulatory pathway and participates directly in homologous pairing and strand exchange in recombination (Walker, 1985). In contrast, our understanding of DNA repair processes in eukaryotes is still in the rudimentary stages.

The yeast *Saccharomyces cerevisiae* provides a good model system for gaining an understanding of the DNA repair processes in higher eukaryotes. The small genome size, ease of genetic analyses, and the feasibility of targeting cloned DNA sequences to homologous sites in the genome are some of the features that make *S. cerevisiae* an excellent organism for the study of DNA repair and other processes. Genetic analyses of DNA repair mutants revealed the existence of three competing pathways for the repair of UV damage (Game and Cox, 1972, 1973; Cox and Game, 1984). These correspond to the *RAD3* pathway, which mediates excision repair, the *RAD6* pathway, which affects postreplication repair and mutagenesis, and the *RAD52* pathway, involved in double strand break repair and genetic recombination. During the past few years, studies of the cloned DNA repair genes have provided some interesting information. The *RAD1*, *RAD2*, *RAD3*, *RAD4*, and *RAD10* genes of *S. cerevisiae* are absolutely required for incision of UV damaged DNA and, in this respect, they resemble the *E. coli uvrA*, *uvrB*, and *uvrC* genes. But in contrast to these *E. coli* genes, whose function is limited to incision, the yeast genes appear to be multifunctional; for example, the *RAD3* gene is essential for cell viability (Higgins, *et al.*, 1983; Naumovski and Friedberg, 1983), and the *RAD1* gene is involved in mitotic recombination (Schiestl and Prakash, 1988). It would not be surprising if future investigations reveal additional roles in other cellular processes for yeast DNA repair genes. The recent observation that RAD6 protein required for postreplication repair, mutagenesis, and sporulation is a ubiquitin conjugating enzyme that attaches ubiquitin to histones H2A and H2B attests to the intricacies involved in DNA repair processes in eukaryotes. It is hoped that the studies of DNA repair in yeast, in addition to providing information on the enzymes and proteins that directly participate in DNA repair, will also reveal pathways of chromatin remodeling during DNA repair and in related processes.

The work in our laboratories has focused primarily on the genes required for excision repair and for postreplication repair and mutagenesis. This review will briefly summarize the information on various cloned genes in the *RAD3* and *RAD6* epistasis groups. The proteins encoded by the *RAD3* and *RAD6* genes have been successfully overproduced and purified from yeast and biochemically characterized. These studies will be discussed. The *RAD50* through *RAD57* genes will not be covered in this review. Various aspects of DNA repair in yeast have been previously reviewed (Haynes and Kunz, 1981; Lawrence, 1982; Game, 1983; Moustacchi, 1987; Friedberg, 1988).

EPISTASIS GROUPS

In *S. cerevisiae*, mutations in genes *RAD1* through *RAD24* cause sensitivity to UV light (Cox and Parry, 1968) and to various other DNA damaging agents depending on the gene, whereas mutations in genes *RAD50* through *RAD57* cause greatly enhanced sensitivity to ionizing radiation and only a weak level of sensitivity to UV light (Game and Mortimer, 1974). A third group of mutants, *mms1* through *mms22*, were isolated on the basis of sensitivity to the alkylating agent methyl methanesulfonate (MMS) (Prakash and Prakash, 1977). Mutants sensitive to other DNA damaging agents have also been identified in yeast (Henriques and Moustacchi, 1980; Ruhland *et al.*, 1981). Some information about the repair pathways in which different genes function was gleaned from a comparison of the sensitivity of the single, double and triple mutant strains. Two genes interact in an epistatic manner if the double mutant between them is no more sensitive to the DNA damaging agent than either single mutant. Epistasis suggests that either the products of these genes form a multiprotein complex, or that they function in a repair pathway in which the action of one gene product is dependent upon a reaction carried out by another gene product. If genes belong to different epistasis groups then, due to impairment of alternate repair pathways, the sensitivity of the double mutant is much greater than that of either single mutant.

For UV repair, yeast genes have been classified into three epistasis groups which are usually referred to by the prominent member in that group, viz., (1) the *RAD3* epistasis group, (2) the RAD6 epistasis group, and (3) the RAD52 epistasis group.

THE *RAD3* EPISTASIS GROUP: ROLE IN EXCISION REPAIR

Ten genes: *RAD1, RAD2, RAD3, RAD4, RAD10, RAD7, RAD23, RAD14, RAD16,* and *MMS19* belong to the *RAD3* epistasis group and these genes are involved in excision repair of DNA damage induced by UV light and by other DNA damaging agents, such as DNA crosslinking agents, that distort the DNA helix (Unrau *et al.* 1971; Waters and Moustacchi, 1974; Prakash, 1975, 1977a, b; Reynolds and Friedberg, 1978; Prakash and Prakash, 1979; Jachymczyk *et al.*, 1981; Miller *et al.*, 1982a, b; Resnick and Setlow, 1982; Miller *et al.*, 1984;). These genes are functionally similar to the *E. coli uvrA, uvrB,* and *uvrC* genes and human xeroderma pigmentosum genes. Based upon their sensitivity to DNA damaging agents and defect in excision repair, these genes can be classified into two groups. Mutations in the *RAD1, RAD2, RAD3, RAD4,* and *RAD10* genes render cells highly sensitive to DNA damage whereas mutations in the *RAD7, RAD23, RAD14, RAD16,* and *MMS19* genes cause only a moderate level of sensitivity. As expected of mutations in genes belonging to the same epistasis group, double mutant combinations between *rad1, rad2, rad3, rad4,* and *rad10* show the same level of sensitivity to UV light as any of these single mutants (Game and Cox, 1972). However, the *rad7 rad14* and *rad7 rad23* double mutants show a higher level of sensitivity to UV light or to crosslinking agents than their single mutant counterparts (Prakash and Prakash, 1979; Miller *et al.*, 1982b; Perozzi and Prakash, 1986). The defect in excision repair in these mutants roughly parallels the level of their sensitivity. The *rad1, rad2, rad3, rad4,* and *rad10* mutants are highly defective in the excision of UV induced pyrimidine dimers and of psoralen plus light induced DNA crosslinks. With the exception of the *mms19* mutation, where the high level of excision defect does not correlate with the moderate level of sensitivity of *mms19* strains, the excision repair defect of the other mutant strains *rad7, rad23, rad14,* and *rad16* corresponds to their sensitivity level. Furthermore, in parallel with enhanced sensitivity of *rad7 rad23* double mutants, one observes a greater excision defectiveness in the double mutant strain than in either of the single mutant strains (Miller *et al.*, 1982b).

Even though the mutants carrying mutations in genes in the *RAD3* epistasis group were known to be defective in the removal of pyrimidine dimers, it was not clear from earlier studies which mutants were specifically defective in the incision of damaged DNA. This ambiguity arises because in a mutant defective in a postincision step of excision repair, DNA ligase may seal the nick, resulting in retention of damage in DNA, thereby making it difficult to distinguish between mutants defective in the incision *vs.* the postincision step. Only a few incision nicks are observed in DNA from UV irradiated *RAD*$^+$ cells, indicating that nicks formed in UV damaged DNA are repaired rapidly, whereas no nicks are observed in DNA obtained from UV irradiated *rad1, rad2, rad3,* and *rad4* mutants, suggesting that these mutants are defective in the incision mechanism (Reynolds and Friedberg, 1981). To circumvent the problem of rapid repair of incision nicks in UV damaged DNA, Wilcox and Prakash (1981) examined nicking of

UV damaged DNA in various double mutant strains in which excision defective *rad* mutations were coupled to the DNA ligase defective *cdc9* mutation. The *cdc9* mutation results in thermolabile DNA ligase activity and temperature sensitive growth. Due to the presence of the *cdc9* mutation, at the restrictive temperature no ligation would occur in mutants defective in a postincision step(s); therefore, one could distinguish between mutants defective in the incision *vs.* the postincision step(s). These studies clearly showed that the *rad1*, *rad2*, *rad3*, *rad4*, and *rad10* mutants are defective in incision of UV irradiated DNA, whereas reduced levels of incision occur in the *rad14-1* and *rad16-1* mutant strains (Wilcox and Prakash, 1981).

In contrast to the short half life of nicks formed in the UV damaged DNA, incision nicks in crosslinked DNA persist for considerably longer periods, making it easier to determine whether a mutant is defective in the incision or postincision step(s). Alkaline sucrose sedimentation of DNA obtained from yeast cells treated with the crosslinking agent 4,5′,8-trimethyl psoralen plus 360 nm light irradiation revealed that the *rad1*, *rad2*, *rad3*, *rad4*, *rad10*, and *mms19* mutants are highly defective in incision of crosslinked DNA, whereas incision was reduced in the *rad14-1* mutant, and the *rad16-1* mutant showed the *RAD*$^{+}$ level of incision proficiency (Miller *et al.*, 1982a).

The fact that incision of damaged DNA is highly defective in the *rad1*, *rad2*, *rad3*, *rad4*, and *rad10* mutants, which are also very sensitive to DNA damage, suggests that the products of these genes may be intimately involved in the incision reaction. The role(s) of the remaining genes, *RAD7*, *RAD23*, *RAD14*, *RAD16*, and *MMS19*, in excision repair is less apparent. Some of these genes may be involved in allowing access of the damaged DNA to the repair complex. In mammalian cells, pyrimidine dimers are removed more efficiently from the actively transcribed genes than from the inactive genes (Bohr *et al.*, 1985; Leadon, 1986; Madhani *et al.*, 1986; Mellon *et al.*, 1986), and the transcribed strand is subject to preferential repair (Mellon *et al.*, 1987). There may be protein factors in the repair complex that interact with components of the transcriptional apparatus. Another possible role for some of these genes could be in repair synthesis or in turnover of the incision complex from the damage site.

STRUCTURE AND FUNCTION OF GENES IN THE *RAD3* EPISTASIS GROUP

The *RAD3* Gene

The *RAD3* gene of *S. cerevisiae* is of particular interest among the excision repair genes because of its multiple roles in excision repair and involvement in the maintenance of cell viability. *RAD3* is the only gene in the excision repair group whose encoded protein has been extensively purified and characterized.

Role in Incision and Postincision Steps of Excision Repair and in Cell Viability

The *rad3* mutations such as *rad3-1* and *rad3-2* that were isolated by screening for UV sensitive mutants of *S. cerevisiae* are unable to carry out incision of DNA damaged by UV light or by DNA crosslinking agents (Reynolds and Friedberg, 1981; Wilcox and Prakash, 1981; Miller *et al.*, 1982a). An unsuspected role of *RAD3* emerged when the cloned *RAD3* gene was used to make genomic rad3Δ mutations and these mutations were shown to cause a recessive lethal phenotype (Higgins, *et al.*, 1983; Naumovski and Friedberg, 1983), indicating that in addition to its role in excision repair, *RAD3* performs an essential function in the maintenance of cell viability. In contrast, deletion mutants of the excision repair genes *RAD1*, *RAD2*, *RAD4*, *RAD10*, *RAD7*, and *RAD23*, and of various other DNA repair genes are viable.

The *RAD3* gene encodes a protein of 778 amino acids with an M_r of 89,779, containing 15.3% acidic and 14.5% basic residues (Naumovski *et al.*, 1985; Reynolds *et al.*, 1985a). Toward the amino terminal region, *RAD3* contains a "Walker type A" consensus amino acid sequence found in proteins that bind and hydrolyze purine nucleotides. The sequence Gly-X-Gly-Lys-Thr present in RAD3 between amino acids 45 and 49 (Reynolds *et al.*, 1985a) represents the highly conserved core of this sequence, and Gly 47 and Lys 48 are invariant residues in this sequence (Walker *et al.*, 1982). Site-directed mutagenesis of Gly-47 to Asp and Lys 48 to Glu inactivates the DNA repair function but not the essential function of *RAD3*, as the mutant cells are UV sensitive (Naumovski and Friedberg, 1986). In our laboratory, we mutated the Lys 48 residue to the similarly charged Arg residue to yield the *rad3 Arg-48* mutation. Interestingly, analysis of excision repair in the *rad3 Arg-48* mutant revealed that *RAD3* is also involved in a step(s) subsequent to the incision of UV damaged DNA. The *rad3 Arg-48* mutant is as UV sensitive as the incision defective *rad3-2* mutant (Higgins, 1987). The incision capacity of the *rad3 arg-48* mutation was examined by coupling it with the *cdc9* mutation, which is defective in DNA ligase activity at the restrictive temperature of 36°C (Sung *et al.*, 1988a). The *rad3 Arg-48 cdc9* double mutants were irradiated with UV light and, following a period of incubation at the restrictive temperature, DNA from the mutant cells was examined on alkaline sucrose gradients. These studies provided evidence for a limited level of incision of UV damaged DNA in the *rad3 Arg-48 cdc9* cells. However, even though incision of damaged DNA occurs in the *rad3 Arg-48* cells, the mutant is totally defective in the removal of pyrimidine dimers. These observations indicate that the *rad3 Arg-48* mutation affects a post incision step in excision repair (Sung *et al.*, 1988a). In the *rad3 Arg-48* cells, in the absence of a post incision step such as the release of the incision enzyme complex from the site of damage in the DNA or strand displacement synthesis occurring after incision, joining of incision nicks by DNA ligase could lead to retention of pyrimidine dimers in DNA.

Genetic analysis of the only available *rad3 ts-1* mutation has provided no clues concerning the nature of the essential function of the *RAD3* gene (Naumovski and Friedberg, 1987). At the restrictive temperature, the *rad3 ts-1* cells undergo 2-4 cell doublings and stop division at different cell cycle stages rather than at a specific stage. DNA, RNA, and protein syntheses at the restrictive temperature, as judged by the incorporation of radiolabeled precursors into these macromolecules, is not affected by the *rad3 ts-1* mutation. It is possible that *rad3 ts-1* is a leaky mutation, and therefore, isolation of other tighter *rad3 ts* mutations might be helpful in unraveling the essential function of *RAD3*.

The *rem1* mutation confers a semidominant, hyperrecombinational and hypermutable phenotype in mitotic cells. This mutation has recently been localized to the *RAD3* gene (Montelone *et al.*, 1988). Double mutants between *rem1* and *rad50* or *rad52* genes, which are both involved in double strand break repair and in genetic recombination, are inviable, whereas the triple mutant *rem1 rad1 rad52* is viable (Malone and Hoekstra, 1984; Montelone *et al.*, 1988). To explain these observations, Malone and Hoekstra (1984) suggested that during DNA replication, rem1 introduces lesions in DNA, which are acted upon by the excision repair proteins, resulting in double strand breaks. An alternative explanation of these observations is that the excision repair complex containing the mutant rem subunit has lost the specificity to bind and incise only the damaged DNA. Incision of undamaged DNA by the rem repair complex could account for these observations and also for the hyperrecombination observed in the *rem* mutant and the lowered recombination in the *rem rad1* mutants (Schiestl and Prakash, 1988).

Purification of the RAD3 *protein*

To facilitate the purification of RAD3 protein for biochemical characterization, the *RAD3* open reading frame (ORF) was placed downstream of the yeast *ADC1* promoter to overproduce the protein, and antibodies were raised against a truncated RAD3 protein obtained from *E. coli* (Sung *et al.*, 1987a). In immunoblotting experiments using affinity purified anti-RAD3 antibodies, yeast strains harboring the chimeric *ADC1-RAD3* plasmid were found to contain 30-50 fold the amount of RAD3 protein detected in strains harboring the *ADC1* vector without the RAD3 insert. RAD3 protein thus overproduced was purified to apparent homogeneity by a combination of ammonium sulfate precipitation and column chromatography on DEAE Sephacel, Bio-Rex 70, DNA agarose, hydroxyapatite, and MonoQ. The M_r of purified RAD3 protein, as judged by $NaDodSO_4$/PAGE is 89 kDa, which is in close agreement with the predicted M_r of 89,779 (Sung *et al.*, 1987a).

RAD3 *protein is a DNA helicase*

Purified RAD3 protein is a single stranded DNA dependent ATPase (Sung *et al.*, 1987a). The *RAD3* ATPase activity coelutes with the RAD3 protein in various column fractions, and ATPase activity is partially inhibited by the anti-RAD3 antibodies raised against a denatured and truncated RAD3 protein. The RAD3 ATPase activity absolutely requires single stranded DNA, and UV irradiation of single-stranded DNA does not affect the ATPase activity. No ATPase activity is seen with double-stranded DNA (Sung *et al.*, 1987a) and with polyribonucleotides (Sung and Prakash, 1988). The ATPase activity requires Mg^{2+} or Mn^{2+} and Ca^{2+} or Zn^{2+} do not support this activity. RAD3 has an approximate K_m for ATP of 67 μM (Sung *et al.*, 1987a).

The RAD3 ATPase activity could supply the energy for translocation of RAD3 on single-stranded DNA and for unwinding the duplex DNA. To examine the possibility of helicase activity in RAD3, we examined the displacement of 3′ end labeled single-stranded DNA fragments that were annealed to circular single-stranded M13 DNA (Figure 1A). The RAD3 protein catalyzes the unwinding of all the examined DNA fragments that varied in length from 71-851 nucleotides (Sung *et al.*, 1987b). The pH optimum for both RAD3 ATPase and helicase activities is about 5.6. Both ATP and dATP support the RAD3 helicase activity. No unwinding occurs with ATP(γ)S or other NTPs or dNTPs. Like the ATPase activity, the helicase activity requires Mg^{2+} or Mn^{2+}. All DNA helicases move unidirectionally on single-stranded DNA and begin the unwinding of duplex DNA from a specific direction. To examine the polarity of unwinding by RAD3, we used a linear DNA substrate that contains terminal duplex regions of 143 bp and 202 bp and a long central single-stranded region of about 7 kb in length (Figure 1B). Only the 202 nucleotide fragment is displaced by RAD3, indicating that RAD3 binds the central single-stranded DNA and moves along it in the 5′→3′ direction, resulting in unwinding of the 202 nucleotide fragment (Sung *et al.*, 1987b).

Role of RAD3 *ATPase and DNA helicase activities in a postincision step*

To understand the biological role of the RAD3 ATPase and helicase activities, we examined the rad3 Arg-48 mutant protein in which the Lys-48 residue in the consensus Walker type A sequence has been changed to arginine (Sung *et al.*, 1988a). The rad3 Arg-48 mutant protein was overproduced by fusion of the rad3 mutant gene with the *ADC1* promoter and the rad3

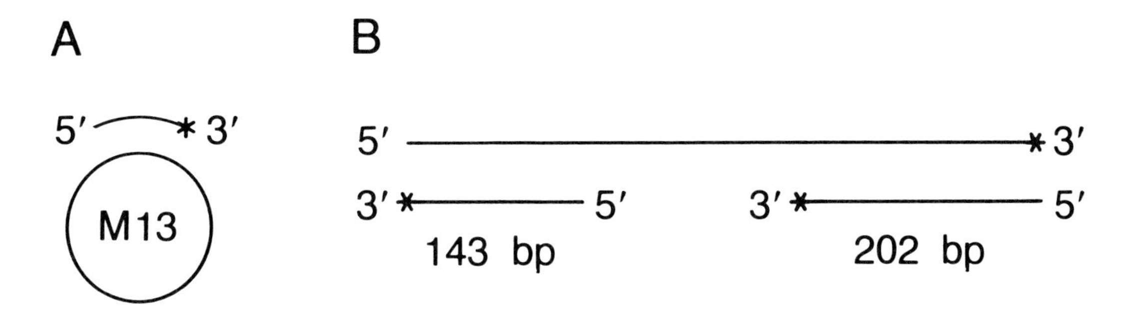

Figure 1. Helicase substrates used in this study. The substrate in (A) is a partial duplex of M13mp7 single-stranded DNA and one of a number of complementary fragments of different lengths. The linear substrate in (B), contains terminal duplex regions of 143 bp and 202 bp and a long central single-stranded region of approximately 7000 nucleotides. Unwinding of duplex DNA in the 3′ to 5′ direction or 5′ to 3′ direction with respect to the bound single-stranded DNA will result in the displacement of the 143 base fragment or 202 base fragment, respectively. Asterisk denotes the position of the ^{32}P label. This figure is taken from Sung, P., L. Prakash, S. W. Matson, and S. Prakash. (1987) *Proc. Natl. Acad. Sci. USA* **84**:8951-8955.

Arg-48 mutant protein purified from yeast cells. The rad3 Arg-48 mutant protein lacks any single-stranded DNA dependent ATPase activity or DNA helicase activity (Sung *et al.*, 1988a), indicating that ATP hydrolysis provides the energy for the RAD3 helicase activity. The rad3 Arg-48 mutant protein, however, binds single-stranded DNA with the same affinity as the RAD3 protein. Thus, the absence of the ATPase and helicase activities in the mutant protein is not due to its inability to bind single-stranded DNA. The rad3 Arg-48 protein retains approximately 40% as much ATP binding activity as the RAD3 protein (Sung *et al.*, 1988a). As has been mentioned previously, since the *rad3 Arg-48* mutation causes a post incision defect in excision repair, the RAD3 ATPase and DNA helicase activities could be needed for the release of the incision complex from DNA and for strand displacement repair synthesis. In this regard, the role of RAD3 could be similar to that of the uvrD helicase of *E. coli* (Caron *et al.*, 1985; Husain *et al.*, 1985).

Our studies with the rad3 Arg-48 mutant protein establish the role of the invariant lysine residue in the Walker type A sequence in ATP binding and hydrolysis. NMR studies with adenylate kinase had indicated that ionic interaction could occur between the invariant lysine residue in the conserved sequence and the α phosphoryl group of the bound nucleotide (Fry *et al.*, 1986). Our studies show that replacement of the invariant lysine residue by arginine abolishes the RAD3 ATPase and DNA helicase activities but not the ability to bind ATP. Replacement of the invariant lysine in the human Ha-ras encoded protein with asparagine reduces the affinity of the protein for GDP and GTP by 100-fold (Sigal *et al.*, 1986). Thus, the lysine residue is specifically required for nucleotide hydrolysis but it can be replaced by the similarly charged arginine residue for nucleotide binding.

Many questions remain about the role of RAD3 in excision repair and in cell viability: for instance, what is the essential role of RAD3? Does RAD3 possess different activity domains and does it interact with different protein complexes in DNA repair and in its role in viability? What is the nature of involvement of RAD3 in the incision and post incision reactions?

The *RAD1* Gene

RAD1 encodes a protein of 1,100 amino acids of M_r 126,360, containing 15.8% acidic and 14.7% basic residues (Reynolds *et al.*, 1987). In addition to its role in excision repair, *RAD1* is also involved in mitotic recombination (Schiestl and Prakash, 1988). Intrachromosomal recombination of a *his3* duplication, in which one *his3* gene is deleted at the 3′ end (*his3*Δ3′) and the other deleted at the 5′ end (*his3*Δ5′), is lowered about four-fold by the *rad1*Δ mutation and about 7-fold by the *rad52*Δ mutation compared to the RAD^+ strain (Schiestl and Prakash, 1988). In the *rad1*Δ *rad52*Δ double mutant, recombination between the *his3* genes is reduced synergistically, almost 200-fold, indicating that the *RAD1* and *RAD52* genes provide alternate pathways for recombination of the *his3* duplication. The *rad1*Δ mutation does not affect the frequency of intrachromosomal and interchromosomal recombination between heteroalleles that differ by point mutations rather than deletions. The *RAD1* gene, however, affects the frequency of integration of linear DNA fragments into homologous sites in the yeast genome. The effect of *RAD1* on recombination is specific rather than a characteristic shared by all other excision repair proteins (Schiestl and Prakash, 1988). Different domains in *RAD1* could account for its involvement in excision repair and recombination.

The *RAD2* Gene

The *RAD2* gene encodes a protein of 1,031 amino acids of M_r 117,847 and it contains 17.8% acidic and 15.3% basic residues (Madura and Prakash, 1986) The carboxyl terminus of RAD2 is basic and its deletion results in loss of RAD2 function. The *RAD2* gene shows regulated expression upon exposure of yeast cells to DNA damaging agents (Madura and Prakash, 1986; Robinson *et al.*, 1986). The only other *RAD* gene known to show induced expression upon DNA damage is *RAD54*, which is involved in double strand break repair and genetic recombination (Cole *et al.*, 1987).

Several other DNA damage inducible genes have been identified in yeast. These include the six *DIN* genes (Ruby and Szostak, 1985) and two *DDR* genes (McClanahan and McEntee, 1984), which were isolated by directly screening for DNA damage inducible sequences. The induction kinetics of two *DDR* genes, *DDRA2* and *DDR48*, has been examined in the *rad3-2*, *rad6*-Δ, and *rad52-1* mutant strains. The level of induction of the *DDRA2* transcript is greatly reduced in the *rad3* mutant strain while the *rad6* and *rad52* mutations lower the induction level of the *DDR48* transcript (Maga *et al.*, 1986). It has been suggested that the *RAD3*, *RAD6*, and *RAD52* genes affect the induction of *DDR* transcripts indirectly by controlling chromatin structure (Maga *et al.*, 1986). The yeast transposable element *TY1* (McClanahan and McEntee, 1984), ubiquitin encoding gene *UBI4* (Treger *et al.*, 1988), and the gene encoding the small subunit of ribonucleotide reductase (Elledge and Davis, 1987; Hurd *et al.*, 1987) are also induced upon DNA damage. The DNA ligase *CDC9* gene and the DNA polymerase I gene show periodic expression during the cell cycle, reaching a peak at the G_1/S phase boundary; both of these genes show induced expression in response to DNA damage (Peterson *et al.*, 1985; Johnson *et al.*, 1986; White *et al.*, 1986; Johnston *et al.*, 1987).

At present, it is not known whether yeast DNA damage-inducible genes are coordinately regulated like the SOS genes of *E. coli*. The role of the *DIN* and *DDR* genes in DNA repair also remains unknown.

The *RAD4* Gene

The *RAD4* gene is toxic to *E. coli*; therefore, when plasmids carrying *RAD4* are propagated in *E. coli*, only a mutationally inactivated *rad4* gene is recovered (Fleer *et al.*, 1987). Because of this problem, the wild type RAD4 gene could not be cloned using yeast genomic libraries that had been propagated in *E. coli*. The *RAD4* gene has been recently cloned directly from the yeast genome (Fleer *et al.*, 1987; Gietz and Prakash, 1988). The RAD4 gene encodes a protein of 754 amino acids of M_r 87,173 (Gietz and Prakash, 1988). In contrast to the RAD1, RAD2, and RAD3 proteins, RAD4 protein is basic, and it contains 18.7% basic and 16.4% acidic residues. Two features in RAD4 suggest that it interacts with DNA. It contains six clusters of 4-7 basic amino acids and an α-helix-turn-α-helix DNA binding motif. The carboxyl terminus of RAD4 is very acidic, there being 19 acidic and 2 basic residues in the last 44 amino acids. The carboxyl terminus of RAD3 is also highly acidic, with 12 acidic and 1 basic residues in the last 20 amino acids. Replacement of the RAD3 acidic sequence with another sequence which is slightly basic does not affect RAD3 function (Reynolds *et al.*, 1985a). It remains to be seen if these acidic sequences are required for interaction with histones and whether simultaneous deletion of these sequences from RAD3 and RAD4 will affect the efficiency of excision repair.

The *RAD10* Gene

RAD10 encodes a protein of 210 amino acids of M_r 24,310, containing 13.3% basic and 11.0% acidic residues (Reynolds *et al.*, 1985b). The middle portion of RAD10, amino acid residues 78-159, possesses features that indicate its involvement in DNA binding. This region is highly basic; it contains 17 basic and only three acidic residues. Eight of the ten tyrosine residues present in RAD10 occur in this region; these tyrosines could be involved in DNA binding by intercalating between the bases, as has been observed for the bacteriophage fd gene 5 and T4 gene 32 proteins (Anderson *et al.*, 1975a,b; McPherson *et al.*, 1979). The RAD10 and RAD4 proteins show considerable sequence similarity between amino acids 33 - 56, and 111 - 135, respectively (Friedberg, 1988). RAD10 shows extensive sequence homology with the human excision repair protein encoded by *ERCC1*, indicating evolutionary conservation of DNA repair genes in eukaryotes (van Duin *et al.*, 1986).

The *RAD7* and *RAD23* Genes

The *RAD7* gene encodes a protein of 565 amino acids with an M_r of 63,705 and it contains 14.9% acidic and 13.8% basic residues (Perozzi and Prakash, 1986). The amino terminus of RAD7 is highly hydrophilic, while the carboxyl terminus is quite hydrophobic (Perozzi and Prakash, 1986). It had been suggested that *RAD7* and *RAD23* genes belong to two different gene clusters that are related by duplication (McKnight *et al.*, 1981). However, the *RAD7* and *RAD23* genes, even in multicopy plasmids, do not show any evidence of complementation of the UV sensitivity conferred by the *rad23*Δ and *rad7*Δ mutations, respectively (Perozzi and Prakash, 1986); furthermore, the nucleotide sequences of these two genes are also quite dissimilar

(Perozzi and Prakash, 1986; Melnick and Sherman, personal communication). Interestingly, a *rad7* gene in which the first 99 amino acids have been deleted, when cloned in a 2 μ multicopy plasmid, elicits wild type level of complementation of the *rad7Δ* mutation, but in the *rad7Δ rad23Δ* double mutant strain, this amino terminally deleted *rad7* gene shows no complementing activity. These observations suggest that the RAD7 and RAD23 proteins interact physically.

THE *RAD6* EPISTASIS GROUP

Mutations in genes belonging to the *RAD6* epistasis group confer sensitivity to a wide variety of DNA damaging agents, including UV, ionizing radiation, and alkylating agents. This epistasis group consists of the *RAD6, RAD18, RAD9, RAD15, REV1, REV2, REV3*, and *MMS3* genes (Haynes and Kunz, 1981; Martin *et al.*, 1981; Lawrence, 1982; Game, 1983; Prakash and Prakash, unpublished observations). Genes in this group are involved in a variety of functions, including postreplication repair of UV damaged DNA, DNA strand break repair, and mutagenesis. *RAD6* is the most pleiotropic among these genes, and is required for DNA repair, mutagenesis, and sporulation. Mutations in the *RAD18* gene affect DNA repair but not mutagenesis. *RAD9* has recently been shown to affect DNA repair indirectly, by controlling the G_2 arrest of cells that occurs in eukaryotic cells in response to DNA damage (Weinert and Hartwell, 1988; Schiestl *et al.*, 1989). The role of *RAD15* is not known. The *REV1, REV2, REV3*, and *MMS3* genes are involved in mutagenic repair. Mutations in these genes lower mutagenesis induced by DNA damaging agents such as UV and ionizing radiation (Lawrence, 1982). Among the three *REV* genes, mutations in *REV3* have the most pronounced influence on mutagenesis (Lemontt, 1971). The *MMS3* gene is required for UV mutagenesis in a/α diploids but not in a/a or α/α diploids, or in *MAT*a or *MAT*α haploids, indicating an involvement of *MMS3* in mating type control of mutagenesis (Martin *et al.*, 1981). UV mutagenesis is also decreased by mutations in the cell division cycle genes *CDC7* and *CDC8* (Prakash *et al.*, 1979; Njagi and Kilbey, 1982). The *CDC7* encoded protein shows homology to protein kinases (Patterson *et al.*, 1986), and *CDC8* encodes thymidylate kinase (Sclafani and Fangman, 1984).

In contrast to UV mutagenesis of nuclear genes, for which there is a strong requirement of the *RAD6* and *REV3* genes, UV mutagenesis of mitochondrial DNA is not dependent on these two genes (Polakowska *et al.*, 1983). The *CDC8* gene, however, is apparently needed for UV mutagenesis of both nuclear and mitochondrial DNA (Prakash *et al.*, 1979; Polakowska *et al.*, 1983). It would be of interest to determine whether yeast mitochondrial DNA polymerase shows an enhanced capacity for mutagenic bypass of UV induced lesions in template DNA.

The *RAD6* Gene

Rad6 mutants are extremely sensitive to a variety of DNA damaging agents, such as UV light, ionizing radiation, and alkylating agents (Cox and Parry, 1968; Game and Mortimer, 1974; Prakash, 1974; Miller *et al.*, 1982a). *rad6* mutants are defective in postreplication repair of damaged DNA (Prakash, 1981) and in mutagenesis induced by UV, γ-rays, and alkylating agents (Prakash, 1974; Lawrence and Christensen, 1976; McKee and Lawrence, 1979; Lawrence, 1982). In its effects on DNA repair and induced mutagenesis, the *rad6* mutation resembles the *recA* mutation of *E. coli*. However, unlike *recA* mutants, *rad6* mutants are not defective in genetic recombination. The frequencies of spontaneous and UV induced mitotic recombination are elevated in *rad6* mutants (Kern and Zimmermann, 1978; Montelone *et al.*, 1981); the *rad6* mutation also affects plating efficiency and growth rate, even in the absence of treatment with a DNA damaging agent. The doubling time of *rad6Δ* strains is about twice that of *RAD6*$^{+}$ strains. In addition, *rad6/rad6* diploids do not undergo sporulation.

Domain Structure of the RAD6 Protein

RAD6 encodes a highly acidic protein (23.3% acidic and 11.6% basic residues) of 172 amino acid residues and M_r 19,704 (Reynolds *et al.*, 1985c). There is a pronounced clustering of acidic residues at the carboxyl end of RAD6 protein, so that 20 of the last 23 residues are acidic, including an uninterrupted stretch of 13 aspartates (Table 1). The RAD6 protein was overproduced in yeast by placing the *RAD6* ORF under the control of the *ADC1* promoter. Due to the acidic nature of the protein, substantial purification of the protein could be obtained on DEAE Sephacel, and further fractionation on hydroxyapatite and Sephadex G75 yielded apparently homogeneous RAD6 protein (Morrison et al., 1988).

In gel filtration experiments using Sephadex G75 and G100, RAD6 protein elutes as a single symmetric peak with an apparent M_r of 29,000. Since both the calculated M_r and the estimated M_r on NaDodS0$_4$/PAGE of RAD6 is near 20,000, it was concluded that native RAD6 protein is monomeric and has an asymmetric shape (Morrison *et al.*, 1988). The RAD6 polyacidic tail sequence is readily cleaved by proteinase K, yielding a protease resistant portion consisting of approximately 154 of the N-terminal residues, which behaves like a typical globular polypeptide in gel filtration studies. This observation suggests that in RAD6 protein, the N-terminal 150 or so residues constitute a globular domain to which the polyacidic sequence appends as a freely-extending tail; a comparison of the Stokes radii of wild type RAD6 and a mutant rad6 protein containing the N-terminal 149 residues lent support to this proposal (Morrison *et al.*, 1988).

Biological Role of the Acidic Domain

To gain insight into the biological role of the polyacidic sequence, we examined mutations in which different portions of the polyacidic tail were deleted (Table I) (Morrison *et al.*, 1988). In the rad6-149, rad6-153, and rad6-164 mutant proteins, the last 23, 19, or 8 amino acid residues, respectively, have been deleted. DNA fragments harboring these *rad6* mutations were introduced into a low copy *CEN* vector and *rad6Δ* yeast cells transformed with the resulting plasmids. The UV sensitivity of the *rad6-153* and *rad6-164* mutants was no different from that of the RAD6 strains, whereas the *rad6-149* mutation increased UV sensitivity only slightly, and UV mutagenesis occurred in all three *rad6* mutant strains, indicating that the *RAD6* polyacidic sequence is not required for the DNA repair and mutagenesis functions (Morrison *et al.*, 1988). The polyacidic sequence, however, plays an essential role in sporulation, because essentially no sporulation occurred in the *rad6-149* mutant. Addition of the first four residues of the polyacidic tail, as in the *rad6-153* mutant, restored substantial sporulation ability (Morrison *et al.*, 1988).

RAD6 Protein is a Ubiquitin Conjugating Enzyme

The biochemical activity of RAD6 was discovered in Alex Varshavsky's laboratory when it was found that the nucleotide sequence of the *S. cerevisiae* gene corresponding to the ubiquitin conjugating enzyme E2$_{20K}$ that they had cloned was identical to the published *RAD6* sequence (Jentsch *et al.*, 1987). RAD6 is one of at least five ubiquitin-conjugating (E2) enzymes that can be isolated from yeast cell extracts by affinity chromatography on ubiquitin-

Table 1. Carboxyl-terminal Residues of Wild Type RAD6 and Mutant rad6 proteins

Protein	Residues
	149 153 172
RAD6	Ser Trp↓Glu $(Asp)_2$ Met $(Asp)_2$ Met $(Asp)_{13}$ Glu Ala Asp
rad6-153	Ser Trp Glu $(Asp)_2$ Met
rad6-149	Ser Trp

The start of the polyacidic sequence in RAD6 protein is indicated by the arrow. The 23-carboxyl-terminal amino acids in RAD6 protein contain 20 acidic residues. In rad6-153 protein, the last 19 residues of the polyacidic tail have been deleted, and in rad6-149 protein, all 23 residues of the acidic tail have been deleted. This table and legend are taken from Sung, P., S. Prakash, and L. Prakash. (1988) *Genes and Development* 2:1476-1485.

Sepharose. The biochemical reactions involved in the attachment of ubiquitin to target proteins have been extensively characterized using various enzyme fractions isolated from rabbit reticulocytes (Pickart and Rose; 1985; Hershko and Ciechanover, 1986). In an ATP dependent reaction, the carboxyl-terminal glycine of ubiquitin is joined *via* a thiolester bond to a cysteine residue of the ubiquitin-activating enzyme, E1. The ubiquitin-conjugating, or E2, enzymes accept the activated ubiquitin from E1. In the E2 enzymes also, ubiquitin is attached to a cysteine residue via a thiolester bond. The E2 enzymes may directly transfer the ubiquitin to target proteins by a reaction in which an isopeptide bond is formed between the carboxyl-terminal glycine of ubiquitin and the ε-amino group of lysines in target proteins, or the E2 enzymes may require another distinct protein, E3, for transfer of ubiquitin to target proteins. Protein substrates requiring E3 for their ubiquitination acquire a multiply-ubiquitinated configuration and are destined for degradation *via* an ATP dependent proteolytic system (Hershko *et al.*, 1984a,b; Hershko and Ciechanover, 1986).

In the work of Jentsch *et al.* (1987), RAD6 protein from yeast was obtained by passing the cell extract through a ubiquitin-Sepharose column, followed by gel electrophoresis of the DTT eluate and elution from the gel of the $E2_{20K}$ corresponding to RAD6 protein. Jentsch *et al.* (1987) also assayed an extract from *E. coli* cells carrying a RAD6-containing plasmid for ubiquitin conjugation activity. RAD6 protein obtained from yeast or present in *E. coli* extracts showed the ability to attach a single molecule of ubiquitin to H2B *in vitro*.

RAD6 Protein Polyubiquitinates Histones

To study the ubiquitin conjugating activity of RAD6, the protein was purified to homogeneity by fractionating yeast cell extracts on columns of ubiquitin-Sepharose and MonoQ, and the purified protein used in ubiquitin conjugation assays containing E1, ^{125}I-ubiquitin, and histone H2A or H2B as substrate. Interestingly, we observed that RAD6 protein attaches multiple molecules of ubiquitin onto both histones, yielding products tagged with one to seven molecules of ubiquitin (Sung *et al.*, 1988b). Histone polyubiquitination has no requirement for E3 and appears to proceed through a processive mechanism. Since acquisition of a multiply-ubiquitinated configuration normally leads to degradation of the protein substrate *via* the ubiquitin-specific proteolytic pathway (Herhsko *et al.*, 1984a,b; Hershko and Ciechanover, 1986), this finding raises the possibility that RAD6 targets histones for degradation *in vivo*.

The role of the acidic tail domain of RAD6 in histone polyubiquitination was investigated using homogeneous rad6-153 and rad6-149 proteins (Table I) obtained from yeast cell extracts. The rad6-153 protein, which carries only 4 residues of the 23-residue polyacidic sequence, possesses a diminished but readily assayable level of histone-polyubiquitinating activity, while the rad6-149 protein, which lacks the entire polyacidic sequence, has lost the ability to multiply ubiquitinate histones (Sung *et al.*, 1988b). These observations highlight the importance of the acidic domain of RAD6 in the recognition of histones during their modification. However, the fact that horse heart cytochrome c, also a highly basic protein, is not modified by RAD6, indicates that the non-acidic, globular domain of RAD6 also plays an essential role in determining substrate specificity (Sung *et al.*, 1988b).

As mentioned in an earlier section, the *rad6-153* and *rad6-149* mutants show somewhat reduced, and little or no sporulation, respectively, but are both normal in DNA repair and induced mutagenesis. Since the rad6-149 protein is devoid of the histone-polyubiquitinating activity, one might infer that this activity is relevant for only the sporulation function of RAD6, and that RAD6 protein ubiquitinates other, presumably non-basic, substrates when carrying out its DNA repair and mutagenesis functions. It remains to be seen if the proteins encoded by genes such as *RAD9*, *RAD15*, *RAD18*, *REV1*, *REV2*, and *REV3*, which are all members of the *RAD6* epistasis group, are substrates for the RAD6 ubiquitin ligase.

The *RAD18* Gene

Like the *RAD6* gene, *RAD18* is required for postreplication repair of UV damaged DNA (Di Caprio and Cox, 1981; Prakash, 1981). However, in contrast to *rad6* mutations, mutations in the *RAD18* gene do not affect mutagenesis (Lawrence, 1982) or sporulation. Spontaneous mutation and mitotic recombination rates are elevated in *rad18* mutants (Boram and Roman, 1976; Quah *et al.*, 1980).

RAD18 encodes a protein of 487 amino acids with an M_r of 55, 512, and it contains 15.8% basic, 13.3% acidic, 37.2% polar, and 33.7% non polar residues (Jones *et al.*, 1988). Three potential zinc binding, DNA binding sequences are present in the RAD18 protein sequence. These occur between residues 28-48, 51-65, and 190-210, and are of the form C-X_2-C-X_{11}-C-X_4-C; C-X_3-H-X_6-C-X_2-C, and C-X_2-C-X_{12}-H-X_3-C. *RAD18* also contains a Walker type A sequence for binding purine nucleotide(s). This sequence occurs between the residues 353 - 373. Thus, RAD18 shows features that suggest that it binds DNA and ATP (Jones *et al.*, 1988).

FUTURE PROSPECTS

The availability of cloned *RAD* genes makes it feasible to overproduce their encoded proteins in yeast for purification and biochemical characterization, individually and in various combinations. In the near future, it should be possible to construct the incision enzyme complex *in vitro*, and to eventually define the role of its protein components in recognition and binding to damage, in incision, and in dissociation of the enzyme complex from the damage site. Isolation of additional *rad3 ts* mutations for viability coupled with their genetic and biochemical analyses will be necessary to elucidate the essential role of *RAD3*. Future studies will also be focused on determining whether different biochemical activities or interactions with different sets of proteins account for the requirement of *RAD3* in excision repair and in cell viability. Some of the proteins in the RAD7, RAD23, RAD14, RAD16, and MMS19 group may affect the accessibility of damaged DNA in chromatin and their characterization may bring to light mechanisms involved in the opening of chromatin during DNA repair.

The discovery of RAD6 as a ubiquitin conjugating enzyme has added a new and exciting dimension to DNA repair studies in eukaryotes. As has been pointed out in this review and elsewhere, RAD6 very likely acts upon substrates other than histones in its roles in DNA repair and mutagenesis (Sung *et al.*, 1988b). The identification of RAD6 substrate(s) and the mechanism(s) by which ubiquitination modulates the activity or stability of these proteins is expected to yield substantive new information. The combination of genetic and biochemical studies should also help in elucidating the role of RAD18 in DNA repair and of RAD9 in G2 arrest of cells in response to DNA damage.

Finally, it should be possible to use *S. cerevisiae* as a vehicle for overproduction of human DNA repair proteins to facilitate their purification and biochemical characterization. The existence of homology between yeast and human DNA repair proteins suggests the possibility of structure-function studies in which domains between yeast and human proteins can be interchanged and their effects examined.

ACKNOWLEDGEMENTS

Research in our laboratories is supported by grants CA35035, CA41261, and GM19261 from the National Institutes of Health, and DE-FG02-88ER60621 from the Department of Energy.

REFERENCES

Anderson, R. A., and Coleman, J. E. (1975a) Physicochemical properties of DNA binding proteins: gene 32 protein of T4 and *Escherichia coli* unwinding protein. *Biochemistry* **14**: 5485-5491.

Anderson, R. A., Y. Nakashima, and J. E. Coleman. (1975b) Chemical modifications of functional residues of fd gene 5 DNA-binding protein. *Biochemistry* **14**: 907-917.

Bohr, V. A., C. A. Smith, D. S. Okomuto, and P. C. Hanawalt. (1985) DNA repair in an active gene: removal of pyrimidine dimers from the DHFR gene of CHO cells is much more efficient than in the genome overall. *Cell* **40**: 359-369.

Boram, W. R., and H. Roman. (1976) Recombination in *Saccharomyces cerevisiae*: a DNA repair mutation associated with elevated mitotic recombination. *Proc. Natl. Acad. Sci. USA* **73**: 2828-2832.

Caron, P. R., S. R. Kushner, and L. Grossman. (1985) Involvement of helicase II (*uvrD* gene product) and DNA polymerase I in excision mediated by the uvrABC protein complex. *Proc. Natl. Acad. Sci. USA* **82**: 4925-4929.

Cole, G. M., D. Schild, S. T. Lovett, and R. K. Mortimer. (1987) Regulation of *RAD54*- and *RAD52-lacZ* gene fusions in *Saccharomyces cerevisiae* in response to DNA damage. *Mol. Cell. Biol.* **7**: 1078-1084.

Cox, B. S. and J. C. Game. (1974) Repair systems in Saccharomyces. *Mutat. Res.* **26**: 257-264.

Cox, B. S. and J. M. Parry. (1968) The isolation, genetics, and survival characteristics of ultraviolet light sensitive mutants in yeast. *Mutat. Res.* **6**: 37-55.

Di Caprio, L., and B. S. Cox. (1981) DNA synthesis in UV-irradiated yeast. *Mutat. Res.* **82**: 69-85.

Elledge, S. J., and R. W. Davis. (1987) Identification and isolation of the gene encoding the small subunit of ribonucleotide reductase from *Saccharomyces cerevisiae*: DNA damage-inducible gene required for mitotic viability. *Mol. Cell. Biol.* **7**: 2783-2793.

Fleer, R., C. M. Nicolet, G. A. Pure, and E. C. Friedberg. (1987) RAD4 gene of *Saccharomyces cerevisiae*: molecular cloning and partial characterization of a gene that is inactivated in *Escherichia coli*. *Mol. Cell. Biol.* **7**: 1180-1192.

Friedberg, E. C. (1988) Deoxyribonucleic acid repair in the yeast *Saccharomyces cerevisiae*. *Microbiol. Rev.* **52**: 70-102.

Fry, D. C., S. A. Kuby, and A. S. Mildvan. (1986) ATP-binding site of adenylate kinase: Mechanistic implications of its homology with *ras*-encoded p21, F1-ATPase, and other nucleotide-binding proteins. *Proc. Natl. Acad. Sci. USA* **83**: 907-911.

Game, J. C. (1983) Radiation-sensitive mutants and DNA repair in yeast, p. 109-137. **In**: J. F. T. Spencer, D. M. Spencer, and A. R. W. Smith (ed.), *Yeast genetics. Fundamental and applied aspects.* Springer-Verlag, New York.

Game, J. C. and B. S. Cox. (1972) Epistatic interactions between four *rad* loci in yeast. *Mutat. Res.* **16**: 353-362.

Game, J. C. and B. S. Cox. (1973) Synergistic interactions between *rad* mutants in yeast. *Mutat. Res.* **20**: 35-44.

Game, J. C. and R. K. Mortimer. (1974) A genetic study of X-ray sensitive mutants in yeast. *Mutat. Res.* **24**: 281-292.

Gietz, R. D., and S. Prakash. (1988) Cloning and nucleotide sequence analysis of the *Saccharomyces cerevisiae RAD4* gene required for excision repair of UV damaged DNA. *Gene* **74**: 535-541.

Haynes, R. H. and B. A. Kunz. (1981) DNA repair and mutagenesis in yeast, p. 371-414. **In**: J. Strathern, E. W. Jones, and J. R. Broach (ed.), *The molecular biology of the yeast Saccharomyces.* Life cycle and inheritance. Cold Spring Harbor Laboratory, Cold Spring Harbor, NY.

Henriques, J. A. P., and E. Moustacchi. (1980) Isolation and characterization of *pso* mutants sensitive to photoaddition of psoralen derivatives in *Saccharomyces cerevisiae. Genetics* **95**: 273-288.

Hershko, A., and A. Ciechanover. (1986) The ubiquitin pathway for the degradation of intracellular proteins. *Prog. Nucl. Acid Res.* **33**: 19-56.

Hersko, A., H. Heller, E. Eytan, G. Kaklij, and I. A. Rose. (1984a) Role of the α-amino group of protein in ubiquitin-mediated protein breakdown. *Proc. Natl. Acad. Sci. USA* **81**: 7021-7025.

Hershko, A., E. Leshinsky, D. Ganoth, and H. Heller. (1984b) ATP-dependent degradation of ubiquitin-protein conjugates. *Proc. Natl. Acad. Sci. USA* **81**: 1619-1623.

Higgins, D. R. (1987) Isolation, structure, and characterization of the *RAD3* gene of the yeast *Saccharomyces cerevisiae. PhD thesis, University of Rochester, Rochester, NY.*

Higgins, D. R., S. Prakash, P. Reynolds, R. Polakowska, S. Weber, and L. Prakash. (1983) Isolation and characterization of the *RAD3* gene of *Saccharomyces cerevisiae* and inviability of *rad3* deletion mutants. *Proc. Natl. Acad. Sci. USA* **80**: 5680-5684.

Hurd, H. K., C. W. Roberts, and J. W. Roberts. (1987) Identification of the gene for the yeast ribonucleotide reductase small subunit and its inducibility by methyl methanesulfonate. *Mol. Cell. Biol.* **7**: 3673-3677.

Husain, L., B. van Houten, D. C. Thomas, M. Abdel-Monem, and A. Sancar. (1985) Effect of DNA polymerase I and DNA helicase II on the turnover rate of UvrABC excision nuclease. *Proc. Natl. Acad. Sci. USA* **82**: 6774-6778.

Jachymczyk, W. J., R. C. von Borstel, M. R. A. Mowat, and P. J. Hastings. (1981). Repair of interstrand cross-links in DNA of *Saccharomyces cerevisiae* requires two systems for DNA repair: the *RAD3* system and the *RAD51* system. *Mol. Gen. Genet.* **182**: 196-205.

Jentsch, S., J. P. McGrath, and A. Varshavsky. (1987) The yeast DNA repair gene *RAD6* encodes a ubiquitin-conjugating enzyme. *Nature (London)* **329**: 131-134.

Johnson, A. L., D. G. Barker, and L. H. Johnston. (1986) Induction of yeast DNA ligase genes in exponential and stationary phase cultures in response to DNA damaging agents. *Curr. Genet.* **11**: 107-112.

Johnston, L. H., J. H. M. White, A. L. Johnson, G. Lucchini, and P. Plevani. (1987) The yeast DNA polymerase I transcript is regulated in both the mitotic cell cycle and in meiosis and is also induced after DNA damage. *Nucl. Acids Res.* **15**: 5017-5030.

Jones, J. S., S. Weber, and L. Prakash. (1988) The *Saccharomyces cerevisiae RAD18* gene encodes a protein that contains potential zinc finger domains for nucleic acid binding and a putative nucleotide binding sequence. *Nucl. Acids Res.* **16**: 7119-7131.

Kern, R., and F. K. Zimmermann. (1978) The influence of defects in excision and error prone repair on spontaneous and induced mitotic recombination and mutation in *Saccharomyces cerevisiae. Mol. Gen. Genet.* **161**: 81-88.

Lawrence, C. W. (1982) Mutagenesis in *Saccharomyces cerevisiae. Adv. Genet.* **21**: 173-254.

Lawrence, C. W., and R. Christensen. (1976) UV mutagenesis in radiation-sensitive strains of yeast. *Genetics* **82**: 207-231.

Leadon, S. A. (1986) Differential repair of DNA damage in specific nucleotide sequences in monkey cells. *Nucl. Acids Res.* **14**: 8979-8995.

Lemontt, J. F. (1971) Mutants of yeast defective in mutation induced by ultraviolet light. *Genetics* **68**: 21-33.

Madhani, H. D., V. A. Bohr, and P. C. Hanawalt. (1986) Differential DNA repair in transcriptionally active and inactive proto-oncogenes: *c-abl* and *c-mos. Cell* **45**: 417-423.

Madura, K., and S. Prakash. (1986) Nucleotide sequence, transcript mapping, and regulation of the *RAD2* gene of *Saccharomyces cerevisiae. J. Bacteriol.* **166**: 914-923.

Maga, J. A., T. A. McClanahan, and K. McEntee. (1986) Transcriptional regulation of DNA damage responsive (*DDR*) genes in different *rad* mutant strains of *Saccharomyces cerevisiae. Mol. Gen. Genet.* **205**: 276-284.

Malone, R. E., and M. F. Hoekstra. (1984) Relationships between a hyper-rec mutation (*rem1*) and other recombination and repair genes in yeast. *Genetics* **107**: 33-48.

Martin, P., L. Prakash, and S. Prakash. (1981) a/α-specific effect of the *mms3* mutation on ultraviolet mutagenesis in *Saccharomyces cerevisiae*. *J. Bacteriol.* **146**: 684-691.

McClanahan, T., and K. McEntee. (1984) Specific transcripts are elevated in *Saccharomyces cerevisiae* in response to DNA damage. *Mol. Cell. Biol.* **4**: 2356-2363.

McKee, R. H., and C. W. Lawrence. (1979) Genetic analysis of gamma-ray mutagenesis in yeast. I. Reversion in radiation-sensitive strains. *Genetics* **93**: 361-373.

McKnight, G. L., T. S. Cardillo, and F. Sherman. (1981) An extensive deletion causing overproduction of yeast iso-2-cytochrome c. *Cell* **25**: 409-419.

McPherson, A., F. A. Jurnak, A. J. Wang, I. Molineux, and A. Rich. (1979) Structure at 2.3Å resolution of the gene 5 product of bacteriophage fd: a DNA unwinding protein. *J. Mol. Biol.* **134**: 379-400.

Mellon, I., V. A. Bohr, C. A. Smith, and P. C. Hanawalt. (1986) Preferential DNA repair of an active gene in human cells. *Proc. Natl. Acad. Sci. USA* **83**: 8878-8882.

Mellon, I., G. Spivak, and P. C. Hanawalt. (1987) Selective removal of transcription-blocking DNA damage from the transcribed strand of the mammalian *DHFR* gene. *Cell* **51**: 241-349.

Miller, R. D., L. Prakash, and S. Prakash. (1982a) Genetic control of excision of *Saccharomyces cerevisiae* interstrand DNA crosslinks induced by psoralen plus near-UV light. *Mol. Cell. Biol.* **2**: 939-948.

Miller, R. D., L. Prakash, and S. Prakash. (1982b) Defective excision of pyrimidine dimers and interstrand DNA crosslinks in *rad7* and *rad23* mutants of *Saccharomyces cerevisiae*. *Mol. Gen. Genet.* **188**: 235-239.

Miller, R. D., S. Prakash, and L. Prakash. (1984) Different effects of *RAD* genes of *Saccharomyces cerevisiae* on incisions of interstrand crosslinks and monoadducts in DNA induced by psoralen plus near UV light treatment. *Photochem. Photobiol.* **39**: 349-352.

Montelone, B. A., M. F. Hoekstra, and R. E. Malone. (1988) Spontaneous mitotic recombination in yeast: the hyper-recombinational rem1 mutations are alleles of the *RAD3* gene. *Genetics* **119**: 289-301.

Montelone, B. A., S. Prakash, and L. Prakash. (1981) Recombination and mutagenesis in *rad6* mutants of *Saccharomyces cerevisiae*: evidence for multiple functions of the *RAD6* gene. *Mol. Gen. Genet.* **184**: 410-415.

Morrison, A., E. J. Miller, and L. Prakash. (1988) Domain structure and functional analysis of the carboxyl-terminal polyacidic sequence of the *RAD6* protein of *Saccharomyces cerevisiae*. *Mol. Cell. Biol.* **8**: 1179-1185.

Moustacchi, E. (1987) DNA repair in yeast: genetic control and biological consequences. *Adv. Radiat. Biol.* **13**: 1-30.

Naumovski, L., G. Chu, P. Berg, and E. C. Friedberg. (1985) *RAD3* gene of *Saccharomyces cerevisiae*: nucleotide sequence of wild-type and mutant alleles, transcript mapping, and aspects of gene regulation. *Mol. Cell. Biol.* **5**: 17-26.

Naumovski, L. and E. C. Friedberg. (1983) A DNA repair gene required for the incision of damaged DNA is essential for viability in *Saccharomyces cerevisiae*. *Proc. Natl. Acad. Sci. USA* **80**: 4818-4821.

Naumovski, L., and E. C. Friedberg. (1986) Analysis of the essential and excision repair functions of the *RAD3* gene of *Saccharomyces cerevisiae* by mutagenesis. *Mol. Cell. Biol.* **6**: 1218-1227.

Naumovski, L., and E. C. Friedberg. (1987) The *RAD3* gene of *Saccharomyces cerevisiae*: isolation and characterization of a temperature-sensitive mutant in the essential function and of extragenic suppressors of this mutant. *Mol. Gen. Genet.* **209**: 458-466.

Njagi, G. D. E., and B. J. Kilbey. (1982) *cdc7-1*, a temperature sensitive cell-cycle mutant which interferes with induced mutagenesis in *Saccharomyces cerevisiae*. *Mol. Gen. Genet.* **186**: 478-481.

Patterson, M., R. A. Sclafani, W. L. Fangman, and J. Rosamond. (1986) Molecular characterization of cell cycle gene *CDC7* from *Saccharomyces cerevisiae*. *Mol. Cell. Biol.* **6**: 1590-1598.

Perozzi, G., and S. Prakash. (1986) *RAD7* gene of *Saccharomyces cerevisiae*: transcripts, nucleotide sequence analysis, and functional relationship between the *RAD7* and *RAD23* gene products. *Mol. Cell. Biol.* **6**: 1497-1507.

Peterson, T. A., L. Prakash, S. Prakash, M. A. Osley, and S. I. Reed. (1985) Regulation of *CDC9*, the *Saccharomyces cerevisiae* gene that encodes DNA ligase. *Mol. Cell. Biol.* **5**: 226-235.

Pickart, C. M., and I. A. Rose. (1985) Functional heterogeneity of ubiquitin carrier proteins. *J. Biol. Chem.* **260**: 1573-1581.

Polakowska, R., L. Prakash, and S. Prakash. (1983) Ultraviolet light induced mutagenesis of mitochondrial genes in the *rad6*, *rev3*, and *cdc8* mutants of *Saccharomyces cerevisiae*. *Mol. Gen. Genet.* **189**: 513-515.

Prakash, L. (1974) Lack of chemically induced mutation in repair-deficient mutants of yeast. *Genetics* **78**: 1101-1118.

Prakash, L. (1975) Repair of pyrimidine dimers in nuclear and mitochondrial DNA of yeast irradiated with low doses of ultraviolet light. *J. Mol. Biol.* **98**: 781-795.

Prakash, L. (1977a) Repair of pyrimidine dimers in radiation-sensitive mutants *rad3*, *rad4*, *rad6*, and *rad9* of *Saccharomyces cerevisiae Mutat. Res.* **45**: 13-20.

Prakash, L. (1977b) Defective thymine dimer excision in radiation-sensitive mutants *rad10* and *rad16* of *Saccharomyces cerevisiae*. *Mol. Gen. Genet.* **152**: 125-128.

Prakash, L. (1981) Characterization of postreplication repair in *Saccharomyces cerevisiae* and effects of *rad6*, *rad18*, *rev3* and *rad52* mutations. *Mol. Gen. Genet* **184**: 471-478.

Prakash, L., D. Hinkle, and S. Prakash. (1979) Decreased UV mutagenesis in *cdc8*, a DNA replication mutant of *Saccharomyces cerevisiae*. *Mol. Gen. Genet.* **172**: 249-258.

Prakash, L., and S. Prakash. (1977) Isolation and characterization of MMS-sensitive mutants of *Saccharomyces cerevisiae*. *Genetics* **86**: 33-55.

Prakash, L., and S. Prakash. (1979) Three additional genes involved in pyrimidine dimer removal in *Saccharomyces cerevisiae*: *RAD7*, *RAD14*, and *MMS19*. *Mol. Gen. Genet.* **176**: 351-359.

Quah, S.-K., R. C. von Borstel, and P. J. Hastings. (1980) The origin of spontaneous mutation in *Saccharomyces cerevisiae*. *Genetics* **96**: 819-839.

Resnick, M. A. and J. K. Setlow. (1982) Repair of pyrimidine dimer damage induced in yeast by ultraviolet light. *J. Bacteriol.* **109**: 979-986.

Reynolds, P., D. R. Higgins, L. Prakash, and S. Prakash. (1985a) The nucleotide sequence of the *RAD3* gene of *Saccharomyces cerevisiae*: a potential adenine nucleotide binding amino acid sequence and a nonessential acidic carboxyl terminal region. *Nucl. Acids Res.* **13**: 2357-2372.

Reynolds, P., L. Prakash, D. Dumais, G. Perozzi, and S. Prakash. (1985b) Nucleotide sequence of the *RAD10* gene of *Saccharomyces cerevisiae*. *EMBO J.* **4**: 3549-3552.

Reynolds, P., L. Prakash, and S. Prakash. (1987) Nucleotide sequence and functional analysis of the *RAD1* gene of *Saccharomyces cerevisiae*. *Mol. Cell.Biol.* **7**: 1012-1020.

Reynolds, P., S. Weber, and L. Prakash. (1985c) *RAD6* gene of *Saccharomyces cerevisiae* encodes a protein containing a tract of 13 consecutive aspartates. *Proc. Natl. Acad. Sci. USA* **82**: 168-172.

Reynolds, R. J. (1978) Removal of pyrimidine dimers from *Saccharomyces cerevisiae* nuclear DNA under nongrowth conditions as detected by a sensitive, enzymatic assay. *Mutat. Res.* **50**: 43-56.

Reynolds, R. J. and E. C. Friedberg. (1981) Molecular mechanisms of pyrimidine dimer removal in excision-defective mutants of *Saccharomyces cerevisiae*: incision of ultraviolet-irradiated deoxyribonucleic acid *in vivo*. *J. Bacteriol.* **146**: 692-704.

Robinson, G. W., C. M. Nicolet, D. Kalainov, and E. C. Friedberg. (1986) A yeast excision-repair gene is inducible by DNA damaging agents. *Proc. Natl. Acad. Sci. USA* **83**: 1842-1846.

Ruby, S. W., and J. W. Szostak. (1985) Specific *Saccharomyces cerevisiae* genes are expressed in response to DNA-damaging agents. *Mol. Cell. Biol.* **5**: 75-84.

Ruhland, A., E. Haase, W. Siede, and M. Brendel (1981) Isolation of yeast mutants sensitive to the bifunctional alkylating agent nitrogen mustard. *Mol. Gen. Genet.* **181**: 346-351.

Sancar, A. and W. D. Rupp. (1983) A novel repair enzyme: UVRABC excision nuclease of *Escherichia coli* cuts a DNA strand on both sides of the damaged region. *Cell* **33**: 249-260.

Schiestl, R. H. and S. Prakash. (1988) *RAD1*, an excision repair gene of *Saccharomyces cerevisiae*, is also involved in recombination. *Mol. Cell. Biol.* **8**: 3619-3626.

Schiestl, R. H., P. Reynolds, S. Prakash, and L. Prakash. (1989) Cloning and sequence analysis of the *Saccharomyces cerevisiae RAD9* gene and further evidence that its product is required for cell cycle arrest induced by DNA damage. *Mol. Cell. Biol.* **9**: 1882-1896

Sclafani, R. A., and W. L. Fangman. (1984) Yeast gene *CDC8* encodes thymidylate kinase and is complemented by herpes thymidine kinase gene TK. *Proc. Natl. Acad. Sci. USA* **81**: 5821-5825.

Sigal, I. S., J. B. Gibbs, J. S. D'Alonzo, G. L. Temeles, B. S. Wolanski, S. H. Socher, and E. M. Scolnick. (1986) Mutant ras-encoded proteins with altered nucleotide binding exert dominant biological effects. *Proc. Natl. Acad. Sci. USA* **83**: 952-956.

Sung, P., D. Higgins, L. Prakash, and S. Prakash. (1988a) Mutation of lysine-48 to arginine in the yeast RAD3 protein abolishes its ATPase and DNA helicase activities but not the ability to bind ATP. *EMBO J.* **7**: 3263-3269.

Sung, P., L. Prakash, S. Weber, and S. Prakash. (1987a) The *RAD3* gene of *Saccharomyces cerevisiae* encodes a DNA-dependent ATPase. *Proc. Natl. Acad. Sci. USA* **84**: 6045-6049.

Sung, P., L. Prakash, S. Matson, and S. Prakash. (1987b) RAD3 protein of *Saccharomyces cerevisiae* is a DNA helicase. *Proc. Natl. Acad. Sci. USA* **84**: 8951-8955.

Sung, P. and S. Prakash. (1988) Purification and biochemical characterization of the *RAD3* protein of *Saccharomyces cerevisiae*. p. 191-199. **In**: E. C. Friedberg and P. C. Hanawalt (ed.) *Mechanisms and consequences of DNA damage processing*. UCLA Symposium on Molecular and Cellular Biology, New Series, Volume 83, Alan R. Liss, Inc., New York.

Sung, P., S. Prakash, and L. Prakash. (1988b) The RAD6 protein of *Saccharomyces cerevisiae* polyubiquitinates histones, and its acidic domain mediates this activity. *Genes Develop.* **2**: 1476-1485.

Treger, J. M., K. A. Heichman, and K. McEntee. (1988) Expression of the yeast *UBI4* gene increases in response to DNA-damaging agents and in meiosis. *Mol. Cell. Biol.* **8**: 1132-1136.

Unrau, P., R. Wheatcroft, and B. S. Cox. (1971) The excision of pyrimidine dimers from DNA of ultraviolet irradiated yeast. *Mol. Gen. Genet.* **113**: 359-362.

Van Duin, M., J. de Wit, H. Odjik, A. Westerveld, A. Yasui, M. H. M. Koken, J. H. J. Hoeijmakers, and D. Bootsma. (1986) Molecular characterization of the human excision repair gene *ERCC1*: cDNA cloning and amino acid homology with the yeast DNA repair gene *RAD10*. *Cell* **44**: 913 923.

Walker, G. C. (1985) Inducible DNA repair system. *Ann. Rev. Biochem.* **54**: 425-457.

Walker, J. E., M. Saraste, M. J. Runswick, and N. J. Gay. (1982) Distantly related sequences in the α- and β-subunits of ATP synthase, myosin, kinases and other ATP-requiring enzymes and a common nucleotide binding fold. *EMBO J.* **1**: 945-951.

Waters, R. and E. Moustacchi. (1974) The disappearance of UV-induced pyrimidine dimers from the nuclear DNA of exponential and stationary phase cells of *S. cerevisiae* following post-irradiation treatments. *Biochim. Biophys. Act.* **353**: 407-419.

Weinert, T. A., and L. H. Hartwell. (1988) The *RAD9* gene controls the cell cycle response to DNA damage in *Saccharomyces cerevisiae*. *Science* **241**: 317-322.

White, J. H. M., D. G. Barger, P. Nurse, and L. H. Johnston. (1986) Periodic transcription as a means of regulating gene expression during the cell cycle: contrasting modes of expression of DNA ligase genes in budding and fission yeast. *EMBO J.* **5**: 1705-1709.

Wilcox, D. R., and L. Prakash. (1981) Incision and postincision steps of pyrimidine dimer removal in excision-defective mutants of *Saccharomyces cerevisiae*. *J. Bacteriol.* **148**: 618-623.

Yeung, A. T., W. D. Mates, E. Y. Oh, and L. Grossman. (1983) Enzymatic properties of purified *Escherichia coli* uvrABC proteins. *Proc. Natl. Acad. Sci. USA* **80**: 6157-6161.

12

DNA METABOLIZING ENZYMES OF *DROSOPHILA*

James B. Boyd, Kengo Sakaguchi and Paul V. Harris

Large macromolecular complexes or "protein machines" have been shown to mediate many forms of prokaryotic DNA metabolism (Alberts, 1985; Echols, 1986). Those observations nearly guarantee the existence of analogous complexes in eukaryotic DNA repair as postulated by Haynes *et al.* (1978). One of the major goals in studies of eukaryotic DNA repair is to identify the participating proteins and to reconstitute functional *in vitro* complexes from those proteins. The recent *in vitro* reproduction of bacterial excision repair with four isolated proteins (uvrABC excision nuclease, helicase II, pol I and DNA ligase) serves as a model for such studies (Husain *et al.*, 1985). This example highlights the necessity for combining molecular and genetic approaches to fully understand such complex macromolecular processes. As the most genetically accessible metazoan organism, *Drosophila* represents a logical focus for such studies in higher organisms. An overview of DNA metabolic enzymes in this organism is presented here together with a description of current studies in our laboratory.

MOLECULAR AND GENETIC ANALYSIS OF DNA REPAIR

Genetic studies of repair deficient mutants in *Drosophila* have revealed an extensive overlap in the functions that participate in DNA repair, recombination and synthesis (see Boyd *et al.*, 1987). Mutants at several gene loci are known, for example, to be deficient in both DNA repair and meiotic recombination. Thus, most enzymes currently under study are potentially involved in more than one metabolic pathway. Since the biological functions of most *Drosophila* enzymes involved in DNA metabolism have not been precisely defined, all DNA metabolizing enzymes are considered for the purposes of this review to be potentially involved in DNA repair. This assumption is necessary since *Drosophila* geneticists and biochemists have generally pursued parallel but independent paths. Recent advances in gene cloning, however, should soon integrate these two areas by permitting geneticists to identify the protein products of genetically characterized genes and by enabling biochemists to establish the biological function of well characterized enzymes. The fusion of these complementary approaches is essential for a complete understanding of DNA repair in this higher eukaryote.

Cloning of repair genes in *Drosophila* benefits from the extensive molecular characterization of the genome of this organism. The presence of giant polytene chromosomes, a small genome size and a low frequency of repeated DNA sequences has greatly facilitated that analysis. Our laboratory is building on that foundation by employing a forward genetic approach to clone the repair related genes *mei-9* and *mei-41* (Boyd *et al.*, 1987). Yamamoto and Mason (personal communication) have succeeded in mutagenizing both loci in crosses that mobilize transposable P elements. A combination of *in situ* hybridization and reversion analysis has established that the recovered mutants were indeed induced by transposon insertions. The method of transposon tagging has recently permitted cloning of the *mei-41* gene with the aid of these mutants (Banga, unpublished observations). Once a coding sequence has been established, a conceptional amino acid sequence will be employed to identify potentially homologous proteins from other organisms. Cloned cDNA sequences will then be employed to over-express the protein products of both these genes in order to permit direct tests for enzymatic function. This approach, and its extension to other genes, is expected to lead to an identification of essential components of DNA repair complexes.

Although an enzymatic analysis of macromolecular complexes has only begun in *Drosophila*, suggestions from the genetic literature point to their importance in DNA repair. The *phr* mutant of *Drosophila*, which is entirely devoid of photorepair (Boyd and Harris, 1987), also exhibits a partial deficiency in excision repair. That observation has been interpreted to reflect an interaction between those two repair systems at the protein level. Observations of synthetic lethality between two otherwise viable repair-related mutants (Smith *et al.*, 1980) also implicates interactions between the normal gene products of those genes. Molecular and genetic analyses are also beginning to reveal levels of complexity beyond those found in prokaryotes. The recent discovery that the *RAD6* gene of yeast encodes a ubiquitin conjugating protein (Jentsch *et al.*, 1987; Prakash, this volume) substantiates suggestions made in earlier studies of *Drosophila* that implicated repair-related genes in chromatin structure (Gatti *et al.*, 1983). Several of the *Drosophila* repair-related mutants exhibit pleiotropisms as complex as those exhibited by the *rad6* mutants (Boyd *et al.*, 1987). The *mei-9* and *mei-41* genes mentioned above are essential for meiosis as well as for resistance to an unusually broad spectrum of mutagenic agents. Mutants at the *mei-9* locus, however, are deficient in excision repair, whereas the *mei-41* mutants are deficient in postreplication repair. A particularly striking example that points to the importance of chromatin structure in repair is found among mutants of the *mus101* locus (see Boyd *et al.*, 1987) which were initially identified by virtue of their hypersensitivity to chemical mutagens. Biochemical studies have identified a defect in postreplication repair (Boyd and Setlow, 1976) and cytogenetic analyses indicate that those mutants are defective in repair of chromosome breaks (Graf *et al.*, 1979; Baker and Smith, 1979). The discovery that the then extant mutants are hypomorphs (Baker and Smith, 1979) led to the recovery of a conditional lethal allele of that gene (Smith *et al.*, 1985). A cytogenetic analysis had revealed that the new mutant exhibits an altered capacity to condense metaphase chromosomes (Gatti *et al.*, 1983). A further intriguing observation has revealed that *mus101* is one of the trans-acting genes that are required for chorion gene amplification (cited in Snyder *et al.*, 1986). Finally, the wild type product of that gene has been shown to be required for the phenomenon of rDNA magnification in which the redundancy of rDNA undergoes rapid alteration (Hawley *et al.*, 1985). Such observations indicate that an extended analysis of DNA repair in *Drosophila* will uncover new levels of complexity among macromolecular complexes that are unique to eukaryotes.

ENZYMOLOGICAL ANALYSIS OF DNA METABOLISM

DNA Polymerases

Because DNA synthesis is an essential feature of most cellular responses to DNA damage, replication complexes undoubtedly play a central role in many DNA repair mechanisms. *Drosophila* provides a number of unique variations in DNA replication which have implications for DNA repair. These include the phenomena of polytenization, rDNA magnification, and chorion gene amplification (Spradling and Orr-Weaver, 1987). The flexibility of the replication process revealed by those phenomena suggests that a limited number of replication enzymes may be able to participate in the varied forms of DNA repair. Before the role of replication in repair can be understood at the molecular level, however, it will first be necessary to identify the basic replication proteins. Toward this end, three distinct DNA polymerases, together with their accessory proteins, are currently under investigation. Since each of these enzymes is potentially involved in some form of DNA repair, current progress in the characterization of all three is summarized here. *Drosophila* reverse transcriptase is not discussed here, because it is probably elicited by retrotransposons (Becker, *et al.* 1987). It is anticipated that the potential of *Drosophila* for genetic analysis will ultimately provide a means to define the role that each DNA polymerase plays in repair.

Although Cohen noted in the early 1970's that *Drosophila* embryos constitute a rich source of DNA polymerases (Cohen *et al.*, 1973), it was not until later in that decade that the major replication complex became the object of intensive investigation (Brakel and Blumenthal, 1978a,b). Early observations of multiple enzyme forms were shown to be due in part to proteolytic degradation during isolation (Brakel and Blumenthal, 1977). That artifact continued to plague the field (Villani *et al.*, 1980; Vilanni *et al.*, 1981) until antibodies to purified α polymerase permitted an identification of the undegraded subunits of that enzyme (Sauer and Lehman, 1982). That information, in turn, made it possible to develop the methods for controlling proteolysis (Kaguni *et al.*, 1983a,b) that have been employed in the purification of each of the enzymes described in this section.

α Polymerase

The extraordinary concentration of α polymerase in *Drosophila* embryos has made this enzyme a prototype in the analysis of the eukaryotic replication complex. The specific activity found in embryos is approximately 30 times that of mammalian tissue culture cells (Banks *et al.*, 1979) and over 150 times that found in adult males (Dusenbery and Smith, 1983). A review of *Drosophila* α polymerase has appeared recently (Lehman *et al.*, 1987) together with an integration of the *Drosophila* observations with studies of other eukaryotic polymerases (Kaguni and Lehman, 1988). Several of the basic features of this protein are listed in Table 1. In agreement with studies of α polymerase from other eukaryotes, the *Drosophila* enzyme is composed of four polypeptides. The largest subunit of 182 kDa carries both the polymerase function (Kaguni *et al.*, 1983a; Cotterill *et al.*, 1987a) and a 3′→5′ exonuclease activity that participates in proofreading (Cotterill, *et al.*, 1987c; Reyland *et al.*, 1988; Cotterill *et al.*, in press). An extended search for a *Drosophila* primase has revealed that it copurifies with α polymerase and, in fact, resides in either or both of the 60 kDa and 50 kDa polypeptides (Conaway and Lehman, 1982; Kaguni *et al.*, 1983b; Cotterill *et al.*, 1987b). Although the fourth subunit of 73 kDa has no detectable enzymatic activity, its influence on the polymerizing subunit suggests that it may play a regulatory role (Cotterill *et al.*, in press).

The recovery of a functional 182 kDa subunit from α polymerase has recently produced several surprising observations (Cotterill *et al.*, 1987c; Cotterill *et al.*, in press). The 3′→5′ exonuclease activity mentioned above is, in fact, only observed in the isolated subunit. In addition, the fidelity of incorporation by the polymerase is increased more than 100 fold (Cotterill *et al.*, 1987c) over that of the complete polymerase-primase (Kaguni *et al.*, 1984). These observations, coupled with a 10-fold preference of the exonuclease activity for mismatched nucleotides, strongly implicates the nuclease in proofreading. The 73 kDa subunit appears to be the factor responsible for masking the proofreading function for reasons that are not yet clear. Since the processivity of the 182 kDa subunit is increased by *E. coli* SSB, it is anticipated that an analogous *Drosophila* protein makes up an additional component of the replication complex. In fact, a protein which stimulates DNA synthesis on single stranded DNA templates without increasing processivity has been identified in embryonic extracts (Kaguni *et al.*, 1983c). If α polymerase participates in DNA repair in *Drosophila*, as is likely in mammalian cells, the 182 kDa peptide plays a key role.

As part of an effort to reconstruct a complete replication complex, a *Drosophila* RNase H with affinity for the polymerase-primase has been purified to near homogeneity (DiFrancesco and Lehman, 1985). With a molecular weight of 180 kDa this enzyme is considerably larger than its mammalian counterparts. The specificity with which *Drosophila* RNase H stimulates polymerization by the polymerase-primase stands in striking contrast to the relatively poor stimulation by RNase H from *E. coli* and strongly supports a role for the homologous nuclease in replication. The stimulation is unique in that it is observed only when the RNA and DNA synthetic capacities of the polymerase are coupled (Lehman *et al.*, 1987). Stimulation of the coupled reaction appears to occur through an increased recycling of the polymerase because the nascent DNA chains are significantly shorter than those produced by the polymerase alone. The production of shorter DNA chains which accompanies stimulation implies that the nuclease increases the rate of initiation by the polymerase. Since RNA intermediates are not invoked in most current repair models, it is possible that the primase and RNase H activities of the replication complex are replaced by alternate accessory proteins in the formation of a repair complex.

β-Polymerase

Extraction conditions that release α polymerase to a soluble fraction leave another polymerase in the nucleus (Sakaguchi and Boyd, 1985). That enzyme has been designated β-polymerase in analogy with the closely related mammalian enzyme. *Drosophila* β-polymerase has been isolated to near homogeneity from embryos by sequential fractionation with DEAE cellulose, phosphocellulose, electrofocusing, and DNA cellulose chromatography. A combination of exclusion chromatography and denaturing gel electrophoresis indicates that this enzyme is composed of a single polypeptide of molecular weight 110 kDa. Although it is more than twice the size of the mammalian enzyme (Zmudzka *et al.*, 1986), like its mammalian counterpart it represents the smallest recognized DNA polymerase of its species (Table 1).

DNA polymerase β is distinguished from the other characterized *Drosophila* enzymes by its insensitivity to NEM (Table 1). Like α polymerase, this enzyme is most readily recovered from embryos, although it has been detected in flies (Furia *et al.*, 1979). Extensive studies, primarily employing polymerase inhibitors, have implicated β polymerase in DNA repair in mammals. In addition, the exclusive appearance of an analogous enzyme at meiosis in plants has implicated β polymerase in DNA recombination (see Sakaguchi and Boyd, 1985). The availability of a homogeneous enzyme from a genetically tractable organism such as *Drosophila* should make it possible to define the precise function of this enzyme class through a combined genetic and molecular analysis. Efforts are therefore currently underway in this laboratory to clone the structural gene of β polymerase in *Drosophila*. Knowledge of the cytogenetic position of that locus will in turn permit the recovery of mutants whose capacity for DNA repair and recombination can potentially define the biological function of this enzyme.

γ-Polymerase

The simplicity of mitochondrial replication provides a unique opportunity in eukaryotes to assemble a model *in vitro* replication system. The unidirectional mode of synthesis, for example, may permit a focused analysis of the continuous mode of replication. It is also possible that this system will shed light on the mechanism of nuclear DNA synthesis, because mitochondrial DNA synthesis may share synthetic factors with replication of the leading strand. Toward this end, a third *Drosophila* DNA polymerase has recently been purified to near-homogeneity from mitochondria (Wernette and Kaguni, 1986). In analogy with related enzymes from vertebrates, this enzyme is designated γ polymerase. This enzyme has again been purified from embryos because its concentration in that tissue exceeds that at other developmental stages by more than an order of magnitude.

The unique response of this enzyme to inhibitors clearly distinguishes it from α and β polymerase (Table 1). Unlike the chick enzyme, which is a homotetramer, the *Drosophila* enzyme is a heterodimer composed of two subunits with molecular weights of 125 kDa and 35 kDa. The polymerization function itself has been associated with the 125 kDa polypeptide. Retention of the molecular mass and K_m for dNTPs of the enzyme throughout purification argues against the artifactual proteolysis that has plagued previous polymerase studies. Because

Table 1. *Drosophila* DNA polymerases.

	Subunit Molecular Weight	S Value	pI	pH Optimum	Mg^{++} Optimum	Optimum Template	NEM	50% Inhibition by: ddTTP	Aphid.	Optimum KCL conc.	Reference
α-Polymerase	182kDa 73kDa 60kDa 50kDa	8.7	≈ 8	7.5	12mM	Activated calf thymus DNA	>90% 0.1mm	Insen	0.3μg/ml	50mM	Banks *et al.*, 1979 Lehman *et al.*, 1987
β-Polymerase	110kDa	≈ 5.0	≈ 5.0	8.4	15mM	Activated calf thymus DNA	Insen.	2μM	Insen.	250mM	Sakaguchi and Boyd, 1985
γ-Polymerase	125kDa 35kDa	7.6		9.0	3mM	poly (dA) p(dT)	0.1mM	≈5μM		120mM	Kaguni *et al.*, 1988 Wernette and Kaguni, 1986

Abbreviations: NEM, N-ethymaleimide; ddTTP, dideoxythymidine triphosphate; Aphid, Aphidicolin.

detergent is required for extraction and maintenance of enzyme activity, this protein is postulated to be associated with the mitochondrial membrane.

The fact that the *in vitro* properties of this enzyme are consistent with those predicted from *in vivo* studies of mitochondrial DNA synthesis suggests that γ polymerase will provide a solid foundation for the construction of a complete *in vitro* replication complex (Wernette *et al.*, in press; Kaguni *et al.*, 1988). In particular, the enzyme efficiently utilizes both single and double strand templates, as is required by the apparent continuous mode of mitochondrial replication. Although it is responsive to the secondary structure of the template, the enzyme exhibits an unusual ability to replicate through secondary structure in the single strand ϕ X174 template. The exceptional efficiency of the enzyme on the synthetic homopolymer poly(dA)·oligo$(dT)_{10}$ correlates well with the extreme A+T content of the *Drosophila* mitochondrial genome. Although γ polymerase incorporates only about 30 nucleotides during each association event, it tends to complete initiated strands because of a preference for previously utilized termini. The existence of ribosubstitution in the mitochondrial genome is apparently not attributable to misincorporation of ribonucleotides by γ polymerase because this enzyme exhibits a strong preference for deoxyribonucleotides; nor is the enzyme stabilized or stimulated by ATP. Since this enzyme is also stimulated by *E. coli* SSB, it is likely that the native replication complex includes a related homologous protein. Finally, the fidelity of the isolated enzyme as assayed by the ϕX174 SS→RF system is equal to that of *Drosophila* α polymerase.

Since γ polymerase is presently the only identified mitochondrial DNA polymerase, it is likely to participate in DNA repair processes similar to those recently implicated in mammalian mitochondria (Tomkinson and Linn, 1986; Domena and Mosbough, 1985). In particular, the *Drosophila* mitochondrial enzyme Nuclease 3, that is described below, may participate with γ polymerase in the response of mitochondria to DNA cross-linking agents.

Topoisomerases

Analyses of excision repair in established cell cultures of *Drosophila* have implicated topoisomerase activity in a preincision event (Harris and Boyd, 1987). In particular, it is postulated that such an activity is required to render the majority of DNA lesions accessible to repair complexes. It is, therefore, possible that one or both of the recognized *Drosophila* topoisomerases participates in repair in this organism. Because the *Drosophila* topoisomerases are considered in detail elsewhere in this volume they are discussed only briefly here.

Topoisomerase I

Topoisomerase I (Hsieh and Brutlag, 1980) was first identified in *Drosophila* as ω protein (Baase and Wang, 1974) (Table 2). As in the case of α polymerase, however, its characterization was initially plagued by proteolytic degradation (Wang *et al.*, 1980). More recently the availability of antibodies raised against this 135 kDa peptide (Javaherian *et al.*, 1982) has permitted a cytological and molecular localization of this protein to transcribed regions of the genome (Fleischmann *et al.*, 1984; Gilmour *et al.*, 1986). In the latter study topoisomerase I was cross linked to DNA in living cells by UV irradiation before the enzyme with its associated DNA was isolated by immunoprecipitation. Nucleic acid hybridization was then employed to identify the recovered sequences. Modifications of that very elegant technology could also prove valuable in identifying proteins that are associated with damaged or newly repaired

Table 2. *Drosophila* topoisomerases.

	Subunit Molecular Weight	**assay pH**	**Mg^{++} requirement**	**ATP requirement**	**Optimum NaCl conc.**	**Inhibition by: novobiocin**	**Reference**
Topo-isomerase I	135kDa	7.5	–	–	200mM	–	Hsieh & Brutlag 1980
			(stimulated)				Javaherian *et al.* 1982
Topo-isomerase II	164kDa	7.9	+	+	125mM	+	Hsieh & Brutlag 1980
	(protein is homodimer)		(7mM optimum)				Sander & Hseih 1983
							Wyckoff *et al.* in press
							Osheroff *et al.* 1983

regions of the genome. The apparent participation of topoisomerase I in transcription does not exclude a further role for this enzyme in DNA repair.

Topoisomerase II

One of the major advantages of employing *Drosophila* to analyze metazoan DNA metabolism is the potential for establishing the biological function of a protein through mutational analysis. That goal is closest to realization with topoisomerase II as a result of the recent cloning of the structural gene (*top2*) for that protein (Nolan *et al.*, 1986; Hsieh *et al.*, 1987; Wyckoff *et al.*, in press). The cloned sequences have been employed to localize *top2* to 37D2-6 on the left arm of chromosome 2 by *in situ* hybridization. With that information it is now possible to mount an intensive genetic analysis of that region of the chromosome. That study will begin with an investigation of the segregation distorter gene *Sd* which currently maps to the same chromosomal region as *top2*. A possible relationship between those genes is particularly intriguing in view of the fact that the *top2* gene of yeast may also be involved in chromosome segregation (see review by Hsieh *et al.*, 1987).

The complete nucleotide sequence of the *top2* gene (Wyckoff *et al.*, in press) has confirmed earlier structural studies that demonstrated a molecular weight of about 170 kDa for the largest active polypeptide (Table 2) (Sander and Hsieh, 1983; Shelton *et al.*, 1983; Osheroff *et al.*, 1983). Even the most rigorous efforts to inhibit proteolysis have failed to eliminate a variable amount of the lower molecular weight polypeptides that retain enzymatic activity and cross react with antibodies raised against the largest polypeptides (Sander and Hsieh, 1983; Hsieh, 1983; Heller *et al.*, 1986; Fairman and Brutlag, 1988). At this point it is tempting to speculate that one or more of those apparent derivatives of the largest polypeptide may be generated *in vivo* to participate in DNA repair. The bulk of topoisomerase II, however, is potentially involved in DNA replication because its levels correlate well with mitotic activity (Fairman and Brutlag, 1988). The fact that its activity is modified by phosphorylation (Ackerman *et al.*, 1985; Sander *et al.*, 1984) suggests that it is under further cellular regulation. Intensive investigation of the mechanism of action of that enzyme has laid a strong foundation for an understanding of its biological function (Osheroff and Brutlag, 1983; Sander and Hsieh, 1983; Osheroff, 1986, 1987; Osheroff and Zechiedrich, 1987).

As one of the major polypeptides of the nuclear matrix which, in contrast to topoisomerase I, is distributed uniformly throughout the genome (Berrios *et al.*, 1985; Heller *et al.*, 1986), topoisomerase II is also likely to play a fundamental role in chromatin structure. In recognition of that role, its nucleotide sequence specificity and its action on chromatin have been the object of intensive study (Sander and Hsieh, 1985; Udvardy *et al.*, 1985; Udvardy *et al.*, 1986; Sander *et al.*, 1987). The genetic analyses, which have been made possible by cloning of the topoisomerase II gene, should prove particularly helpful in sorting out what are clearly multiple functions of this fundamental and versatile protein.

Deoxyribonucleases

Among the nucleases currently recognized in *Drosophila* some, such as AP-endonucleases and a uracil specific nuclease, undoubtedly participate in repair whereas those associated with lysosomes probably do not. The genetic potential of *Drosophila* has yet to be exploited, however, in assigning repair functions to any of these enzymes. In view of that uncertainty, each of the identified *Drosophila* nucleases will be briefly described here.

An early survey of DNases in *Drosophila* employed an *in situ* assay for detecting nucleases following their separation by gel electrophoresis (Boyd and Mitchell, 1965). An analysis of the optimum assay conditions for each separable activity revealed that at least seven distinct enzymes could be detected during the course of development (Boyd, 1969). The relative dearth of nuclease activity that is detectable at pH 7 in early embryonic and late pupal stages is consistent with a predominantly digestive function for that activity (Muhammed *et al.*, 1967; Boyd and Boyd, 1970). In contrast, the striking increase in nucleases active at low pH that is observed just prior to metamorphosis implicates that activity in histolytic functions. Three such non-specific nucleases will be discussed prior to reviewing three alternate enzymes that have been implicated in DNA repair (Table 3).

Non specific nucleases

DNase-l

Among those *Drosophila* enzymes that are directly involved in DNA metabolism, this nuclease is currently the best studied genetically. Naturally occurring electrophoretic variants have been employed to map the genetic locus for this acid-active nuclease to the right arm of chromosome three at 61.8 cM (Grell, 1976; Detwiler and MacIntyre, 1978). Cytogenetic analysis has further localized that gene to a 10 band region between 90C2 and 90E on the polytene chromosomes (Detwiler and MacIntryre, 1978). Electrophoretic analysis of this enzyme in heteroallelic individuals has revealed an additive pattern that is indicative of a monomeric protein. A tissue specific modification of this activity that has been detected in late larval and early pupal stages may explain observations of two classes of acid active nucleases in *Drosophila hydei* (Boyd and Boyd, 1970). Null mutations have been recovered both by screening mutagenized stocks for an enzyme deficiency (Grell, 1976) and by examining extant strains for altered levels of polypyrimidine sequences (Stone *et al.*, 1983). Homozygotes of the null mutations exhibit reduced viability, a small size and disturbed ovarian development (Grell, 1976; Stone *et al.*, 1983). An accumulation of low molecular weight polypyrimidine tracts in ovaries of these strains is correlated with the appearance of abnormal nuclei (Stone *et al.*, 1983). Based on that observation the authors suggest that one function of DNase-1 in ovarian tissue is to complete the degradation of DNA in terminal nurse cells. Since meiotic recombination and mutagen sensitivity are normal in strains homozygous for the null mutations (Grell, 1976; Boyd unpublished observations), it is unlikely that DNase-1 participates in DNA recombination or repair

DNase-l is potentially homologous with a nuclease that has been characterized from salivary glands of *Drosophila hydei* (Table 3) (Boyd, 1970b). Both enzymes are active at acid pH, do not require a divalent cation and increase at puparium formation. If this analogy is sustained, however, these enzymes are not analogues of mammalian lysosomal nucleases as has been assumed. In particular, the *D. hydei* enzyme is an exonuclease that readily degrades only about half of a double stranded substrate. Since much of this *D. hydei* enzyme may be free in the haemolymph, it is important to firmly establish the intracellular location of DNase-1 in *D. melanogaster*. As yet, evidence for a lysosomal origin of that enzyme is circumstantial (Detwiler and MacIntyre, 1978).

Table 3. *Drosophila* deoxyribonucleases. Enzyme sources are *Drosphila melanogaster*, unless stated otherwise.

Enzyme	Source	Assay pH	Mg++ conc. Employed	Substrate Preference	Inhibitors	Map Positon	Mode of Diges-tion	S value/ Molecular Weight	References
DNase 1	Flies	4.5	None Required	Native DNA		61.8			Grell, 1976 Detwiler & Mac-Intyre, 1978
DNase 2	Flies	8.5		Denatured DNA		45.9			Grell, 1976
Nuclease 3	Fly and Embryo Mitochon-dria	5.5	5mM	Native DNA	NaCl above 100mM, Ca^{++}		Endonuclease	Aggregates	unpublished observations
Ap-endo nuclease I	Embryos	6.5	None Required	Depurinated DNA			Exonuclease	66kDa	Spiering & Deutsch, 1981, 1986
AP-endo nuclease II	Embryos	7.0	10mM	Depurinated DNA			Endonuclease	63kDa	Spiering & Deutsch, 1981, 1986
Uracil-specific Nuclease	3rd Instar Larvae	7.5	None Required	Uracil Con-taining DNA			Endonuclease		Deutsch & Spiering, 1982
Salivary nuclease	Drosophila Hydei Salivary Glands	5.4	None Required	Native DNA	NaCl, Cu^{++}		Exonuclease	3.4 S	Boyd, 1970b
"	"	8.6	25mM	Native DNA	NaCl, Ca^{++} Intestinal Factor		Endonuclease	6.3 S	Boyd, 1970a

DNase-2

Unlike the acid active nuclease, DNase-2 has not received any attention since its initial brief description (Grell, 1976). The genetic locus for this activity was mapped to 45.9 cM on the left arm of the third chromosome with the aid of a naturally occurring null mutation. That mutation eliminates the presence of two electrophoretically separable nuclease activities that are active on denatured DNA at pH 8.5 (Table 3). No other phenotype is associated with that null mutation.

Alkaline Active Salivary Nuclease.

An endonuclease that is active at alkaline pH has been characterized from the salivary glands of *Drosophila hydei* (Boyd, 1970a) (Table 3). Since expression of this enzyme and its counterpart in *D. melanogaster* is correlated with metamorphosis (Boyd and Logan, 1972), it is likely that this enzyme has a developmental function rather than a role in DNA repair.

Repair Related Nucleases

AP-endonuclease

An endonuclease activity that specifically cleaves depurinated DNA was initially detected in cultured cells and larval brain ganglia of wild type *Drosophila* (Osgood and Boyd, 1982). Two independently established cell lines of the excision deficient *mei-9*[a] mutation were found to exhibit a 50% reduction in that activity. Since that result could be reproduced with larval brain ganglia, it was possible to analyze other mutations of that gene that had not been introduced into tissue culture. Each of four independently isolated *mei-9* alleles, together with a *mus201* mutant, were also found to exhibit a significant reduction in enzyme activity. These observations parallel those of Kuhnlein *et al.*, (1978) who have identified an AP-endonuclease deficiency in the excision deficient human disorder xeroderma pigmentosum D. The question of whether the reduction in enzymatic activity is a direct or indirect effect of the mutations remains to be resolved. Direct cloning of both the *mei-9* gene (Binninger in this laboratory) and the structural gene for the AP-endonuclease should serve to clarify this question.

The discovery of Spiering and Deutsch (1981) that two fractions of AP-endonuclease activity can be recovered from *Drosophila* (Table 3) opens the possibility that the *mei-9* and *mus 201* mutants may be influencing only one of those two activities. That possibility could explain the elevated levels of AP-endonuclease in *mei-9* embryos relative to brain ganglia and tissue culture cells (Table 5, Osgood and Boyd, 1982). In addition to the fact that the two AP-endonuclease activities are separable by phosphocellulose chromatography, they also exhibit somewhat different physical and enzymological activities (Table 3). Although both enzymes cleave at the 3′-side of an apurinic site, they appear to produce different cleavage products (Spiering and Deutsch, 1986; Deutsch, 1987). Since contaminating enzymes can complicate such studies, a clear distinction between these activities awaits their final purification.

Uracil Specific Nuclease

In searching for a uracil glycosylase in *Drosophila*, Deutsch and Spiering (1982) instead found an endonuclease that acts specifically on uracil-containing DNA. That activity, which was detected only in third instar larvae, does not require a divalent cation (Table 3). Since the

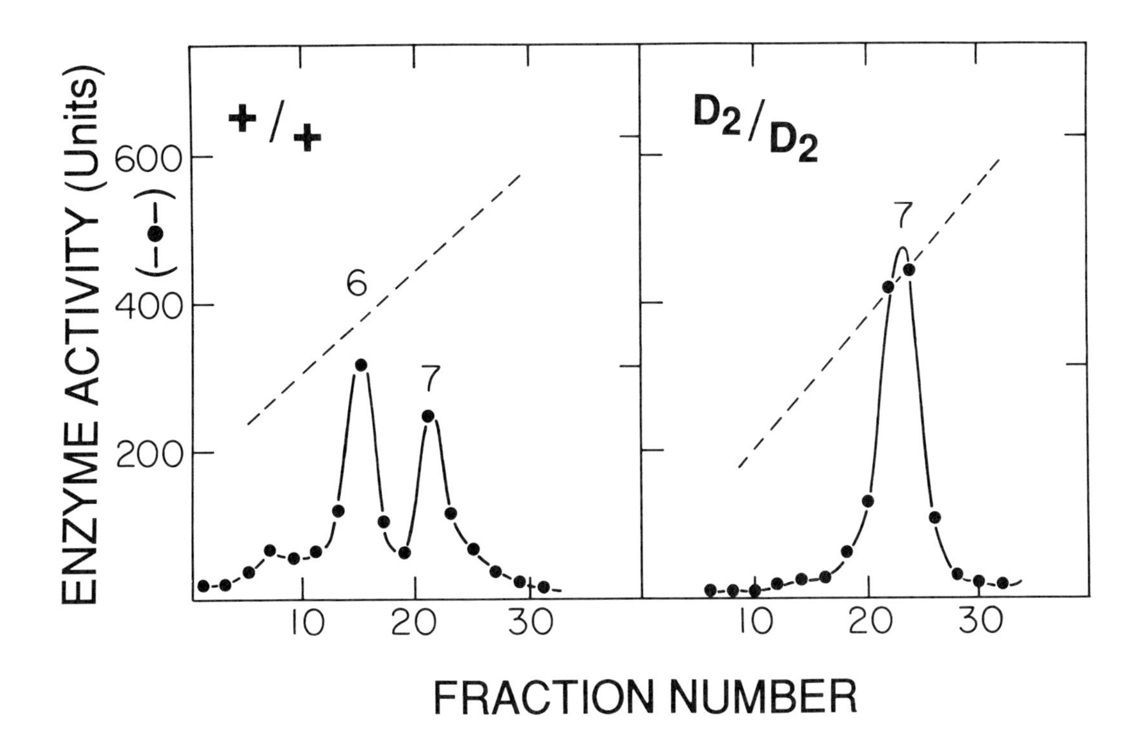

Figure 1. Electrofocusing of deoxyribonucleases from *Drosophila* ovaries. Ovaries from 4-5 day-old females were sonicated in extraction buffer as described (Sakaguchi and Boyd, 1985). Electrofocusing was conducted following centrifugation and Sephadex G-25 chromatography. Nuclease activity was assayed with a modification of the method of Lehman and Nussbaum (1964) in the presence of ATP. +/+ - ovaries from wild type females; D2/D2 - ovaries from females homozygous for the *mus308*^{D2} allele.

enzyme can act on DNA containing either A:U or G:U base pairs, it can potentially participate in repair of mismatches generated either by cytosine deamination or by uracil misincorporation. Its tissue specific appearance, however, suggests that it may serve a developmental function (see Deutsch, 1987)

Nuclease 3

As discussed below, the most consistent enzyme defect detected in a screen of mutagen sensitive mutants occurs in mutant alleles of the *mus308* gene (unpublished observations). Each of six alleles exhibits a change in a mitochondrial nuclease similar to that shown in Figure 1 for the *mus308*^{D2} allele. In that analysis ovaries from wild type (+/+) or mutant (D2/D2) females were sonicated, centrifuged and subjected to isoelectric focusing (see Boyd *et al.*, 1988). Non-specific nuclease activity is assayed with a uniformly labelled substrate. Numbers above the peaks represent the approximate pIs of the associated nuclease activities. The activity focusing at pI 6 has a pH optimum of 5.5, is stimulated by ATP and Mg^{2+} and is associated with ribonuclease and DNA-dependent ATPase activities. Additional studies suggest the pI 6 activity has, in fact, been shifted to the pI 7 region in the mutant tissue. Since the *mus308* mutants are unique among *Drosophila* mutagen-sensitive mutants in their exclusive sensitivity to DNA cross linking agents, this nuclease is potentially involved in resistance of the mitochondrial genome to such agents. The mutagen sensitivity is probably not due to a defect in nuclear repair because *mus308* mutants exhibit a normal response to nuclear DNA cross links. The hypothesis of a mitochondrial defect has been further supported by electron microscopic studies of flies that have been exposed to the DNA cross linking agent nitrogen mustard. Mitochondria in the flight muscle of treated *mus308* flies exhibit a higher frequency of mitochondrial aberrations than are found in treated wild type flies. These observations have stimulated a thorough characterization of the pI 6 nuclease and an analysis of the response of the mitochondrial genome to DNA cross linking agents.

DNA Ligases

DNA ligase I

Since ligation is an essential feature of most repair processes, it is reassuring that the two typical eukaryotic forms of that activity have recently been recovered from *Drosophila* embryos (Takahashi and Senshu, 1987). The strong similarity between the partially purified ligase I of that group and a ligase purified to near homogeneity by Rabin *et al.* (1986) establishes the identity of those two activities (Table 4). Both groups have identified multiple molecular weight forms of that activity which Rabin *et al.* have convincingly demonstrated to be structurally related. The smaller molecular weight forms can, in fact, be derived from the 86 kDa polypeptide by proteolysis with retention of ligase activity by the two largest peptides. In an exhaustive investigation of the mechanism of ligation the latter group has established a mode of action identical to that of the *E. coli* and T4 DNA ligases (Rabin and Chase, 1987). Utilization of a series of substrates has revealed a unique spectrum of ligation reactions that include blunt-end ligation and the ligation of oligonucleotides containing terminal mismatches. The availability of a homogeneous protein places these investigators in a position to explore the function of this enzyme through molecular cloning.

DNA ligase II

An alternate ligase has been recovered from isolated nuclei, in contrast to ligase I which is predominantly cytoplasmic in origin (Takahashi and Senshu, 1987). Several properties, including the elution point from phosphocellulose, the K_m for ATP and peptide molecular weight distinguish this activity from ligase I (Table 4). The concentration of ligase I is correlated with that stage of embryonic development in which DNA synthesis is most intense whereas ligase II is present at more constant levels throughout embryonic development. That observation makes ligase II the form most likely to be involved in DNA repair complexes.

DNA Damage-Specific Enzymes

In addition to the damage-specific nucleases described above, three further enzymatic activities have been detected that act on modified DNA. Although none of these activities have been extensively purified, their identification in *Drosophila* opens the possibility of an eventual genetic analysis of critical metazoan repair functions. A review of the currently recognized damage-specific enzymes in *Drosophila* has recently appeared (Deutsch, 1987)

Table 4. *Drosophila* DNA ligases. Inferred information is included in parentheses. Polypeptides shown in brackets are presumably proteolytic degration products of the larger polypeptides.

	Intracellular Loctation	Elution from Phosphocel lulose	Km for ATP	Requirements			Stimulation KCl mM	Polypeptide MW, kDa	S Value	Inhibi- tion by NEM	Reference
				ATP	Mg++	DTT					
Ligase I	Cytoplasm	0.2M KCl	2.7μm	+	+	+	None	86 [75]		+	Takahasi & Senshu, 1987
(Ligase I)	(Cytoplasm)	0.2M NaCl	1.6μm	+	+	+		83	4.3	+	Rabin *et al.*, 1986
								[75] [64]			Rabin and Chase, 1987
Ligase II	Nuclei	0.6M KCl	30μm	+	+	-	50-100	90 70	4.2	+	Takahasi & Senshu, 1987

DNA Glycosylases

Early failures to detect glycosylases that are specific for uracil, N-7 methylguanine and 3-methyladenine in *Drosophila* (Friedberg *et al.*, 1978, Deutsch and Spiering, 1982; Green and Deutsch, 1983) has supported the concept that glycosylases play a less prominent role in *Drosophila* repair than they do in mammalian repair. Nevertheless, Breimer (1986) has recently detected a glycosylase activity in extracts of *Drosophila* embryos that acts on damaged pyrimidines. Since the molecular weight associated with this activity is much higher than that of the corresponding mammalian enzymes, it will be particularly interesting to compare these two activities in their purified forms.

Methyltransferase

Since a methyltransferase-like activity recently detected in *Drosophila* has a somewhat broader specificity than previously identified enzymes of this class (Green and Deutsch, 1983), this function may partially compensate for the lack of glycosylases in this organism. In that study crude pupal extracts were shown to transfer methyl groups from N-7 methylguanine and 3-methyladenine to protein. Again, enzyme purification is required to substantiate these preliminary observations.

Base Insertion

A genetic analysis of this phenomena is badly needed to establish its significance in eukaryotic DNA repair. An initial step in that direction was taken by Deutsch and Spiering (1985) who detected an activity in *Drosophila* embryos that adds purines to DNA with AP-sites. Since the partially purified activity abolishes the alkali lability of the substrate, the bases are presumed to be added to the free sugars. Synthetic oligonucleotides were employed to demonstrate base specificity in the addition reaction. Unfortunately a thorough analysis of this activity has not been possible because it is extremely labile.

Screening *Drosophila* Mutants for Enzyme Deficiencies

An analysis of DNA repair in *Drosophila* was initiated in this laboratory through the isolation of mutants that are hypersensitive to DNA damaging agents. At present, mutants in 12 of the 30+ identified genes have been shown to exhibit specific repair defects (see Boyd *et al.*, 1987; Henderson *et al.*, 1987). In an initial attempt to identify the protein products of these genes, we have surveyed mutants at 18 different genetic loci for defects in a series of DNA metabolizing enzymes. Those mutants were chosen which yield adequate quantities of homozygous embryos: they represent 6 loci on the X chromosome, 4 on the 2nd chromosome and 8 genes on the 3rd. In selecting the activities to be tested, a list of enzymes was drawn up that are known, primarily from prokaryotic studies, to be potentially involved in DNA repair. A subset of those enzymes that could be readily quantified after minimal purification from *Drosophila* embryos was then selected for the mutant analysis. To compensate for differences in embryo viability, relative activities of different enzymes have been compared within a single extract where possible. A qualitative summary of that screen as presented in Table 5 is summarized below:

Table 5. Assay of DNA metabolizing enzymes in mutagen-sensitive *Drosophila* mutants

Fractionation	Crude Extract	Phosphocellulose		Electrofocusing				
Enzyme Assayed	Topo-isomerase I	Nuclease		DNA polymerase			Nuclease	
		Non-specific	AP-endo-nuclease	α	β	pl 7	# 3	pl 7.5
Mutant Allele								
mei-9[D2]	+	+	(+)	+*	+	+	+	
mei-41[D5]	+	+	+	+	+	+	+	–
mus 101[D1]	+	+	+	+*				
mus 102[D1]	+	+	+	+	+	+	±	±
mus 105[D1]	+	+	+	+	+	+	+	
mus 108[A1]				+	+	+	+	±
mus 201[D1]	+	+	(+)					
mus 205[A1]	+	+	+	+	+	+	+	–
mus 206[A1]	+	+	+	+	+	+	+	–
mus 207[A1]	+	+	+	+	+	+	+	–
mus 302[D1]	+			+	+	+	+	±
mus 304[D3]	+	+	+	+	+	+	+	–
mus 305[D1]		+	+	±	–	+	+	±
mus 306[D1]	+	+	+	+	+	+	+	–
mus 307[D1]	+	+	+	+	+	+	+	–
mus 308[D2]	+	+	+	+	+	+	–	±
mus 310[D1]		+	+	+			+	+
mus 311[D3]	+	±	+	+	+	+	+	+

* Dussenbery and Smith (1983) have previously reported normal levels of polymerase α in mutants of *mus101* and *mei-41.*

( ) See Osgood and Boyd, 1982 for additional results.

Between 0.5-2g of homozygous mutant embryos were sonicated and franctionated as described by Boyd *et al.*, 1988. In this procedure DNA is removed from a high speed supernatant fraction by DEAE cellulose chromatography. Topoisomerase I activity was monitored after removal of NaCl and detergent by Sephadex G-25 chromatography. The above sample was then fractionated either by phosphocellulose chromatography or by electrofocusing (Sakaguchi and Boyd, 1985). The phosphocellulose column was eluted with a gradient of phosphate (0.04M-0.9M) in PEMG buffer containing 0.01% Triton X-100. Topoisomerase I was assayed in a reaction mixture containing 10mM sodium phosphate, pH 7.0, 0.16 M NaCl, 1.0 mM EDTA, 0.5mM dithiothreitol, 0.1 mg/ml BSA, 9μg/ml superhelical DNA, and 0.05 to 0.20 μg extract protein. After incubation at 30°C for 15 min, the reaction was stopped by adding SDS to 0.2%. The extent of the DNA relaxation was quantified on agarose gels. AP-endonuclease and a non-specific nuclease were assayed according to Osgood and Boyd (1982). DNA polymerases were assayed as described by Sakaguchi and Boyd (1985). Nuclease 3 and the PL 7.5 nuclease were assayed in the presence of ATP by a modification of Lehman and Nussbaum (1964). + approximately wild type levels of activity. – major alteration in enzyme activity. ± potential modifications of enzyme activity.

1. Topoisomerase I activity was assayed as the capacity of a DNA-free supernatant fraction to relax superhelical PM2 DNA. In an agarose gel assay that is linear to 200 ng of protein, each of 15 tested mutants exhibit over 85% of control enzyme levels. Since all these values are within two standard errors of the mean, none of these mutants are significantly deficient in topoisomerase I activity.

2. When the above extract is subjected to phosphocellulose chromatography, two separable peaks of nuclease activity are recovered (unpublished observations); an AP-endonuclease and a non-specific nuclease. The unbound AP-endonuclease observed by Spiering and Deutsch (1981, see below) is not observed under these conditions. The ratio of these two activities has

been compared between mutant and wild type extracts. This analysis failed to reveal a partial deficiency of AP-endonuclease activity previously seen in unfractionated extracts of *mei-9* and *mus 201* brain ganglia (Osgood and Boyd, 1982). The defect previously seen in larval tissue is therefore not reflected in the single embryonic activity assayed under the alternate conditions employed here. Mutant *mus311*^{D3} embryos, however, do exhibit a 75% reduction of the non-specific activity.

3. Two ATP-stimulated nucleases can also be separated from the crude extract by electrofocusing (see Figure 1). A striking modification of the activity with pI 6.2 in six tested alleles of the *mus308* gene constitutes the most definitive mutant effect seen in this screen. This enzyme, Nuclease 3, has therefore been further studied as described above. Although the activity of the alternate enzyme with pI 7.5 has not been associated with a specific genetic locus, we are in a position to study its genetic control because it is present in some stocks and not in others.

4. This same fractionation scheme has revealed the presence of three species of DNA polymerase in *Drosophila* embryo extracts (Sakaguchi and Boyd, 1985). The activities with pIs of 5 and 8 have been identified as α and β polymerase respectively, and the activity with pI 7 remains uncharacterized. Among 15 mutants that have been analyzed for the minor polymerases, β polymerase is deficient in *mus305*^{D1}. Since a partial reduction in α polymerase has also been observed in that mutant, neither enzyme can be definitively associated with that locus.

In summary, this screen of *Drosophila* mutants for enzyme defects has revealed a striking modification of an ATP-stimulated nuclease in alleles of the *mus308* gene. Other less extreme alterations have been observed: A non-specific endonuclease is partially reduced in a *mus311* mutant and α and β polymerase levels are altered in a *mus305* mutant. This investigation will also allow us to study the genetic control of the second ATP-stimulated nuclease. A previous study (Osgood and Boyd, 1982) has revealed a developmental alteration in AP-endonuclease activity associated with mutations of the *mei-9* and *mus 201* genes.

SUMMARY

Several genetic observations in *Drosophila* can be interpreted in terms of macromolecular DNA-repair complexes. A prototype for such a complex is currently being reconstructed with the proteins involved in DNA replication. Several enzymes that may participate in analogous repair complexes have been purified and characterized to various extents. These include three DNA polymerases, two topoisomerases, two DNA ligases, two AP endonucleases, a mitochondrial nuclease, a uracil specific endonuclease, a damage specific glycosylase, a base insertase, and a methyltransferase. As more of these proteins are purified to homogeneity, it will be possible to examine their mutual interactions in an attempt to define the structure and function of specific repair complexes. That effort is expected to be accelerated by the application of both forward and reverse genetic approaches. In the forward approach, mutants in the *mei-9* and *mei-41* genes, that are deficient in excision and postreplication repair respectively, have been recovered with transposon insertions. Those mutations are being exploited to clone the corresponding wild type genes in an attempt to identify their protein products. With the recent availability of homogeneous polymerases, ligase and topoisomerase, it is now possible to pursue the reverse genetic approach by recovering clones of the structural genes for those proteins. That effort is now well underway with the topoiomerase II gene which has been cloned and cytogenetically localized. The phenotypes of mutants that can be recovered with such information should eventually reveal the extent to which each of these proteins participates in the processes of DNA repair, recombination and synthesis.

ACKNOWLEDGEMENTS

We are grateful to the many authors mentioned throughout the text who have shared their unpublished manuscripts with us. David M. Binninger, Edith Leonhardt, Andrew W. Singson, Laurie S. Kaguni, Susan Whitehouse-Hills and Douglas L. Wendell provided valuable assistance with the manuscript. This study has been supported by the U.S. Department of Energy (EV70210) and the United States Public Health Service (GM32040) The literature search for this article was complete on June 15, 1988. Since that date a repair-related *Drosophila* gene has been cloned with the aid of an antibody to a human apurinic endonuclease [Kelley *et al.*, 1989, (*Mol. Cell. Biol.* **9**: 965-973)].

REFERENCES

P. Ackerman, C. V. C. Glover and N. Osheroff. (1985) Phosphorylation of DNA topoisomerase II by casein kinase II: Modulation of eukaryotic topoisomerase II activity *in vitro*. *Proc. Natl. Acad. Sci. USA* **82**: 3164-3168

B. M. Alberts. (1985) Protein machines mediate the basic genetic processes. *Trends in Genetics* **1**: 26-30

W. A. Baase, J. C. Wang. (1979) A DNA relaxing enzyme isolated from *Drosophila* embryos. *Biochemistry* **13**: 4299-4303

B. S. Baker and D. A. Smith. (1979) The effects of mutagen-sensitive mutants of *Drosophila melanogaster* in nonmutgenized cells. *Genetics* **92**: 833-847

G. R. Banks, J. A. Boezi and I. R. Lehman. (1979) A high molecular weight DNA polymerase from *Drosophila melanogaster* embryos. *J. Biol. Chem.* **254**: 9886-9892

J.-L. Becker, F. Barre-Sinoussi, D. Dormont, M. Best-Belpomme and J.-C. Chermann. (1987) Characterization of the purified RNA dependent DNA polymerase isolated from *Drosophila*. *Cell. Mol. Biol.* **33**: 225-235

M. Berrios, N. Osheroff and P. A. Fisher. (1985) *in situ* localization of DNA topoisomerase II, a major polypeptide component of the *Drosophila* nuclear matrix fraction. *Proc. Natl. Acad. Sci. USA* **82**: 4142-4146

J. B. Boyd. (1969) *Drosophila* deoxyribonucleases. I. Variation of deoxyribonucleases in *Drosophila melanogaster*. *Biochim. Biophys. Acta* **171**: 103-112

J. B. Boyd. (1970a) Characterization of an alkaline active deoxyribonuclease from the prepupal salivary gland of *Drosophila hydei*. *Biochim. Biophys. Acta* **209**: 339-348

J. B. Boyd. (1970b) Characterization of an acid active deoxyribonuclease from the larval salivary gland of *Drosophila hydei*. *Biochim. Biophys. Acta* **209**: 349-356

J. B. Boyd and H. Boyd. (1970) Deoxyribonucleases of the organs of *Drosophila hydei* at the onset of metamorphosis. *Biochem. Genetics* **4**: 447-459

J. B. Boyd and P. V. Harris. (1987) Isolation and characterization of a photorepair-deficient mutant in *Drosophila melanogaster*. *Genetics* **116**: 233-239

J. B. Boyd, P. V. Harris and K. Sakaguchi. (1988) Use of *Drosophila* to study DNA repair. **In**: *DNA Repair. A laboratory manual of research procedures* **Vol. 3**. pp 99-113. Marcel Dekker, New York.

J. B. Boyd and W. R. Logan. (1972) Developmental variations of a deoxyribonuclease in the salivary glands of *Drosophila hydei* and *Drosophila melanogaster*. *Cell Differentiation* **1**: 107-118

J. B. Boyd, J. M. Mason, A. H. Yamamoto, R. K. Brodberg, S. S. Banga and K. Sakaguchi. (1987) A genetic and molecular analysis of DNA repair in *Drosophila*. *J. Cell Sci.* **Suppl. 6**: 39-60

J. B. Boyd and H. K. Mitchell. (1965) Identification of deoxyribonucleases in polyacrylamide gel following their separation by disk electrophoresis. *Analytical Biochem.* **13**: 28-42

J. B. Boyd and R. B. Setlow. (1976) Characterization of postreplication repair in mutagen-sensitive strains of *Drosophila melanogaster*. *Genetics* **84**: 507-526

C. L. Brakel and A. B. Blumenthal. (1977) Multiple forms of *Drosophila* embryo DNA polymerase: Evidence for Proteolytic Conversion. *Biochemistry* **16**: 3137-3143

C. L. Brakel and A. B. Blumenthal. (1978a) Three forms of DNA polymerase from *Drosophila melanogaster* embryos. *Eur. J. Biochem.* **88**: 351-362

C. L. Brakel and A. B. Blumenthal. (1978b) Replication of poly dA and poly rA by a *Drosophila* DNA polymerase. *Nucleic Acids Res.* **5**: 2565-2575

L. H. Breimer. (1986) A DNA glycosylase for oxidized thymine residues in *Drosophila melanogaster. Biochem. Biophys. Res. Comm.* **134**: 201-204

L. H. Cohen, P. E. Penner and L. A. Loeb. (1973) Multiple DNA polymerases displayed by isoelectric focusing. *Ann. N.Y. Acad. Sci.* **209**: pp 354-362

R. C. Conaway and I. R. Lehman. (1982) A DNA primase activity associated with DNA polymerase α from *Drosophila melanogaster* embryos. *Proc. Natl. Acad. Sci. USA* **79**: 2523-2527

S.M. Cotterill, G. Chui and I. R. Lehman. (1987a) DNA polymerase-primase from embryos of *Drosophila melanogaster*. The DNA polymerase subunit. *J. Biol. Chem.* **262**: 16100-16104

S. M. Cotterill, G. Chui and I. R. Lehman. (1987b) DNA polymerase-primase from embryos of *Drosophila melanogaster*. The primase subunits. *J. Biol. Chem.* **262**: 16105-16108

S. M. Cotterill, M. E. Reyland, L. A. Loeb and I. R. Lehman. (1987c) A cryptic proofreading 3′→5′ exonuclease associated with the polymerase subunit of the DNA polymerase-primase from *Drosophila melanogaster. Proc. Natl. Acad. Sci. USA* **84**: 5635-5639

S. M. Cotterill, M. E. Reyland, L. A. Loeb and I. R. Lehman. Enzymatic activities associated with the polymerase subunit of the DNA polymerase-primase of *Drosophila melanogaster*. Cold Spring Harbor, in press.

C. Detwiler and R. MacIntyre. (1978) A genetic and developmental analysis of an acid deoxyribonuclease in *Drosophila melanogaster. Biochem. Genet.* **16**: 1113-1134

W. A. Deutsch. (1987) Enzymatic studies of DNA repair in *Drosophila melanogaster. Mutation Res.* **184**: 209-215

W. A. Deutsch and A. L. Spiering. (1982) A new pathway expressed during a distinct stage of *Drosophila* development for the removal of dUMP residues in DNA. *J. Biol. Chem.* **257**: 3366-3368

W. A. Deutsch and A. L. Spiering. (1985) Characterization of a depurinated-DNA purine-base-insertion activity from *Drosophila. Biochem. J.* **232**: 285-288

R. A. DiFrancesco and I. R. Lehman. (1985) Interaction of ribonuclease H from *Drosophila melanogaster* embryos with DNA polymerase-primase. *J. Biol. Chem.* **260**: 14764-14770

J. D. Domena and D. W. Mosbaugh. (1985) Purification of nuclear and mitochondrial uracil-DNA glycosylase from rat liver. Identification of two distinct subcellular forms. *Biochemistry* **24**: 7320-7328

R. L. Dusenbery and P. D. Smith. (1983) DNA polymerase activity in developmental stages of *Drosophila melanogaster. Developmental Genetics* **3**: 309-327

H. Echols. (1986) Multiple DNA-protein interactions governing high-precision DNA transactions. *Science* **233**: 1050-1056

R. Fairman and D. L. Brutlag. (1988) Expression of the *Drosophila* type II topoisomerase is developmentally regulated. *Biochemistry* **27**: 560-565

G. Fleischmann, G. Pflugfelder, E. K. Steiner, K. Javaherian, G. C. Howard, J. C. Wang and S. C. R. Elgin. (1984) *Drosophila* DNA topoisomerase I is associated with transcriptionally active regions of the genome. *Proc. Natl. Acad. Sci. USA* **81**: 6958-6962

E. C. Friedberg, T. Bonura, R. Cone, R. Simmons and Corrie Anderson. (1978) Base excision repair of DNA. **In**: *DNA Repair Mechanisms*. (Eds. Hanawalt, Friedberg, and Fox). Academic Press, pp 163-178.

M. Furia, L. C. Polito, G. Locorotondo and P. Grippo. (1979) Identification of a DNA polymerase β-like form in *Drosophila melanogaster* adult flies. *Nucleic Acids Res.* **6**: 3399-3410

M. Gatti, D. A. Smith and B. S. Baker. (1983) A gene controlling condensation of heterochromatin in *Drosophila melanogaster. Science* **221**: 83-85

D. S. Gilmour, G. Pflugfelder, J. C. Wang and J. T. Lis. (1986) Topoisomerase I interacts with transcribed regions in *Drosophila* cells. *Cell* **44**: 401-407

U. Graf, M. M. Green and F. E. Wurgler. (1979) Mutagen-sensitive mutants in *Drosophila melanogaster*. Effects on premutational damage. *Mutation Res.* **63**: 101-112

D. A. Green and W. A. Deutsch. (1983) Repair of alkylated DNA: *Drosophila* have DNA methyltransferases but not DNA glycosylases. *Mol. Gen. Genet.* **192**: 322-325

E. H. Grell. (1976) Genetics of some deoxyribonucleases of *Drosophila melanogaster. Genetics* **83** S28-S29.

P. V. Harris and J. B. Boyd. (1987) Pyrimidine dimers in *Drosophila* chromatin become increasingly accessible after irradiation. *Mutation Research* **183**: 53-60.

R. S. Hawley, C. H. Marcus, M. L. Cameron, R. L. Schwartz and A. E. Zitron (1985) Repair-defect mutations inhibit rDNA magnification in *Drosophila* and discriminate between meiotic and premeiotic magnification. *Proc. Natl. Acad. Sci. USA* **82**: 8095-8099

R. H. Haynes, L. Prakash, M. A. Resnick, B. S. Cox, E. Moustacchi and J. B. Boyd. (1978) DNA repair in lower eucaryotes. **In**: *DNA Repair Mechanisms*. (Eds. Hanawalt, Friedberg, and Fox). pp 405-411. Academic Press, New York.

R. A. Heller, E. R. Shelton, V. Dietrich, S. C. R. Elgin and D. L. Brutlag. (1986) Multiple forms and cellular localization of *Drosophila* DNA topoisomerase II. *J. Biol. Chem.* **261**: 8063-8069

D. S. Henderson, D. A. Bailey, D. A. R. Sinclair and T. A. Grigliatti. (1987) Isolation and characterization of second chromosome mutagen-sensitive mutations in *Drosophila melanogaster. Mutation Res.* **177**: 83-93

T.-s. Hsieh. (1983) Purification and properties of type II DNA topoisomerase from embryos of *Drosophila melanogaster. Methods in Enzymology*. **Vol. 100B**. (Eds. Wu, Grossman and Moldave), pp 161-170. Academic Press, New York.

T.-s. Hsieh and D. Brutlag. (1980) ATP-dependent DNA topoisomerase from *D. melanogaster* reversibly catenates duplex DNA rings. *Cell* **21**: 115-125

T.-s. Hsieh, M. P. Lee, J. M. Nolan and E. Wyckoff. (1987) Molecular genetic analysis of topoisomerase II gene from *Drosophila melanogaster. NCI Monog.* **4**: 7-10

I. Husain, B. Van Houten, D. C. Thomas, M. Abdel-Monem and Aziz Sancar. (1985) Effect on DNA polymerase I and DNA helicase II on the turnover rate of UvrABC excision nuclease. *Proc. Natl. Acad. Sci. USA* **82**: 6774-6778

K. Javaherian, Y-C. Tse and J. Vega. (1982) *Drosophila* topoisomerase I: isolation, purification and characterization. *Nucleic Acids Res.* **10**: 6945-6955

S. Jentsch, J. P. McGrath and A. Varshavsky. (1987) The yeast DNA repair gene *RAD6* encodes a ubiquitin-conjugating enzyme. *Nature* **329**: 131-134

L. S. Kaguni, R. A. DiFrancesco and I. R. Lehman. (1984) The DNA polymerase-primase from *Drosophila melanogaster* embryos. Rate and fidelity of polymerization on single-stranded DNA templates. *J. Biol. Chem.* **259**: 9314-9319

L. S. Kaguni and I. R. Lehman. (1988) Eukaryotic DNA polymerase-primase: Structure, mechanism and function. *Biochim. Biophys. Acta* **950**: 87-101.

L. S. Kaguni, J.-M. Rossignol, R. C. Conaway and I. R. Lehman. (1983a) Isolation of an intact DNA polymerase-primase from embryos of *Drosophila melanogaster. Proc. Natl. Acad. Sci. USA* **80**: 2221-2225

L. S. Kaguni, J.-M. Rossignol, R. C. Conaway G. R. Banks and I. R. Lehman. (1983b) Association of DNA primase with the β and γ subunits of DNA polymerase α from *Drosophila melanogaster* embryos. *J. Biol. Chem.* **258**: 9037-9039

L. S. Kaguni, J.-M. Rossignol, R. C. Conaway and I. R. Lehman. (1983c) The DNA polymerase-primase complex of *Drosophila melanogaster* embryos. **In**: *Mechanisms of DNA Replication and Recombination*. (Ed. C. N. Cozzarelli). pp 495-510. Alan R. Liss, Inc., New York.

L. S. Kaguni, C. M. Wernette, M. C. Conway and P. Yang-Cashman. (1988) Structural and catalytic features of the mitochondrial DNA polymerase from *Drosophila melanogaster* embryos. **In**: *Cancer Cells: Eukaryotic DNA Replication* **Vol. 6** (Eds. T.J. Kelley and B.W. Stillman). pp. 425-432.

U. Kuhnlein, B. Lee, E. E. Penhoet and S. Linn. (1978) Xeroderma pigmentosum fibroblasts of the D group lack an apurinic DNA endonuclease species with a low apparent K_m. *Nucleic Acids Res.* **5**: 951-960

I. R. Lehman, R. A. DiFrancesco, L. S. Kaguni and S. Cotterill. (1987) Assembly of a DNA replication complex from *Drosophila melanogaster* embryos. **In**: *DNA Replication and Recombination*, pp 89-100. Alan Liss, New York.

I. R. Lehman and A. L. Nussbaum. (1964) The deoxyribonucleases of *Escherichia coli*. V. On the specificity of exonuclease I (phosphodiesterase). *J. Biol. Chem.* **239**: 2628-2636

A. Muhammed, J. M. Goncalves and J. E. Trosko. (1967) Deoxyribonuclease and deoxyribonucleic acid polymerase activity during *Drosophila* development. *Devel. Biol.* **15**: 23-32

J. M. Nolan, M. P. Lee, E. Wyckoff and T.-s. Hsieh. (1986) Isolation and characterization of the gene encoding *Drosophila* DNA topoisomearse II. *Proc. Natl. Acad. Sci. USA* **83**: 3664-3668

C.J. Osgood and J.B. Boyd. 1982. Apurinic endonuclease from *Drosphila melanogaster*: Reduced enzymatic activity in excision-deficient mutants of the *mei-9* and *mis(2)201* loci. Mol. Gen. Genet., **186:** 235-239.

N. Osheroff. 1986. Eukaryotic topoisomearse II. Characterization of enzyme turnover. J. Biol. Chem., **261:** 9944-9950.

N. Osheroff. 1987. Role of the divalent cation in topoisomerase II mediated reactions. Biochem., **26:** 6402-6406.

N. Osheroff and D.L. Brutlag. 1983. Recognition of supercoiled DNA *Drosophila* topoisomerase II. In: Mechanisms of DNA Replication and Recombination, pp 55-64. Alan R. Liss, New York.

N. Osheroff, E.R. Shelton and D.L. Brutlag. 1983. DNA Topoisomerase II from *Drosophila melanogaster*. Relaxation of supercoiled DNA. J. Biol. Chem., **258:** 9536-9543.

N. Osheroff and E.L. Zechiedrich. 1987. Calcium-promoted DNA cleavage by eukaryotic topoisomerase II: Trapping the covalent enzyme-DNA complex in an active form. Biochem., **26:** 4303-4309.

B.A. Rabin, R.S. Hawley and J. W. Chase. 1986. DNA ligase from *Drosophia melanogaster* embryos. Purification and physical characterization. J. Biol. Chem., **261:** 10637-10645.

B.A. Rabin and J.W. Chase. 1987. DNA ligase from *Drosophila melanogaster* embryos. Substrate specificity and mechanism of action. J. Biol. Chem., **262:** 14105-14111.

M. E. Reyland, I. R. Lehman and L. A. Loeb. (1988) Specificity of proofreading by the 3′→5′ exonuclease of the DNA polymerase-primase of *Drosophila melanogaster. J. Biol. Chem.* **263**: 6518-6524

K. Sakaguchi and J. B. Boyd. (1985) Purification and characterization of a DNA polymerase β from *Drosophila. J. Biol. Chem.* **260**: 10406-10411

M. Sander and T.-s. Hsieh. 1983. Double strand DNA cleavage by type II DNA topoisomerase from *Drosophila melanogaster*. J. Biol. Chem., **258:** 8421-8428.

M. Sander and T.-s. Hsieh. 1985 *Drosophila* topoisomerase II double-strand DNA cleavage: analysis of DNA sequence homology at the cleavage site. Nucleic Acids Res., **13:** 1057-1072.

M. Sander and T.-s. Hsieh, A. Udvardy and P. Schedl. 1987 Sequence dependence of *Drosophila* topoisomerase II in plasmid relaxation and DNA binding. J. Mol. Biol., **194:** 219-229.

M. Sander, J.M. Nolan and T.-s. Hsieh. 1984. A protein kinase activity tightly associated with *Drosophila* type II DNA topoisomerase. Proc. Natl. Acad. Sci. USA, **81:** 6938-6942.

B. Sauer and I.R. Lehman. 1982. Immunological Comparisons of purified DNA polymerase α from embyos of *Drosophila melanogaster* with forms of the enzyme present *in vivo*. J. Biol. Chem., **257:** 12394-12398.

E.R Shelton, N. Osheroff and D.L. Brutlag. 1983. DNA topoisomerase II from *Drosophila melanogaster*. Purification and physical characterization. J. Biol. Chem., **258:** 9530-9535.

D.A. Smith, B.S. Baker and M. Gatti. 1985. Mutations in genes encoding essential mitotic functions in *Drosophila melanogaster*. Genetics, **110:** 647-670.

P.D. Smith, R.D. Snyder and R. L. Dusenbery. 1980. Isolation and characterization of repair-deficient mutants of *Drosophila melanogaster*. In: DNA Repair and Mutagenesis in Eukaryotes. (Eds. Generoso, Shelby and deSerres). Plenum. Vol. 15 pp 175-188, New York.

P. B. Snyder, V. K. Galanopoulos and F. C. Kafatos. (1986) trans acting amplification mutants and other eggshell mutants of the third chromosome in *Drosophila melanogaster*. *Proc. Natl. Acad. Sci. USA* **83**: 3341-3345

A. L. Spiering and W. A. Deutsch. (1981) Apurinic DNA endonucleases from *Drosophila melanogaster* embryos. *Mol. Gen. Genet.* **183**: 171-174

A. L. Spiering and W. A. Deutsch. (1986) *Drosophila* apurinic/apyrimidinic DNA endonucleases. Characterization of mechanism of action and demonstration of a novel type of enzyme activity. *J. Biol. Chem.* **261**: 3222-3228

A. Spradling and T. Orr-Weaver. (1987) Regulation of DNA replication during *Drosophila* development. *Ann. Rev. Genet.* **21**: 373-403

J.C. Stone, N.A. Dower, J. Hauseman, Y.M.T. Cseko and R. Sederoff. 1983. The characterization of mutant affecting DNA metabolism in the development of *D. melanogaster*. Can. J. Genet. Cytol., **25:** 129-138.

M. Takahashi and M. Senshu. 1987. Two distinct DNA ligases from *Drosophila melanogaster* embyos. FEBS Letters, **213:** 345-352.

A.E. Tomkinson and S. Linn. 1986. Purification and properties of a single strand-specific endonuclease from mouse cell mitochondria. Nucleic Acids Res., 14: 9579-9593.

A. Udvardy. P. Schedl, M. Sander and T.-s. Hsieh. 1985. Novel partitioning of DNA cleavage sites for *Drosophila* topoisomerase II. Cell, **40:** 933-941.

A. Udvardy, P. Schedl, M. Sander and T.-s Hsieh. 1986. Topoisomerase II cleavage in chromatin. J. Mol. Biol., **191:** 231-246.

G. Villani, P.J. Fay, R.A. Bambara and I.R. Lehman. 1981. Elongation of RNA-primed DNA templates by DNA polymerase α from *Drosophila menaogaster* embryos. J. Biol. Chem., **256:** 8202-8207.

G. Villani, B. Sauer and I. R. Lehman. (1980) DNA polymerase α from *Drosophila melanogaster* embryos. Subunit structure. *J. Biol. Chem.* **255**: 9479-9483

J. C. Wang, R. I. Gumport, K. Javaherian, K. Kirkegaard, L. Klevan, M. L. Kotewicz and Y.-C. Tse. (1980) DNA topoisomerases. **In**: *Mechanistic Studies of DNA Replication and Genetic Recombination*, pp 769-784, Academic Press, New York.

C. M. Wernette, M. C. Conway and L. S. Kaguni. (1988) The mitochondrial DNA polymerase from *Drosophila melanogaster* embryos: Kinetics, processivity and fidelity of DNA polymerization. *Biochemistry* in press.

C. M. Wernette and L. S. Kaguni. (1986) A mitochondrial DNA polymerase from embryos of *Drosophila melanogaster*. Purification, subunit structure and partial characterization. *J. Biol. Chem.* **261**: 14764-14770

E. Wyckoff, D. Natalie, J. M. Nolan, M. Lee and T.-s. Hsieh. Structure of the *Drosophila* DNA topoisomerase II gene: Nucleotide sequence and homology among topoisomerases II. *J. Mol. Biol.* In Press.

B. Z. Zmudzka, D. SenGupta, A. Matsukage, F. Cobianchi, P. Kumar and S. H. Wilson. (1986) Structure of rat DNA polymerase β revealed by partial amino acid sequencing and cDNA cloning. *Proc. Natl. Acad. Sci. USA* **83**: 5106-5110

13

FIDELITY OF ANIMAL CELL DNA POLYMERASES

Lawrence A. Loeb and Fred W. Perrino

INTRODUCTION

DNA replication in animal cells is an exceptionally accurate process. High accuracy is required to ensure that the correct sequence of nucleotides is perpetuated from generation to generation. Yet, some limited variation from parental nucleotide sequences is also necessary for evolutionary adaptation to environmental change. Based upon measurements of spontaneous mutation frequencies in eukaryotes, it can be inferred that less than one error is produced for every 10^{10} to 10^{12} nucleotides copied (Drake, *et al.*, 1969). The frequency of errors in the DNA replication process itself could be even lower, since many mutations result from other causes, such as damage to DNA by environmental agents and endogenous reactive metabolites. Conversely, postreplication repair processes may greatly reduce mutation frequencies (Glickman and Radman, 1980).

The primacy of DNA polymerases in ensuring the fidelity of DNA replication seems evident on the basis of their catalytic function. DNA polymerases copy a single-stranded DNA template by sequentially incorporating complementary deoxynucleoside monophosphates. Bacterial and phage DNA polymerases are known to synthesize in conjunction with a proofreading exonuclease which excises incorrect nucleotides prior to further chain elongation.

Incorporation of non-complementary nucleotides, in the absence of exonucleolytic excision would, result in permanent changes in the sequence of newly replicated DNA. Experimental evidence that DNA polymerases are, in fact, an important determinant in the accuracy of DNA replication has been provided by studies of mutant bacteriophage T4 DNA polymerase. Mutations in gene 43 for T4 DNA polymerase can confer a mutator phenotype; *i.e.*, increased frequency of mutations throughout the T4 genome. The alterations in gene 43 may affect either the polymerase activity or the exonuclease activity (*e.g.*, Reha-Krantz and Bessman, 1983; Speyer, 1965). Correlation of altered DNA polymerases with increased mutation frequency has also been documented for Pol III, the major replicating enzyme in *E. coli* (Hall and Brammar, 1973). Mutations in *dna*E or *dna*Q, which code for the DNA polymerase and proofreading exonuclease activities, respectively, can result in mutator phenotypes (Sevastopoulos and Glaser, 1977; Konrad, 1978; Scheuermann and Echols, 1984). Isolated evidence also indicates that mutation in mammalian DNA polymerases can cause a mutator phenotype. Cells containing a mutant DNA polymerase α, selected on the basis of resistance to aphidicolin, exhibit an increased frequency of mutations at three different chromosomal loci (Liu, *et al.*, 1983). In addition, exposure of cells to conditions that increase the error rates of purified DNA polymerases *in vitro* increase mutagenesis *in vivo* (Fry and Loeb, 1986). These conditions include addition of certain divalent metal ions and changes in the concentrations of deoxynucleoside triphosphates (Sirover and Loeb, 1976; Phear, *et al.*, 1987). The evidence that errors made by DNA polymerases *in vivo* can cause mutations does not necessarily imply an absence of post-replication error correcting mechanisms. The increased frequency of errors by DNA polymerases may simply exceed the cell's capacity for correction.

This chapter will focus on the fidelity of purified animal DNA polymerases. First, we will consider the four classes of DNA polymerases and their postulated roles in different DNA synthetic processes, each of which might have different requirements for accuracy. Second, we will summarize current methods for quantitating the fidelity of DNA polymerases. Third, we will describe studies on the fidelity of the different animal DNA polymerases. Last, we will analyze mechanisms by which these enzymes might enhance accuracy of base selection in the absence of a proofreading exonuclease and consider the hypothesis that such an exonuclease must be operative in eukaryotic cells.

ANIMAL DNA POLYMERASES

Four classes of DNA polymerases have been identified in animal cells, and each of these is considered in detail in other chapters in this book. The role of each class in DNA metabolism has been surmised predominantly from correlative observations. Parallels have been drawn between rates of DNA replication or DNA repair and the amount of a specific DNA polymerase in a cell population. The effects of inhibitors which are selective for different DNA polymerases on DNA synthetic processes have been noted and the association of a particular DNA polymerase with a subcellular structure, *i.e.*, nuclei or mitochondria, has also been taken to imply specific function. Ultimately, however, unambiguous demonstration of the role of a particular DNA polymerase in a DNA synthetic process is likely to involve analysis of conditional mutants, and such mutants are only now becoming available.

It is generally believed that DNA polymerase α (Bollum, 1960), the major species of animal DNA polymerase, is central to DNA replication and is also involved in DNA repair (Fry and Loeb, 1986). DNA polymerase α is induced in cells during progression from a quiescent to a proliferative state, and remains relatively constant in cycling cells (Wahl, *et al.*, 1988). A functionally complete replication complex containing DNA polymerase α, reconstituted *in vitro*, would be expected to be exceptionally accurate; the error rate might be similar to the spontaneous mutation rate, *i.e.*, one error in 10^{-10} to 10^{-12} nucleotides polymerized. DNA polymerase β (Weisbach, *et al.*, 1971; Baril *et al.*, 1971, and Chang and Bollum, 1971) is believed to function in DNA repair. Studies with cDNA probes indicate that the level of mRNA for DNA polymerase β varies only slightly during the cell cycle and is similar in non-dividing and dividing cells (Zmudzka *et al.*, 1988), but does increase after exposure of CHO cells to several DNA damaging agents (Fornace *et al.*, 1989). Since there are no *in vivo* studies that quantitate the accuracy of DNA repair processes, it would be difficult to predict the accuracy that one would hope to mimic *in vitro*. DNA polymerase γ (Fridlender *et al.*, 1972) is localized in mitochondria and is assumed to be responsible for mitochondrial DNA replication; the evidence for DNA repair in mitochondria is minimal (Clayton, *et al.*, 1974; Clayton, 1982). Since the mitochondrial genome is small and present in multiple copies, it can be argued that the accuracy of mitochondrial DNA replication need not be as high as that of nuclear DNA replication (Eigen and Schuster, 1977). Lastly, DNA polymerase δ (Byrnes, *et al.*, 1976), an enzyme with many properties similar to DNA polymerase-α, is also considered to function in DNA replication. DNA polymerase δ differs from most preparations of purified DNA polymerase-α in having a 3′→5′ exonuclease activity, which might serve a proofreading function. A comparison of the fidelities of DNA polymerases α and δ might provide an estimate of the contribution of this exonuclease to accuracy. Attractive models of DNA replication involving leading strand synthesis by polymerase δ and lagging strand synthesis by polymerase α have been proposed (Ottiger and Hübscher, 1984; Focher *et al.*, 1988; Prelich and Stillman, 1988). With respect to fidelity, there is no reason to suppose a difference in accuracy between leading and lagging strand synthesis. Thus, the potential for enhanced accuracy implied by the exonuclease activity of DNA polymerase δ is thought-provoking.

In conclusion, when one considers the uncertainties still inherent in assigning functions to the different *E. coli* DNA polymerases, a situation where a multitude of conditional mutants has been analyzed, it seems premature to propose physiological functions for the different animal DNA polymerases. However, these proposed physiological roles are valuable because they lead to models that can be addressed experimentally. There are few conjectures about the role of these enzymes in other important DNA metabolic pathways such as recombination, gene amplification, and resynthesis of DNA during mismatch correction.

METHODS TO QUANTITATE THE FIDELITY OF DNA SYNTHESIS *IN VITRO*

The extensive synthesis carried out by DNA polymerases *in vitro* has made it feasible to assess the accuracy of this process. One can examine a polymerase polypeptide alone, or in combination with various accessory proteins and other factors. Thereafter, one can assess the contribution of each component to the accuracy of DNA synthesis and establish the mechanisms for correct nucleotide addition and for the correction of mistakes once they are made. A major goal of such studies is to assemble a system containing known, pure components that copies DNA with accuracy similar to that prevailing *in vivo*. Current assays to quantitate errors by DNA polymerases and replicating systems are of two general categories, chemical and genetic.

Chemical Assays

Chemical methods for measuring the fidelity of DNA synthesis can be simple and rapid, but sensitivity is limited by the chemical purity of the reaction components. These methods can be subdivided into two groups:

Synthetic Polynucleotide Template

The simplest, most extensively utilized assay for *in vitro* fidelity employs a polynucleotide template containing one or two nucleotide species and measures the frequency of misincorporation of a non-complementary nucleotide (Figure 1). The error rate is the ratio of non-complementary to total nucleotides incorporated. High reproducibility can be achieved using a double-label assay, where the incorrect dNTP contains a radioactive element of high specific activity and a correct dNTP contains a different radioisotope of low specific activity (Agarwal *et al.*, 1979). The major problem with these assays is the requirement for exceptionally pure templates and substrates. The alternating copolymer poly [d(A-T)] can be obtained in high purity by enzymatic synthesis with a very accurate DNA polymerase and minute amounts of poly[d(A-T)] as an initial template. With a poly[d(A-T)] template, dGTP can be used as an incorrect substrate to avoid the use of dCTP which deaminates to dUTP; the latter can substitute for the complementary substrate, dTTP, thus increasing the apparent error rate. Clearly, rigorous analyses are required to document the purity of the template and substrates. Even so, sensitivity in these assays is limited to 10^{-5} by trace contamination of reagents which are below the level of detection in routine analytical procedures. Furthermore, with synthetic homopolymer and copolymer templates, the primer slips on the template during DNA synthesis (Chang, *et al.*, 1972), and this slippage may enhance misincorporation.

Natural DNA Templates

A qualitative assessment of misincorporation opposite multiple template sites on natural DNA (*i.e.*, DNA containing all four nucleotides) can be obtained from gel electrophoresis of "minus" sequencing reactions (Sanger and Coulson, 1975; Hillebrand, *et al.*, 1984) (Figure 2). In this system, a single-stranded DNA template with a known sequence is hybridized to a complementary 5′-[^{32}P]-labeled oligonucleotide. Synthesis in the presence of only three complementary deoxynucleoside triphosphates extends the oligonucleotide primer up to, but not opposite, the template nucleotide that normally base pairs with the absent dNTP. Extension opposite or beyond this template position requires incorporation of one of the three non-complementary nucleotides present in the reaction mixture. The frequency of errors at any position can be estimated from the yield of oligonucleotides extended opposite or beyond that position. By using four separate reactions, each minus a different dNTP, it is possible to assess the frequency of errors at multiple template positions. Even though this assay does not easily yield quantitative information on the types and frequency of errors by DNA polymerases, it facilitates study of the effects of DNA structure and neighboring nucleotides on misincorporation. This assay can be used to assess the relative misincorporation frequencies of different DNA polymerases and to measure the incorporation of nucleotide analogues at multiple template sites, even if the analogues are unlabeled.

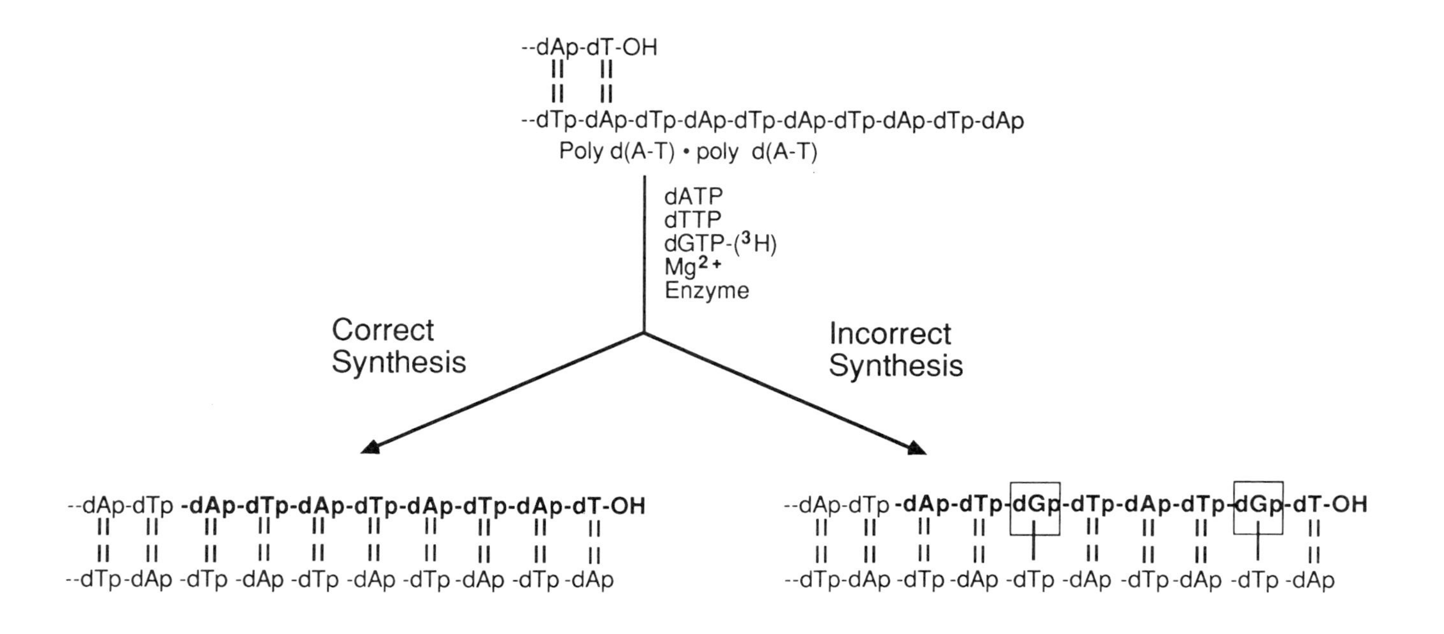

Figure 1. Fidelity of DNA synthesis using poly [d(A-T)] as a template. In this assay, ^{3}H-dGTP is the non-complementary substrate.

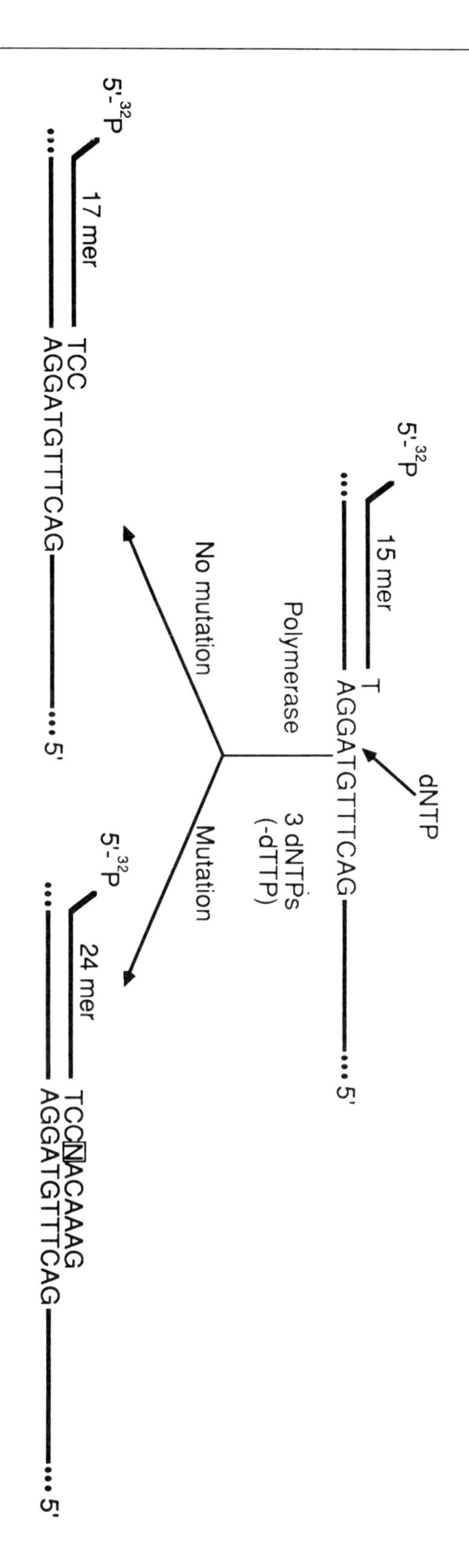

Figure 2. Fidelity of DNA synthesis using the "minus" reaction. In this assay, dTTP is the "minus" substrate.

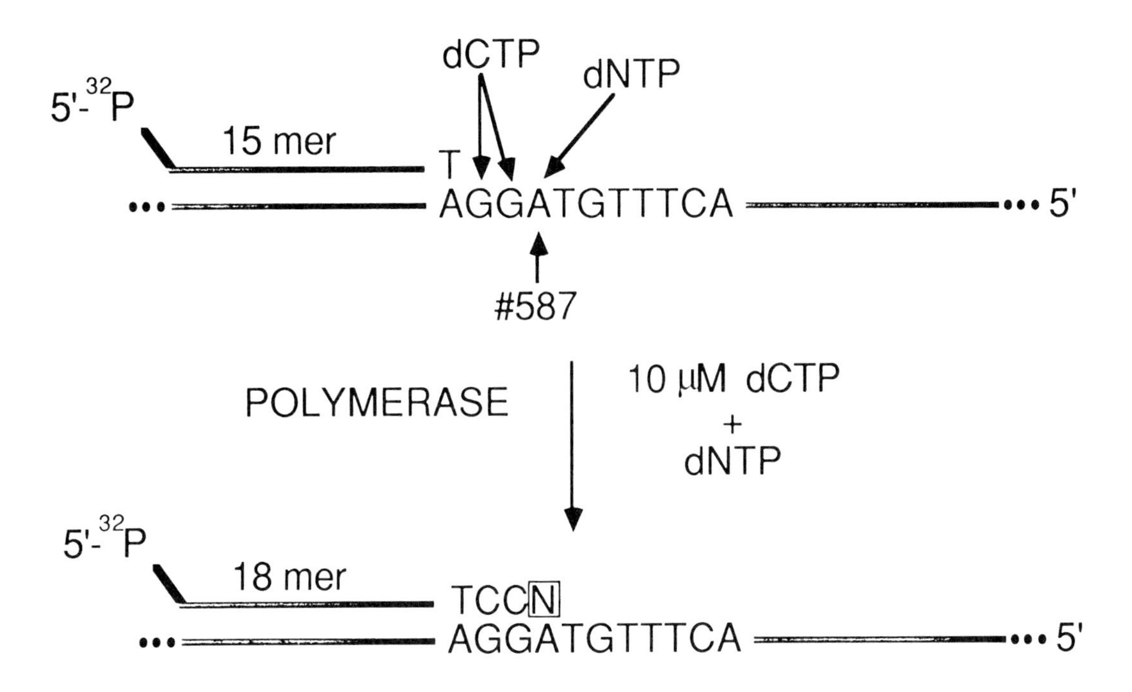

Figure 3. Fidelity of DNA synthesis using site-specific misinsertion of a single non-complementary nucleotide. The assay is modeled after the one described by Boosalis *et al.* (1987), using an oligonucleotide template that corresponds to the sequence of ϕX174 *am*3 DNA (Preston, *et al.*, 1988). The target position 587 is the central position in the *amber*3 codon.

A quantitative gel electrophoresis assay has been developed to investigate the fidelity of DNA synthesis. The rate of nucleotide insertion (or misinsertion) at a targeted template site is determined by quantitating the amounts of [^{32}P]-labeled primer extended by single, successive nucleotide addition steps (Boosalis *et al.*, 1987). By positioning the 3′-terminus of the primer a few nucleotides upstream from the target position, insertion during chain elongation is measured. Kinetic parameters (K_m and V_{max}) for insertion of correct and incorrect nucleotides are obtained by measuring the amounts of "plus one" addition products as a function of the concentration of a single dNTP. In the illustration in Figure 3, the presence of complementary dCTP at a saturating concentration permits synthesis up to the target position; subsequent misincorporation is detected by the inclusion of high concentrations of a non-complementary substrate. Assuming that steady state conditions prevail, and that the rate limiting steps for insertion of complementary and non-complementary nucleotides are the same, then the ratio of V_{max}/K_m for a non-complementary nucleotide to V_{max}/K_m for the complementary nucleotide is equal to the frequency of misinsertion. This method has been used with DNA polymerases lacking a proofreading 3′→5′ exonuclease but may be applicable to DNA polymerases that proofread, since the high concentrations of deoxynucleoside triphosphates could force the reaction in the forward direction. This method has recently been modified to measure the extension of a mispaired nucleotide by using oligonucleotide primers that contain a non-complementary nucleotide at the 3′-terminus (Perrino and Loeb, 1989a).

Genetic Assays for Quantitating Misincorporation

Two different categories of genetic assays, reversion and forward mutation, have been devised to measure the fidelity of DNA polymerases *in vitro*. Reversion assays measure conversion of a nucleotide sequence that codes for an inactive gene product to a sequence that codes for an active gene product. Forward mutation assays measure conversion of a sequence that codes for an active product to a sequence that reduces or abolishes activity. The utility of these assays depends, in part, on whether the alterations in nucleotide sequence are detected by a selection procedure or by a screening procedure. The currently utilized reversion assays for measuring the fidelity of DNA synthesis depend on either selection or screening. Selection procedures are designed so that the original nucleotide sequence fails to code for a required gene product(s), while reversion of the sequence produces active protein(s) and permits phage growth. One can then score for growth in the presence of a large number of original sequences that give no signal. Screening assays are dependent on changes in phenotype that must be distinguished in a large population of plaques. The currently utilized forward mutation assay identifies altered nucleotide sequences based on inactivation of a non-essential gene, and is inherently less sensitive than is a selection assay. However, the special utility of a forward mutation assay is in the wide variety of sequence alterations it reveals and its consequent capacity to delineate the spectrum of errors produced by a DNA polymerase.

ϕX Fidelity Assay

The first genetic method for measuring the fidelity of DNA synthesis uses a single-stranded ϕX174 DNA template containing an amber mutation (Weymouth and Loeb, 1978). The assay, diagrammed in Figure 4, measures the frequency of DNA polymerase errors that revert an amber codon to a codon yielding active proteins required for phage replication. The assay employs an oligonucleotide primer that is hybridized to a designated complementary sequence on the ϕX174 DNA template upstream from the amber mutation. The DNA polymerase to be assayed is used to extend the primer beyond the amber site. The partially double-stranded DNA product is then transfected into *E. coli* spheroplasts to complete replication of the ϕX genome and produce progeny phage. The error rate of the DNA polymerase is estimated from the frequency of revertant progeny phage revealed by plating on two bacterial strains, one permissive and the other nonpermissive for the amber mutation.

The amber codon of ϕX174 *am*3 DNA is located in a segment of the DNA that codes for two proteins, both required for phage production, and punctuated one reading frame apart. All substitutions that code for both active proteins occur at a single nucleotide position on the ϕX genome. Incorporation of any of the three non-complementary nucleotides at position 587 opposite the A in the amber codon results in reversion to wild-type phenotype. The different base substitutions are identified by sequencing the revertant phage DNA. Misincorporation can be detected at a frequency approaching 1 in 10^6 correct nucleotides incorporated when all four dNTP's are present in equal concentrations in the DNA polymerase reaction mixture. Higher sensitivity can be obtained by carrying out reactions under "pool bias" conditions, where the concentration of one or more of the non-complementary dNTP's exceeds that of the com-

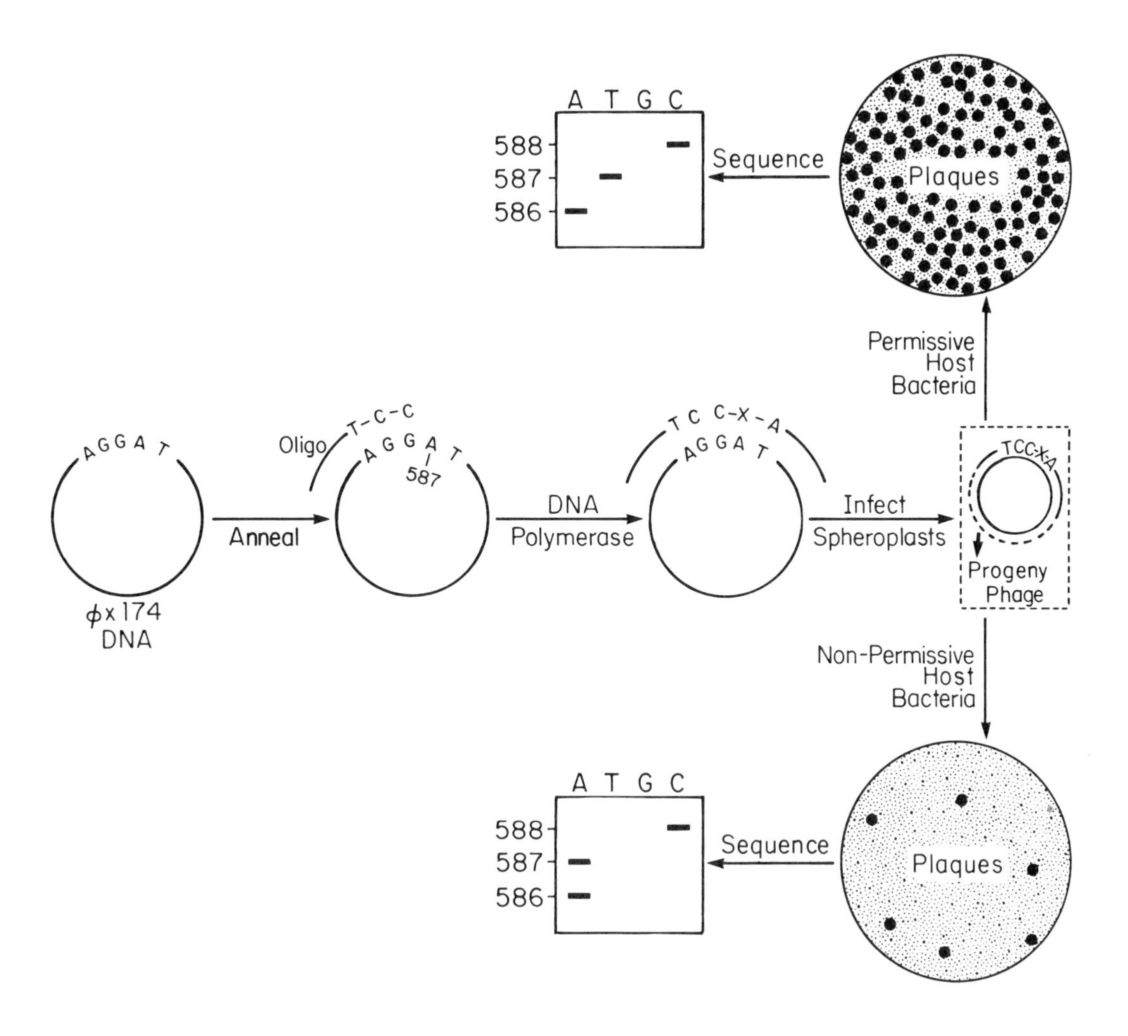

Figure 4. Scheme for the ϕX174 fidelity assay taken from Fry and Loeb, 1986, with permission.

plementary dNTP (Kunkel and Loeb, 1979). The error rate at equal dNTP concentrations is estimated by extrapolating from the mutation frequencies observed under biased conditions. This extrapolation is valid only if the observed error rate is proportional to the ratio of incorrect to correct nucleotides in the reaction mixture.

In contrast to the single site analysis at the *am*3 codon of ϕX174, base substitution errors can be detected at all three positions within the *am*16 codon of ϕX174 on the basis of plaque morphology and growth at different temperatures (Fersht and Knill-Jones, 1983). Eight different base substitution errors are detectable in this reversion assay. If one utilized a series of *amber* mutations, one could quantitate a variety of single-base substitutions. However, assays for reversions at nonsense codons are still unable to detect frameshifts, deletions, and complex mutations.

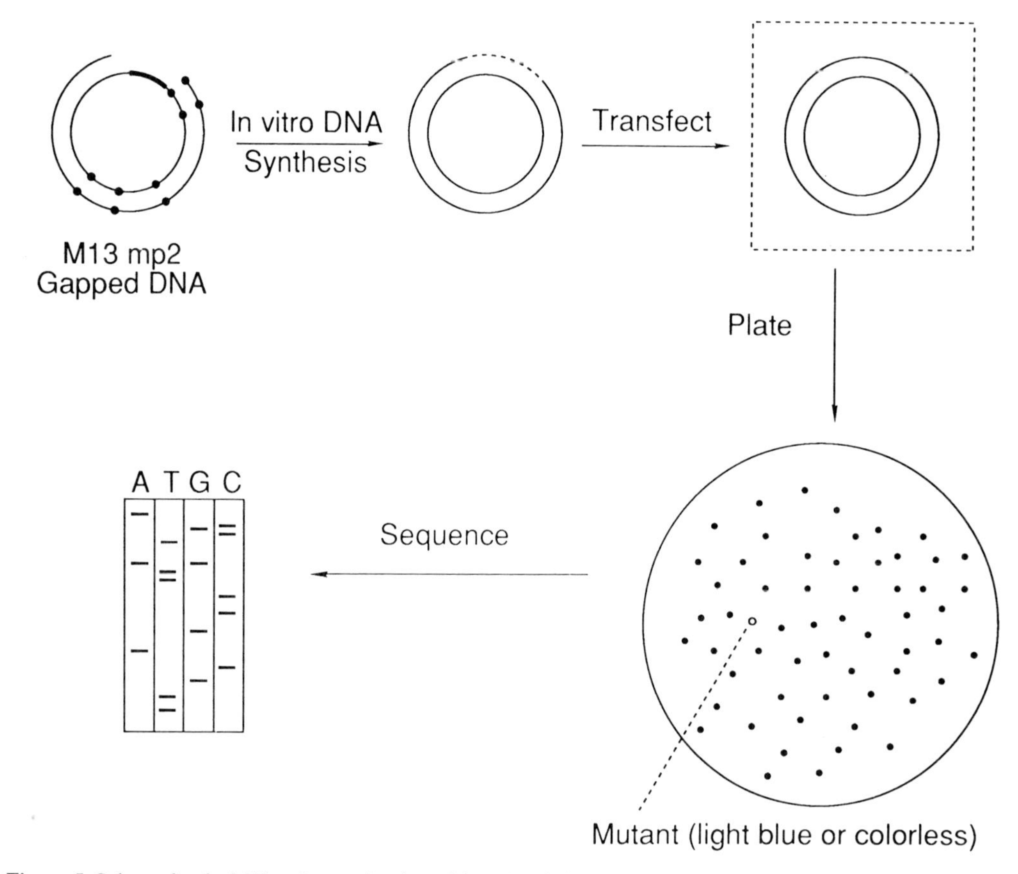

Figure 5. Scheme for the M13mp2 assay is adapted from Kunkel, 1987.

M13 Fidelity Assays

In order to observe a wider spectrum of the mutations produced by a DNA polymerase *in vitro*, a forward mutation assay is desirable. Kunkel (1985a) developed a forward mutation system that measures errors produced during DNA synthesis *in vitro* opposite a single-stranded gap containing a part of the *lac*I and *lac*Z genes in M13mp2 DNA (Figure 5). After DNA polymerization and ligation, the DNA is transfected into an *E. coli* host that has a chromosomal deletion in the *lac*Z operon, and also harbors an episome containing the part of the coding sequence for the *lac*Z gene missing from M13mp2 DNA. The two partial proteins produced within the M13mp2-infected host cell, one coded by M13mp2 DNA and the other by the endogenous episome, reconstitute β galactosidase activity by intracistronic complementation. Errors during DNA synthesis *in vitro* result in a decrease in, or loss of, α-complementation due to mutations in the *lac*Z phage genes. These errors are detected as light blue or colorless plaques following transfection into *E. coli* and growth in the presence of an indicator dye. Because β galactosidase activity is not essential for M13mp2 plaque production, over 200 different base substitution errors at more than 100 sites can be scored within the 250-nucleotide *lac*Zα sequence. Base substitutions usually yield light blue plaques, while deletions and frameshifts yield colorless plaques. However, a definitive description of mutations requires sequencing of individual plaques.

Kunkel *et al.* (1987), have also developed a fidelity assay which measures reversion frequency at opal codons within the *lac*Z-complementation gene of bacteriophage M13mp2. This assay complements the φX174 reversion assay. However, since the detection procedure is non-selective, one may have to plate as many as 10^6 infected *E. coli* to measure rare mutants.

FIDELITY OF ANIMAL DNA POLYMERASES

DNA polymerase α

A large number of studies have been carried out on the *in vitro* error rate of DNA polymerase α purified by different techniques from a variety of animal tissues. Until recently, purification of DNA polymerase α was based on sequential chromatographic columns requiring seven to ten days at 0-4°C and resulting in a heterogeneous array of catalytically active polypeptides (Loeb *et al.*, 1986). Because of the apparent proteolysis occurring during the prolonged procedures, recent work has emphasized rapid purification in "cocktails" of protease inhibitors (Wahl *et al.*, 1984). Whether DNA polymerase α is proteolytically processed *in vivo*, and the physiological significance of its extreme susceptibility to proteolytic degradation *in vitro* remain to be determined.

Most chromatographically purified forms of DNA polymerase α lack 3′→5′ exonuclease activity, and the fidelity of these various forms has been extensively studied. However, preparations of purified DNA polymerase-α from mouse myeloma (Chen *et al.*, 1979), calf thymus (Ottiger and Hübscher, 1984), and HeLa cells (Vishwanatha *et al.*, 1986) have been reported to contain an associated 3′→5′ exonuclease activity; the fidelity of these preparations has not as yet been determined. By the use of protease inhibitors during immunoaffinity chromatography and rapid conventional chromatography, the isolation of a high molecular weight DNA polymerase α (140-180 KDa), tightly associated with DNA primase and devoid of exonuclease activity, has been routinely reported (Fry and Loeb, 1986). The complex from *Drosophila* embryos and calf thymus tissue have been most extensively investigated, and have provided new insights into the fidelity of DNA polymerase α.

Even though extensive purification of DNA polymerase α by conventional chromatography yields a heterogeneous group of polypeptides, it has not been established unambiguously that proteolysis alters the fidelity of DNA polymerase α. Brosius *et al.* (1983), purified 9S, 7S, and 5.7S forms of DNA polymerase α from calf thymus and measured their fidelity in copying poly [d(A-T)]. The results indicate a 15-fold difference in accuracy among the different forms, the smallest species exhibiting the greatest frequency of misincorporation. In contrast, Abbotts and Loeb (1984) observed no difference between various forms of calf thymus DNA polymerase-α using the φX174 system. Interestingly, Krauss and Linn (1982), observed that DNA polymerase α made more mistakes in copying synthetic polynucleotides when purified from cells that had reached confluency than did the same enzyme purified from exponentially dividing cells. It would be fascinating if DNA polymerase α is altered (not necessarily by proteolysis) as cells approach confluency and these alterations are manifested by enhanced mutagenesis, implying that non-replicating cells can tolerate mutations. Alternatively, since DNA polymerase α participates in both DNA replication and DNA repair, it could be that the form responsible for DNA repair in non-dividing cells is the one that is less accurate. Further experiments on the fidelity of DNA synthesis in a variety of tissues undergoing the transition from division to quiescence are needed to confirm and extend these observations.

Using the φX174 *am*3 DNA system, the error rates of conventionally purified forms of DNA polymerase α generally vary from 1/20,000 to 1/40,000 (Kunkel and Loeb, 1981). The nucleotide sequence of DNA from a limited number of revertants was determined and misinsertion of dAMP opposite the template A was the most frequent event, followed by misinsertion of dCMP and dGMP (Kunkel and Loeb, unpublished results). In studies with a φX174 *am*16 DNA template, the most frequent base substitutions within the amber codon are T (template):G (1/13,500), G:G (1/22,700) and G:A (1/35,700) (Grosse *et al.* 1983). All other mispairings were below the level of detection, $<5 \times 10^{-6}$, and thus apparently occurred at a frequency lower than those obtained using the φX174 *am*3 DNA (Kunkel and Loeb, 1981). These differences could be due to differences in nucleotide sequence context in the templates or the result of differences in the forms of the DNA polymerase α studied.

The influence of the nucleotide sequence of the template on the spectrum of errors produced by conventionally purified forms of DNA polymerase α has been extensively documented using the M13mp2 forward mutation system (Kunkel, 1985a). Using preparations of DNA polymerase α from diverse sources, three different classes of errors were most frequent: single-base substitutions, frameshifts, and large deletions. Even though single-base substitutions constituted the most frequently observed errors (approx. 65%), there were a considerable number of frameshifts as well as deletions ranging from 5 to 300 bases in length. With respect to single-base substitutions, transitions occur more frequently than transversions. The frequency of mispairs at template A is similar to that obtained using the φX174 *am*3 fidelity assay (Loeb and Kunkel, 1981).

The studies in prokaryotes indicate that DNA replication is catalyzed by a multisubunit form of the DNA polymerase (Kornberg, 1980). However, prior to delineating such a complex in eukaryotes it was first necessary to define what constituted a catalytically active subunit, and this task was problematic due to extensive proteolysis during the prolonged purification. The first documentation that DNA polymerase α undergoes proteolytic hydrolysis was obtained with DNA polymerase α from *Drosophila* embryos (Kaguni *et al.*, 1983). Conventional purification procedures invariably yielded multiple protein fractions. However, when purification was conducted in buffers that included a "cocktail" of proteolysis inhibitors the final fraction consistently contained a four subunit complex that had both a high molecular weight form of DNA polymerase α and DNA primase. This complex copied φX174 *am*3 DNA with an error rate of 1 in 200,000, approximately five- to ten-fold more accurate than that previously reported with conventional forms of DNA polymerase α. This result is consistent with the correlation by Brosius, *et al.* (1983), that the larger and presumably more native DNA polymerase α polypeptide is indeed more accurate during *in vitro* DNA synthesis.

Using monoclonal antibodies against DNA polymerase α, it is now feasible to rapidly purify the enzyme from a variety of animal cells (Tanaka, *et al.*, 1982; Perrino and Loeb, 1989a). Immunoaffinity chromatography has provided a uniformity of results on the subunit composition of DNA polymerase-primase (Loeb *et al.*, 1986). However, studies of the fidelity of the DNA polymerase-primase complex in the φX174 and M13 assays have not been consistent; error rates have varied from ≈1/500,000 (Reyland and Loeb, 1987) to ≈1/5000 (Kunkel and Bebenek, 1988). Using φX174 *am*3 DNA the error rate opposite the template A at position 587 for DNA polymerase-primase from calf thymus, from a human lymphoblast cell line (TK-6), and from Chinese hamster lung cells (V-79) is, on the average, one in 600,000 nucleotides incorporated (Reyland and Loeb, 1987). Since each of these enzymes exhibits no detectable 3′→5′ exonuclease activity, the high fidelity must be mediated by a mechanism other than exonucleolytic proofreading. The conversion of a high fidelity complex to one of much lower fidelity was observed by sequential sampling of the enzyme during prolonged storage at -70°C. It is possible that the same proteolysis that causes heterogeneity during purification by conventional chromatography might also occur during storage. However, incubation of the purified DNA polymerase-primase with a variety of proteases has so far not resulted in increased misincorporation by the DNA polymerase-primase.

In contrast to the results obtained using the φX174 *am*3 DNA reversion assay, studies with the M13mp2 forward mutation assay have failed to yield a greater accuracy with the human KB cell DNA polymerase-primase than that obtained with conventionally purified forms of DNA polymerase α (Roberts and Kunkel, 1988). The reason for this discrepancy is not known. The principle difference in these two assays is that φX174 *am*3 DNA contains a single target while the M13mp2 system contains a 250 base target. The difference in fidelity could be the result of differences in the target sites, preparations of the DNA polymerase-primase, or in reaction conditions used in the two assays.

The M13mp2 system has yielded an error spectrum for immunopurified DNA polymerase α (Kunkel and Bebenek, 1988). The most frequent errors were C(template):A mispairs followed by G:A>G:G>A:A. In general, transversions were observed at a frequency of approximately 1 in 20,000. Those produced *via* purine:purine mispairs occurred more frequently than those produced *via* pyrimidine:pyrimidine mispairs.

Kinetic investigations on site-specific misincorporation of non-complementary nucleoside triphosphates by DNA polymerase-primase complexes have been carried out first with *Drosophila* DNA polymerase α (Boosalis, *et al.*, 1987) and then with the calf thymus enzyme (Perrino and Loeb, 1989a). The results indicate that the relative insertion frequency for a non-complementary nucleotide is approximately 1/20,000 that of the complementary substrate. However, the relative insertion frequency for the formation of A(template):G and C:C mispairs are $\approx 10^{-6}$, that of the complementary pairings. With both the *Drosophila* and calf thymus DNA polymerase α, the difference in K_m between insertion of a correct and incorrect nucleotide was 10^{-3} to 10^{-4} while the difference in V_{max} was at most 10-fold. Thus, it was concluded that the fidelity of DNA polymerase α at the level of misinsertions is governed by K_m rather than V_{max} effects (Boosalis *et al.*, 1987; Loeb and Perrino, 1989a).

Enhancement in Insertional Fidelity by DNA polymerase α

In all systems that have been investigated, discrimination at the insertion step is much greater than that which one would predict from differences between correct and incorrect base-pairings. Thus, DNA polymerase α must in some way increase the free energy difference between the correct and incorrect base pairs. In the absence of DNA polymerase, the difference in free energy between correct and incorrect Watson-Crick base pairs in aqueous solution is usually estimated to be 1-3 kcal mol^{-1}. This difference would program approximately one mispairing for every 100 nucleotides polymerized. Studies on non-enzyme-mediated polymerization using activated nucleotides and poly dC as a template have substantiated this estimate (Lohrmann and Orgel, 1980). Petruska *et al.* (1988), propose that the polymerase amplifies the free energy difference between correct and incorrect base pairs by reducing entropy differences and increasing enthalpy differences. They envision an active site which excludes water and accommodates best those base pairs which are closest to correct, such that $\Delta\Delta G^o$ is increased beyond the values observed in aqueous solution.

The mechanism for DNA synthesis by DNA polymerase α suggests that the following steps are responsible for nucleotide discrimination. The DNA polymerase binds to the ssDNA template, then to the 3′-OH terminus. The template base determines which nucleotide is then bound and positioned at the active site for phosphodiester linkage (Fisher and Korn, 1981, a and b). The K_m discrimination mechanism postulated for DNA polymerase α has suggested that the information at the active site ensures only correctly paired bases can be positioned prior to catalysis (Boosalis, *et al.*, 1987). Whether the enzyme changes conformation at each nucleotide addition step as directed by the template base remains to be shown. This mechanism is in contrast to the V_{max} discrimination mechanism proposed for *E. coli* DNA polymerase I in which nucleotide binding occurs prior to the conformational change in the DNA polymerase (Kuchta, *et al.*, 1987; Ferrin and Mildvan, 1986; El-Deiry, *et al.*, 1988). The mechanism by which DNA polymerases select, position, and catalyze the phosphodiester linkage needs further investigation. In light of the structural ,and presumed functional, differences as well as the various associated activities, it seems possible that different DNA polymerases mediate these processes by different mechanisms.

Other mechanisms have been invoked for increasing the accuracy of nucleotide insertion without exonucleolytic proofreading. Hopfield (1980) proposed an energy relay model in which the energy released by cleavage of the penultimate phosphate bond is applied by the polymerase to proofread insertion at the next nucleotide addition step. This model predicts that the error rate for the first nucleotide should be greater than that for subsequent nucleotides. This was not the case in studies with DNA polymerase α; the position of the primer was systematically changed with respect to the target sequence and no difference in error frequency was observed (Abbotts and Loeb, 1985). Other evidence against this mechanism is provided in studies on the relative rates of single-nucleotide additions. Studies on misinsertion by *Drosophila* DNA polymerase α were carried out under conditions in which the target for mispairing was three nucleotides from the primer terminus (Boosalis *et al.*, 1987). In the studies with calf thymus DNA polymerase α, the target was the first nucleotide (Perrino and Loeb, 1989a). Yet, the misinsertion frequencies for both enzymes are remarkably similar, suggesting that misincorporation at the first nucleotide addition is not more frequent.

Inefficient Extension of Mismatches by DNA Polymerase α

Stable misincorporation by DNA polymerase requires misinsertion of incorrect nucleotides and then extension of the mismatches by additions of complementary nucleotides. Misinsertions by DNA polymerase-primase complex are much more frequent than the error rate of the enzyme using the ϕX174 *am*3 reversion assay. One mechanism by which frequent misinsertion might not result in mutagenesis would be a lack of extension on the misinserted nucleotide. This mechanism was initially suggested from studies on conventionally purified DNA polymerase α in which the inability to bind to 3′-terminal mispairs (Fisher and Korn, 1981a; Fisher and Korn, 1981b) and reduced rates of mispair extension (Reckman *et al.*, 1983) were detected. Two recent findings suggest that the DNA polymerase-primase complex extends mismatched primer termini with uncommonly low efficiency. First, the error rate of DNA polymerase α in the ϕX174 reversion assay is less than proportional to the ratio of incorrect to correct substrates

(Perrino and Loeb, 1989a) implying that not all misinserted nucleotides produce viable DNA molecules. Second, gel electrophoresis of the products produced in primer extension reactions carried out with DNA polymerase α using reduced levels of one of the four nucleotides revealed an unusual property; enzyme pause sites were detected that corresponded to primers extended with a misincorporated base at the 3′ terminus. No such pause sites were detected with several other DNA polymerases that lack 3′→5′ exonuclease activity. Verification of the inefficient extension of 3′-mispaired termini by DNA polymerase α has been carried out using preformed template-primers containing a single mismatched nucleotide at the 3′-terminus. The frequency of extension of a mispaired primer relative to that of a correctly paired primer with DNA polymerase α is exceptionally low. For calf thymus DNA polymerase α it varies from 1/1800 for a C (template):T mismatch to 2×10^{-6} for an A:G mismatch (Perrino and Loeb, 1989a). These results suggest that inefficient extension of an inserted mismatch could contribute significantly to the fidelity of polymerization by DNA polymerase α. Detailed studies on the interactions between DNA polymerase α and 3′terminal mispairs are now required to determine whether or not misinsertion causes dissociation of the polymerase from the DNA template. The lack of efficient extension of a misinserted nucleotide by DNA polymerase α underscores the involvement of a separate proofreading exonuclease within the DNA replicating complex.

The Possible Association of a 3′→5′ Exonuclease with DNA polymerase α.

Prokaryotic DNA polymerases invariably exploit a 3′→5′ exonuclease for the preferential excision of incorrect nucleotides prior to further elongation during DNA replication. In the case of *E. coli* Pol I this exonuclease is an integral part of the DNA polymerase; with Pol III it is located on a separate subunit of the holoenzyme complex. The contribution of proofreading to the accuracy of prokaryotic DNA polymerases may be as large as 1000-fold (Kunkel, *et al.*, 1984). It is difficult to conceive that eukaryotes would have discarded during evolution such a potent error correcting mechanism. Nevertheless, nearly all purified preparations of DNA polymerase α have been shown to lack any 3′→5′ exonuclease.

There is one tantalizing exception to the generalization that DNA polymerase α lacks a proofreading exonuclease. The purified multisubunit *Drosophila* DNA polymerase α has no detectable exonuclease activity. However, upon dissociation of the subunits by sedimentation through glycerol gradients containing 50% ethylene glycol, a highly active 3′→5′ exonucleolytic activity can be easily detected and this activity cosediments with the 182 kDa subunit (Cotterill *et al.*, 1987). Moreover, the isolated polymerase subunit is at least 100-fold more accurate than the intact complex, suggesting that the uncovered 3′→5′ exonuclease is responsible for the increased accuracy. In the absence of polymerization, the preference for exonucleolytic excision of mispaired primer termini opposite a template A is C>A>>G (Reyla-nd *et al.*, 1988). However, this preference for excision is different under conditions for concomitant DNA synthesis: A = G>C.

The 3′→5′ exonuclease associated with the polymerase subunit of *Drosophila* DNA polymerase-primase has many of the hallmarks of a proofreading activity. It preferentially hydrolyzes mismatched termini and its appearance is associated with an enhancement in the fidelity of the DNA polymerase. The ratio of exonuclease to polymerase activity in the isolated 182 kDa subunit is greater than that of *E. coli* DNA polymerase I, a prototype of an enzyme that proofreads (Brutlag and Kornberg, 1972). It is conceivable that the high accuracy of the intact DNA polymerase-primase (1/200,000), is due to the cryptic exonuclease operating at low efficiency. The unmasking of the 3′→5′ exonuclease when the subunits are separated may reflect a relationship whereby this potent exonuclease is regulated *in vivo*.

The conditions used to demonstrate an exonuclease with *Drosophila* DNA polymerase have not revealed a similar activity using calf thymus DNA polymerase α (Perrino and Loeb, unpublished). So far, cryptic exonucleases have not been reported in DNA polymerase-primase from other sources. Recently, the separate proofreading subunit of *E. coli* DNA polymerase III (ε subunit) has been used to proofread for the calf thymus polymerase-primase complex (Perrino and Loeb, 1989b). Perhaps the *Drosophila* DNA polymerase-primase with an exonuclease may be an example of convergent evolution, or a variant of a yet unrecognized, physiologically significant form of mammalian DNA polymerase α having an exonuclease.

DNA polymerase β

The small size of, and simplicity of catalysis by, DNA polymerase β should simplify elucidation of the structural determinants and catalytic mechanisms which confer accuracy of DNA synthesis in the absence of exonucleolytic proofreading. DNA polymerase β is a single polypeptide of 39,000 daltons, is apparently devoid of any additional catalytic activities, and has been cloned and expressed in milligram amounts in *E. coli* (Abbotts, *et al.* 1988). It seems likely that studies on the three dimensional structure of the enzyme, and the results of traditional structure-function analysis and site specific mutagenesis will be forthcoming. Thus, it should be possible to describe the active site, identify the determinants of accuracy, and correlate amino acid sequence and three-dimensional structure with fidelity.

Initial studies on the fidelity of DNA polymerase β using synthetic polynucleotide templates suggested that this enzyme might be more accurate than DNA polymerase α (Loeb and Kunkel, 1982). However, in studies with ϕX174 DNA and subsequently with M13mp2, the error rates for single-base substititutions were greater than 1/5000 in both systems analyzed (Loeb and Kunkel, 1982; Kunkel and Bebenek, 1988), and were significantly greater than that exhibited by DNA polymerase α. An extensive comparison of errors by DNA polymerase β from four different animal sources yields the following generalizations: DNA polymerase β is error prone; single-base substitutions, particularly involving T:G and C:A mispairings, are very frequent; single-base deletions occur nearly as frequently, and are found predominantly opposite runs of identical template nucleotides; the error spectrum of DNA polymerase β differs greatly from that of DNA polymerase α (Kunkel and Bebenek, 1988).

Even in the case of the most error-prone eukaryotic DNA polymerase, Pol-β, the frequency of misincorporation is less than that calculated from free energy differences between correct and incorrect base pairs, suggesting an active role of Pol-β in ensuring fidelity.

Like DNA replication, DNA repair could be mediated by a multienzyme complex. A DNA polymerase β associated protein, DNAse V, contains both a 3′→5′ and a 5′→3′ exonuclease and could proofread incorrect nucleotides inserted by DNA polymerase β. Addition of DNAse V to reactions catalyzed by DNA polymerase β resulted in a small (*i.e.*, four-fold) increase in fidelity (Probst *et al.*, 1975). Also, DNA polymerase β is non-processive *in vitro* and associated proteins might enhance processivity *in vivo*. Enhanced processivity of DNA polymerase β has been shown to reduce frameshift errors (Kunkel, 1985b). Detailed studies on the relationship of processivity to fidelity are required for DNA polymerase β as well as other eukaryotic DNA polymerases.

The mechanism(s) by which DNA polymerase β increases accuracy beyond that derived from differences in base-pairing energies are not known. The simple demonstration that increasing the concentration of each of the three noncomplementary deoxynucleoside triphosphates results in parallel increases in error rates argues against any proofreading mechanism for removing errors after misinsertion. The finding that the error rate at an initial nucleotide addition step is similar to those at subsequent steps (Abbotts and Loeb, 1984) is not in accord with an energy relay mechanism (Hopfield, 1980). Possibly DNA polymerase β exemplifies the concept of a passive polymerase. Its major function in fidelity is to increase $\Delta\Delta G^0$ by exclusion of water from the active site.

DNA polymerase γ

DNA polymerase γ is purified from isolated mitochondria and is believed to function in mitochondrial DNA replication. Initial studies suggested that this DNA polymerase is highly error prone (Kunkel and Loeb, 1981). However, early preparations of mitochondrial DNA polymerase were devoid of associated exonucleolytic activity. More recent studies show that a 3′→5′ exonuclease copurifies with DNA polymerase activity. Like the proofreading exonuclease activities of *E. coli* DNA Pol I and T4 DNA polymerase, this mitochondrial exonuclease hydrolyzes mismatched terminal nucleotides more rapidly than matched, and hydrolysis can be inhibited by deoxynucleoside monophosphates. Furthermore, conditions that inhibit the exonuclease promote misincorporation (Kunkel and Mosbaugh, 1989). In an M13mp2 reversion assay, porcine mitochondrial DNA polymerase exhibits an error rate for A:C mispair formation of 10^{-6}. However, the error rate in the M13mp2 forward mutation assay is much higher, suggesting that errors not scored in the reversion assay may occur at a much higher frequency. The subunit structure of mammalian mitochondrial DNA polymerases is uncertain and it is not yet known which polypeptides contain polymerase and exonuclease activity. Nonetheless, current studies suggest that exonucleolytic proofreading plays a role in the fidelity of DNA polymerase γ.

DNA polymerase δ

That DNA polymerase δ exists as a distinct entity was initially suggested by its tight association with a 3′→5′ exonuclease activity. Early studies indicated that this exonuclease preferentially excises non-complementary nucleotides from the 3′-primer terminus and that excision is diminished by the addition of deoxynucleoside monophosphates (Byrnes, *et al.*, 1976). Using an M13 reversion assay, Kunkel *et al.*, 1987 demonstrated that DNA polymerase δ is highly accurate, producing less than one error per 10^6 nucleotides polymerized. This error rate was 10- and 500-fold less than those exhibited by DNA polymerases α and β, respectively. A role for the exonuclease in fidelity is indicated by the recent demonstration that inhibition of exonuclease activity, either by addition of deoxynucleoside monophosphates, or by increasing the rate of polymerization with high concentration of dNTPs, increases misincorporation. These effects are exhibited by prokaryotic DNA polymerases and are diagnostic of the role of an exonuclease in proofreading. Further studies on the subunit composition of DNA polymerase δ are required to unambiguously determine whether the exonuclease and DNA polymerase activities reside in the same polypeptide. In fact, proofreading *in vitro* by a 3′→5′ exonuclease may occur independent of a physical association; *i.e.*, a single-strand specific 3′→5′ exonuclease could excise misinsertions by DNA polymerase, particularly if the polymerase dissociates from the template after insertion of non-complementary nucleotides (Perrino and Loeb, 1989b).

FUTURE DIRECTIONS

The Fidelity of DNA Replication *In Vivo*

A major goal of studies on the fidelity of eukaryotic DNA replication is to assemble an *in vitro* system containing only homogeneous components, which can copy DNA with an accuracy similar to that achieved in cells. Will it be necessary to reassemble the entire DNA replicating complex to obtain this accuracy? Is proofreading required for high fidelity in eukaryotes? So far, studies on proofreading in eukaryotes are conflicting. Based on the ratio of incorporation of aminopurine to adenine in HeLa cell nuclei, it has been proposed that proofreading does not occur (Wang *et al.*, 1981) or, if it does occur, aminopurine monophosphates in DNA are not proofread. However, other studies on the spectrum of mutations in the *aprt* gene suggest a next nucleotide effect, which is a hallmark of proofreading (Phear, *et al.*, 1988). Perhaps the fidelity of DNA replication may be surprisingly low, and errors made by DNA polymerases are later removed by a post-replication process such as mismatch correction. Based on the success of studying *E. coli* DNA replication using bacteriophage DNA to define the *E. coli* replisome, investigators have started to measure the fidelity of SV40 DNA replication *in vitro* using soluble eukaryotic extracts (Roberts and Kunkel, 1988) Initial reports suggest this system is highly accurate. The essential components of this highly accurate system must now be defined and purified in order to establish their contribution to fidelity in reconstituted systems. As we use viral DNA probes to reveal the mechanisms of eukaryotic DNA replication, will we find an enzyme system homologous to those in prokaryotes? Are the requirements for accuracy of DNA replication so precise that only minimal alterations were tolerated during evolution, or is the task of replicating the eukaryotic genome so enormous that new mechanisms were evolved to ensure fidelity?

ACKNOWLEDGEMENTS

We thank Mary Whiting for typing and patience. This work was supported in part by NIH grants T32-CA09437 (F.W.P.) and OIG-R-35-CA-39903 (L.A.L.)

REFERENCES

J. Abbotts and L. A. Loeb. (1985) On the fidelity of DNA replication: Use of synthetic oligonucleotide-initiated reactions. *Biochim. Biophys. Acta* **824**: 58-65

J. Abbotts and L. A. Loeb. (1984) On the fidelity of DNA replication: Lack of primer position effect on the fidelity of mammalian DNA polymerases. *J. Biol. Chem.* **259**: 6712-6714

J. Abbotts, D. N. SenGupta, B. Z. Zmudzka, S. G. Widen and S. H. Wilson. (1988) Human DNA polymerase β. Expression in *Escherichia coli* and characterization of the recombinant enzyme. **In**: *DNA Replication and Mutagenesis*, (Eds., R. E. Moses and W. C. Summers) Am. Soc. Microbiol., Washington, DC, pp. 55-65

S. S. Agarwal, D. K. Dube and L. A. Loeb. (1979) On the fidelity of DNA replication. VII. Accuracy of *E. coli* DNA polymerase I. *J. Biol. Chem.* **254**: 101-106

E. F. Baril, O. E. Brown, M. D. Jenkins, and J. Laszlo. (1971) Deoxyribonucleic acid polymerases with rat liver ribosomes and smooth membranes. Purification and properties of the enzymes. *Biochemistry* **10**: 1981-1992

F. J. Bollum. (1960) Calf thymus polymerase. *J. Biol. Chem.* **235**: 2399-2403

M. S. Boosalis, J. Petruska and M. F. Goodman. (1987) DNA polymerase insertion fidelity: Gel assay for site-specific kinetics. *J. Biol. Chem.* **262**: 14689-14696

S. Brosius, F. Grosse and G. Krauss. (1983) Subspecies of DNA polymerase from calf thymus with different fidelity in copying synthetic template-primers. *Nucleic Acids Res.* **11**: 193-202

D. Brutlag and A. Kornberg. (1972) Enzymatic synthesis of deoxyribonucleic acid XXXVI. A proofreading function for the 3'→5' exonuclease activity in deoxyribonucleic acid polymerases. *J. Biol. Chem.* **247**: 241-248

J. J. Byrnes, K. M. Downey, V. L. Blank, V. L. and A. G. So. (1976) A new mammalian DNA polymerase with 3' to 5' exonuclease activity: DNA polymerase δ. *Biochemistry* **15**: 2817-2823

L. M. S. Chang and F. J. Bollum. (1971) Low molecular weight deoxyribonucleic acid polymerase in mammalian cells. *J. Biol. Chem.* **246**: 5835-5837

L. M. S. Chang, G. R. Cassani and F. J. Bollum (1972) Deoxynucleotide-polymerizing enzymes of calf thymus gland. VII. Replication of homopolymers. *J. Biol. Chem.* **247**: 7718-7723

Y-c. Chen, E. W. Bohn, S. R. Planck and S. H. Wilson. (1979) Mouse DNA polymerase α subunit structure and identification of a species with associated exonuclease. *J. Biol. Chem.* **254**: 11678-11687

D. A. Clayton. (1982) Replication of animal mitochondrial DNA. *Cell* **28**: 693-705

D. A. Clayton, J. N. Doda and E. C. Friedberg. (1974) The absence of a pyrimidine dimer repair mechanism in mammalian mitochondria. *Proc. Natl. Acad. Sci. USA* **71**: 2777-2781

S. M. Cotterill, M. E. Reyland, L. A. Loeb and I. R. Lehman. (1987) A cryptic proofreading 3'→5' exonuclease associated with the polymerase subunit of the DNA polymerase-primase from *Drosophila melanogaster*. *Proc. Natl. Acad. Sci. USA* **84**: 5635-5639

J. W. Drake, E. F. Allen, S. A. Forsberg, R. M. Preparata and E. O. Greenberg. (1969) Spontaneous mutation. *Nature* **221**: 1128-1132

M. Eigen and P. Schuster. (1977) The hypercycle, a principle of natural self-organization. *Die Naturwissenschaften*, **64D**, 541-565

W. S. El-Deiry, A. G. So and K. M. Downey. (1988) Mechanisms of error discrimination by *Escherichia coli* DNA polymerase I. *Biochemistry* **27**: 546-553

L. J. Ferrin and A. S. Mildvan. (1986) NMR studies of conformations and interactions of substrates and ribonucleotide templates bound to the large fragment of DNA polymerase I. *Biochemistry* **25**: 5131-5145

A. R. Ferscht and J. W. Knill-Jones. (1983) Kinetics of base misinsertion by DNA polymerase I of *Escherichia coli*. *J. Mol. Biol.* **165**: 655-667

P. A. Fisher and D. Korn. (1981) Ordered sequential mechanism of substrate recognition and binding by KB cell DNA polymerase α. *Biochemistry* **20**: 4560-4569

P. A. Fisher and D. Korn. (1981) Properties of the primer-binding site and the role of magnesium ion in primer-template recognition by KB cell DNA polymerase α. *Biochemistry* **20**: 4570-4578

F. Focher, E. Ferrari, S. Spadari and U. Hübscher. (1988) Do DNA polymerases α and δ act coordinately as leading and lagging strand replicases? *FEBS Lett.* **228**: 6-10

A. J. Fornace, Jr., B. Zmudzka, M. C. Hollander and S. H. Wilson. (1989) Induction of β-polymerase mRNA by DNA damaging agents in Chinese hamster ovary cells. *Mol. Cell. Biol.* **9**: 851-853

B. Fridlender, M. Fry, A. Bolden and A. Weissbach. (1972) A new synthetic RNA-dependent DNA polymerase from human tissue culture cells. *Proc. Natl. Acad. Sci. USA* **69**: 452-455

M. Fry and L. A. Loeb. (1986) *Animal Cell DNA Polymerases*. CRC Press, Boca Raton, FL

B. W. Glickman and M. Radman. (1980) *Escherichia coli* mutator mutants deficient in methylation-instructed DNA mismatch correction. *Proc. Natl. Acad. Sci. USA* **77**: 1063-1067

F. Grosse, G. Krauss, J. W. Knill-Jones and A. R. Fersht (1983) Accuracy of DNA polymerase ???? in copying natural DNA. *EMBO J.* **2**: 1515-1519

R. M. Hall and W. J. Brammar. (1973) Increased spontaneous mutation rates in mutants of *E. coli* with altered DNA polymerase III. *Mol. Gen. Genet.* **121**: 271-276

G. G. Hillebrand, A. H. McClusky, K. A. Abbott, G. G. Revich and K. L. Beattie. (1984) Misincorporation during DNA synthesis, analyzed by gel electrophoresis. *Nucleic Acids Res.* **12**: 3155-3171

J. J. Hopfield. (1980) The energy relay: A proofreading scheme based on dynamic cooperativity and lacking all characteristic symptoms of kinetic proofreading in DNA replication and protein synthesis. *Proc. Natl. Acad. Sci. USA* **77**: 5248-5252

L. S. Kaguni, R. A. DiFrancesco and I. R. Lehman. (1984) The DNA polymerase-primase from *Drosophila melanogaster* embryos. Rate and fidelity of polymerization on single-stranded DNA templates. *J. Biol. Chem.* **259**: 9314-9319

L. S. Kaguni, J-M. Rossignol, R. C. Conaway and I. R. Lehman. (1983) Isolation of an intact polymerase-primase from embryos of *Drosophila melanogaster*. *Proc. Natl. Acad. Sci. USA* **80**: 2221-2225

E. B. Konrad. (1978) Isolation of an *Escherichia coli* K-12 *dna*E mutation as a mutator. *J. Bacteriol.* **133**: 1197-1202

A. Kornberg. (1980) *DNA Replication*. W.H. Freeman and Co.,San Francisco, 1980.

S. W. Krauss and S. Linn. (1982) Changes in DNA polymerases α, β, and γ during the replicative life span of cultured human fibroblasts. *Biochemistry* **21**: 1002-1009

R. D. Kuchta, V. Mizrahi, P. A. Benkovic, K. A. Johnson and S. J. Benkovic. (1987) Kinetic mechanism of DNA polymerase I (Klenow). *Biochemistry* **26**: 8410-8417

T. A. Kunkel. (1985a) The mutational specificity of DNA polymerase β during *in vitro* DNA synthesis: Production of frameshift, base substitution and deletion mutations. *J. Biol. Chem.* **260**: 5787-5796

T. A. Kunkel. (1985b) The mutational specificity of DNA polymerases α and γ during *in vitro* DNA synthesis. *J. Biol. Chem.* **260**: 12866-12874

T. A. Kunkel. (1986) Frameshift mutagenesis by eucaryotic DNA polymerases *in vitro*. *J. Biol. Chem.* **261**: 13581-13-587

T. A. Kunkel and K. Bebenek. (1988) Recent studies of the fidelity of DNA synthesis. *Biochim. Biophys. Acta* **951**: 1-15

T. A. Kunkel, L. A. Loeb and M. F. Goodman. (1984) On the accuracy of T4 DNA polymerases in copying ϕX174 DNA *in vitro*. *J. Biol. Chem.* **259**: 1539-1545

T. A. Kunkel and L. A. Loeb. (1979) On the fidelity of DNA replication: Effect of divalent metal ion activators and deoxyribonucleoside triphosphate pools on *inD vitro* mutagenesis. *J. Biol. Chem.* **254**: 5718-5725

T. A. Kunkel and L. A. Loeb. (1981) Fidelity of mammalian DNA polymerases. *Science* **213**: 765-767

T. A. Kunkel and D. W. Mosbaugh. (1989) Exonucleolytic proofreading by a mammalian DNA polymerase γ. *Biochemistry* **28**: 988-995

T. A. Kunkel, R. D. Sabatino and R. A. Bambara. (1987) Exonucleolytic proofreading by calf thymus DNA polymerase δ. *Proc. Natl. Acad. Sci. USA* **84**: 4865-4869

P. K. Liu, C-c. Chang, J. E. Trosko, D. K. Dube, G. M. Martin and L. A. Loeb. (1983) Mammalian mutator mutant with an altered DNA polymerase α. *Proc. Natl. Acad. Sci. USA* **80**: 797-801.

L. A. Loeb and T. A. Kunkel. (1982) Fidelity of DNA synthesis. *Annu. Rev. Biochem.* **52**: 429-457.

L. A. Loeb, P. K. Liu and M. Fry. (1986) DNA polymerase α: Enzymology, function *in vivo*, fidelity, and mutagenesis. *Prog. Nucleic Acids Res. Mol. Biol.* **33**: 57-110

R. Lohrmann and L. E. Orgel. (1980) Efficient catalysis of polycytidylic acid-directed oligoguanylate formation by Pb^{2+}. *J. Mol. Biol.* **142**: 555-567

H. P. Ottiger and U. Hübscher. (1984) Mammalian DNA polymerase α holoenzymes with possible functions at the leading and lagging strand of the replication fork. *Proc. Natl. Acad. Sci. USA* **81**: 3993-3997

F. Perrino and L. A. Loeb. (1989a) Differential extension of 3′ mispairs is a major contribution to the high fidelity of calf thymus DNA polymerase α. *J. Biol. Chem.* **264**: 2898-2905

F. Perrino and L. A. Loeb. (1989b) Proofreading by the ε subunit of *ED. coli* DNA polymerase III increases the fidelity of calf thymus DNA polymerase α. *Proc. Natl. Acad. Sci. USA* **86**: 3085-3088

J. Petruska, L. C. Sowers and M. F. Goodman. (1986) Comparison of nucleotide interactions in water, proteins, and vacuum: Relevance to DNA polymerase fidelity. *Proc. Natl. Acad. Sci. USA* **83**: 1559-1562

G. Phear, J. Nalbantoglu and M. Meuth. (1987) Next-nucleotide effects in mutations driven by DNA precursor pool imbalances at the *aprt* locus of Chinese hamster ovary cells. *Proc. Natl. Acad. Sci. USA* **84**: 4450-4454

G. Prelich and B. Stillman. (1988) Coordinated leading and lagging strand synthesis during SV40 DNA replication *in vitro* requires PCNA. *Cell* **53**: 117-126

B. D. Preston, B. J. Poiesz and L. A. Loeb. (1988) Fidelity of HIV-1 reverse transcriptase. *Science* **242**: 1168-1171

G. S. Probst, D. M. Stalker, D. W. Mosbaugh and R. R. Meyer. (1975) Stimulation of DNA polymerase by factors isolated from Novikoff hepatoma. *Proc. Natl. Acad. Sci. USA* **72**: 1171-1176

B. Reckmann, F. Grosse and G. Krauss. (1983) The elongation of mismatched primers by DNA polymerase α from calf thymus. *Nucleic Acids Res.* **11**: 7251-7260

L. J. Reha-Krantz and M. J. Bessman. (1981) Studies on the biochemical basis of spontaneous mutation VI. Selection and characterization of a new bacteriophage T4 mutator DNA polymerase. *J. Mol. Biol.* **145**: 677-695

M. E. Reyland and L. A. Loeb. (1987) On the fidelity of DNA replication: XX Isolation of high fidelity DNA polymerase-primase complexes by immunoaffinity chromatography. *J. Biol. Chem.* **262**: 10824-10830

M. E. Reyland, I. R. Lehman and L. A. Loeb. (1988) Specificity of proofreading by the 3′→5′ exonuclease of the DNA polymerase-primase of *Drosophila melanogaster*. *J. Biol. Chem.* **263**: 6518-6524

J. D. Roberts and T. A. Kunkel. (1988) Fidelity of a human cell DNA replication complex. *Proc. Natl. Acad. Sci. USA* **85**: 7064-7068

F. Sanger and A. R. Coulson. (1975) A rapid method for determining sequences in DNA by primed synthesis with DNA polymerases. *J. Mol. Biol.* **94**: 441-448

R. H. Scheuermann and H. A. Echols. (1984) A separate editing exonuclease for DNA replication: The ε subunit of *Escherichia coli* DNA polymerase III holoenzyme. *Proc. Natl. Acad. Sci. USA* **81**: 7747-7751

C. G. Sevastopoulos and D. A. Glaser. (1977) Mutator action by *Escherichia coli* strains carrying *dna*E mutations. *Proc. Natl. Acad. Sci. USA* **74**: 3947-3950

M. A. Sirover and L. A. Loeb. (1976) Infidelity of DNA synthesis *in vitro*: Screening for potential metal mutagens or carcinogens. *Science* **194**: 1434-1436

J. F. Speyer. (1965) Mutagenic DNA polymerase. *Biochem. Biophys. Res. Commun.* *21*: 6-8

S. Tanaka, S-Z. Hu, T. S-F. Wang and D. Korn. (1982) Preparation and preliminary characterization of monoclonal antibodies against human DNA polymerase α. *J. Biol. Chem.* **257**: 8386-8390

J. K. Vishwanatha, S. A. Coughlin, M. Wesolowski-Owen and E. F. Baril. (1986) Multiprotein of DNA polymerase α from HeLa cells. *J. Biol. Chem.* **261**: 6619-6628

A. F. Wahl, S. P. Kowalski, L. W. Harwell, E. M. Lord and R. A. Bambara. (1984) Immunoaffinity purification and properties of a high molecular weight calf thymus DNA polymerase α. *Biochemistry* **23**: 1895-1899

A. F. Wahl, A. M. Geis, B. H. Spain, S. W. Wong, D. Korn and T. S-F. Wang. (1988) Gene expression of human DNA polymerase α during cell proliferation and the cell cycle. *Molec. Cell. Biol.* **8**: 5016-5025

M-l. Wang, R. H. Stellwagen and M. F. Goodman. (1981) Evidence for the absence of DNA proofreading in HeLa cell nuclei. *J. Biol. Chem.* **256**: 7097-7100

A. Weissbach, A. Schlabach, F. Fridlender and A. Bolden (1971) DNA polymerase from human cells. *Nature* **231**: 167-170

L. A. Weymouth and L. A. Loeb. Mutagenesis during *in vitro* DNA synthesis. *Proc. Natl. Acad. Sci. USA* **75**: 1924-1928

B. Z. Zmudzka, A. Fornace, J. Collins and S. H. Wilson. (1988) Characterization of DNA polymerase β mRNA: Cell cycle and growth response in cultured human cells. *Nucleic Acids Res.* **16**: 9587-9596

14

BASE PAIRING AND MISPAIRING DURING DNA SYNTHESIS

Kenneth L. Beattie, Ming-Derg Lai and Rogelio Maldonado-Rodriguez

---

"It has not escaped our notice that the specific pairing we have postulated immediately suggests a possible copying mechanism for the genetic material."

James Watson and Francis Crick, 1953

INTRODUCTION

In their landmark paper on the structure of DNA, Watson and Crick (1953a) made the very perceptive statement quoted above, which has had a lasting influence on our thinking about how fidelity in DNA synthesis is achieved. Since that time a great deal of information regarding the mechanisms of DNA replication has been acquired, through genetic, biochemical and physical experimentation.

Genetic and biochemical evidence suggests that DNA polymerases possess a substrate recognition/discrimination mechanism which minimizes formation of incorrect base pairs during DNA synthesis (Gillin and Nossal, 1976; Engler and Bessman, 1978; Loeb *et al.*, 1980; Loeb and Kunkel, 1982; El-Deiry *et al.*, 1988; Kuchta *et al.*, 1988). It is generally presumed that the specific base pairing (A•T and G•C) postulated by Watson and Crick plays an integral role in the "base selection" function of DNA polymerases. In addition, other factors contribute to the accuracy of DNA replication. A "proofreading" 3′-exonucleolytic activity, associated with many DNA polymerases or replication complexes, is capable of immediately correcting mistakes made during DNA synthesis, before polymerization continues (Brutlag and Kornberg, 1972; Huang and Lehman, 1972; Bessman *et al.*, 1974). A postreplicational mismatch correction mechanism exists, which further decreases the net error rate in DNA replication (Wagner and Meselson, 1976; Radman *et al.*, 1978; Radman and Wagner, 1986; Modrich, 1987).

The precise manner in which the DNA polymerase recognizes Watson-Crick base pairs and discriminates against incorrect base pairs, and the extent to which hydrogen-bonded base pair formation contributes to the specificity of incorporation are unknown. In this chapter this problem will be examined by considering what is known about mispairing and DNA polymerase function. No attempt will be made to review the enormous body of literature on mutagenesis, mispairing and polymerase fidelity. Instead, a representative but diverse set of data will be discussed in order to illustrate the authors' specific perspective on the subject.

BASE PAIRING AND MISPAIRING IN NUCLEIC ACIDS

Possible Mispairing Mechanisms

Traditionally, because of the viewpoint originally forwarded by Watson and Crick (1953a,b,c), the specificity of nucleotide incorporation during DNA synthesis is considered to arise largely from the specific base pairing (A•T and G•T) that exists in the double helical product of polymerization. Furthermore, it is generally assumed that rare mistakes in the polymerase reaction ("misincorporations") occur *via* formation of "mispairs" that involve transient structural forms of the nucleotides that arise by tautomerization, ionization, N-glycosyl rotation, etc. Thus, "mispairing" during DNA synthesis has been perceived in terms of the rare occurrence of a non-Watson-Crick base pair between dNTP and template residues prior to phosphodiester bond formation. These rare events could account for spontaneous base substitution mutagenesis, or mutagenesis induced by chemical damage to nucleic acid components that potentiates "ambiguous" base pairing.

Several ways in which mispairing is envisioned to occur during DNA synthesis are depicted in Figure 1. These specific examples (in which a base in an alternative structural form mispairs with G) are given solely to illustrate structural alternatives that might participate in mispairing events. For example, 5-BrU could mispair with G as the enol tautomer (a), the ionized base (b), or through alteration of the relative orientations of N-glycosyl bonds ("wobbling") (c). Also shown (d) is a mispair involving N-glycosyl bond rotation, in which the *syn* rotamer of A mispairs with G (a "Hoogsteen" base pair).

Base Mispairs Observed in Nucleic Acids

Numerous physical studies have been carried out to determine the structures of double helical nucleic acids containing single base mismatches, and to characterize structural alterations in DNA complexed with other molecules. A rationale of this approach has been that these structures may be relevant to mispairing events that occur during DNA synthesis. Accordingly, this section presents a brief compendium of non-Watson-Crick base pairs that have been observed in nucleic acids and their complexes with other molecules.

The first high resolution crystallographic determination of a nucleic acid structure was for transfer RNA (Quigley and Rich, 1976; Jack *et al.*, 1976; Kim, 1978). Several unusual structural features were found to exist in tRNA, including intercalation of bases and a variety of non-Watson-Crick base pairing. The existence of these unusual structures illustrates the enormous conformational diversity of nucleic acids.

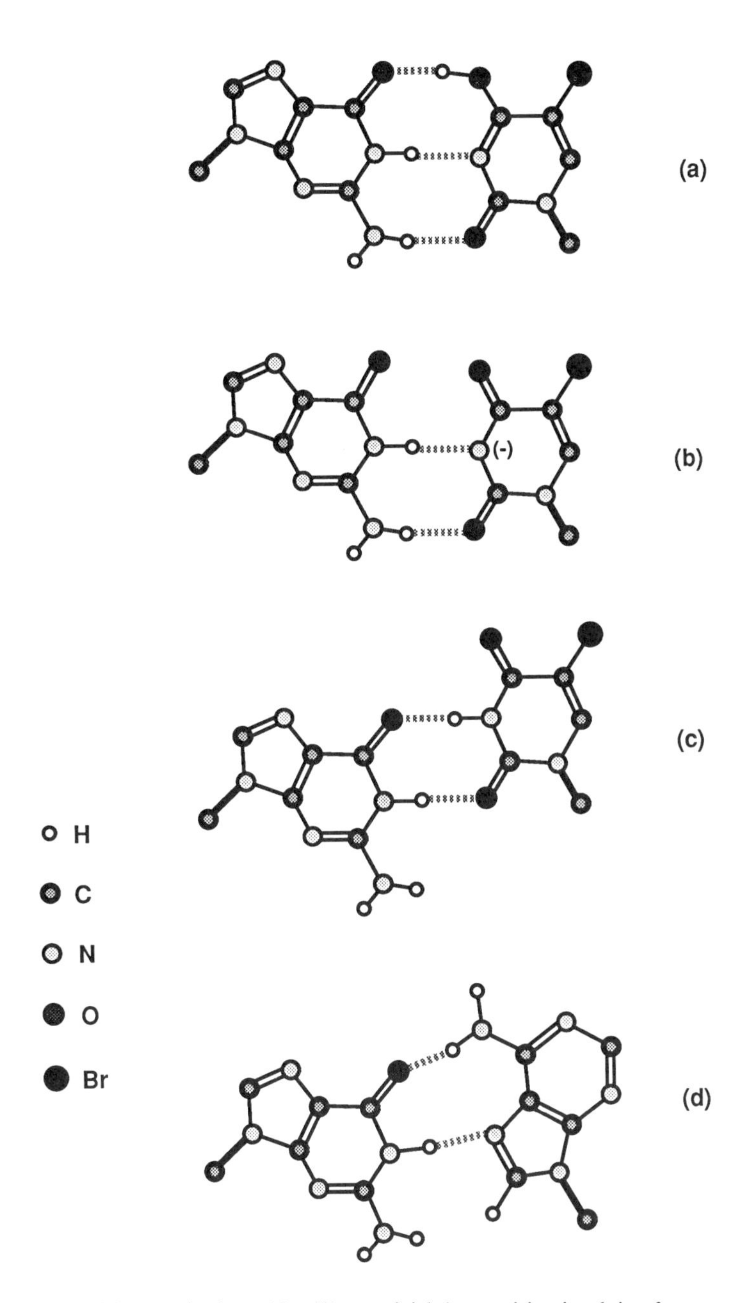

Figure 1. Possible Mispairing Mechanisms. Non-Watson-Crick base pairing involving four structural alterations is depicted here. A guanine residue (on the left) is shown to mispair with 5-BrU *via* (a) tautomerization of BrU (to O^4-enol form), (b) ionization of BrU, (c) wobbling, and to mispair with adenine *via* (d) N-glycosyl rotation of A into the *syn* conformation. These examples, given to point out several structural alterations that may participate in mispairing during DNA synthesis, are based on structures observed in nucleic acid complexes. Although base substitution mutagenesis is traditionally viewed as arising *via* formation of rare hydrogen-bonded base mispairs during DNA synthesis, this notion arose solely from consideration of interactions within duplex DNA. In the context of DNA synthesis protein-nucleic acid interactions may contribute significantly to the specificity of incorporation, and the role of H-bonded base pairing has not yet been determined.

Among the drug-DNA complexes that have been examined at high resolution, some have revealed unusual base pairing induced by the bound drug. For example, in the complex between the self-complementary octamer d(GCGTACGC) and the bis-intercalator antibiotic triostin A (Quigley *et al.*, 1986), both A•T and G•C Hoogsteen base pairs were observed, in which the purine residues were rotated into the *syn* N-glycosyl conformation. In the G•C Hoogsteen pair the C residue appeared to be protonated. The observation that drug-DNA interactions can stabilize unusual base pair structures leads to the prediction that protein-nucleic acid interactions at the active site of DNA polymerases (which might be sequence-dependent and unique for different polymerases) might stabilize specific mispairs during DNA synthesis.

A variety of high resolution X-ray crystallographic and NMR studies have been carried out with DNA duplexes containing mispairs. The results have shown that mispairing within duplex DNA can involve wobbling, ionization and N-glycosyl rotation, the mechanism of mispairing sometimes being influenced by nucleotide sequence. For example, G•T mispairs in all sequence contexts (reported to date) are of the wobble type (Hunter *et al.*, 1987; Kalnik *et al.*, 1988; Kennard, 1987; Patel *et al.*, 1987). However, A•G mispairing within the d(CGCGAAT-TAGCG) duplex occurred by rotation of the N-glycosyl bond of A to the *syn* form (Brown *et al.*, 1986), but without N-glycosyl rotation within d(CCAAGATTGG) (Prive *et al.*, 1987) and within d(CGAGAATTCGCG) (Patel *et al.*, 1984). It appears that B-DNA can accommodate the extra distance between opposing N-glycosyl bonds at the *anti,anti* Pu•Pu mismatch through minor adjustments in backbone torsion angles (Prive *et al.*, 1987). In the A•C mispair within duplex DNA, base pairing appears to involve wobbling as well as protonation of A (Hunter *et al.*, 1986; Sowers *et al.*, 1986). The 2-aminopurine•C base pair was also found to involve base protonation (Eritja *et al.*, 1986b).

There is relatively little evidence supporting mispairing *via* tautomerization, although rare tautomers have been widely presumed to play a role in spontaneous transition mutagenesis as originally suggested by Watson and Crick (1953b,c). Fresco *et al.* (1980) provided spectroscopic evidence that A•C and I•U mispairs involving rare tautomers can exist within double helices dominated by A•U and I•C base pairs, respectively. The only crystallographic study thus far suggesting mispairing *via* tautomerization was in the study of a hairpin loop structure formed by d(CGCGCGTTTTCGCGCG) (Chattopadhyaya *et al.*, 1988). An unusual T•T base pair, with near-Watson-Crick geometry, was observed between the hairpin loops of separate molecules within the crystal, which was proposed to involve one thymine in its O^4-enol form.

MISPAIRING DURING DNA SYNTHESIS

The existence of mispairs in nucleic acid complexes, often involving disfavored structural forms of nucleotides, testifies to the high degree of flexibility and dynamic behavior of nucleic acids, and may be relevant to mechanisms of misincorporation during DNA synthesis. The principal unknown factor with regard to the latter is the additional constraints to mispairing that the DNA polymerase exerts. Certain rare structural forms of the nucleotides may exist within duplex DNA but may not be possible in the template-primer-dNTP-polymerase complex. Consideration of the types of "mispairs" that are permitted or not permitted to occur during DNA synthesis may shed light on the "base selection" and "proofreading" mechanisms of DNA polymerases.

Fidelity of DNA Synthesis

A variety of biochemical assays have been used to assess the frequency of misincorporation during DNA synthesis. These include (i) measurements of incorporation of noncomplementary *vs.* complementary nucleotides during DNA synthesis with synthetic polynucleotides of repeating base sequence (Trautner *et al.*, 1962; Hall and Lehman, 1968; Agarwal *et al.*, 1979; El-Diery *et al.*, 1984); (ii) measurement of the mutation frequency in a transfection assay following *in vitro* primer elongation on bacteriophage templates (Weymouth and Loeb, 1978; Kunkel and Loeb, 1979; Fersht and Knill-Jones, 1983; Kunkel, 1985a; Reckmann and Krauss, 1986); and (iii) gel electrophoretic analysis of primer elongation on bacteriophage templates in the absence of one or more of the four canonical dNTPs (Hillebrand *et al.*, 1984; Hillebrand and Beattie, 1984; Revich *et al.*, 1984; Boosalis *et al.*, 1987). The results of these studies have indicated that the *in vitro* error frequency of purified DNA polymerases is typically 10^{-7} to 10^{-4}, depending on the DNA polymerase studied. This fidelity is much higher than the 10^{-1} to 10^{-2} error frequency predicted from measured free energy differences between correct and incorrect base pairs (1-3 kcal/mol), according to the thermodynamic relationship, $\Delta G=-RT\ln(\text{incorrect/correct})$ (Loeb and Kunkel, 1982). Therefore, since base pairing alone cannot supply sufficient specificity to account for the high fidelity of DNA synthesis, the polymerase must contribute substantially to the accuracy of polymerization, through base selection and proofreading functions.

There is great variability between different polymerases with regard to the relative contributions of base selection and proofreading in the overall accuracy of DNA synthesis. With DNA polymerase encoded by bacteriophage T4, proofreading plays a major role in fidelity (Muzyczka *et al.*, 1972; Bessman *et al.*, 1974; Lo and Bessman, 1976), although base selection is still a significant factor, in light of the existence of strong "mutator" phenotype associated with certain mutations in the T4 DNA polymerase gene that do not diminish 3′-exonuclease activity (Hershfield, 1973; Gillin and Nossal, 1976). In the case of the DNA polymerase I of *E. coli*, base selection plays a much greater role in fidelity than does 3′-exonuclease editing (El-Deiry *et al.*, 1988; Kuchta *et al.*, 1988). Finally, DNA polymerases that lack detectable 3′-exonuclease activity, such as retroviral DNA polymerases (Battula and Loeb, 1976; Roberts *et al.*, 1988) and eucaryotic DNA polymerase beta (Stalker *et al.*, 1976; Loeb and Kunkel, 1982; Weissbach, 1979; Fry, 1983), must rely on base pair recognition to achieve fidelity in DNA synthesis (error frequencies of 10^{-4} to 10^{-3}, still lower than expected if the specificity of incorporation were due to base pairing alone).

A large body of data have now accumulated that indicate that the nucleotide sequence of the template exerts a large influence on the frequency of misincorporation during DNA synthesis. Kunkel (1985a,b) and Roberts and Kunkel (1986) found that the occurrence of any given base substitution during *in vitro* DNA synthesis on M13mp2 templates varied considerably along the region of the *lacZ* gene examined. Pless *et al.* (1981) showed earlier that incorporation of 2-aminopurine opposite template T occurred with a high degree of sequence dependence during *in vitro* DNA synthesis. Similarly, Hillebrand *et al.* (1984) and Hillebrand and Beattie (1984) found that *in vitro* misincorporation on bacteriophage templates exhibited a high degree of sequence dependence. Likewise, Maldonado-Rodriguez *et al.* (1989a,b) found that misincorporation within the *lacI* gene of *E. coli* during *in vitro* DNA synthesis occurred at greatly varying frequencies at different template T residues. The data in the latter study resembled the spectrum of *in vivo* spontaneous mutations (at T residues in the *lacI* gene) in *E. coli* strains deficient in mismatch correction (Schaaper and Dunn, 1987), suggesting that the well known "spectra" of spontaneous and induced mutagenesis *in vivo* (Benzer and Freese, 1958; Freese, 1959a,b; Champe and Benzer, 1962; Foster *et al.*, 1983; Miller, 1983) may largely reflect sequence effects on mispairing events during DNA replication.

Specificity of Misincorporation

A number of experimental approaches have been used to characterize the rare mutational events that occur during *in vitro* DNA synthesis. These have revealed large sequence effects on the specificity of mispairing during DNA synthesis.

Of major significance in this area is Kunkel's extensive characterization of base changes associated with mutations that occur within the *E. coli lacZ*α target sequence during *in vitro* DNA synthesis on bacteriophage M13mp2 template, catalyzed by various eucaryotic DNA polymerases (Kunkel, 1985a,b; Kunkel and Alexander, 1986; Roberts and Kunkel, 1986). A subset of the data obtained by Kunkel and coworkers is shown in Figure 2, which portrays the base substitutions observed at different template positions within the *lacZ* gene, resulting from *in vitro* polymerization catalyzed by eucaryotic DNA polymerase beta (Pol β, above the line) versus eucaryotic DNA polymerase alpha (Pol α, below the line). To minimize the bias inherent to the genetic system, Figure 2 displays only data collected at template positions at which base substitutions of both transition and transversion types are detectable (16 G sites, 8 A sites, 14 T sites and 8 C sites). For this purpose all data for each class of DNA polymerase (Kunkel, 1985a,b) were combined (total of 152 mutants for Pol β and 178 mutants for Pol α).

The results compiled in Figure 2 illustrate the dramatic effect of DNA sequence on both the frequency and kind of misincorporation during DNA synthesis. For example, among the 80 base substitutions seen at 16 template G residues after *in vitro* polymerization catalyzed by Pol α, only one (1.25%) occurred at position 29 and at position 84, whereas 22 (27.5%) occurred at position 149. Additionally, of the 62 base substitutions induced by Pol β at 14 template T residues, none occurred at positions -10, -7 and 121, whereas 25 (40.3%) were detected at position 103. Transition-type base substitutions predominated at most template positions, although at certain positions (eg., G141, G149, T70 and T103) transversions predominated.

The data of Figure 2 also suggest that interactions at the DNA polymerase active site (unique for each polymerase) can influence both the frequency and kind of base substitution errors at certain template positions during DNA polymerization. For example, position G148 represented a "hotspot" for base substitutions induced by Pol α, but not for Pol β. The reverse situation occurred at T70, T87 and T103. Although the data are not extensive, there is a suggestion of different types of misincorporation by the two polymerases at position -11 (template A) and position 118 (template G).

The data of Kunkel suggest that interactions involving both the DNA polymerase and the primer-template strongly influence the nature of mispairing during DNA synthesis. This conclusion was also made by Lai and Beattie (1988a,b), who conducted *in vitro* misincorporation reactions with different DNA polymerases, then determined the identity of misincorporated bases in the newly synthesized DNA.

As illustrated in Figure 3, DNA sequence exerted a major influence on the identity of mispairs formed during *in vitro* misincorporation reactions. The template sequence in the *lacZ* target region is indicated, and vertical bars serve to distinguish between transition-type

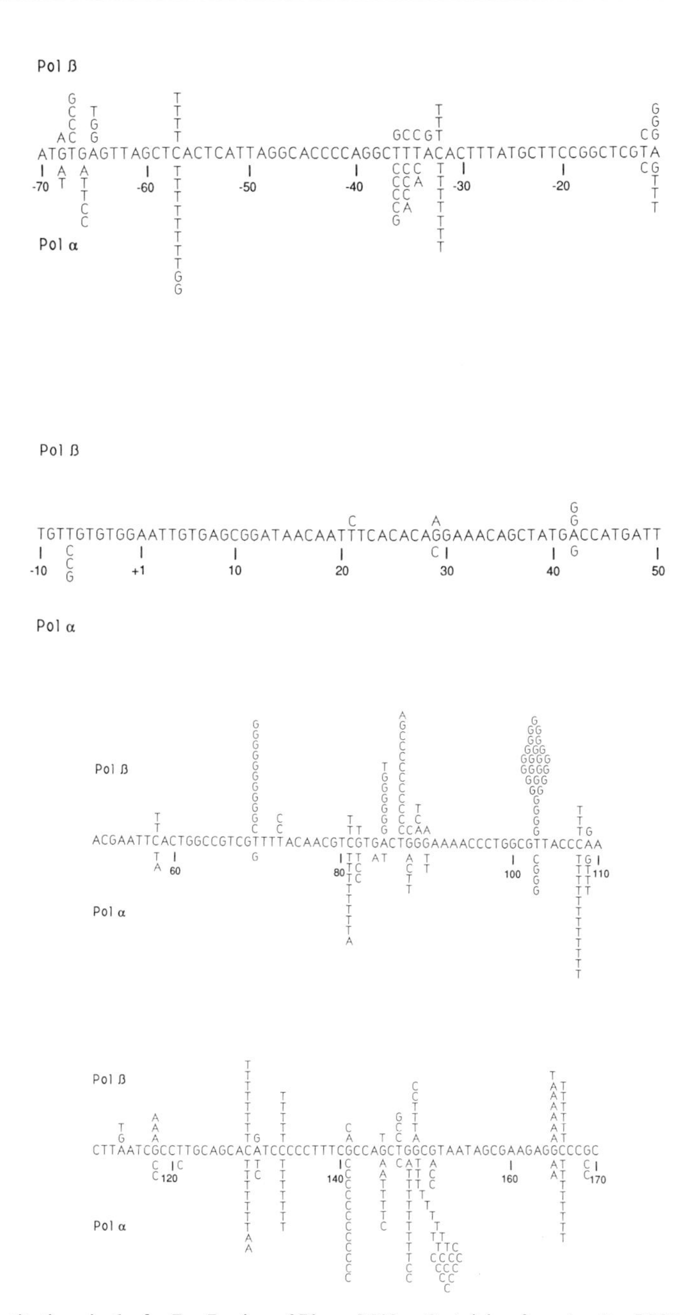

Figure 2. Base Substitutions in the *lacZ*-α Region of Phage M13mp2, Arising from *in vitro* DNA Synthesis Catalyzed by Eucaryotic DNA Polymerases. The template sequence is displayed horizontally, numbered according to transcriptional initiation (+1 being the first transcribed base). The data shown represent a subset of the base substitution mutants collected by Kunkel for eucaryotic DNA polymerase beta (Kunkel, 1985a) and eucaryotic DNA polymerase alpha (Kunkel, 1985b). The letters displayed above (for Pol β) and below (for Pol α) the template sequence represent the bases identified in mutant viral templates (each mutant residue is lined up with the corresponding wild-type residue). The set of base substitutions shown here (152 for Pol β and 178 for Pol α) are restricted to template positions at which both transitions and transversions are detectable, to permit assessment of sequence effects on the kind of misincorporation during DNA synthesis. For each class of polymerase data were combined for enzymes isolated from different sources.

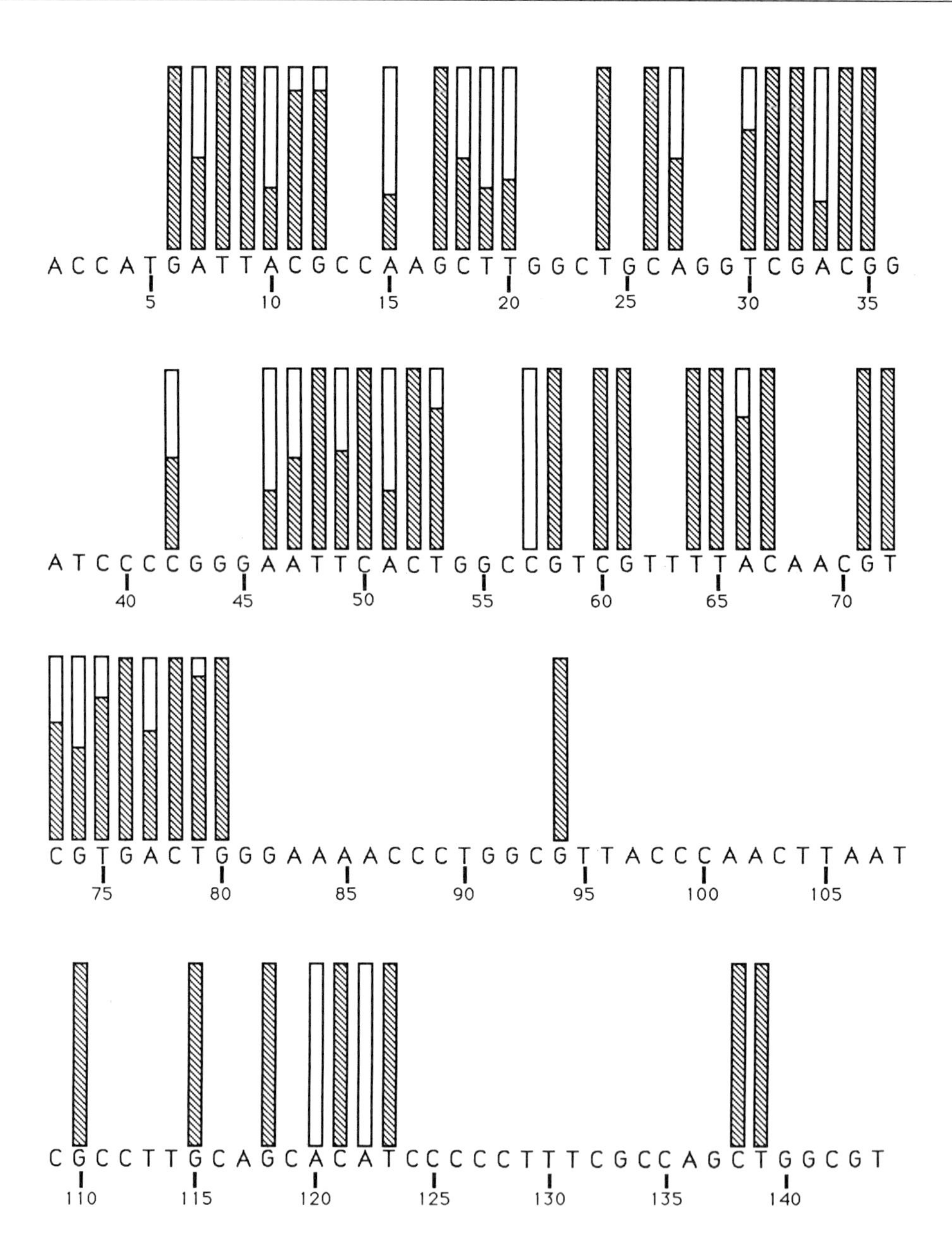

Figure 3. Base Misincorporations Occurring In The Region of Phage M13mp9 During *In Vitro* DNA Synthesis. The template sequence is indicated horizontally, numbered according to translational initiation (base 1 is the first residue in the coding sequence, which corresponds to position +42 in the data of Figure 2). For each vertical bar the kind of misincorporation that occurred during *in vitro* polymerization is represented by the relative size of open bar (tranversion-type mispairing) versus hatched bar (transition-type mispairing). These data were obtained by direct sequencing of DNA synthesized under infinite pool bias conditions (i.e., in the absence of one of the four deoxynucleoside triphosphates). "Minus" reactions were conducted as depicted in Figure 4, then the nascent strand (containing misincorporated bases) was isolated and sequenced as described by Lai and Beattie (1988a). The data shown here, compiled by Lai and Beattie (1989) from a large number of separate misincorporation reactions, do not represent a complete mispairing spectrum, since data were not available at all template positions.

mispairs (Py•Pu, represented by hatched bars) and transversion-type mispairs (Py•Py and Pu•Pu, represented by open bars). These data represent a combination of all data collected with DNA polymerase I Klenow enzyme (Kf pol) and Maloney murine leukemia virus DNA polymerase (MMLV pol) (Lai and Beattie, 1989). It was pointed out previously (Lai and Beattie, 1988a) that the "mispairing spectra" were remarkably similar for different DNA polymerases although, at about 20% of the sites examined, significant differences were seen in the type of mispairing determined for Kf pol and MMLV pol. Inspection of Figure 3 reveals several instances of a major effect of DNA sequence on the type of mispairing. For example, at most template cytosines transition-type mispairing predominated, but at position C57 only transversion-type mispairing was detected. Similarly, at most template adenines, both transitions and transversions were detected, but at positions A120 and A122 only transversions (*via* A•A mispairing) were observed. Using site-directed mutagenesis Lai and Beattie (1989a) confirmed that the identity of nearest neighbors to the misincorporation site can influence the type of misincorporation. Conversion of C121 (see Figure 3) to T121 changed the type of misincorporation that occurred on both sides of position 121. In the original template only A•A mispairing occurred at A120 and A122 (Figure 3), whereas after position 121 was changed from T to C, only A•C mispairing was detected at A120, and a mixture of A•A and A•C mispairing was seen at A122 (Lai and Beattie, 1989). Although these results confirm that the immediate nearest neighbors can contribute to the specificity of misincorporation, we showed previously (Lai and Beattie, 1988a) that longer range sequence effects (greater than two positions away from the misincorporation site) can also affect the type of mispairing during DNA synthesis.

The extent to which DNA sequence effects on misincorporational specificity arise directly from stacking interactions within the DNA, as opposed to indirectly arising through protein-nucleic acid interactions at the polymerase active site is presently unknown. As discussed previously, since certain drug-DNA interactions can stabilize unusual base paired structures, protein-nucleic acid interactions (which may be sequence dependent) could also influence the stability of a non-Watson-Crick base pair during polymerization. These sequence effects may contribute to the well known positional effects on the frequency and type of spontaneous and induced mutation within genes (the so-called "mutational spectra" which are unique for different mutagens) (Miller, 1983).

Incorporation of Modified Nucleotides

Experiments conducted in this laboratory using an electrophoretic assay of misincorporation and related work conducted by others have revealed several important phenomena relevant to substrate recognition/discrimination in DNA synthesis.

As in the aforementioned studies of mispairing of the canonical bases during DNA synthesis, we have observed DNA sequence effects on the base pairing specificity of nucleotide analogs during DNA synthesis. For example, although N^4-methoxy-dCTP was utilized by *E. coli* DNA polymerase I "Klenow fragment" (Kf Pol) most efficiently in place of dTTP, as expected from its known tendency to assume the imino tautomer, the analog replaced dCTP to a measurable extent, more efficiently at certain template positions (Reeves and Beattie, 1985). Also, mispairing of 5-bromo-U in the template with dGTP during DNA synthesis catalyzed by Kf Pol occurred with considerable sequence dependence (Driggers and Beattie, 1988). Again, the precise causes of these sequence effects are unknown, although it seems reasonable that nearest neighbor stacking interactions (which may affect the relative stability of a given non-Watson-Crick base pair within different sequence contexts) may at least partially account for the sequence effects.

Several examples have been found in which there is a lack of correspondence between the stability of a chemically modified base pair within duplex DNA and the formation of the same base pair during DNA synthesis. Although 5-BrU•A base pairs are more stable than T•A base pairs within duplex DNA (Inman and Baldwin, 1964), incorporation of BrdUTP (opposite template A) is much less efficient than that of dTTP during polymerization catalyzed by some DNA polymerases (Driggers and Beattie, unpublished). Studies with synthetic DNA containing the base analog 2-aminopurine (AP) (Eritja *et al.*, 1986b) showed that although the AP•A pair is more stable than the AP•C pair within duplex DNA, during DNA synthesis the formation of AP•C is strongly favored over AP•A. Similar studies with synthetic DNA containing xanthine (X) (Eritja *et al.*, 1986a) showed that although the relative thermodynamic stabilities of base mispairs with X within duplex DNA are T>G>A≈C, the relative incorporation rates opposite template X are T>C>>A≈G during polymerization catalyzed by *Drosophila* DNA polymerase alpha. Thus, even though X•C is less stable than X•G within duplex DNA, X•C is formed during DNA synthesis much more readily than X•G. In addition, although inosine-containing mispairs generally destabilize duplex DNA less than mispairs involving the canonical bases (Martin *et al.*, 1985; Aboul-ela *et al.*, 1985; Kawase *et al.*, 1986), inosine pairs only with cytidine during DNA synthesis (Bessman *et al.*, 1958; Figure 5, below). Finally, benzimidazole, which is incapable of forming H-bonded base pairs, is nevertheless incorporated into nucleic acids in *E. coli* and acts as a base analog mutagen in *Salmonella* (Seiler, 1972; Seiler, 1973).

The results cited above suggest that the stability of hydrogen bonded base pairing cannot be the major determinant of whether a given mispairing event occurs during DNA synthesis. This conclusion is further evidenced by the relative discrimination of different DNA polymerases against modified base pairs during polymerization. The electrophoretic assay of misincorporation, used to assess DNA polymerase fidelity as mentioned previously, offers an extremely sensitive tool for assessing the base pairing specificity of chemically modified nucleotides during DNA synthesis. Figure 4 illustrates the rationale of the assay and its usefulness in assessing the potential of a given nucleotide analog for ambiguous base pairing during DNA synthesis. Figures 5 and 6 show the results of experiments conducted to assess the base pairing specificity of some modified nucleoside residues during primer elongation catalyzed by different DNA polymerases.

Two physical properties of inosine lead to the prediction that this nucleoside would be prone to mispairing during DNA synthesis: (1) Deoxyinosine (dI) generally has greater mispairing potential within duplex DNA than the normal nucleosides, I•A mispairing being especially nondisruptive (Martin *et al.*, 1985; Aboul-ela *et al.*, 1985; Kawase *et al.*, 1986). (2) Physical evidence suggests that inosine exists in aqueous solution as an approximately equimolar mixture of the keto tautomer (having the base pairing specificity of guanine) and the enol tautomer (having the base pairing specificity of adenine) (Wolfenden, 1969). These findings prompted us to test the mispairing potential of inosine during DNA synthesis, using the very sensitive electrophoretic assay of misincorporation. The data of Figure 5 show the base pairing specificity of dI during primer elongation catalyzed by two different DNA polymerases. Enhancement of primer elongation occurred only when dITP was added to the "-G" reaction (lanes 6 and 18), indicating that dITP replaced only dGTP in chain elongation. Thus, none of

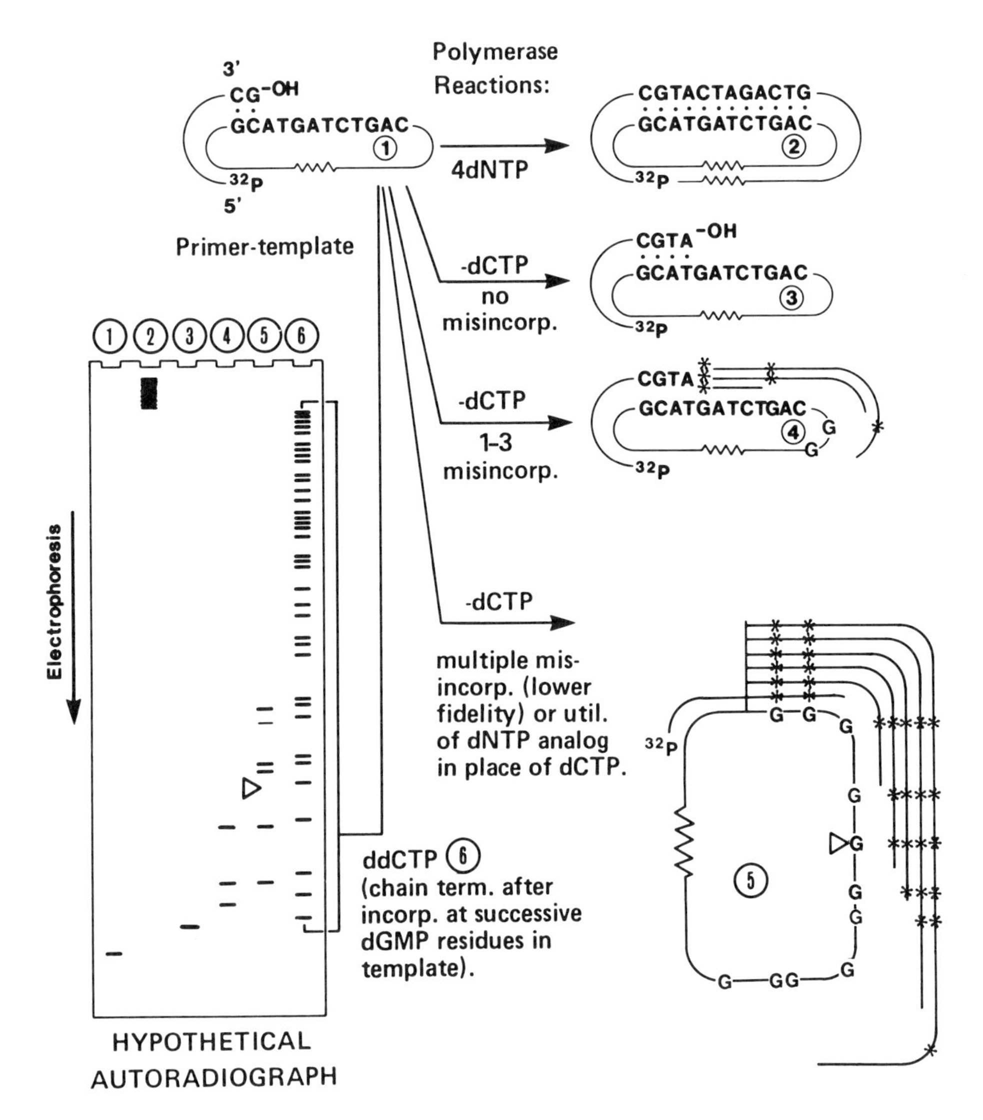

Figure 4. Gel Electrophoretic Assay of Misincorporation. A [5'-^{32}P]-primer, annealed to a DNA template, is elongated in the presence of only 3 of the 4 dNTPs. In such a "minus" reaction, chain elongation past each template position complementary to the missing dNTP depends on incorporation of one of the 3 normal dNTPs present, or (if added) a chemically modified dNTP analog. The relative rate of misincorporation (or incorporation of analog) at different template positions is revealed by the autoradiographic banding patterns obtained after polyacrylamide gel electrophoresis of the denatured products (compared with the "dideoxy" sequencing pattern). A "hotspot" for misincorporation at the 5th G residue in the template (marked in the lower right diagram) is seen in the hypothetical autoradiograph as the absence of a "-C" band (lane 5) corresponding to the 5th "ddC" band (lane 6). The base mispairing capacity of modified nucleotide analogs (in either dNTP or template) can be sensitively determined by carrying out the full set of "minus" reactions with unmodified versus chemically modified dNTP or template and noting whether there is increased chain elongation in the presence of the analog (see examples in Figs. 5 and 6).

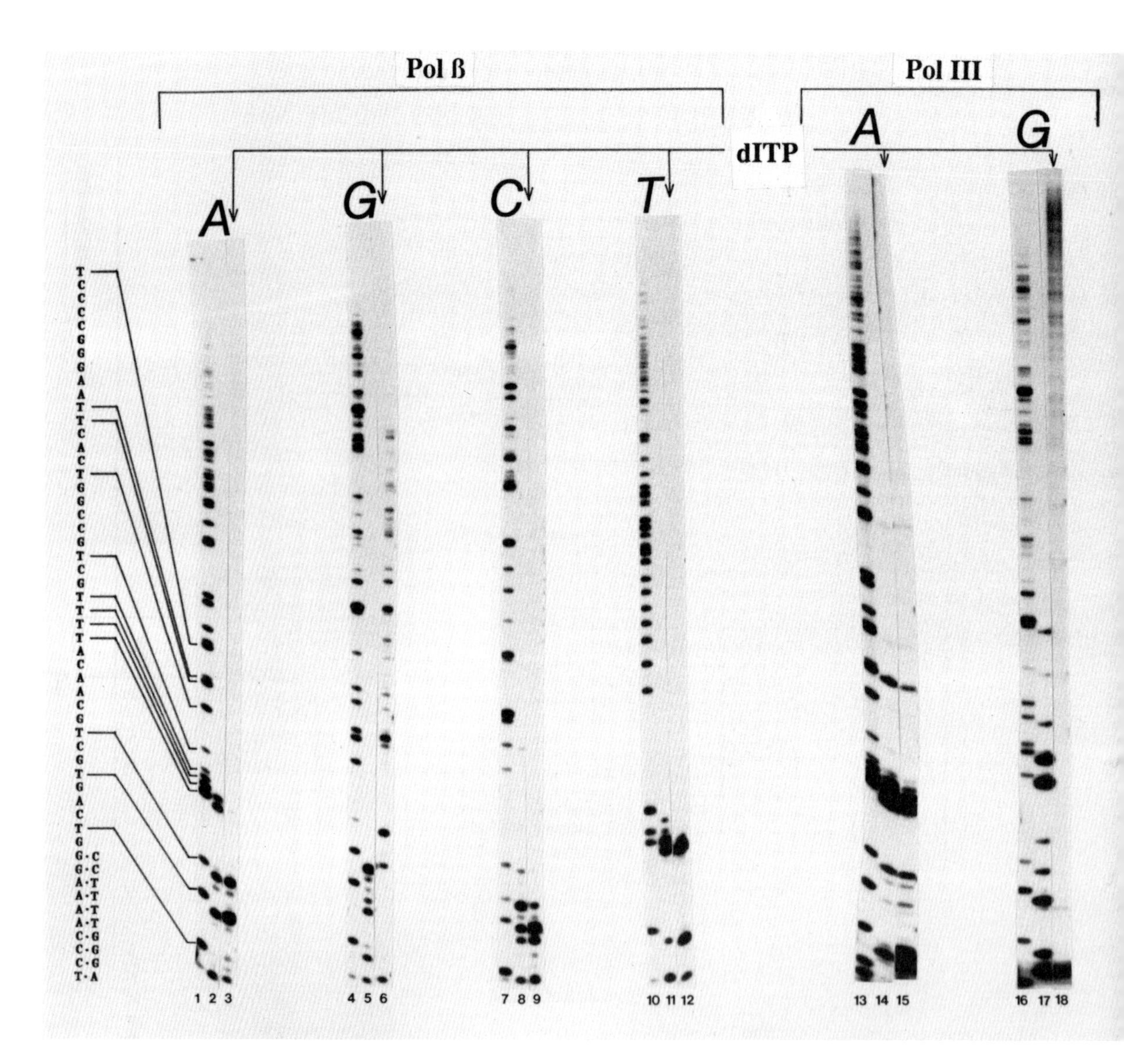
Pol ß
Pol III
dITP
A
G
C
T
A
G
T C C C C G G G A A T T C A C T G G C C G T C G T T T T A C A A C G T C G T G A C T G G·C G·C A·T A·T A·T A·T C·G C·G C·G T·A
1 2 3
4 5 6
7 8 9
10 11 12
13 14 15
16 17 18

the DNA polymerases tested misincorporated dIMP any more frequently than the normal nucleotides, even though mispairing of dI was predicted to occur more readily than that of the normal nucleosides. We also found (data not shown) that replacement of dG with dI in the M13mp9 template led to no greater extent of primer elongation in "-C" reactions conducted with any of the polymerases tested, indicating that dI in the template likewise has no greater mispairing potential than does dG.

Interestingly, two of the polymerases that are considered to be relatively error-prone (MMLV pol and Pol β) utilized dITP in place of dGTP less efficiently than did the more accurate *E. coli* Pol I and Pol III holoenzyme (data not shown), suggesting that these polymerases can exert some discrimination against the normally acceptable I•C base pair.

In similar experiments conducted with the base analog 5-bromouracil, we also noted different behavior by different polymerases (Driggers and Beattie, 1988 and unpublished data). BU•G mispairing was readily detectable with *E. coli* Pol I, MMLV Pol and Pol β by use of the electrophoretic assay (both when BU was in the template and in the dNTP). However, no such BU•G mispairing was detected with *E. coli* DNA polymerase III holoenzyme. The Watson-Crick base pair, BU•A, however, was somewhat discriminated against during DNA synthesis catalyzed by *E. coli* Pol I and Pol β, but not by Pol III holoenzyme.

One additional example to be cited is the analysis of the base pairing specificity of 1,N^6-Etheno-dA during DNA synthesis. The structure of this analog of dA (with hydrogen bonding sites disrupted by the presence of the etheno ring as seen in Figure 6A), together with the known destabilizing effect of its presence within duplex DNA (Lee and Wetmur, 1973), led us to expect that Etheno-dATP would not be detectably incorporated during DNA synthesis. We were thus surprised to discover (Revich *et al.*, 1984; Revich and Beattie, 1986) that the analog is utilized by *E. coli* Pol I in place of dATP, though at relatively low efficiency. The analog also showed low but detectable ambiguous base pairing ability (slight incorporation in place of dGTP). Recent genetic assays conducted with Etheno-dA-containing bacteriophage DNA (Maldonado-Rodriguez *et al.*, 1989c) also suggested that Etheno-dA in a DNA template is normally recognized as dA, and occasionally as G during *in vivo* DNA replication.

Figure 6B shows the results of electrophoretic assays of Etheno-dATP utilization by different DNA polymerases. The incorporation of the analog in place of dATP was readily detected with *E. coli* Pol I (lanes 3 and 4 *vs.* 2) and *E. coli* Pol III holoenzyme (lane 18 *vs.* 17). Surprisingly, Pol β (generally regarded as a relatively error-prone polymerase) discriminated very efficiently against this analog during DNA synthesis. As seen in Figure 6B, Pol β utilized

Figure 5. Electrophoretic Assay of the Base Pairing Specificity of Deoxyinosine Triphosphate During DNA Synthesis. A [5′-^{32}P]-oligonucleotide primer was annealed to M13mp9 viral strand and the gel electrophoretic assay of primer elongation was conducted as illustrated in Figure 4 and as described by Hillebrand *et al.* (1984) and Driggers and Beattie (1988). The nucleotide sequence of the template and 3′-end of the primer is displayed along the left (3′-OH residue of primer pairs with position 81 in the M13mp9 template shown in Figure 3). Lines connect successive T residues in the template with corresponding "dideoxy-A" termination bands in lane 1. For Pol β (lanes 1-12) all four "minus" reactions were carried out and the effect of added dITP on primer elongation was tested. For each "minus" reaction three lanes are shown, displaying the products of the appropriate "dideoxy" sequencing reaction (lanes 1, 4, 7, 10), the "minus" reaction (lanes 2, 5, 8, 11), and the "minus" reaction containing dITP (lanes 3, 6, 9, 12). For *E. coli* Pol III holoenzyme only the results of the "-A" (lanes 13-15) and "-G" (lanes 16-18) reactions are shown. With Pol β and Pol III holoenzyme (and from data not shown, Pol I of *E. coli* and DNA polymerase of Maloney murine leukemia virus) stimulation of primer elongation by added dITP occurred only in the "-G" reaction, indicating that during DNA synthesis dI base pairs only with dC and lacks the mispairing properties that would be predicted from physical studies of tautomeric equilibrium and mispairing of dI within duplex DNA.

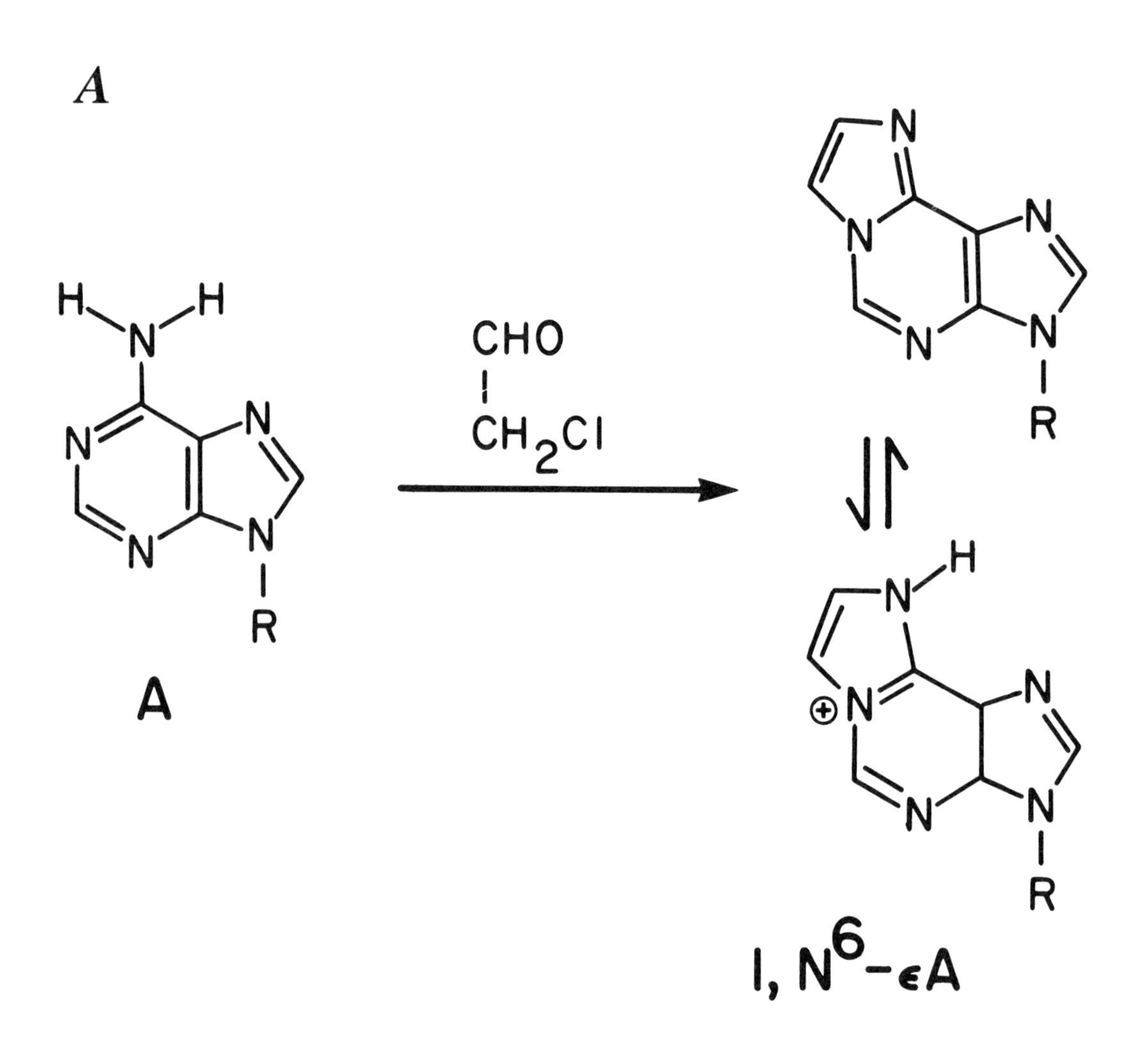

Figure 6. Electrophoretic Assay of the Base Pairing Specificity of 1,N^6-Etheno-dATP During DNA Synthesis. Panel A shows the chemical structure of 1,N^6-Etheno-A, formed by reaction of A with chloroacetaldehyde, a metabolite of vinyl chloride. Panel B shows the results of the gel electrophoretic assay of the specificity of incorporation of the analog during DNA synthesis catalyzed by *E. coli* Pol I (lanes 1-4), eucaryotic Pol β (lanes 5-15) and *E. coli* Pol III holoenzyme (lanes 16-18). The data are displayed for each "minus" reaction as in Figure 5: lane1, "dideoxy-A" sequencing reaction for Pol I and Pol β; lane 7, "ddG"; lane 10, "ddC"; lane 13, "ddT"; lane 16, "ddA" for Pol III; lanes 2, 5 and 17, "-A" reactions for Pol I, Pol β and Pol III, respectively; DNA products from reactions containing Etheno-dATP are displayed immediately following each "minus" reaction. For Pol β the results of all four "minus" reactions are shown, while for Pol I and Pol III the results of only the "-A" reactions are shown. See Revich and Beattie (1986) for details of the Pol I catalyzed reactions. As seen from extensive primer elongation in the "-A" reactions containing Etheno-dATP (lanes 2, 3, 18), incorporation of Etheno-dA in place of dA occurred readily with the *E. coli* polymerases. Greater primer elongation in lane 3 (polymerization carried out at pH 7) compared with lane 4 (polymerization at pH 8, with polymerase activity comparable to that at pH 7) is in accord with the proposal (Revich and Beattie, 1986) that during DNA synthesis the protonated, *syn* form of Etheno-dA pairs with dT. As opposed to Pol I (Revich and Beattie, 1986) Pol β totally discriminated against ambiguous base pairing of the analog during polymerization, and comparison of lanes 5 and 6 shows that Pol β efficiently discriminates against the Etheno-dA•dT base pair during DNA synthesis.

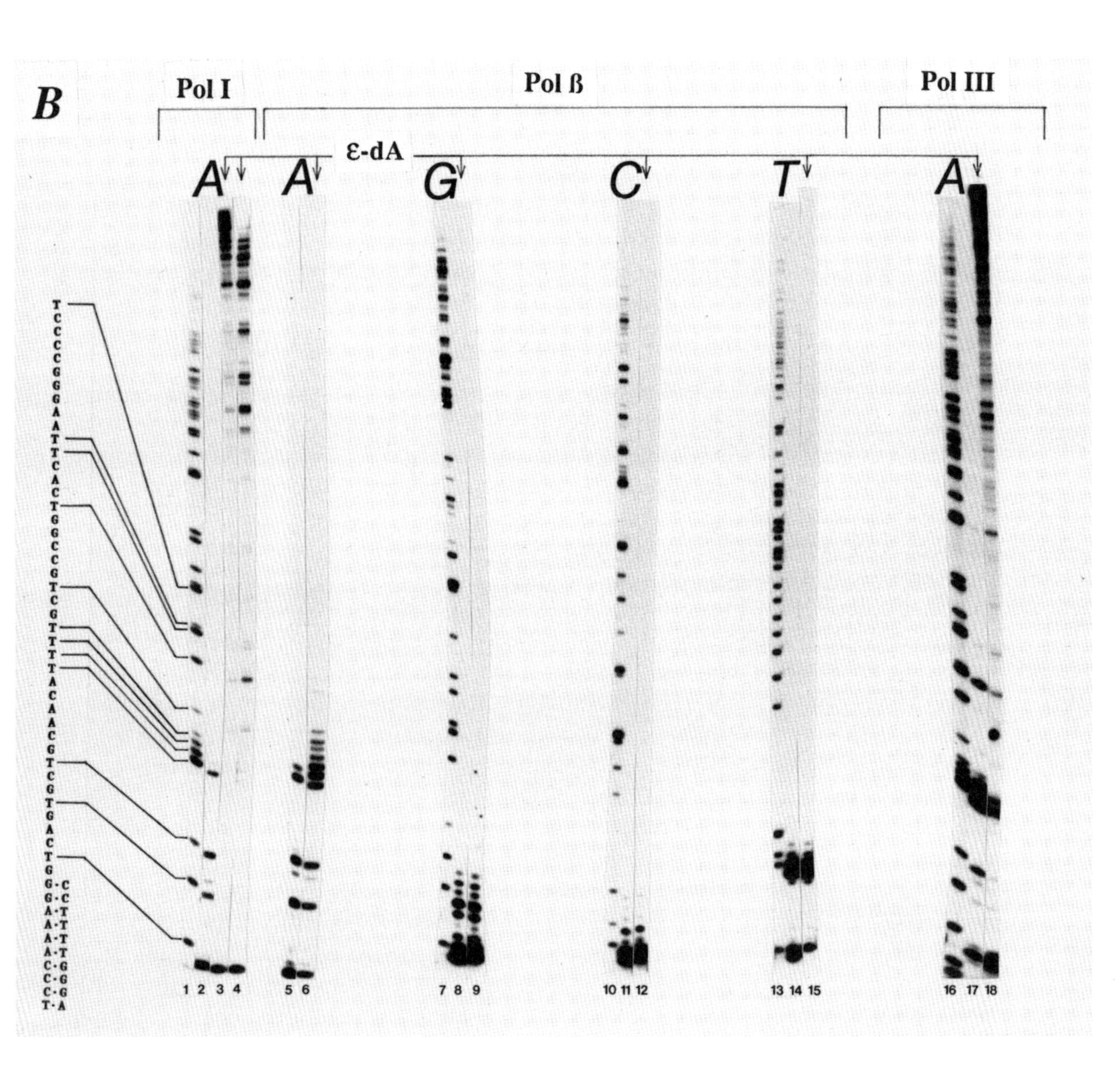
B
Pol I
Pol ß
Pol III
ε-dA
A
A
G
C
T
A
T
C
C
C
C
G
G
G
A
A
T
T
C
A
C
T
G
G
C
C
G
T
C
G
T
T
T
T
A
C
A
A
C
G
T
C
G
T
G
A
C
T
G
G·C
G·C
A·T
A·T
A·T
A·T
C·G
C·G
C·G
T·A
1 2 3 4
5 6
7 8 9
10 11 12
13 14 15
16 17 18

Etheno-dATP in place of dATP at a barely detectable level (lanes 5 and 6) and, in contrast to Pol I (Revich and Beattie, 1986), did not permit ambiguous base pairing of the analog during polymerization. Etheno-A, therefore, represents an analog that from structural considerations should not base pair at all during DNA synthesis, but is nevertheless incorporated with relative ease during primer elongation catalyzed by some but not all DNA polymerases.

The results discussed above evoke the following conclusions pertinent to mechanisms of base substitution mutagenesis: (1) The DNA sequence exerts a large influence on the occurrence of a given non-Watson-Crick base pair during DNA synthesis. (2) The physical properties of certain base analogs and their base pairs within duplex DNA do not yield accurate predictions of the specificity of base pairing during DNA synthesis. (3) The identity of the polymerase has a large influence on the base pairing properties of chemically modified nucleotides, indicating that to a certain extent, interactions at the polymerase active site, unique to each enzyme, can significantly influence the base pairing and mispairing events that occur during DNA synthesis.

ROLE OF DNA POLYMERASE IN THE SPECIFICITY OF INCORPORATION

In considering how substrate specificity is achieved in DNA synthesis, two extreme viewpoints can be proposed: (1) The specificity of nucleotide incorporation during DNA synthesis is determined by H-bonding between the incoming dNTP and the opposing template residue, with the polymerase playing an active role in the chemical step (phosphodiester bond formation) but a passive role in specificity (through template-directed dNTP binding); (2) Phosphodiester bond formation is preceded by a substrate recognition mechanism in which interactions between opposing complementary bases and amino acid residues in the polymerase play the major role, rather than direct hydrogen bonding between complementary bases.

Although there is presently no direct evidence in support of either of these extreme viewpoints, the available evidence, discussed above, suggests that base pairing alone cannot account for the specificity of incorporation. The wide spectrum of polymerase fidelity may reflect the extent to which different DNA polymerases subscribe to these diverse means of discrimination. The widely accepted view that the specificity of nucleotide incorporation is determined by base pairing grew out of the concept of base pairing within the double helix, and has persisted largely because it is conceptually simple and logical. However, as pointed out by Morozov *et al.* (1982), "This concept does not follow directly from the Watson-Crick hypothesis, ... Moreover, there are experimental data that generally contradict the hypothesis of a deciding role of complementarity of the bases in template synthesis."

As discussed above, the most convincing argument against the base pairing model of polymerase fidelity is the lack of correspondence between the base pairing specificity (within duplex DNA) of several nucleotide analogs and their specificity of incorporation during DNA synthesis. Petruska *et al.* (1986) presented a useful hypothesis on DNA polymerase fidelity, wherein exclusion of water at the enzyme's active site greatly magnifies the free energy differences between correctly paired and mispaired bases (H-bonding being the main determinant), such that additional active site constraints are not required to select or reject nucleotides. However, this hypothesis does not seem to explain the unexpected incorporation patterns of certain nucleotide analogs. The hypothesis that free energy differences (involving correct *vs.* incorrect substrates) are amplified by exclusion of water could apply equally to a substrate recognition/discrimination mechanism involving a protein-nucleic acid H-bonding network within the polymerase-DNA-dNTP complex.

The remaining discussion will address the question of how DNA polymerase might utilize protein-nucleic acid interactions to achieve substrate recognition/discrimination prior to phosphodiester bond formation. Whether base pairing also plays a role in substrate specificity prior to the chemical step remains unknown. The exonucleolytic "proofreading" activity associated with some DNA polymerases, which may contribute significantly to the net accuracy of synthesis and may depend more strongly on hydrogen bonded base pairing than the preincorporation base selection process, will not be discussed here.

Several base pair recognition mechanisms can be envisioned, and it is entirely possible that different DNA polymerases achieve base pair selection in different ways.

As a framework for discussion let's begin with a simplified consideration of progression of the enzyme along the reaction coordinate in a single catalytic cycle:

$$E \xrightarrow{(1)} E\bullet DNA_n\bullet dNTP \xrightarrow{(2)} [E\bullet DNA_n\bullet dNTP]^{\ddagger} \xrightarrow{(3)} E\bullet DNA_{n+1}\bullet PP_i \xrightarrow{(4)} E\bullet DNA_{n+1}$$

"Milestones" along the reaction coordinate include (1) formation of polymerase-primer-template-dNTP complex, $E\bullet DNA_n\bullet dNTP$; (2) creation of a transition state for the chemical step, $[E\bullet DNA_n\bullet dNTP]^{\ddagger}$; (3) incorporation (phosphodiester bond formation); and (4) pyrophosphate release. See Kuchta *et al.* (1988) for a more detailed kinetic scheme. In thinking about the catalytic cycle it is useful to keep in mind the dynamic nature of protein conformation and enzyme-substrate interactions. During the course of the reaction, the enzyme and its complexes with substrates and products can be considered (in the simplest sense) to pass through a specific conformational and interactive pathway.

The question arises, at which point(s) along this pathway does discrimination occur? Available evidence suggests that relatively little discrimination derives from the dNTP binding step. This conclusion comes from NMR titration studies (Ferrin and Mildvan, 1986) and kinetic studies (Travaglini *et al.*, 1975) of *E. coli* DNA polymerase I, which indicated that the presence of template or template-primer did not affect the relative affinities of correct *vs.* incorrect dNTPs for the enzyme. Recent kinetic studies of Kuchta *et al.* (1988) also suggested relatively little selectivity in the dNTP binding step, and indicated that the major component of fidelity is a dramatically reduced rate of phosphodiester bond formation for incorrect *vs.* correct dNTPs. On the basis of rapid-quench kinetic studies, Mizrahi *et al.* (1985) proposed that the rate-limiting step in polymerization catalyzed by *E. coli* DNA polymerase I is a conformational change within the E•DNA•dNTP complex immediately preceding the chemical step. Possibly, as suggested by Ferrin and Mildvan (1986), this rate-determining conformational change, which presumably prepares the complex for nucleophilic attack of 3′-OH (primer terminus) on the alpha phosphoryl group (dNTP), may constitute a critical "verification" step which insures a high fidelity of incorporation. Although Ferrin and Mildvan (1986) suggested that the verification step involves base pairing between the incoming dNTP and the complementary template residue, it is also possible that the verification step involves isolated interactions between the enzyme and opposing but non-H-bonded residues in dNTP and template.

Let's return for a moment to the concept that progression along the reaction coordinate involves a specific conformational and interactive pathway of the E•DNA•dNTP complex. The recognition event, positioned at some point along this pathway, may be a prerequisite to the chemical step. The recognition step could involve interactions (between the protein and opposing complementary bases) that stabilize the transition state for phosphodiester bond formation, or could comprise interactions that stabilize a high energy conformation of the enzyme somewhere between E•DNA•dNTP and $[E \bullet DNA_n \bullet dNTP]^{\ddagger}$ along the reaction coordinate (i.e., interactions that lower a conformational energy barrier that must be crossed in order to reach the transition state for phosphodiester bond formation).

Substrate recognition during DNA synthesis may in fact occur at several levels, whether or not the major recognition step involves H-bonded base pairing between dNTP and template. H-bonding of the newly formed base pair might be important for events at the end of the catalytic cycle, such as pyrophosphate release and/or translocation. Kuchta *et al.* (1988) recently provided evidence that 3′-terminal mismatches caused decreased rates of both PP_i release and addition of the subsequent nucleotide, increasing the opportunity for 3′-exonucleolytic editing by *E. coli* DNA polymerase I (Klenow form).

The principal recognition step in DNA synthesis could involve several types of interactions:

(1) Opposing complementary bases could be recognized *via* an intricate H-bonding network, as occurs in enzyme-substrate interactions in tyrosyl-tRNA synthetase (Fersht, 1987), sequence-specific recognition by Eco RI endonuclease (McClarin *et al.*, 1986) and repressor-operator interactions (Jordan and Pabo, 1988; Aggarwal *et al.*, 1988). Notably, disruption of a single hydrogen bond in the foregoing interactions can have a dramatic effect on binding or catalysis. As mentioned previously, exclusion of water from "recognition pocket(s)" within the polymerase could greatly amplify the free energy difference between H-bonding networks associated with correct *vs.* incorrect base pairs.

(2) Steric fit (spatial complementarity) of opposing bases in binding pocket(s) within the polymerase could also be a major component of substrate recognition/discrimination. At the recognition step, progression along the reaction coordinate may be facilitated if opposing template and dNTP residues make precise van der Waals contacts with complementary binding pockets.

(3) The recognition (verification) step could involve actual H-bonded base pairing between incoming dNTP and the complementary template residue, plus additional interactions within the E•DNA•dNTP complex (electrostatic, H-bonding, van der Waals) that are essential to progression along the reaction coordinate.

(4) The laboratory of E. I. Budowsky (Morozov *et al.*, 1986) proposed a very intriguing model for substrate recognition by DNA polymerases, wherein electrostatic interactions play a key role. According to their hypothesis, prior to each incorporation event (and without a requirement for H-bonded base pairing between template and dNTP residues) the DNA polymerase responds to a given template residue (presumably by adopting a specific conformational state), forming a distinct recognition site for the complementary dNTP. Furthermore, these authors proposed that substrate recognition derives from the unique electronic structures of the bases. Specifically, two characteristics of the electronic structure of the bases were found to distinguish the bases, namely the sign of the electrostatic potential around exocyclic substituents at C6 of purines or C4 of pyrimidines (positive for A and C; negative for G and T)

and the magnitude of the negative charges of the N-glycosyl bonds (substantially greater for N1 of pyrimidines than for N9 of purines). Thus, according to this theory, prior to incorporation opposite each template residue, the enzyme would respond to the electronic structure of the template residue, forming a recognition site for the complementary dNTP, in which the spatial distribution of electrostatic potential would permit entry of the complementary dNTP into the recognition site, while preventing entry of noncomplementary dNTPs. The authors found that this model accurately predicted the specificity of incorporation of all the natural nucleotides, as well as a variety of chemically modified nucleotides, including those exhibiting ambiguous base pairing in DNA synthesis and those whose incorporation was not predicted on the basis of H-bonded base pairing.

In conclusion, there is a strong possibility that a key component of substrate selectivity comprises interactions within the polymerase-template-dNTP complex distinct from H-bonded base pairing. Accordingly, "mispairing" events during DNA synthesis, whether involving the canonical bases or chemically modified residues, may be governed by interactions within the complex other than H-bonded base pairing. Certain misincorporation events may exploit intricate protein-nucleic acid interactions at a recognition or verification step along the reaction coordinate, interactions that are not predicted by consideration of H-bonded base pairing.

FUTURE DIRECTIONS

What kind of experimental approaches are needed to further characterize the base pair recognition mechanism(s) of DNA polymerases? This problem could be vigorously addressed by protein engineering. To achieve the full potential of this powerful new technology will require (i) physical studies of polymerase-substrate interactions, using X-ray crystallography and NMR; (ii) enzyme kinetics to further characterize the polymerase reaction pathway, particularly with respect to discrimination against specific mispairs at different template positions; (iii) use of molecular modeling and computational chemistry to visualize protein-nucleic acid interactions and protein dynamics and to explain the observed misincorporation patterns of nucleotide analogs; and (iv) site-directed mutagenesis to test specific models.

It is hoped that the full power of protein engineering can be brought to bear to explain base substitution mutagenesis at the level of DNA polymerase function.

> "Man is privileged to receive truth,
> eager to define knowledge, and
> ignorant of the difference."
>
> Anonymous

ACKNOWLEDGMENTS

This work was supported by Grants GM25530 and GM30590 from the National Institutes of Health and by Grant Q-1006 from the Robert A. Welch Foundation. K.L.B. was recipient of Research Career Development Award CA00891 from the National Cancer Institute. R.M.-R. was recipient of fellowship support from COFAA-IPN, Mexico and from Colgate Palmolive, Mexico.

REFERENCES.

Aboul-ela, D. Koh, I. Tinoco, Fr. and F. H. Martin. (1985) Base-base mismatches. Thermodynamics of double helix formation for $dCA_3XA_3G + dCT_3YT_3G$ (X,Y=A,C,G,T). *Nucl. Acids Res.* **13**: 4811-4824

S. S. Agarwal, D. K. Dube and L. A. Loeb. (1979) On the fidelity of DNA replication. Accuracy of *Escherichia coli* DNA polymerase I. *J. Biol. Chem.* **254**: 101-106

A. K. Aggarwal, D. W. Rodgers, M. Drottar, M. Ptashne and S. C. Harrison. (1988) Recognition of a DNA operator by the repressor of phage 434: A view at high resolution. *Science* **242**: 899-907

N. Battula and L. A. Loeb. (1976) On the fidelity of DNA replication. Lack of exodeoxyribonuclease activity and error-correcting function in avian myeloblastosis virus DNA polymerase. *J. Biol. Chem.* **251**: 982-986

S. Benzer and E. Freese. (1958) Induction of specific mutations with 5-bromouracil. *Proc. Natl. Acad. Sci. U.S.A.* **44**: 112-119

M. J. Bessman, I. R. Lehman, J. Adler, S. B. Zimmerman, E. S. Simms and A. Kornberg. (1958) Enzymatic synthesis of deoxyribonucleic acid. III. The incorporation of pyrimidine and purine analogues into deoxyribonucleic acid. *Proc. Natl. Acad. Sci. U.S.A.* **44**: 633-640

M. J. Bessman, N. Muzyczka, M. F. Goodman and R. L. Schnaar. (1974) Studies on the biochemical basis of spontaneous mutation. II. The incorporation of a base and its analogue into DNA by wild-type, mutator and antimutator DNA polymerases. *J. Mol. Biol.* **88**: 409-421

M. S. Boosalis, J. Petruska and M. F. Goodman. (1987) DNA polymerase insertion fidelity. Gel assay for site-specific kinetics. *J. Biol. Chem.* **262**: 14689-14696

T. Brown, W. N. Hunter, G. Kneale and O. Kennard. (1986) Molecular structure of the G•A base pair in DNA and its implications for the mechanism of transversion mutations. *Proc. Natl. Acad. Sci. U.S.A.* **83**: 2402-2406

D. Brutlag and A. Kornberg. (1972) Enzymatic synthesis of deoxyribonucleic acid. XXXVI. A proofreading function for the 3′→5′ exonuclease activity in deoxyribonucleic acid polymerases. *J. Biol. Chem.* **247**: 241-248

S. P. Champe and S. Benzer. (1962) Reversal of mutant phenotypes by 5-fluorouracil: An approach to nucleotide sequences in messenger-RNA. *Proc. Natl. Acad. Sci. U.S.A.* **48**: 532-546

R. Chattopadhyaya, S. Ikuta, K. Grzeskowiak and R. E. Dickerson. (1988) X-ray structure of a DNA hairpin molecule. *Nature (Lond.)* **334**: 175-179

P. H. Driggers and K. L. Beattie. (1988) Effect of pH on the base-mispairing properties of 5-bromouracil during DNA synthesis. *Biochemistry* **27**: 1729-1735

W. S. El-Deiry, K. M. Downey and A. G. So. (1984) Molecular mechanisms of manganese mutagenesis. *Proc. Natl. Acad. Sci. U.S.A.* **81**: 7378-7382

W. S. El-Deiry, A. G. So and K. M. Downey. (1988) Mechanisms of error discrimination by *Escherichia coli* DNA polymerase I. *Biochemistry* **27**: 546-553

M. J. Engler and M. J. Bessman. (1978) Characterization of a mutator DNA polymerase I from *Salmonella typhimurium*. *Cold Spr. Harb. Symp. Quant. Biol.* **43**: 929-935

R. Eritja, D. M. Horowitz, P. A. Walker, J. P. Ziehler-Martin, M. S. Boosalis, M. F. Goodman, K. Itakura and B. E. Kaplan. (1986a) Synthesis and properties of oligonucleotides containing 2′-deoxynebularine and 2′-deoxyxanthosine. *Nucl. Acids Res.* **14**: 8135-8153

R. Eritja, B. E. Kaplan, D. Mhaskar, L. C. Sowers, J. Petruska and M. F. Goodman. (1986b) Synthesis and properties of defined DNA oligomers containing base mispairs involving 2-aminopurine. *Nucl. Acids Res.* **14**: 5869-5884

L. J. Ferrin and A. S. Mildvan. (1986) NMR studies of conformations and interactions of substrates and ribonucleotide templates bound to the large fragment of DNA polymerase I. *Biochemistry* **25**: 5131-5145

A. R. Fersht. (1987) Dissection of the structure and activity of the tyrosyl-tRNA synthetase by site-directed mutagenesis. *Biochemistry* **26**: 8031-8037

A. R. Fersht and J. W. Knill-Jones. (1983) Fidelity of replication of bacteriophage φX174 DNA *in vitro* and *in vivo*. *J. Mol. Biol.* **165**: 633-654

P. L. Foster, E. Eisenstadt and J. H. Miller. (1983) Base substitution mutations induced by metabolically activated aflatoxin B1. *Proc. Natl. Acad. Sci. U.S.A.* **80**: 2695-2698

E. Freese. (1959a) The specific mutagenic effect of base analogues on phage T4. *J. Mol. Biol.* **1**: 87-105.

E. Freese. (1959b) The difference between spontaneous and base-analogue induced mutations of phage T4. *Proc. Natl. Acad. Sci. U.S.A.* **45**: 622-633

J. R. Fresco, S. Broitman and A.-E. Lane. (1980) Base mispairing and nearest-neighbor effects in transition mutations. **In**: B. Alberts (Ed.). *Mechanistic studies of DNA replication and genetic recombination*. pp. 753-768. Academic Press, New York.

M. Fry. (1983) Eukaryotic DNA polymerases. **In**: S. T. Jacob (Ed.). *Enzymes of DNA synthesis and modification*. Vol. 1, pp. 39-92. CRC Press, Boca Raton, FL.

F. D. Gillin and N. G. Nossal. (1976) Control of mutation frequency by bacteriophage T4 DNA polymerase. II. Accuracy of nucleotide selection by the L88 mutator, CB120 antimutator, and wild type phage T4 DNA polymerases. *J. Biol. Chem.* **251**: 5225-5232

Z. W. Hall and I. R. Lehman. (1968) An *in vitro* transversion by a mutationally altered T4-induced DNA polymerase. *J. Mol. Biol.* **36**: 321-333.

M. S. Hershfield. (1973) On the role of deoxyribonucleic acid polymerase in determining mutation rates. Characterization of the defect in the T4 deoxyribonucleic acid polymerase caused by the ts L88 mutation. *J. Biol. Chem.* **248**: 1417-1423

G. G. Hillebrand and K. L. Beattie. (1984) Template-dependent variation in the relative fidelity of DNA polymerase I of *Escherichia coli* in the presence of Mg^{2+} versus Mn^{2+}. *Nucl. Acids Res.* **12**: 3173-3183

G. G. Hillebrand, A. H. McCluskey, K. A. Abbott, G. G. Revich and K. L. Beattie. (1984) Misincorporation during DNA synthesis, analyzed by gel electrophoresis. *Nucl. Acids Res.* **12**: 3155-3171

W. M. Huang and I. R. Lehman. (1972) On the exonuclease activity of phage T4 deoxyribonucleic acid polymerase. *J. Biol. Chem.* **247**: 3139-3146

W. N. Hunter, T. Brown, N. N. Anand and O. Kennard. (1986) Structure of an adenine-cytosine base pair in DNA and its implications for mismatch repair. *Nature (Lond.)* **320**: 552-555

W. N. Hunter, T. Brown, G. Kneale, N. N. Anand, D. Rabinovich and O. Kennard. (1987) The structure of guanosine-thymidine mismatches in B-DNA at 2.5 Å resolution. *J. Biol. Chem.* **262**: 9962-9970

R. B. Inman and R. L. Baldwin. (1964) Helix-random coil transitions in DNA homopolymer pairs. *J. Mol. Biol.* **8**: 452-469

A. Jack, J. E. Ladner and A. Klug. (1976) Crystallographic refinement of yeast phenylalanine transfer RNA at 2.5 Å resolution. *J. Mol. Biol.* **108**: 619-649

S. R. Jordan and C. O. Pabo. (1988) Structure of the lambda complex at 2.5 Å resolution: Details of the repressor-operator interactions. *Science* **242**: 893-899

M. W. Kalnik, M. Kouchakdjian, B. F. L. Li, P. F. Swann and D. Patel. (1988) Base pair mismatches and carcinogen-modified bases in DNA: An NMR study of G:T and G:O^4meT pairing in dodecanucleotide duplexes. *Biochemistry* **27**: 108-115

Y. Kawase, S. Iwai, H. Inoue, K. Miura and E. Ohtsuka. (1986) Studies on nucleic acid interactions I. Stabilities of mini-duplexes ($dG_2A_4XA_4G_2$:$dC_2T_4YT_4C_2$) and self-complementary d(GGGAAXYTTCCC) containing deoxyinosine and other mismatched bases. *Nucl. Acids Res.* **14**: 7727-7736

O. Kennard. (1987) The molecular structure of base pair mismatches. **In**: F. Eckstein and D. M. J. Lilley (Eds.). *Nucleic acids and molecular biology*. Vol. 1, pp. 25-52. Springer-Verlag, Berlin.

S.-H. Kim. (1978) Three-dimensional structure of transfer RNA, and its functional implications. *Adv. Enzymol.* **46**: 279-315

R. D. Kuchta, P. Benkovic and S. J. Benkovic. (1988) Kinetic mechanism whereby DNA polymerase I (Klenow) replicates DNA with high fidelity. *Biochemistry* **27**: 6716-6725

T. A. Kunkel. (1985a) The mutational specificity of DNA polymerase-β during *in vitro* DNA synthesis. Production of frameshift, base substitution and deletion mutations. *J. Biol. Chem.* **260**: 5787-5796

T. A. Kunkel. (1985b) The mutational specificity of DNA polymerases α and γ during *in vitro* DNA synthesis. *J. Biol. Chem.* **260**: 12866-12874

T. A. Kunkel and P. S. Alexander. (1986) The base substitution fidelity of eucaryotic DNA polymerases. Mispairing frequencies, site preferences, insertion preferences, and base substitution by dislocation. *J. Biol. Chem.* **261**: 160-166

T. A. Kunkel and L. A. Loeb. (1979) On the fidelity of DNA replication. Effect of divalent metal ion activators and deoxyribonucleoside triphosphate pools on *in vitro* mutagenesis. *J. Biol. Chem.* **254**: 5718-5725

M.-D. Lai and K. L. Beattie. (1988a) Influence of DNA sequence on the nature of mispairing during DNA synthesis. *Biochemistry* **27**: 1722-1728

M.-D. Lai and K. L. Beattie. (1988b) Influence of divalent metal activator on the specificity of misincorporation during DNA synthesis catalyzed by DNA polymerase I of *Escherichia coli*. *Mutation Res.* **198**: 27-36

M.-D. Lai and K. L. Beattie. (1989) DNA sequence effects on mispairing during *in vitro* DNA synthesis. *J. Biol. Chem.* submitted.

C. H. Lee and J. G. Wetmur. (1973) Physical studies of chloroacetaldehyde labelled fluorescent DNA. *Biochem. Biophys. Res. Commun.* **50**: 879-885

K.-Y. Lo and M. J. Bessman. (1976) An antimutator deoxyribonucleic acid polymerase. *J. Biol. Chem.* **251**: 2475-2479

L. A. Loeb and T. A. Kunkel. (1982) Fidelity of DNA synthesis. *Ann. Rev. Biochem.* **51**: 429-457

L. A. Loeb, T. A. Kunkel and R. M. Schaaper. (1980) Fidelity of copying natural DNA templates. **In**; B. M. Alberts (Ed.). *Mechanistic studies of DNA replication and genetic recombination*. pp. 735-751. Academic Press, New York.

R. Maldonado-Rodriguez, P. H. Driggers and K. L. Beattie. (1989a) Genetic and biochemical analysis of *in vitro* mutagenesis in the *lacI* gene of *Escherichia coli*. *J. Biol. Chem.* submitted.

R. Maldonado-Rodriguez, J. M. Espinosa-Lara and K. L. Beattie. (1989b) Influence of neighboring base sequence on mutagenesis induced by *in vitro* misincorporation in the lacI gene of *Escherichia coli*. *EMBO J.* submitted.

R. Maldonado-Rodriguez, J. M. Espinosa-Lara and K. L. Beattie. (1989c) Mutations arising from replication of 1,N^6-etheno-A-containing templates *in vivo*. Manuscript in preparation.

F. H. Martin, M. M. Castro, F. Aboul-ela and I. Tinoco. (1985) Base pairing involving deoxyinosine: implications for probe design. *Nucl. Acids Res.* **13**: 8927-8938

J. A. McClarin, C. A. Frederick, B.-C. Wang, P. Greene, H. W. Boyer, J. Grable and J. M. Rosenberg. (1986) Structure of the DNA-Eco RI endonuclease recognition complex at 3 Å resolution. *Nature (London)* **234**: 1526-1541

J. H. Miller. (1983) Mutational specificity in bacteria. *Ann. Rev. Genet.* **17**: 215-238

V. Mizrahi, R. N. Henrie, J. F. Marlier, K. A. Johnson and S. J. Benkovic. (1985) Rate-limiting steps in the DNA polymerase I reaction pathway. *Biochemistry* **24**: 4010-4018

P. Modrich. (1987) DNA mismatch correction. *Annu. Rev. Biochem.* **56**: 435-466

Y. V. Morozov, F. A. Savin and E. I. Budowsky. 1982. Electronic structure, spectroscopic, photochemical, and functional properties of compounds of the vitamin B_6 group and components of nucleic acids and their analogs. **In**: V. P. Skulachev (Ed.). *Biology reviews*. Vol. 3, pp. 167-227. Harwood Academic Publishers GmbH, Chur, Switzerland.

N. Muzyczka, R. L. Poland and M. J. Bessman. (1972) Studies on the biochemical basis of spontaneous mutation. I. A comparison of the deoxyribonucleic acid polymerases of mutator, antimutator, and wild type strains of bacteriophage T4. *J. Biol. Chem.* **247**: 7116-7122

D. J. Patel, S. A. Kozlowski, S. Ikuta and K. Itakura. (1984) Deoxyguanosine-deoxyadenosine pairing in the d(C-G-A-G-A-A-T-T-C-G-C-G) duplex: Conformation and dynamics at and adjacent to the dG•dA mismatch site. *Biochemistry* **23**: 3207-3217

D. J. Patel, L. Shapiro and D. Hare. (1987) Conformation of DNA base pair mismatches in solution. **In**: F. Eckstein and D. M. Lilley (Eds.). *Nucleic acids and molecular biology*. Vol. 1, pp. 70-84. Springer-Verlag, Berlin.

J. Petruska, L. C. Sowers and M. F. Goodman. (1986) Comparison of nucleotide interactions in water, proteins, and vacuum: Model for DNA polymerase fidelity. *Proc. Natl. Acad. Sci. U.S.A.* **83**: 1559-1562

R. C. Pless, L. M. Levitt and M. J. Bessman. (1981) Nonrandom substitution of 2-aminopurine for adenine during deoxyribonucleic acid synthesis *in vitro*. *Biochemistry* **20**: 6235-6244

G. G. Prive, U. Heinemann, S. Chandradegaran, L.-S. Kan, M. L. Kopka and R. E. Dickerson. (1987) Helix geometry, hydration, and G:A mismatch in a B-DNA decamer. *Science* **238**: 498-504

G. J. Quigley and A. Rich. (1976) Structural domains of transfer RNA molecules. *Science* **194**: 796-806

G. J. Quigley, G. Ughetto, G. A. van der Marel, J. H. van Boom, A. H.-J. Wang and A. Rich. (1986) Non-Watson-Crick G:C and A:T base pairs in a DNA-antibiotic complex. *Science* **232**: 1255-1258

M. Radman, G. Villani, S. Boiteux, A. R. Kinsella, B. W. Glickman and S. Spadari. (1978) Replicational fidelity: mechanisms of mutation avoidance and mutation fixation. *Cold. Spr. Harb. Symp. Quant. Biol.* **43**: 937-946

M. Radman and R. Wagner. (1986) Mismatch repair in *Escherichia coli*. *Annu. Rev. Genet.* **20**: 523-538

B. Reckmann and G. Krauss. (1986) Mutations induced by DNA polymerase α upon *in vitro* replication of M13mp8(+) DNA. *Nucl. Acids Res.* **14**: 2365-2380

S. T. Reeves and K. L. Beattie. (1985) Base-pairing properties of N^4-methoxydeoxycytidine 5′-triphosphate during DNA synthesis on natural templates, catalyzed by DNA polymerase I of *Escherichia coli*. *Biochemistry* **24**: 2262-2268

G. G. Revich and K. L. Beattie. (1986) Utilization of 1,N^6-etheno-2′-deoxyadenosine 5′-triphosphate during DNA synthesis on natural templates, catalyzed by DNA polymerase I of *Escherichia coli*. *Carcinogenesis* **7**: 1569-1576

G. G. Revich, G. G. Hillebrand and K. L. Beattie. (1984) High-performance liquid chromatographic purification of deoxynucleoside 5′-triphosphates and their use in a sensitive electrophoretic assay of misincorporation during DNA synthesis. *J. Chromatogr.* **317**: 283-300

J. D. Roberts, K. Bebenek and T. A. Kunkel. (1988) The accuracy of reverse transcriptase from HIV-1. *Science* **242**: 1171-1173

J. D. Roberts and T. A. Kunkel. (1986) Mutational specificity of animal cell DNA polymerases. *Environ. Mut.* **8**: 769-789

R. M. Schaaper and R. L. Dunn. (1987) Spectra of spontaneous mutations in *Escherichia coli* strains defective in mismatch correction: the nature of *in vivo* DNA replication errors. *Proc. Natl. Acad. Sci. USA* **84**: 6220-6224.

J. P. Seiler. (1972) The mutagenicity of benzimidazole and benzimidazole derivatives. I. Forward and reverse mutations in Salmonella typhimurium caused by benzimidazole and some of its derivatives. *Mutat. Res.* **15**: 273-276

J. P. Seiler. 1973. The mutagenicity of benzimidazole and benzimidazole derivatives. II. Incorporation of benzimidazole into the nucleic acids of *Escherichia coli*. *Mutat. Res.* **17**: 21-25

L. C. Sowers, G. V. Fazakerley, H. Kim, L. Dalton and M. F. Goodman. (1986) Variation of nonexchangeable proton resonance chemical shifts as a probe of aberrant base pair formation in DNA. *Biochemistry* **25**: 3983-3988

D. M. Stalker, D. W. Mosbaugh and R. R. Meyer. (1976) Novikoff hepatoma deoxyribonucleic acid polymerase. Purification and properties of a homogeneous β polymerase. *Biochemistry* **15**: 3114-3121

T. A. Trautner, M. N. Swartz and A. Kornberg. (1962) Enzymatic synthesis of deoxyribonucleic acid. X. Influence of bromouracil substitutions on replication. *Proc. Natl. Acad. Sci. U.S.A.* **48**: 449-455

E. C. Travaglini, A. S. Mildvan and L. A. Loeb. (1975) Kinetic analysis of *Escherichia coli* deoxyribonucleic acid polymerase I. *J. Biol. Chem.* **250**: 8647-8656

R. Wagner and M. Meselson. (1976) Repair tracts in mismatched DNA heteroduplexes. *Proc. Natl. Acad. Sci. U.S.A.* **73**: 4135-4139

J. D. Watson and F. H. C. Crick. (1953a) Molecular structure of nucleic acids. A structure for deoxyribose nucleic acid. *Nature (London)* **171**: 737-738

J. D. Watson and F. H. C. Crick. (1953b) Genetical implications of the structure of DNA. *Nature (London)* **171**: 964-967

J. D. Watson and F. H. C. Crick. (1953c) The structure of DNA. *Cold Spring Harbor Symp. Quant. Biol.* **18**: 123-131

A. Weissbach. (1979) The functional roles of mammalian DNA polymerase. *Arch. Biochem. Biophys.* **198**: 386-396

L. A. Weymouth and L. A. Loeb. 1978. Mutagenesis during *in vitro* DNA synthesis. *Proc. Natl. Acad. Sci. U.S.A.* **75**: 1924-1928

R. V. Wolfenden. (1969) Tautomeric equilibria in inosine and adenosine. *J. Mol. Biol.* **40**: 307-310

15

THE MECHANISM OF RNA POLYMERASE

Cheng-Wen Wu and Paul T. Singer

INTRODUCTION

The biochemical basis of life centers on the transfer of information from DNA to RNA to protein, the so-called "central dogma of molecular biology." The first step in this process is gene transcription, the polymerization of ribonucleotide triphosphates into long-chain ribonucleic acid molecules by RNA polymerase under the direction of a DNA template. Due to the importance of this reaction to essentially all forms of life, a clear picture of the molecular mechanism by which it occurs is crucial to the understanding of a great variety of biological processes.

The enzyme responsible for transcription, DNA-dependent RNA polymerase (RPase), is found in all living cells, but varies greatly in structure from species to species. Viral RPases are relatively small (≈100 kDa) proteins, consisting of a single polypeptide. Eubacterial RPases are larger (≈500 kDa) and are made up of multiple subunits, while the eukaryotic enzymes are even larger, requiring multiple ancillary proteins for fully functional transcriptional activity. Furthermore, eukaryotes do not transcribe all RNAs using a single enzyme, but instead use three enzymes, RPase I, II and III, depending on the nature of the genes transcribed. Not surprisingly, there are some rather notable differences between these proteins in terms of structure, DNA-binding characteristics, requirements for protein cofactors and, presumably, mechanism.

It has recently been observed (Sentenac *et al.*, 1983; Allison *et al.*, 1985) that there are significant homologies between all three types of the eukaryotic polymerases from a variety of genera, including *Drosophila* and *Saccharomyces*, and the prokaryotic enzyme from *Escherichia coli*, in terms of both their amino acid sequences and the genes encoding them. It was concluded that the three variants, RPase I, II and III, arose by duplication and divergence of a single ancestral enzyme, which was itself a descendant of a prokaryotic protein. It is reasonable to believe that many characteristics of the enzyme have remained unchanged over time, at least in terms of the mechanism of transcription itself. It is likely that the primary difference between these enzymes is confined to the regulation of transcription, in that eukaryotes have many more genes than prokaryotes. These eukaryotic genes require not only differential expression, but must respond to developmental and environmental stimuli with a degree of subtlety that is not required in prokaryotes.

Scope of this Review

In light of the degree to which the prokaryotic and eukaryotic enzymes are related structurally, the mechanism of transcription in prokaryotes can be used as a model for that in higher organisms. While many interesting and important questions require knowledge of the eukaryotic polymerases, the system is not, as yet, well defined. Indeed, fully functional *in vitro* transcription, using defined molecular components rather than cellular extracts, is still not possible. This is in contrast to the prokaryotic system, where virtually homogeneous enzyme capable of accurate transcription (as observed *in vivo*) has been available for many years. Consequently, the mechanism of the prokaryotic enzyme has been amenable to study using a great variety of biochemical and physical techniques, and is now fairly well understood.

In this article, the overall mechanism by which a eubacterial RPase converts ribonucleoside triphosphates to RNA chains, in a DNA sequence-specific manner, will be reviewed from a physicochemical point of view. Due to the magnitude of the subject, emphasis will be placed on the mechanism of RPase from *E. coli*; in particular, how the enzyme first locates, and then binds to its cognate specific sequence (promoter) on the DNA template, which has been a major area of research in this laboratory. Since the present volume concerns transcription in eukaryotes, parallels will be drawn between the two systems whenever possible. Where appropriate, the reader will be directed to more detailed reviews for in-depth discussions of various topics which will not be duplicated here. Of particular interest is the compendium of articles presented at the Steenbock Symposium in July, 1986 (Reznikoff, *et al.*) which covers a broad spectrum of topics ranging from the genes encoding both the prokaryotic and the eukaryotic RPases, to chromatin structure in eukaryotes.

The Transcription Cycle

The process of transcription was first described as a series of discrete steps by Goldthwait (Anthony *et al.*, 1966) (see Figure 1): binding, or the location of the specific site on the DNA template (promoter) by RPase; initiation, or the binding of the first two ribonucleotides to the enzyme-DNA complex and the formation of the initial phosphodiester bond; elongation of the

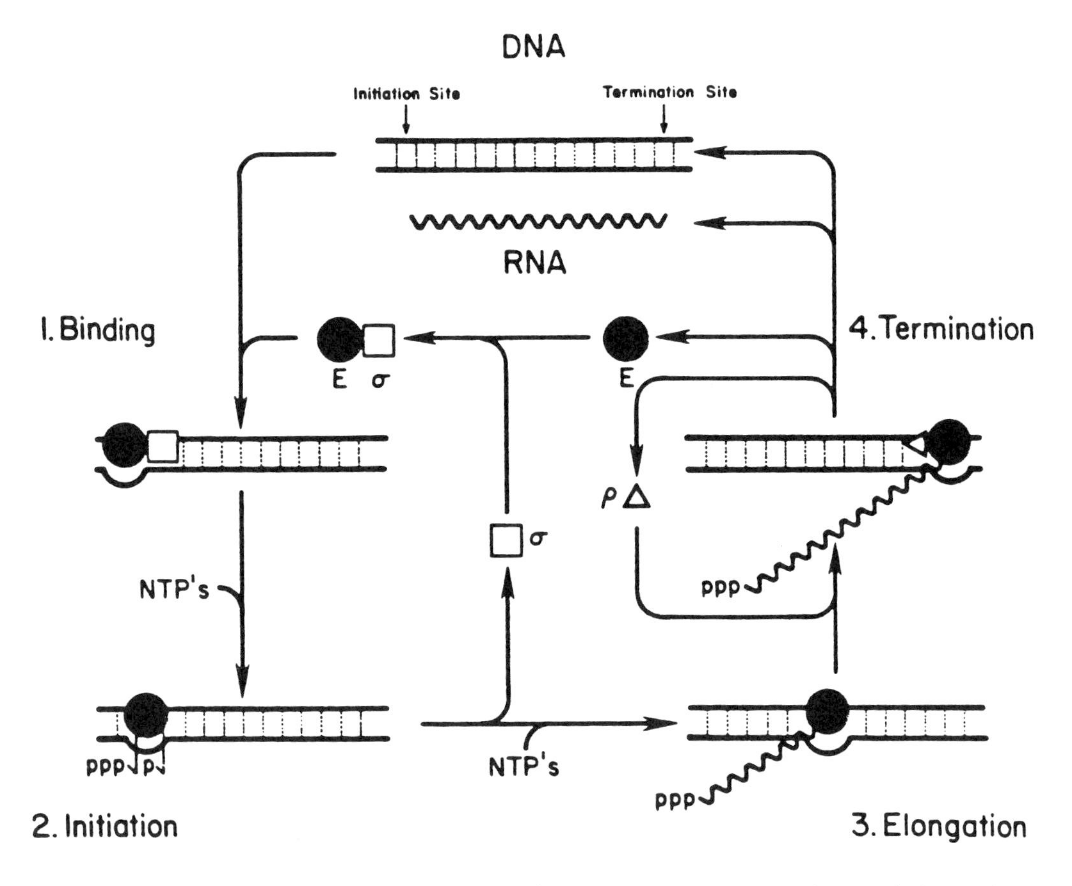

Figure 1. Schematic representation of transcription by *E. coli* RPase. The four major steps in transcription are shown: binding of RPase to the promoter with the melting of about 10 base pairs at the initiation site; initiation, or the formation of the first phosphodiester bond; the elongation of the nascent RNA chain; and termination and the release of both the newly synthesized RNA and the RPase. As shown in this figure, termination also requires a protein factor (ρ), which cycles on and off the polymerase, although ρ-independent termination also occurs.

RNA chain; and finally, termination of RNA synthesis and release of the product (Roberts, 1969). Furthermore, the binding of RPase to the promoter consists of at least two kinetically distinguishable steps (McClure, 1980): the formation of the so-called "closed" complex, and the subsequent isomerization to the "open" complex. This isomerization involves conformational changes in the protein as well as the melting of approximately ten base pairs of the DNA double helix (Hsieh and Wang, 1978), and is a prerequisite for transcriptional activity. Since transcription is effectively the net result of the coupling of each of these reactions (McClure and Chow, 1980), insight into the overall process absolutely requires a clear understanding of each of the constituent reactions.

Subunit Structure and Function of RNA Polymerases

The eubacterial RPases are fairly large, multisubunit metalloproteins which require zinc ions for enzymatic activity (Scrutton *et al.*, 1971). In certain species, such as *E. coli* and *B. subtilis*, a great deal is known not only of the structure of the enzyme, but also the molecular mechanism by which it catalyzes the incorporation of ribonucleotide monomers into long-chain

RNA polymers. Due to the extensive study of the enzyme from *E. coli*, it will be considered a "typical" prokaryotic RPase, and will serve as the focal point for the detailed discussion.

RNA polymerase activity was first discovered in *Escherichia coli* almost three decades ago (Hurwitz *et al.*, 1960; Stevens, 1960). It is a protomer that can exist in either of two forms: core enzyme and holoenzyme (Burgess, 1969). Both contain two α subunits, a β subunit and a β'subunit; holoenzyme contains an additional σ subunit. While both forms are capable of transcription, only the holoenzyme is able to recognize specific sites on the DNA, known as promoters (Hinkle and Chamberlin, 1971, 1972). The σ subunit cycles on and off the core enzyme, its presence being required for the process of promoter recognition and initiation of transcription, but not for that of elongation or termination. Thus, while a typical bacterium contains approximately 3000 RPase molecules (Burgess, 1976), there are only a fraction of that many copies of σ.

E. coli RPase holoenzyme has a molecular weight of about 480 kDa, while the individual polypeptides have molecular weights of: α = 37 kDa, β = 155 kDa, β' = 165 kDa, and σ = 82 kDa (Chamberlin, 1974; Burgess, 1969; Lowe *et al.*, 1979). A variety of structural studies have indicated that the RPase molecule is relatively compact, with none of the subunits separated from the others by more than 15 Å (King *et al.*, 1974; Wu and Wu, 1974; Hillel and Wu, 1977, 1978). Furthermore, RPase is a zinc metalloenzyme, with one metal ion located in the β subunit, and another in the β'subunit (Speckhard *et al.*, 1977; Wu *et al.*, 1977; Miller *et al.*, 1979).

The two largest subunits, which together comprise greater than 60% of the total mass of the holoenzyme, are structurally quite similar to each other as well as to the large subunit of a number of eukaryotic RPases, as evidenced by the homology between the genes for these polypeptides. Regions of each gene show striking similarities; one 126 bp stretch in the gene encoding the large subunit of RPase II in yeast is 62% identical to that encoding the β' subunit of the *E. coli* enzyme. In fact, there are six regions (totalling 324 amino acid residues) that display 36 - 55% homology among the large subunits of yeast RPase II and RPase III and the β' subunit of *E. coli* RPase (Allison *et al.*, 1985). Apparently, the primordial gene was smaller than any of its present-day descendants. In addition, the homology between the prokaryotic polypeptide and each of the eukaryotic gene products is not as great as between the two eukaryotic products themselves (Allison *et al.*, 1985), which is consistent with the further divergence of the eukaryotic genes. It should also be noted that this structural similarity is not restricted to the RPase from yeast, but has also been found for both the protein and genes for RPase II from *Drosophila* and vaccinia virus (Broyles and Moss, 1986), as well as the genes for RPase II from several mammals (Ingles *et al.*, 1983). Lastly, the homology between the eukaryotic and prokaryotic polymerases has been extended to include the 140 kDa subunit of RPase II from yeast and the β subunit from *E. coli* (Sweetser *et al.*, 1987).

The similarity between the prokaryotic and eukaryotic RPases can be summarized as follows: the prokaryotic enzyme consists of two large (β and β') and two small (α_2) subunits, while the eukaryotic counterparts (RPases I-III) each consist of two large and 5-10 smaller subunits. To date, it has been shown that the large subunits of all these enzymes are related; it remains to be seen whether the homology extends to the α subunit (37 kDa, though all characterized RPase II enzymes contain a subunit of approximately 45 kDa). It is possible that eukaryotic RPase, like its prokaryotic counterpart, exists as a core enzyme, with an ensemble of corollary protein factors which are sufficiently tightly bound to the core enzyme as to copurify with it, much as σ can be copurified with the enzyme from *E. coli*.

E. coli RPase has been studied using a variety of means with the goal of assigning the known functions of the holoenzyme to the various constituent subunits. It is known that the β subunit contains the substrate binding site (Wu and Wu, 1974), and that the intrinsic zinc ion contained within the subunit is directly involved in coordinating the nucleoside triphosphate:Mg^{2+} complex (Chatterji *et al.*, 1984). The β subunit is generally believed to be involved in the catalytic process itself, although no isolated subunit, including β, has ever been demonstrated to have any catalytic activity whatsoever. The transcription inhibitors rifamycin (Lowe and Malcolm, 1976; Rabussay and Zillig, 1970) and streptolydigin (Schleif, 1969) have also been shown to interact with β, as mutations conferring resistance to these agents have been identified in that subunit.

The largest subunit of RPase, β′, is involved in DNA template binding (Zillig *et al.*, 1970; Fukuda and Ishihama, 1974), but also binds polyanions, such as heparin (Zillig *et al.*, 1970), which are known to inhibit the binding of RPase to DNA. Using a photocrosslinking assay, it has been demonstrated in our laboratory that the initial binding of RPase to nonspecific DNA occurs primarily *via* the β and β′ subunits (Hillel *et al.*, 1980) which are both positively charged at physiological pH. The role of the α subunit is less well-defined than the others, but specific binding and catalysis depend strictly on the presence of both α subunits (Heil and Zillig, 1970; Zillig *et al.*, 1976; Kawakami and Ishihama, 1980). This polypeptide may serve as the "glue" that holds together the other subunits, performing both a structural role and possibly mediating conformational changes between the various functional locations of the enzyme. Indeed, a model based on the results of a chemical crosslinking study (Hillel and Wu, 1977) places the two α subunits at the juncture of β and β′, while σ contacts both the large subunits and at least one α.

The σ subunit is potentially the most interesting of the constituents of RPase, in that it confers upon the enzyme the ability to recognize particular sequences on the DNA template, and initiate transcription only at those sites. In this regard, the prokaryote *Bacillus subtilis* is more interesting than *E. coli*, and may very well be more instructive of the mechanism of eukaryotic transcription. In *B. subtilis*, no less than five different σ factors have been isolated (see Losick and Youngman, 1984), each of which confers specificity for different classes of promoter. The most abundant form found in the vegetative bacterium, while having a molecular weight of only 55 kilodaltons, has affinity for promoters which are quite similar to those recognized by σ from *E. coli.* (Moran *et al.*, 1982). During sporulation, an alternative σ factor is produced, which is presumably responsible for the transcription of addition proteins involved in the sporulation process (Doi, 1982). In *E. coli*, at least two alternative σ factors have been identified: σ^{32} (Grossman *et al.*, 1984), which was first isolated as a heat shock protein; and σ^{73} (Hirschman *et al.*, 1985), previously known as the *ntrA* (nitrogen fixation) gene product.

PROMOTER BINDING

As mentioned above, the formation of a functional transcription complex in prokaryotes consists of multiple steps. The first of these involves the actual physical encounter of a polymerase molecule with the promoter sequence on the DNA (promoter search), while additional steps involve conformational changes in both the protein and the DNA. The mechanism of promoter search has been a particular interest of this laboratory for several years, for a variety of reasons. The transcriptional activity of any given site is, at least in part, determined by the rate at which the requisite proteins locate and bind to that site. Experimental studies of the specific association of *E. coli* RPase to several strong bacteriophage promoters indicates that the apparent bimolecular association rate constant may exceed 10^{10} $M^{-1}s^{-1}$ (Bujard *et al.* 1982, Chamberlin *et al.*, 1982), a value which exceeds the theoretical diffusion-controlled limit for molecules of their size (von Hippel *et al.*, 1984). An additional factor which should further decrease this rate constant is the fact that there are limited numbers of specific sites amidst a great excess of nonspecific sites on the DNA, each of which is structurally quite similar to the promoters. Since it is known that the binding of RPase to these nonspecific sites is fairly tight ($\approx 10^4$-10^6 M^{-1}; deHaseth *et al.*, 1978), sequestering of polymerase in nonspecific sites lowers the effective concentration of the protein. This also serves to reduce the observed rate of encounter of protein with the specific promoter site.

Since no simple, bimolecular reaction can occur faster than diffusion allows the species to interact, it is of great interest to study the actual mechanism(s) by which these rapid rates can be achieved. Early workers in the field postulated that the observed binding rates could be explained in terms of a "reduction of dimensionality" (Adam and Delbruck, 1968). In essence, this entails the replacement of the simple bimolecular mechanism with one described by multiple steps, one or more of which can occur in either a reduced volume or a reduced dimensionality (one or two dimensions). Due to the increased efficiency of diffusion under these conditions, the overall binding rates can be increased markedly without violating the diffusional limit. The question then becomes "What type of mechanisms are available to a DNA-binding protein that allow such a reduction of either volume or dimensionality, and under what conditions would such processes be expected to play a physiological role?"

Facilitating Mechanisms

For RPase, as well as other similar DNA-binding proteins, there exists a body of theoretical knowledge of mechanisms which have the potential to facilitate the interaction of the protein with a specific binding site on a target DNA molecule (see Berg, Winter and von Hippel, 1981). These mechanisms require only that the binding reaction be diffusion-limited; they follow directly from the geometry of both the binding protein and the nucleic acid chain. Furthermore, Berg and coworkers (Berg and Blomberg, 1976, 1978; Berg *et al.*, 1981) have derived a series of closed-form equations relating the specific association rate to a variety of physical parameters, including nucleic acid chain length, diameter and stiffness, as well as various diffusion coefficients.

Due to the linear nature of nucleic acid polymers, a solution of DNA exists not as a homogeneous spatial distribution of base pairs (each of which is a potential binding site for RPase), but rather as localized regions of high concentration (Jovin, 1976; von Hippel *et al.*, 1978). While the exact dimensions of any given "domain" vary over time, a radius of gyration (r_g) can be calculated as a function of the stiffness of the DNA, which is typically represented by the persistence length (a). Despite the fact that the actual volume of DNA within a domain is two to three orders of magnitude smaller than that of the domain itself, DNA domains are remarkably efficient at trapping protein molecules that enter into the sphere defined by the radius of gyration. The parameter which defines this efficiency, the "absorbancy" of the DNA, has been calculated to range from ≈ 0.4 to 0.9 for chains varying from 2500 to 40000 base pairs in length (Berg and Blomberg, 1978). This efficiency can be intuited by comparison of the ever-moving DNA chain to the tentacles of a jellyfish, which, as any beachgoer can attest, are almost impossible to avoid. Even for short DNA chains (less than the persistence length), rotation is sufficiently rapid to virtually insure the contact of the rigid rod with anything that enters within the sphere of diameter equal to the length of the molecule. It is important to remember that this mechanism applies strictly to binding reactions that are diffusion-limited, as, by definition, such reactions require only a single collision to allow binding.

By measuring the rate at which RPase diffuses out of a domain (Singer and Wu, 1987), an experimental estimate of the absorbancy parameter was made for a plasmid containing about 5000 base pairs. This value (≈0.9) was considerably larger than the calculated value (≈0.5). This difference can be attributed to the treatment of the diffusing ligand species as a dimensionless point in the theoretical derivation, while RPase, with a molecular weight of about 5×10^5 daltons, measures 50 Å in hydrodynamic radius. This dimension makes it even less likely that the protein will avoid contact with the waving "arms" of the DNA chain.

Once the protein enters a domain and binds to the DNA chain, subsequent encounters are much more likely to occur to the same molecule than to another, due simply to effective concentration. In dilute solution, DNA domains may be quite far apart; as a consequence, the effective intermolecular concentration of binding sites (*i.e.* base pairs) can be vanishingly small, while the intramolecular concentration of binding sites is quite large, and independent of overall concentration. Instead, the domain concentration of base pairs depends only on the length of the DNA and the radius of gyration. A 5000 bp chain can be calculated to have an intradomain concentration of about 0.2 mM, based on a radius of ≈0.2 μm (von Hippel *et al.*, 1978). Under most experimental conditions, this value greatly exceeds the global concentration of DNA. This can have important consequences in terms of interpretation of experimental results.

The radius of gyration of a domain is quite large with respect to that of a single base pair. As a consequence, the diffusional limit for the encounter rate between such a domain and freely diffusing protein molecules, which is proportional to the reaction radius, is much faster than for interaction with any single base pair. Furthermore, the intradomain base pair concentration is quite high relative to that calculated for the solution as a whole; not only is the intradomain binding *rate* quite rapid, it is also essentially independent of overall DNA concentration. Thus, the overall binding process has now been broken down into two steps, the first of which is rapid by virtue of large target size, and the second by virtue of its effectively unimolecular nature. Between these two steps, the overall binding rate can be several orders of magnitude faster than predicted without accounting for the tendency of DNA in dilute solution to form domains, without violating diffusional constraints.

While this enhancement of the nonspecific binding rate goes a long way toward explaining the anomalously fast rates observed for the specific binding of RPase to its promoter sites, it does not address the issue that the promoter is located amongst a great excess of similar nonspecific sites. On average, each nonspecific site must be sampled once in order to insure (statistically speaking) contact with the specific site. Thus, one collision of RPase with the DNA chain is not sufficient for specific binding to occur. Furthermore, there are distinct advantages to having these same enzymes bind rather tightly to nonspecific sites on the DNA (von Hippel *et al.*, 1974), including having a significant fraction of the RPase pool sequestered on these inactive sites, from which it can be readily mobilized *via* a slight shift in equilibrium binding constants. It is this nontrivial, nonspecific binding that presents the greatest bottleneck to specific binding, serving as a severe kinetic liability to specific binding, and which requires the most creative solutions on the part of evolution (and enterprising investigators).

To facilitate the sampling of multiple nonspecific sites by a bound protein during a single nonspecific lifetime, Berg and coworkers (Berg *et al.*, 1981) postulated three major mechanisms: intersegment transfer, linear diffusion ("sliding"), and correlated binding ("hopping"). Sliding and intersegment transfer are similar in that they entail direct transfer of a nonspecifically bound protein molecule between two sites: in the former, the two sites are contiguous, while in the latter, they are located far apart (with respect to the linear sequence of bases in a DNA chain). All three, however, turn the rather high affinity of RPase (as well as other proteins, including the *lac* repressor) for nonspecific DNA from a kinetic liability into a kinetic asset, by allowing for intramolecular transfer of bound RPase directly from nonspecific to specific sites.

The intersegment transfer model was suggested by the observations of Breslof and Crothers (1975) that bound ethidium bromide can be rapidly transferred directly from one DNA molecule to another. Due to the flexibility of DNA, short-lived contacts between linearly separated portions of the chain are quite common. In this model, transient ternary complexes, consisting of both chains and the protein, form during these contacts. Upon dissociation of the complex, the bound protein has a fifty percent chance of being transferred to the second chain. It has been determined (Park *et al.*, 1981a) that the different subunits can be photocrosslinked to DNA roughly in proportion to their molecular mass during the first several hundred milliseconds following nonspecific binding. Subsequently, the σ subunit is preferentially crosslinked to DNA, presumably as a consequence of its role in the recognition of promoter sequences. Thus, it is quite possible that large portions of the surface of RPase are capable of interacting nonspecifically with DNA, which would promote the formation of the ternary complex (protein in contact with two linearly separate DNA segments) requisite to the process of intersegment transfer. However, while such a mechanism may indeed play some role in facilitating specific site location, calculations based on both purely theoretical grounds (Berg and Blomberg, 1978; Berg *et al.*, 1981) and the observed rate of ring-closure by DNA ligase (Wang and Davidson, 1966) indicate that intersegment transfer alone cannot account for the experimentally observed specific binding rate.

The linear arrangement of binding sites on the nucleic acid chain has ramifications with respect to the treatment of each site as an independently diffusing species in solution. Once RPase has bound to any given nonspecific site, the remaining sites are no longer equivalent — only two can be adjacent to the occupied site, two are "next-nearest neighbors," and so on. Thus, immediately following dissociation of the protein from the surface of the DNA molecule, it is more likely to rebind to the same, or a nearby, site than it is to a distant site. A protein in this situation is not truly "dissociated" from the site, since implicit in the definition of dissociation is diffusion to a distance sufficient to insure that all sites, including the one just vacated, are equivalent. For clarity, this can be referred to as macroscopic dissociation, as opposed to microscopic dissociation. Because many microscopic dissociations and reassociations can occur per macroscopic event, and there is some probability that the reassociation can occur at a nearby site rather than the original, the net effect can be to sample multiple sites per binding event. It is important to note that this phenomenon of "hopping" does not require any special circumstances or functionality of either DNA or protein to occur; it follows directly as a consequence of the linear geometry of the nucleic acid molecule. Thus, hopping must occur, and, in the absence of any other facilitating mechanisms, would be expected to enhance the specific binding rate considerably (Berg *et al.*, 1981).

The putative facilitating mechanism that has the potential for increasing the specific binding rate to the greatest extent is that of linear diffusion, or sliding. By diffusing laterally along the DNA chain during each nonspecific encounter, the effective size of the ultimate target, the promoter, is increased substantially. This "target expansion" can result in enhancement of the specific association rate by more than three orders of magnitude, which is sufficient to explain the experimentally observed rates.

In order to distinguish between sliding and hopping, which both involve transfer of protein between adjacent binding sites on the DNA molecule, a clearer picture of the molecular nature of binding is required. The binding of RPase, as well as a family of mechanistically similar nucleic acid binding proteins including *lac* repressor and RNase A, is electrostatic in nature. This means that the binding energy is provided almost entirely by the entropy of a small number of monovalent cations which are displaced from the surface of the nucleic acid upon binding of the protein (Record *et al.*, 1977). These counterions are normally condensed into a dense cloud near and on the surface of the highly anionic nucleic acid polymer (Manning, 1969; Manning, 1978). In the presence of even vanishingly small concentrations of Na^+ or K^+, the electric field around the DNA is sufficiently strong to condense enough counterions to shield ≈88% of the total negative charge of the phosphate groups strung along the DNA backbone (Manning, 1977). Upon the approach of RPase, approximately eleven of these counterions are displaced from the DNA surface and diluted into the bulk solvent (deHaseth *et al.*, 1978). It is the energy of this dilution which provides the binding free energy, and since the increase in entropy depends on the concentration of ions already in solution, the equilibrium association constant (K_a) for RPase to DNA varies greatly with cation concentration (Record *et al.*, 1976; deHaseth *et al.*, 1978). Since the association of RPase is essentially diffusion-limited, the changes in K_a are reflected primarily in the nonspecific dissociation rate constant, k_{off} (von Hippel *et al.*, 1974).

The displacement of condensed counterions by RPase underscores the difference between the sliding and hopping mechanisms. In the hopping mechanism, the bound RPase molecule diffuses away from the surface of the DNA chain far enough to allow recondensation of the counterions, however transiently. In the case of the sliding mechanism, however, the ions remain diluted in solution; as one counterion is displaced in front of the sliding protein, a second simultaneously rebinds behind it. This fact has important consequences in terms of the potential efficiency of each mechanism in sampling additional nonspecific sites on the DNA chain; each "hop" requires transient loss of the entire nonspecific binding energy (≈8 kcal mol^{-1}). Depending on the exact nature of the interaction between the sliding protein and DNA, the energy barrier to sliding could in fact be quite nominal. In experiments in our laboratory (Singer and Wu, 1988), the activation energy for the linear diffusion of *E. coli* RPase along relaxed DNA was measured as ≈4.8 kcal mol^{-1}, a value which is sufficiently small so as to be readily supplied by the thermal energy of the solvent at physiological temperatures. Thus, a nonspecifically bound RPase molecule can translocate linearly along the DNA much more efficiently *via* a sliding mechanism than *via* hopping.

Without going into the mathematical consequences of these facilitating mechanisms (see Berg, Winter and von Hippel, 1981), the effects of sliding, if it occurs, can potentially dwarf those of the other mechanisms. In studies employing a variety of DNA binding proteins, including the *lac* repressor (Berg and Blomberg, 1978; Winter and von Hippel, 1981), restriction endonucleases (Jack *et al.*, 1982; Ehbrecht *et al.*, 1985), and RPase (Belintsev, *et al.*, 1980;

Park *et al.*, 1982b; Singer and Wu, 1987), the value of the one dimensional diffusion coefficient, D_1, describing the sliding of bound proteins along various DNA templates has been estimated to be between 5×10^{-10} and 6×10^{-9} cm^2s^{-1}. In all cases but one, the estimated value is below the theoretical upper limit to the diffusion coefficient (4×10^{-9} cm^2s^{-1}), which is almost two orders of magnitude smaller than the three-dimensional diffusion coefficient, due to the frictional resistance of the solvent to the rotation of the protein as it travels along a helical groove of the DNA molecule (Schurr, 1979). In the one exception (Park *et al.*, 1982b), the assumption was made that the entire length of phage T7 DNA was scanned *via* sliding during a single nonspecific lifetime, which was in fact not the case. When multiple sampling of the DNA is accounted for, the resulting estimate of D_1 is more in line with the other results. A one-dimensional diffusion coefficient of this magnitude translates into $\approx 10^6$ steps per second, each resulting in translocation of a single base pair.

Due to the random-walk nature of diffusional processes, a more intuitive parameter, the mean sliding distance, L_S, can be defined to quantitate sliding. This parameter takes into account not only how fast a bound protein slides, but also how long it remains bound:

$$L_s \times 3.4 \times 10^{-8}\ cm/bp = \left(\frac{D_1}{k_{off}}\right)^{1/2}$$

Using a rapid-mixing/photocrosslinking technique developed in this laboratory (Wu *et al.*, 1983), the nonspecific dissociation rate (k_{off}) was measured to be 0.3 s^{-1} at 50 mM KCl. Using this technique, the amount of RPase that binds transiently to nonspecific sites on the DNA can be quantitated and analyzed mathematically. The fact that nonspecific sites display markedly different transient binding behavior as a function of distance from the promoter provides direct evidence that some form of linear translocation of the nonspecifically bound RPase is occurring. The distance over which the promoter exerts its influence is much too great to be explained by only a "hopping" mechanism; in fact, a mathematical model of binding which incorporates only the domain trapping and sliding mechanisms is capable of fitting the data to within experimental error (Singer and Wu, 1987). In the study, the value of the one-dimensional diffusion coefficient was calculated to be 1.5×10^{-9} cm^2s^{-1}, which, together with the previously mentioned value of k_{off}, defines a mean sliding distance of about 2000 bp at 50 mM KCl. Thus, RPase is two thousand times more likely to find the promoter than it would be if each site on the DNA needed to be sampled individually. Since this sampling of sites is rapid compared to the overall process of specific site location, the end result is a tremendous enhancement in the specific binding rate.

Furthermore, a variety of factors have the potential to modulate the degree of enhancement. Since the nonspecific dissociation rate is exquisitely sensitive to salt concentration, the mean sliding distance is strongly dependent on salt concentration (Barkley, 1981; Singer and Wu, 1988). Cellular mechanisms which are capable of regulating the intracellular levels of cations would have the potential of affecting the degree to which sliding plays a role in transcription.

Sliding is likely to play an even more important role in eukaryotic transcription than it does in prokaryotes, for the simple reason that the eukaryotic system is much more complex, requiring the association of as many as ten different protein factors at the initiation site prior to transcription (Lewis and Burgess, 1982). The chances of high order complexes forming in three dimensions can become vanishingly small, whereas if some or all of the proteins first bind nonspecifically to DNA and then locate each other in one dimension, the probability can be increased tremendously. Furthermore, as mentioned previously, the two large subunits of all three eukaryotic polymerases show significant homology to the β and β′ subunits of the *E. coli* enzyme, and might very well share the propensity of this protein to diffuse linearly along the DNA template.

Due to the more complex nature of the transcriptional machinery in eukaryotes, the term "preinitiation complex" has been coined to describe the assembly of all the necessary components (excluding RPase) prior to any actual catalysis. Significantly, these complexes are stable, presumably reflecting the difficulty of assembling the relatively large number of components, and require specific DNA sequences in the region of the gene as well as *trans*-acting transcription initiation factors (Tower *et al.*, 1986; Ciliberto *et al.*, 1983). Furthermore, there is a good deal of evidence that protein-protein interactions play a major role in the assembly of these complexes (Dynan and Tjian, 1985; Takahashi *et al.*, 1986). More importantly, the ultimate binding of RPase I is apparently DNA sequence independent, instead relying entirely on protein contacts with the previously-bound transcription initiation factor (Kownin *et al.*, 1987). This reliance of eukaryotic RPase on the specific binding of trans-acting protein factors to align the enzyme with the desired initiation site has also been described for RPases II (Corden *et al.*, 1980; Takahashi *et al.*, 1986) and III (Sakonju *et al.*, 1980). While this may represent a fundamental difference between the polymerases of prokaryotes and eukaryotes, the ability of the prokaryotic σ subunit to cycle on and off the core enzyme blurs this distinction somewhat.

Open *vs.* Closed Complexes

Once the RPase molecule has encountered the promoter site on the DNA chain, the so-called "closed complex" is free to form (see Chamberlin, 1974). This complex is characterized by the displacement of approximately 12 intimately associated monovalent cations from the surface of the DNA (deHaseth *et al.*, 1978). In the case of the prokaryotic enzyme, polyanions such as heparin (Walter *et al.*, 1967) and Congo Red (Krakow and von der Helm, 1971) can form stable complexes with RPase in the absence of DNA, but cannot actively displace the protein from specific sites. Thus, heparin is used to distinguish complexes of RPase with nonspecific DNA from those at promoters or other tight-binding sites.

This complex undergoes a salt and temperature dependent transition to the "open" complex (Hawley and McClure, 1980), which is characterized by the displacement of approximately five fewer counterions than the closed complex (deHaseth *et al.*, 1978) and, more importantly, the topological unwinding of the DNA to the extent of about 540° (Saucier and Wang, 1972; Gamper and Hearst, 1982). This unwinding results primarily from the melting of approximately 10-15 base pairs of superhelical DNA (Wang *et al.*, 1977). The nature of this unwinding was further elucidated by the observation that the methylating agent dimethylsulfate, which reacts with nitrogen atoms in adenine and cytidine bases in single-stranded but not double-stranded DNA, modified a region containing eleven base pairs straddling the initiation site of the A3 early promoter from phage T7 (Siebenlist, 1979; Kirkegaard *et al.*, 1983). As further evidence of the melting of the base pairs in the region of the transcriptional start site, hyperchromicity of the DNA bases has been observed upon RPase binding (Hsieh and Wang, 1978; Reisbig *et al.*, 1979), due to the loss of stacking of the purine and pyrimidine bases upon disruption of the ordered secondary structure of the double helix.

Rather than being a simple isomerization, there is kinetic evidence that the conversion of closed to open complex involves at least one additional intermediate (Buc and McClure, 1985), though it must be pointed out that under physiological conditions (temperature and/or ionic strength) these intermediates were not observed. The possibility still exists, however, that transcription from different promoters may be rate-limited by different steps along this mechanistic pathway (McClure, 1985), or even that corollary protein factors may influence the lifetime of one or more intermediate. For example, transcription at the *lac* UV5 promoter is rate-limited by a slow isomerization which precedes the rapid ($k = 1\ sec^{-1}$) melting of the DNA (Buc and McClure, 1985).

INITIATION

When initially proposed (Anthony *et al.*, 1966), the process of initiation simply entailed the formation of the first phosphodiester bond. It eventually was discovered, however, that this sharp distinction between initiation and elongation was somewhat blurred. While rifamycin specifically blocks the translocation of the nascent RNA chain when it is only two bases long (Yarbrough *et al.*, 1976), there are certain structural and functional criteria that indicate that a more comprehensive definition of initiation includes the formation of multiple bonds. In a more practical view, the line demarcating elongation from initiation is taken to be the release of σ factor, with a concomitant decrease in the size of the DNase I footprint from about 70 to about 30 base pairs (Rohrer and Zillig, 1977).

Nucleotide Binding Sites and Substrate Specificity

The 5′ terminus of RNA molecules synthesized by *E. coli* RPase is almost exclusively a purine nucleotide (Maitra and Hurwitz, 1965; Maitra *et al.*, 1967), and very often ATP (RPases from bacteriophages, such as T7, prefer GTP). The apparent K_m for the initiating nucleotide has been measured to be 150 μM, while that for each of the three remaining nucleotide triphosphates was approximately ten-fold lower (Wu and Goldthwait, 1969, 1969a). Furthermore, the initiating nucleotide binds as the free nucleotide, while the others bind only in the presence of magnesium. In fact, the substrate for RPase is more accurately referred to as the nucleotide triphosphate-Mg^{2+} complex (Chatterji and Wu, 1982). Thus there are at least two distinct nucleotide binding sites in RPase: a purine-specific, Mg^{2+}-independent initiation site, and a polymerization site which binds all four nucleotide triphosphates (as a complex with magnesium) equally well. By using nucleotide analogs, a great deal about the specificity of the initiation site has been learned (see Krakow *et al.*, 1976). There is an absolute requirement for a free 3′-OH group (Armstrong and Epstein, 1979), but 2′-deoxynucleotides may serve as substrate in the presence of Mn^{2+}, rather than Mg^{2+} (Krakow *et al.*, 1976). While 2′-O-methylribose and arabinose derivatives are inactive in the role of both initiator and polymerization substrate (Gerard *et al.*, 1974), 3′-modified, 2′-deoxynucleotides can be used to affinity label only the initiation site (Armstrong and Epstein, 1979). It has been suggested that this reflects a catalytic role for the 2′ hydroxyl group, possibly in the activation of the 3′ hydroxyl prior to bond formation (Wu and Tweedy, 1982).

Upon binding of nucleotide triphosphates to both the initiation and polymerization sites on RPase, the first phosphodiester bond may be formed. Due to the absolute requirement of the triphosphate moiety in occupants of the polymerization site for phosphodiester bond formation, it was not surprising to find that nucleotide mono- and di-phosphates do not bind to this site

(Wu and Goldthwait, 1969, 1969a). The initiation site, however, will bind these analogs, though the affinity drops markedly as the phosphates are removed. Similarly, nonhydrolyzable ATP analogs, such as those containing a bridging sulfur atom, have been demonstrated to bind to RPase but act as competitive inhibitors, rather than substrate (Stutz and Scheit, 1975). Hydrolyzable, sulfur-containing ATP analogs not only function as substrate for RPase (Eckstein and Grindl, 1970; Reeve *et al.*, 1977; Yee *et al.*, 1979; Armstrong *et al.*, 1979), but because of the introduction of an asymmetric phosphorus atom, can be used to study the stereochemistry of the enzymatic reaction (Burgess and Eckstein, 1978). Based on these studies, in which inversion occurs at the asymmetric phosphorus (Yee *et al.*, 1979), it was inferred that the polymerization reaction occurs *via* a concerted mechanism involving a transient pentavalent phosphorus atom, rather than *via* a covalent enzyme intermediate. Interestingly, studies employing β-S analogs indicate that the S enantiomer is a better initiator than the R enantiomer (Yee *et al.*, 1979), as was the case for the α-S compound, while the R isomer of an initiating dinucleotide was preferred. While this may imply that results obtained in studies employing dinucleotide initiators should be compared to the physiologically more important reaction employing mononucleotide initiators with caution (Tweedy and Wu, 1982), it is also possible that there is a distinct difference between the formation of the very first phosphodiester bond (as observed using the mononucleotide) and the second such bond. In fact, this observed stereochemical difference may reflect a general distinction between initiation and elongation. Such a conclusion would be consistent with the previously mentioned results regarding the ability of rifamycin to interact differently with the complex of RPase with newly-synthesized dinucleotide tetraphosphate than with the preformed dinucleotide (Oen and Wu, 1978).

Role of Zinc in Catalysis

During catalysis, an intrinsic Zn^{2+} ion within the β subunit holds the incoming nucleoside triphosphate in the proper orientation to allow electrophilic attack on the α phosphorus atom by the 3′-hydroxyl group of the nascent RNA chain (Chatterji *et al.*, 1984), with the concomitant release of pyrophosphate. In nuclear magnetic resonance studies employing the paramagnetic Co^{2+} ion in place of Zn^{2+}, the distance from the intrinsic divalent cation to various nuclei in ATP bound to the initiation site was measured in the absence (Chatterji and Wu, 1982) and presence (Chatterji *et al.*, 1984) of template DNA. These results indicated that the metal was directly involved in coordinating the nucleoside triphosphate base, rather than any of the phosphate moieties involved in bond formation. Thus the intrinsic metal ion found in the β subunit may play a role in the recognition and/or orientation of the initiating nucleotide, rather than in rendering the substrate more reactive (Wu and Wu, 1987), although the triphosphate moiety moves >3 Å closer to the metal in the presence of template DNA.

Consistent with the homology between the prokaryotic and eukaryotic RPases at the level of both protein and gene sequence, several RPases isolated from higher organisms have been shown to be zinc metalloenzymes containing approximately two molar equivalents of the metal. The sources identified so far include *E. gracilis* (Falchuk *et al.*, 1976; Falchuk *et al.*, 1977), yeast (Auld *et al.*, 1976); Lattke and Weser, 1976; Wandzilak and Benson, 1977) and wheat germ (Petranyi *et al.*, 1977). RPases I, II and III are included in this list.

Formation of the Initial Phosphodiester Bond

Prior to the initiation step, the antibiotic rifamycin is capable of inhibiting transcription (Kerrich-Santo and Hartman, 1974), although it has been demonstrated, using the abortive initiation technique, that the drug does not interfere with the formation of the first phosphodiester bond (Johnston and McClure, 1976). From additional kinetic results obtained using the same technique (McClure and Cech, 1978) and from physical studies of the ternary complex (Yarbrough *et al.*, 1976), it has been concluded that rifamycin prevents the translocation of RPase along the template concomitant with the movement of the initiating dinucleotide tetraphosphate out of the polymerization site. In fact, by virtue of its binding, the drug induces a conformational change in the enzyme (Yarbrough *et al.*, 1976). Due to this very precise effect of rifamycin on transcription, it has been employed in many studies of the molecular mechanism of transcription.

By simultaneously exposing preformed RPase-promoter complexes to both rifamycin and the four nucleoside triphosphates, the kinetics of the competing initiation and inhibition reactions can be measured (Mangel and Chamberlin, 1974a). Using this technique, the initiation rate was determined to be about 3 s^{-1} at 400 μM nucleotides, and depended almost exclusively on nucleotide concentration, over the range 40 - 800 μM. From further application of the "rifamycin challenge assay" (Rose *et al.*, 1974), it was concluded that initiation proceeds *via* the rapid binding of nucleoside triphosphate to the second, or elongation, site, followed by the rate-limiting association of the initiating purine nucleotide to the initiation site. While rifamycin is apparently capable of blocking the transfer of the initial dinucleoside tetraphosphate such that the second nucleotide occupies the initiation site of RPase, it does not have a similar effect on subsequent translocations, as evidenced by the fact that preformed dinucleotide or trinucleotide initiators circumvent the inhibitory action of the drug (Oen and Wu, 1978). Thus the efficacy of rifamycin as an inhibitor depends on some structural idiosyncrasy of the complex comprising RPase, template DNA, and the newly-formed dinucleoside tetraphosphate; the exact nature of this requirement is not known.

Time Course of Nucleotide Incorporation

In order to study the kinetics of incorporation of the first few nucleotides into RNA, a series of non-steady-state, single-turnover experiments were performed in this laboratory (Shimamoto and Wu, 1980a). By preincubating the RPase-poly(dA-dT) complex with radioactive substrate (either ATP or UTP), and then chasing with a large excess of both (unlabelled) nucleotides, the single nucleotide bound complex partitions between bond formation and dissociation of the radiolabelled nucleoside triphosphate. Due to the presence of large amounts of unlabelled substrate, the RNA chain is elongated sufficiently to allow acid precipitation of the product, and quantitation of the incorporated radiolabel. It was determined that ATP bound first, followed by UTP, which is consistent with results obtained employing other techniques (McClure *et al.*, 1978; Smagowicz and Scheit, 1978).

The above technique was adapted for use with a rapid multimixing apparatus (Shimamoto and Wu, 1980b), and unlike those results obtained using rifamycin challenge protocols (Rhodes and Chamberlin, 1974), the rate-limiting step in productive initiation was found to be independent of concentration for the initiator and second nucleotides, and hence unimolecular in nature. Furthermore, kinetic analysis of the data revealed that there were two distinct relaxations associated with the formation of the original complex, with apparent rate constants of 10 s^{-1} and 25 s^{-1}, respectively. The slower reaction compares well with estimates of other workers for the overall initiation rate. While the exact nature of these intramolecular processes is still not known, the results suggest either the actual formation of the first bond, or the translocation of the nascent chain by one base with respect to the two binding sites on the enzyme. Interestingly, there is some evidence to suggest the presence of an additional binding site for UTP in the initiation complex, and that binding to this site may increase the probability of bond formation.

In a similar experiment employing the A1 promoter on T7 DNA (Shimamoto and Wu, 1981), the actual time course of the incorporation of each of the first four nucleotides into RNA was measured. The order of incorporation followed the known sequence of the 5′ terminus of the transcript from this promoter (pppApUpCpGpA) (Dunn and Studier, 1981). Interestingly, the presence of the third or fourth nucleotide triphosphate stimulated the incorporation of the second (*i.e.* first bond formation) and inhibited commensurately the release of abortive products, possibly *via* stabilization of the transcription bubble by hydrogen bonding to the complementary base in the transiently single-stranded DNA template.

Abortive *vs.* Productive Initiation

Once the initial phosphodiester bond has been formed, the growing oligonucleotide chain can do one of two things: change register with respect to the two nucleotide binding sites, thereby allowing the next required nucleoside triphosphate to bind the elongation site; or release from the RPase:promoter complex as a so-called "abortive" initiation product. The drug streptolydigin inhibits both polymerization (von der Helm and Krakow, 1972) and abortive initiation (McClure, 1980), presumably *via* an interaction with the polymerization site (Cassani *et al.*, 1971). The release of abortive oligonucleotides occurs not only after the formation of the first bond, but after some (small) number of such bonds; the technical distinction between initiation and elongation occurs when abortive products are no longer released. In that the σ subunit dissociates from the holo-RPase complex at about this time, it is logical to conclude that it is involved in this transition as well. The length of the nascent RNA chain at σ release seems to vary: three bonds on the T7 A1 promoter (Kinsella *et al.*, 1982), five or eight bonds at the *gal* promoter (DiLauro *et al.*, 1979), seven or eight bonds using poly[d(A- T)] (Hansen and McClure, 1980), eight bonds at the *lac*UV5 promoter (Carpousis and Gralli, 1980), and fifteen bonds at the λPR′ promoter (Dahlberg and Blattner, 1973). It has been estimated that 50% or more of the complexes that enter the initiation stage abort at a RNA chain length of less than nine bases (Carpousis and Gralli, 1980).

McClure (1985) surmises that abortive initiation cycles may serve to delay the clearance of the promoter by RPase. This is lent further credence by the fact that there are mutationally silent, but nonetheless evolutionarily conserved, base pairs around the transcriptional start site which may "program" this putative time-delay. Perhaps such a mechanism insures that consecutive transcriptional complexes initiating at the same promoter are sufficiently separated linearly along the DNA for a so far undetermined reason. There is also some evidence that the release of σ occurs after a set amount of time for different promoters (Shimamoto, 1987), but different degrees of pausing (or possibly abortive cycling) translate into the observed difference in chain length at the time of release.

ELONGATION OF THE NASCENT RNA CHAIN

Once a sufficient number of phosphodiester bonds have been formed (a number which varies from promoter to promoter) and σ factor has been released from the transcription complex, the polymerization reaction is said to be in the elongation stage. In a manner quite analogous to translation in the ribosome, the growing RNA chain shifts register between the two nonequivalent nucleotide binding sites, such that the most recently added nucleotide moves from the polymerization site to the initiation site, displacing the nucleotide previously occupying that site. As this occurs, however, the DNA duplex is not reformed immediately, but rather a "bubble" of melted DNA moves along with the transcriptional machinery. Subsequently, a new nucleoside triphosphate occupies the polymerization site, as determined by hydrogen bonding with the DNA template, and the next phosphodiester bond is formed. This process repeats itself approximately forty times each second (Chamberlin *et al.*, 1982), resulting in a net chain elongation rate of approximately forty bases per second.

Unlike the similar enzyme, DNA-dependent DNA polymerase, there is little or no editing of the RNA product; the misincorporation rate is one per 10^4 - 10^5 nucleotides (Loeb and Mildvan, 1981), as opposed to less than one per 10^9 for the DNA polymerase. In evolutionary terms, it is more advantageous to generate large amounts of RNA rapidly, even with some errors, than to take a long time (and extra energy) to insure the fidelity of transcripts.

In addition to the release of σ factor, several other changes occur during, and in fact serve to define, the transition from initiation to elongation. Initiating RPase is more susceptible to cleavage by various proteases, presumably as a consequence of protein conformation (Novak and Doty, 1968); the DNase I footprint of RPase is less than half the size (30 *vs.* 70 bp) for the elongating enzyme (Rohrer and Zillig, 1977; Schmitz and Galas, 1979). Most importantly, however, the actively elongating complex proceeds along the DNA template for a great number of base pairs, synthesizing transcripts many thousands of bases long (Dunn and Studier, 1980). This is probably directly related to the release of the σ subunit, as core RPase has a much greater affinity for DNA than does the holoenzyme (although the specificity is absent). For a detailed discussion of the elongation complex, the reader is referred to Yager and von Hippel (1987).

RNA Path Through the Transcription Complex

By selectively preparing 5′ photoaffinity labelled RNA transcripts of various lengths (4-116 nucleotides), Hanna and Meares (1983) were able to map the path of the leading end of the nascent RNA through the transcription complex in *E. coli.* Using the A1 early promoter of bacteriophage T7 D111, they were able to show that transcripts up to twelve bases in length are in intimate contact primarily with the DNA template, and to a minor extent with the β and β′ subunits of RPase. For longer transcripts (up to 94 bases), however, the 5′ terminus of the transcript crosslinked only to the enzyme. Significant amounts of crosslinked σ were detected only for transcripts less than four bases long. These results are consistent with a mechanism wherein σ is released after three phosphodiester bonds have been formed, as previously reported for the T7 A1 promoter (Kinsella, *et al.*, 1982). It further confirms that the "transcription bubble" of melted DNA base pairs is indeed transiently replaced with a DNA:RNA hybrid. It is interesting to note, however, that the "bubble" is approximately 12 base pairs while the enzyme is still in the initiation stage, but this increases to about 18 base pairs once the transition to elongation has occurred (Siebenlist *et al.*, 1980; Gamper *et al.*, 1983). Thus 6 base pairs are melted out during elongation that are not hybridized to RNA; these may interact directly with the RPase or may simply lie at the interface between the two double-stranded structures.

Formation of DNA-RNA Hybrid

In an assay employing 3′-dC tailed sequences (Kadesch and Chamberlin, 1982), not native promoters, RPase II from wheat germ functioned like the prokaryotic (*E. coli*) enzyme, transcribing and releasing free RNA as product. However, the *in vitro* reaction using enzyme from calf thymus, HeLa cells and *Drosophila* is aberrant in that the RNA product is isolated as the hybrid, and the double-stranded DNA template is not reformed (Dedrick and Chamberlin, 1985), although the transcriptional start sites were almost identical for all the enzymes. Such variables as polymerase to template ratio, temperature, pH, and solvent conditions (i.e. concentrations of glycerol, spermidine, ammonium sulfate and magnesium ions) had no effect on this phenomenon. However, the sequence of the 3′ end of the duplex DNA led to significant differences: AT-rich sequences favor formation of the DNA-RNA hybrid, while GC-rich sequences enhance displacement of the growing transcript. This is interesting in that the actual transcriptional start site is approximately six dC residues upstream of the 3′ end of the duplex DNA, within the dC tail, for each of the enzymes examined (Dedrick and Chamberlin, 1985).

While the RPases from wheat germ and calf thymus differ in subunit composition (Hodo and Blatti, 1977), extracts of the former are unable to complement the latter with regard to restoring the ability to release RNA products (Dedrick and Chamberlin, 1985). Thus it is apparently not simply the loss or damage of a subunit during the purification of the calf thymus enzyme that is responsible for this biochemical difference. However, a protein factor which complements this activity in calf thymus RPase II preparations has been purified from HeLa cell extracts (Kane and Chamberlin, 1985), and awaits further characterization.

Dynamic Aspects of Elongation

Following the formation of each phosphodiester bond, both the RNA and the DNA must change register with respect to the initiation and polymerization binding sites on the enzyme, permitting the next nucleoside triphosphate to fill the vacated polymerization site. The nature of the process by which this translocation occurs has never been clearly determined. It is obvious, however, that RPase translates along, and possibly rotates about, the axis of the nucleic acid helix after each bond formed, as the geometry of the 3′-hydroxyl of the RNA with respect to the catalytic site of the enzyme must be maintained.

An intriguing proposal is that of Dennis and Sylvester (1981), who suggested that the β subunit of RPase contains two functionally equivalent active sites which are involved alternately in the formation of successive phosphodiester bonds. In essence, the RPase topologically rolls along the DNA template during elongation. The evidence for such a model is two-fold: appropriate *bis*-3′- 5′ cyclic dinucleotides (Hsu and Dennis, 1982) were effective competitive inhibitors of initiation, and photoaffinity probes incorporated into each of the two sites labelled two distinct sites in the enzyme (Panka and Dennis, 1985). Further work is required to corroborate this hypothesis.

The fact that the DNA-RNA hybrid formed during elongation is approximately twelve base pairs in length has important topological consequences, as this number corresponds to just less than a single turn of the A-type double helix. This allows the nascent RNA strand to be bent away from the DNA, avoiding intertwining of the two (see Figure 2). Furthermore, the process

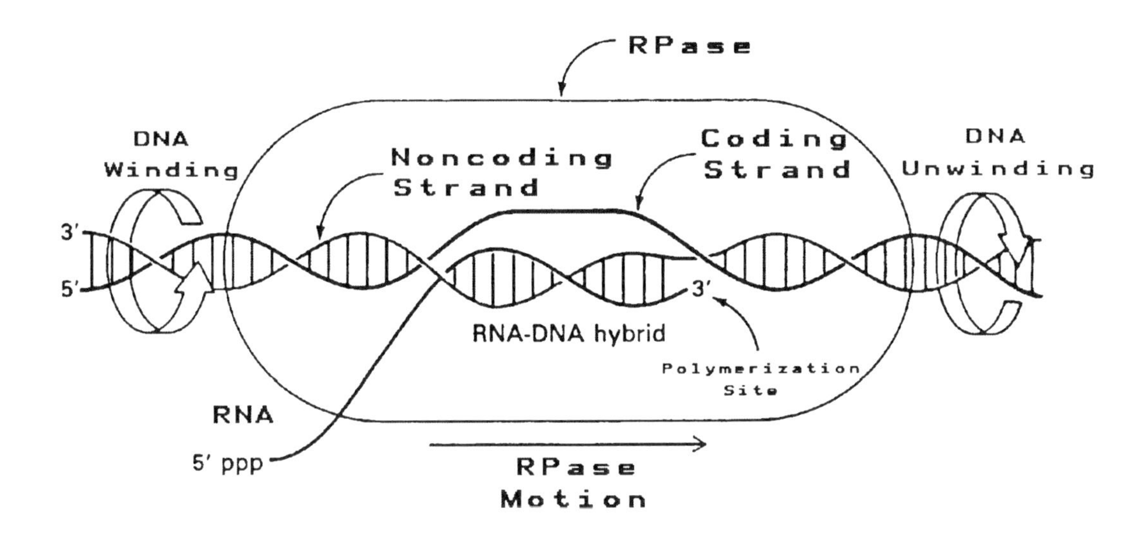

Figure 2. Model of the transcription bubble during elongation.

of DNA unwinding (melting) is dynamic; as the template is traversed, it is constantly being unwound at the leading edge of the transcription bubble and rewound at the trailing edge. Unlike the condition at the leading edge of the replication fork, the unwinding need not involve a topoisomerase, as the topology of the DNA template is unchanged once the transcription bubble has passed through.

Pausing

During the course of elongation, two things can occur at any time: the process can continue smoothly, or the elongation complex can remain associated with its site for some period of time. This poorly understood phenomenon is referred to as "pausing," and has been associated with the premature dissociation of the transcriptional complex (see Andrews and Richardson, 1985). Pausing seems to be DNA sequence dependent, as there are well-known "pause sites" in many genes that have been characterized.

In an attempt to explain pausing, von Hippel and coworkers (Yager and von Hippel, 1987; von Hippel *et al.*, 1987) have evaluated the thermodynamics of the elongating transcription complex as a function of DNA (and hence RNA) sequence, and found that there is good correlation between the degree of pausing at a given site and the free energy of the transcription complex. Changes in the free energy of the transcription complex arise primarily from: the loss and formation of hydrogen bonds between the two strands of the DNA double helix and between the nascent RNA chain and the template strand of the DNA; secondary structure of the nascent RNA chain; and any sequence-dependence in the affinity of RPase for both DNA and RNA. In addition, the presence of a stable hairpin in the RNA transcript within the DNA-RNA hybrid has been associated with the length of pausing (Landick and Yanofsky, 1984).

TERMINATION

Termination of the transcriptional process can be further dissected into four separate processes: the cessation of phosphodiester bond synthesis, dissociation of the DNA-RNA hybrid, unmelting of the transcription bubble, and dissociation of the DNA-RPase complex. It is, in many ways, a special case of pausing; the distinction merely reflects the length of the pause, and whether dissociation of the elongation complex occurs during that period of time. While termination of transcription requires the pausing of RPase (Reisbig and Hearst, 1981; Morgan *et al.*, 1983a, 1983b), pausing is not sufficient to insure that termination occurs.

While scientists would like to believe that a direct cause and effect relationship can be determined for every phenomenon, modern physics has taught us that many events are best explained statistically. Transcription termination is one case where this is especially so. Premature termination occurs at a great many sites along the template, and can be thought of as random, although it is in fact related to DNA sequence. Specific termination, on the other hand, occurs at specific loci on the DNA template, and can be classified into a minimum of two general categories, depending on whether the protein factor ρ is involved or not.

Termination at ρ-independent sites can be both specific and efficient, depending only on the thermodynamics of the interaction of the elongation complex with the particular site on the template (Yager and von Hippel, 1987). A DNA sequence motif associated with transcription termination is the presence of a palindromic, GC-rich region followed by four or more thymidines. The resulting RNA transcript will tend to form into a stable hairpin structure due to the preponderance of rG-rC base pairs in the stem of the hairpin. The presence of such structures promotes pausing at the site, while at the same time a significant portion of the DNA-RNA hybrid is composed of relatively unstable rU-dA base pairs. As a consequence, the hybrid dissociates, releasing the RNA from the DNA template and hence from the RPase. As noted above, termination, and especially ρ-independent termination, is mechanistically very similar to elongation pausing.

While ρ-independent termination can be observed *in vitro* using defined systems, it is still possible that other, as yet uncharacterized, protein factors may be involved in termination at other classes of terminator sequences, especially *in vivo* (Chamberlin *et al.*, 1987). Especially important is the fact that some terminators are markedly more efficient *in vivo* than they are *in vitro*, and the released RNA product can differ in length by several bases (Briat and Chamberlin, 1984).

Rho Factor

A second class of termination events requires not only a pause site at the locus of termination, but a second site located upstream of the pause site on the nascent RNA chain (see von Hippel *et al.*, 1984). This site, which has been found as much as 300 base pairs away from the ultimate termination site (Richardson *et al.*, 1987), has been referred to as a *rut* (**r**ho **ut**ilization) site (Salstrom *et al.*, 1979) and a *fer* (**f**actor **e**ntry **r**egion) site (Platt, 1986). Experimental evidence suggests that the pause site must be sufficiently long-lived (10 sec under *in vitro* assay conditions) for the subsequent binding of ρ to the upstream site. Lastly, the ρ binding site must not have too much secondary structure, as it is the interaction of free cytosine residues with ρ that activates an RNA-dependent ATPase activity (Lowery and Richardson, 1977; Richardson and Conaway, 1980). It is this ATPase activity which is associated with the ultimate release of both ρ and the RPase core enzyme from the DNA template, along with the release of the RNA product, as evidenced by the observation that nucleoside triphosphate analogs which are not cleavable by ρ (but are substrates for RPase) block ρ-dependent termination (Howard and de Crombrugghe, 1976). It is thought that ρ moves along the RNA chain towards the transcription bubble in an energy-dependent process (provided by the hydrolysis of ATP), and disrupts the DNA-RNA hybrid either directly, or indirectly *via* interaction with RPase (McKenney *et al.*, 1981).

NusA and Antitermination

In the case of the λt^{R2} from λ phage (Briat and Chamberlain, 1984) and the *E. coli rrnB* T1 terminators (Chamberlin *et al.*, 1987), the role of an additional protein factor has been demonstrated conclusively. This protein, *nusA*, may in fact be a subunit of RPase, as it has been shown to bind to core, but not holo, enzyme (Greenblatt and Li, 1981). A single *nusA* protein is found in the transcription elongation complex; it essentially takes the place of σ following release of that subunit (Chamberlin *et al.*, 1987). This role of *nus*A in elongation is independent of any role in termination, and may be a general characteristic of transcription. In fact, it may be more accurate to talk of a σ/*nusA* cycle, although more work remains to be done on this subject. There is a body of evidence that the binding of *nusA* to the elongating RPase promotes the pausing of the enzyme at specific sites on the DNA template, ultimately resulting in termination (Greenblatt *et al.*, 1981; Kingston and Chamberlin, 1981; Farnham *et al.*, 1982). In at least one case, σ factor was demonstrated to enhance the rate of release of RPase from the DNA (Chamberlin *et al.*, 1987).

The *nusA* protein has also been implicated in the suppression of termination in λ by the bacteriophage *N* protein (for a review see Friedman and Gottesman, 1983). Apparently, the core RPase-*nusA* elongation complex binds, in a sequence-dependent manner, three additional protein factors (Greenblatt *et al.*, 1987); first the ribosomal protein *S10*, then *nusB*, and finally the phage *N* protein, in the form of a complex with a second copy of *nusA*. From various lines of evidence, it has been concluded that this "elongation control particle" (Greenblatt *et al.*, 1987) is really a ribonucleoprotein complex containing a portion of the transcribed RNA chain (Greenblatt, 1984). This evidence includes the observation that ribosomal translation through the nut_r (*N* **u**tilization) site in the phage DNA disrupts N-mediated termination (Olson *et al.*, 1984), while treatment of the complex with ribonuclease T1 causes the dissociation of N (Greenblatt *et al.*, 1987). Thus it is the presence of this particular sequence in the elongation complex that allows it to "remember" that the site has been traversed. This model may be of more general significance than in simply explaining the activity of a single bacteriophage protein. In fact, the complexity of this system may provide some insight into transcription in eukaryotes, where there are multitudes of protein factors associated with transcription whose functions are unknown.

MECHANISM OF TRANSCRIPTION BY EUKARYOTIC RPASES

Purified RPase II preparations are essentially incapable of recognizing, and/or initiating transcription at, endogenous promoter sequences, but instead act at regions of single-stranded DNA and nicks (see Lewis and Burgess, 1982). Taken with the evidence that the purified enzyme is also deficient with respect to release of RNA product, it appears that while (at least some) purified eukaryotic RPases are capable of transcribing RNA from a DNA template, other properties of the prokaryotic holoenzyme are lacking. By way of comparison, *E. coli* core polymerase, which lacks the σ subunit, behaves essentially the same as RPase II from wheat germ with respect to: initiation at nicks and single-stranded regions of the DNA template, transcriptional start site using poly dC tailed templates, and release of final RNA product. Furthermore, the three enzymes that are deficient in this last property are all from animal sources, and have more in common antigenically with each other than with RPase II from plant sources (Weeks *et al.*, 1982), possibly reflecting more general mechanistic differences between the two eukaryotic kingdoms (Dedrick and Chamberlin, 1985).

By way of analogy to its prokaryotic homologue, some eukaryotic RPases have recently been demonstrated to display discrete, resolvable steps during the process of transcription. Safer *et al.* (1985) were able to distinguish, and isolate, distinct preinitiation, initiation, and elongation complexes of human RPase II. Using HeLa whole cell extracts prepared by the method of Manley *et al.* (1980), they found that by simply omitting those components required for the conversion of one complex to the next, significant amounts of the intermediate complexes could be accumulated. The process of transcription was further elucidated by the observation that different concentrations of the ionic detergent Sarkosyl could arrest the process at new intermediates (Hawley and Roeder, 1985). They proposed that transcription by RPase II could be broken down into a minimum of four steps. The first step involves the association of protein factors with the promoter, which is then committed to transcription. The conversion of this complex to a "rapid start complex," which is functionally equivalent to the prokaryotic open complex, is blocked by 0.015% Sarkosyl. The next step requires two nucleotides and is sensitive to Sarkosyl concentrations above 0.25%. At this point, the first phosphodiester bond is formed, and the reaction proceeds to the elongation of the RNA chain.

Recently, the mechanism of RPase II transcription has been described in greater detail (Van Dyke *et al.*, 1988), as more has been learned of the nature of the protein factors involved. For the human RPase II-major late promoter of adenovirus system, these steps can be summarized as follows. Transcription Factor IID and a gene-specific transcription factor referred to as the upstream stimulatory factor (Sawadogo and Roeder, 1985) bind to the TATA box and the upstream sequence, respectively, in what is apparently the rate-limiting step for the entire process. Following this, the preinitiation complex is assembled upon the addition of RPase II and transcription factors IIB and IIE. At this point, a step analogous to the closed-to-open transition of prokaryotic transcription occurs, although the exact nature of the transition is not known. It is known, however, that it requires energy in the form of the hydrolysis of the β-γ phosphate bond of either ATP or dATP (Cai and Luse, 1987). After the transcription complex moves off the promoter, some factors (minimally TFIID and the upstream stimulatory factor) remain associated with the promoter (Van Dyke *et al.*, 1988). While this might imply maintenance of a stable complex through multiple cycles of transcription at the same promoter, such a phenomenon has yet to be demonstrated *in vitro*.

CONCLUSIONS

As should be evident from the preceding discussion, a great deal is known of the detailed molecular mechanism of transcription in *E. coli*, which is viewed as a paradigm for all prokaryotes. With the recent advances in the development of *in vitro* transcription systems employing the eukaryotic polymerases, a body of knowledge is beginning to be accumulated. Due to the much greater complexity of the process of transcription in eukaryotes, details have been much less forthcoming than was the case for the prokaryotic system. Nevertheless, the more is learned, the more it becomes obvious that there is great similarity between these evolutionarily disparate systems, both in terms of structural homology between the proteins involved, and functional similarities between the individual steps which comprise transcription. While much has been learned of the process of transcription in eukaryotes, there is still a great deal more to be discovered.

REFERENCES

Adam, G. and Delbruck, M. (1968) *Structural Chemistry and Molecular Biology* (A. Rich and N. Davidson, eds.; Freeman, San Francisco) pp. 198-204

Allison, L. A., Moyle, M., Shales, M. and Ingles, C. J. (1985) *Cell* **42**: 599 - 610

Anthony, D. D., Zeszotok, E. and Goldthwait, D. (1966) *Proc. Natl. Acad. Sci. USA* **56**: 1026-1033

Armstrong, V. W. and Eckstein, F. (1979) *Biochemistry* **18**: 5117-5122

Armstrong, V. W., Yee, D. and Eckstein, F. (1979) *Biochemistry* **18**: 4120-4123

Auld, D. S., Atsuya, I., Campino, C. and Valenzuela, P. (1976) *Biochem. Biophys. Res. Commun.* **69**: 548-554

Barkley, M. D. (1981) *Biochemistry* **20**: 3833-3839

Belintsev, B. N., Zavriev, S. K. and Shemyakin, M. F. (1980) *Nucleic Acids Res.* **8**: 1391-1404

Berg, O. G. and Blomberg, C. (1976) *Biophys. Chem.* **4**: 367-381

Berg, O. G. and Blomberg, C. (1978) *Biophys. Chem.* **8**: 271-280

Berg, O. G., Winter, R. B. and von Hippel, P. H. (1981) *Biochemistry* **20**: 6929-6948

Bresloff, H. and Crothers, D. (1975). *J. Mol. Biol.* **95**: 103-123

Briat, J. and Chamberlin, M. J. (1984) *Proc. Natl. Acad. Sci. USA* **81**: 7373-7377

Broyles, S. S. and Moss, B. (1986) *Proc. Natl. Acad. Sci. USA* **83**: 3141-3145

Buc, H. and McClure, W. R. (1985) *Biochemistry* **24**: 2712-2723

Bujard, K., Niemann, A., Breunig, K., Roisch, U., Dresel, A., von Gabain, A., Gentz, R., Stuber, D., and Weiher, H. (1982) **In**: *Promoters, Structure and Function* (Rodriguez, R. L. and Chamberlin, M. J., eds) pp 121-140, Praeger, New York

Burgess, P. M. and Eckstein, F. (1978) *Proc. Natl. Acad. Sci. USA* **75**: 4795-4800

Burgess, R. R. (1969) *J. Biol. Chem.* **244**: 6168-6176

Burgess, R. R., (1976). **In**: *RNA Polymerase*, (R. Losick, M. J. Chamberlin, eds.; New York: Cold Spring Harbor Lab) pp. 69-100

Cai, H. and Luse, D. S. (1987) *Mol. Cell. Biol.* **7**: 3371-3379

Cassani, G., Burgess, R. R., Goodman, H. M. and Gold, L. (1971) *Nature New Biol.* **230**: 197-200

Chamberlin, M. J., (1974). *Ann. Rev. Biochem.* **43**: 721-775

Chamberlin, M. J., Arndt, K. M., Briat, J.-F., Reynolds, R. L. and Schmidt, M. C. (1987) **In**: *RNA Polymerase and the Regulation of Transcription*: A Steenbock Symposium (Elsevier Science Publishing Co., New York, New York) pp. 347-356

Chamberlin, M. J., Rosenberg, S., and Kadesch, T. (1982) **In**: *Promoters, Structure and Function* (Rodriguez, R. L. and Chamberlin, M. J., eds) pp 34-53, Praeger, New York

Chatterji, D. and Wu, F. Y.-H. (1982) *Biochemistry* **21**: 4657-4664

Chatterji, D., Wu, C.-W. and Wu, F. Y.-H. (1984) *J. Biol. Chem.* **259**: 284-289

Ciliberto, G., Castagnoli, L. and Cortese, R. (1983) *Curr. Top. Dev. Biol.* **18**: 59-87

Corden, J., Wasylyk, B., Buchwalder, A., Sassone-Corsi, P., Kedinger, C. and Chambon, P. (1980) *Science* **209**: 1406-1414

Dedrick, R. L. and Chamberlin, M. J. (1985) *Biochemistry* **24**: 2245-2253

Dennis, D. and Sylvester, J. E. (1981) *FEBS Lett.* **124**: 135-139

Dennis, D. D. (1984) *Cell* **37**: 359-365

Doi, R. H. (1982) *Arch. Biochem. Biophys.* **214**: 772-781

Dunn, J. J. and Studier, F. W. (1981) *J. Mol. Biol.* **148**: 303-330

Dynan, W. S. and Tjian, R. (1985) *Nature* **316**: 774-778

deHaseth, P. L., Lohman, T. M., Burgess, R. R., and Record, M. T., Jr. (1978) *Biochemistry* **17**: 1612-1622

Eckstein, F. and Grindl, H. (1970) *Eur. J. Biochem.* **13**: 558-564

Ehbrecht, H.-J., Pingoud, A., Urbanke, C., Maass, G., and Gualerzi, C. (1985) *J. Biol. Chem.* **260**: 6160-6166

Falchuk, K. H., Mazus, B., Ulpino, L. and Vallee, B. L. (1977) *Biochemistry* **15**: 4468-4475

Falchuk, K. H., Ulpino, L., Mazus, B. and Vallee, B. L. (1977) *Biochem. Biophys. Res. Commun.* **74**: 1206-1212

Farnham, P. J., Greenblatt, J. and Platt, T. (1982) *Cell* **29**: 945-951

Fire, A., Samuels, M. and Sharp, P. A. (1984) *J. Biol. Chem.* **259**: 2509-2516
Friedman, D. and Gottesman. M. (1983) **In:** *Lambda II* (Hendrix, R. W., Roberts, J. W., Stahl, F. W. and Weisberg, R. A., Eds.; Cold Spring Harbor Laboratory (1983) pp. 21-52
Fukuda, R. and Ishihama, A. (1974) *J. Mol. Biol.* **87**: 523-540
Gamper, H. B. and Hearst, J. E. (1982) *Cell* **29**: 81-90
Gerard, G. F., Rotman, F. and Boezi, J. A. (1971) *Biochemistry* **10**: 1974-1981
Greenblatt, J. (1984) *Cell Biol.* **62**: 79-88
Greenblatt, J. and Li, J. (1981) *Cell* **24**: 421-428
Greenblatt, J., Horwitz, R. J. and Li, J. (1987) **In:** *RNA Polymerase and the Regulation of Transcription*: A Steenbock Symposium (Elsevier Science Publishing Co., New York, New York) pp. 357-366
Greenblatt, J., McLimont, M. and Hanly, S. (1981) *Nature (London)* **292**: 215-220
Grossman, A. D., Erickson, J. W. and Gross, C. A. (1984) *Cell* **38**: 383-390
Hawley, D. K. and McClure, W. R. (1980) *Proc. Natl. Acad. Sci. USA* **77**: 6381-6389
Hawley, D. K. and Roeder, R. G. (1985) *J. Biol. Chem.* **260**: 8163-8172
Heil, A. and Zillig, W. (1970) *FEBS Lett.* **11**: 165-168
Hillel, Z. and Wu, C.-W. (1977) *Biochemistry* **16**: 3334-3342
Hillel, Z. and Wu, C.-W. (1978) *Biochemistry* **17**: 2554-2561
Hinkle, D. C. and Chamberlin, M. J. (1971) *Cold Spring Harbor Symp. Quant. Biol.* **35**: 65-72
Hinkle, D. C. and Chamberlin, M. J. (1972) *J. Mol. Biol.* **70**: 157-185
Hirschman, J., Wong, P.-K., Sei, K., Keener, J. and Kustu, S. (1985) *Proc. Natl. Acad. Sci. USA* **82**: 7525-7529
Hodo, H. G. and Blatti, S. P. (1977) *Biochemistry* **16**: 2334-2343
Howard, B. and de Crombrugghe, B. (1976) *J. Biol. Chem.* **251**: 2520-2525
Hsieh, T. and Wang, J. C. (1978) *Nucl. Acids Res.* **5**: 3337-3345
Hsu, C.-Y. J and Dennis, D. (1982) *Nucleic Acids Res.* **10**: 5637-5647
Hurwitz, J., Bresler, A. and Diringer, R. (1960) *Biochem. Biophys. Res. Commun.* **3**: 15-19
Ingles, C. J., Biggs, J., Wong, J., K.-C., Weeks, J. R. and Greenleaf, A. L. (1983) *Proc. Natl. Acad. Sci. USA* **80**: 3396-3400
Jack, W. E., Terry, B. J., and Modrich, P. (1982) *Proc. Natl. Acad. Sci. U.S.A.* **79**: 4010-4014
Johnston, D. E. and McClure, W. R. (1976) **In:** *RNA Polymerase*, (R. Losick, M. J. Chamberlin, eds.; New York: Cold Spring Harbor Lab) pp. 413-428
Jovin, T. M. (1976). *Ann. Rev. Biochem.* **45**: 889-920
Kadesch, T. R., Williams, R. C. and Chamberlin, M. J. (1980) *J. Mol. Biol.* **70**: 187-195
Kane, C. M. and Chamberlin, M. J. (1985) *Biochemistry* **24**: 2254-2260
Kawakami, K. and Ishihama, A. (1980) *Biochemistry* **19**: 3491-3495
Kerrich-Santo, R. E. and Hartman, G. R. (1974) *Eur. J. Biochem.* **43**: 521-532
King, A. M. Q., Lowe, P. A. and Nicholson, B. H. (1974) *Biochem. Soc. Trans.* **2**: 76-78
Kingston, R. E. and Chamberlin, M. J. (1981) *Cell* **27**: 523-531
Kirkegaard, K., Buc, H., Spassky, A. and Wang, J. C. (1983) *Proc. Natl. Acad. Sci. USA* **80**: 2544-2548
Kownin, P, Bateman, E. and Paule, M. R. (1987) *Cell* **50**: 693-699
Krakow, J. and von der Helm (1971) *Cold Spring Harbor Symp. Quant. Biol.* **35**: 73-83
Krakow, J. S., Rhodes, G. and Jovin, T. M. (1976) **In:** *RNA Polymerase* (Losick, R. and Chamberlin, M. J., eds.; Cold Spring Harbor Press, Cold Spring Harbor New York) pp. 101-125
Krakow, J., Rhodes, G. and Jovin, T. M. (1976) **In:** *RNA Polymerase*, (R. Losick, M. J. Chamberlin, eds.; New York: old Spring Harbor Lab) pp. 127-157
Lattke, H. and Weser, U. (1976) *FEBS Lett.* **65**: 288-292
Lewis, M. and Burgess, R. R. (1982) *Enzymes, 3rd Ed.* **15**: 110-153
Loeb, A. A. and Mildvan, A. S. (1981) **In:** *Advances in Inorganic Biochemistry, Metal Ions in Genetic Information Transfer* (Eichhorn, G. L. and Marzilli, L. G., Eds.; Elsevier Press, New York) 3, 125-142
Losick, R., Youngman, P. (1984) **In:** *Microbial Development* (R.
Losick, L. Shapiro, eds.; New York: Cold pring Harbor Lab) pp. 63-88
Lowe, P. A. and Malcolm, A. D. B. (1976) *Biochem. Biophys. Acta* **454**: 129-137
Lowe, P. A., Hager, D. A. and Burgess, R. R. (1979) *Biochemistry* **18**: 1344-1352
Lowery, C. and Richardson, J. P. (1977) *J. Biol. Chem.* **252**: 1381-1385
Lyon, M. F. (1974) *Proc. Roy. Soc.* **187**: 243-268
Maitra, U. and Hurwitz, J. (1965) *Proc. Natl. Acad. Sci. USA* **54**: 815-822
Maitra, U., Nakata, Y. and Hurwitz, J. (1967) *J. Biol. Chem.* **242**: 4908-4918
Mangel, W. F. and Chamberlin, M. J. (1974) *J. Biol. Chem.* **249**: 2995-3001
Mangel, W. F. and Chamberlin, M. J. (1974a) *J. Biol. Chem.* **249**: 3002-3006
Manley, J. L., Fire, A., Cano, A., Sharp, P. A. and Gefter, M. L. (1980) *Proc. Natl. Acad. Sci. USA* **77**: 3855-3859

Manning, G. S. (1969) *J. Chem Phys.* **51**: 924-933
Manning, G. S. (1977) *Biophys. Chem.* **7**: 95-99
Manning, G. S. (1978) *Q. Rev. Biophys.* **11**: 179-203
Matsui, T., Segall, J., Weil, P. A. and Roeder, R. G. (1980) *J. Biol. Chem.* **255**: 11992-11996
McClure, W. R. (1980) *J. Biol. Chem.* **255**: 1610-1616
McClure, W. R. and Cech, C. L. (1978) *J. Biol. Chem.* **253**: 8949-8956
McClure, W. R. and Chow, Y. (1980) *Methods in Enzymology* **64**: 277-297
McClure, W. R., 1985. *Ann. Rev. Biochem.* **54**: 171-204
McClure, W. R., Cech, C. L. and Johnston, D. E. (1978) *J. Biol. Chem.* **253**: 8941-8948
McKenney, K., Shimatake, H., Court, D., Schmeissner, V. and Rosenberg, M. (1981) **In**: *Gene Amplification and Analysis* (J. Chirikian and T. Papas, eds., Elsevier/North-Holland, New York, New York) vol. 2 pp. 383-413
Miller, J. A., Serio, G. F., Howard, R. A., Bear, J. L., Evans, J. E. and Kimball, A. P. (1979) *Biochim. Biophys. Acta* **579**: 291-297
Moran, C. P., Jr., Lang, N., LeGrice, S. F. (1982) *Mol. Gen. Genet.* **186**: 339-346
Novak, R. L. and Doty, P. (1968) *J. Biol. Chem.* **243**: 6068-6071
Oen, H. and Wu, C.-W. (1978) *Proc. Natl. Acad. Sci. USA* **75**: 1778-1782
Olson, E. R., Tomich, C. C. and Friedman, D. I. (1984) *J. Mol. Biol.* **180**: 1053-1063
Panka, D. and Dennis, D. (1985) *J. Biol. Chem.* **260**: 1427-1431
Park, C. S., Hillel, Z. and Wu, C.-W. (1982a) *J. Biol. Chem.* **257**: 6944-6949
Park, C. S., Wu, F. Y.-H. and Wu, C.-W. (1982b) *J. Biol. Chem.* **257**: 6950-6956
Petranyi, P., Jendrisak, J. J. and Burgess, R. R. (1977) *Biochem. Biophys. Res. Commun.* **74**: 1031-1038
Platt, T. (1986) *Ann. Rev. Biochem.* **54**: 339-372
Rabussay, D. and Zillig, W., 1970. *FEBS Lett.* **5**: 104-106
Record, M. T., Jr., deHaseth, P. L. and Lohman, T. M. (1977) *Biochemistry* **16**: 4791-4798
Record, M. T., Jr., Lohman, T. M. and deHaseth, P. L. (1976) *J. Mol. Biol.* **107**: 145-152
Reeve, A. E., Smith, M. M., Pigiet, V. and Huang, R. C. C. (1977) *Biochemistry* **16**:, 4464-4469
Rhodes, G. and Chamberlin, M. J. (1974) *J. Biol. Chem.* **249**: 6675-6683
Rhodes, G. and Chamberlin, M. J. (1975) *J. Biol. Chem.* **250**: 9112-9120
Richardson, J. P. and Conaway, R. (1980) *Biochemistry* **19**: 4293-4299
Riesbig, R. R., Woody, A.-Y. M. and Woody, R. W. (1979) *J. Biol. Chem.* **254**: 11208-11217
Roberts, J. W. (1969) *Nature* **224**: 1168-1174
Rohrer, H. and Zillig, W. (1977) *Eur. J. Biochem.* **79**: 401-407
Rose, I. A., O'Connel, E. L., Litwin, S. and Bar Tana, J. (1974) *J. Biol. Chem.* **249**: 5163-5168
Safer, B., Yang, L., Tolunay, H. E. and Andersen, W. F. (1985) *Proc. Natl. Acad. Sci. USA* **82**: 2632-2636
Sakonju, S., Bogenhagen, D. F. and Brown, D. D. (1980) *Cell* **19**: 13-25
Salstrom, J. S., Fiandt, M and Szybalski, W. (1979) *Mol. Gen. Genet.* **168**: 211-230
Samuels, M., Fire, A. and Sharp, P. A. (1982) *J. Biol. Chem.* **257**: 14419-14427
Saucier, J. and Wang, J. C. (1972) *Nature New Biol.* **239**: 167-170
Sawadogo, M. and Roeder, R. G. (1985) *Cell* **43**: 165-170
Schleif, R. (1969) *Nature* **223**: 1068-1069
Schurr, J. M. (1979) *Biophys. Chem.* **9**: 413-414
Scrutton, M. C., Wu, C.-W. and Goldthwait, D. A. (1971) *Proc. Natl. Acad. Sci. USA* **68**: 2497-2501
Sentenac, A., Ruet, A. and Fromageot, P. (1968) *Eur. J. Biochem.* **5**: 385-394
Shimamoto, N. and Wu, C.-W. (1980a) *Biochemistry* **19**: 842-848
Shimamoto, N. and Wu, C.-W. (1980b) *Biochemistry* **19**: 849-856
Shimamoto, N., Kamigochi, T. and Utiyama, H. (1987) **In**: *RNA Polymerase and the Regulation of Transcription*: A Steenbock Symposium (Elsevier Science Publishing Co., New York, New York) pp. 409-412
Shimamoto, N., Wu, F. Y.-H. and Wu, C.-W. (1981) *Biochemistry* **20**: 4745-4755
Siebenlist, U., (1979). *Nature* **279**: 651-652
Singer, P. T. and Wu, C.-W. (1987) *J. Biol. Chem.* **262**: 14178-14189
Smagowicz, J. W. and Scheit, K. H. (1978) *Nucleic Acids Res.* **9**: 2440-2447
Speckhard, D. C., Wu. F. Y.-H. and Wu, C.-W. (1977) *Biochemistry* **16**: 5228-5234
Stevens, A. (1960) *Biochem. Biophys. Res. Comm.* **3**: 92-96
Stutz, A. and Scheit, K. H. (1975) *Eur. J. Biochem.* **50**: 343-349
Sweetser, D., Nonet, M. and Young, R. A. (1987) *Proc. Natl. Acad. Sci. USA* **84**: 1192-1196
Takahashi, K., Vigneron, M., Matthes, H., Wildeman, A., Kenke, M. and Chambon, P. (1986) *Nature* **319**: 121-126
Tower, J., Culotta, V. C. and Sollner-Webb, B. (1986) *Mol. Cell. Biol.* **6**: 3451-3462
Van Dyke, M. W., Roeder, R. G. and Sawadogo, M. (1988) *Science* **241**: 1335-1338
von der Helm, K. and Krakow, J. S. (1972) *Nature New Biol.* **235**: 82-83

von Hippel, P. H., Bear, D. G., Morgan, W. D., and McSwiggen, J. A. (1984) *Ann. Rev. Biochem.* **53**: 389-446
von Hippel, P. H., Rezvin, A., Gross, C. A. and Wang, A. C. (1974) *Proc. Natl. Acad. Sci. USA* **71**: 4808-4812
von Hippel, P. H., Rezvin, A., Gross, C. A. and Wang, A. C. (1978). **In**: *Protein Ligand Interactions* (Sund, H. and Blauer, G., eds; Berlin; Walter de Gruyter) p. 270
Walter, G., Zillig, W., Palm, P. and Fuchs, E. (1967) *Eur. J. Biochem.* **3**: 194-201
Wandzilak, T. M. and Benson, R. W. (1977) *Biochem. Biophys. Res. Commun.* **76**: 247-252
Wang, J. C. and Davidson, N. (1966). *J. Mol. Biol.* **19**: 469-482
Wang, J. C., Jacobsen, J. H. and Saucier, J. (1977). *Nucl. Acids Res.* **4**: 1224-1241
Weeks, J. R., Coulter, D. E. and Greenleaf, A. L. (1982) *J. Biol. Chem.* **257**: 5884-5891
Winter, R. B. and von Hippel, P. H. (1981) *Biochemistry* **20**: 6948-6960
Worcel, A., Kmiec, E. and Shimamura, A. (1987) **In**: *RNA Polymerase and the Regulation of Transcription*: A Steenbock Symposium (Elsevier Science Publishing Co., New York, New York) pp. 209-218
Wu, C.-W. and Goldthwait (1969a) *Biochemistry* **8**: 4450-4458
Wu, C.-W. and Goldthwait (1969b) *Biochemistry* **8**: 4458-4464
Wu, C.-W., Hillel, Z. and Park, C. S. (1983) *Anal. Biochem.* **128**: 481-489
Wu, C.-W., Wu, F. Y.-H. and Speckhard, D. C. (1977) *Biochemistry* **16**: 5449-5454
Wu, F. Y.-H. and Wu, C.-W. (1974) *Biochemistry* **13**: 2562-2566
Wu, F. Y.-H. and Wu, C.-W. (1987) *Ann. Rev. Nutr.* **7**: 251-272
Yarbrough, L. R., Wu, F. Y.-H. and Wu, C.-W. (1976) *Biochemistry* **15**: 2669-2676
Yee, D., Armstrong, V. W. and Eckstein, F. (1979) *Biochemistry* **18**: 4116-4120
Zillig, W., Palm, P. and Heil, A. (1976) **In**: *RNA Polymerase* (Losick, R. and Chamberlin, M. J., eds.; Cold Spring Harbor Press, Cold Spring Harbor New York) pp. 101-125
Zillig, W., Zechez, K., Rabussay, D., Schechner, M., Sethi, V. S., Palm, P., Heil, A. and Seifert, W. (1970) *Cold Spring Harbor Symp. Quant. Biol.* **35**: 47-58

16

THE MECHANISM OF TRANSCRIPTION INITIATION BY RNA POLYMERASE II

Donal S. Luse

INTRODUCTION

It is clear that the regulation of transcription initiation is a key step in the control of eukaryotic gene expression. Since the original development of soluble systems which support accurate initiation of transcription by RNA polymerase II (Weil *et al.*, 1979; Manley *et al.*, 1980) much progress has been made on the identification and characterization of both core transcription factors and regulatory proteins (see, for example, Briggs *et al.*, 1986; Reinberg and Roeder, 1987; Reinberg *et al.*, 1987; Zheng *et al.*, 1987;Burton *et al.*, 1988 for recent work on core transcription factors and Maniatis *et al.* 1987 for a recent review of regulatory elements). I will focus in this article on a subject which has received less attention than factor characterization, namely the molecular mechanisms involved in transcription initiation. By "molecular mechanisms" I mean the pathway by which an initiation-competent RNA polymerase II/factor complex is first assembled and then converted to a transcribing complex committed to RNA chain elongation. Defining the intermediates, and the rate-limiting step(s), in the initiation process will be an essential part of understanding how regulatory molecules intervene to stimulate or repress transcription.

Our expectations concerning the RNA polymerase II initiation process have been strongly influenced by knowledge of the initiation mechanism for prokaryotic RNA polymerase (see von Hippel *et al.*, 1984; McClure, 1985 for recent reviews). At least three steps are involved in initiation in bacteria: *specific binding* of the RNA polymerase to the promoter, *isomerization* of the binary complex to an initiation-competent form (which requires melting of the template and which may involve several intermediates; see Buc and McClure, 1985; Spassky *et al.*, 1985), and *clearance* of the RNA polymerase from the promoter. Given the wide variation in prokaryotic promoter sequences it is not surprising that these promoters differ greatly in the extent to which they support specific binding and efficient isomerization by the polymerase. Both of these steps may be independently influenced by regulatory proteins (reviewed by McClure, 1985). An important and poorly-appreciated feature of initiation is the frequent failure of pre-initiation complexes (at least *in vitro*) to make a successful transition to elongation upon synthesis of the first phosphodiester bond. After one or a few bonds are made, the polymerase often ejects the nascent RNA chain and then reinitiates transcription without releasing from and rebinding to the promoter (Carpousis and Gralla, 1985; Straney and Crothers, 1987). This cycle of *abortive initiation* may continue for many rounds before the RNA polymerase successfully produces a chain of greater than a threshold length (between two and about ten bases, depending on the promoter; reviewed by von Hippel *et al.*, 1984); once this point is passed, the transcription complex is *elongation committed.* Enzymatic and chemical probing of initiation-competent (open) complexes and abortively initiating complexes suggests that the RNA polymerase essentially retains its preinitiation configuration while abortively initiating (Carpousis and Gralla, 1985; Straney and Crothers, 1987). (It should be noted that since the RNA polymerase is cycling between preinitiation and initiation states during abortive initiation, the relative lifetimes of the two states will strongly affect the results of nuclease probe studies. In particular, transcription complex configurations unique to the abortively initiating state could be so short-lived that only the preinitiation configuration can be detected.)

Achieving elongation commitment coincides with the loss of promoter contacts by the polymerase (the upstream footprint boundary shifts from about -50 to -10; Carpousis and Gralla, 1985; Straney and Crothers, 1987) and with acquisition of resistance to "initiation-specific" reagents such as rifampicin (Carpousis and Gralla, 1985). This transition probably also coincides with the loss of the σ initiation factor from the complex (Hansen and McClure, 1980; Straney and Crothers, 1985; but see also Shimamato *et al.*, 1986 and Stackhouse and Meares, 1988 for alternative viewpoints). Abortive initiation is not a minor pathway for the bacterial polymerase. At *lac* UV5, for example, about 98% of the *in vitro* initiations are abortive even in the presence of NTP levels which support efficient elongation (Carpousis and Gralla, 1985). Thus, escape of the polymerase from the promoter is not a trivial process; the efficiency of this step may have a profound effect on overall promoter function. The failure to begin elongation without "false starts" presumably reflects the reluctance of the RNA polymerase to release the very strong promoter contacts that were necessary for its original specific binding to the promoter. The fact that such contacts are missing in the elongation-committed complex prompts the use of the term "clearance" to describe the final step in the initiation process.

Our laboratory approached the problem of transcription initiation by RNA polymerase II with the assumption that the process would, at least generally, resemble the prokaryotic case. We have concentrated on the initiation step itself and on the transition to elongation commitment. We have shown that for one particular RNA polymerase II promoter, a number of the features of bacterial initiation are indeed observed: productive initiation by RNA polymerase II is accompanied by abortive initiation, and the transition to elongation commitment occurs during the addition of the first few bases and not simply with the formation of the first phosphodiester bond (Luse *et al.*, 1987; Cai and Luse, 1987a, b; Jacob and Luse, 1987). We have also studied changes in the protein-DNA interactions within the complex as initiation occurs (Cai and Luse, 1987b). I will discuss these results below; in order to complete the picture of the initiation process I will begin by briefly reviewing how the RNA polymerase reaches an initiation-competent state.

FORMATION OF THE PREINITIATION COMPLEX

One important distinction between bacterial RNA polymerase and RNA polymerase II is that the latter requires a number of additional components to accurately initiate transcription while the core prokaryotic polymerase requires (at most promoters) only the addition of σ factor (von Hippel *et al.*, 1984). An exhaustive discussion of RNA polymerase II transcription factors is beyond the scope of this article. I will concentrate here on those factors required for initiation (as opposed to stimulatory components) at the Adenovirus 2 major late (Ad 2 ML) promoter, since this well-characterized TATA box promoter is the most frequently used template in factor studies and is also the promoter employed in our own experiments. Various members of this group of factors have been described by a number of laboratories. (See, in particular, Reinberg and Roeder, 1987; Reinberg *et al.*, 1987; Zheng *et al.*, 1987; Burton *et al.*, 1988; Flores *et al.*, 1988; in all of these studies, HeLa cells served as the source of the factors.) Recent results indicate that at least five components are required at Ad 2 ML in addition to the RNA polymerase; these are termed TFIIA, TFIIB, TFIID, TFIIE and TFIIF using the nomenclature originally developed by Roeder and colleagues (Matsui *et al.*, 1980; Reinberg and Roeder, 1987; Reinberg *et al.*, 1987; Flores *et al.*, 1988; for an alternative purification scheme for RNA polymerase II factors, see Moncollin *et al.*, 1986 and Zheng *et al.*, 1987). The first step in assembling the preinitiation complex involves the binding of certain transcription factors to the promoter in the absence of RNA polymerase. Those templates that acquire factors are then committed to transcription; the promoter-factor complex is stable upon challenge with additional template DNA (Davison *et al.*, 1983; Fire *et al.*, 1984). The primary factor involved in template commitment is TFIID; TFIIA is probably involved as well (Samuels and Sharp, 1986; Reinberg *et al.*, 1987). The TFIID fraction interacts with the TATA box, protecting this sequence very strongly against nuclease attack and conferring somewhat weaker protection on the remainder of the -45 to +35 region (Sawadogo and Roeder, 1985). After TFIID is bound the RNA polymerase itself can enter the complex, followed by the other factors (Reinberg *et al.*, 1987). The exact number of additional components and the extent to which they interact with one another and with the RNA polymerase remains uncertain. The initiation factor activities in a chromatographic fraction containing TFIIE and TFIIF will co-sediment with RNA polymerase II; furthermore the activities in the TFIIB and TFIIE/TFIIF fractions can interact, since co-incubation of these fractions produces an initiation factor complex that sediments more rapidly than the individual activities (Reinberg and Roeder, 1987). Use of RNA polymerase II affinity columns has identified two RNA polymerase associated proteins, termed RAP 30 (30kDa) and RAP 74 (74kDa), both of which are required for initiation at the Ad 2 ML promoter; these components interact with each other, since mono-specific anti-RAP 30 antibody co-precipitates RAP 74 (Burton *et al.* 1988). Flores *et al.* (1988) have recently shown, using anti-RAP 30 antibodies, that RAP 30 copurifies with TFIIF. Zheng *et al.* (1987) have purified an initiation factor, which they term BTF3, by a different chromatographic procedure from that of Reinberg and Roeder. This activity also binds to RNA polymerase II and shares some chromatographic properties with TFIIF/RAP 30; however, BTF3 appears to correspond to a 27kDa protein. The precise relationship of BTF3 to the other HeLa cell initiation factors remains to be demonstrated.

It is important to emphasize that few of the RNA polymerase II initiation factors have been purified to homogeneity; thus, many of these preparations may contain more than one activity. The nature of the RNA polymerase II that participates in accurate initiation is also uncertain. (The reader is referred to the article in this volume by Corden for a complete treatment of this subject.) In particular, it is known that the largest subunit of RNA polymerase II has a heptapeptide repeat at its C-terminus; the number of repeats ranges from 26 in yeast to 52 in mammalian RNA polymerase II (Allison *et al.*, 1988; Bartolomei *et al.*, 1988). The non-modified largest subunit has a molecular weight of about 220 kDa and is referred to as IIa. The C-terminal repeat of the largest subunit is very sensitive to proteolysis; the proteolyzed largest subunit (apparent molecular weight of about 170 kDa) is referred to as IIb. The heptapeptide repeat may also be heavily phosphorylated, giving an apparent molecular weight for the largest subunit of 240kDa (Cadena and Dahmus, 1987). The phosphorylated subunit is known as IIo. Genetic approaches clearly indicate that the C-terminal "tail" is required for polymerase function at some genes, since cells cannot live when the only RNA polymerase II available has a heptapeptide repeat truncated to less than half of its original length (Allison *et al.*, 1988; Bartolomei *et al.*, 1988; Zehring *et al.*, 1988). Furthermore, photocrosslinking studies by Bartholomew *et al.* (1986) strongly suggest that only polymerase containing the IIo subunit can participate in transcription initiation at the Ad 2 ML promoter. However, recent studies by Zehring *et al.* (1988) show that certain *Drosophila* promoters are accurately utilized *in vitro* by a *Drosophila* RNA polymerase II that apparently contains only the IIb form of the largest subunit. Determination of the exact role of the C-terminal repeat in RNA polymerase II function will require further investigation.

A preinitiation complex may be assembled by incubation of Ad 2 ML promoter DNA with either partially purified transcription factors and RNA polymerase II (Davison *et al.*, 1983; Fire *et al.*, 1984; Hawley and Roeder, 1985) or with unfractionated nuclear extracts (Hawley and Roeder, 1987; Luse *et al.*, 1987). Preinitiation complex assembly at Ad 2 ML is rather slow. (The $t_{1/2}$ is about 8 min at 30°C; Hawley and Roeder, 1987.) This complex is very stable in the absence of NTPs (Cai and Luse, 1987a, b) and will begin transcription with no perceptible lag upon the addition of substrate (Hawley and Roeder, 1985; Luse *et al.*, 1987).

FORMATION OF THE FIRST PHOSPHODIESTER BOND AND ABORTIVE INITIATION

Since we were interested in the initiation process itself we decided to examine the formation of the first few phosphodiester bonds directly, by synthesizing RNA in an *in vitro* transcription system in the presence of limiting levels of NTPs. The source of our RNA polymerase II and transcription factors was a nuclear extract of HeLa cells, prepared essentially as described by Dignam *et al.* (1983). We chose as our template the major late promoter of Adenovirus 2; this extensively studied promoter is strongly utilized *in vitro* and has the further advantage (for our limiting-NTP studies) of initiating at a single base (Weil *et al.*, 1979). The DNA sequence of the RNA-like strand near the initiation site of Ad 2 ML is the following:

```
              +1
...T C C T C A C T C T C T T C C G C A T C G C T...
```

We therefore began our experiments by incubating cloned DNA containing the Ad 2 ML promoter in HeLa nuclear extracts with ATP, CTP and GTP (each at 100 μM) and very low levels (< 1μM) of ^{32}P-labelled UTP (see Coppola *et al.*, 1983 for details). When we purified the RNAs made in this reaction and electrophoresed them on 20% polyacrylamide gels we observed major bands 6, 7, 13 and 17 bases long; these products correspond to pauses before the addition of the third, fourth, fifth and sixth U residues to the growing chain. Most importantly, the RNAs generated by our "short transcription" protocol were present in ternary complexes and were not the result of termination by the RNA polymerases, since addition of excess NTPs after the original reaction chased all of the short RNAs into a large run-off transcript (Coppola *et al.*, 1983). Thus, we learned from this initial experiment that elongation commitment at Ad 2 ML must occur before position +6, since essentially no transcripts of this size were aborted. We also showed that none of the short RNAs, including the 17-mer, are capped, even though this *in vitro* system quantitatively caps longer run-off RNAs (Coppola *et al.*, 1983). Capping is therefore not a participant in the initiation/elongation commitment process.

Unfortunately, our original protocol had several drawbacks which prevented us from examining initiation directly. First, the nuclear extracts retain NTPs at the micromolar level in spite of extensive dialysis (Coppola and Luse, 1984). This substrate concentration is sufficient to support elongation past position +10 (Luse *et al.*, 1987). Second, we originally collected the phenol-extracted short RNAs by precipitation with EtOH; however, NTP-initiated RNAs four bases long or shorter do not precipitate efficiently with EtOH. Finally, in our early work we separated the RNAs by electrophoresis on sequencing-type polyacrylamide gels (*i.e.*, 20% acrylamide and 1% bis-acrylamide). We subsequently discovered that very short (< 5 bases) NTP-initiated RNAs do not resolve well on these gels from the mass of unincorporated ^{32}P-NTPs and the other rapidly migrating labeled species generated by *in vitro* transcription reactions (our unpublished observations).

In light of these difficulties we modified our experimental approach considerably. We first separated preinitiation and initiation steps by assembling preinitiation complexes and then purifying these complexes by gel filtration on Bio Gel A-1.5m (Luse *et al.*, 1987). This method eliminates almost all NTP contamination, allowing us to examine RNA synthesis one base at a time. We also utilized dinucleotide primers as substrates for initiation. It had already been shown (Samuels *et al.*, 1984) that UpC, CpA and ApC are all efficiently utilized in initiation at the Ad 2 ML promoter (note the sequence, above). Since short RNAs initiated with NpN are much less charged than those initiated with pppN, we were able to obtain better resolution of transcripts from unincorporated substrate using dinucleotide primers. Resolution was further enhanced by the use of gels containing a higher concentration of bis (3%, along with 20% acrylamide; Landick *et al.*, 1984).

With these modifications in hand we first wished to determine whether very short transcripts are retained in ternary complex. To do this (Luse *et al.*, 1987 for detailed methods) we incubated Bio Gel-purified preinitiation complexes with the following substrates: either CpA, very low (0.5 μM) levels of ^{32}P-CTP and limiting UTP or CpA and 0.5 μM ^{32}P-CTP alone. (In the former case, elongation can continue until the G residue at +11 must be added; in the latter case, only a single phosphodiester bond can be made (see the transcript sequence, above.)) A similar pair of reactions was carried out with ApC primer and labeled UTP, with or without limiting CTP. In both cases, dATP was also present to satisfy the energy requirement for transcription initiation. (Bunick *et al.* (1982) showed that ATP is needed at RNA polymerase II promoters not only as a substrate for RNA synthesis but also in a second capacity, since ATP analogs with non-cleavable β-γ bonds will not support transcription. Sawadogo and Roeder (1984) subsequently showed that this requirement must be met before ten bonds are made and that dATP can be substituted for ATP.) After incubation for 5 min with the substrates the reactions were rapidly gel filtered by centrifugation through G25 Sephadex. (We have shown that purified RNAs five bases or shorter are included in G25.) The RNAs in the excluded fraction were purified by phenol extraction, concentrated by lyophilization and electrophoresed on a 20% polyacrylamide-3% bis gel. As shown in Figure 1, we obtained a ladder of short RNAs with either primer. (Note that some elongation past position +10 (ApC-primed) or +11 (CpA primed) was observed in spite of the absence of added GTP, presumably due to residual GTP present in the other NTPs.) The minimum size of RNAs in these ladders (≥ 4 bases) shows that ternary complexes stable to brief purification are obtained after the synthesis of as few as two bonds by the RNA polymerase. Conspicuously absent from all of the lanes in Figure 1 is the trimer (CpApC or ApCpU) resulting from the addition of only one base to the primer. These results raised the possibility that CpApC and ApCpU were abortively initiated.

To investigate this we repeated the CpA-primed RNA synthesis reactions in Figure 1 and again chromatographed the reactions directly on G25; in this case we ran the columns in the conventional manner and extracted RNA from all of the fractions (see Luse and Jacob, 1987 for detailed methods). The results are shown in Figure 2. As in the experiment in Figure 1, the large majority of the short RNAs ran in the excluded fraction; however, a very strong band appeared in the included fractions with the mobility expected for CpApC (panel A, arrow). When the only substrates were CpA and CTP, the only product obtained was the putative CpApC (panel B). We extracted the RNA from this band and determined its composition by digestion with various ribonucleases; we also measured its length by HPLC in the presence of appropriate standards. The results of all of these tests confirmed that the labeled product in question is indeed CpApC (Luse and Jacob, 1987). If the CpApC is actually a transcript, its production should require a promoter and be sensitive to α-amanitin. As shown in Figure 3, essentially no CpApC was produced in the presence of α-amanitin, in the absence of any template or in the absence of the major late promoter (pUC18). Thus we can conclude that the CpApC in Figure 2 was a transcript made by RNA polymerase II. Furthermore, since no CpApC was found in the excluded fraction upon gel filtration and since CpA + ^{32}P-CTP reactions did not yield detectable labeled run-off RNA upon chase (Luse *et al.*, 1987) we can conclude that CpApC is synthesized only abortively at Ad 2 ML. It is important to emphasize that the abortive synthesis of CpApC accompanies normal RNA synthesis (panel A of Figure 2) and is not simply the result of the absence of substrates for further elongation. Similar results were obtained with ApC priming at Ad 2 ML; that is, ApCpU was only made abortively whereas essentially all longer RNAs were retained in a ternary complex (Luse and Jacob, 1987). It should be noted that in the ApC primed reaction a considerable amanitin-insensitive background is routinely observed (about 25% of the minus-amanitin signal with the same assay conditions as in Figure 3). This synthesis is not related to transcription since it is not dependent on DNA and is inhibited by reagents that do not affect transcription, as will be shown in Figure 4, below (see Luse and Jacob, 1987).

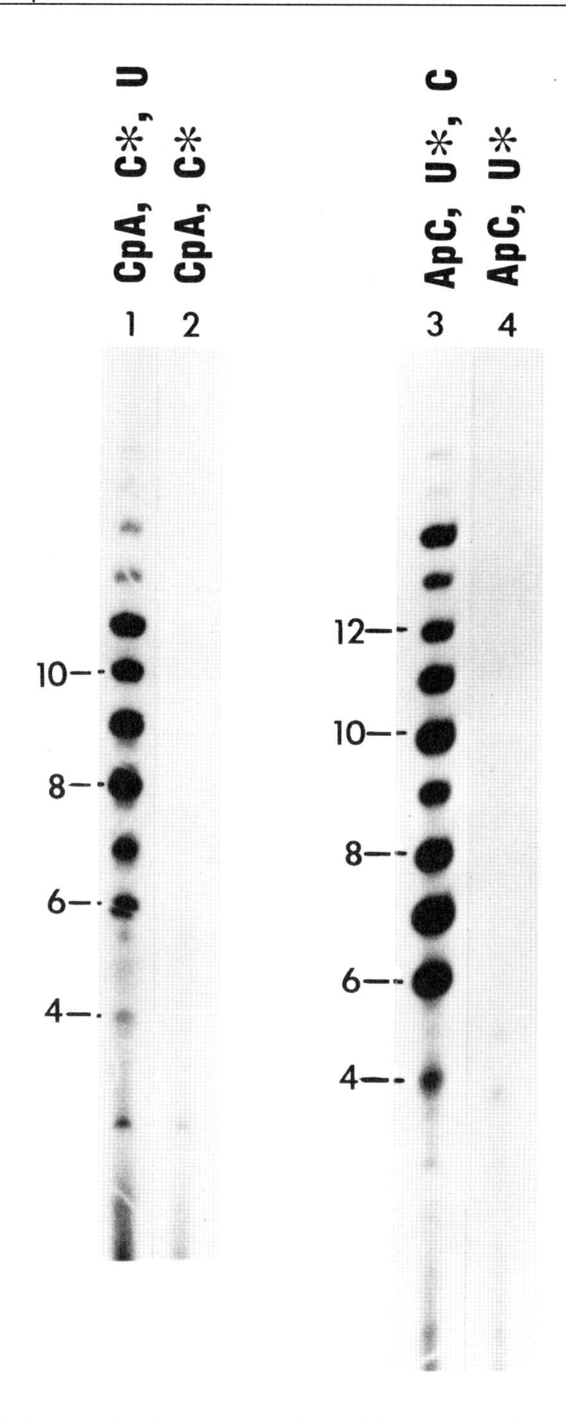

Figure 1. Absence of single-bond product in ternary complexes. RNA was synthesized by incubating preinitiation complex for 5 min with the substrates indicated in the Figure (2000 μM of the dinucleotide primer, 0.5 μM of the labeled NTP and 10 μM of the other NTP, as well as 10 μM dATP to satisfy the energy requirement). The reactions were rapidly chromatographed on G25 Sephadex. RNAs were purified from the excluded fraction and electrophoresed on a 20% polyacrylamide/3% bis gel. Numbers in the margins give the lengths of the indicated RNAs. (Reproduced from Luse *et al.*, 1987 by permission.)

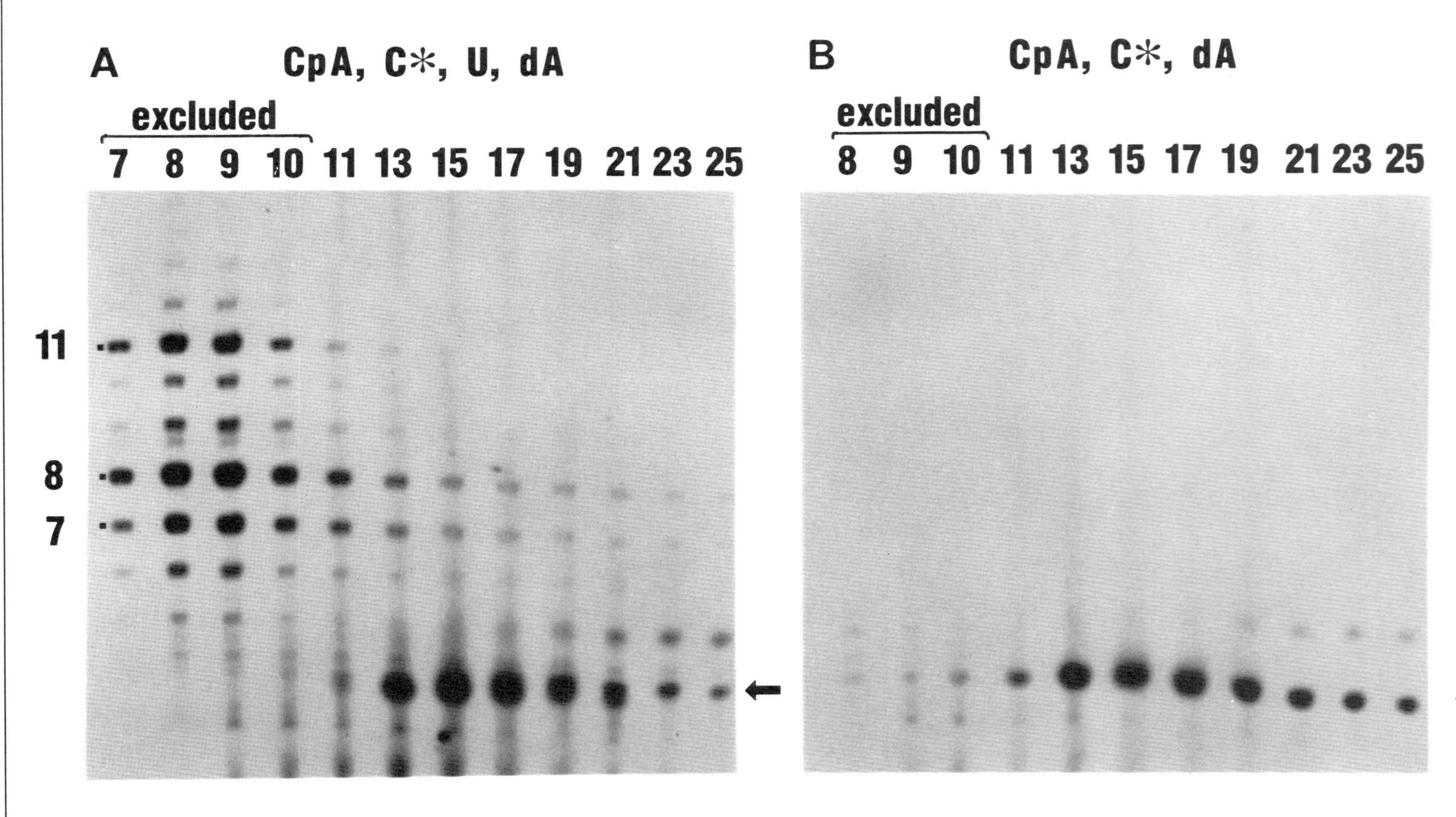

Figure 2. Fractionation of CpA-primed transcription reactions on G25 Sephadex. RNA was synthesized by incubating preinitiation complex with 2 mM CpA, 0.5 μM ^{32}P-CTP and 10 μM dATP, with or without 0.15 μM UTP, for 5 min. The reactions were then fractionated on G25; RNA was purified from each fraction and resolved by electrophoresis as in Figure 1. Fraction numbers are given above the lanes and lengths of certain RNAs are given by numbers on the left of panel A. (Reproduced from Luse and Jacob, 1987 by permission.)

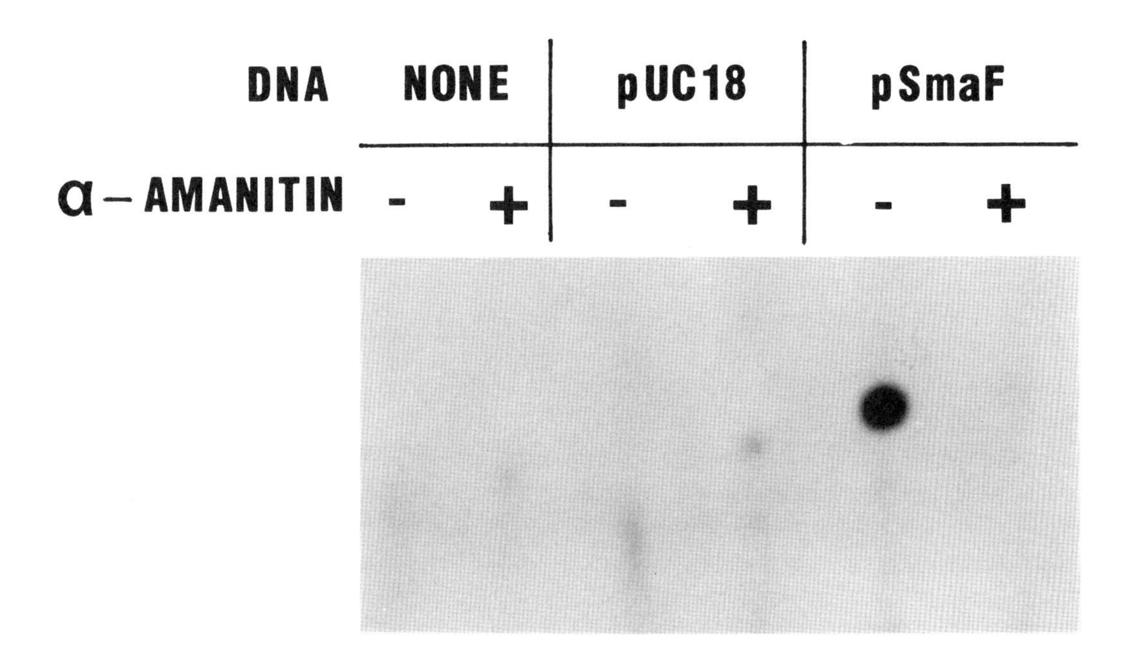

Figure 3. Amanitin sensitivity and promoter dependence of CpApC synthesis. RNA was synthesized using CpA, ^{32}P-CTP and dATP at the same concentrations as in Figure 2B. Three preinitiation complexes were used: one made with the normal Ad 2 ML promoter-containing plasmid (pSmaF), one with plasmid DNA without Ad 2 promoter (pUC18) and one without template. All reactions were fractionated on G25 columns and RNA was extracted from the included fractions, purified and electrophoresed as in Figures 1 and 2. (The mobility of the labeled product here is the same as that of the band marked with the arrow in Figure 2.) (Reproduced from Luse and Jacob, 1987, with permission.)

In prokaryotic systems the transition from abortive to productive transcription coincides with the acquisition of resistance to reagents that are "initiation-specific" (see von Hippel *et al.*, 1984; McClure, 1985). A collection of reagents known to prevent initiation without affecting elongation had also been assembled for RNA polymerase II but the exact point in the eukaryotic transcription process at which sensitivity to these reagents is lost had not been determined. It has been known for some time that elongating RNA polymerase II complexes are not disrupted by high concentrations (0.5%) of the ionic detergent sarkosyl (Gariglio *et al.*, 1974) or by high ionic strength (such as 0.2M $(NH_4)_2SO_4$; Gissinger *et al.*, 1974; Zheng *et al.*, 1987), whereas initiation cannot occur in the presence of either reagent. We had demonstrated (Coppola and Luse, 1984) that RNA polymerase II transcription complexes having nascent chains of 6 bases or longer are resistant to remarkably high levels of proteolysis (200 mg/ml proteinase K for 20 min); these complexes are also relatively resistant to inactivation by DNase I digestion (Coppola and Luse, 1984; Cai and Luse, 1987a). In light of these observations it was

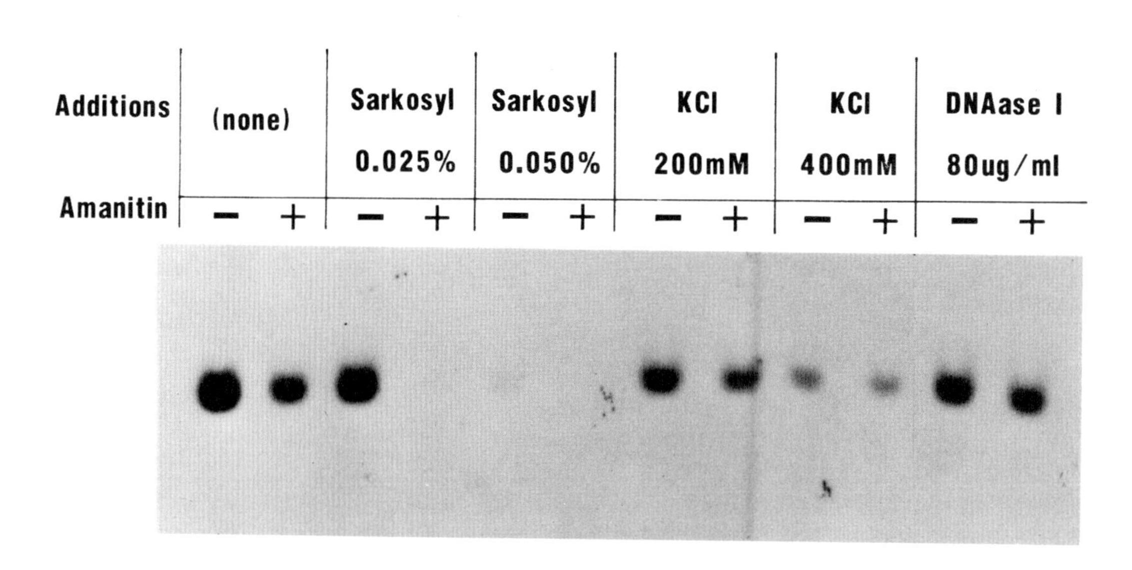

Figure 4. The effect of inhibitors on the synthesis of ApCpU. RNA was synthesized by incubating preinitiation complex with 2 mM ApC, 0.5 μM ^{32}P-UTP and 10 μM dATP, along with the additions or after the pretreatments shown in the Figure. RNAs were purified and resolved by electrophoresis as in Figure 3. (Excision of the bands and scintillation counting gave the following values for cpm above the amanitin control: no addition, 3394 cpm; 0.025% sarkosyl, 2511 cpm; 0.050% sarkosyl, 119 cpm; 200 μM KCl, 272 cpm; 400 μM KCl, 68 cpm; DNase I, 456 cpm.) (Reproduced from Luse and Jacob, 1987 with permission.)

clearly of interest to test the sensitivity of the abortive initiation reaction to reagents which do not inhibit the elongating polymerase. The results of such a test are shown in Figure 4 (for a complete description, see Luse and Jacob, 1987). It is apparent that abortive initiation is completely sensitive to high salt concentrations, to sarkosyl levels above 0.025% and to DNase digestion; it is also sensitive to very low levels of proteolysis (not shown in the Figure). These are exactly the same sensitivities displayed by the preinitiation complex when assayed by non-abortive RNA synthesis (Cai and Luse, 1987a). Thus, the "initiation step" as defined by these reagents can be the formation of the first phosphodiester bond: in addition to possibly inhibiting formation of the preinitiation complex, these treatments also inhibit synthesis of a single bond by a fully assembled preinitiation complex. It is important to mention that no amanitin-sensitive synthesis of CpApC or ApCpU occurs without dATP, indicating that the energy requirement must be fulfilled before a single phosphodiester bond is made (Luse and Jacob, 1987).

It was also of interest to determine the relative extent of abortive and productive transcription supported by the Ad 2 ML promoter. We used CpA-primed reactions for this test. The RNAs were separated by electrophoresis (as in Figure 3) and quantitated by excision and scintillation counting. The results of several measurements of this type are given in Table 1.

TABLE 1. Relative Amounts of Abortive and Productive RNA Synthesis at AD 2 ML

Expt. #1: 2 mM CpA, 0.5 μM ^{32}P-CTP, ± 0.15 μM UTP

Abortive (no UTP)	Abortive (+ UTP)	Productive
82	12	6.9

Expt. #2: 2 mM CpA, 0.5 μM ^{32}P-CTP, ± 0.15 μM UTP

Abortive (no UTP)	Abortive (+ UTP)	Productive
72	11	6.3

Expt. #3: 2 mM CpA, 0.5 μM ^{32}P-CTP, ± 50 μM UTP

Abortive (no UTP)	Abortive (+ UTP)	Productive
63	8	5.6

In all cases, the numbers given are fmole of RNA produced per 25 μl reaction. RNAs were quantitated by excision from gels and scintillation counting. Note that the first two experiments used identical conditions; they were performed using two different batches of preinitiation complex.

These data indicate that abortive initiation occurs more frequently than productive transcription, even though the abortive process is considerably reduced when elongation is possible. If we assume that the same preinitiation complexes can initiate either abortively or productively (a point which will be further discussed below), then clearly multiple rounds of abortive initiation must occur. For the first two sets of reactions in the Table, either 6-7 moles or about 2 moles of CpApC were made per mole of initiation-competent template, depending on whether elongation was possible. Furthermore, it was demonstrated in Figure 4 that abortive initiation is not inhibited by 0.025% sarkosyl, which will inhibit the assembly of new preinitiation complexes from individual components (Hawley and Roeder, 1987). Thus, the preinitiation complex must produce multiple copies of the aborted RNAs while remaining intact, as is the case with abortive initiation by prokaryotic RNA polymerase. (This point was reinforced by the demonstration that dilution of the preinitiation complex ten-fold does not reduce the level of abortive initiation, on a per template basis, as compared to the normal reaction; Luse and Jacob, 1987).

We can conclude based on the data in the above Figures (discussed more extensively in Luse *et al.*, 1987; Cai and Luse, 1987a and Luse and Jacob, 1987) that RNA polymerase II must choose between alternate pathways during initiation at the Ad 2 ML promoter: after forming a single bond, the polymerase can either abort the nascent RNA and reinitiate without disassembly of the initiation complex or it can continue elongation. Several potential objections to this conclusion seem serious enough to warrant further comment. First, abortive initiation might only occur at a hypothetical set of "defective" transcription complexes distinct from those which give rise to full-length RNAs. This seems highly unlikely since abortive initiation is strongly inhibited by productive transcription. Also, abortive and productive transcription both require an energy source and display exactly the same response to inhibitors (Figure 4; Luse and Jacob, 1987), as one might expect if the two reactions were carried out by the same complexes. Another objection to our conclusions could be based on our choice of substrate concentrations. We have used micromolar levels of the ^{32}P-NTP (the first base added) in order to improve the accuracy of our measurements, since the level of amanitin-insensitive, template-independent background increases rapidly with ^{32}P-NTP levels above 2-3 μM. (Increasing the concentration of the first NTP added to the primer to 10 μM leads to an NpNpN signal that is less than twice the background; however, the numbers of moles of NpNpN made per reaction is greater than for 1 μM NTP; Luse and Jacob, 1987). The most important substrate concentration with regard to the abortive formation of the first bond should be the concentration of the second NTP to be added. In most of our experiments we have used very low levels of this second base (typically 0.15 μM; see Figure 2) in order to keep backgrounds low and to keep the elongated RNAs short enough so that they could be resolved along with the shortest aborted products on the same gel. This strategy has clearly not resulted in general failure to retain any short RNA in ternary complex (Figure 2); however, in order to test whether abortive initiation is especially sensitive to levels of the next base added we repeated our CpA-primed short RNA synthesis reaction with 0.5 μM ^{32}P-CTP and 50 μM UTP; the results of such a test are shown in Table 1. It is clear that, while the absolute numbers are somewhat different from those obtained with much lower substrate levels, the overall result is the same. Abortive production of CpApC remains the preferred reaction even in the presence of high levels of the next NTP to be added. (We have also performed the same experiment shown in Table 1 but with 3 μM CTP and 50 μM of both UTP and GTP; the results were essentially the same as those given in the Table for experiment #3.)

A final objection to our approach involves our use of dinucleotide primers. Unfortunately, we have been unable to devise a way of resolving pppNpN products from both the unincorporated pppN precursor and other labeled, rapidly migrating material generated in our *in vitro* transcription reactions. A definitive answer to this objection should be obtained by repeating our experiments with purified transcription factors and RNA polymerase. There are, however, several points worth noting with regard to the behavior of dinucleotide primers in initiation at the Ad 2 ML promoter. First, initiation with these primers retains all of the known properties of initiation with normal substrates, including response to inhibitors and requirement for ATP or dATP; this argues against the idea that we have somehow bypassed the initiation step by using a primer already containing one phosphodiester bond. Secondly, we have indirectly examined the formation of a single bond at Ad 2 ML with ATP and CTP as substrates. We showed that no labeled elongation products can be generated by incubating preinitiation complex for 5 min with ATP and ^{32}P-CTP, followed by chase with an excess of cold CTP and limiting UTP (Luse *et al.*, 1987). Thus, either all ternary complexes containing pppApC are extremely short-lived or the pppApC is produced abortively.

FORMATION OF BONDS 2-10 AND PROMOTER CLEARANCE

Having shown that RNA polymerase II produces RNA abortively at the Ad 2 ML promoter when only one bond can be made, we next asked what would happen if only two bonds could be made. To do this we incubated Bio Gel-purified preinitiation complexes with UpC, ATP and

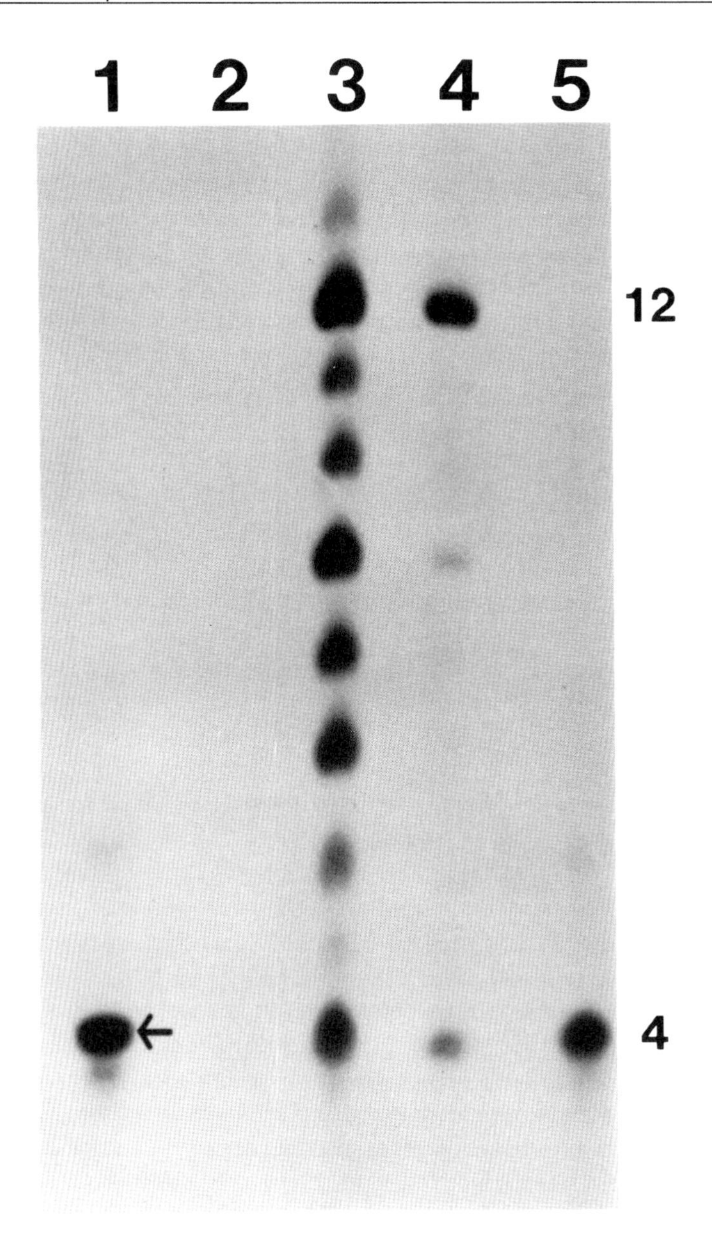

Figure 5. Presence of two-bond RNA in ternary complex. RNA was synthesized by incubating preinitiation complex with 2 mM UpC, 100 μM ATP and 0.5 μM ^{32}P-CTP for 5 min; 1 μg/ml α-amanitin was included in the reaction in lane 2. In lanes 3-5 the reactions were continued for a further 5 min, with the addition of 0.15 μM UTP (lanes 3 and 5) or 0.15 μM UTP and 100 μM CTP (lane 4); in lane 5, 1 μg/ml α-amanitin was also included in the second 5 min incubation. All reactions were fractionated on G25; RNAs from the excluded fractions were purified and resolved by electrophoresis as in Figures 1 and 2. Lengths of selected RNAs are given to the right of the Figure. (Reproduced from Luse *et al.*, 1987 with permission.)

^{32}P-CTP. As in the experiment in Figure 1, we then rapidly gel-filtered the reactions and examined the RNAs present in the excluded fractions. The results of this experiment are shown in Figure 5. In this case RNA was present in the excluded fraction (lane 1, arrow), suggesting that the synthesis of two bonds does allow the formation of a ternary complex. This RNA had the mobility expected for UpCpApC, and composition analysis with various ribonucleases confirmed that it is UpCpApC (Luse *et al.*, 1987). Its production is sensitive to α-amanitin (lane 2). Most importantly, when the original reaction was chased with excess unlabeled CTP and limiting UTP, most of the UpCpApC chased to the G-stop at +12 (lane 4); if α-amanitin was introduced along with the chase substrates, no elongation occurred (lane 5). These observations confirm that the UpCpApC was present in ternary complex and was not abortively initiated (see Luse *et al.*, 1987 for details of the experiment).

We were interested in testing the effects of the initiation inhibitors on the ternary complex which has made two phosphodiester bonds (which we refer to as "complex 2"). We therefore treated complex 2 preparations with sarkosyl, high salt, proteinase K or DNase I and then assayed for the ability of the complexes to elongate their nascent RNAs upon chase with NTPs. The results are summarized in Table 2. We found (Cai and Luse, 1987a) that complex 2 has acquired most of the properties of an elongation complex. It is completely resistant to sarkosyl (up to 2%) and is as resistant to DNase I digestion as more elongated complexes. Complex 2 was not fully resistant to all of the reagents tested, however. It is much more salt resistant than the preinitiation complex but its activity falls off above 0.5 M KCl and is almost lost at 0.8 M KCl. Complex 2 is still relatively sensitive to proteolysis but it is less sensitive than the preinitiation complex. Finally, complex 2 is rather short-lived at 25°C, with a half time for inactivation of about 15 min (Cai and Luse, 1987a) The preinitiation complex (in the *absence* of ATP- see below and Cai and Luse 1987a,b) and the more elongated post-initiation complexes are stable for hours at 25°.

We extended these tests to complexes allowed to elongate to positions 6-13 (which collection of complexes we refer to as "complex 10"). We found that these complexes, as noted in Table 2, are essentially resistant to all of the reagents tested including extensive proteolysis (Cai and Luse, 1987a). However, full elongation commitment probably does not occur until at least 10 phosphodiester bonds have been made. Although all complexes with 6 or more bases are resistant to sarkosyl, salt and proteolysis, complexes with 6 or 7 base chains are quite unstable and fail to elongate after a 5 min incubation at 25°C; complexes with 10 base RNAs are stable for hours at 25° (Cai and Luse, 1987a). The discrimination in stability as a function of length of nascent RNA is quite sharp. We have developed a procedure for extensively purifying elongation complexes by sucrose gradient sedimentation through a zone of sarkosyl (Coppola and Luse, 1984). We have recently found that sarkosyl-sucrose gradient purification of a mixture of complexes containing 10, 11, 12, 13 and 17 base RNAs does lead to recovery of ternary complexes with all lengths (10-17 bases) of RNA; however, the 10 base complexes are inactive when chased with excess NTPs whereas the ≥11 base complexes are active (Linn and Luse, unpublished observations). Thus, the addition of a single base to the growing chain can have a profound effect on the stability of the elongation complex.

CHANGES IN PROTEIN-DNA INTERACTIONS THAT ACCOMPANY INITIATION

To further extend our study of the initiation reaction we investigated changes in the interaction between the polymerase/factor complex and the template as initiation occurs. In light of our results on the consequences of forming one *vs.* two bonds we were particularly interested in how the transcription complex changes as bonds one and two are made. We began this study in a very simple way, by examining the length of run-off RNA made by various

TABLE 2. Resistance of Preinitiation and Early Elongation Complexes to Various Reagents

Complex	Sarkosyl (%)	KCL (M)	Prot. K (0.2mg/ml)	DNase I (μg/ml, 2′)	25 °C
Preinitiation	0.025	0.15	(none)	20	>5hr,-ATP 2min,+ATP
Complex 2	2.0	0.35	<1 min	80	15min,+ATP
Complex 10	2.0	0.80	20 min	80	5hr,+ATP

The values shown are the highest levels tested which did not inhibit the complex in question. All data are from Cai and Luse, 1987a.

transcription complexes which had been trimmed with DNase I and then chased with excess NTPs. We found (Coppola and Luse, 1984; Cai and Luse, 1987a) that increasing digestion of complex 10 led to progressively shorter run-off sizes until an apparent limit distribution of 30-38 bases was reached at about 100 μg/ml of DNase I for 2 min (at 25°C). Performing the same digestion and chase on complex 2 led to 20-23 base run-off RNAs (Cai and Luse, 1987a), which was the expected result since complex 10 has made 8-11 bonds more than complex 2 and the length of the complex 2 transcripts includes the two base UpC primer. When we digested the preinitiation complex with DNase, we expected to see about the same length run-off as for complex 2. However, when complex 2 and preinitiation complex were digested to the equivalent extent and chased, we found to our surprise that the run-offs from the trimmed preinitiation complex were consistently 10 bases *longer* than the chased RNAs produced from trimmed complex 2 (Cai and Luse, 1987a).

This unexpected result prompted us to investigate further the entire question of changes in template protection by the proteins in the complex as initiation proceeds. The simplest approach would have been conventional footprinting. However, this was not possible in our case since only a small fraction of the template DNA is actually assembled into transcription complex. For example, in the experiments described in Table 1, we generated between 6 and 9 fmole of short RNA, and therefore had 6-9 fmole of active transcription complex, in a reaction volume that contained about 40 fmole of template. (Poor efficiency of transcription complex assembly *in vitro* is a generally observed problem with mammalian RNA polymerase II systems; this subject is discussed in more detail in Cai and Luse, 1987b and in Luse and Jacob, 1987). Limited DNase digestion of an end-labeled DNA population of which only 25% bear transcription complexes would produce a band pattern from the protected DNA superimposed on a much stronger pattern from the protein-free DNA; clearly such a result would be very difficult to interpret. However, extensive digestion of the transcription reaction should leave a protected "core" of DNA from the complexes while the noncomplexed DNA should be digested essentially to completion. We reasoned that at least the boundaries of the region protected from

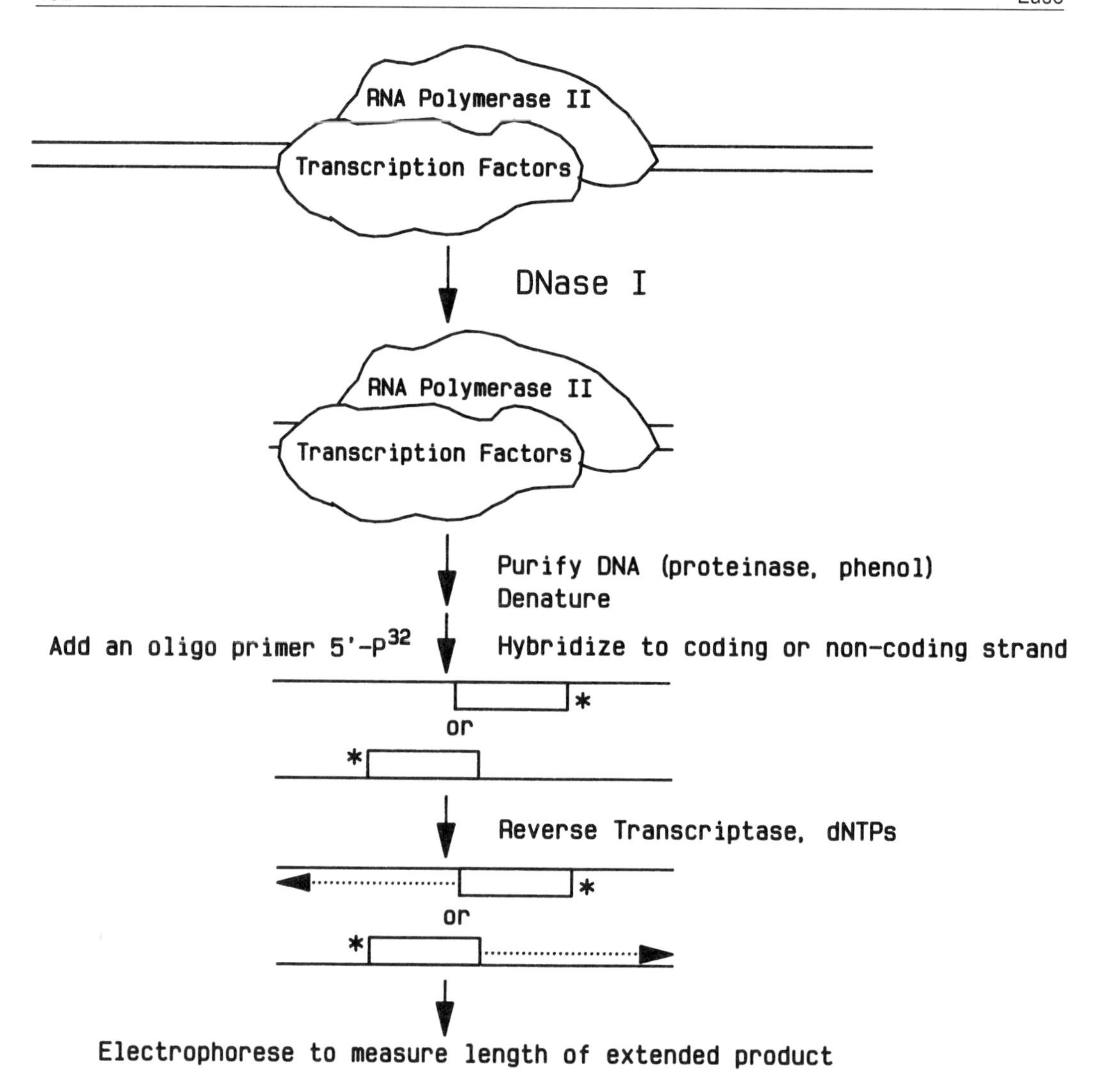

Figure 6. Assessing template protection by DNase digestion and primer extension: a schematic of the experimental approach.

DNase digestion by transcription complex could be investigated by a primer-extension approach. This technique is outlined in Figure 6. Short DNA primers, homologous to the central portion of the anticipated protected region, are synthesized and end-labeled. Transcription complexes are extensively digested with DNase I and the surviving DNAs are purified and hybridized to the primers. The hybrids are then extended with reverse transcriptase. The lengths of the extended products locate the upstream or downstream edge of protection against DNase I conferred by the complex.

Before we could actually employ this approach it was necessary to establish certain experimental parameters. We wished to digest with DNase I as extensively as possible, in order to produce the sharpest possible protection "edge" and also to reduce background (discussed in more detail below). However, disruption of the complexes (particularly the preinitiation complex) by over-digestion was also a concern. Our earlier studies on the lengths of run-off

RNAs made by trimmed complexes had shown that digestion at 80 μg/ml for 2 min gives an essentially limit digest on complexes 2 and 10. Preinitiation complex is inactivated for initiation by this level of digestion; however, preliminary tests indicated that these inactivated complexes nevertheless gave a reproducible and interpretable protection pattern (see Cai and Luse, 1987b). Thus, we decided to compare the extent of protection conferred by all complexes (preinitiation, complex 2 and complex 10) after DNase I digestion for 2 min at 80 μg/ml.

In order to examine both the upstream and the downstream edges of the transcription complex we synthesized primers that would hybridize to either strand of the probable protected region. Here I will discuss results that we obtained with two of these primers: primer 18d, bases -8 to +10 of the coding strand, and primer 20u, bases +15 to -5 of the anticoding strand. (The primer designations consist of the primer length, in bases, followed by a letter indicating whether that primer is elongated in the upstream or downstream direction.) In preliminary tests we found that digestion for 2 min with 80 μg/ml DNase I reduced the size of bulk DNA to 12-25 base pairs. We were therefore concerned that fully digested (non-protein-protected) DNA would nevertheless be large enough to hybridize to our primers and support some minimal extension. To evaluate this directly we digested (pure) template DNA for 2 min at 80 μg/ml DNase, purified the surviving fragments, hybridized them to end-labeled 18d or 20u primer and then elongated the hybrids with reverse transcriptase. The results (see Cai and Luse, 1987b) showed that the pure DNA digest supported elongation with the 18d primer to +25 (*i.e.*, an extension of 15 bases) and with the 20u primer to -15 (*i.e.*, an extension of 10 bases). Thus, to be certain we were examining only protection due to complexes we disregarded any extension products shorter than those produced by naked DNA alone.

We prepared a single batch of preinitiation complex which was either analyzed directly or was used as the source of all other complexes to be analyzed. Complexes were immediately digested with DNase I or incubated at 25°C for various periods before digestion; the surviving fragments were purified, hybridized to either the 18d or the 20u primer and treated with reverse transcriptase as described above (see Cai and Luse, 1987b). The results obtained with the 18d primer are shown in panel A of Figure 7. (Complex "zero", in the Figure, is preinitiation complex; run-off complex has been incubated with all four NTPs for 20 min before digestion.) Primer extension in the downstream direction shows protection (beyond the +25 background produced by naked DNA; see lane 20) only for the preinitiation complex (lanes 1-3) and complex 10 (lanes 6-8). Preinitiation complex protects strongly to +29 and more weakly to about +35; complex 10 protects strongly to +34 and somewhat more weakly to approximately +44. Complex 2 (lanes 4 and 5) does not protect beyond +25, in agreement with our earlier result that trimmed complex 2 produces a shorter run-off RNA than does trimmed preinitiation complex. In data not shown here, we repeated the extensions in lanes 4 and 5 using a different primer whose background extension reached only to +11. These data showed that complex 2 does protect the template out to +25 (Cai and Luse, 1987b). All of the downstream protection patterns (including the complex 2 pattern) are incomplete; that is, nuclease cleavage can occur at many points within the outside edge of the complex as indicated by the strong ladder pattern of bands. The downstream protection patterns for preinitiation complex and complex 10 did not change substantially even when these complexes were incubated at 25°C for hours before DNase digestion (lanes 3 and 8), which is consistent with the retention of transcriptional activity by these complexes upon extended incubation at this temperature (Cai and Luse, 1987a).

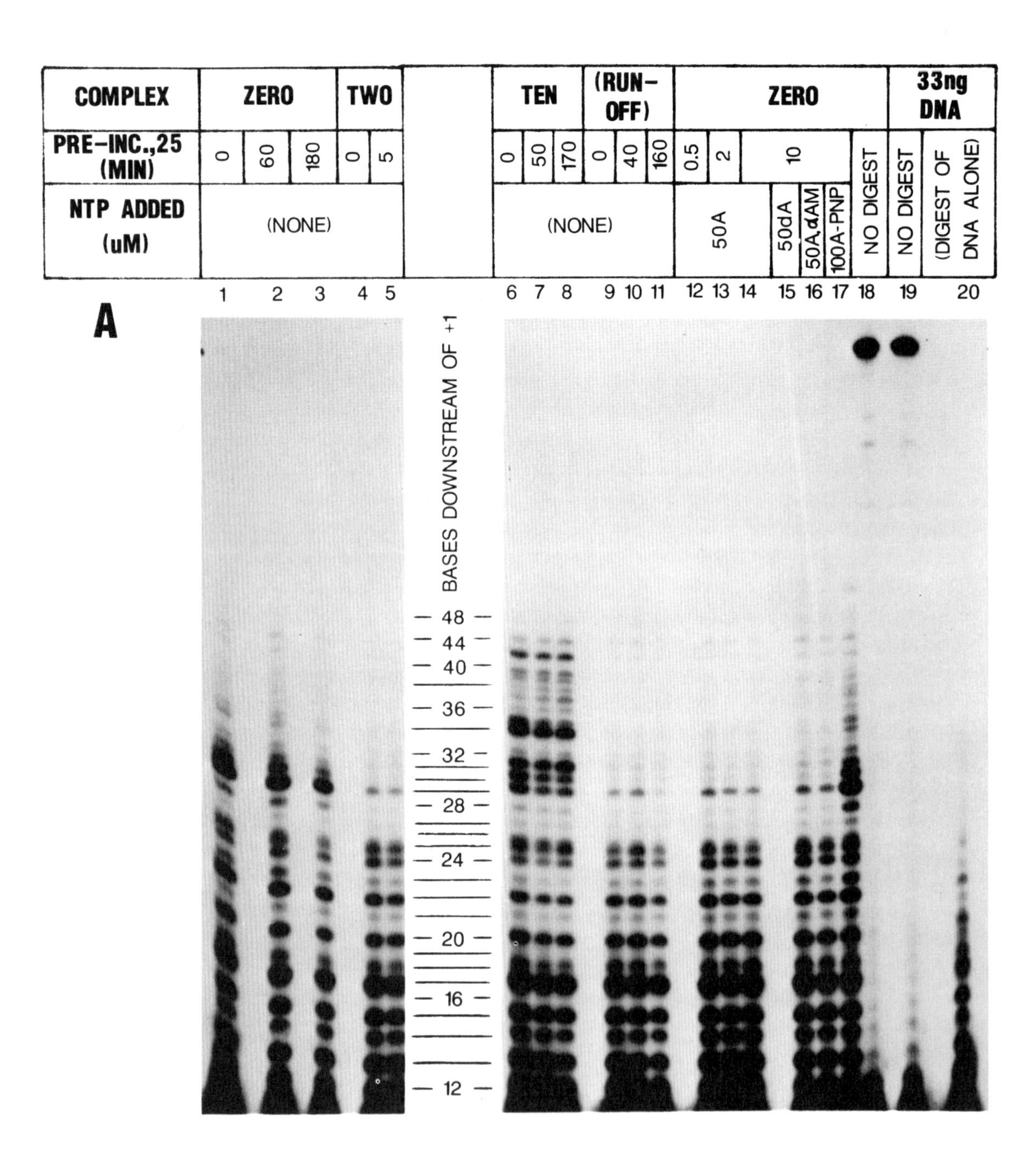

COMPLEX
ZERO
TWO
TEN
(RUN-OFF)
ZERO
33ng DNA
PRE-INC.,25 (MIN)
0
60
180
0
5
0
50
170
0
40
160
0.5
2
10
NO DIGEST
NO DIGEST
(DIGEST OF DNA ALONE)
NTP ADDED (uM)
(NONE)
(NONE)
50A
50dA
50A,dAM
100A-PNP
1 2 3 4 5
6 7 8 9 10 11 12 13 14 15 16 17 18 19 20
A
BASES DOWNSTREAM OF +1
48
44
40
36
32
28
24
20
16
12

Figure 7. Primer extension on DNA prepared from DNase I-treated transcription complexes. A single preparation of preinitiation complex was used as the source of all complexes in both panels. Samples of complexes [preinitiation complex (complex zero), complex 2, complex 10 or a complex chased with excess NTPs (run-off complex)] were preincubated at 25°C, with or without the addition of other NTPs, as indicated in the Figure; all samples were then digested with DNase I (80 μg/ml) for 2 min. The surviving DNAs were purified and hybridized to the 18d (panel A) or the 20u (panel B) primer and the hybrids were extended with reverse transcriptase as diagrammed in Figure 6.

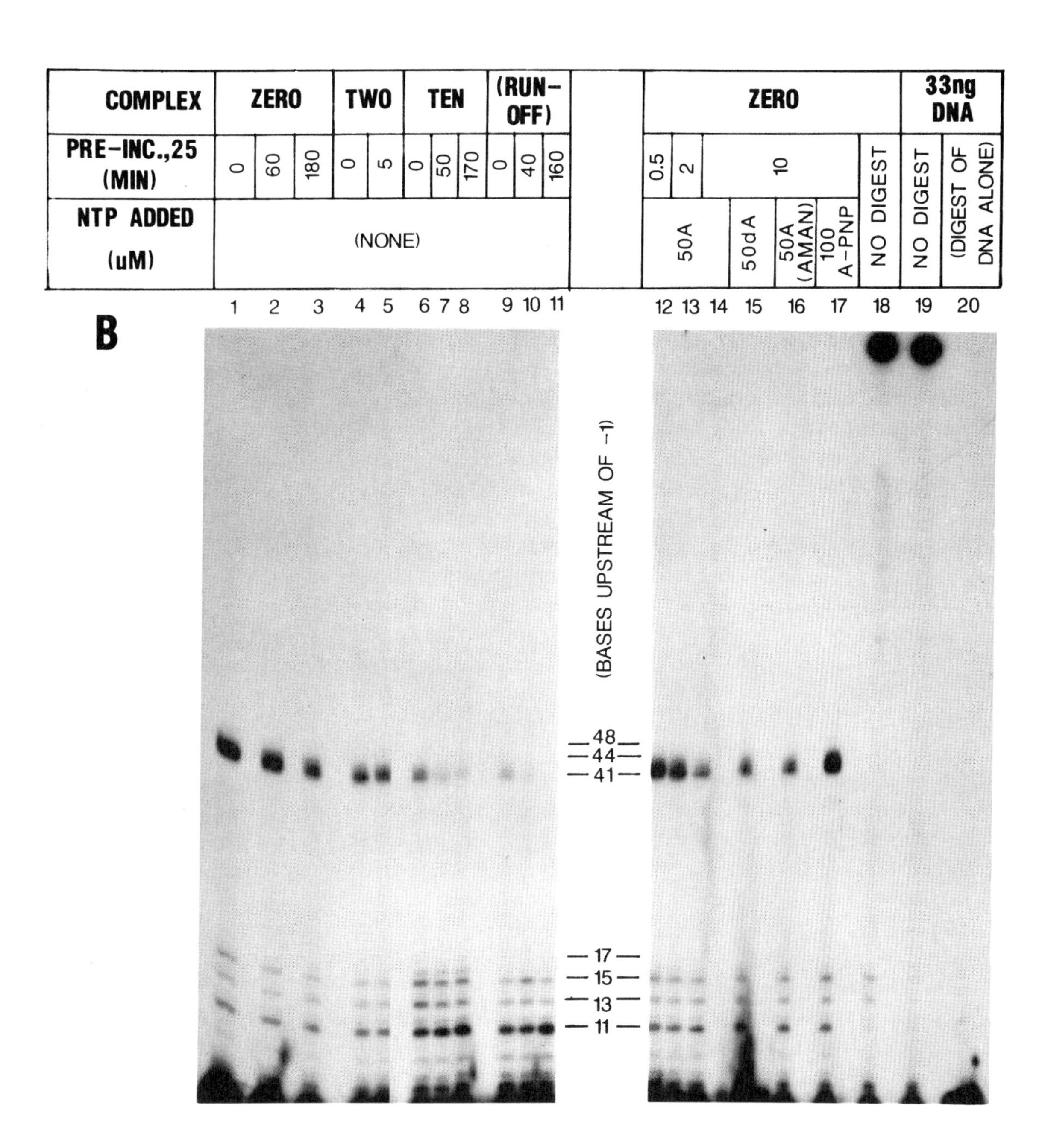

Note that one lane in each panel contained DNA purified from complex without DNase digestion (lane 18), pure (nondigested) template DNA that had never been assembled into transcription complex (lane 19), or pure template DNA that had been DNase digested at 80 mg/ml for 2 min (lane 20). The size markers were generated by extension of the primers (after hybridization to nondigested template DNA) in the presence of dideoxy NTPs. (A-PNP = β-γ imido ATP) (Reproduced from Cai and Luse, 1987b, with permission.)

A striking and unanticipated result was obtained when we examined the effect of satisfying the transcription initiation energy requirement on the downstream protection pattern. When the preinitiation complex was incubated with ATP or dATP, the same inward movement of the downstream protection boundary was obtained as is seen with complex 2 (compare lanes 4-5 and 12-16). Incubations as brief as 30 sec showed the completely shifted pattern (lane 12) and co-incubation with α-amanitin did not block the shift (lane 16). However, no shift was observed if β-γ imido ATP, which does not satisfy the energy requirement, was substituted for ATP (lane 17). Thus, the major initiation-associated change in template protection downstream of the initiation site occurs upon exposure to ATP, independent of the formation of the first bond. Since DNase digestion at 80 μg/ml for 2 min inactivates the preinitiation complex we wondered if this level of digestion also destroyed the ability of the complex to respond to ATP. When preinitiation complex was digested for 2 min, exposed to ATP and then digested for another 2 min, the downstream protection boundary extended only to +25; thus, the digested preinitiation complex remains ATP-responsive (Cai and Luse, 1987b).

When the same DNA fragment preparations used in Figure 7A were hybridized to the 20u primer and extended in the upstream direction, the results in panel B of Figure 7 were obtained. In this case the extension pattern is much simpler. All of the fragments which supported any extension yielded essentially a single band at -42. No ladder of bands was apparent below the -40 position, indicating that template protection over the upstream region (in contrast to the results in panel A) is continuous. This protection pattern is very similar to that described by Sawadogo and Roeder (1985) for the TFIID initiation factor on the Ad 2 ML promoter: a hypersensitive site for DNase I cleavage from -40 to -45 and complete resistance to nuclease cleavage from -40 to -10. Thus, we assume that the 20u extension product at -42 is indicative of the presence or absence of TFIID. The -42 band does not move downstream as transcription proceeds. Instead, the band intensity simply decays; once initiation has occurred (lanes 4-11) *or* once the energy requirement has been met (lanes 12-16) the -42 signal declines. Therefore, as in the case with downstream protection, a major change in the upstream protection pattern can occur upon satisfaction of the energy requirement. However, this change is clearly much slower than the downstream protection change. To determine the lifetime of the complex in the presence of an energy source more precisely, we incubated preinitiation complex with dATP and measured both the transcriptional activity and the reduction in the -42 extension band as a function of incubation time. Half of the upstream protection was lost in about 16 min; activity decayed even faster, with a 50% reduction in about 2 min (Cai and Luse, 1987b). Thus, destabilization of the complex by ATP or dATP involves more changes than the simple loss of TFIID. It is also worth noting that complex 10 does not lose transcriptional activity upon extended incubation at 25°C even though no TFIID remains with the complex under these conditions (lanes 7 and 8 of Figure 7B). Thus, TFIID is not required for further function of the transcription complex once ten bonds are made.

Since the behavior of the transcription complex protection patterns is somewhat unusual, one might be concerned that the protection we observed was not provided by transcription complexes at all but instead by some other DNA binding proteins. In this regard it should be noted that downstream protection, except for the inward movement caused by ATP, has the properties expected for a transcription complex: the boundary moves downstream by about ten bases when ten bases are added to the RNA chain and the protection disappears entirely when the complex has been run off (see lanes 9-11, Figures 7A and 7B). It is also worth noting that the upstream protection patterns allow the estimation of the fraction of template active in transcription, by simply comparing the intensity of extension products produced by DNase-digested complex (for example, the -42 band in lane 1, Figure 7B) with the intensity of the extension product produced by DNA extracted from the same amount of nondigested complex (lane 18 of the same Figure). If the protection that we have studied was produced by transcription complexes one would expect that the amount of DNA active in transcription, as determined by protection, should agree with the number of active templates determined by direct measurement of the amount of RNA synthesized. We made such a pair of measurements for one batch of transcription complex and found that the two estimates agreed within 5% (Cai and Luse, 1987b). Finally, if we assembled preinitiation complex in the presence of 0.025% sarkosyl, which does not allow a functional complex to form, no protection was observed, either upstream or downstream, beyond the background level conferred by naked DNA digests (Cai and Luse, 1987b).

SUMMARY

The molecular dynamics of transcription initiation by RNA polymerase II have only recently become the subject of detailed study. As reviewed above, a pathway for the assembly of the preinitiation complex has been outlined (Reinberg and Roeder, 1987; Reinberg *et al.*, 1987; Flores *et al.*, 1988 and references therein). It is clear from this work and the work of other laboratories that DNA-binding transcription factors must recognize the promoter as the first step in transcription complex assembly, followed by the entry of RNA polymerase II and other factors (some of which may join the complex in direct association with the polymerase). An investigation of the details of preinitiation complex assembly must await the more complete purification and characterization of the components involved.

Our laboratory has begun to investigate the conversion of RNA polymerase II preinitiation complexes assembled at the Ad 2 ML promoter into the elongation-competent form (Luse *et al.*, 1987; Cai and Luse, 1987a,b; Luse and Jacob, 1987). We have found that the initiation step at Ad 2 ML involves a choice between the pathway to productive RNA synthesis (that is, elongation commitment) and an abortive pathway. This choice is required only after the formation of the first phosphodiester bond; once two or more bonds are made the ternary complex is essentially elongation committed. Abortive initiations represent the majority of initiations at Ad 2 ML, even at substrate levels which support efficient elongation. The early elongation complexes (in particular, those that have nascent RNA chains of 3-7 bases) are unstable. Once ten or more bonds have been made the elongation complex is fully stable as assayed by its remarkable resistance to high ionic strength, detergent and proteolysis. We have also investigated the changes in protein-DNA interaction that accompany initiation by measuring changes in the pattern of protection of the template against DNase I digestion. We found to our surprise that the largest change in the transcription complex occurs when the energy requirement for initiation is satisfied. ATP apparently causes a major conformational change in the complex, perhaps as the result of the loss of a subunit, since the downstream boundary of template protection immediately moves inward (that is, towards the initiation site) by at least 5 base pairs. The preinitiation complex is stable for hours without ATP, but transcriptional activity is lost in a few minutes and upstream protection in the TATA box region (and by inference, binding of the TFIID factor) disappears with a half-time of about 15 min after exposure to ATP. Complex 10 incubated for extended periods at 25°C loses all upstream protection but retains transcriptional activity, consistent with our assertion that this complex is fully elongation committed.

It is important to emphasize that all of our work and the large majority of studies on initiation factors have involved a single promoter, the major late promoter of adenovirus. The Ad 2 ML promoter is a relatively simple RNA polymerase II promoter with a single initiation site. Although initiation at Ad 2 ML is stimulated by upstream elements (Sawadogo and Roeder, 1985) this promoter functions well when the only viral sequences present are the TATA box and the initiation site. It is clearly necessary to extend initiation analysis to promoters which are weak or completely dependent on upstream elements, to promoters with multiple initiation sites and to promoters which lack a TATA box.

I would like to conclude by addressing two points concerning our laboratory's recent results on the process of initiation by RNA polymerase II: what are the implications of this work for the understanding of gene regulation and what are the currently unanswered questions that can be approached with our methods? The most important aspect of our work with regard to transcriptional control is the demonstration that promoter clearance is not an automatic process during initiation by RNA polymerase II, even at a strong promoter. This raises the possibility that promoters may differ in their intrinsic efficiencies of clearance and that clearance may be affected by control proteins. Our results on the fate of the transcription complex once initiation has occurred are also important in a regulatory context. These experiments suggest that TFIID does not remain bound to the promoter indefinitely after initiation occurs. The half-life of the postinitiation TFIID-promoter complex (about 15 min) is much longer than the rate of initiation at strongly utilized promoters, which may be as rapid as once every ten seconds (see, for example, Brock and Shapiro, 1983). However, most promoters are probably not strongly utilized in the nucleus. Given that assembly of the TFIID-DNA complex is likely to be the rate-limiting step in initiation (at least at Ad 2 ML; see Hawley and Roeder, 1985), an important function of control proteins will probably be to stabilize the binding of TFIID to certain promoters. This will not only serve to increase the efficiency of transcription complex assembly at those promoters but will also help to retain those genes in a potentially active condition. Two proteins have already been described (Sawadogo and Roeder, 1985; Abmayr *et al.*, 1988) which stabilize TFIID-TATA box interaction at the Ad 2 ML promoter. It is also useful to recall the recent observation of Gilmour and Lis, who showed that RNA polymerases may be "poised" at the beginning of heat shock transcription units prior to heat shock. If the clearance processes at the hsp70 and Ad 2 ML promoters are similar, our results would suggest that the RNA polymerases poised at hsp70 must have made ten or more phosphodiester bonds.

What experiments should now be done to extend the study of initiation at RNA polymerase II promoters? Obviously it will be very important to determine to what extent promoters differ in their efficiency of clearance. It may not be generally true that weak promoters clear less efficiently than strong ones; rapid clearance could result, for example, from weak promoter contacts which are easily broken. Systematic point mutagenesis of a single promoter coupled with clearance assays may allow the identification of sequence features which especially affect clearance. At the very least, such an approach should yield clearance-limited promoters which may serve as tools to assay the effects of known regulatory proteins on the promoter clearance process. Finally, it is important to recall that kinetic analyses of promoter function using the abortive initiation reaction as a assay have been instrumental in prokaryotic systems in measuring the separate contributions of promoter binding and isomerization to the efficiency of assembly of the preinitiation complex (reviewed by McClure, 1985). The availability of an abortive initiation assay for RNA polymerase II will hopefully allow similar determinations to be made for eukaryotic promoters.

ACKNOWLEDGMENTS

I would like to thank David Price for his comments on the manuscript. I would also like to thank the current and former members of my laboratory who made major contributions to the work described here: Haini Cai, Joseph Coppola, George Jacob, Tad Kochel, Eileen Kuempel, Steve Linn and Jon Neumann. Our studies are supported by a grant from the National Institute of General Medical Sciences.

REFERENCES

Abmayr, S. M., J. L. Workman and R. G. Roeder (1988) The Pseudorabies Immediate Early Protein Stimulates *in vitro* Transcription by Facilitating TFIID:Promoter Interactions. *Genes and Devel.* **2**: 542

Allison, L. A., J. K.-C. Wong, V. D. Fitzpatrick, M. Moyle and C. J. Ingles (1988) The C-Terminal Domain of the Largest Subunit of RNA Polymerase II of *Saccharomyces cerevisiae, Drosophila melanogaster*, and Mammals: a Conserved Structure with an Essential Function. *Mol. Cell. Biol.* **8**: 321

Bartholomew, B., M. E. Dahmus and C. F. Meares (1986) RNA Contacts Subunits IIo and IIc in HeLa Cell RNA Polymerase II Transcription Complexes. *J. Biol. Chem.* **261**: 14226

Bartolomei, M. S., N. F. Halden, C. R. Cullen and J. L. Corden (1988) Genetic Analysis of the Repetitive Carboxy-Terminal Domain of the Largest Subunit of Mouse RNA Polymerase II. *Mol. Cell. Biol.* **8**: 330

Briggs, M. R., J. T. Kadonaga, S. T. Bell and R. Tjian (1986) Purification and Biochemical Characterization of the Promoter-Specific Transcription Factor, Sp1. *Science* **234**: 47

Brock, M. L. and D. J. Shapiro (1983) Estrogen Regulates the Absolute Rate of Transcription of the *Xenopus laevis* vitellogenin genes . *J. Biol. Chem.* **258**: 5449

Buc, H. and W. R. McClure (1985) Kinetics of Open Complex Formation between *Escherichia coli* RNA Polymerase and the *lac* UV5 Promoter. Evidence for a Sequential Mechanism Involving Three Steps. *Biochemistry* **24**: 2712

Bunick, D., R. Zandomeni, S. Ackerman and R. Weinmann (1982) Mechanism of RNA Polymerase II-Specific Initiation of Transcription *in vitro*: ATP Requirement and Uncapped Runoff Transcripts. *Cell* **29**: 877

Burton, Z. F., M. Killeen, M. Sopta, L. G. Ortolan and J. Greenblatt (1988) Rap30/74: A General Initiation Factor that Binds to RNA Polymerase II. *Mol. Cell. Biol.* **8**: 1602

Cadena, D. L. and M. E. Dahmus (1987) Messenger RNA Synthesis in Mammalian Cells is Catalyzed by the Phosphorylated Form of RNA Polymerase II. *J. Biol. Chem.* **262**: 12468

Cai, H. and D. S. Luse (1987a) Transcription Initiation by RNA Polymerase II *in vitro*: Properties of Preinitiation, Initiation and Elongation Complexes. *J. Biol. Chem.* **262**: 298

Cai, H. and D. S. Luse (1987b) Variations in Template Protection by the RNA Polymerase II Transcription Complex during the Initiation Process. *Mol. Cell. Biol.* **7**: 3371

Carpousis, A. J. and J. D. Gralla (1985) Interaction of RNA Polymerase with *lac* UV5 Promoter DNA during mRNA Initiation and Elongation. Footprinting, Methylation, and Rifampicin-Sensitivity Changes Accompany Transcription Initiation. *J. Mol. Biol.* **183**: 165

Coppola, J. A., A. S. Field and D. S. Luse (1983) Promoter-Proximal Pausing by RNA Polymerase II *in vitro*: Transcripts Shorter than 20 Nucleotides are not Capped. *Proc. Natl. Acad. Sci. USA* **80**: 1251

Coppola, J. A. and D. S. Luse (1984) Purification and Characterization of Ternary Complexes Containing Accurately Initiated RNA Polymerase II and Less than 20 Nucleotides of RNA. *J. Mol. Biol.* **178**: 415

Davison, B. L., J.-M. Egly, E. R. Mulvihill and P. Chambon (1983) Formation of Stable Preinitiation Complexes between Eukaryotic Class B Transcription Factors and Promoter Sequences. *Nature* **301**: 680

Dignam, J. D., R. M. Lebowitz and R. G. Roeder (1983) Accurate Transcription Initiation by RNA Polymerase II in a Soluble Extract from Mammalian Nuclei. *Nucl. Acids Res.* **11**: 1475

Fire, A., M. Samuels and P. A. Sharp (1984) Interactions between RNA Polymerase II, Factors, and Template Leading to Accurate Transcription. *J. Biol. Chem.* **259**: 2509

Flores, O., E. Maldonaldo, Z. Burton, J. Greenblatt and D. Reinberg (1988) Factors Involved in Specific Transcription by Mammalian RNA Polymerase II. RNA Polymerase II-Associating Protein 30 is an Essential Component of Transcription Factor IIF. *J. Biol. Chem.* **263**: 10812

Gariglio, P., J. Buss and M. H. Green (1974) Sarkosyl Activation of RNA Polymerase Activity in Mitotic Mouse Cells. *FEBS Letts..* **44**: 330

Gilmour, D. S. and J. T. Lis (1986) RNA Polymerase II Interacts with the Promoter Region of the Noninduced *hsp70* Gene in *Drosophila melanogaster* Cells. *Mol. Cell. Biol.* **6**: 3984

Gissinger, F., C. Kedinger and P. Chambon (1974) Animal DNA-dependent RNA polymerases. X. General Enzymatic Properties of Purified Calf Thymus RNA Polymerases A1 and B. *Biochemie* **56**: 319

Hansen, U. M. and W. R. McClure (1980) Role of the σ Subunit of *Escherichia coli* RNA Polymerase in Initiation. II. Release of σ from Ternary Complexes. *J. Biol. Chem.* **255**: 9564

Hawley, D. K. and R. G. Roeder (1985) Separation and Partial Characterization of Three Functional Steps in Transcription Initiation by Human RNA Polymerase II. *J. Biol. Chem.* **260**: 8163

Hawley, D. K. and R. G. Roeder (1987) Functional Steps in Transcription Initiation and Reinitiation from the Major Late Promoter in a HeLa Nuclear Extract. *J. Biol. Chem.* **262**: 3452

Landick, R., D. Maguire and L. C. Lutter (1984) Optimization of Polyacrylamide Gel Electrophoresis Conditions Used for Sequencing Mixed Oligodeoxynucleotides. *DNA* **3**: 413

Luse, D. S. and G. A. Jacob (1987) Abortive initiation by RNA Polymerase II *in vitro* at the Adenovirus 2 Major Late Promoter. *J. Biol. Chem.* **262**: 14990

Luse, D. S., T. Kochel, E. D. Kuempel, J. A. Coppola and H. Cai (1987) Transcription Initiation by RNA Polymerase II *in vitro*: At Least Two Nucleotides Must Be Added to Form a Stable Ternary Complex. *J. Biol. Chem.* **262**: 289

Maniatis, T., S. Goodbourn and J. A. Fischer (1987) Regulation of Inducible and Tissue-Specific Gene Expression. *Science* **236**: 1237

Manley, J. L., A. Fire, A. Cano, P. A. Sharp and M. L. Gefter (1980) DNA-Dependent Transcription of Adenovirus Genes in a Soluble Whole-Cell Extract. *Proc. Natl. Acad. Sci. USA* **77**: 3855

Matsui, T., J. Segall, P. A. Weil and R. G. Roeder (1980) Multiple Factors Required for Accurate Initiation of Transcription by Purified RNA Polymerase II. *J. Biol. Chem.* **255**: 11992

McClure, W. R. (1985) Mechanism and Control of Transcription in Prokaryotes. *Ann. Rev. Biochem.* **54**: 171

Moncollin, V., N. G. Miyamoto, X.-M. Zheng and J.-M. Egly (1986) Purification of a Factor Specific for the Upstream Element of the Adenovirus-2 Major Late Promoter. *EMBO J.* **5**: 2577

Reinberg, D. and R. G. Roeder (1987) Factors Involved in Specific Transcription by Mammalian RNA Polymerase II: Purification and Functional Analysis of Initiation Factors IIB and IIE. *J. Biol. Chem.* **262**: 3310

Reinberg, D., M. Horikoshi and R. G. Roeder (1987) Factors Involved in Specific Initiation in RNA Polymerase II: Functional Analysis of Initiation Factors IIA and IID and Identification of a New Factor Operating at Sequences Downstream of the Initiation Site. *J. Biol. Chem.* **262**: 3322

Samuels, M., A. Fire and P.A. Sharp (1984) Dinucleotide Priming of Transcription Mediated by RNA Polymerase II. *J. Biol. Chem.* **259**: 2517

Samuels, M. and P. A. Sharp (1986) Purification and Characterization of a Specific RNA Polymerase II Transcription Factor. *J. Biol. Chem.* **261**: 2003

Sawadogo, M. and R. G. Roeder (1984) Energy Requirement for Specific Transcription Initiation by the Human RNA Polymerase II System. *J. Biol. Chem.* **259**: 5321

Sawadogo, M. and R. G. Roeder (1985) Interaction of a Gene-Specific Transcription Factor with the Adenovirus Major Late Promoter Upstream of the TATA Box. *Cell* **43**: 165

Shimamato, N., T. Kamigochi and H. Utiyama (1986) Release of the σ Subunit of *Escherichia coli* DNA-dependent RNA Polymerase Depends Mainly on Time Elapsed after the Start of Initiation, Not on Length of Product RNA. *J. Biol. Chem.* **261**: 11859

Spassky, A., K. Kirkegaard and H. Buc (1985) Changes in the DNA Structure of the *lac* UV5 Promoter during Formation of an Open Complex with *Escherichia coli* RNA Polymerase. *Biochemistry* **24**: 2723

Stackhouse, T. M. and C. F. Meares (1988) Photoaffinity Labeling of *Escherichia coli* RNA Polymerase/Poly[d(A-T)] Transcription Complexes by Nascent RNA. *Biochemistry* **27**: 3038

Straney, D. C. and D. M. Crothers (1985) Intermediates in Transcription Initiation from the *E. coli lac* UV5 Promoter. *Cell* **43**: 449

Straney, D. C. and D. M. Crothers (1987) A Stressed Intermediate in the Formation of Stably Initiated RNA Chains at the *Escherichia coli lac* UV5 Promoter. *J. Mol. Biol.* **193**: 267

von Hippel, P. H., D. G. Bear, W. D. Morgan and J. McSwiggen (1984) Protein-Nucleic Acid Interactions in Transcription: A Molecular Analysis. *Ann. Rev. Biochem.* **53**: 389

Weil, P. A., D. S. Luse, J. Segall and R. G. Roeder (1979) Selective and Accurate Initiation of Transcription at the Ad 2 Major Late Promoter in a Soluble System Dependent on Purified RNA Polymerase II and DNA. *Cell* **18**: 469

Zehring, W.A., J. M. Lee, J. R. Weeks, R. S. Jokerst and A. L. Greenleaf (1988) The C-Terminal Repeat Domain of RNA Polymerase II Largest Subunit Is Essential *in vivo* but Is Not Required for Accurate Transcription Initiation *in vitro*. *Proc. Natl. Acad. Sci. USA* **85**: 3698

Zheng, X.-M., V. Moncollin, J.-M. Egly and P. Chambon (1987) A General Transcription Factor Forms a Stable Complex with RNA Polymerase B (II). *Cell* **50**: 361

17

STRUCTURAL AND FUNCTIONAL STUDIES OF *XENOPUS* TRANSCRIPTION FACTOR IIIA

Cheng-Wen Wu, Esteban E. Sierra, Thomas J. Daly, Guang-Jer Wu and Felicia Y.-H. Wu

---

INTRODUCTION

Xenopus transcription factor IIIA (TFIIIA) is a 38.5 kDa protein specifically required for the initiation of 5S rRNA synthesis by RNA polymerase III. Due to its relative abundance in immature *Xenopus* oocytes, it was the first eukaryotic transcription factor purified to homogeneity. This positive transcription factor binds to a specific DNA sequence of approximately 50 base pairs (b.p.) in the center of the 5S rRNA gene (intragenic control region, ICR). In pre-vitellogenic oocytes of *Xenopus*, TFIIIA is found in stoichiometric association with 5S rRNA as part of a 7S storage particle. In addition, TFIIIA has been found in our laboratories to possess at least four other biochemical activities: (1) it exhibits a high affinity for single-stranded (ss) DNA, (2) it promotes the reassociation of ss DNA into duplex form, (3) it is a DNA stimulated ATPase, and (4) it can bind to the flanking regions of the 5S rRNA gene in a highly ordered state. Although the precise function of these activities is not known, it has been proposed that they may be involved in the generation and maintenance of the proper state of the 5S rRNA gene transcription complex throughout multiple rounds of transcription. In this paper we review the present status of the structure and function of TFIIIA. We will focus on how the multiple activities of this protein are involved in 5S rRNA gene transcription.

THE 5S rRNA GENES IN *XENOPUS LAEVIS*

Expression of the genes encoding 5S rRNA in *Xenopus* is developmentally regulated (see review by Wolffe and Brown, 1988). There are two main classes of 5S rRNA genes : somatic and oocyte types. There are approximately 400 copies of somatic-type and 20,0000 copies of oocyte-type 5S rRNA genes per haploid genome of *Xenopus*. While both classes of genes are transcribed in oocytes, the somatic-type genes account for more than 95% of the 5S rRNA synthesis in somatic cells. Both types of genes contain similar cis-acting sequences and appear to be recognized by the same transcription factors. Understanding their developmental regulation can serve as a paradigm for the regulation of other developmentally controlled genes.

The 5S rRNA genes, as well as the tRNA genes and other small RNA genes, are transcribed by RNA polymerase III (see review by Geiduschek and Tocchini-Valentini, 1988). The distinct feature of the 5S rRNA gene is the presence of its regulatory elements, the intragenic control elements, within the structural gene. They are essential for accurate and efficient transcription initiation. The 5S rRNA genes, therefore, present an excellent system to investigate the developmental regulation and mechanisms of eukaryotic gene transcription.

TFIIIA: DISCOVERY AND PURIFICATION

TFIIIA was first described as the protein component of a ribonucleoprotein particle in pre-vitellogenic *Xenopus* oocytes (Picard and Wegnez, 1979). This particle has a sedimentation coefficient of 7S, and serves as one of two storage particles for 5S rRNA during oogenesis. When the 7S particle was first isolated, it was not known that its protein component was TFIIIA.

Taking advantage of the cellular extract prepared from unfertilized eggs, which is deficient in TFIIIA, purification of TFIIIA (Engelke *et al.* , 1980) can be monitored by a complementary *in vitro* transcription assay. The purified protein has a molecular weight (mol. wt.) of approximately 37 kDa and promotes the specific transcription of oocyte and somatic 5S rRNA genes. Furthermore, it was demonstrated (Engelke *et al.*, 1980) that this factor binds specifically to a sequence located within the 5S rRNA gene, as shown by the DNase I footprinting analysis.

TFIIIA was first identified as the protein component in 7S particles by two groups (Pelham and Brown, 1980; Honda and Roeder 1980). Furthermore, Pelham and Brown (1980) found that 5S rRNA specifically inhibited transcription of the 5S rRNA genes *in vitro*, suggesting that the 5S rRNA sequestered a component of the transcriptional machinery before it could be assembled into a transcription complex. Since 5S rRNA in oocytes is associated with proteins in the form of 7S and 42S particles, the possibility that one of the protein components of these complexes might be similar to the transcription factor isolated by Engelke *et al.* (1980) was examined. Because similar limited proteolytic patterns were found for these two proteins, they suggested that the protein in the 7S particles was identical to TFIIIA. In addition, footprint analysis demonstrated that the protein purified from 7S particles could also protect the ICR of the 5S rRNA genes, similar to the purified TFIIIA. Moreover, Honda and Roeder (1980) confirmed these results and further demonstrated the functional identity of these two proteins by showing that both can promote the transcription of 5S rRNA genes.

TFIIIA and 5S rRNA form a ribonucleoprotein complex *in vivo*, and 5S rRNA can inhibit the transcription of 5S rRNA genes. These observations led to a rather appealing hypothesis (Pelham and Brown, 1980; Honda and Roeder, 1980) in which the synthesis of 5S rRNA is auto-regulated: the increasing level of 5S rRNA in the oocyte results in the formation of 7S particles which then deplete TFIIIA and shutdown the transcription of 5S rRNA genes. It is now known that this interpretation is not as simple as it appears, because once TFIIIA is sequestered into a transcription complex, it remains stably bound to the gene, even in the presence of competitors. Furthermore, the transcriptionally repressed 5S rRNA gene complexes still contain TFIIIA (see below). Thus, the role of 7S particle formation in the regulation of transcription remains elusive.

GENERAL STRUCTURE AND AMINO ACID SEQUENCE OF TFIIIA

Significant information regarding the general structural features of TFIIIA were obtained by enzymatic and physical studies before the amino-acid sequence of TFIIIA was determined. Limited proteolysis of TFIIIA with papain and trypsin, respectively, generates 30 and 20 kDa metastable polypeptide fragments that show altered DNA binding and transcriptional activation properties as compared to the native protein (Smith *et al.*, 1984). The 30 kDa fragment protects the same DNA sequence from DNase I digestion as the native TFIIIA, while the 20 kDa fragment protects only the 3′ three fifths portion of the ICR. Transcriptionally, the 30 kDa fragment displays 20% of the activity of the native protein while the 20 kDa fragment is completely inactive. From these experiments it was concluded that TFIIIA is composed of at least three functional domains: a 20 kDa, N-terminal domain that binds to the 3′ end of the ICR, a 10 kDa domain involved in binding to the 5′ end of the ICR, and a C terminal domain that does not bind DNA but is involved in transcriptional activation, perhaps through contacts with other components of the transcriptional machinery.

There is disagreement in the literature as to the exact sizes of the proteolyzed fragments. Miller *et al.* (1985) reported that the papain digest results in a 33 kDa product, while the trypsin digestion produces a 23 kDa fragment which can be further reduced to a 17 kDa fragment, and after prolonged digestion to 6-, 4- and 3 kDa fragments. We have found (Hazuda and Wu, 1986; unpublished results) that digestion with papain results in a 33 kDa fragment that breaks down to 23 kDa. Proteolysis with trypsin, however, results in a 33 kDa fragment that breaks down to a 28 kDa fragment. Since both the native TFIIIA and the 33 kDa papain fragment have a blocked N-terminus while the 28 kDa tryptic fragment has a four amino acid peptide removed from the N-terminus, both fragments are localized in the amino proximal region of the protein. A very recent paper (Xing and Worcel, 1989) reports the sizes of the papain and trypsin generated fragments as 34 kDa and 27 kDa, respectively, in close agreement with our observations. We believe that the sizes of the proteolytic products should be clearly established in order to ensure the validity of comparisons of results from different groups.

Physical characterization of the TFIIIA (Bieker and Roeder, 1984) suggests that it is a highly asymmetric monomer in its native form. Using analytical gel filtration the protein was found to migrate close to bovine serum albumin, with a Stokes radius of 34Å. The frictional coefficient calculated from this value is 1.53, indicative of an asymmetric molecule. If the protein assumes a prolate ellipsoid geometry, its size is calculated to be approximately 135 × 18Å. Protection of approximately 150Å of the 5S rRNA gene by one TFIIIA molecule is consistent with the highly asymmetric shape of the protein.

The amino acid sequence of TFIIIA, as deduced from the cDNA clone (Ginsberg *et al.*, 1984) failed to show any sequence homology to any of the previously determined DNA binding proteins. The most interesting feature of the sequence is the presence of a repeating motif. This motif, which may fold into a DNA-binding unit, is repeated nine times and comprises 80% of the TFIIIA sequence. The consensus sequence for each motif is:

$$(X)_5\text{-Y/F-X-C-}(X)_4\text{-C-}(X)_3\text{-Y/F-}(X)_5\text{-L-}(X)_2\text{-H-}(X)_3\text{-H}$$

where X stands for any amino acid. Based on the discovery that TFIIIA is a zinc metalloprotein (Hanas *et al.*, 1983; Miller *et al.*, 1985; see next section), this motif was designated the *zinc binding finger* (Miller *et al.*, 1985; Brown *et al.*, 1985). The conserved cysteine and histidine residues have been proposed to be involved in binding zinc while the conserved hydrophobic residues nucleate a stable secondary structure. In this proposed model, the loop between the second cysteine and the first histidine forms the DNA binding finger (Miller *et al.*, 1985).

The repeats presumably arose by gene duplication of an ancestral sequence involved in binding half a turn of DNA. Gene duplication then might have led to specialization, with each finger developing sequence-specific binding characteristics (Miller *et al.* , 1985). Structural studies of the genomic DNA encoding TFIIIA (Tso *et al.*, 1986) appear to support the gene duplication hypothesis. The first six exons of the gene correspond closely to the first six fingers and the last exon coincides with the C-terminal domain. Furthermore, finger structures have now been found in a large number of nucleic acid binding proteins (Berg, 1986), showing that this motif has been widely used during the evolution of proteins, especially those involved in eukaryotic transcriptional regulation.

In addition to the repeating motif in TFIIIA, there are some interesting differences between the finger-containing amino terminus and the C-terminus. Table 1 compares the amino acid composition of the N-terminal region (amino acids 1-276) with that of the C-terminal region of the protein (amino acids 277-344).

The most significant difference observed between the N- and C terminus is the large reduction in the number of aromatic amino acids in the latter. This may reflect an intrinsic difference in the secondary structure of the C-terminal domain from that of the N-terminal domain. The presence of aromatic amino acids only in the nucleic acid binding region may also indicates that they play a functional role, perhaps through stacking interactions, in the binding of TFIIIA to nucleic acids, in particular to RNA and ss DNA.

There is also an increase in the amount of positively charged residues in the C-terminus. This is due mostly to the first half of the C-terminus (36.4% basic between amino acids 277-309) which has a high content of arginine and lysine, resembling that of histones, although they share no significant sequence homology (Miller *et al.*, 1985). It is possible that this domain has a yet-to-be-discovered nucleic acid binding function. In fact, a very recent report (Xing and Worcel, 1989) shows that the C-terminal arm of TFIIIA is responsible for the different footprint patterns produced by the protein on somatic and oocyte 5S rRNA genes. Whether this discrimination plays a role in the regulation of transcription of the gene remains to be clarified.

Table 1. Comparison of the Amino Acid Composition of the N-Terminal and C-Terminal Regions of TFIIIA

	TFIIIA	**N-Terminus**	**C-Terminus**
% Acidic (D[a], E)	11.1	10.9	12.1
% Basic (K, R)	18.4	17.0	24.2
% Aromatic (F, W, Y)	8.7	10.5	1.5
% Hydrophobic (Aromatic + I, L, M, V)	23.6	24.6	19.7

[a] One letter amino acid code

TFIIIA IS A ZINC-METALLOPROTEIN

TFIIIA is a zinc-metalloprotein (Hanas *et al.*, 1983b). Atomic absorption measurements of 7S particles which were dialyzed against EDTA revealed the presence of two tightly bound zinc ions which are resistant to metal chelation. Dialysis of TFIIIA against EDTA or 1, 10-phenanthroline results in the removal of the two zinc ions. The resulting apoprotein can no longer bind specifically to the 5S rRNA gene, but retains the ability to interact with non-specific DNA. Upon addition of $CoCl_2$, $NiCl_2$, $MnCl_2$, or $FeCl_2$, the zinc-depleted TFIIIA fails to restore the specific DNA binding activity, but regains this activity only in the presence of exogenous zinc ions (Hanas *et al.*, 1983b) .

Measurements of the zinc content of 7S particles, purified with a different protocol, found 7 and 11 zinc ions per 7S particle in the absence or presence of exogenously added zinc, respectively (Miller *et al.*, 1985). The discrepancy in the previously published data was attributed to the presence of DTT, a chelating agent, in the purification buffer. Since there are 9 repeating units, or fingers, in TFIIIA, it was postulated that each of these mini-domains contains one zinc ion; thus each TFIIIA molecule would contain nine zinc ions. Based on the primary sequence of the repeating domain, it was hypothesized that each zinc ion could be tetrahedrally coordinated to the conserved cysteine and histidine residues of the finger (Miller *et al.*, 1985). EXAFS analysis (Diakun *et al.* , 1986) showed that 2 sulfur and 2 nitrogen atoms are the ligands for the zinc in TFIIIA.

The data concerning the stoichiometry of zinc ions in TFIIIA remains a matter of contention. We believe that it is important to distinguish between tightly and loosely bound zinc ions. Only two zinc ions seem to be tightly bound and become inaccessible to chelation in the 7S particles; the rest of the zinc ions appear to be easily removed by chelation. These two tightly bound zinc in free TFIIIA, however, are easily removed by EDTA or 1, 10 phenanthroline. The resulting apoprotein is inactive in binding specifically to the 5S rRNA genes. We have tested whether it is possible to increase the specific DNA binding activity of TFIIIA by the addition of exogenous zinc. Our data indicate that 2 zinc ions per TFIIIA is all that is required for maximal binding activity (Shang and Wu, unpublished). It is likely that a similar content of zinc ions is required for the transcriptional activity of TFIIIA, which is readily testable. We cannot rule out, however, that the additional zinc ions, which are not essential for specific DNA binding, may stabilize the native conformation of TFIIIA.

Recently, we examined the stoichiometry of zinc ions in TFIIIA by chemical modification of cysteine and histidine residues with p-hydroxymercuriphenylsulfonate and diethyl pyrocarbonate, respectively (Shang *et al.*, 1989). The results indicate that two zinc ions are released from one protein molecule, concomitantly with the modification reactions, regardless of the presence or absence of denaturing agents.

Preliminary circular dichroism (CD) studies of TFIIIA show that the protein is in a predominantly alpha helical conformation (50%, Huang *et al.*, 1989). Interestingly, the CD spectrum of the apoprotein resembles that of the native protein, suggesting that the apparent secondary structure of the native and apo-TFIIIA are similar. However, it is still possible that zinc may be required for the proper folding of the newly synthesized protein.

Evidence for the role of zinc in finger-folding comes from studies where a single finger was generated through the use of recombinant DNA technology (Frankel *et al.*, 1987). The purified peptide, corresponding to the second finger in the TFIIIA sequence, shows a pronounced change in its CD spectrum when it is incubated in the presence of zinc. Also, the zinc-containing peptide shows protection against proteolytic cleavage, analogous to what we have observed in intact TFIIIA. Limited proteolysis of the zinc-depleted TFIIIA results in greatly enhanced degradation by a variety of proteases, as compared to that of the native protein (Hanas *et al.*, 1989; Sierra and Wu, unpublished observation). These results support the notion that zinc plays a role in the stabilizing the structure of TFIIIA and providing specific recognition of DNA sequences by the protein. Since removal of zinc does not change the secondary structure of TFIIIA appreciably, we suggest that zinc may play a direct role in the recognition of the 5S rRNA gene sequence.

NUCLEIC ACID BINDING PROPERTIES OF TFIIIA

Binding to the SS rRNA Genes

The ability of TFIIIA to bind in a sequence-specific manner to the regulatory region of the 5S rRNA genes is a major role of this protein. Early studies (Pelham and Brown, 1980; Engelke *et al.*, 1980) demonstrated that binding of TFIIIA to the 5S rRNA gene protects b.p. +45 to +96 from DNase I digestion, suggesting a novel intragenic binding by regulatory proteins. This finding was consistent with the deletion mutation analysis of the 5S rRNA gene (Sakonju *et al.*, 1980; Bogenhagen *et al.*, 1980) which demonstrated the absolute requirement for a region extending from b.p. +50 to +83 for an accurate initiation of the 5S rRNA transcription. Deletions from the 5′ end of the gene result in transcribed genes initiated approximately 50 b.p. upstream from base pair +50. This suggests that the ICR acts as a molecular ruler that determines the start site of transcription.

Studies using chemical methylation and ethylation of the 5S rRNA gene demonstrated that the strongest contact points between the protein and the gene are at the 3′ end of the gene, on the non-coding strand (Sakonju and Brown, 1982). Methylation of any of 8 guanosine residues clustered around the 3′ end of the ICR, or the ethylation of their surrounding phosphates, results in significant reduction in TFIIIA binding. Together with the evidence that the mutants deleted the 5′-portion of the ICR do not reduce TFIIIA binding significantly, this indicates that TFIIIA binds strongest to the 3′ end of the ICR.

Fine mapping using point mutations and oligonucleotide directed mutagenesis has further defined three major functional elements in the ICR (Pieler *et al.*, 1985a and b, 1987; Bogenhagen 1985). These have been designated as box A, the intermediate element and box C, for the nucleotides +50 - +64, +67 - +72, and +80 - +97, respectively. The latter two regions are specific for 5S rRNA genes and appear to be the main binding sequences for TFIIIA. The location of these two regions correlates with the strong contact points, which were determined from methylation and ethylation interference as described above. Box A is a common sequence for all class III promoters and seems to be involved in TFIIIC binding. However, some mutations in this region have a marked effect on TFIIIA binding. For example, a linker scanner substitution between +45 and +57 b.p. results in decreasing protection from DNase digestion of the 5′ end of the gene (Sands and Bogenhagen, 1987), indicating the loss of specific protein DNA contacting sequences. Mutations in the spacer regions (nucleotides +65 to +66 and +75 to +79) have no discernible effect on TFIIIA binding; however, insertions or deletions that alter the spacing between the boxes impair the TFIIIA binding.

Structural analysis of the 5S rRNA gene, based on its sensitivity to DNase I, has revealed a structural periodicity approximately every 5 b.p. (Rhodes and Klug, 1986). This repeat is found only in the ICR and seems to be determined by a tendency of guanine residues, singly, in pairs or in triplets, to repeat every 5 residues. Since there are nine fingers in the TFIIIA, it was postulated that each finger tip binds approximately 5 b.p. and that the structural periodicity of the DNA is recognized by each finger.

Studies of the 5S rRNA gene-TFIIIA interactions using DNase I, DNase II, micrococcal nuclease and dimethyl sulfate have revealed nine patches of protein-DNA contacts within the ICR, each one approximately 5 b.p. apart (Fairall *et al.*, 1986). These studies also demonstrated that the strength of the interaction between the protein and the DNA is variable, with the weaker contacts at the 5′ end and the stronger ones at the 3′ end. Based on the protection patterns observed, a model for TFIIIA binding was proposed in which the protein lies on one side of the DNA helix, with each finger projecting into the major groove at alternating planes every 5 b.p., acting like a cradle for the DNA (Fairall *et al.*, 1986).

Evidence against this model comes from studies of the binding of TFIIIA-deletion mutants to the 5S rRNA gene (Vrana *et al.*, 1988), described as follows: TFIIIA deletion mutants, containing progressive truncations from both ends, were analyzed for their ability to interact specifically with the 5s rRNA gene using DNase I and hydroxy radical footprinting techniques. N-terminal deletion mutants that remove part of the first finger to half of the second finger result in localized loss of 3 contacts, as well as a drastically reduced affinity of the protein for the gene. Further deletions of the second finger completely abolish specific DNA binding. Deletions from the C terminus are more gradual in their effects, resulting in the progressive loss of 5′ contacts, but not as severe as the effect on affinity observed for the N-terminal deletion mutants. The overall picture which emerges from this study is that each finger protects a specific region in the DNA, but the C-terminal fingers appear to make a major contribution to the extent of the footprint, and the first two fingers provide the largest contribution to the binding energy. The pattern of interaction of the fingers with the DNA, however, is clustered in three regions that closely resemble those defined by using mutant genes to map the interaction of TFIIIA with the gene (Pieler *et al.*, 1987a and b). Fingers 8 and 9 interact primarily with box A, fingers 5 to 7 with the intermediate element, and fingers 1 and 2 with box C. No information is yet available for fingers 3 and 4. Thus, data from DNase I and hydroxy radical footprinting do not support a uniform and equal binding of each finger to the ICR every 5 b.p..

Several studies have been performed to quantitatively determine the interaction of TFIIIA with the ICR of the 5S rRNA gene. DNase I protection experiments, as well as footprint competition assays, have been used to determine the equilibrium binding constant of TFIIIA to the ICR. Hanas *et al.* (1983a) directly determined the equilibrium dissociation constant (K_d) to be 1 nM for the TFIIIA-5S rRNA gene interaction. Under stoichiometric conditions they also observed some degree of cooperative binding and proposed that two TFIIIA molecules were responsible for the complete protection of the ICR from DNase I digestion. Since there is disagreement among different groups as to the stoichiometry of binding of TFIIIA to the 5S rRNA gene (see Smith *et al.*, 1984; Bieker and Roeder, 1984), we have recently addressed this question (Daly and Wu, submitted). Quantitation of the TFIIIA-ICR stoichiometry was achieved through mobility shift experiments using differentially labeled TFIIIA and an ICR containing fragment. Titration of this fragment with TFIIIA resulted in the formation of several complexes; the smallest (and most prevalent) corresponds to a stoichiometry of 1:1. This complex generates a footprint that protects the ICR from position +97 to at least position +58. Whether TFIIIA is able to protect the remainder of this fragment is not clear at present. The observation of multiple complexes in the mobility shift experiment indicates that more than one molecule of TFIIIA is capable of interacting with the 5S rRNA gene *in vitro*, thus explaining our previous results (Hanas *et al.*, 1983a). The major complex formed under the conditions of the study, however, consists of one molecule of TFIIIA per 5S rRNA gene.

Formation of Alternating Patterns of Protection (APP)

TFIIIA can interact with the flanking regions of the ICR in a unique and highly ordered state (Windsor *et al.*, 1988). Footprint titrations, using a 5S rRNA gene containing DNA fragment, show full footprints at a TFIIIA concentration of 5 nM. As the protein concentration is increased to 50 nM or greater, a dramatic change in the binding pattern is observed. At concentrations higher than these the extent of the protection within the ICR is reduced to about only 25 b.p. in the 3′ end. This is accompanied by the appearance of an alternating pattern of protection (APP) every 10 b.p., extending from both sides of the ICR. The periodicity of the repeat suggests that TFIIIA binds only to one side of the helix, perhaps in a manner analogous to histones. Interestingly, the C-terminal domain of TFIIIA is required for the formation of this structure, since neither the 33 kDa fragment, nor the 28 kDa fragment is capable of promoting its formation. This result suggests that strong, cooperative, protein-protein interactions between TFIIIA molecules are mediated by the C-terminal domain, and are required for the formation of the unique APP structure. The APP is reversible by the addition of non-specific DNA. However, this does not imply that the APP results from a non-specific type of interaction, since incubation of non-specific DNA with TFIIIA results in random, non-specific binding. The presence of ICR specifically bound by the TFIIIA is required for the formation of this highly ordered binding structure. Thus, the ICR seems to serve as a nucleation center in this reaction. Interestingly apo-TFIIIA, which is unable to bind to the 5S rRNA gene specifically, is still able to generate the APP. One possibility is that the ICR may contain structural elements which are recognized by the apo TFIIIA.

The concentration of TFIIIA during the early stages of development is much higher than 50 nM; this suggest that APP formation can take place *in vivo*. It has been proposed (Windsor *et al.*, 1988) that polymerization of TFIIIA can be used to direct RNA polymerase III to its entry site, approximately 60-80 b.p. upstream from the ICR, thus providing a mechanism for action-at a-distance. This hypothesis is being investigated.

Non-specific Binding to Double-Stranded DNA

Currently, the ability of TFIIIA to distinguish specific from non-specific DNA receives little attention. Measurement of the non-specific DNA binding to TFIIIA by footprint competition analysis (Hanas *et al.*, 1984b) showed that a concentration of non-specific plasmid DNA of 1×10^{-8} M was required to successfully compete with TFIIIA from a 5S rRNA gene-containing DNA fragment. The assumption of 4000 non-specific binding sites per plasmid DNA sets the K_d for the non-specific DNA in the range of 5×10^{-5} M. However, our direct measurement of non-specific DNA binding by nitrocellulose filter binding techniques shows that the apparent K_d is approximately 5 nM, and little difference exists between specific and non-specific apparent binding affinities (Hanas *et al.*, 1983b). This is presumably due to the presence of strongly cooperative protein-protein interactions in DNA binding by TFIIIA. This TFIIIA binding activity is currently under study. Initial studies suggest a cooperativity value (ω, McGhee and von Hippel, 1974) of at least 1,000 and perhaps as high as 5,000. Although it is unclear what the *in vivo* role of these interactions would be in the immature oocyte, several *in vitro* activities of TFIIIA seem to require protein-protein interactions. These include the ability to reassociate complementary, ss DNA and the formation of alternating patterns of protection on 5S rRNA genes. These activities are all affected by the removal of the carboxy-terminal domain of TFIIIA through limited proteolysis with trypsin. We have shown (Daly and Wu, submitted) that trypsin digestion of TFIIIA correlates with the loss of the ability of the protein to aggregate. It is possible that protein-protein interactions are an intrinsic property of TFIIIA necessary for the maintenance of stable transcription complexes.

Binding to 5S rRNA

TFIIIA is found in the immature oocyte as part of a ribonucleoprotein complex, in a 1:1 stoichiometry with 5S rRNA. This complex is referred to as the 7S particle, on the basis of its sedimentation behavior (Picard and Wagnez, 1979). Major questions about this interaction include: what is the binding affinity of TFIIIA for 5S rRNA? What is the physico-chemical basis for this interaction? How do the properties of this complex determine its function *in vivo*? Our understanding of the TFIIIA-5S rRNA interaction is limited by our ignorance of the structure of 5S rRNA in solution. Interpretation of the results is typically based on a model where the 121 bases of 5S rRNA fold into a structure containing five ds helices and five ss loops (Fox, 1985; Christiansen *et al.*, 1987).

When the binding affinity of TFIIIA for 5S rRNA is measured using a footprint competition assay, an overall binding constant of 3×10^{-8} M is observed (Hanas *et al.*, 1984b). The affinity is similar for *Xenopus*, wheat germ and yeast 5S rRNAs, while no significant binding is observed when either *E. coli* 5S rRNA or wheat germ tRNA are used as competing ligands.

Direct measurements of the TFIIIA-5S rRNA interaction by nitrocellulose filter binding yields significantly different results (Romaniuk, 1985). Tighter binding (an affinity of 2×10^{-9} M) is observed for the interaction of TFIIIA and *Xenopus* 5S rRNA. Furthermore, large differences in the affinities for different RNA species, ranging from 3 × 10^{-10} M for wheat germ 5S rRNA to 1 × 10^{-7} M for yeast $tRNA^{phe}$, are detected. These results should be interpreted cautiously. The binding isotherms from which these values are calculated appear to be sigmoidal in shape, indicative of cooperative interactions. However, there should be no source of cooperativity for a 1:1 stoichiometric complex such as the 7S particle. The shapes of these curves, therefore, suggest that aspects other than simple binding are being observed. The observation of two different dissociation rates also suggests that at least two different phenomena are being measured.

Characterization of the binding reaction (Romaniuk, 1985) shows that TFIIIA has a broad pH range for 5S rRNA binding activity, with similar binding constants observed between pH 6 and 8; above pH 8 the binding is reduced. Variations in the ionic strength of the binding buffer revealed that the TFIIIA-5S rRNA complex is surprisingly non-ionic in nature. At an ionic strength comparable to the physiological conditions, approximately 68% of the free energy of the interaction is derived from non-electrostatic contacts.

Comparisons of chemically and enzymatically cleaved 5S rRNA, complexed and uncomplexed with TFIIIA, have helped in defining the regions of the 5S rRNA that interact closely with TFIIIA. These experiments have shown that TFIIIA interacts with nucleotides +53 to +116 of the 5S rRNA (Anderson *et al.*, 1984; Huber and Wool, 1986; Christiansen *et al.*, 1987). In addition, helices IV and V undergo a conformational change upon interacting with TFIIIA (Anderson *et al.*, 1984). Thus, TFIIIA makes extensive contacts with the 5S rRNA molecule in a region that includes and extends beyond those residues contacted on the 5S rRNA gene.

Another approach to defining the regions of the 5S rRNA molecule important for interaction with TFIIIA is the creation of "mutant" 5S rRNA molecules (Romaniuk *et al.*, 1987). Using truncated and chimeric 5S rRNA molecules, it has been shown that deletions from the 5′ end of the molecule result in a greater decrease in the binding affinity than those from the 3′ end. Deletions that alter the ability of the 5S rRNA to fold into stable helices or loops greatly decrease the affinity of TFIIIA for the molecule. TFIIIA was also shown to have a higher affinity for somatic 5S rRNA than oocyte 5S rRNA. Chimeric molecules constructed from the 5′ half of somatic 5S rRNA and the 3′ half of oocyte 5S rRNA were observed to bind with a greater affinity than the converse chimera, suggesting that the determinants for the increased affinity for somatic 5S rRNA are located in the 5′ end of the molecule.

It is important to compare the binding of TFIIIA to 5S rRNA and to the 5S rRNA gene in order to understand the elements that determine the specificities of the interaction. For this purpose, RNA made from linker substitution mutants of the 5S rRNA gene were compared with the mutant genes in their ability to bind TFIIIA (Sands and Bogenhagen, 1987). These experiments showed that there is little correlation between the sequences required for specific interactions with the 5S rRNA and the 5S genes.

A basic problem in the studies using mutant 5S rRNAs is that it is not always possible to directly relate the interaction to an altered sequence which may also have an altered secondary structure of the RNA. However, it seems that secondary structure is very important for a high affinity and proper interaction with TFIIIA.

There is little information available about what regions of TFIIIA are involved in the interaction with 5S rRNA. Since the TFIIIA seems to have only one nucleic acid binding site (Hanas *et al.*, 1984b), it is very likely that the RNA binding site overlaps with the DNA binding site. Data from limited proteolysis shows that the 28 kDa, trypsin generated fragment can still bind to the 5S rRNA; so does a smaller, 17 kDa, breakdown product. Whether these trypsin-derived peptides contain all of the 5S rRNA binding domains present in the 7S particle is not known.

Binding of Single-Stranded DNA by TFIIIA

One of the most intriguing properties of TFIIIA is its non-specific, high affinity interaction with ss DNA (Hanas *et al.*, 1984b). The strength of this interaction was studied using a footprint competition assay. Single-stranded DNA from M13 was found to be a surprisingly effective competitor for the specific binding of TFIIIA to the 5S rRNA gene. The calculated K_d was, at most, 1×10^{-7} M; this is at least two orders of magnitude higher than that for non-specific, ds DNA. The interaction of TFIIIA with ss DNA seems to be non-specific in nature; this assay could not distinguish between the affinities of TFIIIA for the coding and non-coding strands of the 5S rRNA gene and non-specific, ss DNA fragment. The effectiveness of the ss DNA in competing for TFIIIA binding is dependent on the size of the DNA, with the larger fragments having a higher binding affinity to the TFIIIA. This strongly suggests that positively cooperative protein-protein interactions may be involved. Further evidence of the presence of protein-protein interactions comes from studies on binding of the proteolytic fragments of TFIIIA to ss DNA (Sierra and Wu, unpublished). The 28 kDa, trypsin-generated fragment, impaired in its ability to establish protein-protein contacts, shows very little affinity for ss DNA. We have suggested that the ss DNA binding ability of TFIIIA may be important for its stable association with the 5S rRNA gene during transcription (Hanas *et al.*, 1984b). Since TFIIIA makes its strongest contacts with the non-coding strand of the 5S rRNA genes (Sakonju and Brown, 1982), it is possible that during the passage of the RNA polymerase, TFIIIA remains bound to this strand by virtue of its strong contact points and its affinity to ss DNA.

Binding of ss DNA is potentially important for the transcription of other RNA polymerase III genes. The presence of non-specific, ss DNA interferes with the binding of a yeast transcription factor to the box A of a tRNA gene (Stillman *et al.*, 1985). It is possible that transcription of tRNA genes requires a transcription factor which serves a similar function as TFIIIA in remaining stably bound to ss DNA during transcription.

TFIIIA: ITS ROLE IN TRANSCRIPTION COMPLEX FORMATION

Chromatin isolated from *Xenopus* oocytes was shown to support specific transcription of 5S rRNA genes if supplemented with exogenous RNA polymerase III (Parker and Roeder, 1977) . This suggested that the proteins required for the transcriptional activation of the genes are associated with chromatin.

The indication that transcription complexes formed from 5S rRNA genes, as well as other genes transcribed by RNA polymerase III, remain stably associated with the gene during multiple rounds of transcription came from work using nuclear extract from *Xenopus* oocytes for the specific transcription of cloned 5S rRNA genes (Bogenhagen *et. al.*, 1982). The stability of these transcription complexes was assessed by template exclusion experiments: pre-incubation of a gene with limiting amounts of extract prevents from transcription of a second gene added subsequently. It suggests that once a transcription complex is formed, its components remain associated with the gene during multiple rounds of transcription, thus preventing their exchange with other genes. TFIIIA is required for the formation of these stable complexes on 5S rRNA genes since template exclusion is not observed in extracts that have been immunologically depleted of TFIIIA. Furthermore, chromatin retains the developmental programming of the cells from which it is prepared. Chromatin isolated from somatic cells retains the inactivation of the oocyte-type 5S rRNA genes, while chromatin isolated from germinal vesicles supports the transcription of both oocyte and somatic types of genes. The factors that determine the expression and developmental control of the 5S rRNA genes form part of the chromatin, not of the extracts.

The question of which factors, or fractions, required for transcription are necessary for the formation of stable complexes has been addressed using fractionated extracts from human cells (Lassar *et al.*, 1983). Template exclusion experiments were used to demonstrate that TFIIIA and fraction IIIC are both necessary and sufficient for the formation of a stable transcription complex on the 5S rRNA genes. These factors seem to act together since each one can be independently made rate limiting for the transcription of the second gene. It was also found that if 5S rRNA genes are pre-incubated with TFIIIA and then added to a reaction containing a second gene and the rest of the components of the transcriptional machinery, the 5S rRNA gene is preferentially transcribed. This suggests that TFIIIA forms metastable complexes with the 5S rRNA genes. The transient nature of this complex was demonstrated by showing that if a second gene is added, before the addition of the other factors, there is sufficient dissociation of TFIIIA from the first gene to prevent its transcriptional advantage.

The mechanism by which transcription complexes remain bound during multiple rounds of transcription is not understood, but it is clear that TFIIIA plays an important role. Two hypotheses are proposed to explain this phenomenon: (a) TFIIIA, together with the whole transcription complex, remain bound to the non-coding strand (Sakonju and Brown, 1982). Support for this idea comes from the fact that TFIIIA makes its most numerous and strongest contacts with the non-coding strand of the 5S rRNA gene. (b) Passage of the polymerase is accompanied by alternative dissociation and reassociation of each finger, which keeps the protein anchored to the gene at all times (Miller *et al.*, 1985). The latter possibility appears unlikely for two reasons: (i) RNA polymerase III (mol. wt. 600 kDa) is much larger than TFIIIA, making this mechanism sterically improbable, and (ii) the formation of open complexes would favor the displacement of the transcription complex to one of the two strands. The molecular mechanism by which stable transcription complexes are maintained remains to be elucidated.

THE ROLE OF TFIIIA IN THE FORMATION OF ACTIVE CHROMATIN

The 5S rRNA genes retain their developmental programming when isolated as chromatin (Parker and Roeder, 1977; Bogenhagen *et al.*, 1982; Wormington *et al.*, 1983; Wormington and Brown, 1983). The relationships between TFIIIA binding to the 5S rRNA gene, transcription complex formation and the chromatin structure of the 5S rRNA genes have been investigated. A large part of these studies has concentrated on the effects of TFIIIA binding on active chromatin formation.

Studies of the fate of transcription complexes during chromatin assembly *in vitro* (Gottesfeld and Bloomer, 1982) have shown that formation of transcriptionally active chromatin on the 5S rRNA gene is dependent on the addition of factors present in an oocyte extract prior to nucleosome reconstitution. Addition of purified TFIIIA before nucleosome reconstitution results in the formation of transcriptionally active chromatin. Binding of TFIIIA to the 5S rRNA gene, thus, is an essential (but not sufficient) step in the formation of transcriptionally active chromatin. This is not surprising since TFIIIA binding is the first step in the formation of active transcription complexes.

Further structural characterization of TFIIIA in chromatin requires the determination of its position within the nucleosome. TFIIIA is able to form a stable complex with nucleosome core particles on a somatic 5S rRNA gene from *Xenopus borealis* (Rhodes, 1985). There is no loss of histones during the formation of the complex, nor is there any sliding of the histone octamer. The nucleosome extends from position -75 to +78. This positioning leaves the 3′ end of the ICR outside the nucleosome, allowing TFIIIA to interact with it and form a triple complex with the DNA and the nucleosome. This suggests a mechanism where partial binding of TFIIIA to the 3′ end of the ICR displaces the rest of the ICR from the surface of the nucleosome resulting in complete TFIIIA binding. H1 binding to the outside of the nucleosome could repress gene activation by preventing TFIIIA from binding to the 3′ side of the ICR (Rhodes, 1985). This may only reflect a partial picture since additional factors may affect the binding of TFIIIA to the DNA organized into nucleosomes.

The location of the nucleosome varies among different 5S rRNA genes. Studies involving nucleosome reconstitution using a somatic 5S rRNA gene from *Xenopus laevis* (Gottesfeld, 1987) showed that the reconstitution is directed by the DNA sequence. The location of the nucleosome in this gene is from position +20 to 80 b.p. downstream of the gene. Formation of this nucleosome, unlike that on the somatic gene from *Xenopus borealis*, excludes binding of TFIIIA to the 5S rRNA gene . The DNA sequences responsible for directing nucleosome positioning, in this case, lie in the 5′ flanking regions of the gene. The pattern of nucleosome positioning observed *in vitro* is the same as observed *in vivo* for the oocyte-type 5S rRNA genes (Gottesfeld and Bloomer, 1980; Young and Carroll, 1983). The pattern of nucleosome positioning for the somatic-type genes *in vivo* is not available for comparison.

Chromatin structure seems to play a role in the developmental regulation of 5S rRNA synthesis. Inactive, oocyte type 5S rRNA genes derived from somatic cell chromatin can be derepressed by treating the chromatin with 0.6 M NaCl, which removes and releases histones (Schlissel and Brown, 1984). This salt-stripped chromatin requires RNA polymerase III and all the other transcription factors necessary for transcription, indicating the absence of transcription complexes in the repressed state. Addition of histone H1 to these genes results in a selective re-repression of the oocyte type genes, even in the presence of excess TFIIIA. Since the binding of TFIIIA is required prior to the formation of active complexes, these results suggest that developmental control of the 5S rRNA genes is not determined solely by the TFIIIA occupancy.

Binding of TFIIIA to the 5S rRNA genes in chromatin is not sufficient to ensure the transcriptional state of the gene. For example, extracts made from mature oocytes can transcribe the somatic 5S rRNA genes 40 to 100 fold more efficiently than the oocyte-type genes (Millstein et. al., 1987; Peck et. al., 1987). Under conditions in which the oocyte-type 5S rRNA genes are transcriptionally inactive, TFIIIA is still associated with them , as shown by footprinting (Peck *et al.*, 1987). These results seem to suggest that additional factor(s) determine(s) the developmental regulation of the genes.

On the other hand, it was found that chromatin assembly in oocyte S-150 results in the deposition of phased nucleosomes on a *Xenopus borealis* somatic 5S rRNA gene (Shimamura *et al.*, 1988). This process inhibits transcription and is independent of the presence of histone H1. Interestingly, the positioning of the nucleosome is identical to the results obtained by

Rhodes (1985). However, footprints of the 5S gene assembled in the S-150 do not show the binding by TFIIIA. This does not rule out the possibility that TFIIIA, and/or other transcription factors, are in limiting amounts under the assay conditions. Thus, the formation of transcriptionally active chromatin may depend on the prior formation of active transcription complexes on the genes, before nucleosome deposition.

TRANSCRIPTIONAL ACTIVATION BY TFIIIA

Two TFIIIA activities, specific binding to the 5S rRNA gene and transcriptional activation, can be uncoupled. Studies with proteolytic fragments of the protein have shown that a papain-generated 33 kDa fragment can bind to the 5S rRNA gene in a manner virtually identical to the native protein, yet it retains only 20% of the transcriptional activity. The 28 kDa tryptic fragment, which retains part of the specific binding activity, is completely inactive in transcription (Smith *et al.* , 1984) . These experiments define the C-terminal domain as the transcriptional activator region of the protein. Two main mechanisms for the activation of transcription are conceivable: (a) interaction with other proteins of the transcription apparatus and/or (b) induction of changes in the structure of the DNA which favor the formation of open complexes during transcription.

Very little is known about the interaction of TFIIIA with other proteins involved in transcription because other purified components are not available. Circumstantial evidence suggests that TFIIIA may interact directly, or may be in very close contact with, TFIIIC. TFIIIA has been shown to form contacts all over the ICR, including box A (Fairall *et al.*, 1986) . The A box, however, has been shown to be the main determinant for TFIIIC binding, which in turn is dependent on prior TFIIIA binding in order to form a stable complex (Pieler *et al.*, 1987). The sharing of part of the binding site, as well as the stabilization of TFIIIC binding by TFIIIA and *vice versa*, suggest that these two proteins may interact with each other. Interestingly, the C-terminus of TFIIIA, which has been shown to be important for protein-protein interactions, is positioned close to the A-box (Vrana *et al.*, 1988; Xing and Worcel, 1989). It has been shown recently that the papain-generated fragment of TFIIIA cannot form a stable complex on a 5S rRNA gene upon the addition of TFIIIC (Hayes *et al.*, 1989). This strongly suggests that the C-terminal end of TFIIIA mediates the protein-protein interactions with TFIIIC essential for the formation of stable transcription complexes. There is no evidence, however, for direct protein-protein contacts.

TFIIIA has been examined for its ability to alter the structure of the ICR. Unwinding studies (Hanas *et al.*, 1984a; Shastry, 1986) have shown that only about one b.p. per 5S gene is unwound by TFIIIA, suggesting that this is not an important property of the protein. We have used CD to study the interaction of TFIIIA with a synthetic DNA fragment containing the sequence of the ICR (Shang, Huang, Wu and Wu, in preparation). The CD spectrum of the ICR containing fragment is characteristic of that for B-form DNA, in agreement with a previous report (Gottesfeld *et al.*, 1987). When TFIIIA interacts with this fragment, it induces a change in the CD spectrum of the DNA, suggesting an alteration in the conformation of the fragment. When TFIIIA interacts with a non-specific DNA fragment, no signal change in the CD spectrum is observed, indicating that conformational change is a sequence-specific event. Furthermore, the mobility of the TFIIIA-ICR complex in a gel is different from that of a complex with non-specific DNA of the same size. These results are contradictory to the previous observations (Gottesfeld *et al.*, 1987) in which changes in the CD signal from an ICR containing fragment were not detectable upon binding to TFIIIA. The reason for the discrepancy is not understood.

The change observed in the CD signal from the ICR, as well as the anomalous mobility in a gel, are consistent with bending of the DNA structure induced by TFIIIA. In fact, a very recent report demonstrates, using electron microscopy, that TFIIIA bends the 5S rRNA gene (Bazett-Jones and Brown, 1989) .

Studies involving the use of TFIIIA deletion mutants have permitted the mapping of the transcriptional activator domain (Vrana *et al.*, 1988). Deletion of the last 31 amino acids from the C-terminus does not affect either binding of TFIIIA to the 5S rRNA gene or the transcription activity of TFIIIA. Further deletion of an additional 19 amino acids results in the loss of 90% of the transcriptional activity of the protein.

Transcriptional activation can be mapped to a region between amino acids 293 and 313, although it remains possible that this deletion affects the conformation of a second site which is the actual activating region. It will be of interest to determine whether this mutant can induce bending of the ICR, thus establishing a correlation between this activity and transcriptional activation. Another very important question that should be addressed with this mutant is whether it is able to form stable transcription complexes. This can serve as an index of its capacity to interact with other components of the transcriptional apparatus, most likely TFIIIC.

ATPase ACTIVITY OF TFIIIA

Highly purified preparations of TFIIIA have been shown to possess a DNA-stimulated ATPase activity (Hazuda and Wu, 1986). This activity is associated with both TFIIIA and the 7S particle throughout purification, and can be abolished by affinity purified anti-TFIIIA antibodies. Kinetic analysis has shown that TFIIIA hydrolyzes ATP with a V_{max} of 1.7 nmol/min/mg of protein and a K_m of 5.0×10^{-5} M, while 7S particles hydrolyze ATP with a V_{max} of 2.7 nmol/min/mg of protein and a K_m of 1.4×10^{-4} M. These results suggest that the free protein has a higher affinity for the substrate, but a reduced catalytic capacity as compared to 7S particles. The turnover number, although rather low, is comparable to that of other DNA-stimulated ATPases.

The ATPase activity of TFIIIA can be stimulated approximately three-fold by the addition of double-stranded DNA. It does not appear to require a specific DNA sequence, since no differences could be detected between the effects of pBR322 and a plasmid containing the 5S rRNA gene. TFIIIA shows a high affinity for ss DNA; neither synthetic polydeoxynucleotides nor 5S rRNA stimulates the ATPase activity, however.

The ATP binding site and the catalytic center of TFIIIA seem to be located in the C-terminal domain of the protein. The 33 kDa, papain-generated fragment possesses approximately 50% of the enzymatic activity, while the 28 kDa, trypsin-generated, fragment exhibits only approximately 10% of the ATPase activity of the intact TFIIIA. Furthermore, while [α-^{32}P]-ATP can be photocrosslinked to TFIIIA, the efficiency of photocrosslinking to the 33 kDa fragment is markedly reduced and no [α-^{32}P]-ATP can be detected to cross-link to the 28 kDa fragment.

We have also used the ATP analog, 5′-p fluorosulfonylbenzoyladenosine (FSBA) as an affinity label of the ATP binding site of TFIIIA (Hazuda, Sierra and Wu, submitted). Using this reagent, TFIIIA can be labeled stoichiometrically, resulting in the abolition of its ATPase activity. We have identified a single lysine residue as the target of the FSBA. This residue is located in the C-terminus, since limited proteolysis of [^{3}H]-FSBA-labeled TFIIIA with trypsin generates a 28 kDa fragment that has lost the label.

Although the ATP binding site of TFIIIA is located in the C terminus, removal of the zinc ions, which are located in the N terminus, abolishes the ATPase activity (Wu, 1987). Moreover, we observe that in the presence of ATP, the N-terminus of TFIIIA is more sensitive to proteolysis than in its absence (Sierra and Wu, unpublished), suggesting that ATP induces a conformational change in the TFIIIA. It is possible that ATP hydrolysis mediates the communication between different domains of TFIIIA.

The role of the ATPase activity of TFIIIA in transcription is not clear. It has been shown (Bieker and Roeder, 1986) that ATP hydrolysis is not required for transcription of 5S rRNA genes. In addition, FSBA-modified TFIIIA, which has no detectable ATPase activity, is transcriptionally active (unpublished observation). In summary, we have observed three clear effects of ATP on TFIIIA: (a) it induces the formation of TFIIIA multimers, (b) it makes the N-terminus more sensitive to proteolysis, and, (c) it affects the binding of TFIIIA to ss DNA, resulting in the formation of complexes that show increased S1 resistance.

TFIIIA PROMOTES DNA REASSOCIATION

The finding that TFIIIA is a ss DNA binding protein (Hanas *et al.*, 1984b) led to the question of whether it could also carry out biochemical activities similar to those of other ss DNA binding proteins (Coleman and Oakley, 1980) . Hanas *et al.* (1985) demonstrated that TFIIIA will promote the reassociation of complementary, ss DNA strands. The 33 kDa papain-generated fragment retains the ability to reassociate the DNA. However, the 28 kDa trypsin-generated fragment is unable to promote reassociation. The most likely explanation for this result is that the C-terminal domain, which by footprinting is known not to be significantly involved in specific DNA binding, is responsible for protein-protein interactions required for the reassociation reaction. It was also found that the zinc ions in TFIIIA are not required for its non-specific interactions with DNA nor for DNA reassociation. These results are consistent with a model in which TFIIIA promotes DNA reassociation through covering the ss DNA by virtue of its strong cooperative binding, thus relieving the kinetic barrier that prevents DNA reassociation.

TFIIIA is required in a stoichiometric manner for the DNA reassociation reaction (Fiser-Littell and Hanas, 1988). The low protein to DNA ratio required for this reaction is surprising because one TFIIIA molecule can reassociate approximately 150 bases. However, data from our laboratory indicates that each TFIIIA molecule binds only approximately 20 bases of ss DNA (Sierra and Wu, unpublished). Direct comparison with other ss DNA binding proteins showed that the mechanism by which TFIIIA promotes DNA reassociation is different from DNA reassociation mechanisms mediated by other ss DNA binding proteins (Fiser Littell and Hanas, 1988) ; however, the mechanism remains unclear.

The best evidence supporting that DNA reassociation may play a role in transcription comes from transcription studies of 5S rRNA genes containing base mismatches in different positions (Sullivan and Folk, 1987). Genes containing mismatches within the ICR, in positions that do not normally affect transcription, are transcriptionally inactive. It is possible that renaturation of the genes following the passage of the polymerase is prevented by the destabilizing effect of the mismatch.

CONCLUSIONS

TFIIIA is a multi-functional transcriptional activator whose structure has evolved to carry out its specialized functions in a modular fashion. The protein is composed of 3 major domains, as defined by limited proteolysis: an N-terminal, 28 kDa, nucleic acid binding domain and two 5 kDa, intermediate and C-terminal domains, which are involved in mediating protein-protein interactions and in ATP hydrolysis. The nucleic acid binding domain can itself be subdivided into 9 mini-domains, called fingers. The modular design of TFIIIA has allowed us to establish the functions of the different domains as well as to investigate the role of these multiple activities in transcription.

We envision the role of TFIIIA in transcription as follows: TFIIIA specifically binds to the ICR of the 5S rRNA genes and induces a conformational change (possibly bending) of the DNA. This is followed by TFIIIC recognition and binding to this structure, forming a stable transcription complex. Upon binding of TFIIIB and RNA polymerase III the DNA must form an "open complex" (by analogy with prokaryotic transcription) for transcription to start. TFIIIA remains associated with the non coding strand by virtue of its high affinity to ss DNA and its multiple contact points with this strand. The remainder of the transcription complex is also displaced to the non-coding strand, possibly by strong protein-protein interactions among its various components. After one round of transcription TFIIIA reassociates the two strands, leading to the priming of the gene for a new round of transcription.

There are still many unanswered questions about the activities of TFIIIA. The roles of the ATPase activity, the tight binding to ss DNA and the APP formation are far from clear. It is possible that they are involved in the developmental regulation of 5S rRNA synthesis, perhaps by keeping chromatin in the proper conformation. Since the 5S rRNA genes are found in clusters (most notably, the oocyte-type genes are located at the telomeres of most chromosomes), the APP structure could be an ideal solution for keeping extensive regions of a chromosome in a special, active conformation. At present, however, this is purely speculative.

Many of the biochemical properties of TFIIIA are still poorly understood. The physico-chemical principles that determine the remarkable variety of interactions of this protein with nucleic acids are currently being studied. Another question that deserves special attention involves the role of zinc ions in the specific recognition of the ICR. Do zinc ions make specific contacts with the bases, as those zinc ions in *E. coli* RNA polymerase (Chatterji and Wu, 1982;, Chatterji *et al.*, 1984) or do they play a merely structural role?

An area that has received little attention so far, due to the absence of purified components, is that of protein-protein interactions between TFIIIA and the various components of the transcriptional machinery. As discussed above, it is very likely that TFIIIA interacts directly with TFIIIC. Binding of TFIIIC to the TFIIIA-ICR complex increases the affinity of the complex to the ICR several-fold and leads to the formation of a stable complex. It is possible that this tight binding is mediated, at least in part, by protein-protein interactions, since no significant changes are observed in the footprint of the ternary complex as compared to that of the TFIIIA-ICR complex. Protein-protein interactions may play a role in the mechanism by which TFIIIA activates transcription. Since deletion of a small part of the protein leads to a 90% decrease in transcription, it has been proposed that this region is involved in contacts with other factors and/or the polymerase. The characterization of these contacts will provide information about the mechanisms of transcriptional enhancement.

TFIIIA exhibits an extraordinary number of activities for a protein of its size. The complexity of these activities is probably a reflection of the intricate mechanisms of transcriptional regulation in eukaryotes. As long as the roles of so many TFIIIA-associated activities remain unclear the study of structure-function relationships in this protein will continue to be an area of great interest.

REFERENCES

Anderson, J., Delihas, Hanas, J. S. and Wu, C.-W. (1984) 5S RNA structure and interaction with transcription factor A. 2. Ribonuclease probe of the 7S particle form Xenopus laevis immature oocytes and RNA exchange properties of the 7S particle. *Biochemistry* **23**: 5752-5759

Baxett-Jones, D. P. and Brown, M. L. (1989) Electron microscopy reveals that transcription factor IIIA bends 5S DNA. *Mol. Cell. Biol.* **9**: 336-341

Berg, J. M. (1986) Potential metal-binding domains in nucleic acid binding proteins. *Science* **232**: 485-487

Beiker, J. J. and Roeder, R. G. (1984) Physical properties and DNA binding stoichiometry of a 5S gene-specific transcription factor. *J. Biol. Chem.* **259**: 6158-6164

Beiker, J. J. and Roeder, R. G. (1986) Characterization of the nucleotide requirement for elimination of the rate-limiting step in 5S RNA gene transcription. *J. Biol. Chem.* **261**: 9732-9738

Bogenhagen, D. F., Sakonju, S. and Brown, D. D. (1980) A control region in the center of the 5S RNA gene directs specific initiation of transcription II. The 3′ border region. *Cell* **19**: 27-35

Bogenhagen, D. F., Wormington, W. M. and Brown, D. D. (1982) Stable transcription complexes of *Xenopus* 5S RNA genes : a means to maintain the differentiated state. *Cell* **28**: 413-421

Bogenhagen, D. F. (1985) The intragenic control region of the *Xenopus* 5S RNA gene contains two factor A binding domains that must be aligned properly for efficient transcription initiation *J. Biol. Chem.* **260**: 6466-6471

Brown, R. S., Sander C. and Argos, P. (1985) The primary structure of the transcription factor TFIIIA has 12 consecutive repeats. *FEBS Letts.* **186** 271-274

Chatterji, D. and Wu, F. Y.-H. (1982) Selective substitution *in vitro* of an intrinsic zinc of *E. coli* RNA polymerase with various divalent metals. *Biochemistry* **21**: 4651-4656

Chatterji, D., Wu, C.-W and Wu, F. Y.-H. (1984) Nuclear magnetic resonance studies on the role of intrinsic metals in *E. coli* RNA polymerase. Effect of the DNA template on the nucleotide-enzyme interaction. *J. Biol. Chem.* **259**: 284-289

Christiansen J., Brown, R. S., Sproat, B. S. and Garrett, R. A. (1987) *xenopus* transcription factor IIIA binds primarily at junctions between double helical stems and internal loops in oocyte 5S RNA *EMBO J.* **6**: 453-460

Coleman, J. E. and Oakley, J. L. (1980) Physical chemical studies of the structure and function of DNA-binding (helix-destabilizing) proteins. *Crit. Rev. Biochem.* **7**: 247-289

Diakun, G. P., Fairall, L. and Klug, A. (1986) EXAPO study of the zinc-binding sites in the protein transcription factor IIIA. *nature* **324**: 698-699

Engelke, D. R., Ng, S.-Y., Shastry, B. S. and Roeder, R. G. (1980) Specific interaction of a purified transcription factor with an internal control region of 5S RNA genes. *Cell* **19**: 717-728

Fairall, L., Rhodes, D. and Klug, A. (1986) Mapping the site of protection on a 5S RNA gene by the *Xenopus* transcription factor IIIA *J. Mol. Biol.* **192**: 577-591

Fiser-Littell, R. M. and hanas, J. S. (1988) *Xenopus* transcription factor IIIA-dependent DNA renaturation. *J. Biol. Chem.* **263**: 17136-17141

Fox, G. E. (1985) **In**: *The Bacteria* (Woese, C. and Wolfe, R., Eds.) **Vol 8**, pp. 257-316, Academic Press, New York

Frankel, A. D., Berg, J. M. and Pabo, C. O. (1987) Metal-dependent folding of a single zinc finger from transcription factor IIIA. *Proc, natl. Acad. Sci. USA* **84**: 4841-4845

Geiduschek, E. P. and Tocchini-Valentini, G. P. (1988) Transcription by RNA polymerase III. *Ann. Rev. Biochem.* **57**: 873-914

Ginsberg, A. M., King, B. O. and Roeder, R. G. (1984) *Xenopus* 5S gene transcription factor, TFIIIA : characterization of a cDNA and measurement of RNA levels throughout development. *cell* **39**: 479-489

Gottesfeld, J. M. (1987) DNA sequence-directed nucleosome reconstitution on 5S RNA genes of *Xenopus laevis*. *Mol. Cell. Biol.* **7: 1612-1622**

Gottesfeld, J. M. and Bloomer, L. S. (1982) Assembly of transcritpionally active 5S RNA gene chromatin *in vitro*. *Cell* **28**: 781-791

Gottesfeld, J. M., Blanco, J. and Tennant, L. L. (1987) The 5S gene internal control region is B-form both free in solution and in a complex with TFIIIA. *Nature* **329**: 460-462

Hanas, J. S., Bogenhagen D. F. and Wu, C.-W. (1983a) Cooperative model for the binding of *xenopus* transcription factor A to the 5S RNA gene. *Proc. Natl. Acad. Sci USA* **80**: 2142-2145

Hanas, J. S., Hazuda D.J., Bogenhagen, D. F., Wu, F. Y.-H. and Wu, C.-W. (1983b) *Xenopus* transcription factor A requires zinc for binding to the 5S RNA gene *J. Biol. Chem.* **258**: 14120-14125

Hanas, J. S., Bogenhagen, D. F. and Wu, C.-W. (1984a) DNA unwinding ability of *Xenopus* transcription factor A. *Nucleic Acids Res.* **12**: 1265-1276

Hanas, J. S., Bogenhagen, D. F. and Wu, C.-W. (1984b) Binding of the *Xenopus* transcription factor A to 5S RNA and to single-stranded DNA. *Nucleic Acids Res.* **12**: 2745-2758

Hanas, J. S., Hazuda, D. J. and Wu, C.-W. (1984b) *Xenopus* transcription factor A promotes DNA reassociation. *J. Biol. Chem.* **260**: 13316-13320

Hanas, J. S., Duke, A. L. and Gaskins, C. J. (1989) Conformational states of *Xenopus* transcription factor IIIA. *Biochemistry* **28**: 4083-4088

Hayes, J., Tullius, T. D. and Wolffe, A. P. (1989) A protein-protein interaction is essential for stable complex formation on a 5S RNA gene. *J. Biol. Chem.* **264**: 6009-6012

Hazuda, D. J. and Wu, C.-W. (1986) DNA-activated ATPase activity associated with *Xenopus* transcription factor A. *J. Biol. Chem.* **261**: 12202-12208

Honda, B. M. and Roeder, R. G. (1980) Association of a 5S RNa gene transcription factor with 5S RNA and altered levels of the factor during cell differentiation. *Cell* **22**: 119-126

Huang, W.-J., Shang, Z., Wu, C.-W. and Wu, F. Y.-H. (1989) Circular dichroism (CD) studies of TFIIIA and its nucleic acids interactions. *J. Cell. Biol.* **107**: 845a

Huber, P. W. and Wool, F. G. (1986) Identification of the binding site on 5S RNA for the transcription factor IIIA : proposed structure of a common binding site on 5S RNA and on the gene. *Proc, Natl. Acad. Sci. USA* **83**: 1593-1597

Lassar, A. B., Martin, P. L. and Roder, R. G. (1983) Transcription of class III genes : formation of preinitiation complexes. *Science* **222**: 740-748

McGhee, J. D. and von Hippel, P. H. (1974) Theoretical aspects of DNA-protein interactions : cooperative binding and non-cooperative binding of large ligands to a one-dimensional homogeneous lattice. *J. Mol. Biol.* **86**: 469-489

Miller, J., McLachlan, A. D. and Klug, A. (1985) Repetitive zinc-binding domains in the protein transcription factor IIIA from*Xenopus* oocytes. *EMBO J.* **4**: 1609-1614

Millstein, L., Eversole-Cire P., Blanco, J. and Gottesfeld, J. M. (1987) Differential transcription of *Xenopus* oocyte and somatic-type 5S genes in a *Xenopus* oocytes extract. *J. Biol. Chem.* **262**: 1-11.

Parker, C. S. and Roeder, R. G. (1977) Selective and accurate transcription of the *Xenopus laevis* 5S RNA genes in isolated chromatin by purified RNA polymerase III. *Proc. Natl. Acad Sci. USA* **74**: 44-48

Pelham, H. R. and Brown, D. D. (1980) A specific transcription factor that can bind either the 5S RNA gene or the 5S RNA. *Proc. Natl. Acad Sci. USA* **77**: 4170-4174

Peck, L. J., Millstein L., Eversole-Cire P., Gottesfeld J. M. and Varshavsky A. (1987) Transcriptionally inactive oocyte-type 5S RNA genes of *Xenopus levis* are complexed with TFIIIA *in vitro*. *Mol. Cell. Biol.* **7**: 3503-3510

Picard, B. and Wegnez, M. (1979) Isolation of a 7S-particle from *Xenopus levis* oocyte : A 5S RNA protein complex. *Proc. Natl. Acad Sci. USA* **76**: 241-245

Pieler, T., Appel, B., Oei, S. L., Mentzel, H. and Erdmann, V. A. (1985a) Point mutational analysis of the *Xenopus laevis* 5S gene promoter. *EMBO J.* **4**: 3751-3756

Pieler, T., Oei, S. L., Hamm, J., Engelke, U. and Erdmann, V. A. (1985b) Functional domains of the *Xenopus laevis* 5S gene promoter. *EMBO. J.* **4**: 3751-3756

Pieler, T., Hamm, J., and Roeder, R. G. (1987) The 5S gene internal control region is composed of three distinct sequence elements, organized as two functional domains with variable spacing. *Cell* **48**: 91-100

Rhodes, D. (1985) Structural analysis of s triple complex between the histone octamer, a *Xenopus* gene for 5S RNA and transcription factor IIIA. *EMBO. J.* **4**: 3473-3482

Rhodes, D. and Klug A. (1986) An underlying repeat in some transcriptional control sequences corresponding to half a double-helical turn of DNA *Cell* **46**: 123-132

Romaniuk, P. J. (1985) Characterization of the RNA binding properties of transcription factor IIIA of *Xenopus laevis* oocytes. *Nucleic Acids Res.* **13**: 5369-5387

Romaniuk, P. J., de Stevenson, I. L. and Wong, H.-H.A (1987) Defining the binding site of *Xenopus* transcription factor IIIA on 5S RNA using truncated and chimeric 5S RNA molecules. *Nucleic Acids Res.* **15**: 2737-2755

Sakonju, S., Bogenhagen, D. F. and Brown, D. D. (1980) A control region in the center of the 5S RNA gene directs specific initiation of transcription I. The 5′ border of the region. *Cell* **19**: 13-25

Sakonju, S. and Brown, D. D. (1982) Contact points between a positive transcription factor and the *Xenopus* 5S RNA gene. *Cell* **31**: 395-405

Sands, M. S. and Bogenhagen, D. F. (1987) TFIIIA binds to different domains of 5S RNA and the *Xenopus borealis* 5S RNA gene. *Mol. Cell. Biol.* **7** 3985-3993

Schlissel. M. S. and Brown, D. D. (1984) The transcriptional regulation of *Xenopus* 5S RNA genes in chromatin; the roles of active stable transcription complexes and histone H1. *Cell* **37**: 903-913

Shang, Z., Liao, Y.-D., Wu, F. Y.-H. and Wu, C.-W. (1989) Zn^{2+} release from *Xenopus* transcription factor IIIA induced by chemical modification. *Biochemistry* (submitted)

Shastry, B. S. (1986) 5S RNA gene specific transcription factor (TFIIIA) changes the linking number of the DNA. *Biochem. Biophys. Res. Comm.* **134**: 1086-1092

Shimamura, A., Tremethick, D. and Worcel, A. (1988) Characterization of the repressed 5S DNA minichromosomes assembled *in vitro* with a high speed supernatant of *Xenopus laevis* oocytes. *Mol. Cell. Biol.* **8**: 4257-4269

Smith, D. R., Jackson, I. J. and Brown, D. D. (1984) Domains of the positive transcription factor specific for the *Xenopus* 5S RNA genes. *Cell* **37**: 645-652

Stillman, D. J., Caspers, P. and Geiduschek, E. P. (1985) Effect of temperature and single-stranded DNA on the interaction of an RNA polymerase III transcription factor with a tRNA gene. *Cell* **40**: 311-317

Sullivan, M. A. and Folk, W. R. (1987) Transcription of eukaryotic tRNA1?^{met} and 5s RNA genes by RNA polymerase III is blocked by base mismatches in the intragenic control regions. *Nucleic Acids Res.* **15**: 2059-2068

Tso, J. Y., Van Den Berg D. J. and Korn, L. J. (1986) Structure of the gene for *Xenopus* transcription factor TFIIIA. *Nucleic Acids Res.* **14**: 2187-2200

Vrana, K. E., Churchill, M. E. A., Tullius, T. D. and Brown, D. D. (1988) Mapping functional regions of transcription factor TFIIIA. *Mol. Cell. Biol.* **8**: 1684-1696

Windsor, W. T., Lee, T.-C., Daly, T. J. and Wu, C.-W. (1988) *Xenopus* transcription factor IIIA binds to the flanking regions of the 5S RNA gene intragenic control region in a unique and highly ordered state. *J. Biol. Chem.* **263**: 10272-10277

Wolffe, A. P. and Brown, D. D. (1988) Developmental regulation of two 5S ribosomal RNA genes. *science* **241**: 1626-1632

Wormington, W. M., Schlissel, M. and Brown, D. D. (1983) Developmental regulation of *Xenopus* 5S RNA genes. *Cold Spring Harbor Symp. Quant. Biol.* **47**: 879-884

Wormington, W. M. and Brown, D. D. (1983) Onset of 5S RNA gene regulation during *Xenopus* embryogenesis. *Dev. Biol.* **99**: 248-257

Wu, C.-W. (1987) *Xenopus* transcription factor A: a zinc metalloenzyme with multiple functions. **In**: *Zinc Enzymes* (Bertini, I., Luchinat, C., Maret, W. and Zeppenzauer, M., Eds) pp. 563-573, Birkhauser, Boston

Xing, Y. Y. and Worcel, A. (1989) The C-terminal domain of transcription factor IIIA interacts differently with different 5S RNA genes. *Mol. Cell. Biol.* **9**: 499-514

Young, D. and Carroll, D. (1983) Regular arrangement of nucleosomes on 5S RNA genes in *Xenopus laevis*. *Mol. Cell. Biol.* **3**: 720-730